AF761675

AN INNOVATIVE APPROACH TO UNDERSTANDING AND TREATING CANCER: TARGETING pH

From Etiopathogenesis to New Therapeutic Avenues

AN INNOVATIVE APPROACH TO UNDERSTANDING AND TREATING CANCER: TARGETING pH

From Etiopathogenesis to New Therapeutic Avenues

TOMAS KOLTAI, MD, PhD

Former Director of The Hospital del Centro Gallego de Buenos Aires and Former Oncologist and Director of Centro Médico del Sindicato de la Alimentación, Buenos Aires, Argentina

STEPHAN J. RESHKIN, PhD

Associate Professor, Department of Bioscience, Biotechnology and Biopharmaceutics, University of Bari, Italy

SALVADOR HARGUINDEY, MD, PhD

Director, Institute for Clinical Biology and Metabolism, Vitoria, Spain

ELSEVIER

ACADEMIC PRESS

An imprint of Elsevier

Academic Press is an imprint of Elsevier
125 London Wall, London EC2Y 5AS, United Kingdom
525 B Street, Suite 1650, San Diego, CA 92101, United States
50 Hampshire Street, 5th Floor, Cambridge, MA 02139, United States
The Boulevard, Langford Lane, Kidlington, Oxford OX5 1GB, United Kingdom

© 2020 Elsevier Inc. All rights reserved.

No part of this publication may be reproduced or transmitted in any form or by any means, electronic or mechanical, including photocopying, recording, or any information storage and retrieval system, without permission in writing from the publisher. Details on how to seek permission, further information about the Publisher's permissions policies and our arrangements with organizations such as the Copyright Clearance Center and the Copyright Licensing Agency, can be found at our website: www.elsevier.com/permissions.

This book and the individual contributions contained in it are protected under copyright by the Publisher (other than as may be noted herein).

Notices

Knowledge and best practice in this field are constantly changing. As new research and experience broaden our understanding, changes in research methods, professional practices, or medical treatment may become necessary.

Practitioners and researchers must always rely on their own experience and knowledge in evaluating and using any information, methods, compounds, or experiments described herein. In using such information or methods they should be mindful of their own safety and the safety of others, including parties for whom they have a professional responsibility.

To the fullest extent of the law, neither the Publisher nor the authors, contributors, or editors, assume any liability for any injury and/or damage to persons or property as a matter of products liability, negligence or otherwise, or from any use or operation of any methods, products, instructions, or ideas contained in the material herein.

Library of Congress Cataloging-in-Publication Data
A catalog record for this book is available from the Library of Congress

British Library Cataloguing-in-Publication Data
A catalogue record for this book is available from the British Library

ISBN: 978-0-12-819059-3

For information on all Academic Press publications
visit our website at https://www.elsevier.com/books-and-journals

Publisher: Stacy Massuci
Acquisition Editor: Rafael Teixeira
Editorial Project Manager: Sam Young
Production Project Manager: Kiruthika Govindaraju
Cover Designer: Alan Studholme

Typeset by SPi Global, India

ELSEVIER Book Aid International
Working together
to grow libraries in
developing countries
www.elsevier.com • www.bookaid.org

Contents

Contributors

Julieta Afonso
Life and Health Sciences Research Institute (ICVS), School of Medicine, University of Minho, Braga; ICVS/3B's—PT Government Associate Laboratory, Braga /Guimarães, Portugal

Fátima Baltazar
Life and Health Sciences Research Institute (ICVS), School of Medicine, University of Minho, Braga; ICVS/3B's—PT Government Associate Laboratory, Braga /Guimarães, Portugal

Stefano Fais
Department of Oncology and Molecular Medicine, Istituto Superiore di Sanità, Rome, Italy

Sara Granja
Life and Health Sciences Research Institute (ICVS), School of Medicine, University of Minho, Braga; ICVS/3B's—PT Government Associate Laboratory, Braga /Guimarães, Portugal

Enrico Pierluigi Spugnini
UOSD SAFU, Department of Research, Diagnosis and Innovative Technologies, IRCCS Regina Elena National Cancer Institute, Rome, Italy

Odília Queirós
IINFACTS—Institute of Research and Advanced Training in Health Sciences and Technologies, Department of Sciences, University Institute of Health Sciences (IUCS), CESPU, CRL, Gandra, Portugal

Diana Valente
Life and Health Sciences Research Institute (ICVS), School of Medicine, University of Minho, Braga; ICVS/3B's—PT Government Associate Laboratory, Braga /Guimarães; IINFACTS—Institute of Research and Advanced Training in Health Sciences and Technologies, Department of Sciences, University Institute of Health Sciences (IUCS), CESPU, CRL, Gandra, Portugal

Preface

There is nothing more difficult to take in hand, more perilous to conduct, or more uncertain in its success, than to take the lead in the introduction of a new order of things. For the reformer has enemies in all those who profit by the old order, and only lukewarm defenders in all those who would profit by the new order, this lukewarmness arising partly from fear of their adversaries … and partly from the incredulity of mankind, who do not truly believe in anything new until they have had actual experience of it.

Niccolo Machiavelli

The longest journey starts with the first step

Attributed to Lao Tzu

This book is the first step in the association of basic and clinical research. Starting with molecular mechanisms and ending at practical bedside experience, it tries to elucidate the intricate activity of metabolism and pH in cancer.

Since the 1920s with Otto Warburg's discoveries on cancer metabolism, the pH alterations of tumors were simply considered a by-product of metabolic changes.

Through different routes, many researchers, including the authors arrived at the conclusion that altered pH is not simply an accidental and harmless consequence of cancer, but an initiator and key player in the whole process of malignization and tumor evolution. What is even more important is that altered pH and metabolism are targetable characteristics and they can be attacked with relatively unsophisticated drugs, many of which are already being used in clinical situations other than cancer.

Reversing the tumor's pH and metabolic alterations have important clinical implications that go from slowing tumor growth and achieving stable disease to apoptosis and tumor reduction. Most of the drugs working against the altered pH and metabolism can be used along with standard chemo- and radio-therapeutic protocols without interfering with and, in most cases, improving their results.

All the chapters related to the mechanisms of these cancer alterations start with the basic molecular description of the phenomenon and end with practical clinical guidelines for treating the problem. Molecular and clinical evidence is highlighted in each of these chapters.

For the sake of a better understanding, we avoided descriptions of very complex molecular issues that have no direct practical applications. With that same purpose, we used a considerable amount of figures to illustrate each chapter. We believe that a good drawing is much better than a long text.

This book describes proven facts, but does not fall short in provocative ideas, not with the intention of generating controversies, but to create a framework for a different way of thinking of cancer.

Chapter 1 starts with a brief introduction on the history of the research of cancer metabolism and pH, and how this knowledge evolved since the original discoveries of Otto Heinrich Warburg, to the present. This description spans almost 100 years of changing concepts. Warburg's research opened the door to the development of a whole new world: proton dynamics as a fundamental participant in the origin and development of cancer. Later research, showed in a step by step process all the implications of this proton dynamics in the life of the cancer cell.

Chapter 2 describes the peculiar metabolism of the cancer cells as compared with that of normal cells. This chapter also highlights the vulnerabilities of this metabolism and how it can be used against the tumor.

Chapter 3 introduces us to the world of the pH paradigm in cancer, showing a different perspective of the disease. Basically, the pH paradigm does not change anything in cancer, what changes is the way we perceive it. Looking at cancer from this angle we will understand that even if there are more than 200 or so different cancers, pathways and oncogenes, the mechanism of proliferation, migration, invasion and metastasis is always the same. This is the pH-centered perception of cancer. It also aims at explaining the molecular basis of the pH alteration.

Chapters 4–12 separately describe each cellular component and compound that participates in the pH alteration. These includes the lactic acid peculiarities (Chapter 4) the sodium hydrogen exchanger (Chapter 5), the voltage gated sodium channels (Chapter 6), the carbonic anhydrases (Chapter 7), the proton pumps (Chapter 8), the sodium/bicarbonate transporter (Chapter 9), the aquaporins (Chapter 10), the invadopodia (Chapter 11) and the Sp1 transcription factor (Chapter 12) as part of the master regulator system headed by the hypoxia inducible factor (HIF).

Chapters 13–16 discuss possible treatments for the pH alterations in cancer, with special emphasis on already available drugs. Nutraceuticals will also be discussed in these chapters. Chapter 17 by Fatima Baltazar and her team describes cancer markers in the frame of the pH-centered approach and recent developments in treatment.

Chapters 18 and 19 are mainly oriented toward a clinical framework and new protocols to deal with the problem, while an offspring of these protocols is discussed in Chapter 20 oriented to the prevention of metastasis.

Chapter 21 describes the clinical experience with metabolic and pH oriented treatments in pets and human patients. This chapter was authored by Drs Enrico Pierluigi Spugnini and Stefano Fais.

Chapter 22 discusses the opportunities that the altered pH gives to nanomolecular particles in order to deliver drugs in a specific manner into the tumor minimizing damage to non cancerous tissues. These particles are also able to modify the abnormal pH situation and they seem to be the next breakthrough in cancer treatment.

The conclusions and future perspectives are placed in Chapter 23.

We have sacrificed erudition for the sake of clarity. This is a scientific sin, but hopefully our readers will forgive us.

When we decided to write this book we had only two goals in mind: make our colleagues aware of the importance of the metabolic and pH paradigm in cancer and also show a different perception of cancer that will have direct consequences in future treatments. If we achieve the purposes we set out for this book, means that in the future oncologist and medical practitioners in general will be conscious of the importance of treating the pH and metabolic problem in their cancer patients. And in certain cases this consciousness will achieve better results in their patients by the use of simple medication mentioned in these pages.

We hope to achieve these goals.

We are grateful to our Publisher for helping us bring this important subject to the attention of the medical community. Our thanks to the collaborators Drs Spugnini, Fais and Baltazar for their experienced insights. We also thank Julia Weiss for her manuscript revision and correction.

The Authors

PART I

Metabolism and pH physiopathology of cancer

CHAPTER 1

Introduction

Introductory words

Here in this book, we shall climb the cancer mountain. And we are going to do it from the metabolic side. But, we will have to use an innovative instrument for this climb: the pH-centered approach. The pH-centered approach enables us to understand that the striking findings of acid-base alterations in cancer are not collateral damage of the disease. From Warburg and up to the late fifties, the metabolic approach led to the idea that cancer was a "mitochondrial disease". However, the explosive development of molecular biology showed that the mitochondrial origin of cancer was a flawed concept. The new culprit was the gene. And a gene-centered approach era started. We are living now in this "genetic era". Cancer is the product of a change in one or many genes, whether mutations, epigenetic silencing, translocations, etc. While there is no doubt that the gene (gene-centered approach) is the root of the malignant process; something is missing for a real understanding of the reasons that lead to cancer development. Genes "speak", but they do not do things by themselves. Genes have a code that they express through proteins. These proteins are the real instruments that make things happen. Different tumors make things happen in different ways. Malignant tumors have different phenotypes and genotypes. A mesothelioma has very little to do with a chronic myeloid leukemia. Both are malignant tumors, both grow, and both may eventually kill their host, and yet, they are both cancers. And of course, their genotypes and phenotypes are strikingly different. This has led scientists to say that cancer is not a disease but a multiplicity of diseases. Melanoma seems to be one disease. For example, when we penetrate into the intricate world of melanoma, we find that melanomas with certain mutations, such as the V600 mutation, behave differently than melanomas with a different kind of mutation. Thus the concept that cancer is many diseases seems appropriate. Furthermore, it is known that the same tumor is genetically and phenotypically heterogeneous. This would mean that one tumor is many diseases. As we said at the beginning, something is still missing in the quest for an integrated vision of cancer. This is the point where the pH-centered approach comes to our aid. All different tumors, no matter their genotype or phenotype or driving pathway, share one feature: an inverted pH gradient in which as the cell increases its alkalinity the surrounding matrix becomes acid while in normal cells it is the opposite. When analyzed in depth, this abnormality leads to the conclusion that the unifying characteristic of all malignant tumors is this pH alteration. It does not matter if it is a glioblastoma, a renal cell carcinoma, or a lymphoma. They all invert their pH gradient.

An Innovative Approach to Understanding and Treating Cancer: Targeting pH
https://doi.org/10.1016/B978-0-12-819059-3.00001-0
© 2020 Elsevier Inc.
All rights reserved.

This pH inversion could seem to be a collateral development of the many metabolic changes that take place in cancer. Is it so?

For an answer based on sheer logic let us analyze a few facts:

(1) The pH inversion occurs very early in the transformation process, even before the metabolic changes ensue (metabolic switch).
(2) Inhibiting the pH inversion in the early stage of transformation, aborts the malignization.
(3) Reversing the pH inversion slows down cancer proliferation and growth.
(4) There are no cancers without the inverted pH gradient.
(5) The most active parts of a tumor (where invasion is going on) show a more pronounced inversion of the pH gradient.
(6) Metastatic cells may be genetically different from the primary tumor, but the pH inversion is equally present.
(7) When cells are under hypoxic conditions (that induce a pH inversion) transformation may develop.
(8) Once transformation starts in these hypoxic cells, the restoration of normal oxygen levels does not restore normality.
(9) pH inversion creates a very harsh environment that jeopardizes the survival of normal cells, and in spite of this, malignant cells thrive and grow in it.

Many diseases but one common pathway: inversion of the pH gradient. Cells, whether normal or malignant do not have a thinking apparatus. But for a few seconds let's suppose the contrary. Then:

(1) Why would a cell invert its pH gradient if this would create an "impossible" environment?
(2) Why would a cell that is transforming invert its pH as the first step towards cancer?
(3) Why would the inhibition of the pH abnormality impede carcinogenesis at that early stage?
(4) Why do all malignancies have an inverted pH?

Researchers have slowly but steadily found answers to these intriguing questions. These answers are the fundamentals of the pH-centered approach to cancer.

We believe that pH inversion is not a collateral consequence of malignization. It is center stage. And the emerging view is that the genetic alterations that give way to cancer are strongly associated with the pH inversion, which is probably the first step in malignization.

GENE MODIFICATION ----------→ pH GRADIENT INVERSION ----------→ CANCER

The starting point of the pH-centered approach is to consider pH inversion as a basic and fundamental behavior in the etiopathogenesis of cancer.

And yet up to now, this phenomenon that is well-known to basic researchers has had no influence in the clinical approach to cancer. There is a gap between basic science and applied science in oncology. This divergence between hard science and practical use of the new findings may last for some time, but sooner or later they will come together.

The purpose of this book is to try to fill the gap and bring together what basic science has discovered with its practical use at bedside. The knowledge of how the pH inversion develops (first part of the book) is the first step towards an understanding of how it can be targeted (second part of the book).

In the first part we discuss the mechanisms that invert the pH and every important protein, enzyme, channel, transporter and exchanger that actively participates in the inversion process. In the second part, the discussion is focused on every drug that may be of use to alter or eliminate the pH abnormality. The discussion is mainly restricted to existing drugs and nutraceuticals that practicing oncologists can find on the market. Most of them are FDA approved, a few are not. Experimental drugs will also be mentioned but without entering into excessive details.

Our own experiences treating cancer patients have shown how useful this approach may be. Therefore, it is difficult to understand why, for example, an extracellular alkalinizer is not used when the patient is receiving medication such as doxorubicin. A simple proton pump inhibitor such as omeprazole or pantoprazole can modify extracellular acidity and increase drug delivery to the tumor.

It is also difficult to understand the reasons why after a patient has finished standard protocols, no metastasis-prevention drugs are used. And without going too far, even aspirin would make a difference. It is also surprising that with all the non-toxic drugs available to reduce migration and invasion, no serious and well planned prospective clinical trials are being conducted.

If we only look at the medical costs aspect, any treatment of the kind discussed in this book has a negligible cost, much less than any of the new "targeted" treatments and also of the old classical protocols. The cost/benefit ratio is highly in favor of any pH-based treatments.

We wish to stress that there is an issue that will be brought up repeatedly throughout this book: the pH-based treatments are complementary to any standard therapies. They do not replace them, they only produce better and more lasting results in conjunction with the usual treatments. pH-centered treatments may reverse drug resistance in many advanced tumors. This is not an insignificant achievement.

Regarding the schedules for the prevention of metastasis the classical approaches do not contemplate any further therapy after surgery and chemoradiotherapy except some hormonal treatments in breast and prostate cancer. Would it not be advisable to maintain a hostile environment that can reduce the risk of metastasis? pH-based treatments can do it. A metronomic cellular acidification would reduce the metastatic risk. And only a few, low cost medicines would be needed!

Throughout this journey into cancer metabolism and pH abnormalities, hard scientific facts will be basically preserved, even when occasionally some knowledge will be sacrificed on the altar of a better understanding. The basic issue discussed here is understanding what role pH inversion has in cancer insurgence and progression, and what it does to the patient and finally how to reverse and resolve it.

Cancer metabolism and pH management are at a fork in the road, like the poem by Robert Frost.

Two roads diverged in a yellow wood,
And sorry I could not travel both
And be one traveler, long I stood
And looked down one as far as I could
To where it bent in the undergrowth.
Then took the other, as just as fair,...

It might seem that oncologists must choose between the standard road and the pH road and that they cannot travel both. However, this is not so. We are not in a yellow wood. pH targeting can be part of any conventional treatment of cancer. Actually, it should be part of it.

Furthermore, in addition to the lack of toxicity and negligible costs, the pH-centered treatments provide a physiologic approach to cancer: they restore an altered pH to what should be normal. Simply said: it does not affect normal cells, because normal cells have a normal pH.

Pharmacological research invests billions in developing new and more efficient drugs, while simple and time-proved pharmaceuticals are being neglected. We wonder why.

If something can benefit the cancer patient, no matter how simple and unsophisticated that "something" may be, does it not deserve well-planned clinical trials?

Unfortunately, there are almost no clinical trials going on in this area.

Maybe, and only maybe, this book will serve the purpose of arousing the interest of those oncologists that are disappointed with the outcomes of many of their patients.

It also may be, that this book will bring what basic researchers have discovered in the last 20 years closer to bedside medical practice.

The basics

The importance of pH homeostasis in normal and malignant cells and tissues lies in the fact that all the biological processes are pH sensitive. Most enzymes and the reactions they catalyze have a very narrow optimum pH range, and many protein interactions such as ligand-receptor association/dissociation and protein-DNA binding, require a proper pH in order to occur.

Thus, pH, a sophisticated way of expressing the concentration of H^+ or "acidity" of a substance, cell or tissue, is an invisible presence in all the biological processes. This "invisibility" started to change in the 1920s when Otto Warburg, a German chemist and medical doctor, found that malignant tissues had a low pH, which meant that they were acidic. He entered a terrain that is still being explored even today.

The importance of pH in cancer

A few examples will clarify the important relationship between cancer and pH.

(1) The primary aim of chemotherapy and radiotherapy is to kill as many malignant cells as possible. The main mechanism involved in this therapeutic death is apoptosis, although in some cases necrosis, necroptosis and ferroptosis also play significant roles.

The first step in the apoptotic process is a decrease in the intracellular pH (pHi) since the endonucleases that breakup DNA and are involved in the process require an acidic environment.[1–3]

(2) Exosomes secreted by cells are though to be a mechanism of communication with other cells. In malignant cells, exosomes seem to participate in tumor progression. Extracellular acidity, which is characteristic of cancer, increases the production of exosomes, to the point that when malignant cells are incubated in alkaline solutions, no exosomes are produced.[4]

(3) Many of the enzymes that participate in cellular metabolism are pH-sensitive.[5] Phosphofructokinase (PFK) an essential enzyme of the glycolytic pathway, shows a direct correlation between activity and increased pH (within physiologic ranges).[6] At a pH of 7.3 ATP acts as inhibitor of PFK blocking glycolysis, but at a pH of 7.6 ATP has no effect on the enzyme unless its concentration is doubled. The malignant cells' cytoplasm is more alkaline than their normal counterparts, thus, PFK's activity is enhanced.

(4) DNA polymerases have an increased activity during proliferation which leads to a higher level of DNA synthesis. DNA polymerase activity is enhanced with a higher intracellular pH.[7]

(5) pH may act as a signal transducer. This is the case of insulin that increases glycolytic activity, but also increases intracellular pH.[8] Actually, it is the increased pH that induces the glycolytic activity. It is sufficient that insulin increase pHi by 0.12 for this to happen.

(6) When a lymphocyte is stimulated with a mitogenic compound the intracellular pH shows a double, both early and late, increasing response. The early response, occurring within minutes, raises pH from 7.18 to 7.35; the second rise occurs 12 h later.[9] The initial pH increase seems to be related to the initiation of mitosis.

(7) The extracellular matrix (ECM) around the tumor cell is acidic and influences the glycolytic metabolism. When the pH of this ECM becomes less acidic the glycolytic metabolism increases and the opposite happens when the pH becomes more acidic.[10]

(8) In vivo experiments with humans also provide evidence of how important pH is in metabolism. A group of volunteers doing physical exercises first received substances that increased acidosis and then substances that increased alkalosis. In the first part of the experiment, the lactic acid increase in blood and muscle was very moderate compared with the second part. This shows that there is a decrease of the glycolytic pathway when the cells are acidified.[11]

(9) In 1961 an experiment was performed in which yeast was exposed to high concentrations of nystatin, which blocked the glycolytic pathway. But when ammonium chloride that increased intracellular pH was added to the medium, the glycolytic pathway was reactivated.[12]

(10) pH is not only important in determining the glycolytic metabolism, it also intervenes in cellular carbohydrate transport: high pH increases glucose transport while low pH decreases it.[13]

(11) In a unicellular system like the streptococcus culture, the addition of glucose to the medium increased intracellular pH (pHi) through proton extrusion and kept it above the extracellular pH (pHe). When the glucose load was consumed, conditions returned to the original state.[14]

(12) The dietary manipulation of pH in mice could achieve full regression of sarcoma 180. This 1979 experiment has shown how an important issue is pH in cancer.[15]

(13) When the oncogenic ras p21 protein was injected to quiescent NIH 3T3 cells the intracellular pH increased 0.17. Amiloride, an inhibitor of the sodium/hydrogen exchanger 1 (NHE1) impeded this pH modification.[16]

(14) In 1987 Wakelam et al.[17] proposed what we may call the oncogenic-pH pathway that clearly shows the relation of pH to cancerogenesis. It starts with growth factors activating their receptors which lead to inositol breakdown and the generation of two second messengers:

- **(a)** inositol 1, 4, 5-trisphosphate that increases intracellular free calcium levels[18];
- **(b)** 1,2 diacylglycerol that activates Protein C kinase. Protein C kinase increases NHE1 activity with the consequent intracellular pH elevation.

The association of increased free calcium and augmented alkaline cytoplasm promotes proliferation.

(15) When the oncogenes v-mos and activated H-ras were introduced into an experimental cell like NIH3T3, pHi was raised due to NHE1 activation and the cell entered S phase of mitosis. The proto-oncogene form of H-ras had only a mild effect or no effect at all.[19]

(16) When the HPV-16 E7 oncogene was expressed in keratinocytes in a time-dependent inducible system, the NHE1-driven intercellular alkalinization was the first event and this alkalinization was necessary for the development of all subsequent tumor hallmarks, such as increased growth, anchorage-independent growth and the Warburg increase in glycolysis and lactate production.[20]

Finally, the best example of the pH-cancer relationship is that almost all the tumors show the inversion of the pH gradient (intracellular alkalosis and extracellular acidosis; see Second phase below for explanation). We say almost, because there might be some exceptions, but to the best of our knowledge all the tumors invert the pH gradient at a certain point of their evolution.

The evolving concept of pH in cancer

First phase

In the 1920s Otto Heinrich Warburg showed for the first time that cancer cells used aerobic glycolysis as their main energy-generating mechanism, instead of the oxidative phosphorylation used by most normal adult cells.[21,22] This was an important breakthrough into the enigmatic world of tumor metabolism. While oxidative phosphorylation produces a considerable amount of energy, aerobic glycolysis is much less efficient, generating only 5% of the energy obtainable through the first process. Both pathways breakdown glucose as an energy source. In the oxidative pathway this breakdown is complete and releases 18 times more energy than aerobic glycolysis where the breakdown is incomplete and ends at the level of lactic acid. This also means that cancer cells that preferentially use the glycolytic pathway need much more glucose in order to fulfill their energy needs. Warburg also found that cancer cells employed a much larger amount of glucose than non cancerous cells. The increased utilization of glucose through the aerobic glycolytic pathway, correspondingly engendered a considerable increment of lactic acid production. Warburg found that cancers were acidic. This actually meant that the interstitial tissues of cancer were acidic, because in those days there were no adequate instruments to measure intracellular acidity. The assumption was, that if the interstitial tissue or ECM (extracellular matrix) of cancers was acidic, then, cancer cells were also acidic.

In this first phase of the cancer pH concept, pH alterations were considered a mere consequence of the excessive lactic acid generation.

For more than 50 years the scientific community believed that cancer cell's cytosol was more acidic than normal cells. This idea even had some support based on the very elementary instruments that were available for measuring intracellular pH and the highly artificial experimental conditions.

Second phase

By the end of the 1970s more accurate systems for pH measurement had been developed. A particularly important tool for non-invasive intracellular pH measurement, was the Nuclear Magnetic Resonance Spectrometer (NMRS) which makes it possible to study living tumors in situ, in experimental animals and in patients. This new instrument opened a window onto a totally different picture: the inside of the cancer cell (cytoplasm) was alkaline while the exterior ECM was highly acidic.[23,24] This came as a total surprise: the cytoplasm had a slightly alkaline pHi, no matter how low the ECM pH was. The difference between the intracellular pH (pHi) and the extracellular pH (pHe) or pH gradient was inverted in relation to normal cells. Part of this long held mistake about intracellular pH was also due to the fact that the earlier microelectrode measurements of pHi always show more acidity than NMRS studies.[25]

At the same time, scientists became aware that keeping different pH levels inside and outside required specialized cell mechanisms. Buffer systems alone were not able to maintain an alkaline intracellular pH when a high production of lactate was present, which is what happens in cancers. In the 50 years between Warburg's discovery and the finding of an alkaline pHi in cancers, scientists felt at ease with the idea that the increased lactate production due to enhanced glucose uptake in a glycolytic cell acidified both the inside and the outside milieu. By the beginning of the 80s, the pH balance in cancer cells had to be fully reviewed and new ideas were urgently needed.

In 1952 before the discovery of the alkalinity of pHi, Hodgkin and Huxley described the voltage gated sodium channel, a membrane protein that regulates the entrance of Na^+ into the cell. It became evident that in addition to the buffer systems, there were ion transport regulators that could also modify pH.

By the end of the 1950s there was a clear distinction between channels and carriers. While a channel is located at the cell membrane and is open simultaneously on both sides (intra and extracellular), a carrier is not simultaneously open at both ends. Furthermore, a channel may allow the traffic of millions of ions, while a carrier has to bind to the transported substance and thus its performance is much slower and limited to specific substances.

Although voltage gated sodium channels respond very quickly to stimulation, permitting a speedy movement of Na^+, the proteins that transport lactate (monocarboxylate transporters) are proportionately much slower.

By the end of this second phase the predominant knowledge was that pHe in cancer was acid and pHi alkaline, the gradient maintained with channels and carriers. pH changes were a consequence of metabolic changes triggered by cancer.

Third phase

Shortly after the discovery of the existence of a pH gradient with an intracellular alkalinity, it was also found that the manipulation of cancer's pH could be a useful therapeutic tool.[26] All the efforts were targeted against the extracellular acidity, disregarding the intracellular alkalinity.[27] pH anomalies were still considered a consequence of cancer without direct participation in the initial oncogenic process.

Fourth phase

We arrive to the present day approach in which extracellular and intracellular pH do not have the same level of importance, with the intracellular pH (pHi) alterations being considered the primary cause and the extracellular pH (pHe) a consequence of pHi. Furthermore, intracellular alkalinity is a very early manifestation of malignant transformation and a basic component and cause of the malignant phenotype.[20] Therefore, therapeutic interventions are aimed to modify the pHi, by acidifying it.[28]

As we shall see throughout this book, the extracellular acidity is a consequence of the intracellular alkalinity. Therefore, if we want to treat cancer successfully, the primary target should be the intracellular alkalinity. In any case, we have to address both, but considering that our primary target should be the intracellular alkalinity.[29]

Another historical change is that Warburg considered the glycolytic metabolism as the cause of cancer. He thought that cancer was the product of a "mitochondrial disease" that induced glycolytic behavior. We shall also show that glycolytic metabolism is partly a consequence of intracellular alkalinization rather than the cause.

Clinical implications

Cancer-related pH changes have clinical implications, because besides creating an adequate environment for growth, proliferation, invasion and metastasis, these changes hinder chemotherapeutic treatments and decrease natural immunologic defenses against the malignancy.[30] At the same time, they offer a target for new treatment modalities. Some of them have already been introduced in practice, like pH sensitive nanoparticle carriers that release a chemotherapeutic drug when they meet an acidic extracellular milieu. This technique allows a major concentration of drug to be delivered at the tumor site without an equivalent release into normal tissue.

The pH changes are absolutely essential for cancer and it cannot develop without them. Therefore, if we can reverse them, tumors are unable to survive.

pH changes are unique to cancer and normal tissues do not express any of those pH changes. This means that targeting pH should not have deleterious effects on nonmalignant tissues.

However, the treatment of the pH abnormalities in cancer is a fully neglected issue among practicing oncologists.

The objectives of this book

In October 2000 in London a Symposium[31] took place in which Robert Gillies' opening speech asked:

> *We know that pH is altered in tumours, but does this have anything to do with tumour biology?*

Now we have a clear answer to his question, and it is: "Absolutely, yes".

The aim of the various chapters in this book is to show all the known consequences of the altered pH dynamics taking place in cancer, and the mechanisms that lead to them.

And we also hope to change the neglected situation of pH in oncological practice and make the medical community aware that targeting the altered pH dynamics as part of conventional therapies will finally get better results.

We shall try to show not only the importance of pH in cancer but also the mechanisms of how pH works from the very beginning of malignization, its progression and evolution and eventually in the apoptosis, when this is achieved. And most important of all, we shall also describe the multiple ways in which the therapeutic manipulation of pH may stop or at least slow down the malignant process and decrease the metastatic risk.

References

1. Barry MA, Reynolds JE, Eastman A. Etoposide-induced apoptosis in human HL-60 cells is associated with intracellular acidification. *Cancer Res* 1993;**53**(10):2349–57.
2. Zanke BW, Lee C, Arab S, Tannock IF. Death of tumor cells after intracellular acidification is dependent on stress-activated protein kinases (SAPK/JNK) pathway activation and cannot be inhibited by Bcl-2 expression or interleukin 1β-converting enzyme inhibition. *Cancer Res* 1998;**58**(13):2801–8.
3. Li J, Eastman A. Apoptosis in an interleukin-2-dependent cytotoxic T lymphocyte cell line is associated with intracellular acidification role of the Na/H-antiport. *J Biol Chem* 1995;**270**(7):3203–11.
4. Ban J-J, Lee M, Im W, Kim M. Low pH increases the yield of exosome isolation. *Biochem Biophys Res Commun* 2015;**461**(1):76–9.
5. Madshus IH. Regulation of intracellular pH in eukaryotic cells. *Biochem J* 1988;**250**(1):1.
6. Ui M. A role of phosphofructokinase in pH-dependent regulation of glycolysis. *Biochim Biophys Acta Gen Subj* 1966;**124**(2):310–22.
7. Gerson DF. In: Nuccitelli R, Deamer DW, editors. *Intracellular pH: its measurement, regulation, and utilization in cellular functions*. New York: Alan R. Liss; 1982. p. 375–83.
8. Fidelman ML, Seeholzer SH, Walsh KB, Moore RD. Intracellular pH mediates action of insulin on glycolysis in frog skeletal muscle. *Am J Physiol Cell Physiol* 1982;**242**(1):C87–93.
9. Gerson DF, Kiefer H, Eufe W. Intracellular pH of mitogen-stimulated lymphocytes. *Science* 1982; **216**(4549):1009–10.
10. Erecińska M, Deas J, Silver IA. The effect of pH on glycolysis and phosphofructokinase activity in cultured cells and synaptosomes. *J Neurochem* 1995;**65**(6):2765–72.
11. Sutton JR, Jones NL, Toews CJ. Effect of pH on muscle glycolysis during exercise. *Clin Sci* 1981; **61**(3):331–8.
12. Sutton DD, Arnow PM, Lampen JO. Effect of high concentrations of nystatin upon glycolysis and cellular permeability in yeast. *Proc Soc Exp Biol Med* 1961;**108**(1):170–5.
13. Kaminskas ED. The pH-dependence of sugar-transport and glycolysis in cultured Ehrlich ascites-tumour cells. *Biochem J* 1978;**174**(2):453–9.
14. Iwami Y, Hata S, Schachtele CF, Yamada T. Simultaneous monitoring of intracellular pH and proton excretion during glycolysis by *Streptococcus mutans* and *Streptococcus sanguis*: effect of low pH and fluoride. *Oral Microbiol Immunol* 1995;**10**(6):355–9.
15. Harguindey S, Henderson ES, Naeher C. Effects of systemic acidification of mice with Sarcoma 180. *Cancer Res* 1979;**39**(11):4364–71.
16. Hagag N, Lacal JC, Graber M, Aaronson S, Viola MV. Microinjection of ras p21 induces a rapid rise in intracellular pH. *Mol Cell Biol* 1987;**7**(5):1984–8.
17. Wakelam MJ, Houslay MD, Davies SA, Marshall CJ, Hall A. The role of N-ras p21 in the coupling of growth factor receptors to inositol phospholipid turnover, In: Biochemical Society Transactions. 619th Meeting. Cambridge; 1987. p. 45–7.

18. Berridge MJ, Irvine RF. Inositol phosphates and cell signaling. *Nature* 1989;**341**:197.
19. Doppler W, Jaggi R, Groner B. Induction of v-mos and activated Ha-ras oncogene expression in quiescent NIH 3T3 cells causes intracellular alkalinisation and cell-cycle progression. *Gene* 1987;**54**(1):147–53.
20. Reshkin SJ, Bellizzi A, Caldeira S, Albarani V, Malanchi I, Poignee M, et al. Na+/H+ exchanger-dependent intracellular alkalinization is an early event in malignant transformation and plays an essential role in the development of subsequent transformation-associated phenotypes. *FASEB J* 2000;**14**(14):2185–97.
21. Warburg OH, editor. *The metabolism of tumours: investigations from the Kaiser Wilhelm Institute for Biology*. Berlin-Dahlem: Constable & Company; 1930.
22. Warburg O. Über den stoffwechsel der carcinomzelle. *Naturwissenschaften* 1924;**12**(50):1131–7.
23. Griffiths JR, Stevens AN, Iles RA, Gordon RA, Shaw D. 31P NMR investigation of solid tumours in the living rat. *Biosci Rep* 1981;**1**:319.
24. Griffiths JR, Cady E, Edwards RHT, McCready VR, Wilkie DR, Wiltshaw E. 31P NMR studies of a human tumour in situ. *Lancet* 1983;**1**:1435.
25. Griffiths JR. Are cancer cells acidic? *Br J Cancer* 1991;**64**(3):425–7.
26. Harguindey S, Orive G, Pedraz JL, Paradiso A, Reshkin SJ. The role of pH dynamics and the Na+/H+ antiporter in the etiopathogenesis and treatment of cancer. Two faces of the same coin—one single nature. *Biochim Biophys Acta* 2005;**1756**(1):1–24.
27. Koltai T. Cancer: fundamentals behind pH targeting and the double-edged approach. *Onco Targets Ther* 2016;**9**:6343.
28. Harguindey S, Stanciu D, Devesa J, Alfarouk K, Cardone RA, Polo Orozco JD, et al. Cellular acidification as a new approach to cancer treatment and to the understanding and therapeutics of neurodegenerative diseases. *Semin Cancer Biol* 2017;**43**:157–79.
29. Keitaro H, Lee C, Tannock IF. Antitumor activity of nigericin and 5-(N-ethyl-N-isopropyl) amiloride: an approach to therapy based on cellular acidification and the inhibition of regulation of intracellular pH. *Oncol Res* 1994;**6**(6):259–68.
30. Kouidhi S, Ben Ayed F, Benammar Elgaaied A. Targeting tumor metabolism: a new challenge to improve immunotherapy. *Front Immunol* 2018;**9**:353.
31. Novartis Foundation. In: Goode SJ, Chadwick DJ, editors. *Symposium on the tumour microenvironment: causes and consequences of hypoxia and acidity*. John Wiley and Sons, Ltd; 2001.

CHAPTER 2

Cancer metabolism

Introduction

Metabolism is the mechanism that cells use to breakdown nutrients like glucose, fatty acids, and amino acids in order to produce energy and building blocks needed for life. Cells accumulate this energy in ATP molecules (high energy molecules). Two different mechanisms for glucose (the main source of energy) breakdown are employed: aerobic and anaerobic metabolism. At the beginning of life on earth, there was no fully developed atmosphere with adequate oxygen supply. Therefore, the first living organisms could only use the anaerobic pathway that did not require oxygen. When photosynthetic organisms appeared, this low oxygen atmosphere slowly changed. In this way, as the earth's atmosphere acquired a stable oxygen level, new organisms containing mitochondria developed through natural selection. These probably appeared through a process of endobiotic association (mitochondria were originally independent organisms that established a symbiotic association within host cells).[1–3] Organisms with mitochondria could generate energy in a more efficient way through aerobic metabolism (oxidative phosphorylation metabolism or OXPHOS). The initial molecule of the metabolic pathway in both cases is glucose, which is degraded in 10 enzymatic steps to pyruvate (the process is known as glycolysis). Each step needs a specific enzyme. Nine of these 10 enzymes are controlled by HIF-1α (Hypoxia Inducible Factor 1α).[4, 5] From pyruvate on, the two metabolisms follow different routes. If there is adequate oxygen supply the pathway is the aerobic (oxidative) pathway. If the supply is inadequate, pyruvate is converted into lactic acid (anaerobic pathway).

Since the pioneering work of Otto Warburg in the 1920s,[6–8] we know that the metabolism of tumors is quite different from that of normal cells: malignant cells preferentially break down glucose to lactic acid (glycolytic metabolism) instead of metabolizing it to CO_2 and H_2O (oxidative phosphorylation) as normal cells usually do. The consequence of this behavior is that significantly less energy (2 molecules of ATP) is obtained in comparison to normal cells (36 ATPs per each molecule of glucose). This low production of ATP in malignant cells is compensated by an heterogeneously increased (2 to 20 fold) uptake of glucose[9] in order to produce the same amount of energy as a normal cell. It is quite frequent that they uptake glucose beyond their metabolic needs.[10] Warburg, studying the highly proliferative malignant ascites cells, thought that this phenomenon was due to decreased mitochondrial efficiency and that this was the cause of cancer. Now, we know that it is not so.[11] However glycolytic cancer cells do not fully lose their ability to use the oxidative pathway.[12]

An Innovative Approach to Understanding and Treating Cancer: Targeting pH
https://doi.org/10.1016/B978-0-12-819059-3.00002-2
© 2020 Elsevier Inc.
All rights reserved.

The Warburg effect

Glycolytic metabolism also takes place in normal tissues when the oxygen supply is low, but when the oxygen levels return to normal, non malignant cells return to oxidative phosphorylation (the mechanism is known as anaerobic glycolysis). This does not happen in malignant cells, since they continue with their glycolytic metabolism even under normal oxygen supply. This "abnormality" was called the Warburg effect or aerobic glycolysis or glycolytic metabolism of cancer cells.[13] The Warburg effect implies a high production of lactic acid (60 times more than in normal cells) because much more glucose is used. When the lactic acid is extruded from the cells, it creates a highly acidic micro-environment. Now, we also know that lactic acid is not the main cause of extracellular acidosis[14, 15] (Fig. 1).

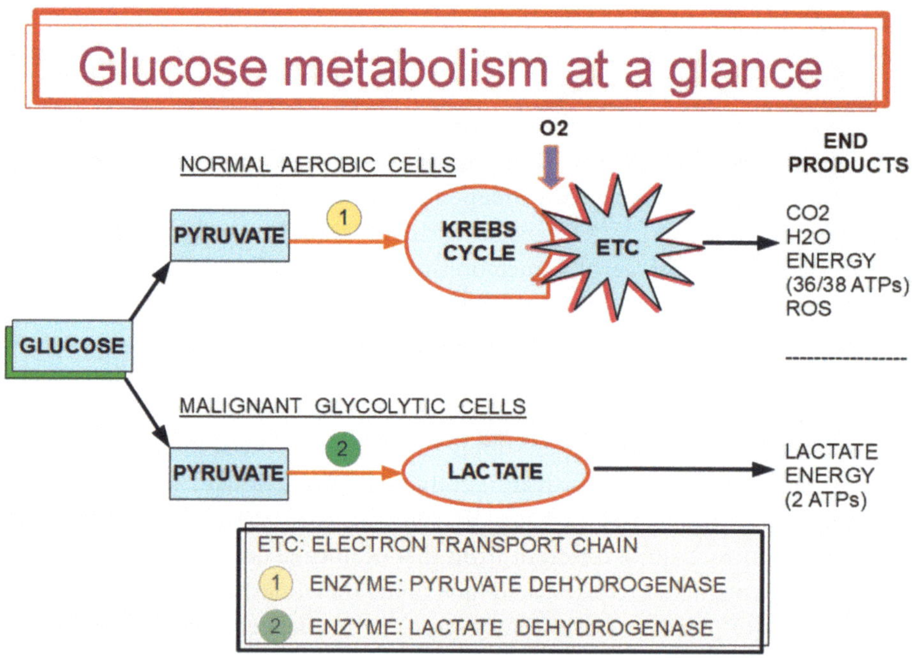

Fig. 1 Shows the difference of glucose metabolism in normal aerobic cells and malignant glycolytic cells. While aerobic metabolism needs oxygen and produces ROS (reactive oxygen species), anaerobic metabolism does not require oxygen and does not produce ROS.

For the next 50 years this was the mainstream knowledge we had on tumor metabolism.

During the last 30 years, it was found that the metabolic differences between normal and malignant cells were not only in carbohydrate metabolism, but went far beyond those described by Warburg: lipids, amino acids and proteins, enzymes, nucleic acids, water, ROS (reactive oxygen species), ions, and, most of it all, acid-base balance behaved differently. The glycolytic phenotype was only the tip of the metabolic iceberg. The changes showed by malignant cells were called the "metabolic switch".

While Warburg described the acidity of tumors, he actually could only measure the extracellular acidity but not intracellular. Mistakenly, he considered that both intra and extracellular pH were acidic in cancer. It took 50 years and new technology to discover that the intracellular pH was alkaline in malignant cells.

Even though we have known that glycolysis depends on a high pH,[16–18] only recently has an integral idea been advanced: that the Warburg effect can be simply explained by an increase in the intracellular pH.[19, 20] Finally, it has also been shown that malignant intracellular alkalinization drives the initial activation of aerobic glycolysis.[21]

Cancer development is an extraordinary metabolic challenge where the biomass needs to double with each mitotic cycle. This requires profound energetic changes and nutritional resources. Therefore, until quite recently, the scientific community wondered why a tumor that needed an important amount of energy for an accelerated growth and proliferation was using an inefficient metabolism like aerobic glycolysis.

From all the previous concepts and experience the following question rises. What would happen if a tumor did not develop a different metabolism of its own?

Since the energy requirements are high because the tumor needs to duplicate its biomass in a short time, it would increase glucose uptake and the mitochondrial oxidative pathway's activity would be correspondingly increased. Energy production would be augmented. And now two vital problems threatening the tumor's life would come into play:

(1) The oxygen requirements for oxidative metabolism would be increased. But the tumor is embedded in a hypoxic environment due to excessive proliferation without a correspondingly efficient vascular supply. This means that there is no enough oxygen available for oxidative metabolism.[22]

(2) A highly increased oxidative metabolism would produce an important amount of ROS (reactive oxygen species), which are highly toxic for the cell.[23, 24]

Thus, the tumor cells need to develop a different metabolism if they are to grow and proliferate in an accelerated manner: a metabolism that needs little or no oxygen at all, and also produces low amounts of ROS. The answer to these needs is aerobic glycolysis, that requires no oxygen and that, by avoiding or reducing mitochondrial activity, reduces ROS production.

And what about the low energy production of aerobic glycolysis? An important increase of glucose uptake compensates the low energetic efficiency of glycolysis. Increased glucose uptake, predominance of glycolytic metabolism, increased fatty acid synthesis, increased role of glutamine, and pH gradient inversion are the main characteristics of these changes. The metabolic abnormalities (metabolic switch) are among the hallmarks of cancer[25] and are essential for the tumors survival. This makes malignant cells vulnerable to drugs that target their peculiar metabolism. Even if the metabolic changes are a highly integrated process, for the sake of better understanding, we will analyze them separately.

The glycolytic phenotype

Many mechanisms are involved in the metabolic switch. The three most important are:

(A) Oncogenic mutations.

(B) Hypoxia.

(C) Intracellular alkalosis with increased activity of proton transporters.

Alkalosis stimulates glycolysis and hinders the Krebs cycle.[26]

Oncogenic mutations. One or more oncogenic mutations are the starting point of the malignant process. Proto-oncogenes, like PI3K, AKT, RAS, c-Myc, promote glycolysis while tumor suppressors tend to inhibit glycolysis.[27] The frequent up-regulation of the PI3K/Akt pathway in many cancers promotes glycolysis by diverse mechanisms:

(a) The activation of Akt phosphorylation increases the expression of GLUTs (glucose membrane transporters) and of glycolytic enzymes like phospho-fructokinase and hexokinase II.

(b) Akt increases hypoxia inducible factor-1 (HIF-1α) activation thus inducing the over-expression of glycolytic enzymes[28, 29] and it represses pro-oxidative enzymes like PDH (pyruvate dehydrogenase).[30]

Hypoxia stabilizes HIF-1α enabling its dimerization with HIF-1β in the nucleus and acting as a transcription factor (Fig. 2).

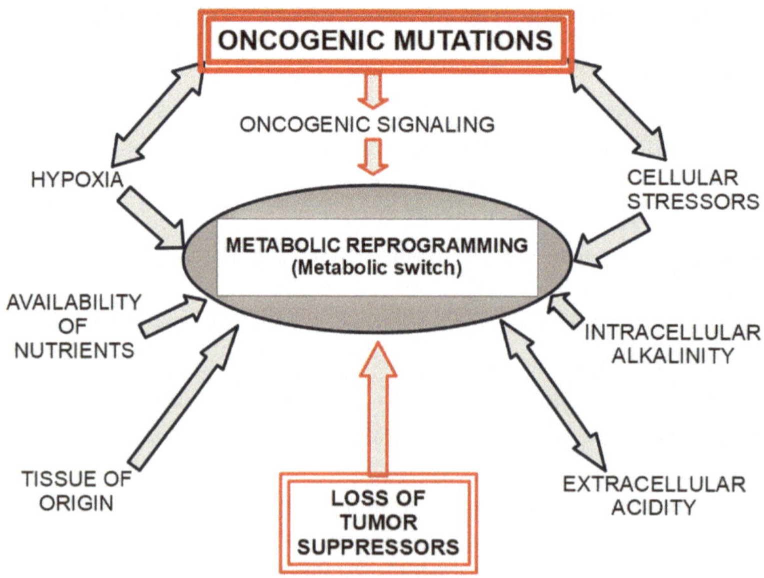

Fig. 2 A general view of multiple factors that lead to metabolic reprogramming. The most important one is oncogenic signaling.

Almost all the enzymes of the glycolytic pathway are regulated by proto-oncogenes, tumor suppressor genes and are strongly affected by hypoxia (Fig. 3).

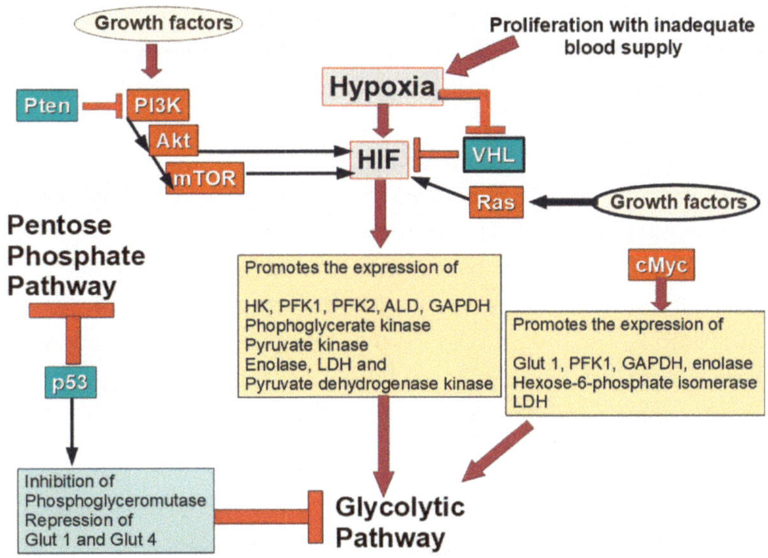

Fig. 3 Relation among proto-oncogenes, tumor suppressor genes, metabolic enzymes, hypoxia, glycolysis and the pentose phosphate pathway.[31–39] Proto-oncogenes are in *red*, tumor suppressor genes in *light blue*. HK, hexokinase; PFK1, phosphor-fructokinase 1; PFK2, phospho-fructokinase 2; ALD, aldolase; GAPDH, glyceraldehydes-3-phospho-dehydrogenase; LDH, lactate dehydrogenase; Glut, glucose transporter.

The same relationship is shown in Fig. 4 referring to the metabolic switch.

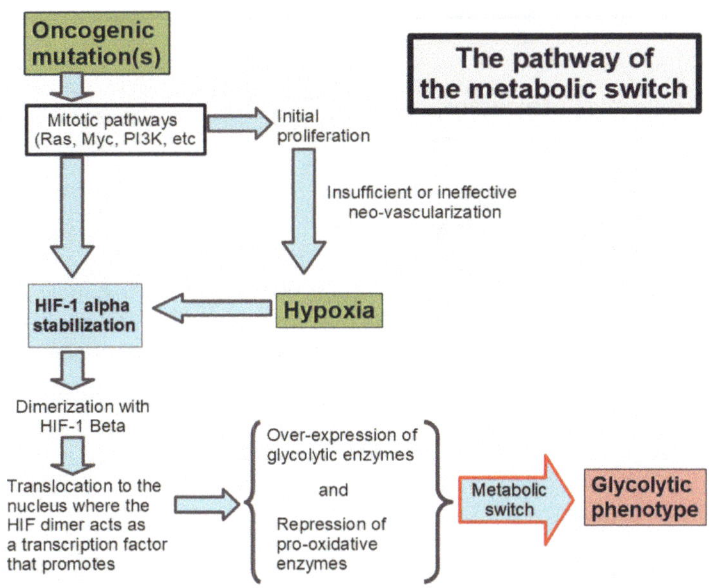

Fig. 4 Cancer's glycolytic phenotype is the consequence of oncogenic mutations that initiate proliferation and hypoxia. The latter develops due to excessive growth that is not compensated by an increased and effective vascular supply (neovascularization).

Over-expression of cMyc in cancer cells enhances the activity of pyruvate kinase isoform M2 (PKM2) whereas in normal tissues the PKM1 isoform predominates.[40] This modification delays the glycolytic process allowing an increased deviation towards the PPP (pentose phosphate pathway) that is necessary for increased synthesis of the nucleotide sugars ribose and deoxyribose and ROS scavengers.

Another situation leading to the glycolytic phenotype is the loss of tumor suppressor genes. For example, mTOR (mammalian target of rapamycin) is negatively regulated by AMPK (AMP activated kinase) which is positively activated by Liver kinase B1 (LKB1) and the increase of the AMP/ATP ratio. AMPK promotes oxidative phosphorylation. Loss of LKB1 tumor suppressor promotes tumorigenesis, decreases AMPK signaling, enhances mTOR signaling and promotes a glycolytic phenotype[41] (Fig. 5). On the other hand, p53 down-regulates the glycolytic phenotype[42, 43] (Fig. 3).

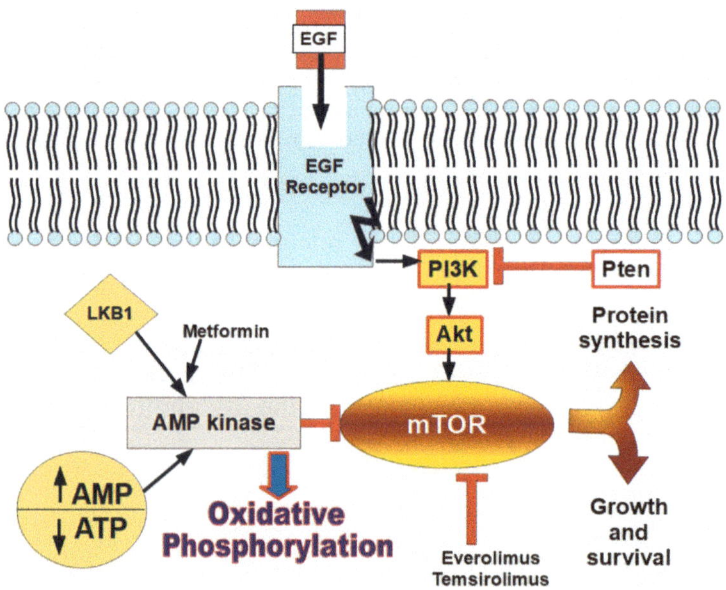

Fig. 5 The loss of LKB1 induces a glycolytic phenotype. AMPKinase is a metabolic checkpoint that is activated by LKB1and/or an increase in the AMP/ATP ratio. Activation of AMPKinase induces oxidative phosphorylation[44] and blocks mTOR. Metformin activates AMPK. Everolimus and Temsirolimus block mTOR.

In general, the metabolic changes are essential partners for cancer's progression. These metabolic instruments may vary according to type of cancer, driver gene, environment, degree of undifferentiation and tumor' evolution.

Four issues should be kept in mind when dealing with the study of cancer metabolism

1. The predominant oncogenic pathway of a tumor and the tissue type determine the metabolic phenotype. Yuneva et al.[45] found that when the driver gene was cMyc the metabolic behavior was different from that found with cMet as a driver. They also described a different metabolic behavior with the same driver but in a different cancer cell line.
2. Tumors are metabolically heterogeneous.[46–48] Sonveaux et al.[49] reported that in most tumors they found coexisting cells with different metabolic profiles, (glycolytic hypoxic areas coexisted with oxygenated oxidative areas). Hypoxic areas produced lactate that served as fuel for normoxic cells implying a symbiotic metabolism where "waste" (lactate) produced by hypoxic cells fuelled normoxic malignant cells. MCT-1 (monocarboxylate transporter-1) played an essential role in lactate shuttle. The inhibition of MCT-1 produced a switch in lactate-fuelled cells to glycolysis
3. The metabolic switch usually gives a competitive advantage because:
   - **(a)** Glycolytic behavior provides the basic bricks for building other necessary compounds like nucleic acids, proteins and fatty acids.
   - **(b)** Energy loss of glycolytic behavior is compensated by a strong increase in glucose consumption.
4. Metabolic change is a progressive process: not all the cells in a cancer tissue adopt the same metabolic changes at the same time, but as tumor progresses the metabolic changes become more evident and increase their modification relative to normal.

Therefore, oncogenic pathways and hypoxia lay the groundwork for the metabolic switch (Fig. 6).

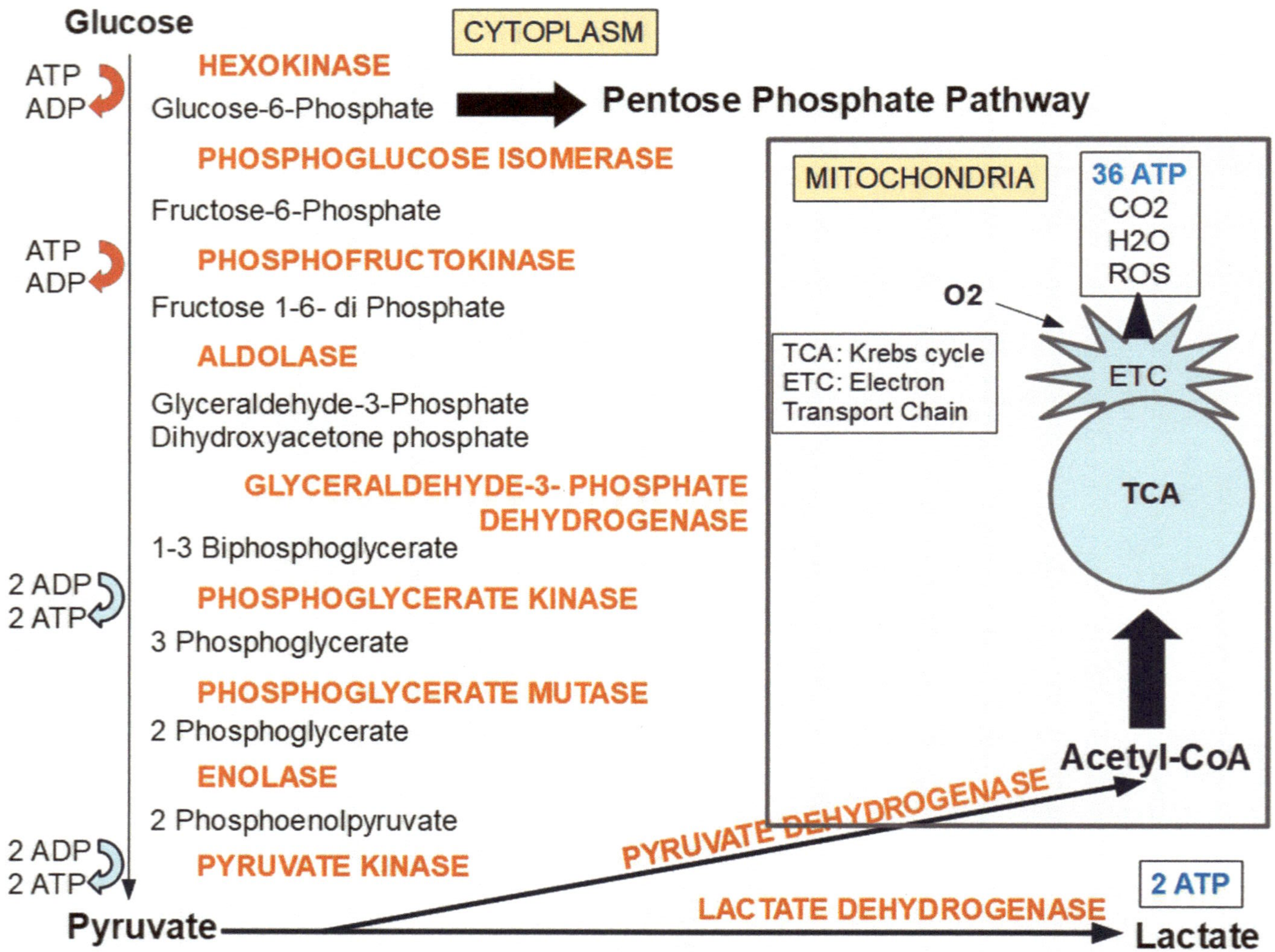

Fig. 6 The glycolytic pathway in cancer cells ends in lactic acid that is extruded from the cell through the action of monocarboxylate transporters (MCTs). On the other hand, normal cells transform pyruvate into acetyl-CoA that enters the tricarboxylic cycle (TCA) and the electron transport system generating energy, ROS, and the end products: CO_2 and H_2O. The "metabolic decision point" is located at the level of pyruvate and depends on PDH (pyruvate dehydrogenase). This enzyme is usually inhibited in cancer cells. Therefore, pyruvate does not enter into the mitochondria, but is transformed into lactic acid.

The metabolic portfolio that may be partially or completely present in cancer consists of:

1. Hypoxia
2. High energy requirements.
3. Increased nutrients consumption.
4. Aerobic glycolysis (Warburg effect) (glycolytic phenotype).
5. Increased fatty acids synthesis (lipogenic phenotype).
6. Increased glutamine consumption (glutaminolytic phenotype).
7. Symbiotic and/or parasitic behavior.

The direct consequence of the metabolic switch is the increased acidity of the extracellular matrix (ECM). For many years, ECM's acidity was considered the natural effect of the important extrusion of lactic acid from the cell. This classical view seems to be partially wrong, because ECM in tumors is acidic even when lactic acid is experimentally eliminated. Increased CO_2 production is probably one of the main culprits of the abnormal ECM's pH through the production of CO_3H_2 (carbonic acid) mediated by the presence of carbonic anhydrases on the cell surface. Of course, lactic acid is still a contributor to acidity.

The ECM's acidity is a very important factor in tumor migration, invasion and eventual metastases,[50, 51] because it activates proteolytic enzymes that degrade surrounding tissues and allow migration. It also acts as a barrier against immuno-defensive systems and impedes the activity of certain chemotherapeutic drugs like doxorubicin and weak bases in general.[52–55]

The utilization of the aerobic glycolytic pathway has been frequently found during embryonic development.[56] After birth, probably due to adequate availability of oxygen, a metabolic change is triggered in the hepatocytes that adopt oxidative phosphorylation. The glycolytic pathway found in cancer cells maintains marked differences with the glycolytic pathway sometimes found in normal cells. Certain glycolytic enzyme isoforms are over expressed only in cancer cells. This is the case of pyruvatekinase isoform M2 (PKM2) (Fig. 5). PKM2 is less active than PKM1, and this slower activity down-regulates phosphoenolpyruvate catabolism to pyruvate. This results in accumulation of intermediate products of glycolysis upstream of phosphoenolpyruvate making it possible to use them for the production of other necessary molecules for proliferation, like amino acids, nucleic acids and lipids.[57–59]

What is the goal of tumor cells when they adopt the energetically inefficient glycolytic phenotype?

Possible answers:

(a) To down-regulate PK's activity as explained above leading to an increased use of the PPP, producing the accumulation of intermediaries needed to synthesize other molecules (ribose for nucleotides, lipids for membranes and other uses, amino acids, ROS scavengers, *etc.*). PKM2 is pro-glycolytic while PKM1 is pro-oxidative.[28]

(b) To thrive in an extremely hypoxic environment.
(c) To increase resistance to permanent or transitory hypoxia;
(d) To activate enzymes that take part in the ECM remodeling, migration, and invasion, by creating an acidic extracellular matrix.[60–73]
(e) To activate translocation of Glut1 and Glut4 to the plasmatic membrane accomplished by increased lactic acid.[74]
(f) To decrease ROS generation through the downregulation of OXPHOS.[75, 76]
(g) To "protect" cancer cells when they are in areas of high oxygen level like the bloodstream.[77]
(h) To increase the uptake of glucose and other nutrients compensating the lower energetic efficiency of the glycolytic phenotype.

In summary, reduced OXPHOX capacity is favorable for rapid growth and increased invasiveness with low ROS production.

Targeting glycolytic enzymes in cancer

Normal cells use the glycolytic pathway in very limited circumstances (during embryonic development,[78] muscle efforts, ischemia and T cell activation and proliferation.[79, 80] But under normoxia, normal tissues do not employ the glycolytic pathway. Therefore, at least theoretically, the glycolytic pathway may be targeted with a very low cost for normal cells.

The Warburg effect is addictive for cancer cells and they depend on it for generating ATP and biosynthetic building blocks. Inhibition of the glycolytic pathway results in slowing down proliferation and invasion and eventually in cell death.[81, 82]

The multiple enzymes involved in the glycolytic pathway highlight many different targets to be considered and the possibility of combining drugs and attacking more than one target simultaneously.

GLUTS are a family of membrane proteins that transport glucose from the extracellular to the intracellular matrix. Fourteen isoforms have been identified and many isoforms are over-expressed in cancers with bad prognoses. Glut1 seems to be the most important in the group and a potential target for anti-cancer therapy. Inhibition of Gluts, particularly Glut1 and Glut2 with phloretin[83, 84] has shown apoptosis in breast, colon and hepatic cancer cell lines. GLUTS are also down-regulated by other natural compounds like resveratrol and quercetin,[85] antidiabetics like thiazolidenediones, caffeine, theophiline and pentoxifilline.[86]

Hexokinase: Cancer cells preferentially express Hexokinase II (HK-II) mainly due to gene amplification,[87] which catalyzes irreversible glucose phosphorylation to glucose-6-phosphate. HK-II is bound to the mitochondrial outer membrane, decreasing apoptosis through the inhibition of cytochrome *c* release. HK II over-expression increases the glycolytic rate.[88–91] HIF-1 and cMyc induce HK-II over-expression.[92] Krushna et al., showed that HK-II deletion targets cancer cells.[93]

HK-II has been targeted with an alkylating agent like 3-bromopyruvate (3-BP) with interesting results.[94–96] 3-BP is undoubtedly a glycolitic inhibitor, but the exact mechanism of action is difficult to identify.[97] Glyceraldehyde-3-phosphate dehydrogenase seems to be the primary target of 3-BP. Lonidamine, an anti-spermatogenic with low human toxicity is an inhibitor of HK-II that has been tested in cancer patients; one third of the patients obtained benefits with lonidamine used alone. However, this drug has not entered into practice.[98, 99] Lonidamine is potentiated by diazepam in human glioblastoma cells.[100]

2 deoxy-glucose (2DG) is an inhibitor of hexokinase through a competitive mechanism with glucose. Thus, glucose deprivation reduces ATP production and interfere with the glycolytic pathway.[101] The good results as an anticancer molecule found in the laboratory[102–105] could not be reproduced in the limited clinical trials performed with this drug. When 2DG was associated with certain other drugs, like metformin or autophagy inhibitors, cytotoxicity became more relevant.[106]

PFK (Phosphofructokinase): This enzyme irreversibly commits the glucose molecule to the glycolytic pathway. Increased PFK-1 activity in cancer is a frequent finding in glycolytic cells and is the consequence of direct oncogenes signaling or through HIF-1 activity.[107, 108] PFK phosphorylation is a highly controlled step that depends on fructose 2–6 biphosphate. It is known as the key enzyme of glycolysis. Its activity shows a strong dependence on an elevated intracellular pH.[109]

PGM (Phosphoglucomutase) isoform 1: Is up-regulated in cancer, due to loss of p53. An experimental drug has been developed under the name PGMI-004A that inhibits this enzyme limiting cancer growth by decreasing glycolysis, the pentose phosphate pathway and biosynthesis in general.[110]

PGK1 (Phosphoglycerate kinase 1): Increased expression of PGK1 is a signal of bad prognosis.[111, 112] The knockdown of PGK1 reduced invasiveness and EMT phenotype.[113] As many other enzymes of the glycolytic chain, PGK1 is under HIF-1α control.

Enolase: Isoform 1 is increased in multiple cancer tissues. In glioblastoma, isoform 2 is up-regulated because enolase 1 gene is usually deleted.[114] Phosphonoacetate-hydraxamate has shown selective toxicity against enolase 1-deleted cells.

Pyruvate kinase (PK): Myc induces over-expression of isoform M2. PKM2 slows down the phosphoenolpyruvate to pyruvate transformation allowing the build up of intermediate products needed for anabolic processes. So it makes sense to activate this rate-limiting enzyme. This theory has been experimentally demonstrated: activation of PKM2 inhibits growth of xenograft experimental models.[115] In the absence of serine, sulfonamide-quinoline activators decrease tumor growth. PKM2 maintains a close relationship with HIF-1α where PKM2 is a HIF-1α target gene, interacting with it, and regulating HIF-1α transcriptional activity through a feedback loop. The interaction between these two proteins plays a fundamental role in the glycolytic phenotype[116] (Fig. 7).

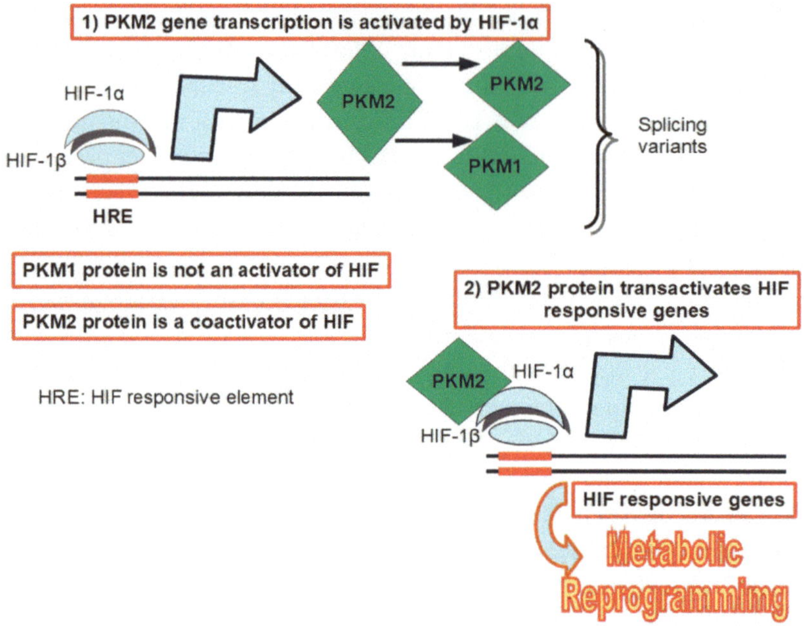

Fig. 7 PKM2 as transactivator of HIF responsive genes stimulates the glycolytic phenotype.

Lactate dehydrogenase (LDH): LDH-A plays a role in cancer cell survival under severe hypoxia. Isoform A favors the pyruvate to lactate pathway and LDH-B, the reverse process. Using RNA interference to knock down LDH-A, cancer cell death was achieved independently of p53 status.[117] Gossypol and sodium oxamate have been tested as LDH inhibitors but their use is limited to the experimental setting.[118] No adequate inhibitor has yet been found.

Pyruvate dehydrogenase (PDH): PDH is a cross-road enzyme complex that is under indirect control of proto-oncogenes. When this enzyme is inhibited, metabolism becomes glycolytic because pyruvate cannot enter into the Krebs cycle (TCA) and is metabolized to lactic acid by lactate dehydrogenase. PDH is inhibited through phosphorylation by PDK (pyruvate dehydrogenase kinase). PDK's activity is strongly increased in glycolytic malignant cells. Thus PDH is inhibited and oxidative metabolism is impeded.

CPI-613 is a novel anti-cancer agent described as a PDH inhibitor, decreasing pyruvate access to mitochondrial metabolism. It is in phase II clinical trials (NCT01832857 and NCT00741403).[119] Finally dichlorocetate is an inhibitor of pyruvate dehydrogenase kinase that activates PDH and is currently used to treat lactic acidosis. It introduces modifications in tumor metabolism by "forcing" pyruvate into the mitochondria, increasing oxidative phosphorylation. Dichloroacetate alone had only modest anti-tumor effects. In the experimental setting with Lewis lung carcinoma, it did not show growth arrest, but there was significant anti-metastatic activity,[120] as well as synergy with radiotherapy in glioblastoma[121] (Fig. 8).

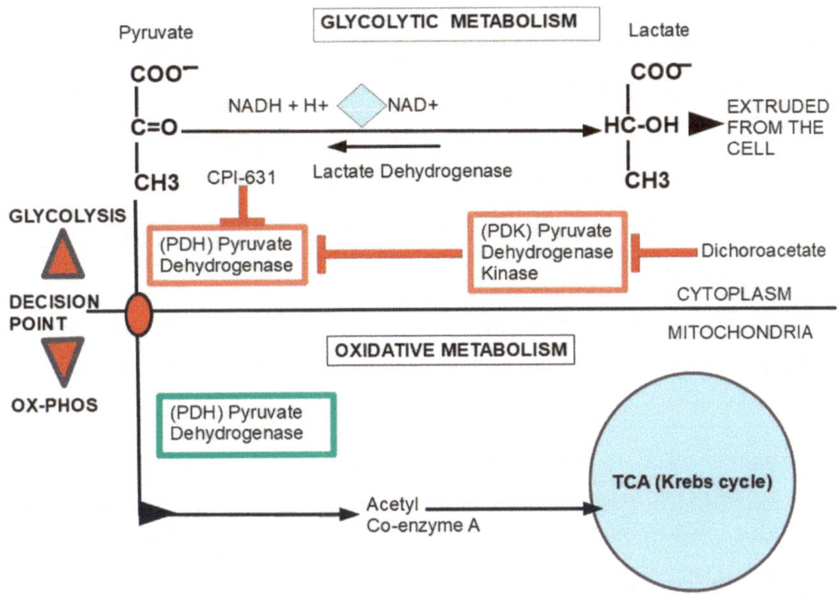

Fig. 8 According to inhibition or lack of inhibition of PDH the cell enters either the glycolytic or the oxidative pathway respectively.

The lactate shuttle[122]

It is well known that advanced tumors are very heterogeneous, not only morphologically but also in their genotype and metabolism. Regarding glucose metabolism, this heterogeneity shows the presence of two different types of malignant cells: glycolytic and oxidative. The important issue is that these oxidative cells, usually located in more oxygenated areas of the tumor, are able to use the lactic acid produced by glycolytic cells as fuel for their own metabolism.[123, 124] Many tumors develop such a shuttle system where the uptake of lactate by the oxidative cells becomes an important source of energy (reversed Warburg effect).[125] We have called these cells lactophagic cells. What we do not know is if all oxidative cells have the ability to become lactophagic. This mechanism can be considered a symbiotic system in which a group of cells produce lactic acid that serves as nourishment for lactophagic cells. Furthermore, it is quite frequent to find stromal cells surrounding the tumor that are "enslaved" by the malignant cells and induced to produce lactic acid to feed the lactophagic cells.[126] This resembles a parasitic system. Monocarboxylate transporters (MTCs) are over-expressed in cancer cells, and also in the symbiotic and parasitic systems.[127–129] The presence of the lactate shuttle is a sign of poor prognosis.[130, 131]

The fact that lactate is not the main cause of extracellular acidity in spite of its enhanced production and cellular extrusion, suggests that lactate clearing is also highly increased. The lactate shuttle issue will be further discussed in Chapter 4.

The pentose phosphate pathway

The glycolytic switch was the first metabolic change in cancer that was discovered in the 1920s. However, the last 30 years have shown that there is an equally important switch of a parallel and strongly connected pathway: the pentose phosphate pathway (PPP). PPP is up-regulated in many cancers.[132–134] As glucose enters the cell through GLUTS's activity, it is immediately phosphorylated by hexokinase, generating glucose-6-phosphate. This molecule is the crossroad where the molecule can follow one of two paths: glycolysis or PPP (Fig. 9).

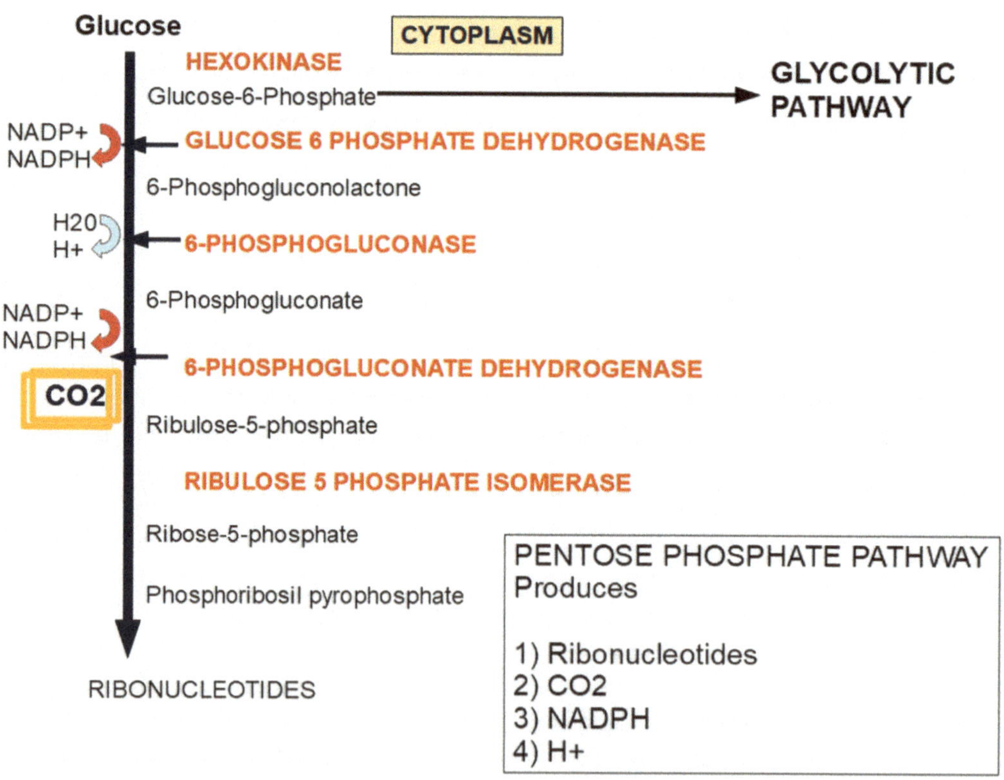

Fig. 9 The pentose phosphate pathway, starting from glucose to ribonucleotides. There are other branches of PPP (pentose phosphate pathway) not shown in the figure.

As shown in Fig. 9 the first step of PPP is at the glucose-6 phosphate. The cancer cell intensively uses PPP to produce two essential molecules for proliferation: pentose sugars for nucleotides and NADPH for ROS scavenging.[135–137] This PPP up-regulation is particularly evident in the G1-S phase of the cell cycle in order to increase the production of pentose for DNA and RNA.[138]

Broadly speaking we may consider the PPP as a detoxification (antioxidant) circuit and a provider of building blocks. At the same time it produces two other molecules that the cell needs to extrude: protons and CO_2. The first is extruded by proton transporters and the second by diffusion. The CO_2 of the PPP explains the reason why a cell that down-regulates oxidative phosphorylation is still a producer of increased amounts of CO_2. And also it shows the participation of PPP in the increased proton production of carbohydrate metabolism.

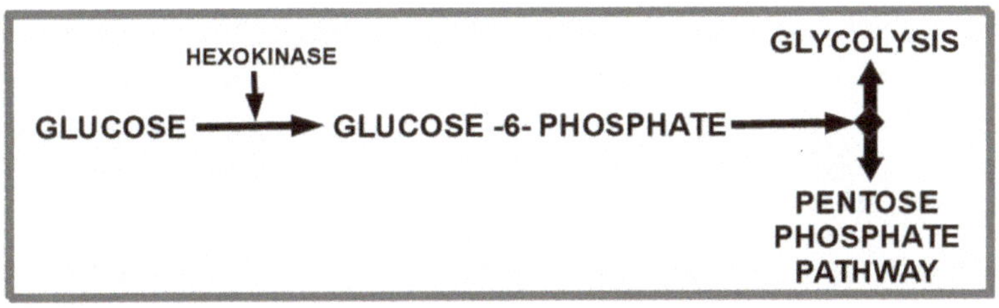

The PPP is regulated by p53 and $NADP^+$/NADPH by binding glucose-6-phosphate dehydrogenase (G-6-PD), the limiting enzyme of the pathway. p53 down-regulates while $NADP^+$ stimulates the pathway.[139] Loss of function of p53 not only increases glycolytic activity, but also augments PPP's activity.[140]

Importantly, robustness of cancer cells under chemotherapy is strongly related to ROS scavengers. Therefore, inhibition of the PPP decreases tumor proliferation and increases the effects of chemotherapeutic drugs through a lower possibility of neutralizing ROS.[141]

G-6-PD is the key enzyme in the PPP and should be considered as an oncogene, because its up-regulation produces malignant transformation in NIH 3T3 cells.[142]

2-deoxyglucose is an inhibitor of hexokinase and gallated cathechins are competitive inhibitors of glucose-6-phosphate dehydrogenase.[143] It has been known for more than 50 years that dehydroepiandrosterone (DHEA) is an inhibitor of G-6-PD.[144, 145]

There are potential inhibitors of the PPP that in theory should enhance the effects of chemotherapy by decreasing ROS detoxification. In the experimental setting this theory has been confirmed.[146]

The lipogenic phenotype

In 1979 Stutowicz et al.[147] found that by acting as an acetyl-CoA donor, ATP-citrate lyase, a lipogenic enzyme involved in fatty acid (FA) synthesis, was markedly increased in breast carcinoma. This increase was even more significant than those of glycolytic enzymes. In 1994 Kuhajda et al.[148] identified OA-519 antigen, a marker of poor prognosis in breast and prostate cancer, as Fatty Acid Synthase (FAS). Abnormal FA synthesis and FAS up-regulation are common in cancer.[149–151]

FA originates mainly from external sources (nutrition), but *de novo* synthesis also becomes very important in many types of cancer.[152, 153]

The regulators of this synthesis are the master transcriptional factors Sterol Regulatory Element-Binding Protein 1 (SREBP-1).[154] (Fig. 10).

Fatty acid synthase

FAS is a key enzymatic complex in FA synthesis from acetyl-CoA and malonyl-CoA, in the presence of NADPH, into long-chain saturated FAs[155] and it is expressed at high levels in liver and adipose tissue, but at low levels in other tissues. FAS is not a single enzyme but a whole enzymatic system composed of two identical 272kDa multifunctional polypeptides, with seven functional domains each, in which substrates are passed from one functional domain to the next.[156–158] Its main function is to catalyze the synthesis of palmitate.[159]

Under a diet with a relatively high fat content (40%), and even after a carbohydrate-rich diet for 3 days, FAS activity remained low in humans. Under normal dietary conditions, *de novo* lipogenesis in man is negligible.[160] The source of long chain saturated FAs is either *de novo* synthesis (mediated by FAS), ingested food, or both.

FAS is the single human enzymatic complex responsible for the conversion of dietary carbohydrate to fat and it is the only eukaryotic enzyme capable of synthesizing palmitate, the precursor of the majority of nonessential FAs. It is down-regulated in most normal cells, except in lipogenic tissues such as liver, lactating breast, fetal lung, and adipose tissue. Conversely, several human cancers, over-express FAS, which has been associated with poor prognosis.[161–163]

FAs are aliphatic acids fundamental for energy production and storage, building blocks for cellular structures as membranes and intermediates in the biosynthesis of hormones and other biologically important molecules.

The metabolism and homeostasis of FAS are transcriptionally regulated by upstream stimulatory factors and sterol regulatory element binding protein-1c (SREBP-1c) in response to feeding and insulin in living animals[164, 165] (Fig. 10).

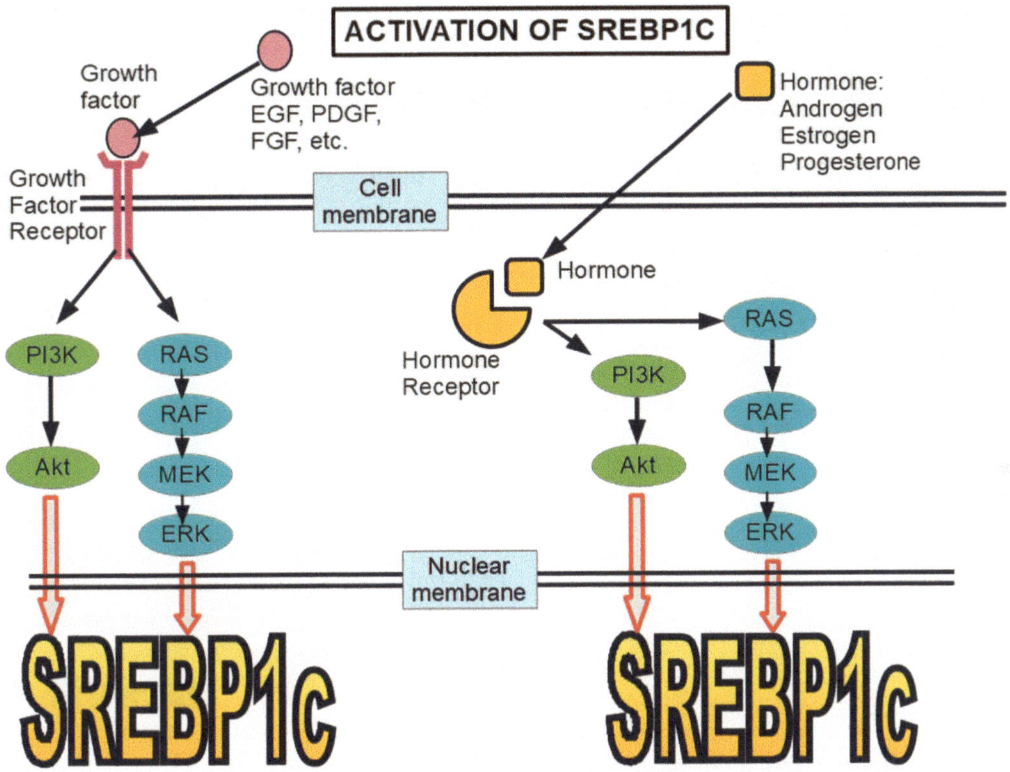

Fig. 10 Left side: growth factors, membrane receptors for growth factors and the pathway that increases the activity of SREBP1c. Right side: activation of SREBP1c by hormones with intracellular receptors. SREBP1c binds the FAS promoter and induces the expression of the gene.

SREBP-1c is a transcription factor regulated by environmental factors, external and internal signaling pathways and hormones and it can be modified by different pharmaceuticals (Fig. 11).

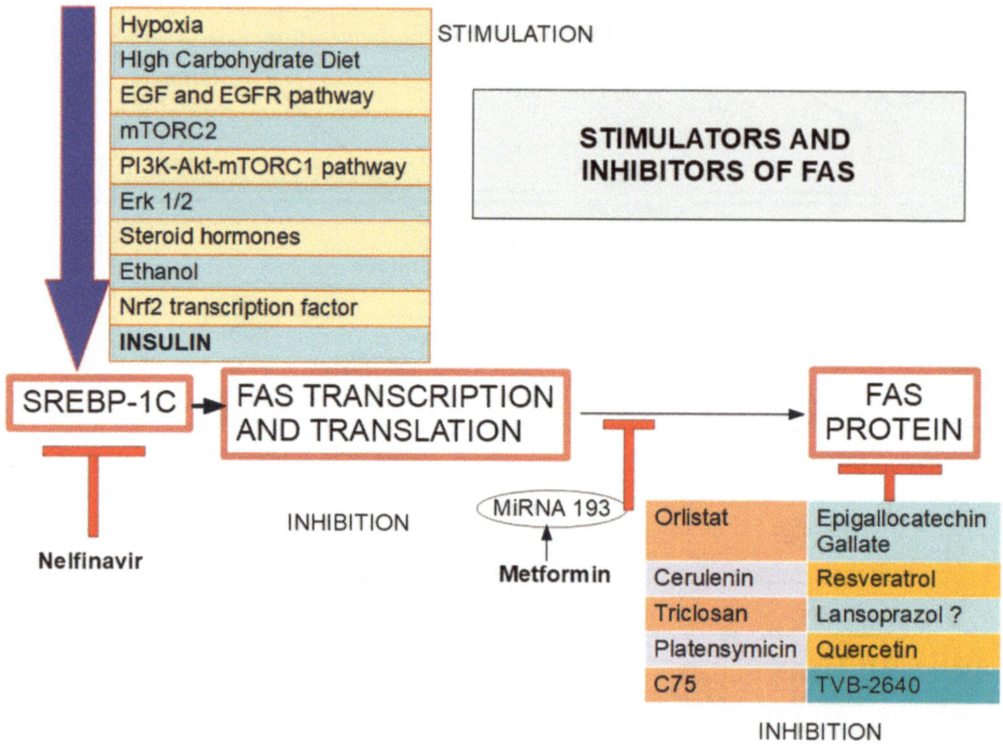

Fig. 11 Factors influencing FAS production and inhibition.

FAS has been investigated as a possible oncogene.[166] Increased retinoblastoma aggressiveness was associated with FAS activation and Camassei postulated that FAS inhibition could represent an alternative treatment strategy in advanced and resistant retinoblastoma.[167]

A molecular link was identified between FAS and the HER2 oncogene, a marker for poor prognosis that is over-expressed in 30% of breast and ovarian cancers. Pharmacologic FAS inhibitors were found to suppress p185(HER2) oncoprotein expression and tyrosine kinase activity in breast and ovarian cancers over-expressing HER2.[168] Similar suppression was observed when FAS gene expression was silenced by RNA interference.

Metformin is not usually considered a FAS inhibitor, however it blocks the stimulation of a high energy diet on colon carcinoma growth *in vivo* and is associated with a reduction of FAS expression.[169] Energy deficiency induced by metformin inhibits lipogenesis in prostate cancer cells.[170] Wahdan-Alaswad et al.[171] found that MIRNA 193b is induced by metformin and it inhibits FAS through a postranslational block of the 3 UTR region of the transcript. We believe that part of the anti-tumoral behavior of metformin is linked to its ability to acidify the intracellular milieu.

The metabolism of invadopodia

Invadopodia, the basic mechanism used for local invasion and migration, require a key role from the plasmatic membrane, which is rich in lipids. Caveolin-1 and cholesterol rich lipid rafts generate the scaffold for the formation of invadopodia.[172, 173] A big proportion of the *de novo* FA synthesized is incorporated into the phospholipids of the new membrane.[174] The caveolin-1 and cholesterol rich lipid rafts are the usual site where NHE1 and NaV1.5 ion channels are located. Although the exact mechanisms remain elusive, we do know that all these structures cooperate together in the formation and functioning of invadopodia.[175] The reduction of cholesterol with statins was proposed as one of the mechanisms that may decrease migration and metastasis by interfering with the lipid rafts at the invadopodial level.[176, 177]

Glutaminolytic phenotype

In 1955, while studying the nutritional requirements of cells in culture, Eagle[178] found that many cells exhibited high glutamine consumption significantly surpassing the consumption of other amino acids. These cells could not maintain their growth in the absence of glutamine and some lost viability. This was surprising because glutamine i not an essential amino acid, a finding that had no explanation at that time.

Glutamine is a major energy source through the production of alpha-ketoglutarate, which enters into the Krebs cycle contributing to the production of citrate. Now, there is some evidence indicating that glutamine may be an "essential" amino acid in certain cancer cells when they become "addict" to it[179] and synthesis cannot keep pace with demand.[180] Not all cancer cells show glutamine dependency,[181] however the addicted cells cannot survive without an exogenous supply of glutamine.

Glutamine in cancer cells is a[182]:

1. Fuel for the TCA cycle. Fig. 12.
2. Nitrogen donor for nucleotide synthesis.
3. Nitrogen donor for protein synthesis.
4. Activator of mTORC1 signaling.
5. Precursor of gluthatione.
6. Suppressor of oxidative stress.

Glutamine analogues have been tested as inhibitors of the glutaminolytic pathway, but their toxicity was too high.[183] The glutaminolytic pathway has permitted the development of a probe to determine glutaminolysis in cells and test the response to different drugs like hyperpolarized NMR with [5-13C]glutamine.[184] Another way to decrease the availability of glutamine is the so called "glutamine trap": phenylbutyrate "captures" glutamine decreasing its availability to cells.[185]

Corbet et al.[186] have shown that "long-term exposure of cancer cells to acidic pH leads to a metabolic reprogramming towards glutamine metabolism. This switch is triggered by the need to reduce the production of protons from glycolysis and further maintained by the NAD+-dependent increase in SIRT1 deacetylase activity to ensure intracellular pH homeostasis".

Interestingly, metformin increases the cancer cells' dependence on glutamine by decreasing glucose oxidation.[187] This may suggest the possibility of synergy between metformin and glutamine metabolism inhibitors.

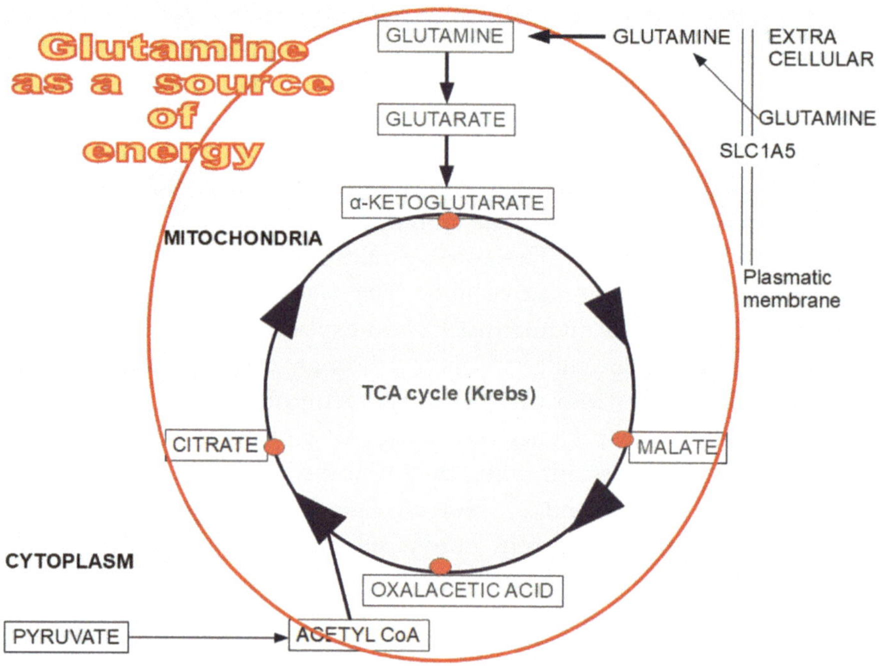

Fig. 12 Glutamine is a source of energy that can replace glucose as the main nutrient.

Serine phenotype

A special case of nutrient deprivation is serine starvation. Maddocks et al.[188] reported that depletion of serine in the diet of mice with HCT116 tumors showed impaired tumor growth. This finding was more accentuated in p53 null mice and there also was a significant loss of cell viability. One of the reasons for this behavior is that p53 null mice produced less serine than WT mice. The prevailing idea is that p53 helps cells to adapt to low serine conditions. Serine also is an allosteric activator of PKM2.[189]

p53 mutations are frequent in cancer, therefore, serine starvation may induce a return to oxidative phosphorylation and decrease tumor growth. The novo synthesis of serine does not seem to satisfy the demand in cancer, so exogenous uptake becomes necessary in certain tumors.[190] Possemato et al.[191] described high expression of PHGDH (phosphoglycerate dehydrogenase) in 70% of cases of ER negative breast cancers. PHGDH has an essential role in serine synthesis and there was increased flux of serine synthesis in these cases. Reducing PHGDH in patients with high expression of this enzyme decreased proliferation. Serine starvation deserves further research.

The acid-base balance in malignant tumors

This subject is one of the central issues of this book and it will be discussed in varying detail in all the chapters. Therefore, only the main points will be mentioned here.

Normal cells have a slightly alkaline intracellular pH (pHi = 7–7.2) and a more alkaline extracellular matrix (pHe = 7.3–7.4). The difference between these two pHs constitutes the gradient[192] and is the foundation of the cancer pH paradigm.

The first step in the malignant transformation process is an increase in the intracellular pH (pHi = 7.4–7.7) to the point that the intracellular milieu becomes more alkaline than the extracellular.[21, 193]

The extracellular matrix receives an acidic burden that is extruded from the malignant cells and it becomes acidic (pHe less than 7).[194–196] Proton production in malignant cells is 65 to 100% higher compared to normal cells.[197] This process is known as the inversion or reversal of the pH gradient and it represents a very important step in the development and progression of cancer.[48] Intracellular alkalinity is essential for the proliferative phenotype and resistance to apoptosis.[198] Extracellular acidity is absolutely necessary for the activation of proteolytic enzymes that allow the degradation of the extracellular matrix, migration, invasion, metastasis and the impairment of the host's immunological defenses against malignant cells. It also implies the overexpression/activation of proton extrusion mechanisms. Intracellular alkalosis not only stimulates glycolysis but at the same time hinders the entrance of pyruvate into the Krebs Cycle.

The increased intracellular pH and the development and maintenance of the inverted pH gradient is the result of the activity of a series of membrane-bound proton transporters (MBPT), mainly the Na+/H+ exchangers but also carbonic anhydrases (CAs, mainly CA IX and XII), vacuolar H+-ATPases, the H+/Cl- symporter, the monocarboxylate transporters (MCT, mainly MCT1 and MCT4) also known as the lactate-proton symporter, the Na+ − dependent Cl-/HCO3– exchangers (NBC) and ATP synthase. Each has its specific inhibitors. More specifically, the NHE1 (Na+/H+ Exchanger isoform 1) extrudes protons and imports sodium in an equimolecular fashion, while carbonic anhydrases 9 and 12 (CAIX and CAXII) catalyze the conversion of the excessive CO_2 production of cancer cells into carbonic acid, which then ionizes into bicarbonate and a proton at the cell surface. The sodium/bicarbonate co-transporter (NBC) imports bicarbonate ions into the cell ion participating in the alkalization of the cytosol.

Other oncogenic-metabolic-pH relations: a few examples

(A) In 1987 Wakelam et al.[199] proposed what we may call the oncogenic-pH pathway that clearly shows the relation of pH to cancerogenesis. It starts with growth factors activating their receptors which leads to inositol breakdown and the generation of two second messengers:

(a) inositol 1, 4, 5-trisphosphate that increases intracellular free calcium levels[200];

(b) 1,2 diacylglycerol that activates protein C kinase. This kinase increases NHE1 activity with the consequent elevation of intracellular pH.

The association of increased free calcium and augmented alkaline cytoplasm promotes proliferation.

(B) CO_2 production from glutamine is increased under alkaline conditions,[201] which means that an alkaline cytoplasm increases the use of glutamine as a source of energy for the Krebs Cycle.

(C) Caloric restriction is a known mechanism for decreasing tumor proliferation.[202, 203] One of the consequences of a ketogenic and calorie restricted diet is the acidification of the cytoplasm.[204]

(D) Glutamine can stimulate glycolysis in the cytoplasm.[205]

(E) Cardiolipin is the only lipid synthesized by mitochondria and is frequently altered in cancer, producing impairment of oxidative phosphorylation.[206] This would explain some therapeutic failures of drugs like dichloroacetate (DCA) that may restore oxidative phosphorylation. Table 1 summarizes the pecularities of cancer metabolism.

The ROS problem

Intense anabolism, as in cancer tissues, means high ROS production. ROS (particularly H_2O_2 but not O_2^-[207]) are cytotoxic for normal and cancer cells. The over-production of ROS by chemotherapeutic agents is precisely one of the mechanisms used by radiotherapy and chemotherapy to achieve apoptosis and/or necrosis. Cancer cells defend themselves from ROS through different mechanisms:

1. Adopting the glycolytic phenotype that produces less ROS than OXPHOS,
2. Importing glutamine as a glutathione precursor.
3. Increasing the PPP (Pentose Phosphate Pathway) that produces NADPH.

Glutathione and thioredoxin are ROS scavengers that require NADPH for their activity.

Part of the NADPH neutralizing ROS comes from the PPP (Fig. 13).

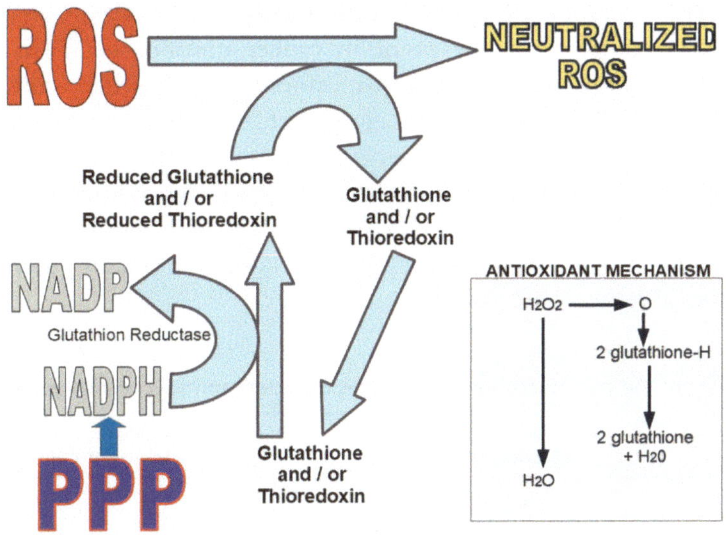

Fig. 13 Mechanism of action of ROS scavengers. Hydrogen peroxide as an example.

ROS also serves as a metabolic signal. When ROS levels are elevated it decreases PKM2 activity, the enzyme that catalyzes the step from phosphoenolpyruvate to pyruvate (step number 10 of the glycolytic chain). This diminished activity allows the accumulation of substrates upstream of phosphoenolpyruvate and a deviation of substrates towards PPP. In this way NADPH production is increased but building blocks, namely amino acids and ribose production, are also increased (Fig. 14).

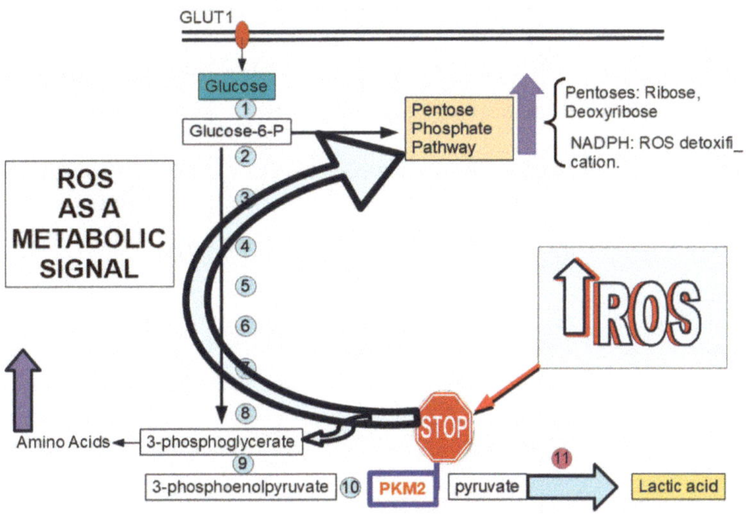

Fig. 14 Increased ROS slows down PKM2 increasing substrates towards PPP and 3-phosphoglycerate. Increased PPP allows better ROS detoxification and more building blocks for other compounds. *Pale blue arrows* show the effects of the PKM2 slowdown produced by ROS accumulation.[208]

The high production of ROS makes NADPH an essential molecule for cancer cells so that it becomes another targetable point in cancer metabolism.[209] In this respect, 6-Amino-nicotinamide (6-AN) is an anti-metabolite that has been known for over 60 years and is one of the most potent antagonists of NADPH.[210] In experimental conditions, 6-AN has shown strong anti-cancer effects in mice with mammary carcinoma, rat lymphosarcoma, mouse melanoma, Walker carcinoma, and mouse glioblastoma. The problem with 6-AN was that its anti-tumor properties were lost at clinically tolerable dosage (0.2 mg/Kg of body weight).[211] When it was used alone in different advanced tumors the clinical results were poor.[212]

Table 1 Summary of the main metabolic differences between normal and malignant cells as described above.

	Normal cell	Malignant cell
Glucose uptake	Standard	10 to 20 fold higher than normal cells
Glucose transporters	Standard	Increased
Glutamine uptake	Standard	Higher in glutaminolytic phenotype
Warburg effect	Absent	Present in almost all advanced tumors. There is usually correlation with degree of malignancy
Anabolic activity	According to functional state of the cell. Increased in embryonic cells and in highly proliferative normal tissues	Increased in highly proliferating cells
Lactic acid in tumoral microenvironment	Low	Increased, creating an acid microenvironment.
pH of the microenvironment	Within physiologic standards	Decreased
Lactate transport	Negligible in most cells except in muscle where it may be increased during exercise.	Increased in the symbiotic and parasitic phenotypes through MCT-1 and MCT-4
Oxygen in the microenvironment	Normoxia	Hypoxia
Symbiotic metabolism (lactate shuttle)	Only in muscle during exercise.	Frequent
PKM isoforms	PKM1 is the standard PKM isoform	Over-expression of PKM-2 is a frequent finding.
Isocitrate dehydrogenase I and II	Wild type	Frequently mutated with increased activity

Table 1 Summary of the main metabolic differences between normal and malignant cells as described above—cont'd

	Normal cell	Malignant cell
PDK activity	Low in normoxia, increased in hypoxia.	Increased in both normoxia and hypoxia.
CO2 hydration on the cell surface	Normal	Highly increased
Oncometabolites	2HG (hydroxyglutarate almost absent)	Accumulation of 2HG
Hexokinase	Isoform II is found only in skeletal muscle and adipose tissue. In other tissues isoform I predominates	Isoform II is the predominant form in cancer cells.
HIF 1α	Low expression in normoxia	Over-expression in both normoxia and hypoxia. It is a powerful inducer of glycolytic enzymes expression.
LDH	Standard	Over-expressed in glycolytic phenotype
FAS expression	Low expression under normal dietary conditions	Over-expression of FAS
Membrane carbonic anhydrases IX and XII	Low expression	Highly expressed
pH gradient	pHe higher than pHi	pHe lower than pHi
CO2 production	Standard level	Increased
Glucose-6-phosphate dehydrogenase (PPP)	Standard activity	Increased activity
Transketolase L1 (PPP)	Standard activity	Increased activity

The pH paradigm and metabolic changes in cancer

We have looked at metabolism with a biased view: carbohydrates, lipids, glutamine, *etc.* as separate entities. But the entire metabolic process is a highly integrated and interwoven mechanism. Furthermore, extracellular acidosis and intracellular alkalosis are powerful drivers of many of the metabolic differences shown between cancer cells and their normal counterparts. This is a phenotype that some authors have called "chronic acidity adapted" cells.[213]

The main objective of this book is the description and analysis of this pH-metabolism relationship. Therefore, we will not go further into it in this chapter.

Multitargeted attack on tumor metabolism

Fig. 14 illustrates a proposed multi-targeted strategy against tumor metabolism.

The first step of a "metabolic" treatment of cancer should be the determination of the tumor's metabolic profile (glycolytic, glutaminolytic, lipogenic). This is not always possible. Evidence shows that inhibiting only one step in cancer metabolism is insufficient for achieving good results. More than one step or even better, more than one pathway needs to be addressed. Furthermore, the simultaneous targeting of the three main metabolic pathways of cancer (glycolysis, glutaminolysis and *de novo* fatty acid synthesis) has been tested with good experimental results (lonidamine, diazonorleucine and orlistat were used) and no toxicity found in normal cells or mice.[214] The combination of three compounds, like 2 deoxyglucose, biguanides and 3-BP (3 bromopyruvate) showed additive experimental results.[215] α-lipoic acid associated with hydroxycitrate (known as Metabloc) was used in patients with advanced metastatic drug-resistant disease. Eleven patients were treated with five partial regressions and three stable disease. α-lipoic acid co-activates PDH through the inhibition of PDK, inducing the oxidative pathway, while hydroxycitrate is an inhibitor of ATP-citrate lyase decreasing the availability of cytoplasmic citrate for lipid synthesis.[216]

The combination of disulfiram with auranofin is other example of a combined attack on the cancer cell redox system. While auranofin, used for the treatment of rheumatoid arthritis inhibits the thioredoxin system, disulfiram, used for the treatment of alcoholic addiction targets superoxide dysmutase (another antioxidant enzyme). The combination increases oxidative stress.[217] (Disulfiram has many other anti-cancer properties).

Metformin, an inhibitor of oxidative phosphorylation, and 2-deoxyglucose, an inhibitor of glycolysis exert synergistic anti-tumor activity.[218] Something similar happens when metformin is combined with DCA, but in this case DCA increases oxidative phosphorylation. The combined effect of stimulation and inhibition of oxidative phosphorylation produces an oxidative stress that induces apoptosis.[219, 220]

Fig. 15 shows possible drug associations for the inhibition of metabolic pathways in cancer.

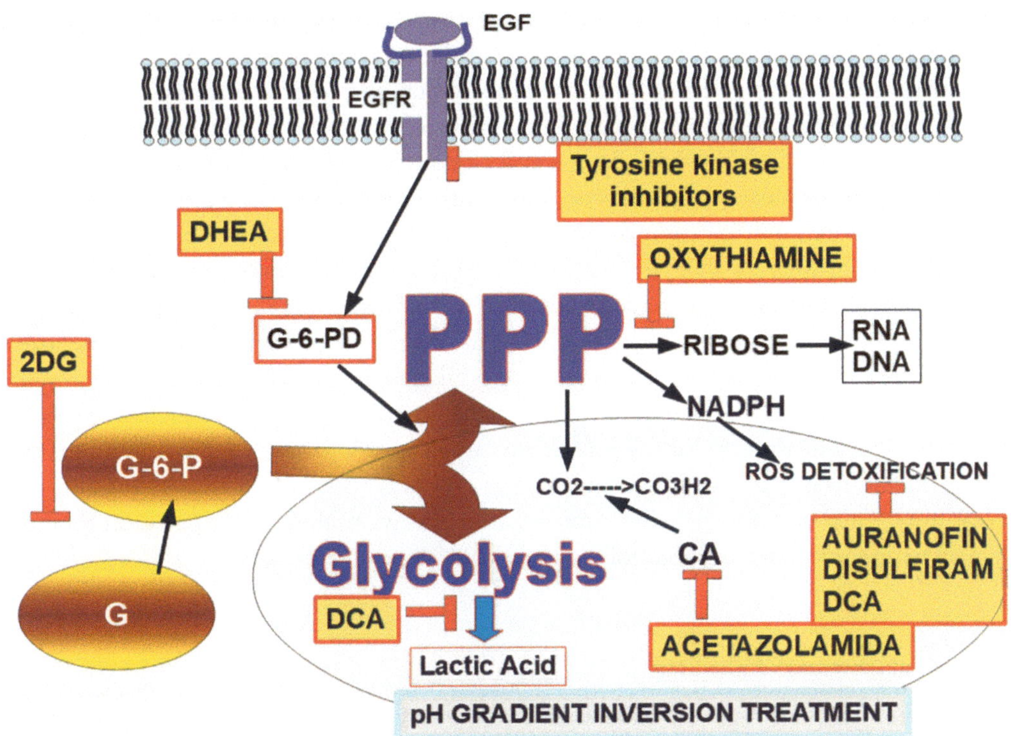

Fig. 15 Multi-targeted metabolic and integrated attack is possible.

Conclusions

There are very important differences among the more than 200 known cancers, but most of them share metabolic "abnormalities".[221] Metabolic targeting is just starting to be a viable strategy in cancer therapy. Ninety years have elapsed since the pioneering investigations of Warburg, who established that blocking one of the two possible metabolic pathways of energy (aerobic glycolysis or oxidative respiration) was not enough to kill tumor cells. It is necessary to block both. This concept seems accurate today and it goes even further: multiple pathways have to be inhibited in order to achieve visible results, because cancer cells can overcome the inhibition of a single pathway. Indeed, if we intend to target cancer metabolism we must first determine which metabolic pathways are preferentially used by the tumor and how the tumor may adapt to defend itself, due to its metabolic heterogeneity, robustness and flexibility. Metabolic reprogramming makes cancer cells more aggressive. This fact fully justifies targeting metabolism as part of an integrated therapeutic approach because metabolic reprogramming is absolutely necessary for cancer to evolve. The role of a tumors' metabolic reprogramming in drug resistance is one of the main problems that oncologists face nowadays and, therefore, targeting the metabolic peculiarities of the disease and its continuous adaptation to new environmental conditions would help prevent cancer drug resistance.

In this context, modification of the characteristic pH imbalance of tumors is a powerful weapon against cancer's peculiar metabolism. There are some repurposed pharmaceuticals like metformin in advanced trials. Others such as CPI-613 are in phase I/II trials. In spite of interesting results in laboratory tests, natural substances such as resveratrol, genistein and curcumin have been unable to establish their places in the metabolic arsenal. There is, however, an unexplored wealth of repurposed drugs that have great potential as complementary treatments for the practicing oncologist. We believe that drugs targeting cancer metabolism have an important role as adjuncts to standard treatments and that because of their non toxicity and non-interference with chemo and/or radiotherapy, should be included in the protocols sooner than later.

References

1. Martin W, Hoffmeister M, Rotte C, Henze K. An overview of endosymbiotic models for the origins of eukaryotes, their ATP-producing organelles (mitochondria and hydrogenosomes), and their heterotrophic lifestyle. *Biol Chem* 2001;**382**(11):1521–39.
2. Gray MW, Burger G, Lang BF. The origin and early evolution of mitochondria. *Genome Biol* 2001;**2**(6) reviews1018-1.
3. Lane N, Martin W. The energetics of genome complexity. *Nature* 2010;**467**:929–34.
4. Levine AJ, Puzio-Kuter AM. The control of the metabolic switch in cancers by oncogenes and tumor suppressor genes. *Science* 2010;**330**(6009):1340–4.
5. Semenza GL. HIF-1: upstream and downstream of cancer metabolism. *Curr Opin Genet Dev* 2010; **20**(1):51–6.
6. Warburg O, Wind F, Negelein E. The metabolism of tumors in the body. *J Gen Physiol* 1927; **8**(6):519–30.
7. Warburg O. *The metabolism of tumors*. London: Arnold Constable; 1930.
8. Burk Dean note's on the conference by Warburg O. *The prime cause and prevention of cancer*. [Revised Lindau Lecture]. English ed; 1966.
9. Kallinowski F, Vaupel P, Runkel S, Berg G, Fortmeyer HP, Baessler KH, et al. Glucose uptake, lactate release, ketone body turnover, metabolic micromilieu, and pH distributions in human breast cancer xenografts in nude rats. *Cancer Res* 1988;**48**(24 Part 1):7264–72.
10. Bauer DE, Harris MH, Plas DR, Lum JJ, Hammerma PS, Rathmell JC, et al. Cytokine stimulation of aerobic glycolysis in hematopoietic cells exceed proliferative demand. *FASEB J* 2004;**18**:1303–5.
11. Koppenol WH, Bounds PL, Dang CV. Otto Warburg's contributions to current concepts of cancer metabolism. *Nat Rev Cancer* 2011;**11**(5):325.
12. Hsu PP, Sabatini DM. Cancer cell metabolism: warburg and beyond. *Cell* 2008;**134**(5):703–7.
13. Vander Heiden MG, Cantley LC, Thompson CB. Understanding the Warburg effect: the metabolic requirements of cell proliferation. *Science* 2009;**324**(5930):1029–33.
14. Newell K, Franchi A, Pouyssegur J, Tannock I. Studies with glycolysis-deficient cells suggest that production of lactic acid is not the only cause of tumor acidity. *Proc Natl Acad Sci* 1993;**90**(3):1127–31.

15. Yamagata M, Hasuda K, Stamato T, Tannock IF. The contribution of lactic acid to acidification of tumours: studies of variant cells lacking lactate dehydrogenase. *Br J Cancer* 1998;**77**(11):1726.
16. Wilhelm G, Schulz J, Hofmann E. pH-dependence of aerobic glycolysis in ehrlich ascites tumour cells. *FEBS Lett* 1971;**17**:158.
17. Kaminskas E. The pH-dependence of sugar-transport and glycolysis in cultured Ehrlich ascites-tumour cells. *Biochem J* 1978;**174**:453–9.
18. Relman AS. Metabolic consequences of acid-base disorders. *Kidney Int* 1972;**1**:347–59.
19. Nagata H, Che XF, Miyazawa K, Tomoda A, Konishi M, Ubukata H, et al. Rapid decrease of intracellular pH associated with inhibition of Na+/H+ exchanger precedes apoptotic events in the MNK45 and MNK74 gastric cancer cell lines treated with 2-aminophenoxazine-3-one. *Oncol Rep* 2011;**25**:341–6.
20. Che X-F, Zheng C-L, Akiyama S-I, Tomoda A. 2-Aminophenoxazine-3-one and 2-amino-4,4α-dihydro-4α,7-dimethyl-3H-phenoxazine-3-one cause cellular apoptosis by reducing higher intracellular pH in cancer cells. *Proc Japan Acad, Series B Phys Biol Sci* 2011;**87**:199–213.
21. Reshkin SJ, Bellizzi A, Caldeira S, Albarani V, Malanchi I, Poignee M, et al. Na+/H+ exchanger-dependent intracellular alkalinization is an early event in malignant transformation and plays an essential role in the development of subsequent transformation-associated phenotypes. *FASEB J* 2000; **14**:2185–97.
22. Brahimi-Horn MC, Chiche J, Pouysségur J. Hypoxia and cancer. *J Mol Med* 2007;**85**(12):1301–7.
23. Sies H. Oxidative stress: introductory remarks. *Oxid stress* 1985;**501**:1–8.
24. Apel K, Hirt H. Reactive oxygen species: metabolism, oxidative stress, and signal transduction. *Annu Rev Plant Biol* 2004;**55**:373–99.
25. Hanahan D, Weinberg RA. Hallmarks of cancer: the next generation. *Cell* 2011;**144**(5):646–74.
26. JSJ C-M, Robles A, Lopez Lazaro M. Role of the intracellular pH in the metabolic switch between oxidative phosphorylation and aerobic glycolysis—relevance to cancer. WebmedCentral, *Cancer* 2011;**2**:3. WMC001716.
27. Muñoz-Pinedo C, El Mjiyad N, Ricci JE. Cancer metabolism: current perspectives and future directions. *Cell Death Dis* 2012;**3**(1):e248.
28. Cairns RA, Harris IS, Mak TW. Regulation of cancer cell metabolism. *Nat Rev Cancer* 2011;**11**:85–95.
29. Cairns RA. Drivers of the Warburg phenotype. *Cancer J* 2015;**21**(2):56–61.
30. Kim JW, Tchernyshyov I, Semenza GL, Dang CV. HIF-1-mediated expression of pyruvate dehydrogenase kinase: a metabolic switch required for cellular adaptation to hypoxia. *Cell Metab* 2006;**3**(3):177–85.
31. Elstrom RL, Bauer DE, Buzzai M, Karnauskas R, Harris MH, Plas DR, et al. Akt stimulates aerobic glycolysis in cancer cells. *Cancer Res* 2004;**64**:3892–9.
32. Düvel K, Yecies JL, Menon S, Raman P, Lipovsky AI, Souza AL, et al. Activation of a metabolic gene regulatory network downstream of mTOR complex 1. *Mol Cell* 2010;**39**(2):171–83.
33. Dang CV, Semenza GL. Oncogenic alterations of metabolism. *Trends Biochem Sci* 1999;**24**:68–72.
34. Minchenko O, Opentanova I, Caro J. (2003) Hypoxic regulation of the 6-phosphofructo-2-kinase/fructose-2- 6-bisphosphatase gene family (PFKFB-1–4) expression in vivo. *FEBS Lett* 2003; **554**:264–70.
35. Simon MC. Coming up for air: HIF-1 and mitochondrial oxygen consumption. *Cell Metab* 2006;**3**:150–1.
36. Hu Y, Lu W, Chen G, Wang P, Chen Z, Zhou Y, et al. K-ras G12V transformation leads to mitochondrial dysfunction and a metabolic switch from oxidative phosphorylation to glycolysis. *Cell Res* 2012;**22**(2):399.
37. Dang CV, Lewis BC, Dolde C, Dang G, Shim H. Oncogenes in tumor metabolism, tumorigenesis, and apoptosis. *J Bioenerg Biomembr* 1997;**29**:345–54.
38. Dang CV. c-Myc target genes involved in cell growth, apoptosis, and metabolism. *Mol Cell Biol* 1999;**19**(1):1.
39. Osthus RC, Shim H, Kim S, Li Q, Reddy R, Mukherjee M, et al. Deregulation of glucose transporter 1 and glycolytic gene expression by c-Myc. *J Biol Chem* 2000;**275**:21797–800.

40. David CJ, Chen M, Assanah M, Canoll P, Manley JL. HnRNP proteins controlled by c-Myc deregulate pyruvate kinase mRNA splicing in cancer. *Nature* 2010;**463**(7279):364–8.
41. Shaw RJ. LKB1 and AMPK control mTOR signaling and growth. *Acta Physiol (Oxford)* 2009;**196**:65–80.
42. Yeung SJ, Pan J, Lee MH. Roles of p53, MYC and HIF-1 in regulating glycolysis—the seventh hallmark of cancer. *Cell Mol Life Sci* 2008;**65**(24):3981.
43. Kawauchi K, Araki K, Tobiume K, Tanaka N. p53 regulates glucose metabolism through an IKK-NF-κ B pathway and inhibits cell transformation. *Nat Cell Biol* 2008;**10**(5):611.
44. Marsin AS, Bertrand L, Rider MH, Deprez J, Beauloye C, Vincent MF, et al. Phosphorylation and activation of heart PFK-2 by AMPK has a role in the stimulation of glycolysis during ischaemia. *Curr Biol* 2000;**10**(20):1247–55.
45. Yuneva MO, Fan TW, Allen TD, Higashi RM, et al. The metabolic profile of tumors depends on both the responsible genetic lesion and tissue type. *Cell Metab* 2012;**15**(2):157–70.
46. Sottoriva A, Spiteri I, Piccirillo SG, Touloumis A, Collins VP, Marioni JC, et al. Intratumor heterogeneity in human glioblastoma reflects cancer evolutionary dynamics. *Proc Natl Acad Sci* 2013; **110**(10):4009–14.
47. Hensley CT, Faubert B, Yuan Q, Lev-Cohain N, Jin E, Kim J, et al. Metabolic heterogeneity in human lung tumors. *Cell* 2016;**164**(4):681–94.
48. Cantor JR, Sabatini DM. Cancer cell metabolism: one hallmark, many faces. *Cancer Discov* 2012; **2**(10):881–98. Sep 25.
49. Sonveaux P, Végran F, Schroeder T, Wergin MC, Verrax J, Rabbani ZN, et al. Targeting lactate-fueled respiration selectively kills hypoxic tumor cells in mice. *J Clin Invest* 2008;**118**(12):3930–43.
50. Alfarouk KO, Verduzco D, Rauch C, Muddathir AK, Adil HB, Elhassan GO, et al. Glycolysis, tumor metabolism, cancer growth and dissemination. A new pH-based etiopathogenic perspective and therapeutic approach to an old cancer question. *Oncoscience* 2014;**1**(12):777.
51. Alfarouk KO, Muddathir AK, Shayoub ME. Tumor acidity as evolutionary spite. *Cancer* 2011; **3**(1):408–14.
52. Gerweck LE, Vijayappa S, Kozin S. Tumor pH controls the in vivo efficacy of weak acid and base chemotherapeutics. *Mol Cancer Ther* 2006;**5**(5):1275–9.
53. Rauch C, Blanchard A, Wood E, Dillon E, Miriam L, Wahl ML, et al. Cell membranes, cytosolic pH and drug transport in cancer and MDR: physics, biochemistry and molecular biology. In: Meszaros A, Balogh G, editors. *Multiple drug resistance*. Nova Science Publishers, Inc; 2009. p. 1–100 [chapter 3].
54. Raghunand N, Gillies RJ. pH and drug resistance in tumors. *Drug Resist Updat* 2002;**3**:39–47.
55. Raghunand N, Mahoney BP, Gillies RJ. Tumor acidity, ion trapping and chemotherapeutics: II. pH-dependent partition coefficients predict importance of ion trapping on pharmacokinetics of weakly basic chemotherapeutic agents. *Biochem Pharmacol* 2003;**66**(7):1219–29.
56. Cuezva JM, Ostronoff LK, Ricart J, López de Heredia M, Di Liegro CM, Izquierdo JM. Mitochondrial biogenesis in the liver during development and oncogenesis. *J Bioenerg Biomembr* 1997;**29**:365–77.
57. Mazurek S, Boschek CB, Hugo F, Eigenbrodt E. Pyruvate Kinase type M2 and its role in tumor growth and spreading. *Semin Cancer Biol* 2005;**15**:300–8.
58. Kroemer G, Pouyssegur J. Tumor cell metabolism: cancer´s achilles heel. *Cancer cell* 2008;**13**:472–82.
59. Christofk HR, Vander Heiden MG, Harris MH, Ramanathan A, Gerszten RE, Wei R, et al. The M2 splice isoform of pyruvate kinase is important for cancer metabolism and tumour growth. *Nature* 2008;**452**(7184):230.
60. Harguindey S, Arranz JL, Washl ML, Orive G, Reshkin JS. Proton transport inhibitors as potentially selective anticancer drugs. *Anticancer Res* 2009;**29**:2127–36.
61. Mahoney BP, Raghunand N, Baggett B, Gillies RJ. Tumor acidity, ion trapping and chemotherapeutics: I. Acid pH affects the distribution of chemotherapeutic agents in vitro. *Biochem Pharmacol* 2003; **66**(7):1207–18.
62. Cosse JP, Michiels C. Tumour hypoxia affects the responsiveness of cancer cells to chemotherapy and promotes cancer progression. *Anti Cancer Agents Med Chem* 2008;**8**(7):790–7. Formerly Current Medicinal Chemistry-Anti-Cancer Agents.

63. Federici C, Petrucci F, Caimi S, Cesolini A, Logozzi M, Borghi M, et al. Exosome release and low pH belong to a framework of resistance of human melanoma cells to cisplatin. *PLoS One* 2014;**9**(2).
64. Rozhin J, Sameni M, Ziegler G, Sloane BF. Pericellular pH affects distribution and secretion of cathepsin B in malignant cells. *Cancer Res* 1994;**54**:6517–25.
65. Webb SD, Sherratt JA, Fish RG. Alterations in proteolytic activity at low pH and its association with invasion: a theoretical model. *Clin Exp Metastasis* 1999;**17**:397–407.
66. Fukumura D, Xu L, Chen Y, Gohongi T, Seed B, Jain RK. Hypoxia and acidosis independently up-regulate vascular endothelial growth factor transcription in brain tumors in vivo. *Cancer Res* 2000;**61**:6020–4.
67. Xu L, Fukumura D, Jain RK. Acidic extracellular pH induces VEGF in human glioblastoma cells via ERK1/2 MAPK signaling pathway. Mechanism of low pH induced VEGF. *J Biol Chem* 2002; **277**(13):11368–74.
68. Lardner A. The effects of extracellular pH on immune function. *J Leukoc Biol* 2001;**69**(4):522–30.
69. Bohloli M, Atashi A, Soleimani M, Kaviani S, Anbarlou A. Investigating effects of acidic pH on proliferation, invasion and drug-induced apoptosis in lymphoblastic leukemia. *Cancer Microenviron* 2016; **9**(2-3):119–26.
70. Peppicelli S, Bianchini F, Calorini L. Extracellular acidity, a "reappreciated" trait of tumor environment driving malignancy: perspectives in diagnosis and therapy. *Cancer Metastasis Rev* 2014; **33**(2-3):823–32.
71. Calorini L, Peppicelli S, Bianchini F. Extracellular acidity as favouring factor of tumor progression and metastatic dissemination. *Exp Oncol* 2012;**34**(2):79–84.
72. Rofstad EK, Mathiesen B, Kindem K, Galappathi K. Acidic extracellular pH promotes experimental metastasis of human melanoma cells in athymic nude mice. *Cancer Res* 2006;**66**(13):6699–707.
73. Cardone RA, Casavola V, Reshkin SJ. The Role Of Disturbed pH Dynamics and The N+/H+ Exchanger In Metastasis. *Nat Rev Cancer* 2005;**5**:786–95.
74. Medina RA, Southworth R, Fuller W, Garlick PB. Lactate-induced translocation of GLUT1 and GLUT4 is not mediated by phosphatidylinositol-3-kinase pathway in the rat heart. *Basic Res Cardiol* 2002;**97**:168–76.
75. Weinberg F, Hamanaka R, Wheaton WW, Weinberg S, Joseph J, Lopez M, et al. Mitochondrial metabolism and ROS generation are essential for Kras-mediated tumorigenicity. *Proc Natl Acad Sci* 2010;**107**(19):8788–93.
76. Rigoulet M, Yoboue ED, Devin A. Mitochondrial ROS generation and its regulation: mechanisms involved in H2O2 signaling. *Antioxid Redox Signal* 2011;**14**(3):459–68.
77. Porporato PE, Dhup S, Dadhich RK, Copetti T, Sonveaux P. Anticancer targets in the glycolytic metabolism of tumors: a comprehensive review. *Front Pharmacol* 2011;**2**:49.
78. Clough JR, Whittingham DG. Metabolism of [14C] glucose by postimplantation mouse embryos in vitro. *Development* 1983;**74**(1):133–42.
79. Wang R, Dillon CP, Shi LZ, Milasta S, Carter R, Finkelstein D, et al. The transcription factor Myc controls metabolic reprogramming upon T lymphocyte activation. *Immunity* 2011;**35**(6):871–82.
80. Pearce EL, Poffenberger MC, Chang CH, Jones RG. Fueling immunity: insights into metabolism and lymphocyte function. *Science* 2013;**342**(6155):12424–54.
81. Pelicano H, Martin DS, Xu RA, Huang P. Glycolysis inhibition for anticancer treatment. *Oncogene* 2006;**25**(34):4633.
82. Gatenby RA, Gillies RJ. Glycolysis in cancer: a potential target for therapy. *Int J Biochem Cell Biol* 2007;**39**(7–8):1358–66.
83. Wu CH, Ho YS, Tsai CY, et al. In vitro and in vivo study of phloretin-induced apoptosis in human liver cancer cells involving inhibition of type II glucose transporter. *Int J Cancer* 2009;**124**: 2210–9.
84. Kim MS, Kwon JY, Kang NJ, Lee KW, Lee HJ. Phloretin induces apoptosis in H-Ras MCF10A human breast tumor cells through the activation of p53 via JNK and p38 mitogen-activated protein kinase signaling. *Ann N Y Acad Sci* 2009;**1171**:479–83.
85. Martin HJ, Kornman F, Huhrmann GF. The inhibitory effects of flavonoids and antiestrogens on the Glut1 glucose transportes in human erythrocytes. *Chem Biol Interact* 2003;**146**:225–35.

86. Granchi C, Fancelli D, Minutolo F. An update on therapeutic opportunities offered by cancer glycolytic metabolism. *Bioorg Med Chem Lett* 2014;**24**:4915–25.
87. Rempel A, Mathupala SP, Griffin CA, Hawkins AL, Pedersen PL. Glucose catabolism in cancer cells: amplification of the gene encoding type II hexokinase. *Cancer Res* 1996;**56**:2648–71.
88. Palmieri D, Fitzgerald D, Shreeve SM. Analyses of resected human brain metastases of breast cancer reveal the association between up-regulation of hexokinase 2 and poor prognosis. *Mol Cancer Res* 2009;**7**(9):1438–45.
89. Pedersen PL, Mathupala S, Rempel A, Geschwind JF, Ko YH. Mitochondrial bound type II hexokinase: a key player in the growth and survival of many cancers and an ideal prospect for therapeutic intervention. *Biomed Biochim Acta* 2002;**1555**:14–20.
90. Pastorino JG, Shulga N, Hoek JB. Mitochondrial binding of hexokinase II inhibits Bax-induced cytochrome c release and apoptosis. *J Biol Chem* 2002;**277**(9):7610–8.
91. Mathupala SP, Rempel A, Pederen PL. Aberrant glycolytic netabolism of cancer cells: a rematkable coordination of genetic, transcriptional, post-translational and mutational events that lead to a critical role for typeII hexokinase. *J Bioenerg Biomembr* 1997;**29**:339–43.
92. Pastorino JG, Hoek JB. Hexokinase II: the integration of energy metabolism and control of apoptosis. *Curr Med Chem* 2003;**10**(16):1535–51.
93. Patra KC, Wang Q, Bhaskar PT, Miller L, Wang Z, Wheaton W, et al. Hexokinase 2 is required for tumor initiation and maintenance and its systemic deletion is therapeutic in mouse models of cancer. *Cancer Cell* 2013;**24**(2):213–28.
94. Marrache S, Dhar S. The energy blocker inside the power house: mitochondria targeted delivery of 3-bromopyruvate. *Chem Sci* 2015;**6**(3):1832–45.
95. Nilsson H, Lindgren D, Forsberg AM, Mulder H, Axelson H, Johansson ME. Primary clear cell renal carcinoma cells display minimal mitochondrial respiratory capacity resulting in pronounced sensitivity to glycolitic inhibition by 3-Bromopyruvate. *Cell Death Dis* 2015;**6**:e1585.
96. Gong L, Wei Y, Yu X, Peng J, Leng X. 3-Bromopyruvic acid, a hexokinase II inhibitor, is an effective antitumor agent on the hepatoma cells : in vitro and in vivo findings. *Anti Cancer Agents Med Chem* 2014;**14**(5):771–6.
97. Ganapathy-Kanniappan S, Geschwind JF, Kunjithapatham R, Bujis M, Vossen JA, Tchernyshyov I, et al. Glyceraldehyde-3-phosphate dehydrogenase (GAPDH) is pyruvylated during 3-bromopyruvate mediated cancer cell death. *Anticancer Res* 2009;**29**(12):4909–18.
98. Mansi JL, de Graeff A, Newell DR, Glaholm J, Button D, Leach MO, et al. A phase II clinical and pharmacokinetic study of Lonidamine in patients with advanced breast cancer. *Br J Cancer* 1991; **64**(3):593–7.
99. Floridi A, Paggi MG, D'Atri S, De Martino C, Marcante ML, Silvestrini B, et al. Effect of lonidamine on the energy metabolism of Ehrlich ascites tumor cells. *Cancer Res* 1981;**41**:4661–6.
100. Miccoli L, Poirson-Bichat F, Sureau F, Goncalves RB, Bourgeois Y, Dutrillaux B, et al. Potentiation of lonidamine and diazepam, two agents acting on mitochondria in human glioblastoma treatment. *J Natl Cancer Inst* 1998;**90**(18):1400–6.
101. Woodward GE, Cramer FB. 2-desoxy-D-glucose as an inhibitor of anaerobic glycolysis in tumor tissue. *J Frankl Inst* 1952;**254**:259–60.
102. Kern KA, Norton JA. Inhibition of established rat fibrosarcoma growth by the glucose antagonist 2-deoxy-d-glucose. *Surgery* 1987;**102**:380–5.
103. Ciavardelli D, Rossi C, Barcaroli D, Volpe S, Consalvo A, Zucchelli M, et al. Breast cancer stem cells rely on fermentative glycolysis and are sensitive to 2-deoxyglucose treatment. *Cell Death Dis* 2014;**5**(7).
104. Ramirez-Peinado S, Alcazar-Limones F, Lagares-Tena L, El Mjiyad N, Caro-Maldonado A, Tirado OM, et al. 2-deoxyglucose induces Noxa-dependent apoptosis in alveolar rhabdomyosarcoma. *Cancer Res* 2011;**71**(21):6796–806.
105. Coleman MC, Asbury CR, Daniels D, Du J, Aykin-Burns N, Smith BJ, et al. 2-deoxy-D-glucose causes cytotoxicity, oxidative stress, and radiosensitization in pancreatic cancer. *Free Radic Biol Med* 2008;**44**(3):322–31.
106. Sahra IB, Laurent K, Giuliano S, Larbret F, Ponzio G, Gounon P, et al. Targeting cancer cell metabolism: the combination of metformin and 2-deoxyglucose induces p53-dependent apoptosis in prostate cancer cells. *Cancer Res* 2010;**70**(6):2465–75.

107. Moreno-Sanchez R, Rodrıguez-Enrıquez S, Marın-Hernandez A, Saavedra E. Energy metabolism in tumor cells. *FEBS J* 2007;**274**:1393–418.
108. Yalcin A, Clem BF, Simmons A, Lane A, Nelson K, Clem AL, et al. Nuclear targeting of 6-phosphofructo-2-kinase (PFKFB3) increases proliferation via cyclin-dependent kinases. *J Biol Chem* 2009; **284**(36):24223–32.
109. Ui M. A role of phosphofructokinase in pH-dependent regulation of glycolysis. *Biochim Biophys Acta Gen Subj* 1966;***124***(2):310–22.
110. Hitosugi T, Zhou L, Elf S, Fan J, Kang HB, Seo JH, et al. Phosphoglycerate mutase 1 coordinates glycolysis and biosynthesis to promote tumor growth. *Cancer Cell* 2012;**22**(5):585–600.
111. Guo S, Xiao Y, Li D, Jiang Q, Zhu L, Lin D, et al. PGK1 and GRP78 overexpression correlates with clinical significance and poor prognosis in Chinese endometrial cancer patients. *Oncotarget* 2018; **9**:680–90.
112. Zieker D, Konigsrainer I, Tritschler I, Loffler M, Beckert S, Traub F, et al. Phosphoglycerate kinase 1 a promoting enzyme for peritoneal dissemination in gastric cancer. *Int J Cancer* 2010;**126**:1513–20.
113. Fu D, He C, Wei J, Zhang Z, Luo Y, Tan H, Ren C. PGK1 is a potential survival biomarker and invasion promoter by regulating the HIF-1α–mediated epithelial-mesenchymal transition process in breast cancer. *Cell Physiol Biochem* 2018;**51**(5):2434–44.
114. Muller FL, Colla S, Aquilanti E, Manso VE, Genovese G, Lee J, et al. Passenger deletions generate therapeutic vulnerabilities in cancer. *Nature* 2012;**488**(7411):337–42.
115. Anastasiou D, Yu Y, Israelsen WJ. Pyruvate kinase M2 activators promote tetramer formation and suppress tumorigenesis. *Nat Chem Biol* 2012;**8**(10):839–47. Erratum in: *Nat Chem Biol.* 2012;8 (12):1008.
116. Luo W, Hu H, Chang R, Zhong J, Knabel M, O'Meally R, et al. Pyruvate kinase M2 is a PHD3-stimulated coactivator for hypoxia-inducible factor 1. *Cell* 2011;**145**:732–44.
117. Allison SJ, Knight JRP, Granchi C, Rani R, Minutolo F, Milner J, et al. Identification of LDH-A as a therapeutic target for cancer cell killing via (1) p53/NAD(H)-dependent and (2) p53-independent pathways. *Oncogenesis* 2014;**3**:e102.
118. Zhao Z, Han F, Yang S, Wu J, Zhan W. Oxamate-mediated inhibition of lactate dehydrogenase induces protective autophagy in gastric cancer cells: involvement of the Akt-mTOR signaling pathway. *Cancer Lett* 2015;**358**(1):17–26.
119. Lee KC, Maturo C, Shorr R, Rodriguez R. Uncommon toxicologic profile at toxic doses of CPI-613 (an agent selectively alters tumor energy metabolism) in rats and minipigs reflects novel mechanism. *Am J Pharmacol Toxicol* 2010;**5**:183–208.
120. Kolesnik DL, Pyaskovskaya ON, Boychuk IV, Dasyukevich OI, Melnikov OR, Tarasov AS, et al. Effect of dichloroacetate on lewis lung carcinoma growth and metastasis. *Exp Oncol* 2015;**37**(2):126–9.
121. Shen H, Hau E, Joshi S, Dilda PJ, McDonald KL. Sensitization of glioblastoma cells to irradiation by modulating the glucose metabolism. *Mol Cancer Ther* 2015;**8**:1794–804.
122. Brooks GA. The science and translation of lactate shuttle theory. *Cell Metab* 2018;**27**(4):757–85.
123. Bonuccelli G, Tsirigos A, Whitaker-Menezes D, Pavlides S, Pestell RG, Chiavarina B, et al. Ketones and lactate "fuel" tumor growth and metastasis: evidence that epithelial cancer cells use oxidative mitochondrial metabolism. *Cell Cycle* 2010;**9**(17):3506–14.
124. Draoui N, Feron O. Lactate shuttles at a glance: from physiological paradigms to anti-cancer treatments. *Dis Model Mech* 2011;**4**(6):727–32.
125. Whitaker-Menezes D, Martinez-Outschoorn UE, Lin Z, Ertel A, Flomenberg N, Witkiewicz AK, et al. Evidence for a stromal-epithelial "lactate shuttle" in human tumors: MCT4 is a marker of oxidative stress in cancer-associated fibroblasts. *Cell Cycle* 2011;**10**(11):1772–83.
126. Sanità P, Capulli M, Teti A, Galatioto GP, Vicentini C, Chiarugi P, et al. Tumor-stroma metabolic relationship based on lactate shuttle can sustain prostate cancer progression. *BMC Cancer* 2014; **14**(1):154.
127. Pavlides S, Whitaker-Menezes D, Castello-Cros R, Flomenberg N, Witkiewicz AK, Frank PG, et al. The reverse Warburg effect: aerobic glycolysis in cancer associated fibroblasts and the tumor stroma. *Cell Cycle* 2009;**8**(23):3984–4001.
128. Marchiq I, Pouysségur J. Hypoxia, cancer metabolism and the therapeutic benefit of targeting lactate/H+ symporters. *J Mol Med* 2016;**94**(2):155–71.

129. Diehl K, Dinges LA, Helm O, Ammar N, Plundrich D, Arlt A, et al. Nuclear factor E2-related factor-2 has a differential impact on MCT1 and MCT4 lactate carrier expression in colonic epithelial cells: a condition favoring metabolic symbiosis between colorectal cancer and stromal cells. *Oncogene* 2018; **37**(1):39.
130. Pértega-Gomes N, Vizcaíno JR, Attig J, Jurmeister S, Lopes C, Baltazar F. A lactate shuttle system between tumour and stromal cells is associated with poor prognosis in prostate cancer. *BMC Cancer* 2014;**14**(1):352.
131. Balliet RM, Capparelli C, Guido C, Pestell TG, Martinez-Outschoorn UE, Lin Z, et al. Mitochondrial oxidative stress in cancer-associated fibroblasts drives lactate production, promoting breast cancer tumor growth: understanding the aging and cancer connection. *Cell Cycle* 2011;**10**(23):4065–73.
132. Jiang P, Du W, Wu M. Regulation of the pentose phosphate pathway in cancer. *Protein Cell* 2014; **5**(8):592–602.
133. Riganti C, Gazzano E, Polimeni M, Aldieri E, Ghigo D. The pentose phosphate pathway: an antioxidant defense and a crossroad in tumor cell fate. *Free Radic Biol Med* 2012;**53**(3):421–36.
134. Board M, Humm S, Newsholme EA. Maximum activities of key enzymes of glycolysis, glutaminolysis, pentose phosphate pathway and tricarboxylic acid cycle in normal, neoplastic and suppressed cells. *Biochem J* 1990;**265**(2):503–9.
135. Patra KC, Hay N. The pentose phosphate pathway and cancer. *Trends Biochem Sci* 2014;**39**(8):347–54.
136. Cosentino C, Grieco D, Costanzo V. ATM activates the pentose phosphate pathway promoting antioxidant defence and DNA repair. *EMBO J* 2011;**30**(3):546–55.
137. Stincone A, Prigione A, Cramer T, Wamelink MM, Campbell K, Cheung E, et al. The return of metabolism: biochemistry and physiology of the pentose phosphate pathway. *Biol Rev* 2015;**90**(3): 927–63.
138. Vizán P, Alcarraz-Vizán G, Díaz-Moralli S, Solovjeva ON, Frederiks WM, Cascante M. Modulation of pentose phosphate pathway during cell cycle progression in human colon adenocarcinoma cell line HT29. *Int J Cancer* 2009;**124**:2789–96.
139. Jiang P, Du W, Wang X, Mancuso A, Gao X, Wu M, et al. p53 regulates bioyntgesis through direct inactivation of glucose-6-phosphate dehydrogenase. *Nat Cell Biol* 2011;**13**:310–6.
140. Bensaad K, Tsuruta A, Selak MA, Vidal MN, Nakano K, Bartrons R, et al. TIGAR, a p53-inducible regulator of glycolysis and apoptosis. *Cell* 2006;**126**(1):107–20.
141. Ramos-Montoya A, Lee WNP, Bassilian S, Lim S, Trebukhina RV, Kazhyna MV, et al. Pentose phosphate cycle oxidative and nonoxidative balance: a new vulnerable target for overcoming drug resistance in cancer. *Int J Cancer* 2006;**119**(12):2733–41.
142. Kuo WY, Lin JY, Tang TK. Human glucose-6-phosphate dehydrogenase (G6PD) gene transforms NIH 3T3 cells and induces tumors in nude mice. *Int J Cancer* 2000;**85**(6):857–64.
143. Shin ES, Park J, Shin JM, Cho D, Cho SY, Shin DW, et al. Catechin gallates are NADP+-competitive inhibitors of glucose-6-phosphate dehydrogenase and other enzymes that employ NADP+ as a coenzyme. *Bioorg Med Chem* 2008;**16**(7):3580–6.
144. Rainieri R, Levy HR. On the specificity of steroid interaction with mammary glucose-6-phosphate dehydrogenase. *Biochemistry* 1970;**9**:2233–43.
145. Mark PA, Banks J. Inhibition of mammalian glucose-6-phosphate dehydrogenase by steroids. *Proc Natl Acad Sci U S A* 1960;**46**:447–52.
146. Raiss B, Comin B, Puigjaner J, Brandes JL, Creppy E, Saboureau D, et al. Oxythiamine and dehydroepiandrosterone induce a G1 phase cycle arrest in Ehrlich's tumor cells through inhibition of the pentose cycle. *FEBS Lett* 1999;**456**:113–8.
147. Stutowicz A, Kwiatkowsky J, Angielski S. Lipogenetic and glucolytic enzyme activities in carcinoma and non-malignant diseases of the human breast. *Br J Cancer* 1979;**38**:681–7.
148. Kuhajda FP, Jenner K, Wood FD, Hennigar RA, Jacobs LB, et al. Fatty acid synthesis: a potential selective target for antineoplastic therapy. *Proc Natl Acad Sci U S A* 1994;**91**(14):6379–83.
149. Kuhajda FP. Fatty acid synthase and cancer: new application of an old pathway. *Cancer Res* 2006; **66**(12):5977–80.
150. Menendez JA, Lupu R. Fatty acid synthase and the lipogenic phenotype in cancer pathogenesis. *Nat Rev Cancer* 2007;**7**(10):763.

151. Flavin R, Peluso S, Nguyen PL, Loda M. Fatty acid synthase as a potential therapeutic target in cancer. *Future Oncol* 2010;**6**(4):551–62.
152. Medes G, Thomas A, Weinhouse S. Metabolism of neoplastic tissue. IV. A study of lipid synthesis in neoplastic tissue slices in vitro. *Cancer Res* 1952;**13**:27–9.
153. Zaidi N, Lupien L, Kuemmerle NB, Kinlaw WB, Swinnen JV, Smans K. Lipogenesis and lipolysis: the pathways exploited by the cancer cells to acquire fatty acids. *Prog Lipid Res* 2013;**52**(4):585–9.
154. Horton JD, Goldstein JL, Brown MS. SREBPs: activators of the complete program of cholesterol and fatty acid synthesis in the liver. *J Clin Invest* 2002;**109**:1125–31.
155. Wakil SJ. Fatty acid synthase, a proficient multifunctional enzyme. *Biochemistry* 1989;**28**:4523–30.
156. Alberts AW, Strauss AW, Hennessy S, Vagelos PR. Regulation of synthesis of hepatic fatty acid synthetase: binding of fatty acid synthetase antibodies to polysomes. *Proc Natl Acad Sci U S A* 1975; **72**(10):3956–60.
157. Stoops JK, Arslanian MJ, Oh YH, Aune KC, Vanaman TC, Wakil SJ. Presence of two polypeptide chains comprising fatty acid synthase. *Proc Natl Acad Sci U S A* 1975;**72**(5):1940–4.
158. Smith S, Witkowski A, Joshi AK. Structural and functional organization of the animal fatty acid synthase. *Prog Lipid Res* 2003;**42**(4):289–317.
159. Chirala SS, Wakil SJ. Structure and function of animal fatty acid synthase. *Lipids* 2004;**39**(11):1045–53.
160. Weiss L, Hoffmann GE, Schreiber R, Andres H, Fuchs E, et al. Fatty-acid biosynthesis in man, a pathway of minor importance. Purification, optimal assay conditions and organ distribution of fatty acid synthase. *Biol Chem Hoppe Seyler* 1986;**367**:905–12.
161. Alo PL, Visca P, Marci A, Mangoni A, Botti C, et al. Expression of fatty acid synthase (FAS) as a predictor of recurrence in stage I breast carcinoma patients. *Cancer* 1996;**77**:474–82.
162. Milgraum LZ, Witters LA, Pasternack GR, Kuhajda FP. Enzymes of the fatty acid synthesis pathway are highly expressed in in situ breast carcinoma. *Clin Cancer Res* 1997;**3**:2115–20.
163. Epstein JI, Carmichael M, Partin AW. OA-519 (fatty acid synthase) as an independent predictor of pathologic stage in adenocarcinoma of the prostate. *Urology* 1995;**45**:81–6.
164. Paulauskis JD, Sul HS. Hormonal regulation of mouse fatty acid synthase gene transcription in liver. *J Biol Chem* 1989;**264**(1):574–7.
165. Latasa MJ, Griffin MJ, Moon YS, Kang C, Sul HS. Occupancy and function of the -150 sterol regulatory element and -65 E-box in nutritional regulation of the fatty acid synthase gene in living animals. *Mol Cell Biol* 2003;**23**(16):5896–907.
166. Baron A, Migita T, Tang D, Loda M. Fatty acid synthase: a metabolic oncogene in prostate cancer? *J Cell Biochem* 2004;**91**(1):47–53.
167. Camassei FD, Cozza R, Acquaviva A, Jenkner A, Rava L, Gareri R, et al. Expression of the lipogenic enzyme fatty acid synthase (FAS) in retinoblastoma and its correlation with tumor aggressiveness. *Invest Ophthalmol Vis Sci* 2003;**44**:2399–403.
168. Menendez JA, Vellon L, Mehmi I, Oza BP, Ropero S, Colomer R, et al. Inhibition of fatty acid synthase (FAS) suppresses HER2/neu (erbB-2) oncogene overexpression in cancer cells. *Proc Natl Acad Sci U S A* 2004;**101**(29):10715–20.
169. Algire C, Amrein L, Zakikhani M, Panasci L, Pollak M. Metformin blocks the stimulative effect of a high energy diet on colon carcinoma growth in vivo and is associated with reduced expresión of fatty acid synthase. *Endocr Relat Cancer* 2010;**17**:351–60.
170. Loubière C, Goiran T, Laurent K, Djabari Z, Tanti JF, Bost F. Metformin-induced energy deficiency leads to the inhibition of lipogenesis in prostate cancer cells. *Oncotarget* 2015;**6**(17):15652–61.
171. Wahdan-Alaswad RS, Cochrane DR, Spoelstra NS, Howe EN, Edgerton SM, Anderson SM, et al. Metformin-induced killing of triple-negative breast cancer cells is mediated by reduction in fatty acid synthase via miRNA-193b. *Horm Cancer* 2014;**5**(6):374–89.
172. Caldieri G, Giacchetti G, Beznoussenko G, Attanasio F, Ayala I, et al. Invadopodia biogenesis is regulated by caveolin-mediated modulation of membrane cholesterol levels. *J Cell Mol Med* 2009;**13**:1728–40.
173. Yamaguchi H, Takeo Y, Yoshida S, Kouchi Z, Nakamura Y, et al. Lipid rafts and caveolin-1 are required for invadopodia formation and extracellular matrix degradation by human breast cancer cells. *Cancer Res* 2009;**69**:8594–602.

174. Swinnen JV, Van Veldhoven PP, Timmermans L, De Schrijver E, Brusselmans K, et al. Fatty acid synthase drives the synthesis of phospholipids partitioning into detergent-resistant membrane microdomains. *Biochem Biophys Res Commun* 2003;**302**:898–903.
175. Busco G, Cardone RA, Greco MR, Bellizzi A, Colella M, Antelmi E, et al. NHE1 promotes invadopodial ECM proteolysis through acidification of the peri-invadopodial space. *FASEB J* 2010; **24**(10):3903–15.
176. Glynn SA, O'Sullivan D, Eustace AJ, Clynes M, O'Donovan N. The 3-hydroxy-3-methylglutaryl-coenzyme A reductase inhibitors, simvastatin, lovastatin and mevastatin inhibit proliferation and invasion of melanoma cells. *BMC Cancer* 2008;**8**(1):9.
177. Collisson EA, Kleer C, Wu M, De A, Gambhir SS, Merajver SD, Kolodney MS. Atorvastatin prevents RhoC isoprenylation, invasion, and metastasis in human melanoma cells. *Mol Cancer Ther* 2003; **2**(10):941–8.
178. Eagle H. Nutrition needs of mammalian cells in tissue culture. *Science* 1955;**22**:501–14.
179. Wise DR, Thompson CB. Glutamine addiction: a new therapeutic target in cancer. *Trends Biochem Sci* 2010;**35**(8):427–33.
180. Lacey JM, Wilmore DW. Is glutamine a conditionally essential amino acid? *Nutr Rev* 1990; **48**(8):297–309.
181. Hensley CT, Wasti AT, DeBernadinis RJ. Glutamine and cancer: cell biology, physiology, and clinical opportunities. *J Clin Invest* 2013;**123**(9):3678–84.
182. Kim HK, Kim H. Oncogenes and tumor suppressors regulate glutamine metabolism in cancer cells. *J Cancer Prev* 2013;**18**(3):221–6.
183. Ramsay EE, Hogg PJ, Dilda PJ. Mitochondrial metabolism inhibitors for cancer therapy. *Pharm Res* 2011;**28**(11):2731.
184. Canapé C, Catanzaro G, Terreno C, et al. Probing treatment response of glutaminolytic prostate cancer cells to natural drugs with hyperpolarized [5-13C]glutamine. *Magn Reson Med* 2015;**73**(6):2296–305.
185. Darmaun D, Welch S, Rini A, Sager BK, Altomare A, Haymond MW. Phenylbutyrate-induced glutamine depletion in humans: effect on leucine metabolism. *Am J Physiol Endocrinol Metab* 1998;**274**(5): E801–7.
186. Corbet C, Draoui N, Polet F, Pinto A, Drozak X, Riant O, Feron O. The SIRT1/HIF-2α axis drives reductive glutamine metabolism under chronic acidosis and alters tumor response to therapy. *Cancer Res* 2014;canres–0705.
187. Fendt SM, Bell EL, Keibler MA, Davidson SM, Wirth GJ, Fiske B, et al. Metformin decreases glucose oxidation and increases the dependency of prostate cancer cells on reductive glutamine metabolism. *Cancer Res* 2013;**73**(14):1–10.
188. Maddocks OD, Berkers CR, Mason SM, Zheng L, Blyth K, Gottlieb E, Vousden KH. Serine starvation induces stress and p53-dependent metabolic remodelling in cancer cells. *Nature* 2013;**493**(7433):542–6.
189. Chaneton B, Hillmann P, Zheng L, et al. Serine is a natural ligand and allosteric activator of pyruvate kinase M2. *Nature* 2012;**491**(7424):458–62.
190. Labuschagne CF, van den Broek NJ, Mackay GM, Vousden KH, Maddocks OD. Serine, but not glycine, supports one-carbon metabolism and proliferation of cancer cells. *Cell Rep* 2014;**7**(4): 1248–58.
191. Possemato R, Marks KM, Shaul YD, Pacold ME, Kim D, Birsoy K, et al. Functional genomics reveal that the serine synthesis pathway is essential in breast cancer. *Nature* 2011;**476**(7360):346.
192. Gerweck LE, Seetharaman K. Cellular pH gradient in tumor versus normal tissue: potential exploitation for the treatment of cancer. *Cancer Res* 1996;**56**(6):1194–8.
193. Rich IN, Worthington-White D, Garden OA, Musk P. Apoptosis of leukemic cells accompanies reduction in intracellular pH after targeted inhibition of the Na+/+ exchanger. *Blood* 2000;**95**(4): 1427–34.
194. Harguindey S, Gillis M. Evolution and cancer: Possible relationships to changes in environmental hydrogen ion concentration. *J Theor Biol* 1980;**86**(3):487–92.

195. Harguindey S, Stanciu D, Devesa J, Alfarouk KO, Cardone RA, Polo Orozco JD, et al. Cellular acidification as a new approach to cancer treatment and to the understanding and therapeutics of neurodegenerative diseases. *Semin Cancer Biol* 2017;**43**:157–79.
196. Harguindey S, Arranz JL, Orozco JD, Rauch C, Fais S, Cardone RA, et al. Cariporide and other new and powerful NHE1 inhibitors as potentially selective anticancer drugs–an integral molecular/biochemical/metabolic/clinical approach after one hundred years of cancer research. *J Transl Med* 2013;**11**(1):282.
197. Martin GR, Jain RK. Noninvasive measurement of interstitial pH profiles in normal and neoplastic tissue using fluorescence ratio imaging microscopy. *Cancer Res* 1994;**54**(21):5670–4.
198. Reshkin SJ, Greco MR, Cardone RA. Role of pHi, and proton transporters in oncogene-driven neoplastic transformation. *Philos Trans R Soc B: Biol Sci* 2014;**369**(1638):20130100.
199. Wakelam MJ, Houslay MD, Davies SA, Marshall CJ, Hall A. The role of N-ras p21 in the coupling of growth factor receptors to inositol phospholipid turnover. *Biochem Soc Trans* 1987;**15**:45–7. 619th Meeting, Portland Press Limited, Cambridge.
200. Berridge MJ, Irvine RF. Inositol phosphates and cell signaling. *Nature* 1989;**341**:197.
201. Kamm DE, Strope GL. The effects of acidosis and alkalosis on the metabolism of glutamine and glutamate in renal cortex slices. *J Clin Invest* 1972;**51**(5):1251–63.
202. Kritchevsky D. Caloric restriction and cancer. *J Nutr Sci Vitaminol (Tokyo)* 2001;**47**(1):13–9.
203. Mc Neil C. Caloric restriction research come of age. *J Natl Cancer Inst* 1996;**88**(16):1105–7.
204. Seyfried TN, Shelton LM. Cancer as a metabolic disease. *Nutr Metab* 2010;**7**:7.
205. Baggetto LG. Deviant energetic metabolism of glycolytic cancer cells. *Biochimie* 1992;**74**(11):959–74.
206. Kiebish MA, Han X, Cheng H, Chuang JH, Seyfried TN. Cardiolipin and electron transport chain abnormalities in mouse brain tumor mitochondria: lipidomic evidence supporting the Warburg theory of cancer. *J Lipid Res* 2008;**49**(12):2545–56.
207. Clément M.V., Akram S., Kumar A.P., Chang M.K., Pervaiz S. 3. Reactive oxygen species, intracellular pH, and cell fate. In: Lagadic-Gossmann D, editor. Proton homeostasis in tumorigenesis and cell death; 2011. p. 49–64, ISBN: 978-81-308-0429-3.
208. Anastasiou D, Poulogiannis G, Asara JM, Boxer MB, Jiang JK, Shen M, et al. Inhibition of pyruvate kinase M2 by reactive oxygen species contributes to cellular antioxidant responses. *Science* 2011;**334**(6060):1278–83.
209. Chakrabarti G, Gerber DE, Boothman DA. Expanding antitumor therapeutic windows by targeting cancer-specific nicotinamide adenine dinucleotide phosphate-biogenesis pathways. *Clin Pharmacol* 2015;**7**:57–68.
210. Johnson WJ, McCoLL JD. 6-Aminonicotinamide- A potent nicotinamide antagonist. *Science* 1955;**122**:834.
211. Herter FP, Weissman SG, Thompson HG, Hyman G, Martin DS. Clinical experience with 6-Aminonicotinamide. *Cancer Res* 1961;**21**:31–7.
212. Perlia CP, Kofman S, Sky-Peck H, Taylor SG. Clinical use of 6- Aminonicotinamide in patients with disseminated neoplastic disease. *Cancer* 1961;**14**:644–8.
213. Corbet C, Pinto A, Martherus R, de Jesus JP, Polet F, Feron O. Acidosis drives the reprogramming of fatty acid metabolism in cancer cells through changes in mitochondrial and histone acetylation. *Cell Metab* 2016;**24**(2):311–23.
214. Cervantes-Madrid D, Dueñas-Gonzalez A. Antitutumor effects of a drug combination targeting glycolysis, glutaminolysis and de novo synthesis of fatty acids. *Oncol Rep* 2015;**34**(3):1533–42.
215. Lea MA, Qureshi MS, Buxhoeveden M, et al. Regulation of the proliferation of colon cancer cells by compounds that affect glycolysis, including 3-bromopyruvate, 2-deoxyglucose and biguanides. *Anticancer Res* 2013;**33**(2):401–7.
216. Schwartz L, Buhler L, Icard P, Lincet H, Steyaert JM. Metabolic treatment of cancer: intermediate results of a prospective case series. *Anticancer Res* 2014;**34**(2):973–80.
217. Brüning A, Kast RE. Oxidizing to death:disulfiram for cancer cell killing. *Cell Cycle* 2014;**13**(10):1513–4.

218. Cheong JH, Park ES, Liang J, Dennison JB, Tsavachidou D, Nguyen-Charles C, et al. Dual inhibition of tumor energy pathway by 2-deoxy glucose and metformin is effective against a broad spectrum of preclinical cancer models. *Mol Cancer Ther* 2011;**2**:molcanther–0497.
219. Choi YW, Lim IK. Sensitization of metformin-cytotoxicity by dichloroacetate via reprogramming glucose metabolism in cancer cells. *Cancer Lett* 2014;**346**(2):300–8.
220. Li B, Li X, Ni Z, Zhang Y, Zeng Y, Yan X, et al. Dichloroacetate and metformin synergistically suppress the growth of ovarian cancer cells. *Oncotarget* 2016;**7**(37):59458.
221. Currie E, Schulze A, Zechner R, Walther TC, Farese Jr RV. Cellular fatty acid metabolism and cancer. *Cell Metab* 2013;**18**(2):153–61.

CHAPTER 3

The pH-centered paradigm in cancer

Part I: The pH paradigm in cancer. Introduction

Cancer is like the mythological figure represented by René Lalique's jewel: a multi-headed monster (Fig. 1).

Fig. 1 René Lalique's multi-headed serpent.

If we want to kill a creature like this, it is not enough to cut off one head, but all of them must be destroyed. Even if just one head survives, the monster can continue to grow and flourish.

It has taken a long time to understand that cancer has many heads, and no matter how hard we hit one of them, in the end the monster will survive because it has many others.

An Innovative Approach to Understanding and Treating Cancer: Targeting pH
https://doi.org/10.1016/B978-0-12-819059-3.00003-4
© 2020 Elsevier Inc.
All rights reserved.

In spite of this knowledge, we are still trying to perform our medical task by attacking two or at most three of these heads, but the others remain intact. To make things worse, the heads can grow back, and will be even stronger than the ones we destroyed and these new heads will be immune to our attacks.

Traditionally we target the mitotic apparatus destroying cancer cells but at a very high price because we also destroy the reproduction of normal and absolutely necessary cells. Alternatively we restrain certain enzymatic circuits, at a very high price as well because these circuits are also used by normal cells; or we restore immunocompetence with antibodies against PD-1 and CTLA-4, but with poor patient tolerance…and so on. Every new target needs toxic and sometimes unbearable treatments. And sooner or later, most of our treatments, with the exception of initial surgery, are rendered useless because malignant tumors always find the way to develop resistance. The Darwinian evolution of cancer works against our therapeutic resources: the more we attack, the more resistant cancer becomes.[1–3]

Today, we are fully aware that there will be no silver bullet to cure cancer. When the first targeted treatments appeared, we had the sensation that the magic bullet was finally within reach. Now we know that it was not so. Advances in molecular biology are creating a new illusion: that genetically tailored treatments for each patient may do the trick. And sometimes they do. Unfortunately, cancer cells always find new ways to circumvent our treatments.

Common view points or perceptions of cancer origin and progression

The gene-centric perception of cancer is based on the concept that different tumors express the same or several different genes. For the sake of simplicity, osteosarcoma over-expresses gene A, small-cell lung carcinoma over-expresses gene B, and nonsmall-cell lung carcinoma over-expresses gene C. Researches work to find the way to repress or counteract the expression of gene A, B or C according to the tumor they intend to treat with specifically tailored drugs. But now, gene A has mutated and new clones appear that over-express gene D. In other words, the tumor has become resistant to treatment. Besides, all advanced tumors are very heterogeneous and in spite of its monoclonal origin, after some time they become polyclonal and resistant to any tailored treatment addressed to the driver gene.

Therefore, from the genetic point of view cancer, is not one disease but a multiplicity of diseases. As a consequence, each of "these (more than 200) different diseases" will require a specifically tailored variety of therapeutic approaches in order to treat a specific tumor. Difficult, but not impossible. A more rational alternative is to search and find for a few common denominators that can apply to a multiplicity of cancers. Here is where the new pH-centric paradigm comes into play.

The pH-centered perception of cancer. Regardless of their origin and genetic background and as opposed to normal cells cancer cells have an aberrant regulation of pH, leading to an inverted intracellular to extracellular pH gradient (pHi to pHe). This disturbance in pH arises very early in carcinogenesis and is one of the most common hallmarks of malignancy. Independently of the mutated gene or genes, the pH-centered concept addresses the mechanism or mechanisms by which the mutated genes produce the malignant phenotype: the inversion of the pH gradient in general and the intracellular alkalosis in particular. The driver gene or genes, the oncogenic mutations, or the pathways involved make no major difference. At the end of the road all the malignant tumors "need" to alter their pH balance in a way that lets them proliferate, migrate, invade and metastasize. By tackling these pH alterations we can make it very difficult for the tumor to continue with its evolution.

Recently, there has been a significant increase in our knowledge concerning the importance and role of pHi and pHe in the etiopathogenesis and progression of cancer. This inverted pH gradient is driven by a series of proton (H^+) export mechanisms that underlie the initiation and further development of the neoplastic process. Indeed, the intracellular acidification induced by proton transport inhibitors (PTI) and/or other drugs has been repeatedly suggested as a potential treatment of cancer.[4] In this context, one of the primary and best studied regulators of both pHi and pHe in tumors is the Na^+/H^+ exchanger isoform 1 (NHE1) (which will be analyzed in depth in this book).

Tumors have a unique metabolic system and a concerted strategy to survive, grow and metastasize.[5] (Fig. 2).

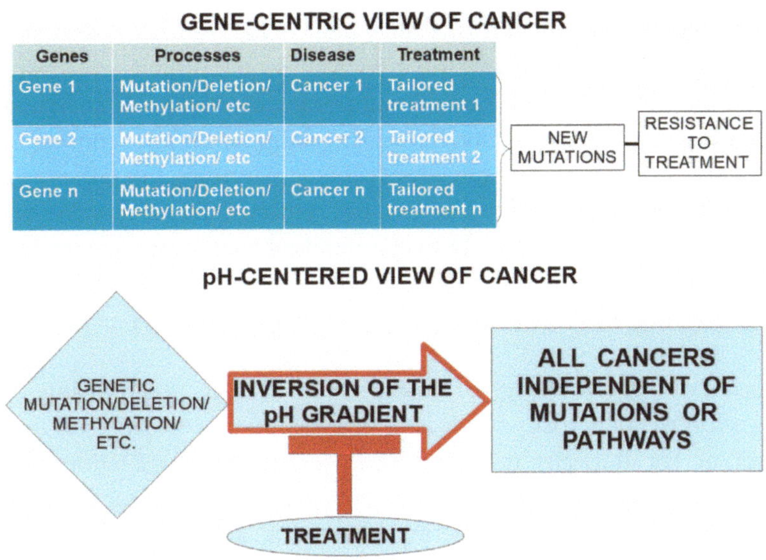

Fig. 2 Two different perceptions of cancer.

Warburg discovered that even in the presence of oxygen malignant cells preferentially degraded glucose to lactic acid (glycolytic metabolism) instead of metabolizing it into CO_2 and H_2O (oxidative phosphorylation), as normal cells usually do.

This glycolytic metabolism of cancer was under-appreciated for almost a century until a recent revival of interest in the fundamental role of the tumor microenvironment and glycolysis in cancer growth and progression.[6–8] This has led the scientific community to adopt the differential tumor metabolism as an additional hallmark of cancer.[9]

Thus, a major shift is taking place from the gene-centric view that has dominated cancer research for the past 20 years, towards the search for the fundamental principles that all tumors share. The gene-centric approach has produced a perception of cancer as a complex group of unrelated diseases and this has led to the idea of a tailored therapy for each patient, based on the tumor's pattern of gene expression. On the other hand, the pH-centered approach allows us to consider cancer as a single disease process (Fig. 2).

Conventional chemotherapies, including the targeted treatments, produce extensive damage to normal cells with serious and sometimes intolerable side effects. Cancer cells and normal cells share the same mitotic apparatus, so that normal and pathologic proliferation are equally handicapped. Thus, if we go against the mitotic apparatus, we will slow down the proliferation of cancer cells, but at the same time we are constrained by toxicity to hematopoietic cells and the intestinal mucosa. This adverse phenomenon on normal tissues is a constant feature with many cancer treatments.

We must select and target those characteristics which are unique to malignant cells. But there is one characteristic of cancer that is not shared with normal tissues: the inversion of the pH gradient. Therefore, modifying the altered intracellular and extracellular acid-base balance in cancerous tissues has negligible effects on normal cells because the attack only targets glycolytic cells. The pH abnormalities in cancer offer an easily treatable soft spot, that can be attacked with existing drugs, with very low toxicity, adequate efficiency, low cost and high tolerability. At the same time, they can be used along with well established treatment protocols that may blast the monster's other heads. Attacking cancer's pH abnormalities is by no means a replacement for other scientifically accepted treatments, but rather a new and synergistic approach that will hopefully improve the results of cancer therapeutics.

Table 1 compares the side effects between standard chemotherapeutic drugs and pH-centered treatments.

Table 2 compares the characteristics of conventional and pH-centered therapeutic strategies.

At this point of our knowledge of cancer's biology, it is quite possible that only the coordinated effort to cut as many heads as possible is the appropriate path for a change in cancer treatment. By all means, it is much simpler than creating a tailored treatment for each patient. A rational approach to targeting the weak spots of cancer should be highly rewarding, if at the same time this targeting is not toxic, and is well tolerated by patients. Furthermore, if repurposed drugs can be used for attacking these weak links in the cancer chain, new approaches could be developed.

The pH paradigm in cancer: Extracellular acidity, intracellular alkalinity and the pH gradient

Søren Sørensen developed a formula based on negative logarithms for expressing proton—or hydrogen ion concentrations, which he called pH, *pondus hydrogenii* (or hydrogen power). pH is a highly compartmentalized phenomenon in multicellular

Table 1 A comparison of side effects between standard chemotherapeutic drugs and pH-centered treatments.

Treatment	Mechanism	Adverse effects
Alkylating agents	DNA damage	Mutations and damage to normal cells. Immunosuppression
Cytotoxic antibiotics	DNA damage and inhibition of RNA synthesis. Inhibition of topoisomerase and increased ROS production	Damage to normal cells
Mitotic disrupters (anti-microtubule drugs)	Inhibit the mitotic apparatus	Damage to normal cells
Intercalating agents	DNA damage	Mutations and damage to normal cells
Antimetabolites	Inhibition of DNA and RNA synthesis	Damage to normal cells
Topoisomerase inhibitors	DNA damage	Damage to normal cells
MABS (monoclonal antibodies): Bevacizumab	Targets VEGF	High blood pressure, bleeding, poor wound healing, blood clots, kidney damage.
MABS: Cetuximab	Targets EGFR	Serious skin rashes, diarrhea, vomiting, weakness, weight loss, etc.
Immune checkpoint inhibitors (PD-1 and CTLA-4)	Increase immune defense against cancer cells	Fatigue, coughing, nausea, loss of appetite, skin rash and itching, life threatening side effects
CAR-T cell immunotherapy	Immune destruction of cancer cells	Cytokine release syndrome and neurotoxicity
Radiotherapy	Damage to DNA and other structures of the cell producing increased ROS	Destruction of normal tissues
Kinase inhibitors	Inhibits certain signaling pathways	Drug reductions and interruptions are common, definitive drug withdrawals as well as toxic deaths in rare cases.
Hormone therapies	Inhibits certain hormones that stimulate cancer growth, e.g., tamoxifen or the anti-androgen abiraterone	Tamoxifen predisposes to uterine cancer. Hormone treatment for prostate cancer produces erectile dysfunction, fatigue, peripheral edema, joint pain, high blood sugar, etc.
Anti-pH gradient inversion	Modification of intra and extracellular pH. Damage limited to glycolytic cells	None, when it is achieved with repurposed drugs. No damage to normal non- glycolytic cells.

Damage to normal cells includes a long list of adverse events that goes beyond the scope of this summary.

Table 2 Conventional vs. pH-centered treatments.

Conventional treatments	pH-centered treatments
Usually target cancer cells but also have undesirable consequences on normal cells.	Only target cancer cells with minor or no consequences on normal cells.
Patient tolerance is low.	Well tolerated.
Resistance to treatment is a frequent limitation.	Low risk of resistance.
Negative effect on quality of life.	Have almost no effect on quality of life.
High cost.	Low cost.
Main objective is to destroy as many cancer cells as possible.	Main objective is to block tumor evolution in order to achieve a stable disease and selectively destroy all cancer cells.
Require association of multiple toxic drugs.	May be associated with conventional treatments in most of the cases.
Most of the drugs need to be administered intravenously.	Most of these drugs can be administered orally.

organisms. Each compartment has a different pH which is the most appropriate for the biochemical and molecular reactions that take place in that compartment. Homeostasis is mainly the name applied to the mechanisms that organisms employ to keep the pH at the optimal level for the respective compartment. The difference in pH between compartments is the pH gradient (Box 1).

BOX 1

The pH behavior in nature and the cancer pH paradigm

pH is a highly compartmentalized feature in multicellular organisms.

The pH in each compartment can be accurately tuned to optimize cellular functions and is tightly controlled by buffers, proton transporters, and specific channels.

Biochemical reactions have an optimal pH condition in which the process is efficiently carried out. The modification of this optimal pH may hinder the reactions.

Most enzymes are pH-dependent.

A gradient is established between different compartments.

Homeostasis consists essentially in keeping these gradients in a narrow range that allows efficient and optimal functioning of the organism.

Chemical reactions rate is often a function of pH.

Multicellular organisms have developed a complex net of mechanisms to keep pH homeostasis. Frequently these mechanisms are overlapping.

Finally, the pH-centered paradigm in cancer explains the fundamental differences between normal and malignant cell physiology.

In cancer we have to consider the pH levels of two very different compartments, the intracellular compartment and the extracellular (Fig. 3). The pH of the extracellular compartment (pHe) is relatively easy to measure, and it has been universally accepted that pHe in cancerous tissues is lower (more acid) than in their normal counterparts.

Research on tumor pH began in the 1920s when Otto Warburg first suggested that aerobic glycolysis was the main metabolic feature of malignant cells and the primary cause of cancer.[10] His hypothesis was that this preference for glycolysis in cancer, also in the presence of oxygen (aerobic glycolysis), was secondary to mitochondrial damage and produced large quantities of lactic acid. Since then much work has been done and has confirmed cancer's extracellular acidity.[11,12]

On the other hand, measuring intracellular pH (pHi) has been a more complicated task, and it had to wait until adequate technical developments came into the field. It was mainly for this reason that Warburg believed that cancer cells were acidic, and this fundamental concept explains the confusion that dominated the field for decades.[13–15]

While at first sight the differences in pHi values between normal and malignant cells may not seem very great, it should be realized that even small variations in pHi mean a much bigger difference in proton concentrations due to the logarithmic nature of pH expression.

The acid-base balance in cells has long been recognized (since the 1930s) as the main mechanism for maintaining homeostasis. Cells can only survive within a very narrow range of pH (less than one unit). It becomes essential to recognize that the pH of normal cells and cancer cells drastically deviate towards opposite ends. This abnormality represents the greatest known difference between normal cellular physiology and cancer pathophysiology.

Maintaining the different physiological pHs within the body is of vital importance for the organism to stay alive. While blood has a pH between 7.35 and 7.45 that is strictly maintained, almost all the chemicals (mainly nutrients) entering the human organism are very far from the blood pH. For instance, small pH variations beyond the range of 7.35 and 7.45 in body fluids can be corrected, but greater alterations can lead to metabolic acidosis or alkalosis events that may even cause death. The intracellular pH is in a constant state of dynamic equilibrium, which is efficiently maintained by ion fluxes through specific membrane-bound proton pumps, transporters, ion channels, ion exchangers, and buffer systems with some overlapping functions among them.

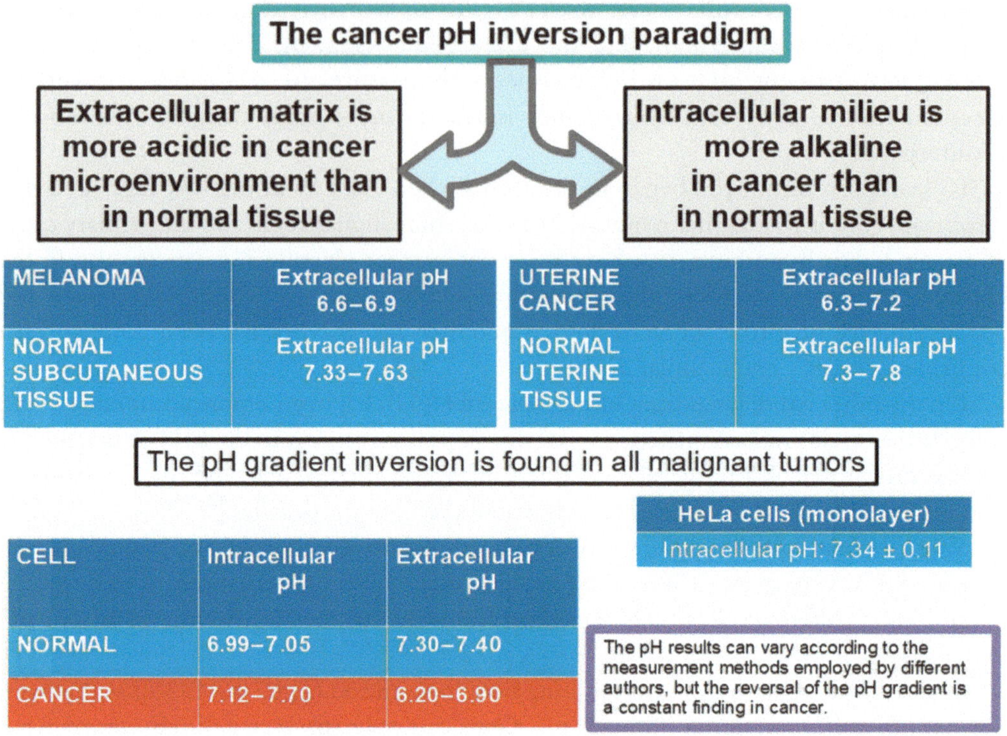

Fig. 3 Inversion of the pH gradient in cancer cells. The differences between normal and cancer tissues are very clear-cut in the extracellular matrix. In the intracellular space, the differences are more subtle and sometimes overlapping values may be found, but in general pHi in cancer is slightly alkaline in comparison with normal counterparts.[16–20]

Effects of pH gradient inversion in cancer

Intracellular alkalinity

The alkaline intracellular milieu represents a proliferative advantage for the malignant cell. Reshkin et al.[21] have shown that cytoplasmic alkalinization is a key transformation-specific process at the very beginning of the oncogenic transformation of a normal cell. In 1982, Winkler et al.[22] found a sharp increase in protein synthesis when pHi reached 6.9 with an optimum at pHi 7.4. However, there has been some controversy about the relationship between proliferation and alkaline intracellular pH. In 1988, Madshus[23] wrote:

> *"Oscillations in intracellular pH have been postulated to be of importance in the control of the cell cycle and cell division in several cell types. Low intracellular pH is common both to prokaryotic and eukaryotic resting cells. This is one reason that these cells have low metabolic activities. A rapid increase in intracellular pH may be important to bring cells from G0/GI and into S phase."*

Similar findings have been reported by Rubin and Fodge[24] and Engstrom[25] stressing the direct relationship between an increase in pH, glycolytic stimulation and DNA biosynthesis.

Indeed, there are pHi oscillations during the different phases of the cell cycle even in normal cells.

Steinhardt & Morisawa[26] showed that, in a normal cell cycle, the pHi was increased to higher alkaline levels within 1 h prior to mitosis. However, when pHi was artificially lowered, mitosis was delayed and even stopped. Lee and Steinhardt[27] observed that meiosis in sea urchins was delayed or inhibited if the intracellular pH became acidic. In 1985 Pouyssegur et al.[28] showed that at pHi below 7.2 growth factors could not induce the G0 to S phase transition. However, at the same time, Mills et al.[29] found that a pHi increase was not essential for proliferation, at least in the case of lymphocytes. Other authors suggested that it was not the pHi that regulated proliferation but that the NHE1 exchanger was the factor responsible for this control. The conclusion is that intracellular alkalosis and extracellular acidosis represent a necessary framework for malignant transformation and tumor development, progression and the activation of metastasis[30,31] (Fig. 4). Whether intracellular pH increase is a necessary "permissive" factor or a determinant issue for proliferation is still being a matter of discussion, the need of NHE1 activity in this sense is beyond controversy.

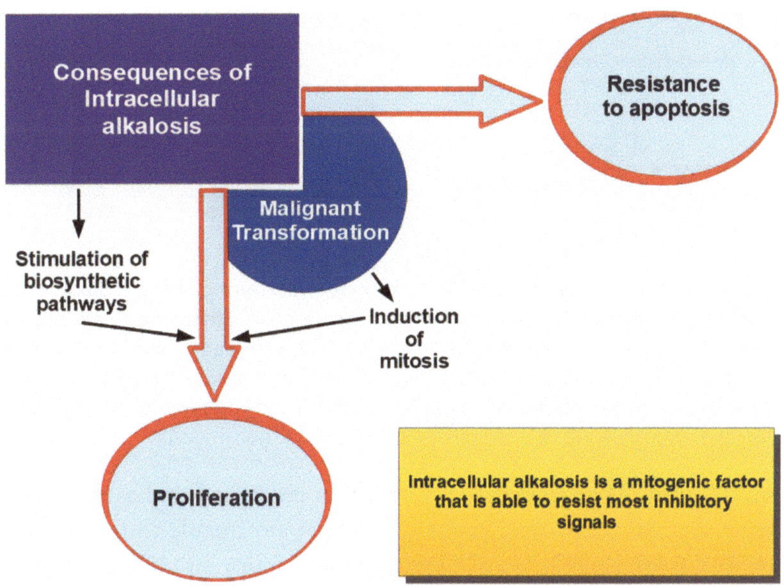

Fig. 4 Pro-tumoral effects of intracellular alkalosis.

Extracellular acidity

Extracellular acidity seems to be a regulator of cellular phenotype.

It is the consequence of:

(1) the metabolic changes produced in malignant cells;

(2) the increased proton production and extrusion from malignant cells that simultaneously produce intracellular alkalinity;

(3) hypoxia;

(4) apoptosis and or necrosis of cells not adapted to new microenvironmental conditions.

The details of how these processes take place will be analyzed in the second part of this chapter.

Extracellular acidity is a key feature that allows cellular migration and invasion through the degradation of the extracellular matrix and promotion of the invadopodial process.

The consequences of extracellular acidity are described in the next three Figs. 5–7.

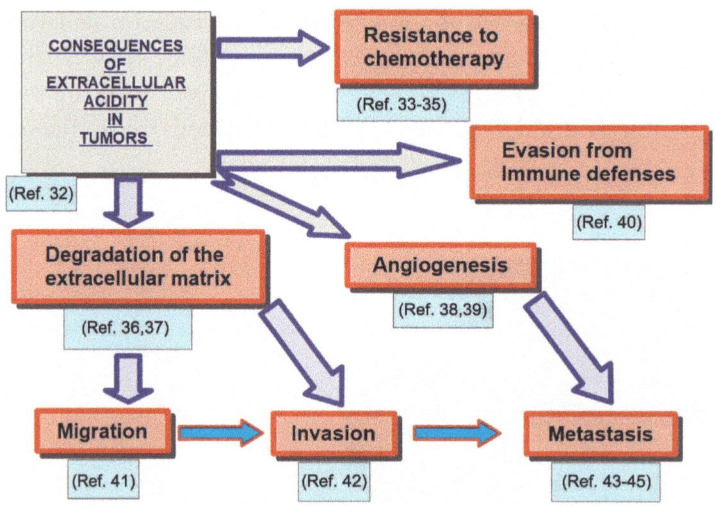

Fig. 5 Consequences of extracellular acidity in cancer development: the main direct consequence of extracellular acidity is the activation of proteolytic enzymes (like cathepsine B, metalloproteases and heparanase) which degrade the extracellular matrix (ECM). This degradation of the ECM is coordinated with the process of migration and invasion, even if all the involved mechanisms are not clearly identified.[32–45] There is no possibility of migration without degradation of the extracellular matrix (see chapter on invadopodia). Without migration, there is neither invasion nor metastasis.

Traditionally, since Warburg, lactic acid has been considered as the main source of extracellular acidity. But experiments with cells that are unable to follow the glycolytic pathway have cast doubts on the issue. Increased production of CO_2 through the pentose phosphate pathway which diffuses freely to the ECM where it is hydrated to CO_3H^- and H^+ (with the intervention of carbonic anhydrase IX and XII) has been recognized as a very important source of protons.[58] Hydrolysis of ATP is possibly another source of

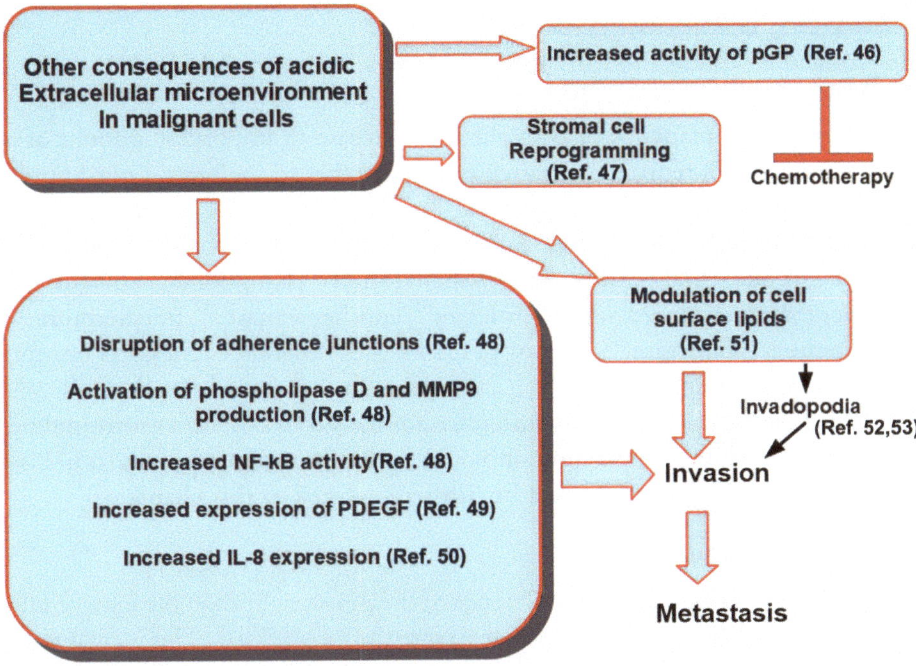

Fig. 6 More consequences of the increased expression of extracellular acidity.[46–53] The mechanism of cell surface lipid modulation by extracellular acidity is not exactly known. In glioblastoma, extracellular acidity produces conformational changes of the cell membrane modifying its rigidity. Extracellular acidity increases GM3 glycosphingolipids in the cell membrane. In addition to acting as a functional barrier to excessive extracellular acidity, GM3 glycosphingolipids also have a role in signal transduction through interaction with cell surface receptors.[54]

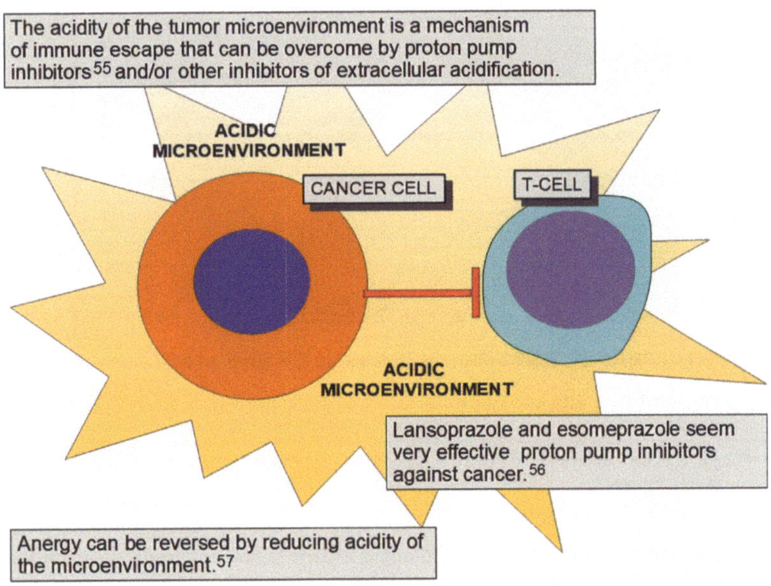

Fig. 7 Immune escape due to extracellular acidity is a process that can be reversed by increasing pH.[55–57]

proton production. Cellular death, whether apoptotic or necrotic, has been neglected as a source of extracellular acidity. However, in tumors that undergo extensive apoptosis/necrosis, whether spontaneous or provoked, the release of important amount of acidic cellular content should represent a source of protons that to the best of our knowledge, has never been measured.

Extracellular acidity is a requisite for the activation of the proteolytic enzymes that digest the ECM like metalloproteases, urokinase form of plasminogen activator, cathepsins[59,60] membrane anchored serine proteases[61] and heparanase.[62] Furthermore, it is a basic requirement for invasion[63] and low pHe increases adherence junction's disruption[64] and stimulates metastatic behavior.[65]

The low extracellular pH creates difficulties for the survival of other surrounding cells and for the activity of anti-tumor immunological cells.[66,67] Low extracellular pH is also a major player in drug resistance as will be further analyzed in this chapter.

The pH gradient inversion

The pH gradient inversion is the consequence of the pH alterations in the extracellular and intracellular compartments, where as we have seen, the intracellular pH is increased and the extracellular pH is decreased in relation to normal cells. All this process is known as the dysregulated proton dynamics and has profound effects on the cancer's behavior.[68–70] Proton extrusion is sufficient to induce this anomaly with all its consequences. Inhibiting proton efflux produced cell death in invasive tumors.[71] Therefore, we may assume that proton extrusion is the *primum movens* that leads to the pH gradient inversion. Fig. 8 summarizes the issue.

Fig. 8 The perfect storm: reversal of the pH gradient is the most accurate definition coined by the team led by Barber[72] of the events leading to malignization.

The hypoxia-pH gradient inversion relationship

There is no doubt that hypoxia, which is almost a constant feature of tumors, is a key promoter of the disease's progression, its resistance to treatment and poor patient outcome.[73]

Hypoxia is a condition where a cell or tissue is deprived of adequate oxygen supply to support a physiological metabolism. This happens when the vascular supply is interrupted or diminished, or when a tumor outgrows its vascular supply. Malignant tumors are usually hypoxic because they grow at a faster pace than their normal counterparts and newly formed vessels are usually inefficient and insufficient to provide the required nutrients and oxygen. Oxygen is delivered along a concentration gradient that is highest at the site of vessels and gradually diminishes until it reaches its final destination cell where it is used by mitochondria to generate high energy molecules. The diffusion distance of oxygen reaches approximately 70–100 μm from blood vessels, so that cells located beyond this distance are chronically exposed to hypoxia.[74]

Hypoxia represents a unique environmental stress that induces multiple changes in cellular physiology through diverse instruments like transcription factors, signaling proteins, epigenetic and metabolic changes, in order to adapt cell function and survival to the new micro-environmental conditions.

In malignant tumors, chronic hypoxia may alternate with periods of reoxygenation at the time new vessels are formed.[75] But normal tissues also may suffer periods of hypoxia and posterior reperfusion.

Therefore, this phenomenon is not unique to cancer. Normal and malignant cells react to hypoxia with profound metabolic changes.

In order to cope with the hypoxia problem, tumors undergo a metabolic reprogramming that permits their accelerated growth and better utilization of available nutrients in conditions of less oxygen and better utilization of available nutrients. Hypoxia is also a stimulus for the up-regulation of proteins with pH regulatory functions like the NHE1 exchanger and membrane carbonic anhydrases (CA IX and CA XII).[76–78] Finally, it has been well established that tumor progression is correlated with genetic instability and hypoxia[79,80] (Table 3).

Table 3 Evidence showing the role of hypoxia in genetic instability

Reference	Findings
Mihaylova et al.[81]	Hypoxia reduced the expression of DNA mismatch repair gene Mlh1, probably by histone deacetylation.
Keysar et al.[82]	Chinese hamster ovary cells under hypoxia/reoxygenation showed increased mutations similar to those obtained with other mutagens. Reducing ROS with DMSO significantly reduced hypoxia toxicity and mutagenesis. The mutations were primarily large DNA deletions.
Rodríguez-Jimenez et al.[83]	Prolonged hypoxia produced genomic instability in human and murine stem cells, particularly in the DNA mismatch repair system (MMR). Usually stem cells are located in hypoxic regions, meaning that they are highly exposed to genomic instability. Trichostatin A, a histone deacetylase inhibitor also enhanced MMR expression.
Bouquet et al.[84]	Hypoxia produced chromatin modifications and activated DNA-PK (DNA-dependent protein kinase) which is involved in DNA double-strand break signaling and repair.
Rice et al.[85]	Exposing CHO cells to hypoxia enhanced the frequency of methotrexate resistance and dihydrofolate reductase gene amplification.
Koshiji et al.[86]	HIF-1α generates genetic instability at the nucleotide level by inhibiting MSH2 and MSH6. HIF-1α displaces the transcriptional activator Myc from Sp1 binding to repress MutSalpha expression, Sp1 recruits HIF-1α to the gene promoter under hypoxia. The experiments were carried out in HCT 116 and HepG2 cell lines (originated in malignant colon and hepatic tumors). But similar results were obtained with a MRC-5 fibroblast cell line (originated in normal human fetal lung).
Yuan et al.[87]	Low pH and hypoxia increased mutagenesis and damaged cellular repair mechanisms in mammalian cells.
Lee et al.[88]	Human lymphocytes obtained from healthy donors were exposed to a hypoxic environment. They showed enhanced sister chromatid exchange compared with their normal counterparts, but no microsatellite instability.
Kumareswaran et al.[89]	DNA double breaks of irradiated cells were tested under hypoxia and normoxia. There was an increase in unrepaired double breaks under hypoxic conditions with increased chromosomal instability.
Bindra et al.[90]	Hypoxia down-regulated the BRCA1 gene through E2F binding to the promoter.
Lu et al.[91]	Hypoxia silenced the BRCA1 gene by epigenetic control of the promoter. It was reversed with a histone deacetylase inhibitor. BRCA1 silencing persisted when the oxygen level returned to normal.
Bindra and Glazer[92]	RAD51 is a critical gene in homologous recombination. Hypoxia decreased its expression in a HIF-independent manner.
Russo et al.[93]	An anoxia-induced endonuclease was identified in rat embryo fibroblasts. This endonuclease was held responsible for enhanced DNA breakage and genomic instability.

Can hypoxia act as a carcinogenic agent by itself independently of genetic instability?

In 1953 Goldblatt and Cameron[94] published an experiment in which they exposed cultured rat heart fibroblasts to intermittent anaerobiosis. Under these conditions, a set of morphologically transformed cells appeared. When the transformed cells were implanted in rats they developed fibrosarcomas. Negative results were obtained with fibroblasts not exposed to intermittent anaerobic conditions. The authors' conclusion was "*that fibroblasts grown in vitro and submitted to repeated periods of anaerobiosis, sometimes become neoplastic*".

At that time, just when the structure of DNA was being explored, Goldblatt and Cameron did not have our sophisticated technology so there was no way of knowing if the cells exposed to intermittent anaerobiosis showed any pro-oncogenic mutations prior to the experiment. Therefore, we can only presume that the developed tumors were a consequence of the intermittent hypoxia. Twenty years later, using mixed cultures of epithelial cells and fibroblasts from the skin of rat embryos, Goldblatt et al.[95] found that when cells were cultivated under hypoxic conditions, they produced malignant tumors when they were transplanted into rats of the same strain.

An important point to highlight is that Goldblatt's first publication was based on intermittent hypoxia and the second, on chronic hypoxia. Goldblatt proposed hypoxia as the cause of malignant transformation, and as a proof of concept, adding oxyhemoglobin to the culture medium prevented malignant transformation by avoiding hypoxia.[96] We must distinguish between acute hypoxia (often anoxia) that produces cell death and chronic relative and/or intermittent hypoxia, with or without re-oxygenation, that is capable of producing DNA damage[97] and therefore, genomic instability, but not death (Table 3).

Further, the well-known phenomenon of some "scar tumors" was suggested to be a consequence of intermittent or relative hypoxia of certain tissues.[98]

Hypoxia and pH gradient inversion

The etiological role of relative hypoxia as a direct etiological factor in oncogenesis (the so called, Goldblatt-Warburg effect), as compared to cellular alkalosis (para-Warburg effect), were further stressed in 1995.[99]

An active proliferative stage such as the one found in cancer requires a slightly alkaline intracellular medium for mitosis and a markedly acid extracellular environment for migration and invasion. Hypoxia can induce both changes because under its influence we find an increased expression of carbonic anhydrase II, IX and XII (CAIX II-CAIX–CAXII) genes and also the expression of the gene for NHE-1[100] (Figs. 9 and 10).

Hypoxia and extracellular acidity are different phenomena, but there is a strong, intricate, and not yet fully understood, relationship between them in the cellular

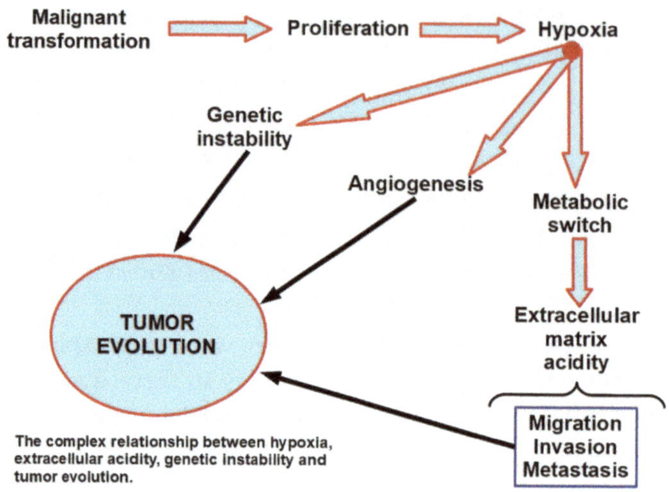

Fig. 9 Hypoxia and its role in the evolution of tumors.

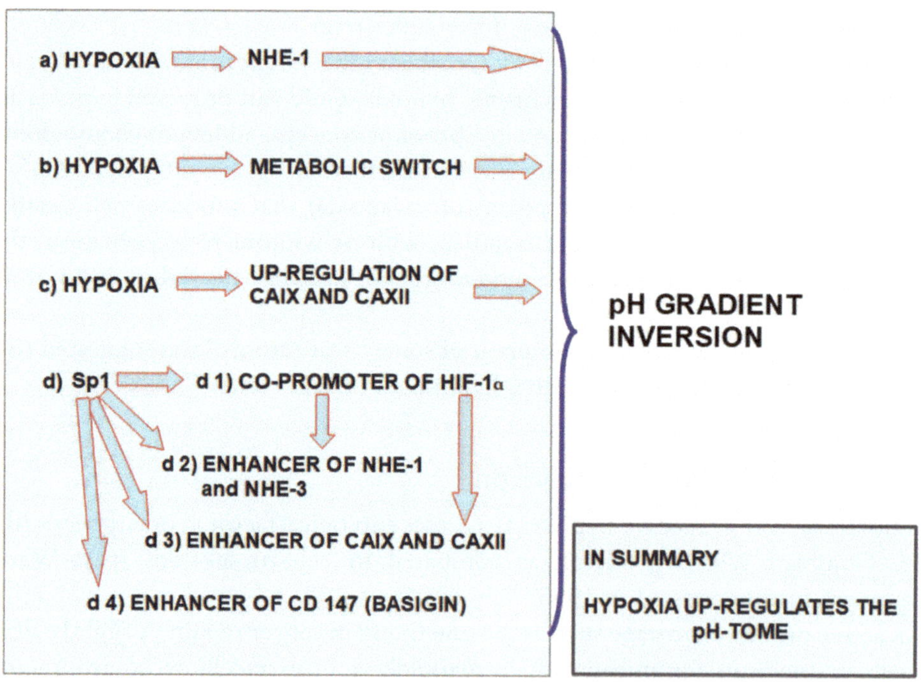

Fig. 10 Interaction between hypoxia and the constituents of the pH gradient inversion.

microenvironment. Also, hypoxia is a well known cause of NHE-1 up-regulation, and together they stimulate proton extrusion.[101]

The metabolic switch (adopting the glycolytic phenotype instead of the oxidative phenotype) is strongly dependent on hypoxia and its agent HIF-1α. While oncogenic transformation already sets the stage for the alterations in pHi and pHe and the first appearance of Warburg metabolism, it is the metabolic switch which is mainly responsible for the later alterations of the pH in progression[102–105] (Fig. 11). For a review of hypoxia as a key regulator mechanism in cancer and its clinical impact, see reference.[106]

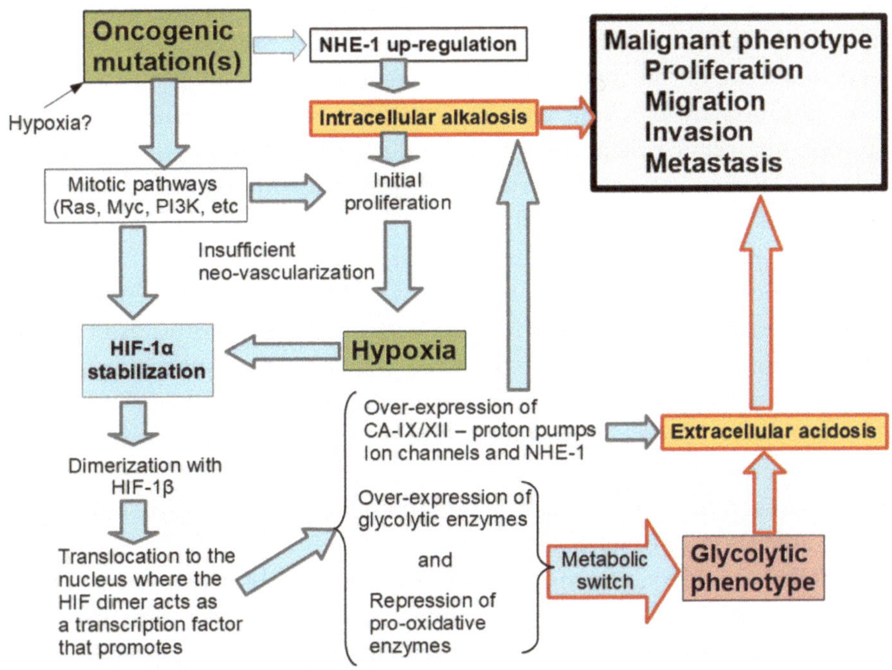

Fig. 11 A full view of the malignization process and the pH centered paradigm.

Clinical impact of the pH centered paradigm in cancer therapeutics

(a) Intracellular acidification as a new pH-centered approach to cancer treatment

The abnormalities of intracellular alkalinization along with extracellular acidification in all types of solid tumors and leukemias have rarely, if ever, been described in any other disease.[107,108] The increasing attempts to induce cellular acidification using proton transport inhibitors (PTIs) and other intracellular acidifiers of different origins is becoming a new therapeutic concept. Cellular acidification is also a tool that may improve selective targeting in cancer treatment, either as a metabolic mediator of apoptosis, to overcome multiple drug resistance (MDR) and/or to restore immune function.

Even though classic medicine recognized the systemic acid-base balance as the fundamental parameter for defining the general concept of homeostasis more than a century ago, this concept has finally been extended to the cellular level (intracellular homeostasis).

The marked deviation of pH occurring in cancer cells and tissues represents the most constant characteristic that differentiates cancer from normality.

The pH-centered approach permits a better understanding of the many faces and stages of malignant disease, as a single and well organized and integrated process driven in one direction by a severely altered acid-base regulation[109,110] (Fig. 11).

Regarding the etiopathogenesis and treatment of cancer, the primary aim of this pH-centered approach is to manipulate the dysregulated abnormal pH in order to stop tumor growth, control local invasion and deactivate the metastatic process, all at once. The main therapeutic objective of this approach is the induction of a pro-apoptotic selective intracellular acidification by blocking proton efflux from the malignant cells.[111,112] This would reverse intracellular alkalinity and extracellular acidity. At the same time, it would also inhibit the Warburg effect and act as an anti-glycolytic measure, thus limiting cancer growth and metastasis. This is a rational approach that can be exploited in the treatment of many, if not all, malignant solid tumors and leukemias.

Since cancer cells use different proton pumps and transporters to protect themselves from an intracellular accumulation of hydrogen ions –in other words, from intracellular acidification, it is most likely that an integrated and concerted utilization of all known inhibitors of proton extrusion in pharmacological dosages will be necessary to treat human malignancies. This is because the multiple pH regulators at the membrane of cancer cells can be co-expressed at the same time (the many heads of the serpent), even in hostile conditions. We have to realize that inhibiting one transporter will be compensated by the over-activation of other/s.

(b) Extracellular acidity, chemoresistance and multiple drug resistance (MDR)

Chemoresistance of tumors is usually a multifactorial event where gene mutations and epigenetic changes seem to be the main culprits through the over-expression of specific proteins that clear the chemotherapeutic drugs. As yet little appreciated microenvironmental factors, like hypoxia and extracellular acidity play an important role in limiting the drugs' abilities to reach the tumor cells in lethal concentrations.[113–115] Ion trapping is one of the reasons why certain chemotherapeutic drugs (weak bases) remain in the highly acidic extracellular environment and the drug does not reach its intracellular destination[116] with the substantial reduction of the cytotoxicity of the trapped drug.[117] This is the case of doxorubicin, daunorubicin, mitoxantrone that are weak bases. Weak acids, on the other hand, show a higher cytotoxicity with an acidic pHe. This would be the case of cyclophosphamide and 5-fluorouracil. However, paclitaxel is not influenced by pHe because it does not dissociate in solution, so that it is neither a basic nor an acidic compound.[118] In spite of this, an increased pHi probably has some effect on paclitaxel's anti-mitotic activity because alkaline pHi increases the assembly of microtubules while paclitaxel inhibits their depolimerization, also acting as a mitotic spindle poison.[119] Furthermore, sustained intracellular acidification decreases Pgp (P glycoprotein) expression[120,121] (Fig. 12).

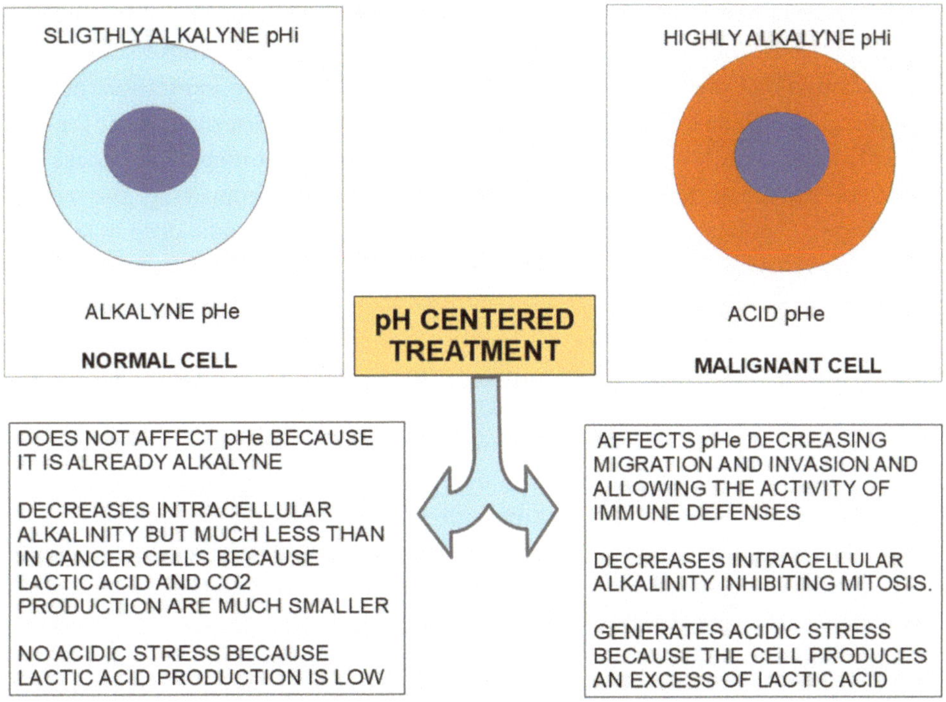

Fig. 12 Different effects of the pH centered treatments in normal and malignant cells. The fundamentals of the pH centered treatments are based on blocking proton extrusion from the cell (H^+). Cancer cells produce much more protons than normal cells, therefore the accumulation of these protons in the cytoplasm create an acidic stress. Normal cells, on the other hand, with a much lower production of CO_2 and almost no production of lactic acid are not affected by the inhibition of proton extrusion.

Some examples of how increasing pHe improves therapeutic results

Raghunand et al.[122] have shown that reducing extracellular acidity by the chronic administration of sodium bicarbonate increased the cytotoxicity of a weakly basic chemotherapeutic drug like mitoxantrone, but not of doxorubicin.

Proton pump inhibitors (PPI) may also reduce pHe acidity improving the activity of weak-base chemotherapeutics.[123–125] Clinical trial NCT01069081 showed that intermittent administration of high doses of a proton pump inhibitor (esomeprazol) increased the response rate to docetaxel-cisplatin treatment in metastatic breast carcinoma patients.[126]

Extracellular acidity also increases the activity of pGP (P glycoprotein) without increasing its expression.[127] Microenvironmental tumor acidity has also been shown to diminish the effects of immunotherapy with anti-CTLA-4 and anti-PD-1 compounds. Indeed, increasing pHe with sodium bicarbonate improves the effects of immunotherapy.[128] Reducing extracellular acidity also enhances the activity of tumor infiltrating T lymphocytes[129] and improves the overall results of cancer immunotherapy.[66] LDHA-associated lactic acid increase decreases the activity of T and NK cells.[130]

The acidic microenvironment is an ideal site for dormant (or residual) cancer cells that are slightly (or not at all) affected by conventional chemotherapeutic drugs.[131] The renewed malignant activity of these cells is strongly associated with the expression of TGFβ2 (transforming growth factor beta 2).[132] Extracellular acidity is a strong inducer for TGFβ2 and TGFβ1 expression,[133] and a proton pump inhibitor like esomeprazole blocks this.[134] Esomeprazol has been found to be particularly useful in improving the results of chemotherapy,[135] either as a chemosensitizer[136] or alone as a cytotoxic drug in rabdomyosarcoma,[137] esophageal cancer,[138] triple negative breast cancer,[139] human osteosarcoma[140] and melanoma.[141] In cervical cancer, esomeprazol showed synergy with paclitaxel.[142]

This knowledge has led to the development of new drug delivery mechanisms that are able to circumvent the extracellular acidity barrier and improve cytotoxicity with less damage to normal cells.

Autophagy is a mechanism that malignant cells use for survival under stress conditions and for recycling molecules. It consists of degrading intracellular structures that are not immediately necessary for survival, and using these materials as a source of nutrients or for the building of new and necessary molecules. Chloroquine has been used as an autophagy inhibitor in cancer with the goal of increasing apoptosis. However, low pHe inhibits chloroquine's autophagy-inhibiting effects.[143]

Quite frequently, cancer patients receive drugs for cardiovascular, neurological and many other diseases and conditions, along with their oncological treatments. These drugs may affect the cancer medications by modifying the pH and other pathways. For example, vasodilators like captopril, hydralazine, nifedipine, prazosine, etc., decrease pHe even further, interfering with weak bases effects.[144] On the other hand, intracellular acidification increases the cytotoxic effects of hyperthermia in glioma.[145]

Finally as a proof of concept, we must mention that the chronic administration of drugs that inhibited the pHi regulation resulted in high cytotoxicity in murine tumors both in vitro and in vivo.[146] That experiment showed that pH manipulation is not only necessary for the correct delivery of chemotherapeutics but is also a powerful tool in cancer treatment.

Conclusions Part I

The pH-centered paradigm is the more rational approach to cancer therapeutics encompassing all stages of cancer development at the same time. It should, and will be soon exploited in the treatment of many, if not all, malignant solid tumors and leukemias.

The initial and "basic" abnormality of the pH-cancer perspective is a strong tendency of cancer cells towards intracellular alkalosis. Increased intracellular alkalosis is a strong mitogenic inducer,[147] and an anti-apoptotic factor, as well as the primary cause of all the further abnormalities in hydrogen ion dynamics.[148] This intracellular alkalosis could not be appreciated in Otto Warburg's times since the discovery of the membrane-bound proton transporters took place well after Warburg's death. This helps us to understand why Warburg thought that cancer cells were acidic. He found a high production of lactic acid but intracellular pH measurement was not yet available. Thus he thought that the same acidic pH of the extracellular matrix was also present in the cytoplasm.[149]

Interstitial acidification negatively impacts the results of chemotherapy,[150,151] decreases immune defenses against tumors[26,152,153] drives tumor progression and activates the entire metastatic process.

Secondarily to the seminal acid-base intracellular abnormality of cancer cells, an extracellular acidity is induced by transmembrane acid extrusion, which then becomes a key factor in tumor progression and metastasis on its own, while tumor hypoxia and extracellular acidity independently influence tumor progression.[154,155]

Intracellular acidification is a new treatment option that reverses the pH paradigm and has major effects on the evolution of the disease.

Part II: Why and how does the pH-centered paradigm develop in cancer?

Enhanced growth and proliferation of any cell requires adequate environmental conditions, like a nutrient- and oxygen-rich environment in-order-to fuel anabolic processes that increase metabolism, as well as an intracellular pH suited for bio-synthesis. These conditions are usually found in normal cells and at the beginning of malignant proliferation. However, as soon as accelerated proliferation starts, such an ideal environment disappears. Increased nutrient consumption and inefficient or insufficient vascular supply transform the cancer environment into highly hypoxic and nutrient poor conditions.[156,157] Therefore, malignant cells need to modify their internal conditions and the external environment in-order-to develop.

The first step of the malignant transformation process are the genetic changes that induce the activity of oncogenic genes and/or the inactivation of suppressor genes. One or more genes suffer alterations in order to start the malignant behavior. Whether it is one gene or a multiplicity of genes may vary from tumor to tumor. But a characteristic of these genetic alterations (mutation, loss of function, transposition, amplification, etc.) is that they produce early alkalinization of the intracellular milieu through an increased activity of the sodium hydrogen exchanger 1 (NHE1) and/or other proton transporters (PTs)[21,158] The early intracellular alkalinization is independent of hypoxia and HIFs and it seems to be a requisite for transformation-driven characteristics including proliferation, anchorage-independent and growth-factor-independent growth and the initiation of the Warburg Metabolism (e.g., glycolysis and lactate production in the presence of oxygen).[159–161] A study on early malignization with a transfected oncogene showed that the first step, even before proliferation, was the increase in NHE1 activity that increased intracellular pH (pHi). When NHE1 was pharmacologically inhibited, no changes in pHi occurred and the cells did not adopt a malignant phenotype (Fig. 13).

The adaptations of the proliferating tumor to the environmental conditions lead to the pH gradient inversion and the metabolic switch[162] (Fig. 14).

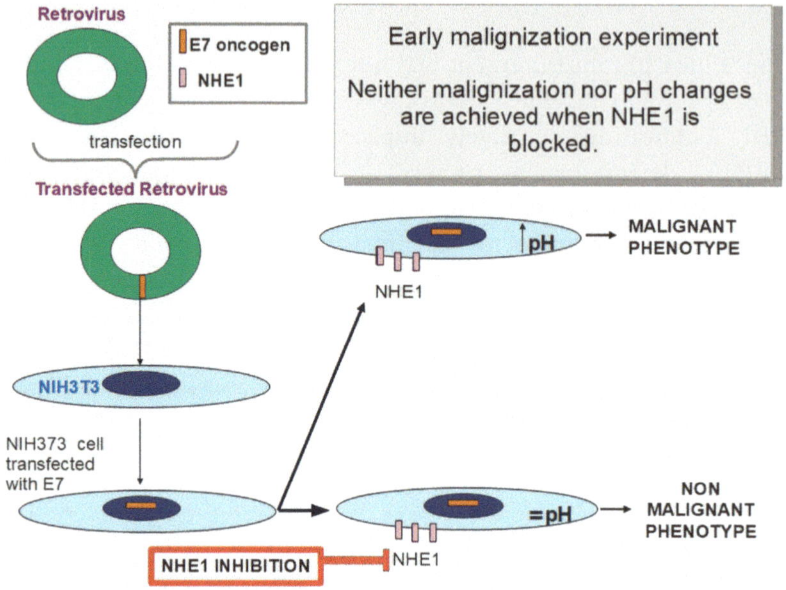

Fig. 13 Early malignization (see explanation in text).[21]

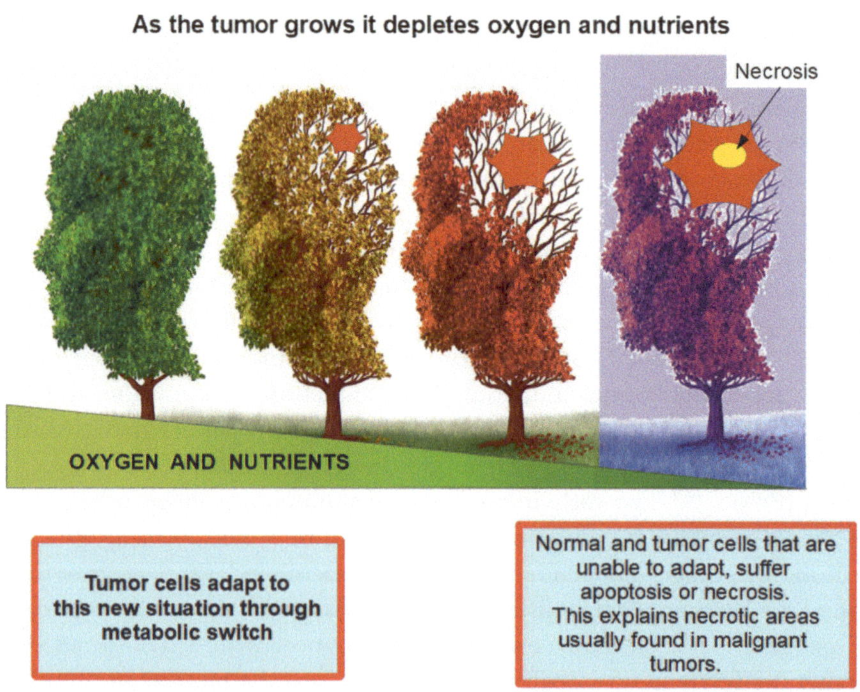

Fig. 14 Tumors permanently develop and evolve on a Darwinian selection model. Tumor cells and surrounding nonmalignant cells that do not adapt to the new micro-environmental conditions are eliminated and replaced by those that adapt better. These new breed of malignant cells thrive and proliferate. As micro-environmental conditions are not the same throughout the tumor, due to distance from feeding vessels, tumors adopt a very heterogeneous phenotype and genotype. A multi-clonal evolution, as observed in advanced tumors, is also a strong contributory factor to their heterogeneity.

The role of hypoxia in cancer metabolism and growth

The direct consequence of the hypoxic state leads to the over-expression of Hypoxia Inducible Factor 1 alpha (HIF-1α), which may be already increased due to oncogenic mutations. For instance, Akt over-expression due to oncogenic mutations is a powerful driver for HIF-1α expression. The HIFs are transcription factors that respond to modifications in oxygen availability[163,164] with the following mechanism: under normal oxygen conditions HIF-1α is bound to the Von Hippel Lindau protein (VHL) that carries HIF-1α to the proteasome where it is degraded. On the other hand, unbound stable HIF-1β is present in the cytoplasm. When oxygen conditions change, and hypoxia develops, HIF-1α is released from VHL and is now ready to bind to HIF-1β.[165] This heterodimer formed by HIF-1α and HIF-1β migrates to the nucleus, where they bind to hypoxia responsive element (HRE) segments of promoter regions of the DNA of genes inducing the transcription of more than 60 genes related to metabolism and mechanisms of adaptation to hypoxia.[166] These genes are known as hypoxia responsive genes.

Mechanisms involved in up-regulating the carbonic anhydrase IX (CA IX) gene is shown in Fig. 15. In the specific case of CA IX the Specificity protein 1 (Sp1) is a

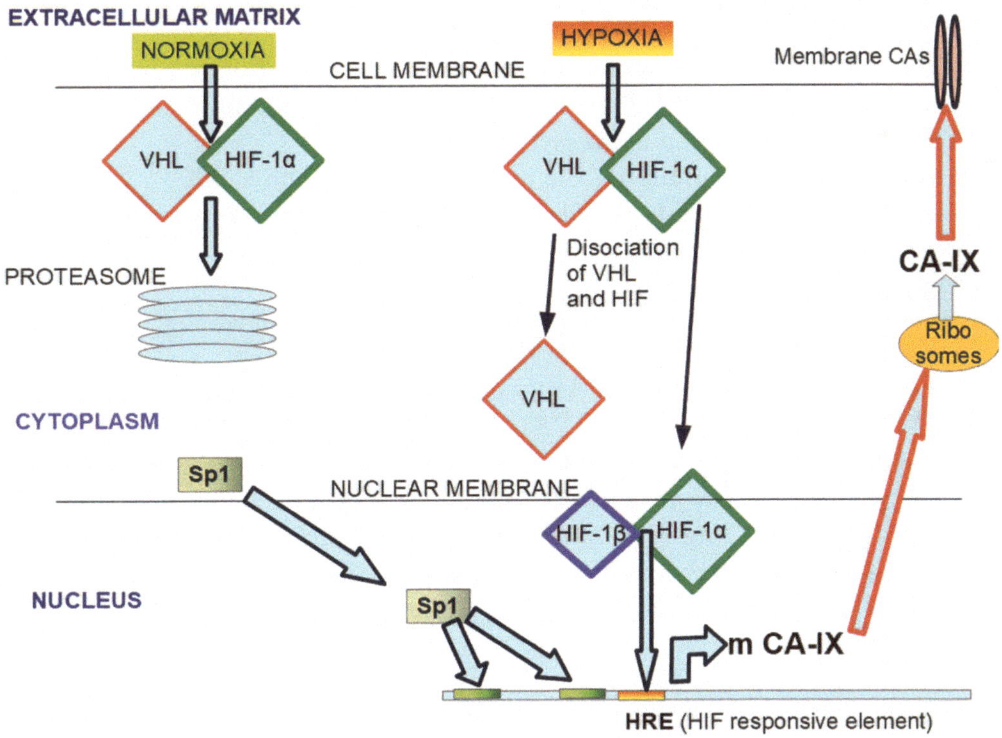

Fig. 15 Activation of HIF-1 alpha (hypoxia inducible factor 1 alpha) and its actions on hypoxia inducible genes.[167] In the figure, the up-regulation of the carbonic anhydrase IX gene is shown as an example. Sp1 is also a promoter of this gene. Both HIF-1α and Sp1 are necessary for promoting this gene's transcription.

transcription factor that enhances/promotes HIF1α and the carbonic anhydrase genes IX and XII (CAIX and CAXII).

The expression of multiple genes by the HIF system, are mostly related to a metabolic switch towards a low or nil oxygen consumption/metabolism, and is also directed towards the stimulation of new vessel formation (neoangiogenesis). The production of new vessels, achieved through the stimulation of HIF/VEGF axis in cancer cells and in peritumoral cells like fibroblasts, macrophages, myofibroblasts, and endothelial cells, is an imperfect process, because the new vessels usually do not become fully developed and functional. Therefore, the hypoxic environment becomes a permanent feature in cancer in spite of these new vessels.

In summary, HIF is a very important transcription factor in cancer because it promotes essential genes activation for cancer proliferation by:

- Upregulating pro-angiogenic molecules such as VEGF (Vascular endothelial growth factor),[168] Semaphorin D4,[169] calcitonin receptor-like receptor (CRLR), SCF (stem cell factor)[170] and angiopoietin II.[171]
- Upregulating pro-metastatic Twist.[172]
- Upregulating MMPs (matrixmetalloproteases) that are essential for ECM degradation, invasion and metastasis.[173–176]
- Downregulating E-cadherin facilitating metastasis.
- Upregulating PDK-1(pyruvate dehydrogenase kinase 1) that decreases oxidative phosphorylation (OXPHOS) metabolism and increases glycolytic metabolism.[177] This is a fundamental step in the glycolytic switch.
- Increasing expression of the glucose transporters, Glut1 and Glut3.[178]
- Increasing expression of 6-phosphofructo 2 kinase/fructose 2-6 bisphosphatase-3 (PFKFB3) gene, a key regulator of the glycolytic flux.[179]
- Upregulating MCT4, a lactate extruder.[180,181]
- Increasing the expression of membrane carbonic anhydrases (CAIX and CAXII).[182,183]

These above mentioned actions of HIFs are only a few examples of its multi-target activities.

Inhibition of the HIF system for interrupting the pathway shown in Fig. 15, has not been successfully developed yet but, as suggested by Semenza,[184] there are FDA (Food and Drug Administration) approved drugs and nutraceuticals that when adequately combined may partially down-regulate the hypoxia-dependent pathway.

Metabolic switch and pH abnormalities

The metabolic switch, which is a consequence of oncogenic mutations[185] and HIF-1α over-expression,[186,187] leads towards a glycolytic phenotype (Warburg effect). The tendency to abandon OXPHOS and adopt the aerobic glycolytic pathway as the main energy source and, in spite of this pathway's low energetic efficiency, it has two important advantages for the cancer cells:

(1) It decreases ROS (reactive oxygen species) production that is much higher under OXPHOS metabolism, and is potentially dangerous for cancer cell survival.

(2) It generates biologic building blocks for other molecules needed in the anabolic process of biomass building.

At the same time, the metabolic switch has an apparent disadvantage: an excess of protons is produced through different mechanisms. Three of these mechanisms are:

Excess of lactic acid production, that dissociates into lactate plus a proton.

ATP hydrolysis that also generates an excess of protons.[188]

Hydration of CO_2 forming CO_3H^- through the activity of membrane CAs mainly CAIX and CAXII.[189]

Cancer cells adapt to this excessive intracellular acidity by increasing the expression and/or activity of several proton (H^+) exporting mechanisms. This adaptation keeps intracellular pH at alkaline levels while the extracellular matrix receives the burden of the exported acidity.[190] In this way a highly abnormal pH gradient is quickly established with an alkaline interior and an acidic exterior.[191] This is the exact opposite to normality.[192]

This picture favors cancer growth and progression through diverse mechanisms (as it has been previously explained:

(1) Intracellular alkaline pH induces cell transformation and stimulates proliferation.[193]

(2) Extracellular acidity is a necessary feature for the activation of matrix degrading enzymes like cathepsin and metalloproteases (metalloproteinases). The matrix degradation is necessary for migration, invasion and eventual metastasis.[194]

(3) Extracellular matrix acidity blocks immunologic attacks against malignant cells and decreases the access of certain chemotherapeutic drugs to the tumor.[66]

The metabolic switch can be experimentally reversed and this reversal decreases proliferation in a very significant manner.[195,196]

Hypoxia plus extracellular acidity contribute to tumor progression and Darwinian selection of resistant cells that can survive in this harsh environment. This means that those malignant cells that do not acquire the adaptations to the environment die. Those who went through the adaptation process survive and evolve. These later cells form the bulk of the tumor and have been called "chronic acidity-adapted cells" by Corbet et al.[197] will. We prefer to rename them as "chronic hypoxia-acidity adapted cells".

The highly regulated pH of cells and the less regulated pH of the surrounding extracellular matrix is the result of a delicate balance between lack of oxygen, scarcity of nutrients, metabolic processes, proton production, proton transport and extrusion, chemical buffering (inside and outside the cell), apoptotic and necrotic by-products and vascular removal of waste products. The balance in this complicated environment is supported by sophisticated mechanisms.

Importantly, pHi is not uniformly distributed in malignant cells nor pHe in tumors. The migrating/ invading cells generate pseudopodia, and proton extrusion at the pseudopodia (front part of the migrating cell) is substantially higher than at the rear end. This creates a higher extracellular acidity at the pseudopodia and an intracellular pH gradient where pHi is higher in the front end and lower in the rear end.

Furthermore, pHe gradients are formed within the tumor by cells with different types of metabolism that is dependent upon their location within the tumor (Fig. 16). Those cells that are closer to vessels obtain energy through OXPHOS. The more distant ones adopt the glycolytic phenotype. Some OXPHOS cells use the lactate produced by the hypoxic cancer cells as an energy source (lactate shuttle, also known as Reverse Warburg effect).

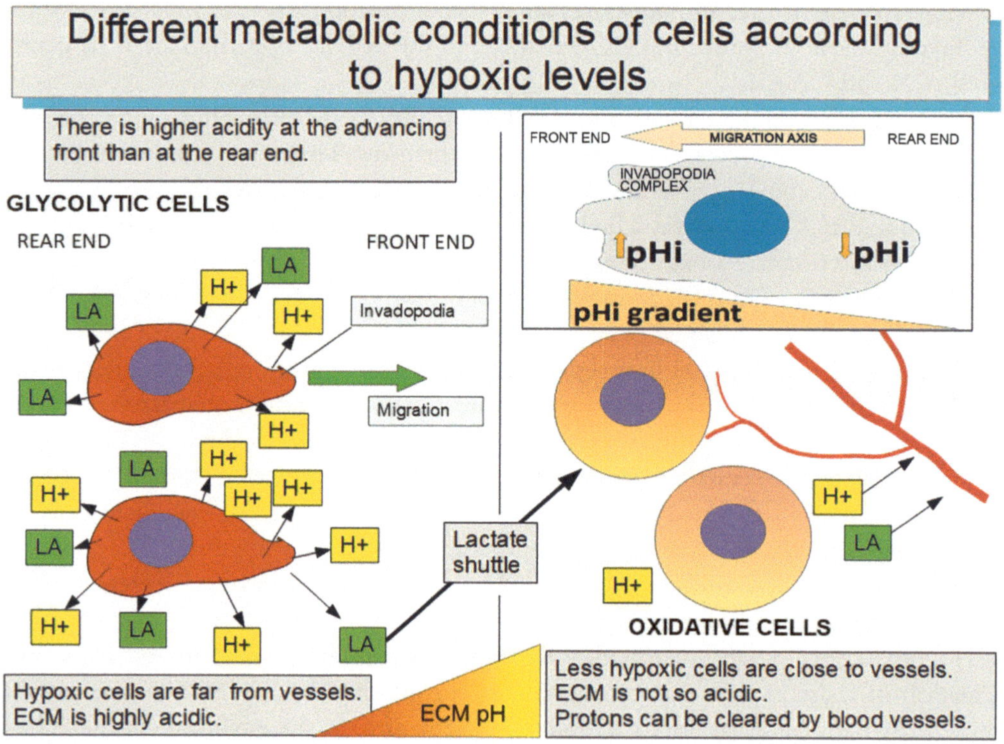

Fig. 16 Hypoxia creates different metabolic conditions and extracellular acidity.

Mechanisms that create an acidic extracellular matrix and an alkaline intracellular milieu

Microenvironmental and intracellular pHs are major issues that influence processes like proliferation, differentiation, metastasis, and angiogenesis, as we have seen in the first part of this chapter. Malignant cells show a pronounced increase in metabolic processes where a 10- to 15-fold rise in glucose consumption is only the tip of a metabolic iceberg in which all the metabolic systems of the cell are involved (including lipids, amino acids, proteins, water and ions). Aerobic glycolysis (Warburg effect) is one of the hallmarks of cancer metabolism, and it implies an excessive production of protons, which if remaining inside the cell would result in fatal intracellular acidosis. In the hypoxic areas of the tumor anaerobic glycolysis if further exacerbated resulting in an even higher production of lactate and protons. Furthermore, alkaline pHi also increases protein synthesis.[198] Increased DNA and RNA synthesis with increasing pHi may be explained by the optimum pH for DNA polymerases between 7 and 8.[199] Phosphofructokinase is the rate-limiting enzyme of glycolysis and its activity is pH sensitive increasing with pHi.[200,201] Seminal research in this area by Tannock and Rotin[202] showed that cytoplasmic alkalization preceded proliferation in certain cells. In the same sense, Reshkin et al., discovered that the first step of the malignant/neoplastic process was precisely the increase of NHE1 activity with the consequent increase in pHi.[21]

Therefore, tumor cells are in a permanent fight against excessive acid load, and they solve this chronic problem on a short and a long term basis. In the short term, malignant cells use chemical buffering and transfer of acids into organelles like lysosomes. The long term solution comes through a series of proton transport and extrusion mechanisms that expel the excess acidity (Fig. 17). This allows cancer cells to maintain a normal intracellular pH or even overshoot this mechanism to produce a tendency towards a slightly alkaline intracellular cytosol.[203]

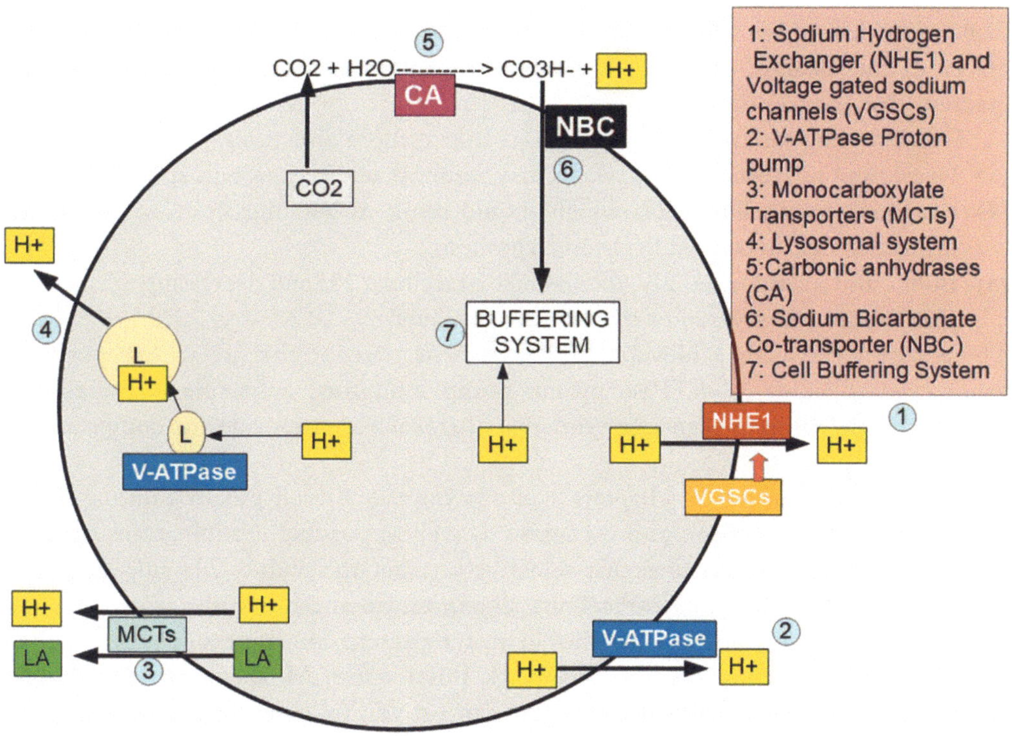

Fig. 17 Cancer creation of an alkaline pHi and an acidic pHe ("proton reversal"). The mechanisms involved are mainly the extrusion of protons by NHE1 (Sodium Hydrogen Exchanger1 also known as Na^+/H^+ antiporter), V-ATPase proton pumps, lactate extrusion through MCTs (monocarboxylate transporters), the hydration of CO_2 by membrane carbonic anhydrases[204] (CAs) and the import of bicarbonate by NBC (sodium bicarbonate co-transporter).

To generate the inverted pH gradient, as is found in cancer, it is necessary the overexpression and/or increased activity of ion channels and transporters, buffer systems and organelles like lysosomes (Fig. 6). Frequently their activity overlaps. Each of them will be discussed in the following chapters. The main participants besides buffers and lysosomes are NHE1, voltage gated sodium channels (VGSCs), V-ATPase proton pumps, carbonic anhydrases IX and XII, sodium bicarbonate cotransporter and monocarboxylate transporters 1 and 4 (Fig. 17).

New therapeutic directions derived from the pH-centered paradigm

Not all cancer cells express or overexpress the same combination of proton transporters and sometimes the differences among different tumors may be very important,[205] this means that there is a significant heterogeneity of proton extruders between different tumors. For example, in three different melanoma cell lines it was found that NHE1 was not up-regulated and furthermore it was lower than in normal or other kind of tumor cells. MCT1 on the contrary seemed to be highly up-regulated but not at the transcription level. The conclusion was that pHi in these cells depended from MCT1 and not NHE1.[206] On the other hand, it was found that in bladder cancer cells and murine mammary sarcoma cells the NHE1 was the driver of the alkaline pHi.[207] Multidrug resistant cancer cells show a higher pH gradient than nonresistant cells.[208]

Thus targeting tumors' pH should go in at least three directions:

(1) Decreasing intracellular pH, which may result in acidic stress and apoptosis.[209]

(2) Increasing extracellular pH, which should result in less migration, invasion, and metastasis (acid-mediated invasion hypothesis).[210]

(3) Both 1 and 2 simultaneously: increasing extracellular pH and decreasing intracellular pH (therapeutic reversal of inverted pH gradient).[211,212]

These objectives seem achievable at the present state of the art with repurposed drugs (acetazolamide, V-ATPase proton pump inhibitors, amiloride, voltage gated sodium channel inhibitors) and experimental drugs such as cariporide and compound 9t, among others.

This, and the following chapters will analyze the altered pH of tumors and its implications for progression and metastasis as well as possible combinations of drugs of different origins and nature that selectively target this vulnerable side of cancer development. Targeting the intracellular alkalinization and the acidic microenvironment of cancer cells and tissues should not represent an aggression against their normal counterparts because the latter lack this feature. More recent studies have emphasized the role of extracellular pH in cancer, yet, to a certain extent neglecting the importance of intracellular pH in cancer growth and progression. We think that the intracellular alkalinity represents the root of the problem, while extracellular acidity is its more damaging consequence, because:

(1) Cellular alkalinization is a very early step in the neoplastic process, even before proliferation.[21]
(2) Mitosis and biosynthesis of complex molecules require an alkaline intracellular milieu.
(3) High pHi is a mediating factor in the expression of some oncogenes like ras and v-mos.[213]
(4) Low pHe is a consequence of the intracellular processes that increase pHi, namely proton (H^+) extrusion.
(5) Acidifying the cell by blocking proton extrusion should immediately increase pHe.[214]
(6) Acidifying the cell restrains proliferation and stimulates apoptosis.

From the point of view of therapy, both pHi and pHe should be addressed simultaneously in-order-to achieve better therapeutic results. This is what we have called "double edged pH targeting".[215] Modifying the inverted pH gradients of tumors should not be harmful for normal cells, because, as we shall explain in more detail, the gradient inversion is a characteristic of tumors where glycolytic cells predominate, while normal tissues do not show this inversion.

Why is pHe so acid and pHi alkaline in cancer cells?

The main reasons for the pH gradient reversal are:
- Over-expression and/or over-activity of proton exporters.
- Over-expression and/or over-activity of membrane CAs.
- Over-expression and/or over-activity of bicarbonate importers (NBCs).
- Hypoxia and HIF-1.
- Increased glucose uptake and metabolism.
- Excessive lactic acid production due to increased glycolysis.
- Increased metabolism with the production of other acid metabolites like CO_2 which is hydrated to CO_3H^- and H^+.
- Increased ATP hydrolysis.
- Death of stromal and tumor cells probably contribute to ECM acidity (apoptosis is initiated with cellular acidification, and to this the liberation of lysosomal contents must be added.
- Decreased clearance of substances accumulated in the ECM.

Tumor heterogeneity and the lactate shuttle in cancer

An important question that comes up from the analysis of cancer metabolism is: do all cancer cells in a tumor have the same type of metabolism?

The answer is no.

In a malignant tumor there are usually areas that use a high proportion of OXPHOS (in the vicinity of functional blood vessels and less hypoxic areas) and other areas where aerobic glycolysis is predominant. This is reflected by the different pHe found in different parts of the tumors. The phenotypical and genotypical heterogeneity of tumors is widely known. Metabolic heterogeneity is also part of this picture. And this goes to the extreme when we realize that there are malignant cells that thrive on lactic acid.

From the perspective of carbohydrate metabolism, we may find four types of cells. Cells that:

(i) Take up glucose and produce CO_2 and H_2O: OXPHOS or oxidative phosphorylation.

(ii) Take up glucose and produce lactic acid that is extruded from the cell: aerobic glycolysis.

(iii) Take up lactic acid and produce CO_2 and H_2O: OXPHOS or oxidative phosphorylation metabolism (malignant lactophagic cells).

(iv) Theoretically at least, there should be also cells that take up glucose and produce lactic acid because they are in a very hypoxic area, but when hypoxia is reduced they might return to CO_2 and H_2O production instead of lactic acid. These would be tumor cells under anaerobic glycolysis (not experimentally confirmed).

Lactic acid has been and is still considered to be the main source of acidification of the ECM, but there are studies that raise doubts about this concept (see Fig. 18). Newell et al.[216] compared pHe in two lines of Ras transfected fibroblasts of the same origin implanted in nude mice. One was glycolysis deficient and had a lower lactic acid content in the ECM, while the other had normal glycolytic function and a much higher lactic acid production. But both had similar pHe, which was more acid than the non-tumor counterpart. The conclusion was that lactic acid is not the major cause of acidic environment in solid tumors in vivo. Similarly, using lactate dehydrogenase deficient cells, Yamagata et al.[217] obtained the same results.

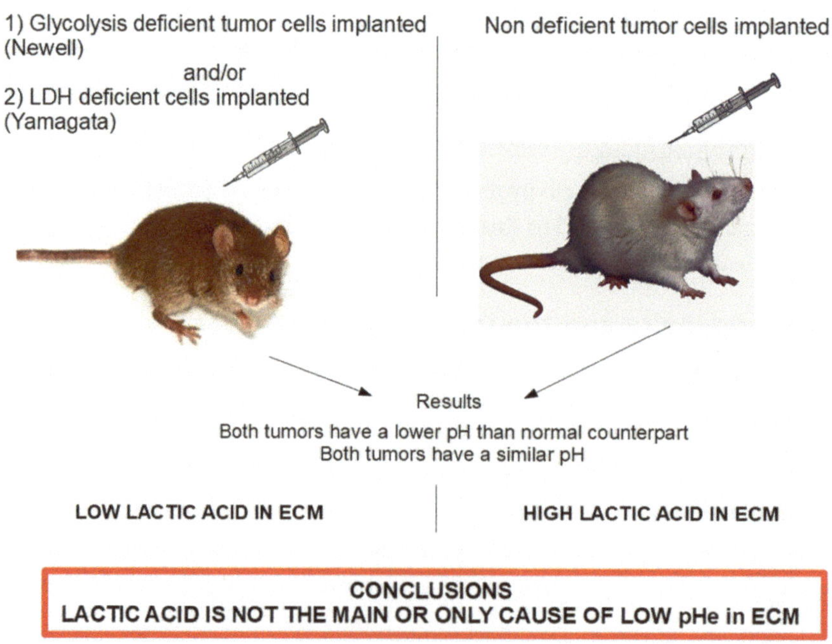

Fig. 18 Experiments of Newell et al.,[216] and Yamagata et al.[217]

This leaves us with the unanswered question: why is the tumor ECM so acid?

The possible answer is that lactic acid is one of the causes, but not the only or main cause. Other possible causative factors could be:

- increased proton exportation due to overexpression of NHE1, VGSCs, and/or proton pump (H^+ ATPase)[218];
- accumulation of CO_2 due to decreased vascular supply. This leads to CO_3H_2 production, which dissociates in CO_3H^- and H^+ increasing protons in the ECM. Gullino et al.[219] described increased PCO_2 and dissolved CO_2 in solid tumors. The parts of the tumor that are far from vascular structures show a lower pHe.[220] The excess of CO2 production may also originate from the PPP (pentose phosphate pathway) which is highly active in tumors[221];
- increased activity of membrane carbonic anhydrases CAIX and CAXII;
- hypoxia;
- other metabolites (Newell et al.[216] suggested that alternative acid-producing metabolic pathways which have not yet been detected, are present during the growth of the tumor in vivo).

At this point we can assume that increased lactic acid and CO_2 are the main causes of acidic ECM in cancer tissues.

Cell death and pH

An issue that seems to have been neglected is cell death in the tumor bulk and its contribution to extracellular acidity. Apoptosis is the main source of cell free circulating DNA (cfDNA)[222–224] and cfDNA is significantly increased in patients with localized cancer and further increased in metastatic cancer. This happens in both early and late stages of cancer.[225] Therefore, we may assume that the natural evolution of cancer goes along with apoptosis of a certain proportion of its cells. The apoptotic process liberates protons and should contribute to ECM acidity without producing modifications of the intracellular pH of the rest of the malignant cells of a tumor. At this point, we cannot establish quantitatively the importance of this process in the determination of pHe. We cannot exclude either that part of the circulating DNA is a product of the apoptosis of circulating cancer cells. However, as histopathology frequently shows necrotic areas in tumors, we must assume that tumor apoptosis/necrosis are contributors to ECM acidity in advanced disease. Normal cells have usually a pHe of 7.4. When pHe decreases below 7.1, as happens in cancer environment, their viability is significantly decreased.[226] Indeed, at a lower pHe apoptosis and/or necrosis will take place.

Clinical implications of the acid-base regulation in cancer

Malignant tumors need to regulate their particular intra and extracellular pH in order to proliferate and invade. The medical practitioner has a series of simple, nontoxic and inexpensive ways to interfere with this regulation. Proton pump inhibitors, carbonic anhydrase inhibitors (CAI) and/or NHE1 blockers are part of daily prescriptions in diseases other than cancer. These are Food and Drug Administration (FDA) approved drugs for certain health problems like peptic ulcer, glaucoma and hypertension. They can easily be added to cancer treatment without interfering with chemotherapeutic drugs (see chapters on pharmacological interventions). This kind of drugs have minimal side effects, but they are powerful modifiers of the malignant cell's intra and extracellular acid-base balance. A full description will be given in the chapter dedicated to pharmacological interventions.

As an example we shall consider here one type of cancer: glioblastoma multiforme (GBM).

Glioblastoma as an ideal test of concept

Treatment of cancer has been progressing slowly but steadily in the last 50 years. Many high mortality tumors of the past are fully curable nowadays. However, there is one striking exception: glioblastoma. No real breakthrough has been achieved with this tumor with ominous prognosis and very short survival. Glioblastoma, being a highly glycolytic malignancy is also strongly pH dependent and NHE1 (sodium hydrogen exchanger 1) driven.[227,228] Therefore, this is one of those pathologies where the lessons learnt on the pH paradigm might bring a promising change.

This tumor is highly dependent on the glycolytic pathway. All the experiments in which the glycolytic pathway was inhibited, or the intracellular pH was decreased, resulted in either a slower growth or increased apoptosis of glioma cells. This is the case with dichloroacetate (DCA: an anti-glycolytic and cell acidifier compound),[229] topiramate (cell acidifier),[230] MCT inhibitors like quercetin that decrease lactate efflux[231,232] and proton pump inhibitors.[233,234]

The prognosis of this tumor is still dismal. Glioblastoma shows the worst 5-year survival rates among all cancers.[235] Therefore, in theory, adding a pH centered approach to the usual treatment protocols may improve the results without further loss of quality of life.

pH evolution in cancer: Its relation with cellular metabolism

Changes in the pH of extracellular matrix are a consequence of events happening inside the malignant cell (excessive production and extrusion of protons). However external events have important participation; such as obliteration of capillaries due to tumor's growth and the distance to an alternative vascular supply in the early stages of neoplastic progression (Fig. 19).

During the tumor's growth and proliferation the pH imbalance becomes progressively more accentuated (Fig. 20). In certain areas of the tumor, the environmental

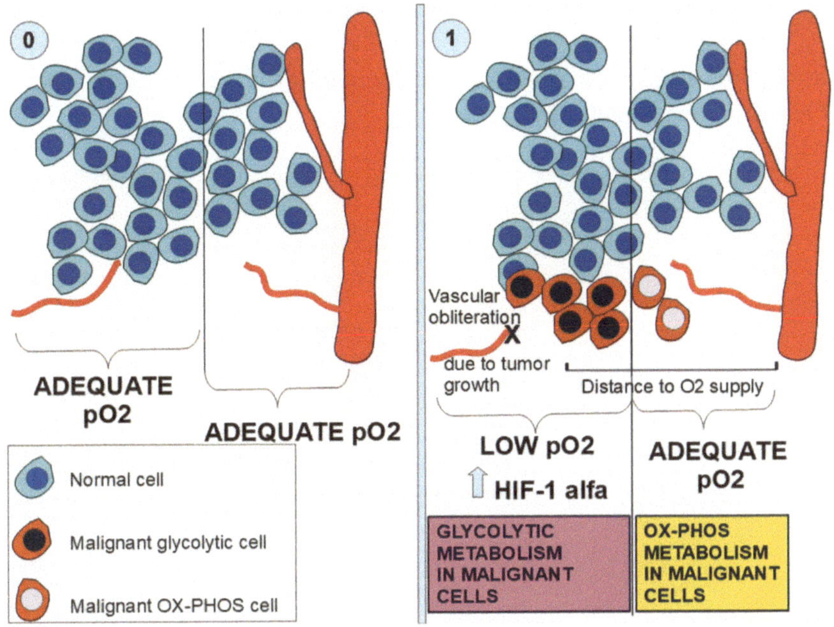

Fig. 19 High production of protons, glycolytic metabolism, and inadequate vascular supply co-participate in the appearance of metabolic and pH heterogeneity. Stage 0 is normal tissue with no transformation. In stage 1 part of the cells remain with oxidative metabolism and those that lack nutrients and oxygen develop glycolytic metabolism creating a lower ECM pH.

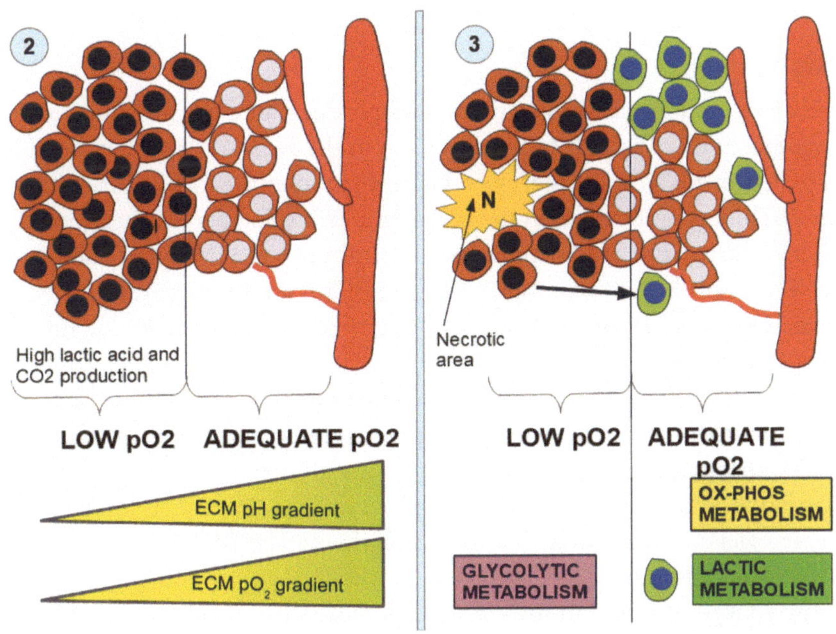

Fig. 20 Stages 2 and 3. In stage 3 there are necrotic areas. Lactic metabolism and the lactate shuttle may be present at any stage of the development. The arrow in stage 3 indicates the lactate shuttle. The highest pHi and lowest pHe is found in cells near necrotic and hypoxic areas.

conditions become so harsh that necrosis is frequently found. Necrotic areas show the lowest pHe in the tumor bulk (Fig. 21).

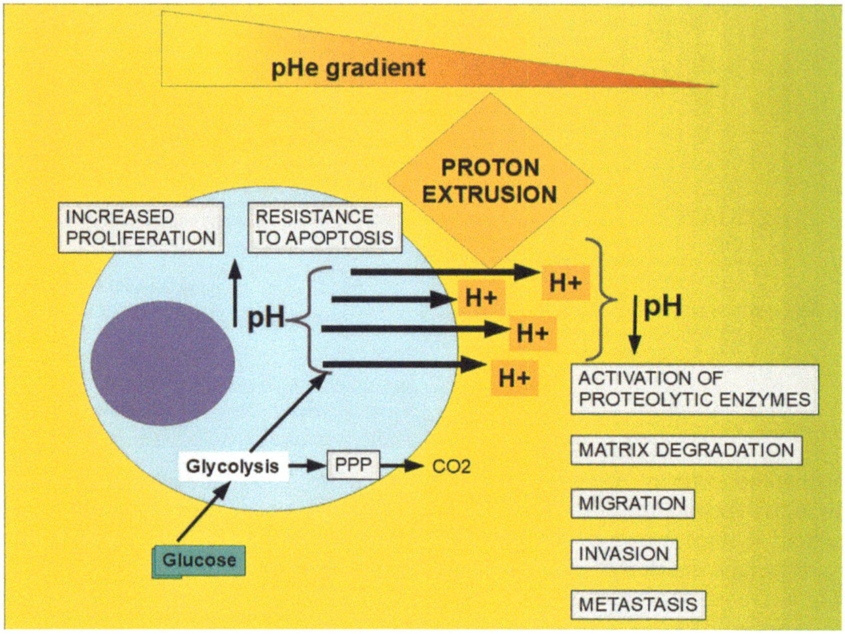

Fig. 21 The pH paradigm: causes and consequences at a glance. *PPP*, pentose phosphate pathway.

Conclusions Part 2

Tumor cells are predominantly glycolytic and produce an excess of lactic acid.

The hyperactivity of the PPP increases CO_2 in the ECM. A deficient vascular supply increases pCO_2 in ECM with increased CO_3H^- production, while an efficient proton extruding machinery is activated in the malignant cells since very early stages of transformation. These three associated mechanisms contribute to ECM acidity and intracellular alkalinity. The contribution of apoptosis to the ECM acidity requires further investigation. There is also a bicarbonate importing activity that is essential for the functioning of this pH neohomeostasis.

OXPHOS cells may also be present in malignant tissues and, in these cases, the lactate shuttle becomes an important issue (see chapter on lactic acid). The lactate shuttle feeds the oxidative malignant cells and saves glucose for glycolytic cells.

Summarizing in a few lines what is the pH paradigm in cancer and its causes:

(1) Increased lactate production due to glycolytic metabolism.

(2) Increased glucose uptake.

(3) Increased CO2 production through the pentose phosphate pathway due to a high metabolic rate and glycolytic metabolism.

(4) Increased ATP hydrolysis due to increased synthesis producing excessive phosphate.
(5) Increased extrusion of protons (H^+).
(6) The net result of 1, 2, 3, 4, and 5 is a higher pHi and a lower pHe: pH gradient inversion.
(7) Behind all this mechanism there is an overexpressed/overactive proton extrusion and bicarbonate import systems that we have called the pH-tome.

References

1. Alfarouk KO, Shayoub MEA, Muddathir AK, Elhassan GO, Bashir AHH. Evolution of tumor metabolism might reflect carcinogenesis as a reverse evolution process (dismantling of multicellularity). *Cancers (Basel)* 2011;**3**:3002–17.
2. Gatenby RA, Gillies RJ, Brown JS. The evolutionary dynamics of cancer prevention. *Nat Rev Cancer* 2010;**10**(8):526–7.
3. Harguindey S, Gillis M. Evolution and cancer: possible relationships to changes in environmental hydrogen ion concentration. *J Theor Biol* 1980;**86**(3):487–92.
4. Harguindey S, Stanciu D, Devesa J, Alfarouk KO, Cardone RA, Polo Orozco JD, et al. Cellular acidification as a new approach to cancer treatment and to the understanding and therapeutics of neurodegenerative diseases. *Semin Cancer Biol* 2017;**43**:157–79.
5. Harguindey S, Reshkin SJ, Orive G, Arranz JL, Anitua E. Growth and trophic factors, pH and the Na +/H + exchanger in Alzheimer's disease, other neurodegenerative diseases and cancer: new therapeutic possibilities and potential dangers. *Curr Alzheimer Res* 2007;**4**(1):53–65.
6. Cardone RA, Casavola V, Reshkin SJ. The role of disturbed pH dynamics and the Na+/H+ exchanger in metastasis. *Nat Rev Cancer* 2005;**5**(10):786–95.
7. Jang M, Kim SS, Lee J. Cancer cell metabolism: implications for therapeutic targets. *Exp Mol Med* 2013;**45**:e45.
8. Alfarouk KO, Verduzco D, Rauch C, Muddathir AK, Adil HB, Elhassan GO, et al. Glycolysis, tumor metabolism, cancer growth and dissemination. A new pH-based etiopathogenic perspective and therapeutic approach to an old cancer question. *Oncoscience* 2014;**1**(12):777.
9. Hanahan D, Weinberg RA. Hallmarks of cancer: the next generation. *Cell* 2011;**144**(5):646–74.
10. Warburg O, Posener K, Negelein E. Uber den Stoffwechsel der Carcinomzelle. *Biochem Z* 1924;**152**:309–44.
11. Wolpert PW, Noller K, Shaughnessy D, Houchin DN, Baccari ME, Miller FA. Tissue pH: a new clinical tool. *Arch Surg* 1970;**101**:308–13.
12. Wike-Hooley JL, Haveman J, Reinhold HS. The relevance of tumour pH to the treatment of malignant disease. *Radiother Oncol* 1984;**2**(4):343–66.
13. House SW, Warburg O, Burk D, Schade AL. On respiratory impairment in cancer cells. *Science* 1956;**124**(3215):267–72.
14. Hulikova A, Harris AL, Vaughan-Jones RD, Swietach P. Regulation of intracellular pH in cancer cell lines under normoxia and hypoxia. *J Cell Physiol* 2013;**228**(4):743–52.
15. McIntyre A, Harris AL. The role of pH regulation in cancer progression. In: *Metabolism in cancer*. Cham: Springer; 2016. p. 93–134.
16. Gerweck LE, Seetharaman K. Cellular pH gradient in tumor versus normal tissue: potential exploitation for the treatment of cancer. *Cancer Res* 1996;**56**(6):1194–8.
17. Ashby BS. pH studies in human malignant tumours. *Lancet* 1966;**288**(7458):312–5.
18. Naeslund J, Swenson KE. Investigations on the pH of malignant tumours in mice and humans after the administration of glucose. *Acta Obstet Gynecol Scand* 1953;**32**:359–67.
19. Harguindey S, Arranz JL, Orozco JD, Rauch C, Fais S, Cardone RA, et al. Cariporide and other new and powerful NHE1 inhibitors as potentially selective anticancer drugs–an integral molecular/biochemical/metabolic/clinical approach after one hundred years of cancer research. *J Transl Med* 2013;**11**(1):282–99.

20. Shirmanova MV, Druzhkova IN, Lukina MM, Matlashov ME, Belousov VV, Snopova LB, et al. Intracellular pH imaging in cancer cells in vitro and tumors in vivo using the new genetically encoded sensor SypHer2. *Biochim Biophys Acta Gen Subj* 2015;**1850**(9):1905–11.
21. Reshkin SJ, Bellizzi A, Caldeira S, Albarani V, Malanchi I, Poignee M, et al. Na+ /H+ exchanger-dependent intracellular alkalinization is an early event in malignant transformation and plays an essential role in the development of subsequent transformation-associated phenotypes. *FASEB J* 2000;**14**:2185–97.
22. Winkler MM, Matson GB, Hershey JW, Bradbury EM. 31P-NMR study of the activation of the sea urchin egg. *Exp Cell Res* 1982;**139**(1):217–22.
23. Madshus IH. Regulation of intracellular pH in eukaryotic cells. *Biochem J* 1988;**250**(1):1–8.
24. Rubin H, Fodge D. Clarkson B, Baserga R, editors. *Control of proliferation in animal cells*. vol. 1. Cold Spring Harbor Laboratory; 1974. p. 801–16.
25. Engström W. Intracellular pH during mitogenic stimulation induced by alkaline pH. *Cell Biol Int Rep* 1981;**5**(11):1063–70.
26. Steinhardt RA, Morisawa M. Nuccitelli R, Deamer DW, editors. *Intracellular pH: its measurement, regulation, and utilization in cellular functions*. New York: Alan R. Liss; 1982. p. 361–74.
27. Lee SC, Steinhardt RA. pH changes associated with meiotic maturation in oocytes of *Xenopus laevis*. *Dev Biol* 1981;**85**(2):358–69.
28. Pouyssegur J, Franchi A, L'Allemain G, Paris S. Cytoplasmic pH, a key determinant of growth factor-induced DNA synthesis in quiescent fibroblasts. *FEBS Lett* 1985;**190**:115–9.
29. Mills GB, Cragoe EJ, Gelfand EW, Grinstein S. Interleukin 2 induces a rapid increase in intracellular pH through activation of a Na+/H+ antiport. Cytoplasmic alkalinization is not required for lymphocyte proliferation. *J Biol Chem* 1985;**260**(23):12500–7.
30. Flinck M, Kramer SH, Pedersen SF. Roles of pH in control of cell proliferation. *Acta Physiol (Oxf)* 2018;**223**(3):e13068.
31. Flinck M, Kramer SH, Schnipper J, Andersen AP, Pedersen SF. The acid-base transport proteins NHE1 and NBCn1 regulate cell cycle progression in human breast cancer cells. *Cell Cycle* 2018; **17**(9):1056–67.
32. Harguindey S, Arranz JL, Washl ML, Orive G, Reshkin JS. Proton transport inhibitors as potentially selective anticancer drugs. *Anticancer Res* 2009;**29**:2127–36.
33. Mahoney BP, Raghunand N, Baggett B, Gillies RJ. Tumor acidity, ion trapping and chemotherapeutics: I. Acid pH affects the distribution of chemotherapeutic agents in vitro. *Biochem Pharmacol* 2003; **66**(7):1207–18.
34. Cosse JP, Michiels C. Tumour hypoxia affects the responsiveness of cancer cells to chemotherapy and promotes cancer progression. *Anticancer Agents Med Chem* 2008;**8**(7):790–7.
35. Federici C, Petrucci F, Caimi S, Cesolini A, Logozzi M, Borghi M, et al. Exosome release and low pH belong to a framework of resistance of human melanoma cells to cisplatin. *PLoS One* 2014;**9**(2):e88193.
36. Rozhin J, Sameni M, Ziegler G, Sloane BF. Pericellular pH affects distribution and secretion of cathepsin B in malignant cells. *Cancer Res* 1994;**54**:6517–25.
37. Webb SD, Sherratt JA, Fish RG. Alterations in proteolytic activity at low pH and its association with invasion: a theoretical model. *Clin Exp Metastasis* 1999;**17**:397–407.
38. Fukumura D, Xu L, Chen Y, Gohongi T, Seed B, Jain RK. Hypoxia and acidosis independently up-regulate vascular endothelial growth factor transcription in brain tumors in vivo. *Cancer Res* 2000;**61**:6020–4.
39. Xu L, Fukumura D, Jain RK. Acidic extracellular pH induces VEGF in human glioblastoma cells via ERK1/2 MAPK signaling pathway. Mechanism of low pH induced VEGF. *J Biol Chem* 2001; **277**(13):11368–74.
40. Lardner A. The effects of extracellular pH on immune function. *J Leukoc Biol* 2001;**69**(4):522–30.
41. Bohloli M, Atashi A, Soleimani M, Kaviani S, Anbarlou A. Investigating effects of acidic pH on proliferation, invasion and drug-induced apoptosis in lymphoblastic leukemia. *Cancer Microenviron* 2016; **9**(2–3):119–26.
42. Peppicelli S, Bianchini F, Calorini L. Extracellular acidity, a "reappreciated" trait of tumor environment driving malignancy: perspectives in diagnosis and therapy. *Cancer Metastasis Rev* 2014;**33**(2–3): 823–32.
43. Calorini L, Peppicelli S, Bianchini F. Extracellular acidity as favouring factor of tumor progression and metastatic dissemination. *Exp Oncol* 2012;**34**(2):79–84.

44. Rofstad EK, Mathiesen B, Kindem K, Galappathi K. Acidic extracellular pH promotes experimental metastasis of human melanoma cells in athymic nude mice. *Cancer Res* 2006;**66**(13):6699–707.
45. Cardone RA, Casavola V, Reshkin SJ. The role of disturbed pH dynamics and the NA+/H+ exchanger in metastasis. *Nat Rev Cancer* 2005;**5**:786–95.
46. Thews O, Gassner B, Kelleher DK, Schwerd G, Gekle M. Impact of extracellular acidity on the activity of P-glycoprotein and the cytotoxicity of chemotherapeutic drugs. *Neoplasia* 2006;**8**(2):143–52.
47. Som A, Bloch S, Ippolito JE, Achilefu S. Acidic extracellular pH of tumors induces octamer-binding transcription factor 4 expression in murine fibroblasts in vitro and in vivo. *Sci Rep* 2016;**6**:27803.
48. Kato Y, Ozawa S, Miyamoto C, Maehata Y, Suzuki A, Maeda T, et al. Acidic extracellular microenvironment and cancer. *Cancer Cell Int* 2013;**13**(1):89.
49. Griffiths L, Dachs GU, Bicknell R, Harris AL, Stratford IJ. The influence of oxygen tension and pH on the expression of platelet-derived endothelial cell growth factor/thymidine phosphorylase in human breast tumor cells grown in vitro and in vivo. *Cancer Res* 1997;**57**(4):570–2.
50. Shi Q, Abbruzzese JL, Huang S, Fidler IJ, Xiong Q, Xie K. Constitutive and inducible interleukin 8 expression by hypoxia and acidosis renders human pancreatic cancer cells more tumorigenic and metastatic. *Clin Cancer Res* 1999;**5**:3711–21.
51. John S, Sivakumar KC, Mishra R. Extracellular proton concentrations impacts LN229 glioblastoma tumor cell fate via differential modulation of surface lipids. *Front Oncol* 2017;**7**:20.
52. Yamaguchi H, Oikawa T. Membrane lipids in invadopodia and podosomes: key structures for cancer invasion and metastasis. *Oncotarget* 2010;**5**:320–8.
53. Yamaguchi H, Takeo Y, Yoshida S, Kouchi Z, Nakamura Y, Fukami K. Lipid rafts and caveolin-1 are required for invadopodia formation and extracellular matrix degradation by human breast cancer cells. *Cancer Res* 2009;**69**(22):8594–602.
54. Russo D, Parashuraman S, D'Angelo G. Glycosphingolipid-protein interaction in signal transduction. *Int J Mol Sci* 2016;**17**(10):1732.
55. Bellone M, Calcinotto A, Filipazzi P, De Milito A, Fais S, Rivoltini L. The acidity of the tumor microenvironment is a mechanism of immune escape that can be overcome by proton pump inhibitors. *Oncoimmunology* 2013;**2**(1):e22058.
56. Lugini L, Federici C, Borghi M, Azzarito T, Marino ML, Cesolini A, et al. Proton pump inhibitors while belonging to the same family of generic drugs show different anti-tumor effect. *J Enzyme Inhib Med Chem* 2016;**31**(4):538–45.
57. Calcinotto A, Filipazzi P, Grioni M, Iero M, De Milito A, Ricupito A, et al. Modulation of microenvironment acidity reverses anergy in human and murine tumor-infiltrating T lymphocytes. *Cancer Res* 2012;**72**(11):2746–56.
58. Helmlinger G, Sckell A, Dellian M, Forbes NS, Jain RK. Acid production in glycolysis-impaired tumors provides new insights into tumor metabolism. *Clin Cancer Res* 2002;**8**(4):1284–91.
59. Duffy MJ. The role of proteolytic enzymes in cancer invasion and metastasis. *Clin Exp Metastasis* 1992;**10**(3):145–55.
60. Palermo C, Joyce JA. Cysteine cathepsin proteases as pharmacological targets in cancer. *Trends Pharmacol Sci* 2008;**29**(1):22–8.
61. Netzel-Arnett S, Hooper JD, Szabo R, Madison EL, Quigley JP, Bugge TH, et al. Membrane anchored serine proteases: a rapidly expanding group of cell surface proteolytic enzymes with potential roles in cancer. *Cancer Metastasis Rev* 2003;**22**(2–3):237–58.
62. Nakajima M, Irimura T, Di Ferrante D, Di Ferrante N, Nicolson GL. Heparan sulfate degradation: relation to tumor invasive and metastatic properties of mouse B16 melanoma sublines. *Science* 1983;**220**(4597):611–3.
63. Martinez-Zaguilan R, Seftor EA, Seftor RE, Chu YW, Gillies RJ, Hendrix MJ. Acidic pH enhances the invasive behavior of human melanoma cells. *Clin Exp Metastasis* 1996;**14**(2):176–86.
64. Chen Y, Chen CH, Tung PY, Huang SH, Wang SM. An acidic extracellular pH disrupts adherens junctions in HepG2 cells by Src kinases-dependent modification of E-cadherin. *J Cell Biochem* 2009;**108**(4):851–9.
65. Kato Y, Nakayama Y, Umeda M, Miyazaki K. Induction of 103-kDa gelatinase/type IV collagenase by acidic culture conditions in mouse metastatic melanoma cell lines. *J Biol Chem* 1992;**267**(16):11424–30.
66. Huber V, Camisaschi C, Berzi A, Ferro S, Lugini L, Triulzi T, et al. Cancer acidity: an ultimate frontier of tumor immune escape and a novel target of immunomodulation. *Semin Cancer Biol* 2017;**43**:74–89.

67. Fischer K, Hoffmann P, Voelkl S, Meidenbauer N, Ammer J, Edinger M, et al. Inhibitory effect of tumor cell-derived lactic acid on human T cells. *Blood* 2007;**109**(9):3812–9.
68. White KA, Grillo-Hill BK, Barber DL. Cancer cell behaviors mediated by dysregulated pH dynamics at a glance. *J Cell Sci* 2017;**130**(4):663–9.
69. Amith SR, Fliegel L. Na^{+}/H^{+} exchanger-mediated hydrogen ion extrusion as a carcinogenic signal in triple-negative breast cancer etiopathogenesis and prospects for its inhibition in therapeutics. *Semin Cancer Biol* 2017;**43**:35–41.
70. Harguindey S, Koltai T, Reshkin SJ. Curing cancer? Further along the new pH-centric road and paradigm. *Oncoscience* 2018;**5**(5–6):132–3.
71. Grillo-Hill BK, Choi C, Jimenez-Vidal M, Barber DL. Increased H+ efflux is sufficient to induce dysplasia and necessary for viability with oncogene expression. *Elife* 2015;**4**:e03270.
72. Webb BA, Chimenti M, Jacobson MP, Barber DL. Dysregulated pH: a perfect storm for cancer progression. *Nat Rev Cancer* 2011;**11**(9):671.
73. Vaupel P. The role of hypoxia-induced factors in tumor progression. *Oncologist* 2004;**9**(Suppl. 5):10–7.
74. Corbet C, Feron O. Tumor acidosis: from the passenger to the driver's seat. *Nat Rev Cancer* 2017; **17**(10):573–93.
75. Rofstad EK, Galappathi K, Mathiesen B, Ruud EB. Fluctuating and diffusion-limited hypoxia in hypoxia-induced metastasis. *Clin Cancer Res* 2007;**13**(7):1971–8.
76. Chiche J, Ilc K, Laferriere J, Trottier E, Dayan F, Mazure NM, et al. Hypoxia-inducible carbonic anhydrase IX and XII promote tumor cell growth by counteracting acidosis through the regulation of the intracellular pH. *Cancer Res* 2009;**69**(1):358–68.
77. Orive G, Reshkin SJ, Harguindey S, Pedraz JL. Hydrogen ion dynamics and the Na+/H+ exchanger in cancer angiogenesis and antiangiogenesis. *Br J Cancer* 2003;**89**:1395–9.
78. Parks SK, Pouysségur J. Targeting pH regulating proteins for cancer therapy—progress and limitations. *Semin Cancer Biol* 2017;**43**:66–73.
79. Bindra RS, Glazer PM. Genetic instability and the tumor microenvironment: towards the concept of microenvironment-induced mutagenesis. *Mutat Res Fundam Mol Mech Mutagen* 2005;**569**(1):75–85.
80. Harris AL. Hypoxia—a key regulatory factor in tumour growth. *Nat Rev Cancer* 2002;**2**:38–47.
81. Mihaylova VT, Bindra RS, Yuan J, Campisi D, Narayanan L, Jensen R, et al. Decreased expression of the DNA mismatch repair gene Mlh1 under hypoxic stress in mammalian cells. *Mol Cell Biol* 2003; **23**(9):3265–73.
82. Keysar SB, Trncic N, LaRue SM, Fox MH. Hypoxia/reoxygenation-induced mutations in mammalian cells detected by the flow cytometry mutation assay and characterized by mutant spectrum. *Radiat Res* 2010;**173**(1):21–6.
83. Rodríguez-Jiménez FJ, Moreno-Manzano V, Lucas-Dominguez R, Sánchez-Puelles JM. Hypoxia causes downregulation of mismatch repair system and genomic instability in stem cells. *Stem Cells* 2008;**26**(8):2052–62.
84. Bouquet F, Ousset M, Biard D, Fallone F, Dauvillier S, Frit P, et al. A DNA-dependent stress response involving DNA-PK occurs in hypoxic cells and contributes to cellular adaptation to hypoxia. *J Cell Sci* 2011;**124**:1943–51.
85. Rice GC, Hoy C, Schimke RT. Transient hypoxia enhances the frequency of dihydrofolate reductase gene amplification in Chinese hamster ovary cells. *Proc Natl Acad Sci U S A* 1986;**83**:5978–82.
86. Koshiji M, To KK, Hammer S, Kumamoto K, Harris AL, Modrich P, et al. HIF-1alpha induces genetic instability by transcriptionally downregulating MutSalpha expression. *Mol Cell* 2005;**17**:793–803.
87. Yuan J, Narayanan L, Rockwell S, Glazer PM. Diminished DNA repair and elevated mutagenesis in mammalian cells exposed to hypoxia and low pH. *Cancer Res* 2000;**60**(16):4372–6.
88. Lee JH, Choi IJ, Song DK, Kim DK. Genetic instability in the human lymphocyte exposed to hypoxia. *Cancer Genet Cytogenet* 2010;**196**:83–8.
89. Kumareswaran R, Ludkovski O, Meng A, Sykes J, Pintilie M, Bristow RG. Chronic hypoxia compromises repair of DNA double-strand breaks to drive genetic instability. *J Cell Sci* 2012;**125**:189–99.
90. Bindra RS, Gibson SL, Meng A, Westermark U, Jasin M, Pierce AJ, et al. Hypoxia-induced downregulation of BRCA1 expression by E2Fs. *Cancer Res* 2005;**65**(24):11597–604.
91. Lu Y, Chu A, Turker MS, Glazer PM. Hypoxia-induced epigenetic regulation and silencing of the BRCA1 promoter. *Mol Cell Biol* 2011;**31**:3339–50.

92. Bindra RS, Glazer PM. Repression of RAD51 gene expression by E2F4/p130 complexes in hypoxia. *Oncogene* 2007;**26**(14):2048.
93. Russo CA, Weber TK, Volpe CM, Stoler DL, Petrelli NJ, Rodriguez-Bigas M, et al. An anoxia inducible endonuclease and enhanced DNA breakage as contributors to genomic instability in cancer. *Cancer Res* 1995;**55**(5):1122–8.
94. Goldblatt H, Cameron G. Induced malignancy in cells from rat myocardium subjected to intermittent anaerobiosis during long propagation in vitro. *J Exp Med* 1953;**97**(4):525–52.
95. Goldblatt H, Friedman L, Cechner RL. On the malignant transformation of cells during prolonged culture under hypoxic conditions. *Biochem Med* 1973;**7**(2):241–52.
96. Goldblatt H, Friedman L. Prevention of malignant change in mammalian cells during prolonged culture in vitro. *Proc Natl Acad Sci U S A* 1974;**71**(5):1780–2.
97. Pires IM, Bencokova Z, Milani M, Folkes LK, Li J-L, Stratford MR, et al. Effects of acute versus chronic hypoxia on DNA damage and genomic instability. *Cancer Res* 2010;**70**(3):926–35.
98. Harguindey S, Galdós I, Takita H. Lung cancer as a scar tumor: report of a case and possible etiopathogenic mechanisms. *Ann Intern Med* 1994;**2**:285–7.
99. Harguindey S, Pedraz JL, Garcia Canero R, Perez de Diego J, Cragoe Jr. EJ. Hydrogen ion-dependent oncogenesis and parallel new avenues to cancer prevention and treatment using a H^+-mediated unifying approach: pH-related and pH-unrelated mechanisms. *Crit Rev Oncog* 1995;**6**(1):1–34.
100. Parks SK, Cormerais Y, Durivault J, Pouyssegur J. Genetic disruption of the pHi-regulating proteins Na+/H+ exchanger 1 (SLC9A1) and carbonic anhydrase 9 severely reduces growth of colon cancer cells. *Oncotarget* 2017;**8**(6):10225.
101. Parks SK, Cormerais Y, Pouysségur J. Hypoxia and cellular metabolism in tumour pathophysiology. *J Physiol* 2017;**595**(8):2439–50.
102. Kim JW, Tchernyshyov I, Semenza GL, Dang CV. HIF-1-mediated expression of pyruvate dehydrogenase kinase: a metabolic switch required for cellular adaptation to hypoxia. *Cell Metab* 2006; **3**(3):177–85.
103. Semenza GL. Regulation of cancer cell metabolism by hypoxia-inducible factor 1. *Semin Cancer Biol* 2009;**19**(1):12–6.
104. Brahimi-Horn MC, Chiche J, Pouyssegur J. Hypoxia signalling controls metabolic demand. *Curr Opin Cell Biol* 2007;**19**(2):223–9.
105. Chiche J, Brahimi-Horn MC, Pouysségur J. Tumour hypoxia induces a metabolic shift causing acidosis: a common feature in cancer. *J Cell Mol Med* 2010;**14**(4):771–94.
106. Vaupel P, Mayer A. Hypoxia in cancer: significance and impact on clinical outcome. *Cancer Metastasis Rev* 2007;**26**(2):225–39.
107. Chen Z, Liu M, Li L, Chen L. Involvement of the Warburg effect in non-tumor diseases processes. *J Cell Physiol* 2018;**233**(4):2839–49.
108. Burns JS, Manda G. Metabolic pathways of the Warburg effect in health and disease: perspectives of choice, chain or chance. *Int J Mol Sci* 2017;**18**(12):2755.
109. Harguindey S, Orive G, Pedraz JL, Paradiso A, Reshkin SJ. The role of pH dynamics and the Na+/H+ antiporter in the etiopathogenesis and treatment of cancer. Two faces of the same coin—one single nature. *Biochim Biophys Acta* 2005;**1756**:1–24.
110. Daniel C, Bell C, Burton C, Harguindey S, Reshkin SJ, Rauch C. The role of proton dynamics in the development and maintenance of multidrug resistance in cancer. *Biochim Biophys Acta* 2013; **1832**(5):606–17.
111. Albatany M, Alex L, Meakin S, Bartha R. Dichloroacetate induced intracellular acidification in glioblatoma: in vivo detection using AACID-CEST MRI at 9.4 Tesla. *J Neurooncol* 2018;**136**:255–62.
112. Swietach P, Patiar S, Supuran CT, Harris AL, Vaughan-Joness RD. The role of carbonic anhydrase 9 in regulating extracellular and intracellular pH in three-dimensional tumor cell growths. *J Biol Chem* 2009;**284**(30):20299–310.
113. Mahoney BP, Raghunand N, Baggett B, Gillies RJ. Tumor acidity, ion trapping and chemotherapeutics: I. Acid pH affects the distribution of chemotherapeutic agents in vitro. *Biochem Pharmacol* 2003; **66**(7):1207–18.
114. Gerweck LE, Vijayappa S, Kozin S. Tumor pH controls the in vivo efficacy of weak acid and base chemotherapeutics. *Mol Cancer Ther* 2006;**5**(5):1275–9.

115. Alfarouk KO, Stock CM, Taylor S, Walsh M, Muddathir AK, Verduzco D, et al. Resistance to cancer chemotherapy: failure in drug response from ADME to P-gp. *Cancer Cell Int* 2015;**15**(1):71.
116. Rauch C, Blanchard A, Wood E, Dillon E, Wahl ML, et al. Cell membranes, cytosolic pH and drug transport in cancer and MDR: physics, biochemistry and molecular biology. In: Meszaros A, Balogh G, editors. *Multiple drug resistance*. Nova Science Publishers, Inc; 2009. p. 1–100. Chapter 3.
117. Raghunand N, Gillies RJ. pH and drug resistance in tumors. *Drug Resist Updat* 2002;**3**:39–47.
118. Vukovic V, Tannock IF. Influence of low pH on cytotoxicity of paclitaxel mitoxantrone and topotecan. *Br J Cancer* 1997;**75**:1167–72.
119. Donato R. Calcium-independent, pH-regulated effects of S-100 proteins on assembly-disassembly of brain microtubule protein in vitro. *J Biol Chem* 1988;**263**:106–10.
120. Lu Y, Pang T, Wang J, Xiong D, Ma L, Li B, et al. Down-regulation of P-glycoprotein expression by sustained intracellular acidification in K562/Dox cells. *Biochem Biophys Res Commun* 2008; **377**(2):441–6.
121. Omran Z, Scaife P, Stewart S, Rauch C. Physical and biological characteristics of multi drug resistance (MDR): an integral approach considering pH and drug resistance in cancer. *Semin Cancer Biol* 2017;**43**:42–8.
122. Raghunand N, Mahoney BP, Gillies RJ. Tumor acidity, ion trapping and chemotherapeutics: II. pH-dependent partition coefficients predict importance of ion trapping on pharmacokinetics of weakly basic chemotherapeutic agents. *Biochem Pharmacol* 2003;**66**(7):1219–29.
123. De Milito A, Fais S. Tumor acidity, chemoresistance and proton pump inhibitors. *Future Oncol* 2005; **1**(6):779–86.
124. De Milito A, Canese R, Marino ML, Borghi M, Iero M, Villa A, et al. pH-dependent antitumor activity of proton pump inhibitors against human melanoma is mediated by inhibition of tumor acidity. *Int J Cancer* 2010;**127**(1):207–19.
125. Taylor S, Spugnini EP, Assaraf YG, Azzarito T, Rauch C, Fais S. Microenvironment acidity as a major determinant of tumor chemoresistance: proton pump inhibitors (PPIs) as a novel therapeutic approach. *Drug Resist Updat* 2015;**23**:69–78.
126. Wang BY, Zhang J, Wang JL, Sun S, Wang ZH, Wang LP, et al. Intermittent high dose proton pump inhibitor enhances the antitumor effects of chemotherapy in metastatic breast cancer. *J Exp Clin Cancer Res* 2015;**34**(1):85.
127. Thews O, Gassner B, Kelleher DK, Schwerd G, Gekle M. Impact of extracellular acidity on the activity of P-glycoprotein and the cytotoxicity of chemotherapeutic drugs. *Neoplasia* 2006;**8**(2):143–52.
128. Pilon-Thomas S, Kodumudi KN, El-Kenawi AE, Russell S, Weber AM, Luddy K, et al. Neutralization of tumor acidity improves antitumor responses to immunotherapeutic interventions. *Cancer Res* 2015;**76**:1381. canres-1743.
129. Calcinotto A, Filipazzi P, Grioni M, Iero M, De Milito A, Ricupito A, et al. Modulation of microenvironment acidity reverses anergy in human and murine tumor-infiltrating T lymphocytes. *Cancer Res* 2012;**72**:2746–56. canres1272.
130. Brand A, Singer K, Koehl GE, Kolitzus M, Schoenhammer G, Thiel A, et al. LDHA-associated lactic acid production blunts tumor surveillance by T and NK cells. *Cell Metab* 2016;**24**(5):657–71.
131. Peppicelli S, Andreucci E, Ruzzolini J, Laurenzana A, Margheri F, Fibbi G, et al. The acidic microenvironment as a possible niche of dormant tumor cells. *Cell Mol Life Sci* 2017;**74**(15):2761–71.
132. Bragado P, Estrada Y, Parikh F, Krause S, Capobianco C, Farina HG, et al. TGFβ2 dictates disseminated tumour cell fate in target organs through TGF-β-RIII and p38α/β signalling. *Nat Cell Biol* 2013;**15**:1351–61.
133. Ahmed S, Tsuchiya T. Novel mechanism of tumorigenesis: increased transforming growth factor-beta 1 suppresses the expression of connexin 43 in BALB/cJ mice after implantation of poly-l-lactic acid. *J Biomed Mater Res A* 2004;**70**:335–40.
134. Peppicelli S, Bianchini F, Toti A, Laurenzana A, Fibbi G, Calorini L. Extracellular acidity strengthens mesenchymal stem cells to promote melanoma progression. *Cell Cycle* 2015;**14**:3088–100.
135. Oral AY, Oral HB, Sarimahmut M, Cevatemre B, Özkaya G, Kormaz Ş, et al. Combination of esomeprazole with chemotherapeutics results in more pronounced cytotoxic effect via apoptosis on A549 nonsmall-cell lung cancer cell line. *Turk J Biol* 2017;**41**(1):231–41.

136. Matuszcak C, Lindner K, Haier J, Hummel R. Proton pump inhibitors as chemosensitizer: new indication for a well-known medication. *Cancer Cell Microenviron* 2015;**2**(1):e667.
137. Perut F, Avnet S, Fotia C, Baglìo SR, Salerno M, Hosogi S, et al. V-ATPase as an effective therapeutic target for sarcomas. *Exp Cell Res* 2014;**320**(1):21–32.
138. Lindner K, Borchardt C, Schöpp M, Bürgers A, Stock C, Hussey DJ, et al. Proton pump inhibitors (PPIs) impact on tumour cell survival, metastatic potential and chemotherapy resistance, and affect expression of resistance-relevant miRNAs in esophageal cancer. *J Exp Clin Cancer Res* 2014;**33**(1):73.
139. Goh W, Sleptsova-Freidrich I, Petrovic N. Use of proton pump inhibitors as adjunct treatment for triple-negative breast cancers. An introductory study. *J Pharm Pharm Sci* 2014;**17**(3):439–46.
140. Ferrari S, Perut F, Fagioli F, Brach Del Prever A, Meazza C, Parafioriti A, et al. Proton pump inhibitor chemosensitization in human osteosarcoma: from the bench to the patients' bed. *J Transl Med* 2013;**11**:268.
141. De Milito A, Canese R, Marino ML, Borghi M, Iero M, Villa A, et al. pH-dependent antitumor activity of proton pump inhibitors against human melanoma is mediated by inhibition of tumor acidity. *Int J Cancer* 2010;**127**(1):207–19.
142. Song T, Jeon HK, Hong JE, Choi JJ, Kim TJ, Choi CH, et al. Proton pump inhibition enhances the cytotoxicity of paclitaxel in cervical cancer. *Cancer Res Treat* 2017;**49**(3):595.
143. Pellegrini P, Strambi A, Zipoli C, Hägg-Olofsson M, Buoncervello M, Linder S, et al. Acidic extracellular pH neutralizes the autophagy-inhibiting activity of chloroquine: implications for cancer therapies. *Autophagy* 2014;**10**:562–71.
144. Adachi E, Tannock IF. The effects of vasodilating drugs on pH in tumors. *Oncol Res* 1999;**11** (4):179–85.
145. Kitai R, Kabuto M, Kubota T, Kobayashi H, Matsumoto H, Hayashi S, et al. Sensitization to hyperthermia by intracellular acidification. *J Neurooncol* 1998;**39**(3):197–203.
146. Yamagata M, Tannock IF. The chronic administration of drugs that inhibit the regulation of intracellular pH: in vitro and anti-tumour effects. *Br J Cancer* 1996;**73**(11):1328–34.
147. Grinstein S, Rotin D, Mason MJ. Na+/H+ exchange and growth factor-induced cytosolic pH changes. Role in cellular proliferation. *Biochim Biophys Acta Rev Biomembr* 1989;**988**(1):73–97.
148. Reshkin SJ, Greco MR, Cardone RA. Role of pHi, and proton transporters in oncogene-driven neoplastic transformation. *Philos Trans R Soc Lond B Biol Sci* 2014;**369**:20130100.
149. Weinhouse S, Warburg O, Burk D, Schade AL. On respiratory impairment in cancer cells. *Science* 1956;**124**(3215):267–72.
150. Shirmanova MV, Druzhkova IN, Lukina MM, Dudenkova VV, Ignatova NI, Snopova LB, et al. Chemotherapy with cisplatin: insights into intracellular pH and metabolic landscape of cancer cells in vitro and in vivo. *Sci Rep* 2017;**7**(1):8911.
151. Wojtkowiak JW, Verduzco D, Schramm KJ, Gillies RJ. Drug resistance and cellular adaptation to tumor acidic pH microenvironment. *Mol Pharm* 2011;**8**(6):2032–8.
152. Lacroix R, Rozeman EA, Kreutz M, Renner K, Blank CU. Targeting tumor-associated acidity in cancer immunotherapy. *Cancer Immunol Immunother* 2018;**5**:1–8.
153. Ohashi T, Akazawa T, Aoki M, Kuze B, Mizuta K, Ito Y, et al. Dichloroacetate improves immune dysfunction caused by tumor-secreted lactic acid and increases antitumor immunoreactivity. *Int J Cancer* 2013;**133**(5):1107–18.
154. Parks S, Chiche J, Pouysségur J. Disrupting proton dynamics and energy metabolism for cancer therapy. *Nat Rev Cancer* 2013;**13**:611–21.
155. Bindra RS, Schaffer PJ, Meng A, Woo J, Maseide K, Roth ME, et al. Alterations in DNA repair gene expression under hypoxia: elucidating the mechanisms of hypoxia-induced genetic instability. *Ann N Y Acad Sci* 2005;**1059**:184–95.
156. Vaupel P, Kallinowski F, Okunieff P. Blood flow, oxygen and nutrient supply, and metabolic microenvironment of human tumors: a review. *Cancer Res* 1989;**49**(23):6449–65.
157. Mbeunkui F, Johann DJ. Cancer and the tumor microenvironment: a review of an essential relationship. *Cancer Chemother Pharmacol* 2009;**63**(4):571–82.
158. Wakabayashi S, Fafournoux P, Sardet C, Pouyssegur J. The Na+/H+ antiporter cytoplasmic domain mediates growth factor signals and controls "H (+)-sensing". *Proc Natl Acad Sci U S A* 1992; **89**(6):2424–8.

159. Pouysségur J, Sardet C, Franchi A, L'Allemain G, Paris S. A specific mutation abolishing Na+/H+ antiport activity in hamster fibroblasts precludes growth at neutral and acidic pH. *Proc Natl Acad Sci U S A* 1984;**81**(15):4833–7.
160. Parks SK, Chiche J, Pouyssegur J. pH control mechanisms of tumor survival and growth. *J Cell Physiol* 2011;**226**(2):299–308.
161. Paris S, Pouyssegur J. Growth factors activate the Na+/H+ antiporter in quiescent fibroblasts by increasing its affinity for intracellular H+. *J Biol Chem* 1984;**259**(17):10989–94.
162. Calderón Montaño JM, Burgos Morón E, Pérez Guerrero C, Salvador Bofill FJ, Robles Frías A, López LM. Role of the intracellular pH in the metabolic switch between oxidative phosphorylation and aerobic glycolysis-relevance to cancer. *WebmedCentral* 2011;**2**(3):1–10.
163. Wilkins SE, Abboud MI, Hancock RL, Schofield CJ. Targeting protein–protein interactions in the HIF system. *ChemMedChem* 2016;**11**(8):773–86.
164. Wang GL, Jiang BH, Rue EA, Semenza GL. Hypoxia-inducible factor 1 is a basic-helix-loop-helix-PAS heterodimer regulated by cellular O2 tension. *Proc Natl Acad Sci U S A* 1995;**92**:5510–4.
165. Maxwell PH, Wiesener MS, Chang GW, Clifford SC, Vaux EC, et al. The tumour suppressor protein VHL targets hypoxia-inducible factors for oxygen-dependent proteolysis. *Nature* 1999;**399**:271–5.
166. Semenza GL. Targeting HIF-1 for cancer therapy. *Nat Rev Cancer* 2003;**3**(10):721–32.
167. Semenza GL. HIF-1: mediator of physiological and pathophysiological responses to hypoxia. *J Appl Physiol* 2000;**88**(4):1474–80.
168. Tsuzuki Y, Fukumura D, Oosthuyse B, Koike C, Carmeliet P, Jain RK. Vascular endothelial growth factor (VEGF) modulation by targeting hypoxia-inducible factor-1α → hypoxia response element → VEGF cascade differentially regulates vascular response and growth rate in tumors. *Cancer Res* 2000; **60**(22):6248–52.
169. Sun Q, Zhou H, Binmadi NO, Basile JRJ. Hypoxia-inducible factor-1-mediated regulation of semaphorin 4D affects tumor growth and vascularity. *Biol Chem* 2009;**284**(46):32066–74.
170. Han ZB, Ren H, Zhao H, Chi Y, Chen K, Zhou B, et al. Hypoxia-inducible factor (HIF)-1 alpha directly enhances the transcriptional activity of stem cell factor (SCF) in response to hypoxia and epidermal growth factor (EGF). *Carcinogenesis* 2008;**29**(10):1853–61.
171. Kelly BD, Hackett SF, Hirota K, Oshima Y, Cai Z, Berg-Dixon S, et al. Cell type-specific regulation of angiogenic growth factor gene expression and induction of angiogenesis in non-ischemic tissue by a constitutively active form of hypoxia-inducible factor 1. *Circ Res* 2003;**93**(11):1074–81.
172. Yang MH, Wu MZ, Chiou SH, Chen PM, Chang SY, Liu CJ, et al. Direct regulation of TWIST by HIF-1alpha promotes metastasis. *Nat Cell Biol* 2008;**10**(3):295–305.
173. Choi JY, Jang YS, Min SY, Song JY. Overexpression of MMP-9 and HIF-1α in breast cancer cells under hypoxic conditions. *J Breast Cancer* 2011;**14**(2):88–95.
174. Petrella BL, Lohi J, Brinckerhoff CE. Identification of membrane type-1 matrix metalloproteinase as a target of hypoxia-inducible factor-2α in von Hippel–Lindau renal cell carcinoma. *Oncogene* 2005; **24**(6):1043.
175. Jing SW, Wang YD, Kuroda M, Su JW, Sun GG, Liu Q, et al. HIF-1α contributes to hypoxia-induced invasion and metastasis of esophageal carcinoma via inhibiting E-cadherin and promoting MMP-2 expression. *Acta Med Okayama* 2012;**66**(5):399–407.
176. Zhu S, Zhou Y, Wang L, Zhang J, Wu H, Xiong J, et al. Transcriptional upregulation of MT2-MMP in response to hypoxia is promoted by HIF-1α in cancer cells. *Mol Carcinog* 2011;**50**(10):770–80.
177. Kim JW, Tchernyshyov I, Semenza GL, Dang CV. HIF-1-mediated expression of pyruvate dehydrogenase kinase: a metabolic switch required for cellular adaptation to hypoxia. *Cell Metab* 2006; **3**(3):177–85.
178. Iida T, Yasuda M, Miyazawa M, Fujita M, Osamura RY, Hirasawa T, et al. Hypoxic status in ovarian serous and mucinous tumors: relationship between histological characteristics and HIF-1α/GLUT-1 expression. *Arch Gynecol Obstet* 2008;**277**(6):539–46.
179. Minchenko A, Leshchinsky I, Opentanova I, Sang N, Srinivas V, Armstead V, et al. Hypoxia-inducible factor-1-mediated expression of the 6-phosphofructo-2-kinase/fructose-2, 6-bisphosphatase-3 (PFKFB3) gene. Its possible role in the Warburg effect. *J Biol Chem* 2002;**277**(8):6183–7.
180. Ullah MS, Davies AJ, Halestrap AP. The plasma membrane lactate transporter MCT4, but not MCT1, is up-regulated by hypoxia through a HIF-1α-dependent mechanism. *J Biol Chem* 2006; **281**(14):9030–7.

181. Doherty JR, Cleveland JL. Targeting lactate metabolism for cancer therapeutics. *J Clin Invest* 2013; **123**(9):3685–92.
182. Kaluz S, Kaluzová M, Liao SY, Lerman M, Stanbridge EJ. Transcriptional control of the tumor-and hypoxia-marker carbonic anhydrase 9: a one transcription factor (HIF-1) show? *Biochim Biophys Acta Rev Cancer* 2009;**1795**(2):162–72.
183. Swietach P, Vaughan-Jones RD, Harris AL. Regulation of tumor pH and the role of carbonic anhydrase 9. *Cancer Metastasis Rev* 2007;**26**(2):299–310.
184. Semenza GL. Hypoxia-inducible factors: mediators of cancer progression and targets for cancer therapy. *Trends Pharmacol Sci* 2012;**33**(4):207–14.
185. Levine AJ, Puzio-Kuter AM. The control of the metabolic switch in cancers by oncogenes and tumor suppressor genes. *Science* 2010;**330**(6009):1340–4.
186. Lu CW, Lin SC, Chen KF, Lai YY, Tsai SJ. Induction of pyruvate dehydrogenase kinase-3 by hypoxia-inducible factor-1 promotes metabolic switch and drug resistance. *J Biol Chem* 2008; **283**(42):28106–14.
187. Vasseur S, Tomasini R, Tournaire R, Iovanna JL. Hypoxia induced tumor metabolic switch contributes to pancreatic cancer aggressiveness. *Cancer* 2010;**2**(4):2138–52.
188. Walsh M, Fais S, Spugnini EP, Harguindey S, Izneid TA, Scacco L, et al. Proton pump inhibitors for the treatment of cancer in companion animals. *J Exp Clin Cancer Res* 2015;**34**:93.
189. Chiche J, Ilc K, Laferrière J, Trottier E, Dayan F, Mazure NM, et al. Hypoxia-inducible carbonic anhydrase IX and XII promote tumor cell growth by counteracting acidosis through the regulation of the intracellular pH. *Cancer Res* 2009;**69**(1):358–68.
190. Griffiths JR, McIntyre DJ, Howe FA, Stubbs M. Why are cancers acidic? A carrier-mediated diffusion model for H+ transport in the interstitial fluid. *Novartis Found Symp* 2001;**240**:46–62. discussion 62–7,152–3.
191. Song CW, Griffin R, Park HJ. Influence of tumor pH on therapeutic response. In: *Cancer drug resistance.* Totowa, NJ: Humana Press Inc; Springer; 2006. p. 21–42.
192. Sharma M, Astekar M, Soi S, Manjunatha BS, Shetty DC, Radhakrishnan R. pH gradient reversal: an emerging hallmark of cancers. *Recent Pat Anticancer Drug Discov* 2015;**10**(3):244–58.
193. Pouyssegur J, Sardet C, Franchi A, L'Allemain G, Paris S. A specific mutation abolishing Na+/H+ antiport activity in hamster fibroblasts precludes growth at neutral and acidic pH. *Proc Natl Acad Sci U S A* 1984;**81**:4833–7.
194. Stubbs M, McSheehy PM, Griffiths JR, Bashford CL. Causes and consequences of tumour acidity and implications for treatment. *Mol Med Today* 2000;**6**(1):15–9.
195. Fantin VR, St Pierre J, Leder P. Attenuation of LDH-A expression uncovers a link between glycolysis, mitochondrial physiology, and tumor maintenance. *Cancer Cell* 2006;**9**:425–34.
196. Sun RC, Fadia M, Dahlstrom JE, Parish CR, Board PG, Blackburn AC. Reversal of the glycolytic phenotype by dichloroacetate inhibits metastatic breast cancer cell growth in vitro and in vivo. *Breast Cancer Res Treat* 2010;**120**(1):253–60.
197. Corbet C, Pinto A, Martherus R, de Jesus JP, Polet F, Feron O. Acidosis drives the reprogramming of fatty acid metabolism in cancer cells through changes in mitochondrial and histone acetylation. *Cell Metab* 2016;**24**(2):311–23.
198. Winkler MM. Nuccitelli R, Deamer OW, editors. *Intracellular pH: its measurement, regulation and utilization in cellular functions*. New York: Alan R. Liss; 1982. p. 325–40.
199. Gerson DF. Nuccitelli R, Deamer DW, editors. *Intracellular pH: its measurement, regulation, and utilization in cellular functions*. New York: Alan R. Liss; 1982. p. 375–83.
200. Fidelman ML, Seeholzer SH, Walsh KB, Moore RD. Intracellular pH mediates action of insulin on glycolysis in frog skeletal muscle. *Am J Physiol* 1982;**242**:87–93.
201. Ui M. A role of phosphofructokinase in pH-dependent regulation of glycolysis. *Biochim Biophys Acta Gen Subj* 1966;**124**(2):310–22.
202. Tannock IF, Rotin D. Acid pH in tumors and its potential for therapeutic exploitation. *Cancer Res* 1989;**49**(16):4373–84.
203. Madshus IH. Regulation of intracellular pH in eukaryotic cells. *Biochem J* 1988;**250**(1):1.
204. Potter CPS, Harris AL. Diagnostic, prognostic and therapeutic implications of carbonic anhydrases in cancer. *Br J Cancer* 2003;**89**(1):2.

205. Lee AH, Tannock IF. Heterogeneity of intracellular pH and of mechanisms that regulate intracellular pH in populations of cultured cells. *Cancer Res* 1998;**58**(9):1901–8.
206. Wahl ML, Owen JA, Burd R, Herlands RA, Nogami SS, Rodeck U, et al. Regulation of intracellular pH in human melanoma: potential therapeutic implications. *Mol Cancer Ther* 2002;**1**(8):617–28.
207. Boyer MJ, Tannock IF. Regulation of intracellular pH in tumor cell lines: influence of microenvironmental conditions. *Cancer Res* 1992;**52**(16):4441–7.
208. Belhoussine R, Morjani H, Sharonov S, Ploton D, Manfait M. Characterization of intracellular pH gradients in human multidrug-resistant tumor cells by means of scanning microspectrofluorometry and dual-emission-ratio probes. *Int J Cancer* 1999;**81**(1):81–9.
209. Harguindey S, Pedraz JL, Garcia Cañero R, Perez de Diego J, Cragoe EJ. Hydrogen ion-dependent oncogenesis and parallel new avenues to cancer prevention and treatment using a H^+-mediated unifying approach: pH-related and pH-unrelated mechanisms. *Crit Rev Oncog* 1995;**6**:1–33.
210. Gatenby RA, Gawlinski ET, Gmitro AF, Kaylor B, Gillies RJ. Acid-mediated tumor invasion: a multidisciplinary study. *Cancer Res* 2006;**66**(10):5216–23.
211. Harguindey S, Arranz JL, Wahl ML, Orive G, Reshkin SJ. Proton transport inhibitors as potentially selective anticancer drugs. *Anticancer Res* 2009;**29**(6):2127–36.
212. Koltai T. Triple-edged therapy targeting intracellular alkalosis and extracellular acidosis in cancer. *Semin Cancer Biol* 2017;**43**:139–46.
213. Doppler W, Jaggi R, Groner B. Induction of v-mos and activated Ha-ras oncogene expression in quiescent NIH 3T3 cells causes intracellular alkalinisation and cell-cycle progression. *Gene* 1987; **54**(1):147–53.
214. Harguindey S, Stanciu D, Devesa J, Alfarouk K, Cardone RA, Orozco JD, et al. Cellular acidification as a new approach to cancer treatment and to the understanding and therapeutics of neurodegenerative diseases. *Semin Cancer Biol* 2017;**43**:157–79.
215. Koltai T. Cancer: fundamentals behind pH targeting and the double-edged approach. *Onco Targets Ther* 2016;**9**:6343.
216. Newell K, Franchi A, Pouyssegur J, Tannock I. Studies with glycolysis-deficient cells suggest that production of lactic acid is not the only cause of tumor acidity. *Proc Natl Acad Sci U S A* 1993; **90**(3):1127–31.
217. Yamagata M, Hasuda K, Stamato T, Tannock IF. The contribution of lactic acid to acidification of tumours: studies of variant cells lacking lactate dehydrogenase. *Br J Cancer* 1998;**77**(11):1726.
218. Izumi H, Torigoe T, Ishiguchi H, Uramoto H, Yoshida Y, Tanabe M, et al. Cellular pH regulators: potentially promising molecular targets for cancer chemotherapy. *Cancer Treat Rev* 2003;**29**(6):541–9.
219. Gullino PM, Grantham FH, Smith SM, Haggerty AC. Modifications of the acid-base balance of the internal milieu of tumors. *J Natl Cancer Inst* 1965;**34**:857–69.
220. Helmlinger G, Sckell A, Dellian M, Forbes NS, Jain RK. Acid production in glycolysis-impaired tumors provides new insights into tumor metabolism. *Clin Cancer Res* 2002;**8**(4):1284–91.
221. Helmlinger G, Yuan F, Dellian M, Jain RK. Interstitial pH and pO, gradients in solid tumors in vivo: high resolution measurements reveal a lack of correlation. *Nat Med* 1977;**3**:177–82.
222. Heitzer E, Auer M, Hoffmann EM, Pichler M, Gasch C, Ulz P, et al. Establishment of tumor specific copy number alterations from plasma DNA of patients with cancer. *Int J Cancer* 2013;**133**:346–56.
223. Thierry AR, Mouliere F, Gongora C, Ollier J, Robert B, Ychou M, et al. Origin and quantification of circulating DNA in mice with human colorectal cancer xenografts. *Nucleic Acids Res* 2010;**38**:6159–75.
224. Mouliere F, El Messaoudi S, Pang D, Dritschilo A, Thierry AR. Multi-marker analysis of circulating cell-free DNA toward personalized medicine for colorectal cancer. *Mol Oncol* 2014;**8**:927–41.
225. Bettegowda C, Sausen M, Leary RJ, Kinde I, Wang Y, Agrawal N, et al. Detection of circulating tumor DNA in early-and late-stage human malignancies. *Sci Transl Med* 2014;**6**(224):224ra24.
226. Gatenby RA, Gawlinski ET. A reaction-diffusion model of cancer invasion. *Cancer Res* 1996; **56**(24):5745–53.
227. Wolf A, Agnihotri S, Micallef J, Mukherjee J, Sabha N, Cairns R, et al. Hexokinase 2 is a key mediator of aerobic glycolysis and promotes tumor growth in human glioblastoma multiforme. *J Exp Med* 2011;**208**:313–26. jem-20101470.

228. Zhou Y, Zhou Y, Shingu T, Feng L, Chen Z, Ogasawara M, et al. Metabolic alterations in highly tumorigenic glioblastoma cells: preference for hypoxia and high dependency on glycolysis. *J Biol Chem* 2011;**286**:32843–53 jbc-M111.
229. Michelakis ED, Sutendra G, Dromparis P, Webster L, Haromy A, Niven E, et al. Metabolic modulation of glioblastoma with dichloroacetate. *Sci Transl Med* 2010;**2**(31):31ra34.
230. Marathe K, McVicar N, Li A, Bellyou M, Meakin S, Bartha R. Topiramate induces acute intracellular acidification in glioblastoma. *J Neurooncol* 2016;**130**(3):465–72.
231. Colen CB, Shen Y, Ghoddoussi F, Yu P, Francis TB, Koch BJ, et al. Metabolic targeting of lactate efflux by malignant glioma inhibits invasiveness and induces necrosis: an in vivo study. *Neoplasia* 2011;**13**(7):620–32.
232. Miranda-Goncalves V, Honavar M, Pinheiro C, Martinho O, Pires MM, Pinheiro C, et al. Monocarboxylate transporters (MCTs) in gliomas: expression and exploitation as therapeutic targets. *Neuro Oncol* 2013;**15**(2):172–88.
233. Di Cristofori A, Ferrero S, Bertolini I, Gaudioso G, Russo MV, Berno V, et al. The vacuolar H+ ATPase is a novel therapeutic target for glioblastoma. *Oncotarget* 2015;**6**(19):17514.
234. Graham RM, Thompson JW, Webster KA. Inhibition of the vacuolar ATPase induces Bnip3-dependent death of cancer cells and a reduction in tumor burden and metastasis. *Oncotarget* 2014;**5**(5):1162.
235. Krex D, Klink B, Hartmann C, von Deimling A, Pietsch T, Simon M, et al. Long-term survival with glioblastoma multiforme. *Brain* 2007;**130**(10):2596–606.

CHAPTER 4

Lactic acid and its transport system

Introduction

Lactic acid (LA) is the final product of the glycolytic pathway in the cell. But it is not the end of the multiple activities of LA in the life of cancer, as we shall see here.

As we have shown in previous chapters, the glycolytic pathway is highly up-regulated in all cancer cells. And so is the uptake of glucose. Both up-regulations lead to a lactic acid production that Warburg calculated as an approximately 60 fold increase compared with normal cells. The intracellular burden represented by such an amount of lactic acid would be highly toxic for the cell if it were to remain inside. LA does not remain in the cell. It is swiftly exported by specialized carriers of the monocarboxylate family (MCTs). This means that LA enters into the ECM (extracellular matrix) and from there it can follow different routes. It can:

(1) remain in the ECM cooperating in the acidity of the ECM.
(2) be cleared through blood vessels.
(3) be used as fuel for oxidative malignant cells that have developed the skill to uptake LA reconverting it to pyruvate that is metabolized through the Krebs cycle (lactophagic cells). This process is known as the inverse Warburg effect.

The proportion in which each of these events occurs has not been determined up to now (Fig. 1). In an over-simplified way this figure shows that the first step of the glycolytic chain is performed by GLUTs (glucose transporters) and the last step is the lactate and proton extrusion through MCTs (monocarboxylate transporters). There are 11 enzymatic steps in the middle since a molecule of glucose enters the cells, produces two molecules of lactate (see chapter on metabolism) and finally is extruded to the ECM.

MCTs are transporters of monocarboxylates and protons (H^+) in which the proton binds to the MCT transferring its energy. In the next step, lactate is bound to the transporter and to the proton and both are carried through the membrane[1] (Fig. 2). The extruded lactate and protons are partly the cause of extracellular acidity, however, the main cause is the hydration of CO2 to carbonic acid, as has been discussed in previous chapters.

Almost a hundred years ago, Warburg found that cancer cells showed increased glycolysis with increased LA as its direct consequence. This picture was present with and without lack of oxygen.[2–4] Half a century later, this phenomenon was named the Warburg effect[5] (or aerobic glycolysis).

An Innovative Approach to Understanding and Treating Cancer: Targeting pH
https://doi.org/10.1016/B978-0-12-819059-3.00004-6
© 2020 Elsevier Inc.
All rights reserved.

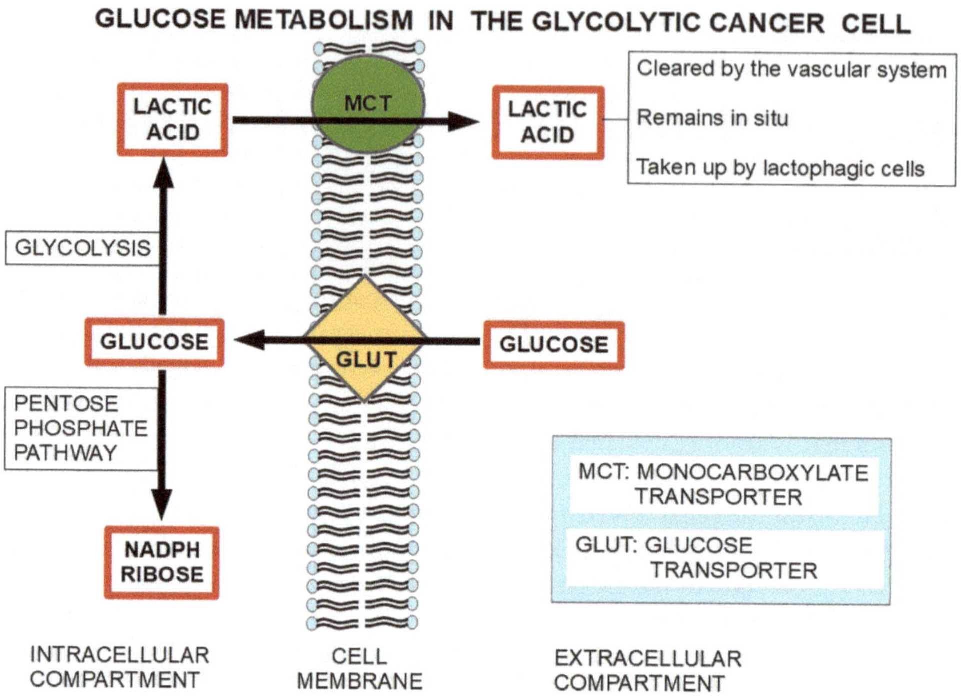

Fig. 1 Origin and destiny of lactic acid in cancer cells.

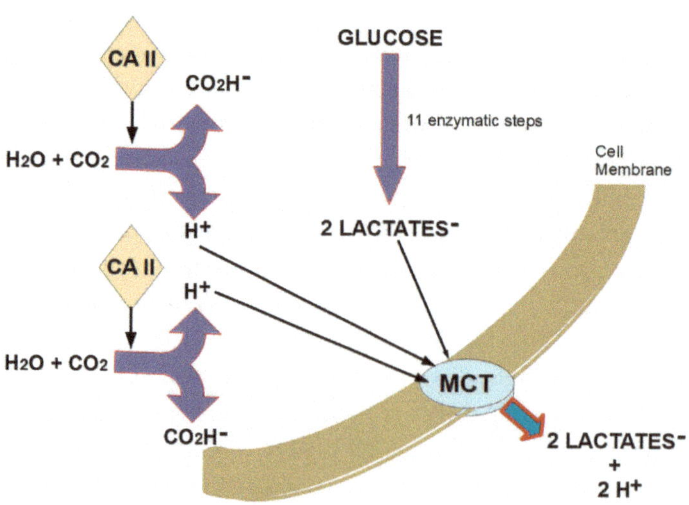

Fig. 2 Lactate and proton extrusion through MCT's activity. CA II: carbonic anhydrase 2.

Fig. 3 shows the increased lactic acid production and low oxygen consumption of cancer cells when compared with normal tissues. It also shows that not all the ATP generated in cancer cells comes from aerobic glycolysis.

As discussed in the preceding chapters, ECM is highly acidic in cancer cells, and lactic acid is an important component of this ECM. Lactic acid concentration in the ECM surrounding tumors is up to 40 fold higher as compared with normal tissues (tumors' concentration of lactic acid = 40 mM/ normal tissues concentration = 1–2 mM).[6] The high concentration of lactates is correlated with high incidence of metastasis[7] and poor prognosis.

These facts seem to show three things:

(1) that lactic acid is a terminal metabolite, and therefore

(2) it is a waste product, and

(3) it is the main factor for ECM acidity in tumors.

Fig. 3 shows lactic acid and ATP production in normal and cancer cells with and without oxygen.

COMPARATIVE TABLE OF ATP PRODUCTION IN DIFFERENT TISSUES ACCORDING TO WARBURG

Cells of	O_2 Consumption	Lactic Acid Produced	ATP produced with O_2	ATP produced without O_2	Total ATP
LIVER	15	1	105	1	106
KIDNEY	15	1	105	1	106
EMBRYO	15	25	105	25	130
CANCER	7	60	49	60	109

Cancer cells produce 60 times more lactic acid than normal cells

45% of ATP is produced by OXPHOS in cancer cells

O_2 consumption per mg of tissue
Lactic acid in cubic millimiter per hour
ATP per cubic millimeter

Fig. 3 Warburg's findings in tumors. Mathematical simulations showed that in glycolytic cancer cells 85% of the glucose was metabolized through the glycolytic pathway and 15% through mitochondrial oxidative phosphorylation (OXPHOS). These proportions may dynamically change according to the amount of available glucose and the concentration of lactate.[8]

In words of San-Millan and Brooks[9] "*Lactate has been the great unknown in human metabolism*" and we may add, the "ideal waste product". But not any longer, because all these assumptions are incorrect.

In this chapter we will discuss the role of lactic acid and its transport system in cancer, which should lead us to understand that lactic acid is neither a terminal nor a waste product nor the main culprit of extracellular acidosis and that it is an important player in the cancer concert.[10]

The experiments by Newell et al.[11] and Yamagata et al.[12] have shown that lactic acid is not such an important contributor to ECM acidity as previously thought (see previous chapter). This may lead to the erroneous concept that lactic acid is not important in cancer. However, lactate is not an innocent bystander because:

(1) Even if it is not the main cause of ECM acidity, lactate still contributes to it.
(2) Lactate is a source of energy for OXPHOS cells (lactate shuttle).
(3) Lactate enhances tumor cell motility,[13] migration and growth.[14]
(4) It is a signaling molecule.[15]
(5) It has immunosuppressive effects in tumors.[16–18]
(6) Lactate induces pro-angiogenic effects.[19–21]
(7) It has a high correlation with metastasis and even promotes it.[22]
(8) It increases cancer cell stemness.[23]
(9) Lactate activates HIF-1α.[24]
(10) Lactate has pro-inflammatory effects with increased NF-kB activation.[25]
(11) There is a positive correlation between lactate and radioresistance.[26]
(12) Lactate increases hyaluronan production involved in migration and growth.[27]
(13) It is not only produced by malignant cells, but also by stromal cells.
(14) Lactate up-regulates aquaporin1 in tumors.[28]

After going through different enzymatic steps (10 steps), one molecule of glucose, produces two molecules of pyruvic acid. Depending on oxygen availability, the next steps lead to OXPHOS or to the glycolytic pathway. In cancer cells the glycolytic pathway is the preferential one.

Malignant tissues show the presence of OXPHOS cells, but there is a high predominance of glycolytic cells that explains the lactic acid overload. Since the cancer microenvironment is very hypoxic, the Warburg effect predominates and lactic acid production is highly increased (Fig. 3).

The coexistence of the glycolytic and oxidative pathways in tumors has been a known fact since Warburg's research, because he determined that tumor cells metabolized 66% of the glucose to lactate. Nevertheless, malignant glycolytic cells retain their OXPHOS ability.[29] The adoption of the aerobic glycolytic pathway by cancer cells is only a part of a major metabolic change that goes well beyond carbohydrate metabolism and involves lipids and proteins and this is known as the metabolic switch.

This metabolic switch becomes active very early in the cancerization process.[30] However this mechanism is also found in nonmalignant cells such as proliferating primary lymphocytes.[31] This means that aerobic glycolysis and its increased production of lactic acid, is a mechanism that is sometimes adopted in situations of high metabolic activity and proliferation. Finally, all glycolytic enzymes are overexpressed in cancer cells, as explained in the chapter on metabolism.

Pro-tumoral activities of lactic acid

Lactate contributes to extracellular space (ECS) acidity

Nowadays, we know that under experimental conditions, malignant cells that are disabled for lactate production, have ECS (extracellular space) pH (pHe) similar to those that can produce lactate without restrictions. In spite of this, lactate extrusion cooperates in ECS acidification because tumors have highly increased lactic acid content in ECS when compared with normal counterparts.

Lactate is a source of energy for OXPHOS cells: The lactate shuttle

For many years, the prevailing concept of glucose metabolism in cancer was that the glycolytic pathway that degrades glucose to lactate was the predominant pathway for the production of energy. Researchers did not understand why a cell that needed a big amount of energy, like a cancer cell, used such an inefficient mechanism, where 36 molecules of ATP were squandered for each molecule of glucose.[32]

Then, it was discovered that these glycolytic cells were not alone in the tumor, because there were other cells that kept OXPHOS (oxidative phosphorylation) as their main metabolic pathway. A third discovery was that some or all the oxidative cells had the ability to use lactate as an energy source. There was no squandering.

Another prevailing concept was that lactic acid in glycolytic cells was confined to the cytoplasm and had no entry into the mitochondria. This is changing too. There is evidence showing that glycolytic cells may fuel their mitochondria with lactate through its oxidation to pyruvate by mitochondrial LDH (lactate dehydrogenase).[33]

The lactate shuttle concept is based on the fact that cells that produce high amounts of lactate through glycolytic metabolism extrude this lactate to the extracellular matrix, after replenishing the NAD^+ pool, from where it is taken up by other malignant oxidative cells that use this lactate as a source of energy.

It is a well known process in normal muscle where white glycolytic fibers produce lactate that feeds red oxidative fibers in the same muscle. There is another example of the lactate shuttle in the brain, between astrocytes and neurons.[34] This cell to cell shuttle (intercellular shuttle) is present even under aerobic conditions.[35] There is yet another form of lactate shuttle which occurs inside the same cell (intracellular shuttle) where there is an uptake of lactate by mitochondria. This is accomplished thanks to mitochondrial lactate dehydrogenase (LDH) that oxidizes lactate.[36]

There have also been some disagreements concerning the real existence of the intracellular shuttle.[37] Yoshida et al. have shown strong evidence against the existence of mitochondrial LDH and the ability of mitochondria to convert lactate into pyruvate.[38] However, they also found that very high levels of lactate (more than a 50 to 150 times higher than pyruvate levels) could "force" an intracellular shuttle (a malignant glycolytic cell produces at least 60 times more lactate than any nonglycolytic cell). We have to add that this does not happen in normal tissues, but it may well be possible in cancer cells.

Without entering into the complex arguments of intracellular shuttle, from this point on, our description will be restricted to the intercellular lactate shuttle.

Advanced tumors are very heterogeneous, not only morphologically but also in their genotype and metabolism. Regarding glucose metabolism, this heterogeneity shows the presence of two different types of malignant cells: glycolytic and oxidative. The oxidative cells are usually located in better oxygenated areas of the tumor, and have the ability to use the lactic acid produced by glycolytic cells as fuel for their own metabolism.[39,40] Many tumors develop a shuttle system where the uptake of lactates by the oxidative cells becomes an important source of energy (reversed Warburg effect).[41] We have called these cells lactophagic cells. What we do not know is if all oxidative cells have the ability to become lactophagic. This mechanism can be considered as a symbiotic system in which a group of cells produces lactic acid that serves for the nutrition of lactophagic cells. Furthermore, it is quite usual to find stromal cells surrounding the tumor that are "enslaved" by the malignant cells and induced to produce lactic acid to nourish the lactophagic cells.[42] This resembles a parasitic system. Monocarboxylate transporters are over-expressed in cancer cells, but in the symbiotic and parasitic systems they are of paramount importance.[43–45] The considerable distance between vessels and acidic areas of the tumor could also suggest that the lactate shuttle is a major mechanism in lactate clearance from the ECS (Fig. 4). The lactate shuttle also works between glycolytic cells and tumor endothelial cells and has a stimulatory effect on neoangiogenesis.[46] In the astrocyte-neuron lactate shuttle Garcia-Espinosa et al.[47] found that these cells can choose between glucose and lactate for their oxidative metabolism, according to their availability. This has not been tested in cancer, but the system probably works in the same way (Box 1).

BOX 1

SYMBIOTIC SYSTEM: MALIGNANT GLYCOLYTIC CELLS PRODUCE LACTIC ACID THAT IS TAKEN UP BY MALIGNANT OXIDATIVE CELLS (Reverse Warburg symbiotic effect).

PARASITIC SYSTEM: STROMAL AND ENDOTHELIAL CELLS ARE "ENSLAVED" BY THE TUMOR AND INDUCED TO ADOPT A GLYCOLYTIC METABOLISM PRODUCING LACTIC ACID THAT IS TAKEN UP BY MALIGNANT OXIDATIVE CELLS (Reverse Warburg parasitic effect).

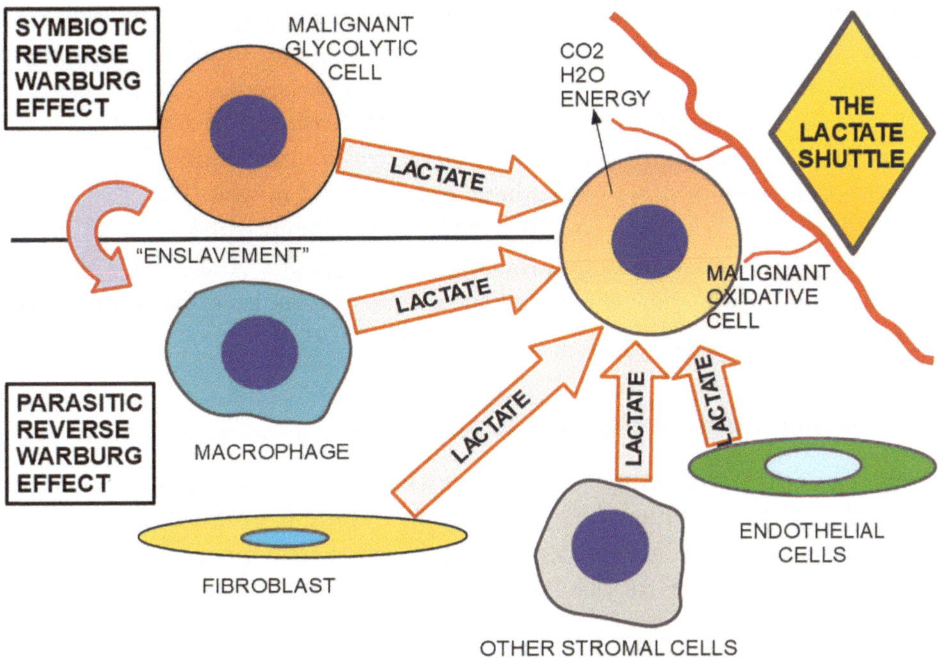

Fig. 4 Glycolytic cancer cell (upper left side) extruding lactic acid that is taken up by an oxidative cancer cell on the right side (Symbiotic reverse Warburg effect). Stromal noncancer cells (lower half) "enslaved" by the tumor also produce lactic acid that feeds oxidative cancer cells (Parasitic reverse Warburg effect).[48–51]

Lactate shuttle and pH

Roth and Brooks[52] showed that lactate transport is a pH sensitive process. They wrote:

> *"Lactate (La^-) transport is pH-sensitive; stimulations to fivefold overshoot above equilibrium values were observed both directly by a proton gradient directed inward, and indirectly by a monensin- or nigericin-stimulated exchange of Na^+ or K^+ for H^+ across the sarcolemma."*
>
> *And: "We conclude that the majority of La^- crosses the skeletal muscle sarcolemma by a specific carrier-mediated process that is saturable at high La^- concentrations, but flux is passively augmented at low intracellular pH by undissociated lactic acid. In addition, a Na^+ H^+ exchange mechanism confirmed in skeletal muscle sarcolemma, does affect both lactate and proton flux, and is potentially an important coregulator of intracellular pH and thus, cellular metabolism".*

Although, Roth and Brooks's experiment was performed on normal rat muscle, we can extract an interesting assumption: that lactophagic cells may have a slightly lower intracellular pH than the glycolytic ones and that this difference stimulates lactate uptake. This hypothesis has not been experimentally confirmed.

The favorable effects of metformin in cancer can also be explained by their report:

(1) Metformin increases lactic acid production through mitochondrial complex I inhibition.

(2) This excess of LA saturates the transport system (Roth and Brooks).

(3) Part of the LA that remains inside the cell decreases intracellular pH.

(4) Cellular acidification slows biosynthetic processes and mitosis.[53–54]

(5) Cellular acidification down-regulates the Wnt/B catenin pathway.

There is indirect evidence sustaining this hypothesis. Melnik et al.[55] have shown that metformin acidifies the intracellular milieu and down-regulates the Wnt/B catenin pathway.

Clinical implications of the lactate shuttle

The presence of the lactate shuttle and high levels of lactic acid are signs of poor prognosis in many different types of tumors.[56–58] The lactate shuttle is a mechanism associated to tumor progression. Therefore, it is a potential target for treatment.

In summary, there are basically two regulators of the lactate shuttles: LDH and MCTs.

Oxidative cells can use lactate and/or glucose as energy sources.

The lactate shuttle is the best proof of the high metabolic flexibility of cancer. It explains why certain anti-glycolytic treatments like 2DG (2-deoxyglucose) or DCA (dichloroacetate) have failed as stand-alone treatments. It is also the proof of concept of the idea first sustained by Warburg, that to "cure" cancer we need to inhibit both metabolisms: glycolytic and oxidative. Lee et al.[59] have shown experimental evidence that targeting both, glycolytic and oxidative metabolism yields better results. The addition of metformin as a mitochondrial poison to 2DG[60,61] or to dichloroacetate (DCA)[62] has produced a significant improvement of anti-cancer results. The reasons for metformin's synergy with the above mentioned drugs may differ, but in any case the down-regulation of the lactate shuttle is one of them. On the one hand, metformin hinders the oxidative metabolism and therefore impedes lactate from becoming a source of energy, while on the other it increases lactic acid production.

MCT can be blocked by monocarboxylic acids that play as lactate-competitors.

DCA (dichloroacetate) is a small-molecule drug that restores the glycolytic metabolism to the oxidative pathway. Using DCA or 2 deoxyglucose seems to inhibit the metabolic reprogramming between malignant and stromal cells.[63] This inhibition slows down tumor growth.

Table 1 shows the effects of the combination of a biguanide (mitochondrial poison) and DCA.

Another way to target the lactate shuttle is through LDH inhibition or down-regulation.

Table 1 Association between DCA and a mitochondrial "poison" (Metformin; Met).

Reference	Findings
Litchfield et al.[64]	In prostate cancer cells the Met-DCA association increased anti-tumor effects.
Hong et al.[65]	There is a synergy of the Met-DCA association in cancer cell killing, but HIF-1α mediates resistance to this activity. Therefore, down-regulation of HIF-1α may be a prerequisite for this type of treatment.
Hong et al.[66]	In hypoxia, TRAIL increased sensitivity to Met-DCA cell-killing in breast cancer cells.
Blackburn et al.[67]	In breast cancer cell lines DCA (1–5 mM), metformin (30–300 μM) or the combination of the two significantly inhibited tumor growth over 4 days of treatment in BALB/c-*Trp53*+/− mice.
Shen et al.[68]	The Met- DCA association synergistically elevated oxidative stress in diffuse intrinsic pontine glioma cells, with growth arrest, mitochondrial depolarization and apoptosis. Also increased radio-sensitivity.
Martinkute et al.[69]	The Met-DCA association sensitized breast cancer cells to apoptosis.
Kiliccioglu et al.[70]	The Met- DCA association produced growth inhibition and increased apoptosis in PC-3 prostate cancer cells. The association allowed for lower doses of each component to achieve the results.
Lea et al.[71]	In bladder cancer cells the association of phenformin with DCA produced enhanced growth inhibition.

Lactate dehydrogenase catalyzes the conversion of pyruvate to lactate, with NADH as a co-factor. It is a tetrameric enzyme composed of two subunits, M and H, whose association can generate five different isoforms. Isoform 5 has the highest activity in converting pyruvate to lactate under anaerobic conditions, such as those found in hypoxic tumors and for this reason it is up-regulated in tumor tissues. LDH inhibition affects the tumor's glycolytic metabolism but does not interfere with oxidative nonmalignant cells.[72]

There is active research going on in this sense. Gossipol, a polyphenolic compound, is being experimentally used for this purpose.[73–75] Gossipol has other anti-cancer effects (like inhibition of the anti-apoptotic proteins), therefore it is difficult to determine how much of its action is related with LDH downregulation.

Galloflavin, synthesized from gallic acid, and known for more than 150 years,[76] is another compound with LDH inhibitory properties (blocks both isoforms LDHA and LDHB[77]) that in the experimental setting has shown anti-tumor activity.[78–81] It acts as a competitive inhibitor of NADH for LDH. Due to it ability to inhibit the B isoform of LDH as well, galloflavin should also interrupt the intracellular lactate shuttle that relies mainly on the B isoform.

Another possible way to interrupt the lactate shuttle is through the inhibition of lactate transporters, particularly monocarboxylate-1 and 4 (MCT-1/ MCT-4). There are no clinically available MCT inhibitors. Although there are a few MCT-1 inhibitors under active research, like AR-C155858,[82] and AZD-3965,[83] none have reached the clinical stage.

Among nutraceuticals there are some compounds like phloretin, sylibin[84] and quercetin with MCT inhibitory effects. Quercetin's inhibition of MCTs has been found in human lung cancer cells.[85] Niflumic acid, an analgesic and anti-inflammatory drug is also an MCT inhibitor.[86] Wang and Morris found that MCT activity was significantly decreased "in the presence of the flavonoids apigenin, biochanin A, chrysin, diosemin, fisetin, genicitin, hesperitin, kaempferol, luteolin, morin, narigenin, phloretin and quercetin".[87] Of the 3 repurposed drug/nutraceuticals, phloretin, quercetin and niflumic acid, quercetin is the most potent in MCT1 inhibition while phloretin is the most potent against MCT4.[88]

The inhibition of MCTs is an important mechanism for achieving intracellular acidification in malignant cells.[89,90]

Lactate enhances tumor cell motility and migration

Epithelial mesenchymal transition is a precondition for migration. Li et al.[91] found that lactate induced Snail and epithelial-mesenchymal transition (EMT) by releasing activated TGF-β1 (Box 2).

BOX 2

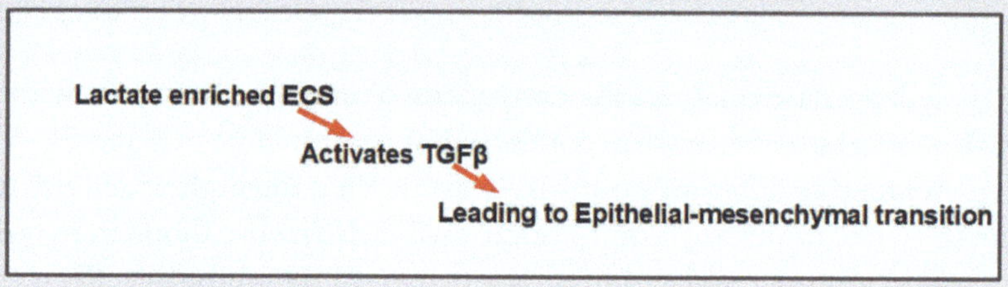

Lactate is a signaling molecule

Lactate has a receptor that has been identified as a G-protein coupled receptor: GPR81 (also called HCA1 that stands for hydrocarboxylic receptor 1). GPR81 expression is increased in most solid malignant tumors and its inhibition impairs growth and invasion.[92] The binding of lactate to GPR81 inhibits intracellular signaling related to cAMP in adipose tissue,[93] whereas in the nervous system, LA binding to GPR81 has an activating effect on cAMP.[94] These contradictory findings hint at a tissue-specific effect of LA-GPR81 binding. The connections between these inhibitory/stimulatory actions of LA and cancer have not been investigated, however those between cAMP and cancer are better known.[95,96] GPR81 is related to glucose and lactic sensing and PKA (phosphokinase A) activation in yeasts[97] and humans. GPR81 signaling increases the expression of MCTs. PKA is a cAMP-dependent kinase, so we may assume that in those tissues where the LA-GPR81 binding activates cAMP accumulation, there is PKA activation. PKA activation has pro-tumor effects.[98]

The known pro-tumor lactic acid pathways where LA is acting as a signaling molecule are the following (Box 3):

BOX 3

Lactic acid-GPR81 → cAMP → PKA-I → Pro-tumoral effects

Lactic acid-GPR81 → MCT1 and MCT4

Lactic acid-GPR81 → PD-L1*

*In this case the up-regulation of PD-L1 is related to decreased cAMP activity.[99]

The tissue specific effects of LA binding to GPR81 seem to originate in the fact that there are two different isoforms of PKA: PKA-I and PKA-II.[100,101] The first is pro-proliferative while the second is pro-differentiation. Different tissues show different proportions of PKA-I and PKA-II, hence the different effects.[102]

The lactate receptor GPR81 is currently considered a pro-oncogenic receptor. Many breast cancer cell lines show an over-expressed GPR81. The knockdown of GPR81 decreased growth and angiogenesis and induced apoptosis[103] (Fig. 5).

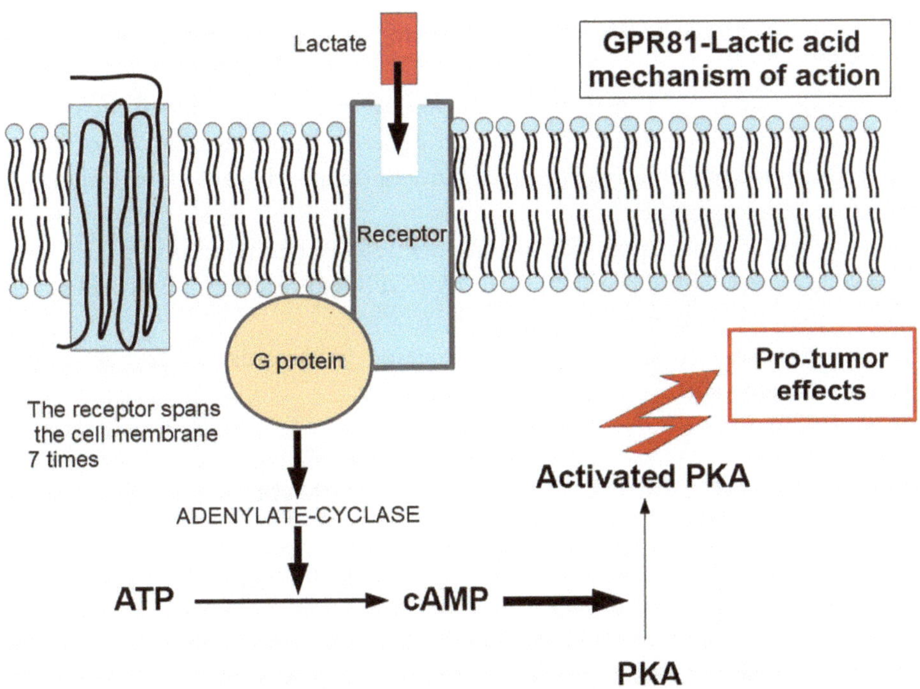

Fig. 5 Activation of GPR81 signaling through lactic acid binding. This may lead to two different responses according to tissue specificity: (1) down-regulation of adenyl-cyclase or (2) up-regulation of adenyl-cyclase. In the latter case cAMP is formed activating PKA with its pro-tumor effects.

Lactate has immunosuppressive effects in tumors

Choi et al.[104] maintain that lactic acid is an immunosuppressive metabolite in tumors. They base this assumption on the acidity generated by lactic acid. Acidity is partially the cause of loss of immunocompetence, but lactic acid has effects on the immune system on its own and independent from acidity. Lactic acid decreases immunosurveillance.[105,106] Fisher et al.[107] found that lactic acid had direct effects on the activity of human cytotoxic lymphocytes, decreasing their proliferation and cytotoxic activity up to 95%. When these cells were relocated in a lactic acid-free medium the cytotoxic activity was restored. As cytotoxic T lymphocytes also have a glycolytic metabolism, the increased lactic acid in the culture medium blocked the extrusion of lactic acid from the lymphocytes and this interfered with their metabolism and function. In this case, the immunosuppressive activity of lactic acid seems independent from acidity. Reinforcing this concept, Gottfried et al.[108] found that lactic acid interfered in the activation of dendritic cells and antigen expression. Macrophages that were pretreated with lactic acid inhibited CD8+ T-cell proliferation and this situation was restored to normal when lactic acid production was blocked with dichloroacetate.[109] Dendritic cells treated with lactic acid could not develop anti-tumor responses in vivo.[110]

Lactate induces pro-angiogenic effects

Angiogenesis is a fundamental process for tumor development. Lactate is an angiogenic inducer in nonmalignant healing tissues[111,112] and is also a driver for angiogenesis in cancer.[113,114]

Macrophages cultured in an acidic and lactate-rich medium produced angiogenic factors. This did not happen when the medium contained pyruvate instead of lactate.[115] MCT1 is connected with the angiogenic activity of lactates, because targeting this transporter reduces angiogenesis.[116]

There are many explanations for the angiogenesis-lactate relationship:

(a) Lactate increases VEGF production and VEGFR expression in endothelial cells.[117]
(b) Lactate increases endothelial cell migration.[118]
(c) Lactate increases HIF-1α effects by decreasing its degradation and activating the PI3K/Akt pathway.[119]

Lactate levels have a high correlation with metastasis and also promote metastasis

A high correlation between elevated levels of lactic acid and metastasis was found in head and neck cancers[120,121] cervical cancer,[122] pancreatic cancer,[123] breast cancer,[124] among others. LDH5 over-expression also shows correlation with metastasis.[125,126] Metastatic tumors have a higher level of lactate than nonmetastatic cancers.[127]

Lactate has pro-inflammatory effects

Lactate increased the transcription of IL-23 which enhanced IL17 and MMP-9 secretion in tumor- associated macrophages, starting a pro-inflammatory pathway that promoted tumor development. This promoter activity of lactate was regulated by pHe: low pHe increased lactate enhancer activity on IL-23 while higher pHe inhibited it.[128,129]

The lactate transport system

Lactate transport into and out of the cells requires specialized proteins. This process is carried out by a family of monocarboxylate transporters (MCTs) with different isoforms. There are 14 known members of the MCT family, but only MCT1, MCT2, MCT3 and MCT4 catalyze the bidirectional proton-linked transport of monocarboxylates such as lactate and pyruvate across the plasma membrane. MCT2 and MCT3 are limited to only a few tissues, therefore our description will be limited to MCT1 and MCT4 that are widely distributed and over-expressed in many tumors. MCT1 and MCT4 are H^+/lactate symporters with bidirectional transport of lactic acid through the plasma membrane. While MCT1 is mainly an importer of lactate into the cell, MCT4 acts as an exporter and is induced by hypoxia. It is precisely MCT4 that is responsible for lactate extrusion in cancer. These two transporters require the expression of a transmembrane protein that acts as a chaperone molecule: CD147, also known as basigin.

The MCTs are plasma membrane transporters that can carry lactate and pyruvate forming part of the transportome (see notes at the end).

MCT4 is regulated by HIF-1 while MCT1 is regulated by cMyc and p53. Extracellular lactate is also an up-regulator of MCT4 but not of MCT1.[130]

MCT1 and MCT 4 are over-expressed in many tumors. The over-expression of these transporters and their chaperone protein basigin (CD147) is usually associated with poor prognoses.[131–133]

p53 is classically known as an anti-cancer gene/protein. This is very evident in its interrelation with MCT1. Boidot et al. reviewed needle biopsies of breast cancer and found that in wild-type p53 tumors, the expression level of MCT1 was not influenced by hypoxia, but in mutated p53 tumors, MCT1 expression was significantly greater in the hypoxia group. Those patients with a higher expression of MCT1 and p53 mutation had worse outcomes than patients with lower MCT1. p53 is also a regulator of glycolysis and apoptosis amid many other anti-cancer activities through the induction of TIGAR (TP53-induced glycolysis and apoptosis regulator).[134]

CD147/basigin/EMMPRIN/gp42

Basigin, (synonyms CD147 and EMMPRIN), is a transmembrane and cell-surface glycoprotein of the immunoglobulin superfamily that plays an important role in intercellular communication exerting pleiotropic functions[135] (Fig. 6).

There is a tight association between basigin and MCTs (MCT1, MCT3 and MCT4), where basigin acts as an ancillary protein. Basigin forms a dimer when it associates with MCT1[136] and serves as a chaperone required for the plasma membrane translocation of MCTs[137] (Fig. 6).

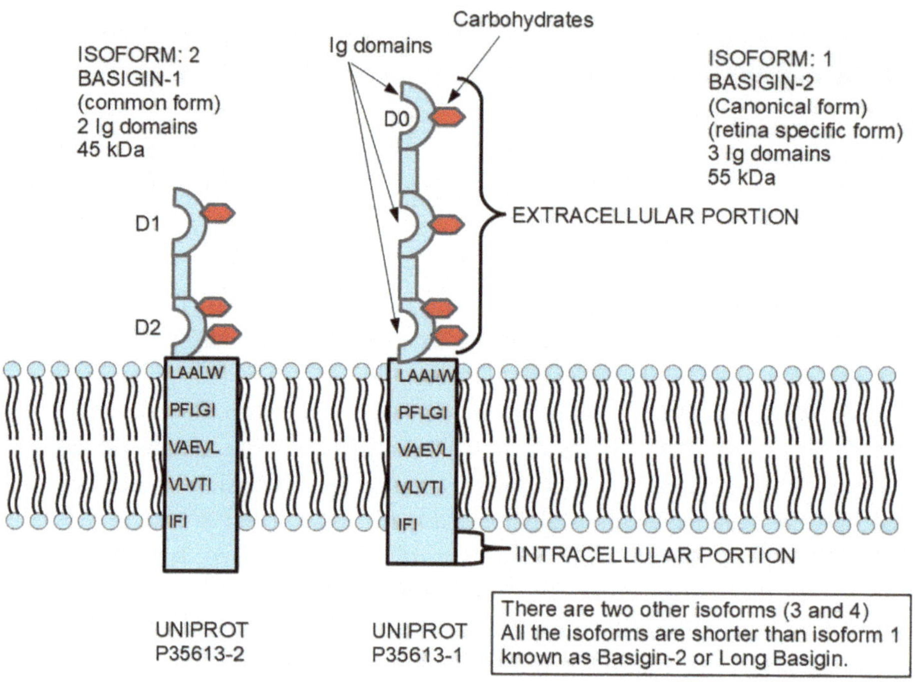

Fig. 6 Basigin 1 and basigin 2 structure. (The letters in basigin are amino acids.)

The most important pro-tumor activity of CD147/Basigin is to control the energetics of glycolytic tumors via MCT1/MCT4.[138] There are also other pro-tumoral activities of basigin, such as stimulation of matrix metalloproteinase and vascular endothelial growth factor expression,[139] participation in the migration/invasion/metastasis process, and the control of malignant signaling pathways. Knutti et al.[140] found that CD147 modulated the Wnt and Jak/STAT pathways in breast cancer cells. CD147 expression in triple-negative breast cancer patients was associated with poor survival.[141]

β1 integrin can interact with CD147 and it seems that the activation of FAK/cortactin signaling induced by β1 integrin can be blocked by a CD147 antibody blockade.[142]

A pro-metastatic pathway in which basigin plays a role has been reported: TRAF6/basigin/MMP9.

(TRAF6 is the TNF receptor associated factor 6). This pathway can be down-regulated with emodine.[143] The RNA interference of CD147 decreases metastasis and invasion of human breast cancer cells.[144] In bladder cancer, CD147 expression was correlated with the stage and grade of the tumor. The same correlation was seen when the association of CD147 and MCT1 was studied. This associated over-expression was also related to cisplatin resistance, shorter overall survival and shorter disease free survival.[145] There is further evidence on a possible a role for CD147 in multidrug resistance.[146] Finally, basigin expression is negatively correlated with genes that participate in the immunological defenses. It has been postulated as a master regulator of these genes.[147]

Although it might seem tempting to target basigin in cancer, there could be serious drawbacks, because by down-regulating this transmembrane protein, severe damage to the retina may ensue,[148,149] unless the inhibitor is specific for Basigin 1. In basigin-null mice there was no expression of MCT1, MCT3 and MCT4, in the neural retina while normal counterparts showed a very rich expression of the three MCTs.[150] Furthermore, the inhibition of basigin increases production of β amyloid peptides. Basigin down regulates the enzymatic activity of the γ secretase complex.[151,152]

There is a direct interaction between basigin and carbonic anhydrase IX (CAIX)[153,154] where CAIX binds to the first globular domain of basigin. The knockdown of CAIX decreased MCT-mediated lactate flux in hypoxic cancer cells.[155] This can lead to the erroneous idea that it would be better to down-regulate MCTs by inhibiting CAs instead of trying to decrease basigin expression; the problem in this case is that anhydrase carbonic enzymatic activity is not related to basigin activation. This means that the usual inhibitors of CAIX enzymatic activity would not inactivate basigin interaction.[156]

Metabolism and transport of lactate is a complex process that involves nonenzymatic proteins, enzymes and membrane transporters forming a network that some authors have called the "transport metabolon".[157] The relationship between MCTs and CAs is isoform-specific and not related to CA catalytic activity.[158]

In summary, CD147 (basigin) is involved in pro-tumoral activities that are dependent and independent from MCTs.

Therapeutic considerations

Lactic acid accumulation in the ECM and its transport metabolon are valid targets in order to decrease tumor proliferation, invasion, migration and metastasis.

Decrease the production of lactic acid. In 2007 Bonnet et al.[159] showed that dichloroacetate (DCA), an orally available small molecule, inhibited the mitochondrial pyruvate dehydrogenase kinase (PDK) and decreased lactic acid levels.[160] The inhibition of PDK increases the oxidative pathway over the glycolytic one. Through this mechanism DCA achieves mitochondrial apoptosis in vivo and in vitro.[161]

DCA has a bioavailability of 100%. The required concentration for pharmacological effects is between 0.20 and 40 µg/mL and it is clinically achievable. DCA will be discussed in length in the chapter on pharmacology.

MCTs' inhibition. There are no inhibitors of MCT available in the clinical setting apart from quercetin. A few experimental drugs are in an advanced phase of research. This is the case of AZD3965 which is specific for MCT1 but not for MCT4. AR-C117977 is another experimental inhibitor of MCT1 with strong immunosuppressant effects; it has not been tested against cancer cells.[162] AR-C155858 is an MCT1 and MCT2 inhibitor,[163] but has no effect on MCT4. None of these compounds have gone beyond the laboratory, or been tested in cancer treatment.

Compounds with MCT inhibitory activity without or with low toxicity, albeit with low potency, are:

- Flavonoids like phloretin (a polyphenol present in apples and pears, inhibits glucose uptake up to 60%) apigenin, biochanin A, chrysin, diosemin, fisetin, genistein, hesperitin, kaempferol, luteolin, morin, narigenin, and quercetin reduce the activity of MCT1.[164]
- NSAIDs like diclofenac and flurbiprofen.[165]

Inhibition of basigin. The inhibition of this chaperone decreases lactic acid export and also sensitizes cancer cell to biguanids' cytotoxicity.[166] This finding confirms our idea that metformin's main anti-cancer mechanism works through cellular acidification.[167] There are no clinically available basigin inhibitors, and we foresee great difficulties in the development of this kind of compound, due to the adverse events related to its inhibition: mainly retinal deterioration and beta amyloid accumulation.

The most promising treatment should involve more than one drug, like a combination of DCA, metformin, diclofenac and phloretin just to mention an example. Emodine is a drug that deserves further research as one of the down-regulators of a pro-metastatic pathway that involves basigin.

Fig. 7 summarizes the pro-tumoral activities of lactic acid.

Conclusions

Normal cells mainly rely on mitochondrial oxidative phosphorylation (OXPHOS) as their source of energy. There are some exceptions, like muscle tissue and heart muscle under certain circumstances, but even in these tissues, OXPHOS is the main energy source.

In cancer cells glycolysis is enhanced and OXPHOS capacity is reduced in almost all rapid growing tumors. Warburg believed that this was due to impairment of mitochondrial OXPHOS. We now know that mitochondrial OXPHOS is conserved in most cancers, although repressed. Aerobic glycolysis (Warburg effect) is the result of various

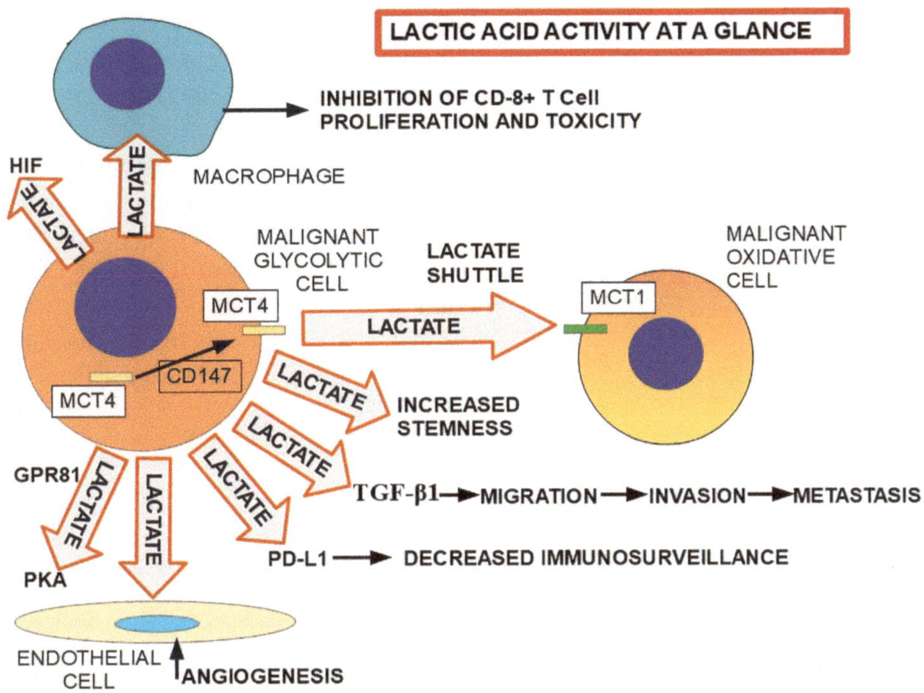

Fig. 7 A summary of the pro-tumoral effects of lactic acid.

factors such as mutations of oncogenes or tumor suppressors, down-regulation of tumor suppressor genes by epigenetic modifications, hypoxic microenvironment, genetic background and others. The lactic acid increase is a direct consequence of glycolytic metabolism. It is not a waste product of metabolism, but a very active metabolite that fuels non glycolytic malignant cells and shows clear pro-cancer activities. Lactic acid has a profound influence on invasion and metastasis and therefore it should be a target for treatment. The association of DCA with phloretin or with other flavonoids could be a viable combination for reducing the pro-tumor effects of lactic acid. Phloretin has its own anti-cancer effects, independent from MCT inhibition. Thwarting lactic acid's effects in cancer has shown an increase in malignant cell apoptosis without substantially affecting normal cells.

Lactic acid is not alone in this pro-tumoral activity: a lactate transport system is an absolutely essential participant. Aside from their roles in lactate transport, basigin (CD147) and the monocarboxylate transporters MCT1 and MCT4 have other prominent pro-tumoral effects. Although there are no clinically approved MCT inhibitors, quercetin (a nutraceutical) and the anti-inflammatory diclofenac may represent an interesting option for reducing lactate extrusion.

Notes: Some neologisms used in this chapter require definitions.

(1) A **metabolon** has been defined as a "temporary, structural-functional, supramolecular complex of sequential metabolic enzymes and cellular structural elements, in which metabolites are passed from one active site to another without complete equilibration with the bulk cellular fluids". This substrate channeling should improve the whole process of moving lactate through the plasmatic membrane. Metabolons have been found in many metabolic pathways like glycolysis and the tricarboxylic acid cycle, as well as in the biosynthesis of DNA, RNA, and proteins[168,169]

(2) **Transportome** Membrane transporters and channels (collectively the transportome) govern cellular influx and efflux of ions, nutrients, and drugs.

References

1. De Bruijne AW, Vreeburg H, Van Steveninck J. Kynetic analysis of L-lactate transport in human erythrocytes via the monocarboxylate-specific carrier system. *Biochim Biophys Acta* 1983;**732**:562–8.
2. Warburg O. Über den stoffwechsel der carcinomzelle. *Naturwissenschaften* 1924;**12**(50):1131–7.
3. Warburg O, Wind F, Negelein E. *The metabolism of tumors in the body*. *J Gen Physiol* 1927;**8**(6):519. http://jgp.rupress.org/content/jgp/8/6/519.full.pdf.
4. Warburg O. *On the origin of cancer cells*. *Science* 1956;**123**(3191):309–14. https://www.drkarafitzgerald.com/wp-content/uploads/2016/08/Warburg1956.pdf.
5. Racker E. Bioenergetics and the problem of tumor growth: an understanding of the mechanism of the generation and control of biological energy may shed light on the problem of tumor growth. *Am Sci* 1972;**60**(1):56–63.
6. Walenta S, Mueller-Klieser WF. Lactate: mirror and motor of tumor malignancy. *Semin Radiat Oncol* 2004;**14**:267–74.
7. Walenta S, Wetterling M, Lehrke M, Schwikert G, Sandfor K, Rofstad EK, et al. High lactate levels predict likelihood of metastases, tumor recurrence, and restricted patient survival in human cervical cancers. *Cancer Res* 2000;**60**(4):916–21.
8. Mendoza-Juez B, Martinez-Gonzalez A, Calvo GF, Perez-Garcia VM. A mathematical model for the glucosa-lactate metabolism of in vitro cancer cells. *Bull Math Biol* 2012;**74**:1125–42.
9. San-Millan I, Brooks GA. Reexamining cancer metabolism: lactate production for carcinogenesis could be the purpose and explanation of the Warburg effect. *Carcinogenesis* 2017;**38**(2):119–33.
10. Hirschhaeuser F, Sattler UG, Mueller-Klieser W. Lactate: a metabolic key player in cancer. *Cancer Res* 2011;**71**(22):6921–5.
11. Newell K, Franchi A, Pouyssegur J, Tannock I. Studies with glycolysis-deficient cells suggest that production of lactic acid is not the only cause of tumor acidity. *Proc Natl Acad Sci* 1993;**90**(3):1127–31.
12. Yamagata M, Hasuda K, Stamato T, Tannock IF. The contribution of lactic acid to acidification of tumours: studies of variant cells lacking lactate dehydrogenase. *Br J Cancer* 1998;**77**(11):1726.
13. Goetze K, Walenta S, Ksiazkiewicz M, Kunz-Schughart LA, Mueller-Klieser W. Lactate enhances motility of tumor cells and inhibits monocyte migration and cytokine release. *Int J Oncol* 2011; **39**(2):453–63.
14. Baumann F, Leukel P, Doerfelt A, Beier CP, Dettmer K, Oefner PJ, et al. Lactate promotes glioma migration by TGF-β2–dependent regulation of matrix metalloproteinase-2. *Neuro-Oncology* 2009; **11**(4):368–80.
15. Hashimoto T, Hussien R, Oommen S, Gohil K, Brooks GA. Lactate sensitive transcription factor network in L6 cells: activation of MCT1 and mitochondrial biogenesis. *FASEB J* 2007; **21**(10):2602–12.

16. Choi SY, Collins CC, Gout PW, Wang Y. Cancer-generated lactic acid: a regulatory, immunosuppressive metabolite? *J Pathol* 2013;**230**(4):350–5.
17. Gottfried E, Kunz-Schughart LA, Ebner S, Mueller-Klieser W, Hoves S, Andreesen R, et al. Tumor-derived lactic acid modulates dendritic cell activation and antigen expression. *Blood* 2006;**107**(5):2013–21.
18. Fischer K, Hoffmann P, Voelkl S, Meidenbauer N, Ammer J, Edinger M, et al. Inhibitory effect of tumor cell–derived lactic acid on human T cells. *Blood* 2007;**109**(9):3812–9.
19. Constant JS, Feng JJ, Zabel DD, Yuan H, Suh DY, Scheuenstuhl H, et al. Lactate elicits vascular endothelial growth factor from macrophages: a possible alternative to hypoxia. *Wound Repair Regen* 2000;**8**(5):353–60.
20. Milovanova TN, Bhopale VM, Sorokina EM, Moore JS, Hunt TK, Hauer-Jensen M, et al. Lactate stimulates vasculogenic stem cells via the thioredoxin system and engages an autocrine activation loop involving hypoxia-inducible factor 1. *Mol Cell Biol* 2008;**28**(20):6248–61.
21. Hunt TK, Aslam RS, Beckert S, Wagner S, Ghani QP, Hussain MZ, et al. Aerobically derived lactate stimulates revascularization and tissue repair via redox mechanisms. *Antioxid Redox Signal* 2007;**9**(8):1115–24.
22. Bonuccelli G, Tsirigos A, Whitaker-Menezes D, Pavlides S, Pestell RG, Chiavarina B, et al. Ketones and lactate "fuel" tumor growth and metastasis: evidence that epithelial cancer cells use oxidative mitochondrial metabolism. *Cell Cycle* 2010;**9**(17):3506–14.
23. Martinez-Outschoorn UE, Prisco M, Ertel A, Tsirigos A, Lin Z, Pavlides S, et al. Ketones and lactate increase cancer cell "stemness," driving recurrence, metastasis and poor clinical outcome in breast cancer: achieving personalized medicine via metabolo-genomics. *Cell Cycle* 2011;**10**(8):1271–86.
24. De Saedeleer CJ, Copetti T, Porporato PE, Verrax J, Feron O, Sonveaux P. Lactate activates HIF-1 in oxidative but not in Warburg-phenotype human tumor cells. *PLoS One* 2012;**7**(10).
25. Samuvel DJ, Sundararaj KP, Nareika A, Lopes-Virella MF, Huang Y. Lactate boosts TLR4 signaling and NF-κB pathway-mediated gene transcription in macrophages via monocarboxylate transporters and MD-2 up-regulation. *J Immunol* 2009;**182**(4):2476–84.
26. Sattler UG, Meyer SS, Quennet V, Hoerner C, Knoerzer H, Fabian C, et al. Glycolytic metabolism and tumour response to fractionated irradiation. *Radiother Oncol* 2010;**94**(1):102–9.
27. Stern R. Hyaluronidases in cancer biology. In: *Hyaluronan in cancer biology*; 2009. p. 207–20.
28. Hayashi Y, Edwards NA, Proescholdt MA, Oldfield EH, Merrill MJ. Regulation and function of aquaporin-1 in glioma cells. *Neoplasia* 2007;**9**(9):777–87.
29. Moreno-Sanchez R, Rodriguez-Enriquez S, Marin-Hernandez A, Saavedra E. Energy metabolism in tumor cells. *FEBS J* 2007;**274**(6):1393–418.
30. DeBerardinis RJ, Lum JJ, Hatzivassiliou G, Thompson CB. The biology of cancer: metabolic reprogramming fuels cell growth and proliferation. *Cell Metab* 2008;**7**(1):11–20.
31. Hedeskov CJ. Early effects of phytohaemagglutinin on glucose metabolism of normal human lymphocytes. *Biochem J* 1968;**110**:373–80.
32. Brooks GA. The science and translation of lactate shuttle theory. *Cell Metab* 2018;**7**(4):757–85.
33. Chen YJ, Mahieu NG, Huang X, Singh M, Crawford PA, Johnson SL, et al. Lactate metabolism is associated with mammalian mitochondria. *Nat Chem Biol* 2016;**12**(11):937–43.
34. Hertz L. The astrocyte-neuron lactate shuttle: a challenge of a challenge. *J Cereb Blood Flow Metab* 2004;**24**(11):1241–8.
35. Brooks GA. Lactate shuttles in nature. *Biochem Soc Trans* 2002;**30**(2):258–64.
36. Brooks GA, Dubouchaud H, Brown M, Sicurello JP, Butz CE. Role of mitochondrial lactate dehydrogenase and lactate oxidation in the intracellular lactate shuttle. *Proc Natl Acad Sci* 1999; **96**(3):1129–34.
37. Gladden LB. Is there an intracellular lactate shuttle in skeletal muscle? *J Physiol* 2007;**582**(3):899.
38. Yoshida Y, Holloway GP, Ljubicic V, Hatta H, Spriet LL, Hood DA, et al. Negligible direct lactate oxidation in subsarcolemmal and intermyofibrillar mitochondria obtained from red and white rat skeletal muscle. *J Physiol* 2007;**582**(3):1317–35.
39. Bonuccelli G, Tsirigos A, Whitaker-Menezes D, Pavlides S, Pestell RG, Chiavarina B, et al. Ketones and lactate "fuel" tumor growth and metastasis: evidence that epithelial cancer cells use oxidative mitochondrial metabolism. *Cell Cycle* 2010;**9**(17):3506–14.

40. Draoui N, Feron O. Lactate shuttles at a glance: from physiological paradigms to anti-cancer treatments. *Dis Model Mech* 2011;**4**(6):727–32.
41. Whitaker-Menezes D, Martinez-Outschoorn UE, Lin Z, Ertel A, Flomenberg N, Witkiewicz AK, et al. Evidence for a stromal-epithelial "lactate shuttle" in human tumors: MCT4 is a marker of oxidative stress in cancer-associated fibroblasts. *Cell Cycle* 2011;**10**(11):1772–83.
42. Sanità P, Capulli M, Teti A, Galatioto GP, Vicentini C, Chiarugi P, et al. Tumor-stroma metabolic relationship based on lactate shuttle can sustain prostate cancer progression. *BMC Cancer* 2014;**14**(1):154.
43. Pavlides S, Whitaker-Menezes D, Castello-Cros R, Flomenberg N, Witkiewicz AK, Frank PG, et al. The reverse Warburg effect: aerobic glycolysis in cancer associated fibroblasts and the tumor stroma. *Cell Cycle* 2009;**8**(23):3984–4001.
44. Marchiq I, Pouysségur J. Hypoxia, cancer metabolism and the therapeutic benefit of targeting lactate/H+ symporters. *J Mol Med* 2016;**94**(2):155–71.
45. Diehl K, Dinges LA, Helm O, Ammar N, Plundrich D, Arlt A, et al. Nuclear factor E2-related factor-2 has a differential impact on MCT1 and MCT4 lactate carrier expression in colonic epithelial cells: a condition favoring metabolic symbiosis between colorectal cancer and stromal cells. *Oncogene* 2018;**37**(1):39.
46. Vegran F, Boidot R, Michiels C, Sonveaux P, Feron O. Lactate influx through the endothelial cell monocarboxylate transporter MCT1 supports an NFkappaB/IL-8 pathway that drives tumor angiogenesis. *Cancer Res* 2011;**71**:2550–60.
47. Garcia-Espinosa MA, Rodrigues TB, Sierra A, Benito M, Fonseca C, Gray HL, et al. Cerebral glucose metabolism and the glutamine cycle as detected by in vivo and in vitro 13C NMR spectroscopy. *Neurochem Int* 2004;**45**:297–303.
48. Semenza GL. Tumor metabolism: cancer cells give and take lactate. *J Clin Invest* 2008;**118**(12):3835–7.
49. Whitaker-Menezes D, Martinez-Outschoorn UE, Lin Z, Ertel A, Flomenberg N, Witkiewicz AK, et al. Evidence for a stromal-epithelial "lactate shuttle" in human tumors: MCT4 is a marker of oxidative stress in cancer-associated fibroblasts. *Cell Cycle* 2011;**10**(11):1772–83.
50. Martinez-Outschoorn UE, Pavlides S, Howell A, Pestell RG, Tanowitz HB, Sotgia F, et al. Stromal–epithelial metabolic coupling in cancer: integrating autophagy and metabolism in the tumor microenvironment. *Int J Biochem Cell Biol* 2011;**43**(7):1045–51.
51. Sonveaux P, Copetti T, De Saedeleer CJ, Vegran F, Verrax J, Kennedy KM, et al. Targeting the lactate transporter MCT1 in endothelial cells inhibits lactate-induced HIF-1 activation and tumor angiogenesis. *PLoS One* 2012;**7**(3).
52. Roth DA, Brooks GA. Lactate and pyruvate transport is dominated by a pH gradient-sensitive carrier in rat skeletal muscle sarcolemmal vesicles. *Arch Biochem Biophys* 1990;**279**(2):386–94.
53. Harguindey S, Stanciu D, Devesa J, Alfarouk K, Cardone RA, Orozco JDP, et al. Cellular acidification as a new approach to cancer treatment and to the understanding and therapeutics of neurodegenerative diseases. *Semin Cancer Biol* 2017;**43**:157–79.
54. Suhaimi NAM, Phyo WM, Yap HY, Choy SHY, Wei X, Choudhury Y, et al. Metformin inhibits cellular proliferation and bioenergetics in colorectal Cancer patient-derived xenografts. *Mol Cancer Ther* 2017;**16**(9):2035–44. https://doi.org/10.1158/1535-7163.MCT-16-0793.
55. Melnik S, Dvornikov D, Müller-Decker K, Depner S, Stannek P, Meister M, et al. Cancer cell specific inhibition of Wnt/β-catenin signaling by forced intracellular acidification. *Cell Discov* 2018;**4**(1):37.
56. Pértega-Gomes N, Vizcaíno JR, Attig J, Jurmeister S, Lopes C, Baltazar F. A lactate shuttle system between tumour and stromal cells is associated with poor prognosis in prostate cancer. *BMC Cancer* 2014;**14**(1):352.
57. Balliet RM, Capparelli C, Guido C, Pestell TG, Martinez-Outschoorn UE, Lin Z, et al. Mitochondrial oxidative stress in cancer-associated fibroblasts drives lactate production, promoting breast cancer tumor growth: understanding the aging and cancer connection. *Cell Cycle* 2011;**10**(23):4065–73.
58. Yokota H, Guo J, Matoba M, Higashi K, Tonami H, Nagao Y. Lactate, choline, and creatine levels measured by vitro 1H-MRS as prognostic parameters in patients with non-small-cell lung cancer. *J Magn Reson Imaging* 2007;**25**:992–9.
59. Lee ZW, Teo XY, Song ZJ, Nin DS, Novera W, Choo BA, et al. Intracellular hyper-acidification potentiated by hydrogen sulfide mediates invasive and therapy resistant cancer cell death. *Front Pharmacol* 2017;**8**:763.

60. Sahra IB, Laurent K, Giuliano S, Larbret F, Ponzio G, Gounon P, et al. Targeting cancer cell metabolism: the combination of metformin and 2-deoxyglucose induces p53-dependent apoptosis in prostate cancer cells. *Cancer Res* 2010;**70**(6):2465–75.
61. Ben Sahra I, Tanti JF, Bost F. The combination of metformin and 2 deoxyglucose inhibits autophagy and induces AMPK-dependent apoptosis in prostate cancer cells. *Autophagy* 2010;**6**(5):670–1.
62. Li B, Li X, Ni Z, Zhang Y, Zeng Y, Yan X, et al. Dichloroacetate and metformin synergistically suppress the growth of ovarian cancer cells. *Oncotarget* 2016;**7**(37):59458.
63. Fiaschi T, Marini A, Giannoni E, Taddei ML, Gandellini P, De Donatis A, et al. Reciprocal metabolic reprogramming through lactate shuttle coordinately influences tumor-stroma interplay. *Cancer Res* 2012;**72**(19):5130–40. https://doi.org/10.1158/0008-5472.CAN-12-1949.
64. Litchfield LM, Mukherjee A, Eckert MA, Johnson A, Mills KA, Pan S, et al. Hyperglycemia-induced metabolic compensation inhibits metformin sensitivity in ovarian cancer. *Oncotarget* 2015;**6**(27):23548.
65. Hong SE, Jin HO, Kim HA, Seong MK, Kim EK, Ye SK, et al. Targeting HIF-1α is a prerequisite for cell sensitivity to dichloroacetate (DCA) and metformin. *Biochem Biophys Res Commun* 2016;**469**(2):164–70.
66. Hong SE, Kim CS, An S, Kim HA, Hwang SG, Song JY, et al. TRAIL restores DCA/metformin-mediated cell death in hypoxia. *Biochem Biophys Res Commun* 2016;**478**(3):1389–95.
67. Blackburn AC, Rooke M, Li Y, Dahlstrom JE, Board PG. Chemoprevention with the metabolism modifying drugs dichloroacetate and metformin in Trp53+/−mice. *Hered Cancer Clin Pract* 2012;**10**(2):A62.
68. Shen H, Tsoli M, Chang C, Chitranjan A, Liu J, Hau E, et al. Hg-04 dichloroacetate and metformin combine to modulate glucose metabolism and potently sensitise dipg cells to radiation therapy. *Neuro-Oncology* 2016;**18**(Suppl. 3):iii48.
69. Martinkute L, Stankevicius E. *2nd International conference in pharmacology: from cellular processes to drug targets*. Poster A2–19 Available at http://www.intrinsicactivity.org/2017/5/S2/A2.19/IA201705S2-A2.19.pdf; 2017.
70. Kiliccioglu I, Konac E, Albayrak G, Donmez C, Bilen CY. Combination of metformin and dichloroacetate inhibits proliferation and induce intrinsic pathway of apoptosis in PC-3 human prostate cancer cells. *Gazi Med J* 2015;**26**(4A):212–4.
71. Lea MA, Altayyar M. Inhibition of growth of bladder cancer cells by 3-(3-Pyridinyl)-1-(4-pyridinyl)-2-propen-1-one in combination with other compounds affecting glucose metabolism. *Anticancer Res* 2015;**35**(11):5889–99.
72. Granchi C, Minutolo F. Anticancer agents that counteract tumor glycolysis. *ChemMedChem* 2012;**7**(8):1318–50.
73. Vander Jagt DL, Deck LM, Royer RE. Gossypol: prototype of inhibitors targeted to dinucleotide folds. *Curr Med Chem* 2000;**7**(4):479–98.
74. Liu G, Kelly WK, Wilding G, Leopold L, Brill K, Somer B. An open-label, multicenter, phase I/II study of single-agent AT-101 in men with castrate-resistant prostate cancer. *Clin Cancer Res* 2009;**15**(9):3172–6.
75. Baggstrom MQ, Qi Y, Koczywas M, Argiris A, Johnson EA, Millward MJ, et al. A phase II study of AT-101 (gossypol) in chemotherapy-sensitive recurrent extensive-stage small cell lung cancer. *J Thorac Oncol* 2011;**6**(10):1757–60.
76. Bohn R, Graebe C. Űber galloflavin. *Ber Dtsch Chem Ges* 1887;**20**(2):2327–31.
77. Manerba M, Vettraino M, Fiume L, Di Stefano G, Sartini A, Giacomini E, et al. Galloflavin (CAS 568-80-9): a novel inhibitor of lactate dehydrogenase. *ChemMedChem* 2012;**7**(2):311–7.
78. Farabegoli F, Vettraino M, Manerba M, Fiume L, Roberti M, Di Stefano G. Galloflavin, a new lactate dehydrogenase inhibitor, induces the death of human breast cancer cells with different glycolytic attitude by affecting distinct signaling pathways. *Eur J Pharm Sci* 2012;**47**(4):729–38.
79. Han X, Sheng X, Jones HM, Jackson AL, Kilgore J, Stine JE, et al. Evaluation of the anti-tumor effects of lactate dehydrogenase inhibitor galloflavin in endometrial cancer cells. *J Hematol Oncol* 2015;**8**(1):2.
80. Vettraino M, Manerba M, Govoni M, Di Stefano G. Galloflavin suppresses lactate dehydrogenase activity and causes MYC downregulation in Burkitt lymphoma cells through NAD/NADH-dependent inhibition of sirtuin-1. *Anti-Cancer Drugs* 2013;**24**(8):862–70.

81. Fiume L, Vettraino M, Carnicelli D, Arfilli V, Di Stefano G, Brigotti M. Galloflavin prevents the binding of lactate dehydrogenase A to single stranded DNA and inhibits RNA synthesis in cultured cells. *Biochem Biophys Res Commun* 2013;**430**(2):466–9.
82. Ovens MJ, Davies AJ, Wilson MC, Murray CM, Halestrap AP. AR-C155858 is a potent inhibitor of monocarboxylate transporters MCT1 and MCT2 that binds to an intracellular site involving transmembrane helices 7–10. *Biochem J* 2010;**425**(3):523–30.
83. Polański R, Hodgkinson CL, Fusi A, Nonaka D, Priest L, Kelly P, et al. Activity of the monocarboxylate transporter 1 inhibitor AZD3965 in small cell lung cancer. *Clin Cancer Res* 2014;**20**(4):926–37.
84. Shim CK, Cheon EP, Kang KW, Seo KS, Han HK. Inhibition effect of flavonoids on monocarboxylate transporter 1 (MCT1) in Caco-2 cells. *J Pharm Pharmacol* 2007;**59**(11):1515–9.
85. Izumi H, Takahashi M, Uramoto H, Nakayama Y, Oyama T, Wang KY, et al. Monocarboxylate transporters 1 and 4 are involved in the invasion activity of human lung cancer cells. *Cancer Sci* 2011;**102**(5):1007–13.
86. Halestrap AP, Meredith D. The SLC16 gene family—from monocarboxylate transporters (MCTs) to aromatic amino acid transporters and beyond. *Pflugers Arch* 2004;**447**(5):619–28.
87. Wang Q, Morris ME. Flavonoids modulate Monocarboxylate transporter-1 mediated transport of γ-hydroxybutyrate in vitro and in vivo. *Drug Metab Dispos* 2007;**35**(2):201–8.
88. Halestrap AP. Monocarboxylic acid transport. *Compr Physiol* 2011;**3**(4):1611–43.
89. Jackson VN, Halestrap AP. The kinetics, substrate, and inhibitor specificity of the monocarboxylate (lactate) transporter of rat liver cells determined using the fluorescent intracellular pH indicator, 2′, 7′-bis(carboxyethyl)-5 (6)-carboxyfluorescein. *J Biol Chem* 1996;**271**(2):861–8.
90. Carpenter L, Halestrap AP. The kinetics, substrate and inhibitor specificity of the lactate transporter of Ehrlich-Lettre tumour cells studied with the intracellular pH indicator BCECF. *Biochem J* 1994;**304**(3):751–60.
91. Li X, Zhang Z, Zhang Y, Cao Y, Wei H, Wu Z. Upregulation of lactate-inducible snail protein suppresses oncogene-mediated senescence through p16 INK4a inactivation. *J Exp Clin Cancer Res* 2018;**37**(1):39.
92. Roland CL, Arumugam T, Deng D, Liu SH, Philip B, Gomez S, et al. Cell surface lactate receptor GPR81 is crucial for cancer cell survival. *Cancer Res* 2014;**74**:18.
93. Cai TQ, Ren N, Jin L, Cheng K, Kash S, Chen R. Role of GPR81 in lactate-mediated reduction of adipose lipolysis. *Biochem Biophys Res Commun* 2008;**377**:987–91.
94. Mosienko V, Teschemacher AG, Kasparov S. Is L-lactate a novel signaling molecule in the brain? *J Cereb Blood Flow Metab* 2015;**35**(7):1069–75.
95. Cho-Chung YS. Hypothesis. Cyclic AMP and its receptor protein in tumor growth regulation in vivo. *J Cyclic Nucleotide Res* 1980;**6**(3):163–77.
96. Pastan IH, Johnson GS, Anderson WB. Role of cyclic nucleotides in growth control. *Annu Rev Biochem* 1975;**44**:491–522.
97. Röck R, Bachmann V, Bhang HE, Malleshaiah M, Raffeiner P, Mayhofer JE, et al. In-vivo detection of binary PKA network interactions upon activation of endogenous GPCRs. *Sci Rep* 2015;**5**:11133.
98. Naviglio S, Caraglia M, Abbruzzese A, Chiosi E, Di Gesto D, et al. Protein kinase A as a biological target in cancer therapy. *Expert Opin Ther Targets* 2009;**13**(1):83–92.
99. Feng J, Yang H, Zhang Y, Wei H, Zhu Z, Zhu B, et al. Tumor cell-derived lactate induces TAZ-dependent upregulation of PD-L1 through GPR81 in human lung cancer cells. *Oncogene* 2017;**36**(42):5829.
100. Cho-Chung YS. Role of cyclic AMP receptor proteins in growth, differentiation, and suppression of malignancy: new approaches to therapy. *Cancer Res* 1990;**50**(22):7093–100.
101. Tortora G, Ciardiello F. Protein kinase a as target for novel integrated strategies of cancer therapy. *Ann N Y Acad Sci* 2002;**968**:139–47.
102. Cheadle C, Nesterova M, Watkins T, Barnes KC, Hall JC, Rosen A, et al. Regulatory subunits of PKA define an axis of cellular proliferation/differentiation in ovarian cancer cells. *BMC Med Genet* 2008;**1**(1):1755.
103. Lee YJ, Shin KJ, Park SA, Park KS, Park S, Heo K, et al. G-protein-coupled receptor 81 promotes a malignant phenotype in breast cancer through angiogenic factor secretion. *Oncotarget* 2016;**7**(43):70898–911.

104. Choi SY, Collins CC, Gout PW, Wang Y. Cancer-generated lactic acid: a regulatory, immunosuppressive metabolite? *J Pathol* 2013;**230**(4):350–5.
105. Brand A, Singer K, Koehl GE, Kolitzus M, Schoenhammer G, Thiel A, et al. LDHA-associated lactic acid production blunts tumor immunosurveillance by T and NK cells. *Cell Metab* 2016;**24**(5):657–71.
106. Pucino V, Bombardieri M, Pitzaliz C, Mauro C. Lactate at the crossroad of metabolism, inflammation, and autoimmunity. *Eur J Immunol* 2017;**47**(1):14–21.
107. Fischer K, Hoffmann P, Voelkl S, Meidenbauer N, Ammer J, Edinger M, et al. Inhibitory effect of tumor cell-derived lactic acid on human T cells. *Blood* 2007;**109**(9):3812–9.
108. Gottfried E, Kunz-Schughart LA, Ebner S, et al. Tumor-derived lactic acid modulates dendritic cell activation and antigen expression. *Blood* 2006;**107**(5):2013–21.
109. Ohashi T, Akazawa T, Aoki M, et al. Dichloroacetate improves immune dysfunction caused by tumor-secreted lactic acid and increases antitumor immunoreactivity. *Int J Cancer* 2013;**133**(5):1107–18.
110. Caronni N, Simoncello F, Stafetta F, Guarnaccia C, Ruiz Moreno JS, Opitz B, et al. *Downregulation of membrane trafficking proteins and lactate conditioning determine loss of dendritic cell function in lung cancer. Cancer Res* 2018. https://doi.org/10.1158/0008-5472.CAN-17-1307. https://www.ncbi.nlm.nih.gov/pubmed/29363545.
111. Porporato PE, Payen VL, De Saedeleer CJ, Préat V, Thissen JP, Feron O, et al. Lactate stimulates angiogenesis and accelerates the healing of superficial and ischemic wounds in mice. *Angiogenesis* 2012;**15**(4):581–92.
112. Constant JS, Feng JJ, Zabel DD, Yuan H, Suh DY, Scheuenstuhl H, et al. Lactate elicits vascular endothelial growth factor from macrophages: a possible alternative to hypoxia. *Wound Repair Regen* 2000;**8**:353–60.
113. Dhup S, Kumar Dadhich R, Ettore Porporato P, Sonveaux P. Multiple biological activities of lactic acid in cancer: influences on tumor growth, angiogenesis and metastasis. *Curr Pharm Des* 2012;**18**(10):1319–30.
114. Polet F, Feron O. Endothelial cell metabolism and tumour angiogenesis: glucose and glutamine as essential fuels and lactate as the driving force. *J Intern Med* 2013;**273**(2):156–65.
115. Jensen JA, Hunt TK, Scheuenstuhl H, Banda MJ. Effect of lactate, pyruvate, and pH on secretion of angiogenesis and mitogenesis factors by macrophages. *Lab Invest* 1986;**54**(5):574–8.
116. Sonveaux P, Copetti T, De Saedeleer CJ, Végran F, Verrax J, Kennedy KM, et al. Targeting the lactate transporter MCT1 in endothelial cells inhibits lactate-induced HIF-1 activation and tumor angiogenesis. *PLoS One* 2012;**7**(3).
117. Kumar VB, Viji RI, Kiran MS, Sudhakaran PR. Endothelial cell response to lactate. Implication of PAR modification of VEGF. *J Cell Physiol* 2007;**211**:477–85.
118. Beckert S, Farrahi F, Aslam RS, Scheuenstuhl H, Königsrainer A, Hussain MZ, et al. Lactate stimulates endothelial cell migration. *Wound Repair Regen* 2006;**14**(3):321–4.
119. Ruan G-X, Kazlauskuas A. Lactate engages receptor tyrosine kinases Axl, Tie2 and vascular endothelial factor receptor 2 to activate phosphoinositide 3-kinase/Akt and promote angiogenesis. *J Biol Chem* 2013;**288**:21161.
120. Walenta S, Salameh A, Lyng H, Evensen JF, Mitze M, Rofstad EK, et al. Correlation of high lactate levels in head and neck tumors with incidence of metastasis. *Am J Pathol* 1997;**150**(2):409.
121. Brizel DM, Schroeder T, Scher RL, Walenta S, Clough RW, Dewhirst MW, et al. Elevated tumor lactate concentrations predict for an increased risk of metastases in head-and-neck cancer. *Int J Radiat Oncol Biol Phys* 2001;**51**(2):349–53.
122. Schwickert G, Walenta S, Sundfør K, Rofstad EK, Mueller-Klieser W. Correlation of high lactate levels in human cervical cancer with incidence of metastasis. *Cancer Res* 1995;**55**(21):4757–9.
123. Anderson M, Marayati R, Moffitt R, Yeh JJ. Hexokinase 2 promotes tumor growth and metastasis by regulating lactate production in pancreatic cancer. *Oncotarget* 2017;**8**(34):56081.
124. Chen P, Zuo H, Xiong H, Kolar MJ, Chu Q, Saghatelian A, et al. Gpr132 sensing of lactate mediates tumor–macrophage interplay to promote breast cancer metastasis. *Proc Natl Acad Sci* 2017;**114**(3):580–5.
125. Wu XZ, Ma F, Wang XL. Serological diagnostic factors for liver metastasis in patients with colorectal cancer. *World J Gastroenterol: WJG* 2010;**16**(32):4084.

126. Koukourakis MI, Giatromanolaki A, Sivrides E, Bougioukas G, Didlis V, Gatter KC, et al. Lactate dehydrogenase-5 (LDH-5) overexpression in non-small-cell lung cancer tissues is linked to tumour hypoxia, angiogenic factor production and poor prognosis. *Br J Cancer* 2003;**89**(5):877.
127. Walenta S, Chau TV, Schroeder T, Kunz-Schughart LA, Fuerst A, Mueller-Klieser W. Metabolic classification of human rectal adenocarcinomas: a novel guideline for clinical oncologists? *J Cancer Res Clin Oncol* 2003;**129**:321–6.
128. Shime H, Yabu M, Akazawa T, Kodama K, Matsumoto M, Seya T, et al. Tumor-secreted lactic acid promotes IL-23/IL-17 proinflammatory pathway. *J Immunol* 2008;**180**(11):7175–83.
129. Yabu M, Shime H, Hara H, Saito T, Matsumoto M, Seya T, et al. IL-23-dependent and-independent enhancement pathways of IL-17A production by lactic acid. *Int Immunol* 2010;**23**(1):29–41.
130. Boidot R, Vegran F, Meulle A, LeBreton A, Dessy C, Sonveaux P, et al. Regulation of monocarboxylate transporter MCT1 expression by p53 mediates inward and outward lactate fluxes in tumors. *Cancer Res* 2012;**72**(4):939–48.
131. Kennedy KM, Dewhirst MW. Tumor metabolism of lactate: the influence and therapeutic potential for MCT and CD147 regulation. *Future Oncol* 2010;**6**(1):127–48.
132. Pinheiro C, Albergaria A, Paredes J, Sousa B, Dufloth R, Vieira D, et al. Monocarboxylate transporter 1 is up-regulated in basal-like breast carcinoma. *Histopathology* 2010;**56**(7):860–7.
133. Pértega-Gomes N, Vizcaíno JR, Miranda-Gonçalves V, Pinheiro C, Silva J, Pereira H, et al. Monocarboxylate transporter 4 (MCT4) and CD147 overexpression is associated with poor prognosis in prostate cancer. *BMC Cancer* 2011;**11**(1):312.
134. Bensaad K, Tsuruta A, Selak MA, Vidal MN, Nakano K, Bartrons R, et al. TIGAR, a p53-inducible regulator of glycolysis and apoptosis. *Cell* 2006;**126**(1):107–20.
135. Muramatsu T. Basigin (CD147), a multifunctional transmembrane glycoprotein with various binding partners. *J Biochem* 2016;**159**(5):481–90.
136. Wilson MC, Meredith D, Halestrap AP. Fluorescence resonance energy transfer studies on the interaction between the lactate transporter MCT1 and CD147 provide information on the topology and stoichiometry of the complex in situ. *J Biol Chem* 2002;**277**(5):3666–72.
137. Kirk P, Wilson MC, Heddle C, Brown MH, Barclay AN, Halestrap AP. CD147 is tightly associated with lactate transporters MCT1 and MCT4 and facilitates their cell surface expression. *EMBO J* 2000;**19**(15):396–904.
138. Le Floch R, Chiche J, Marchiq I, Naiken T, Ilc K, Murray CM, et al. CD147 subunit of lactate/H+ symporters MCT1 and hypoxia-inducible MCT4 is critical for energetics and growth of glycolytic tumors. *Proc Natl Acad Sci U S A* 2011;**108**(40):16663–8.
139. Belton RJ, Chen L, Mesquita FS, Nowak RA. Basigin-2 is a cell surface receptor for soluble basigin ligand. *J Biol Chem* 2008;**283**:17805–14.
140. Knutti N, Huber O, Friedrich K. CD147 (EMMPRIN) controls malignant properties of breast cancer cells by interdependent signaling of Wnt and JAK/STAT pathways. *Mol Cell Biochem* 2019;**451**(1–2):197–209.
141. Liu M, Tsang JY, Lee M, Ni YB, Chan SK, Cheung SY, et al. CD147 expression is associated with poor overall survival in chemotherapy treated triple-negative breast cancer. *J Clin Pathol* 2018;**71**(11):1007–14.
142. Li L, Dong X, Peng F, Shen L. Integrin β1 regulates the invasion and radioresistance of laryngeal cancer cells by targeting CD147. *Cancer Cell Int* 2018;**18**:80.
143. Shi G-H, Zhou L. Emodin suppresses angiogenesis and metastasis in anaplastic thyroid cancer by affecting TRAF-6-mediated pathways in vivo and in vitro. *Mol Med Rep* 2018;**18**(6):5191–7.
144. Li F, Zhang J, Guo J, Jia Y, Han Y, Wang Z. RNA interferente targeting CD147 inhibits metastasis and invasión of human breast cancer MCF-7 cells by downregulating MMP-9/VEGF expresión. *Acta Biochim Biophys Sin* 2018;**50**(7):676–84.
145. Afonso J, Santos LL, Miranda Goncalves V, Morais A, Amaro T, Longato-Filho A, et al. CD147 and MCT1-potential partners in bladder cancer aggressiveness and cisplatin resistance. *Mol Carcinog* 2015;**54**(11):1451–66.
146. Li HX, Tang F, Wang WJ, Li QQ, Bao Y, Chen Q, et al. Effects of multidrug resistance reversing agent on prglycoprotein, EMMPRIN and MMP expression in human breast cancer cells. *Fudan Uni J Med Sci* 2008;**35**(4):498–503.

147. Shang Y-K, Li C, Liu Z-K, Kong L-M, Wei D, Xu J, et al. System analysis of the regulation of the immune response by CD147 and FOXC1 in cancer cell lines. *Oncotarget* 2018;**9**(16):12918–31.
148. Ati-Ali N, Fridlich R, Millet-Puel G, Clerin E, Delalande F, Jaillard C, et al. Rod-derived cone viability factor promotes cone survival by stimulating aerobic glycolysis. *Cell* 2015;**161**(4):817–32.
149. Hori K, Katayama N, Kachi S, Kondo M, Kadomatsu K, Usukura J, et al. Retinal dysfunction in basigin deficiency. *Invest Ophthalmol Vis Sci* 2000;**41**(10):3128–33.
150. Philp NJ, Ochrietor JD, Rudoy C, Muramatsu T, Linser PJ. Loss of MCT1, MCT3, and MCT4 expression in the retinal pigment epithelium and neural rethina of the 5A11/Basigin-null mouse. *Retin Cell Biol* 2003;**44**(3):1305–11.
151. Zhou S, Zhou H, Walian PJ, Jap BK. CD147 is a regulatory subunit of the γ-secretase complex in Alzheimer's disease amyloid β-peptide production. *Proc Natl Acad Sci* 2005;**102**(21):7499–504.
152. Zhou S, Zhou H, Walian PJ, Jap BK. The discovery and role of CD147 as a subunit of gamma-secretase complex. *Drug News Perspect* 2006;**19**(3):133.
153. Ames S, Pastorekova S, Becker HM. The proteoglycan-like domain of CA-IX mediates non catalytic facilitation of lactate transport in cancer cells. *Oncotarget* 2018;**9**(46):27940–57.
154. Klier M, Andes FT, Deitmer JW, Becker HM. Intracellular and extracellular carbonic anhydrases cooperate non-enzymatically to enhance activity of monocarboxylate transporters. *J Biol Chem* 2014;**289**(5):2765–75.
155. Ames S, Jamali S, Becker HM. Extracellular carbonic anydrase IX supports transport activity of MCTs by direct interaction. In: *Presented at the 4th annual meeting of the International Society of Cancer Metabolism 19–21 October 2017. Bertinoro, Italy*; 2017.
156. Jamali S, Klier M, Ames S, Barros LF, McKenna R, Deitmer JW, et al. Hypoxia-induced carbonic anhydrase IX facilitates lactate flux in human breast cancer cells by non-catalytic function. *Sci Rep* 2015;**5**:13605.
157. Deitmer JW, Theparambil SM, Ruminot I, Becker HM. The role of membrane acid/base transporters and carbonic anhydrases for cellular pH and metabolic processes. *Front Neurosci* 2014;**8**:430.
158. Becker HM, Klier M, Deitmer JW. Carbonic anhydrases and their interplay with acid/base-coupled membrane transporters. *Subcell Biochem* 2014;**75**:105–34.
159. Bonnet S, Archer SL, Allalunis-Turner J, Haromi A, Beaulieu C, Thompson R, et al. A mitochondria-K+ channel axis is suppressed in cancer and its normalization promotes apoptosis and inhibits cancer growth. *Cancer Cell* 2007;**11**(1):37–51.
160. Stacpoole PW, Nagaraja NV, Hutson AD. Efficacy of dichloroacetate as a lactate-lowering drug. *J Clin Pharmacol* 2003;**43**:683–91.
161. Michelakis ED, Webster L, Mackey JR. Dichloroacetate (DCA) as a potential metabolic-targeting therapy for cancer. *Br J Cancer* 2008;**99**(7):989–94.
162. Bueno V, Binet I, Steger U, Bundick R, Ferguson D, Murray C, et al. The specific monocarboxylate transporter (MCT1) inhibitor, AR-C117977, a novel immunosuppressant, prolongs allograft survival in the mouse. *Transplantation* 2007;**84**(9):1204–7.
163. Ovens MJ, Davies AJ, Wilson MC, Murray CM, Halestrap AP. AR-C155858 is a potent inhibitor of monocarboxylate transporters MCT1 and MCT2 that binds to an intracellular site involving transmembrane helices 7–10. *Biochem J* 2010;**425**(3):523–30.
164. Wang Q, Morris ME. Flavonoids modulate monocarboxylate transporter-1-mediated transport of gamma-hydroxybutyrate in vitro and in vivo. *Drug Metab Dispos* 2007;**35**(2):201–8.
165. Vellonen KS, Häkli M, Merezhinskaya N, Tervo T, Honkakoski P, Urtti A. Monocarboxylate transport in human corneal epithelium and cell lines. *Eur J Pharm Sci* 2010;**39**(4):241–7.
166. Granja S, Marchiq I, Le Floch R, Moura CS, Baltazar F, Pouyssegur J. Disruption of BASIGIN decreases lactic acid export and sensitizes non-small cell lung cancer to biguanides independently of the LKB1 status. *Oncotarget* 2015;**6**(9):6708.
167. Koltai T. Triple-edged therapy targeting intracellular alkalosis and extracellular acidosis in cancer. *Semin Cancer Biol* 2017;**43**:139–46.
168. Srere PA. The metabolon. *Trends Biochem Sci* 1985;**14**:313–4.
169. Srere PA. Complexes of sequential metabolic enzymes. *Annu Rev Biochem* 1987;**56**:89–124.

CHAPTER 5

The sodium hydrogen exchanger 1 (NHE1)

pH-tome: Introduction

pH-tome is a neologism we have coined in order to describe in one word all the multiple structures and functions that intervene in the pH gradient inversion which is always found in all malignant tissues. Essential parts of the pH-tome are glycolytic metabolism and its enzymes, NHE1, lactic acid, monocarboxylate transporters (MCTs), voltage gated sodium channels (VGSCs), V-ATPase proton pumps, carbonic anhydrases, sodium/bicarbonate cotransporters (NBCs), the buffering system, aquaporins, and other transporters and channels whose precise participation has not been fully disclosed such as potassium channels, calcium channels, etc. pH-tome is the machinery that intervenes in the aberrant regulation of the pH.

pH is one parameter in a group of highly scrutinized and controlled signals in multicellular organisms. Homeostasis means mainly the tight regulation of cellular pH, and its possible modifiers, within physiological ranges. There are two different compartments that interest us in cancer: the intracellular compartment and the extracellular microenvironment surrounding the malignant cells. The pH levels of both compartments are altered in cancer as we have described in previous chapters. Actually, in cancer, there is an inversion of the normal pH scheme, where the intracellular compartment becomes slightly more alkaline while the extracellular space (ECS) becomes strongly acidic.

This inversion of the pH gradient has profound effects on proliferation, resistance to apoptosis, angiogenesis, genomic instability, invasion, migration and metastasis.[1–3]

The main mechanisms responsible for pH homeostasis were mentioned in previous chapters. Now is the time to describe the structural and functional units that form these mechanisms. If we are obliged to be very synthetic on the issue and express the concept in two words, we would say: metabolism and NHE1. We cannot underestimate the importance of all the other components of the pH-tome, but metabolism and NHE1 are still the key players. Metabolism and pH are different faces of the same coin.

Actually the first line management of pH is in the hands of the buffering systems inside the cells, but they cannot fully handle the acidic overload produced by the predominantly glycolytic metabolism usually found in tumors. The consequence of this is the overexpression of most of the proton extrusion systems of the cells. NHE1, NBC and MCTs are the most important proton extruders that are overexpressed in cancer but NHE1 is actually considered the main pH regulator in cancer cells.[4–6] NHE1 achieves particular

An Innovative Approach to Understanding and Treating Cancer: Targeting pH
https://doi.org/10.1016/B978-0-12-819059-3.00005-8
© 2020 Elsevier Inc.
All rights reserved.

importance in areas involved in migration and invasion, like invadopodia. These characteristics have accentuated the interest in NHE1 as a drug target.

NHE1

NHE stands for sodium hydrogen exchanger or Na^+/H^+ exchanger or sodium-hydrogen-antiporter (gene SLC9A1). It belongs to a family in which the NHE1 is the main concern in cancer. They perform the electroneutral transport of Na^+ and H^+ in opposite directions on a one for one basis.

Function and structure

The activation of NHE has many functions beyond the extrusion of hydrogen and the incorporation of sodium into the cell, like adhesion, migration and proliferation, cell volume and pHi.[7] Nine isoforms have been identified, but the most important in the cancer cell is NHE1.

NHE1 structure (Fig. 1).

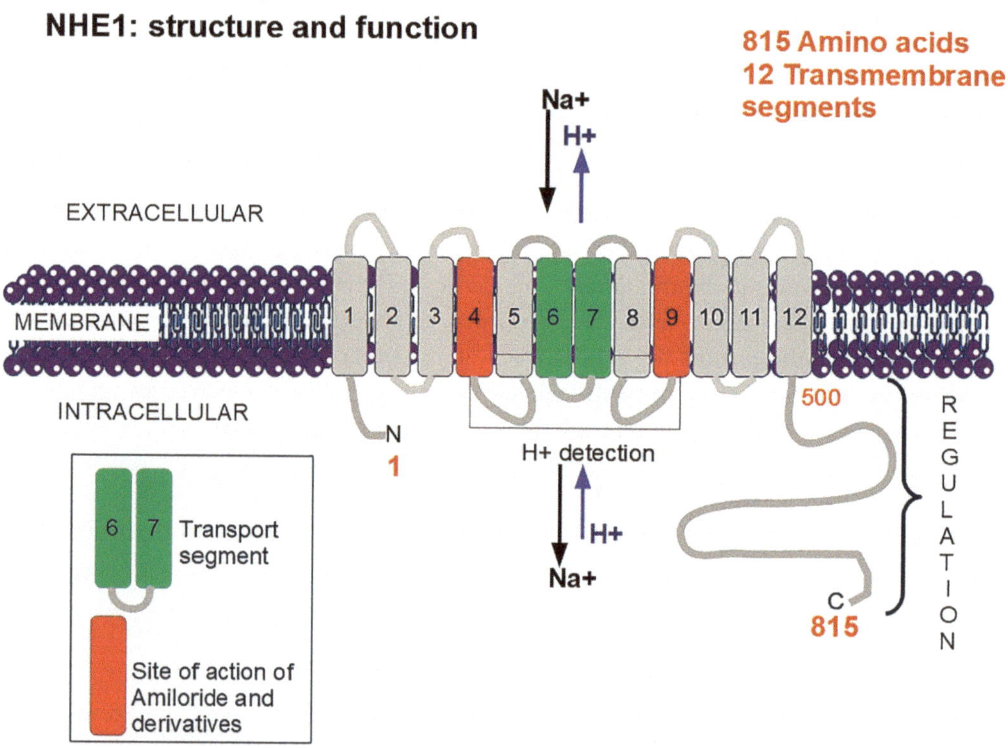

Fig. 1 Structure and function of NHE1. Intracellular pH is the main regulator of NHE1's activity. The antiporter has 12 transmembrane regions (M) an intracytoplasmic (N) amino terminus and a large (C) carboxy terminus. The M6 and M7 regions are highly conserved and are thought to be the ion transport portion of the structure.

NHE1 is almost quiescent when intracellular pH is in normal ranges, but as soon as it goes below this range, it becomes active. While NHE1 is found in almost all cells, the other isoforms are more tissue specific.

The primary regulator of NHE1 functioning is intracellular acidity, which is detected by the intracellular portions of the molecule between segments 4–5, 6–7 and 8–9 (see Fig. 1). When the intracellular pH decreases, the pH sensor of NHE1 induces an allosteric activation of the exchanger. There is an interaction between the pH sensors and the carboxyterminal intracellular portion, which may modify the sensor's sensitivity. This last step has not been fully elucidated. There is evidence showing that in breast cancer, VGSCs, particularly NaV1.5, induces an allosteric change of NHE1 modifying the sensitivity of pH detection and inducing NHE1 activity at a higher pH.

NHE1 inhibitors, like amiloride and its derivatives act on segments 4 and 9 (in red in Fig. 1). Many of the mechanisms that regulate NHE1 activity act upon the carboxy terminal intra cytoplasmic segment (AA 500–815). Experiments performed by Wakabayashi, et al.[8] showed that eliminating this segment reduced the affinity of the internal H^+ sensor and abolished growth factor-induced cytoplasmic alkalinization, while the exchange activity and activation by intracellular H^+ remained unchanged.

These findings led them to propose that the cytoplasmic C terminal segment is the part of the molecule that establishes the pH threshold for the sensor segments (Table 1).

Three different functional areas can be found in NHE1:

(a) The transporter segment that extrudes H^+ from the cell and incorporates Na^+ to the cell.[9]

(b) The sensor area that detects intracellular pH.[10]

(c) The regulator area that determines the threshold of the sensor area.[11]

Table 1 A summary of these concepts.

Functional areas	Possible location	Function
Transporter	Segment 6-Segment 7 (M6-M7)	Na^+ in; H^+ out
Sensor	Portions of the molecule between segment 4–5, 6–7 and 8–9	Intracellular pH sensor.
Regulator	Beyond AA 501 to 815	Determines the threshold of the sensor.

Besides pH, NHE1's activity is also regulated by the phosphorylation of specific amino acids in its C-terminal cytoplasmic segment, as well as by interactions of the C-terminal tail with intracellular proteins and lipids without phosphorylation.[12] We think that these secondary mechanisms of regulation may be important in malignant cells, because many of them are oncogenic signals.

Factors that modify NHE1 activity

Fig. 2 shows the many activators and inducers of NHE1 activity.

The factors increasing NHE1 activity include

- VGSCs which have the ability to increase NHE1 activity.[13]
- Growth factors.[8]

- Thrombin.[14]
- Angiotensin II.[15,16]
- Aldosterone.[17,18]
- Cathecolamines may have a similar effect to angiotensin and aldosterone.
- Oxidative stress.
- Osmotic stress.
- Carbonic anhydrase II.[19]
- Calmodulin.[20]
- PIP2 (phosphatidilinositol biphosphate).[21]
- Akt (in all tissues except heart muscle).[22,23]
- PI3 Kinase.[24]
- Serum deprivation[a]

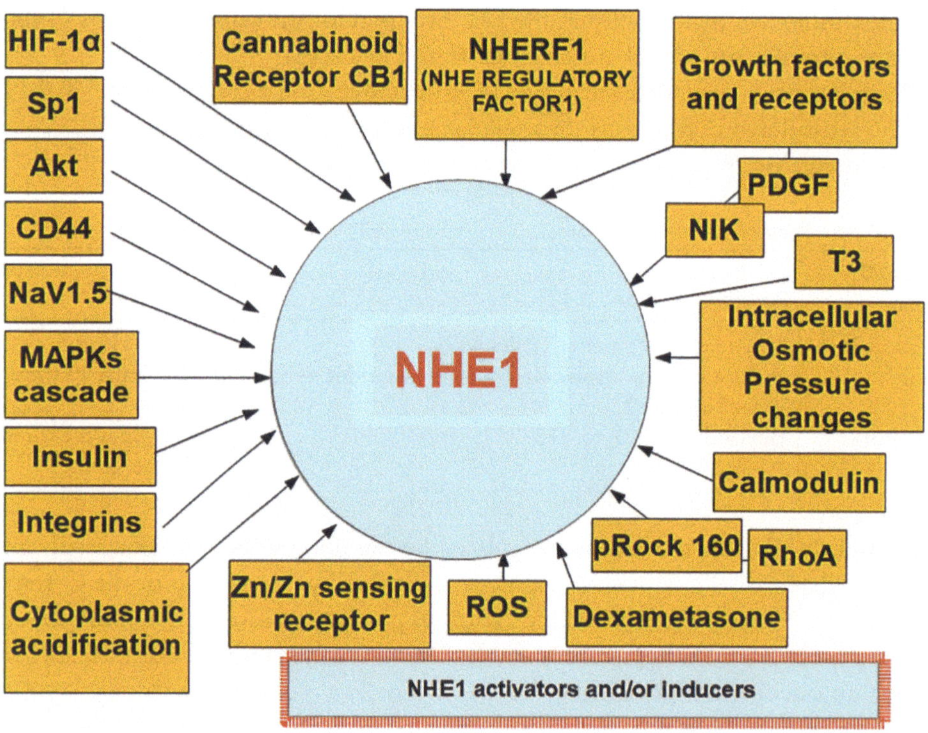

Fig. 2 Activators and/or inducers of NHE1 See references in the text, plus NHERF1,[25] RhoA-pRock160,[26] Mapkinases[27–29] cytoplasmic acidification,[30] cannabinoid receptor,[31] HIF-1α,[32] Zn,[33] growth factors,[34,35] ROS (reactive oxygen species)[36,37] dexamethasone,[38] calmodulin,[39] insulin,[40] CD44.[41]

Conditions that DECREASE NHE1 activity

(1) Depletion of cellular ATP decreases NHE1 activity.

(2) Akt in heart muscle.[42]

(3) Serum deprivation decreases NHE1 activity in non tumor cells.[25]

Stimulation of NHE1 activity: Mechanism of action

Stimulation of NHE1 increases its sensitivity to pH by augmenting the affinity of its allosteric proton binding site. This means that it acts at a higher pH than non-stimulated NHE1. Indeed, Fig. 3 shows that activation of NHE1 shifts the curve of its activity vs intracellular pH (pHi) to the right.

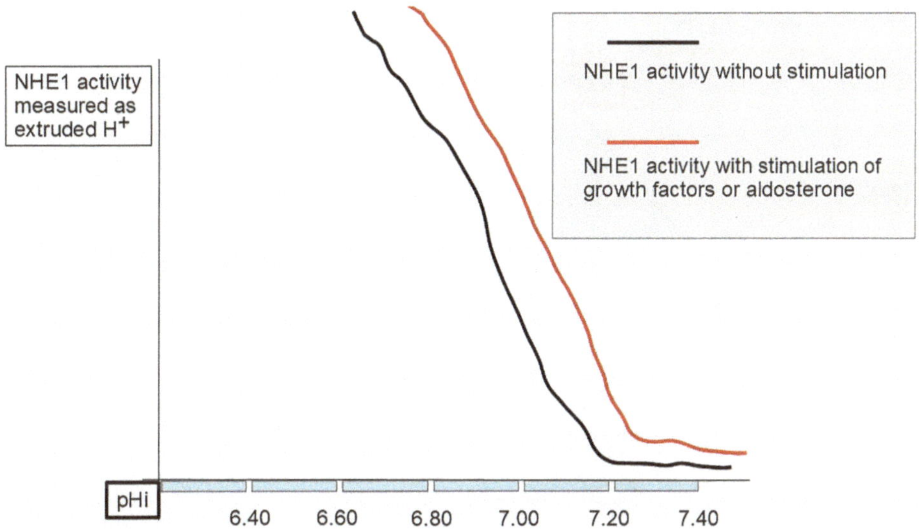

Fig. 3 Stimulating NHE1 activity with growth factors or any other known stimulus changes the pH threshold that activates NHE1. In the example above, we can see that NHE1 shows minimal baseline activity at pHi 7.40 and becomes weakly active at pHi of 7.00 if there is no special stimulation (black line). With stimulation, NHE1 activity starts at 7.20 and increases rapidly. So that at pHi of 7.00 it is much higher than in its non-stimulated counterpart. Thereafter, both curves increase swiftly but at any given pHi the activity of the stimulated NHE1 is higher than the non-stimulated NHE1. The current explanation is that the sensor threshold to pH is modified by stimulators probably through interaction with the cytoplasmic C-terminal segment that introduces allosteric changes in the pH sensor.

While NHE-deficient cells were unable to grow at acidic pH levels, all three isoforms (NHE1, 2 and 3) supported proliferation under acidic conditions.[43] The creation of the

[a] Tumors usually are in a hypoxic and serum deprived microenvironment. Under these conditions lactic acid production usually decreases but extracellular acidity is maintained by NHE1 activity.

highly acid extracellular environment is not only due to the increased activity of each NHE1 molecule, but it is also to an increased number of these membrane molecules.

NHE1 and the metabolic hallmarks of cancer

Hypoxia and acidosis of the ECS are early events in carcinogenesis, as is the metabolic switch towards glycolysis, and as Gillies and Gatenby[44] wrote "*adaptation to hypoxia and acidosis must be a major component of the carcinogenic sequence.*" Cytoplasmic alkalinization, which is the natural counterpart and cause of extracellular acidification is also an early event and the responsibility for this inversion of the pH gradient has been attributed to NHE1 activity (Fig. 4).[45]

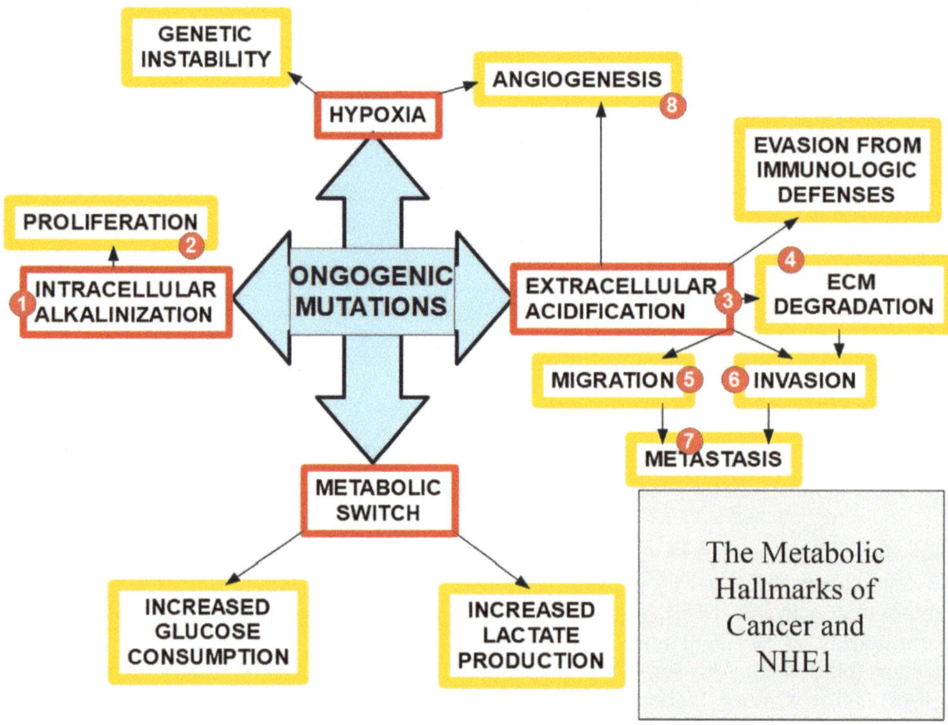

Fig. 4 Four of the most important hallmarks of metabolic differences between normal and malignant cells (in *red* frames). Relationship between NHE1 activity and these hallmarks.

1. Intracellular alkalinization

In the hypothetical case that the NHE1 and MCTs would not exist, the metabolic switch in cancer, would produce an acidic cytoplasm. This would be due to the 60-fold increase of lactic acid production, which, sooner or later, would saturate the lactate export system. NHE1 and MCTs are the tools for avoiding an intracellular lactic acidosis that would be fatal for the malignant cell. There is also another important source of acid inside the cell:

increased CO_2 that produces CO_3H_2 in contact with water. Ionization of CO_3H_2 releases a proton that would participate in decreasing intracellular pH unless it is exported by NHE1. Therefore, NHE1 is a key figure for cancer cell survival. NHE1 is not only overexpressed in many malignant cells, it is also overactive. Overactivity in this case means that NHE1 becomes active at a higher pH as shown in Fig. 3.

2. Proliferation

Approximately 50 years ago, the intracellular concentration of H^+ was considered to be a mitotic control agent.[46]

It has been shown that intracellular alkalinization is a fundamental and necessary condition for proliferation.[47–53] Rotin et al.[54] obtained mutant cells with deficient NHE1 expression. The parent cells with normal NHE1 showed no problems in growing at pH 6.6 or higher, while the mutant offsprings could not grow in the absence of bicarbonate at pH lower than 7. While transplanted parental cells could grow in immunodeficient mice, the mutant cells were unable to develop or grow.

There is an association between mitosis/meiosis and alkaline intracellular milieu. For example mice oocytes grow inside follicles under standard pH conditions, but before acquiring meiotic conditions, pH in the oocyte increases by 0.25 pH due to activation of NHE1 and bicarbonate/chlorine exchanger.[55]

The strong relationship between intracellular alkalinity and cell mitosis has been confirmed by many authors that, *brevitatis causa*, we will not repeat. NHE1 plays a pivotal role in alkalinizing the intracellular milieu and therefore, in proliferation. During the cell cycle, pHi oscillates between 6.8 and 7.3 [56] with maximum alkalinity before mitosis. The oscillation has been attributed to ATP hydrolysis when energy is required for synthesis before mitosis takes place.[57] The pHi oscillations in cancer cells are somewhat different and they are in an alkaline range between 7.2 and 7.5. According to Schwartz et al.,[58] the intracellular alkalosis is due to decreased OXPHOS and decreased CO_2 production. We do not share this idea. We believe that increased NHE1 activity, and probably other proton exporters are the main culprits of pHi alkalinity. Furthermore, growth factors increase NHE1 activity.[49–52]

What is not well known is the molecular mechanism by which increased intracellular pH and NHE1 activity increase proliferation. However, there are some clues. NHE1 is the key figure in cytoplasmic alkalinization and in yeasts this alkalinization activates mTORC1 and PKA (phosphokinase A),[59,60] and there is unpublished evidence that NHE1 and/or intracellular alkalinity promote G1/S transition. It has been suggested that cytosolic pH acts as a cellular signal regulating cell cycle progression and tumorigenesis.

3. Extracellular acidification

Extracellular acidification is the direct consequence of the removal of H^+ by NHE1 from the cell and its extrusion to the ECM. However, even if NHE1 is a very important mechanism in this sense, it is not unique. Other membrane and transport proteins also intervene in the process. In 1987 Hagag et al.[61] found that the microinjection of p21 ras protein in quiescent mouse NIH3T3 cells increased intracellular pH in a quick and sustained manner. This increase did not happen if the cells were treated with amiloride (an NHE1 inhibitor). Further, Reshkin et al.[45] showed that intracellular alkalinization is the first step in oncogene-driven malignant transformation. So the relationship between an oncogenic gene/protein and NHE1 has been clearly established; that is oncogene activation increases intracellular pH through NHE1 activity and this alkalinization drives malignant transformation.

The concept of pHi variations during the mitotic cycle is not universally accepted. Karagiannis and Young[62] working with yeast (*Saccharomyces pombe*), did not find pHi modifications during the cycle. But the pHi during mitosis was always in the range of 7.30 (which means alkaline). These findings are important because there is similarity between *S. pombe* and higher eukaryotes cells.

One very interesting finding in these experiments with *Saccharomyces pombe* was that pH homeostasis during mitosis was heavily dependent on the presence of an energy source (glucose), because glucose starved cells could not regulate pHi in response to minor changes in pHe. Non-starved cells, on the other hand, showed a stern homeostasis of pHi even when there was a strong acid overload in the medium. In the absence of glucose, intracellular H^+ ion concentration showed a three fold increase over a ten fold rise in external H^+ ion concentration. (The caloric restriction treatment of cancer may have a ground in these findings). The authors conclusion was that: "*intracellular pH is tightly regulated during logarithmic growth independently of cell-cycle-related parameters. It is thus unlikely to play any signaling or regulatory role.*"[62]

Experimenting with *Saccharomyces cerevisae* yeast, Gonzalez et al.[63] found that when yeast was subjected to a switch from aerobic to anaerobic conditions, that in addition to an important acidification of pHi (7.5–6.8), both cytoplasmic ATP and phosphoenolpyruvate also decreased markedly.

Therefore, at this metabolic cross-road we need to make some considerations regarding malignant cells:

- Cancer cells are in a highly acidic, hypoxic and nutrient deprived environment.
- They partially cope with this problem by switching to aerobic and anaerobic glycolysis, which will lower pHi.
- In spite of these "unfavorable" factors they manage to create an alkaline intracellular milieu, that allows them to synthesize all the molecules required to produce new cells and to proliferate in an accelerated manner.

Then, what or who allows them to modify their intracellular conditions in the way they do?

The answer is: a very efficient proton extruder mechanism.

NHE1 is one of these proton extruders, and as we shall see below, it is probably one of the key players in proton extrusion (but not the only one).

NHE1 is highly activated and overexpressed in cancer.

Mechanisms of NHE1 expression

The NHE1 gene has in the first exon the entire 5′ noncoding region (786 bases long), and 352 bases of the coding sequence. The noncoding enhancer/promoter region has a powerful TATA box (70% of promoter activity), 4 Sp1 consensus sequences, 3 AP sites, a cyclic AMP response element, and four partial glucocorticoid response elements.[64] We shall come back to this region, specially regarding the enhancer Sp1 consensus sequences in the pharmacology and therapeutic chapter.

NHE1 activity in cancer

To understand the importance of NHE1 in cancer, let us analyze what happens if we inhibit NHE1 in cancer cells. Working with NIH3T3 cells transformed by oncogenic E7 of HPV16, Reshkin et al.[45] found that blocking cytoplasmic alkalinization by inhibiting NHE1, or by acidifying the culture medium prevented the development of transformed phenotypes like increased growth rate and glycolytic metabolism. In the same study, they found that treating nude mice grafted with E7 generated malignancy, delayed the tumor development.

Furthermore, Reshkin et al.[65] found that serum deprivation produced a decrease of lactic acid and it inhibited NHE1 activity in non-tumor cells but increased NHE1 activity in cancer cells, and even more in metastatic cancer cells. Furthermore, inhibition of NHE1 has been identified as one of the apoptotic mechanisms of paclitaxel.[66,67]

Harguindey[68] and Reshkin[69] proposed NHE1 inhibition as an important target in cancer.

Table 2 *provides additional evidence.*

Table 2 Evidence of NHE1 inhibition as an antiproliferative mechanism.

Reference	Findings
Li et al.[70]	A recombinant NHE1 antisense gene was transfected into drug-resistant human SCLC. pHi decreased from 7.25 to 6.86 (averages) and apoptosis increased 6 fold in transfected cells.
Rebillard et al.[71]	The chemotherapeutic drug cisplatin produced apoptosis through a pathway that led to very early NHE1 inhibition with intracellular acidification and increase in membrane fluidity. These events were independent of cisplatin-induced DNA adducts formation.
Manzella et al.[72]	When cultured in 10% serum compared with 0.5% serum, ovarian cancer cell growth was stimulated, and this growth-stimulation was blocked by caripode (an NHE1 inhibitor).
Kooiker et al.[73]	Cariporide inhibition of NHE1 led to a decrease in cellular proliferation in lung squamous cell carcinoma.
Miraglia et al.[74]	EIPA (an amiloride derivative that inhibits NHE1) significantly reduced the pH*i* and increased the intracellular accumulation of doxorubicin in doxorubicin resistant human colon carcinoma cells.

Continued

Table 2 Evidence of NHE1 inhibition as an antiproliferative mechanism.—cont'd

Reference	Findings
McLean et al.[75]	Blocking NHE1 activity with amiloride in malignant glioma reversed both pH alkalinization and the transformed phenotype.
Flinck et al.[76]	Cell cycle progression was dependant on NHE1 and NBC in breast cancer cells. The knockdown of any of them reduced proliferation and delayed cell cycle progression. NHE1 reached its maximum level in the S phase while NBC did so in the G2/M phase. The authors propose a phase specific regulation.
Andersen et al.[77]	Breast cancer proliferation was promoted by MCTs, NBC and NHE1 but through different overlapping mechanisms.
Guan et al.[78]	Elevated NHE1 in gliomas is a marker of poor prognosis.
Guan et al.[79]	Blocking NHE1 in glioma cells increased immunogenecity and increased sensitivity to temozolomide and PD-1 inhibitors.

4. Migration

This will be discussed in depth in the chapters on invadopodia and aquaporins. NHE1 forms part of a complex mechanism that includes VGSCs, Src, cortactin, lipid rafts and caveolin that form the migratory machinery. There is much evidence showing that NHE1 is a fundamental protein in normal and cancer cell migration.[80–82] Inhibition or knockdown of NHE1 impedes the migratory process or, at least, slows it down very much.[83] The relationship between NHE1 and aquaporins has not been elucidated, but migration is strongly associated with cellular volume changes.[84,85] Thus, from a functional point of view this relation may exist. On a speculative basis we assume that pHi and/or pHe is the signaling feature between NHE1 and aquaporins, because migration is a finely tuned and complex process.[86] NHE1 creates the necessary intracellular and extracellular pH conditions at the leading edge of migration, but also participates in actin filament anchoring.[87]

5. Invasion

Degradation of the ECM is a required step for invasion. This is performed by cathepsin and metalloproteases that require an acidic extracellular medium for their activation. NHE1 creates this acidic extracellular environment, particularly at invadopodial level through proton export.

6. ECM degradation

The ECM degradation necessary for invasion occurs mainly through the activity of the invadopodia and will be explained in the chapter on invadopodia. ECS acidity created by proton extruders is essential for creating an adequate pH for proteolytic enzyme activity.

7. Metastasis

Requires migration and invasion as part of the process.

There is no invasion nor migration without degradation of the ECM. The activity of specific proteases centered on the invadopodia is essential for achieving degradation. This activity at the invadopodia is dependent on NHE1 driven extracellular acidification. Research by Greco et al.[88] showed "*a concordance between NHE1 activity, extracellular acidification and protease activity at invadopodia to finely regulate ECM digestion.*"

Lin et al.,[89] by inhibiting NHE1 activity with cariporide, suppressed migration and invasion produced by membrane type 1 metalloprotease (MT1-MMP) and found that MT1-MMP expression at both the RNA and protein levels was regulated by NHE1.

8. Angiogenesis

Angiogenesis is a hallmark of most tumors as has been shown by the seminal works of Judah Folkman[90] and it takes place in a hostile tumor microenvironment with pronounced hypoxia, and low nutrient and glucose levels, while lactate is elevated and pHe is strongly acidic. While the new vessels are functionally defective and inefficient to decrease hypoxia, they are very important for the tumor's development and progression, as has been proved by the benefits achieved with antiangiogenic treatments (see Table 3).

Table 3 Evidence of NHE1 involvement in angiogenesis.

Reference	Findings
Mo et al.[91]	HIF-1α increased overexpression of NHE1 and calpains. When NHE1 was knocked down calpain-2 expression was reduced. Interfering NHE1 with specific RNA inhibited HUVECs proliferation/viability, and migration.
Gao et al.[92]	Cariporide, an NHE1 inhibitor decreased pHi, proliferation, and migration of endothelial cells by decreasing VEGF in leukemia cells.
Sills et al.[93]	Squalamine was an NHE3 inhibitor and a strong antiangiogenic factor.
Zasloff et al.[94]	Squalamine was useful in the treatment of colon cancer.
Akhter et al.[95]	Squalamine is an antibiotic with strong NHE3 inhibitory activity.
Bhargava et al.[96]	A phase I study of squalamine in colon cancer.
He et al.[97]	Targeting NHE1 produced a reduction of pHi and the expression of VGEF in leukemia cells (K562).

Therapeutic implications

NHE1 inhibition in cancer in general and NHE3 inhibition in colorectal cancer decreases intracellular pH and reduces proliferation. There is enough experimental evidence showing the beneficial effects of NHE1/NHE3 inhibition and cancer cell acidification as a therapeutic goal. (See the review by Harguindey et al.[98]).

Amiloride is the basic NHE1 inhibitor, but this inhibition is weak compared with other drugs like cariporide or compound 9t. New derivatives have been synthetized, cariporide being the best known, but they have never been tested in cancer in clinical settings.

Sildenafil is an NHE1 inhibitor[99] and it has also various anticancer effects. Cimetidine has been investigated as an inhibitor of NHE1.[100–103] Similarities between cimetidine and amiloride inhibition of the exchanger were found. Expression of NHE1 can also be decreased by blocking the enhancer sites on its gene with tolfenamic acid and other

Sp1 blockers. Some statins, such as lovastatin, are involved in intracellular acidification due to a partial inhibition of NHE1.[104]

Squalamine is a steroid conjugate that was found in liver and cartilages of the dogfish shark.[105] It only inhibits NHE3 isoform[106] and works as a strong antiangiogenic molecule. Squalamine decreased intracellular alkalinity in tumor cells in spite of not inhibiting NHE1.[107]

NHE1 inhibition is one of the main weapons to be used in the reversal of the pH gradient inversion. This reversal is the core of the pH oriented treatment of cancer (to be discussed in Part II of this book).

Conclusions

NHE1 plays a significant role in at least 2 of the 4 metabolic hallmarks of cancer.

NHE1 has at least four known pro-tumor mechanisms:

- Lowering the pHe that creates the ideal condition for proteolytic enzymes activity.
- Increasing pHi which is a necessary condition for mitotic activity.
- Direct stimulation for metalloproteinase expression.
- Stimulation of angiogenesis.

NHE1 inhibition decreases migration, invasion and metastasis. This means that inhibition of this antiporter should have a place in cancer treatment. Furthermore, there are drugs that inhibit NHE1 and have very low toxicity, like cariporide. Unfortunately, cariporide has only been clinically tested in cardiovascular prevention but never in cancer. In the meantime, amiloride may be used while awaiting clinical cancer-specific trials.

References

1. Martinez-Zaguilan R, Seftor EA, Seftor RE, Chu YW, Gillies RJ, Hendrix MJ. Acidic pH enhances the invasive behavior of human melanoma cells. *Clin Exp Metastasis* 1996;**14**:176–86.
2. Huber V, De Milito A, Harguindey S, Reshkin SJ, Wahl ML, Rauch C, et al. Proton dynamics in cancer. *J Transl Med* 2010;**8**(1):57.
3. Alfarouk KO, Verduzco D, Rauch C, Muddathir AK, Adil HB, Elhassan GO, et al. Glycolysis, tumor metabolism, cancer growth and dissemination. A new pH-based etiopathogenic perspective and therapeutic approach to an old cancer question. *Oncoscience* 2014;**1**(12):777.
4. Boyer MJ, Tannock IF. Regulation of intracellular pH in tumor cell lines: influence of microenvironmental conditions. *Cancer Res* 1992;**52**:4441–7.
5. Slepkov E, Fliegel L. Structure and function of the NHE1 isoform of the Na+/H+ exchanger. *Biochem Cell Biol* 2002;**80**(5):499–508.
6. Spugnini EP, Sonveaux P, Stock C, Perez-Sayans M, De Milito A, Avnet S, et al. Proton channels and exchangers in cancer. *Biochim Biophys Acta Biomembr* 2015;**1848**(10):2715–26.
7. Koliakos G, Paletas K, Kaloyianni M. NHE1: a molecular target for signalling and cell matrix interactions. *Connect Tissue Res* 2008;**49**(3–4):157–61.
8. Wakabayashi S, Fafournoux P, Sardet C, Pouyssegur J. The Na+/H+ antiporter cytoplasmic domain mediates growth factor signals and controls "H(+)-sensing" *Proc Natl Acad Sci U S A* 1992;**89**(6):2424–8.
9. Dutta D, Fliegel L. Molecular modeling and inhibitor docking analysis of the Na+/H+ exchanger isoform one. *Biochem Cell Biol* 2018;**30**(999):1.

10. Wong KY, McKay R, Liu Y, Towle K, Elloumi Y, Li X, et al. Diverse residues of intracellular loop 5 of the Na+/H+ exchanger modulate proton sensing, expression, activity and targeting. *Biochim Biophys Acta Biomembr* 2019;**1861**(1):191–200.
11. Dutta D, Fliegel L. Structure and function of yeast and fungal Na+/H+ antiporters. *IUBMB Life* 2018;**70**(1):23–31.
12. Amith SR, Vincent KM, Wilkinson JM, Postovit LM, Fliegel L. Defining the Na+/H+ exchanger NHE1 interactome in triple-negative breast cancer cells. *Cell Signal* 2017;**29**:69–77.
13. Brisson L, Driffort V, Benoist L, Poet M, Counillon L, Antelmi E, et al. NaV1. 5 Na+ channels allosterically regulate the NHE1 exchanger and promote the activity of breast cancer cell invadopodia. *J Cell Sci* 2013;**126**(21):4835–42.
14. Putney LK, Denker SP, Barber DL. The changing face of the Na+/H+ exchanger, NHE1: structure, regulation, and cellular actions. *Annu Rev Pharmacol Toxicol* 2002;**42**:527–52.
15. da Costa-Pessoa JM, Figueiredo CF, Thieme K, Oliveira-Souza M. The regulation of NHE1 and NHE3 activity by angiotensin II is mediated by the activation of the angiotensin II type I receptor/phospholipase C/calcium/calmodulin pathway in distal nephron cells. *Eur J Pharmacol* 2013;**721**(1–3):322–31.
16. Eguti DM, Thieme K, Leung GP, Mello-Aires M, Oliveira-Souza M. Regulation of Na+/H+ exchanger isoform 1 (NHE1) by calmodulin-binding sites: role of angiotensin II. *Cell Physiol Biochem* 2010;**26**(4–5):541–52.
17. Casey JR. Why bicarbonate? *Biochem Cell Biol* 2006;**84**(6):930–9.
18. Karmazyn M, Liu Q, Gan XT, Brix BJ, Fliegel L. Aldosterone increases NHE1 expression and induces NHE1-dependent hypertrophy in neonatal rat ventricular myocytes. *Hypertension* 2003;**42**:1171–6.
19. Li X, Alvarez B, Casey JR, Reithmeier RAF, Fliegel L. Carbonic anhydrase II binds to and enhances activity of the Na+/H+ exchanger. *J Biol Chem* 2002;**277**:36085–91.
20. Wakabayashi S, Ikeda T, Iwamoto T, Pouyssegur J, Shigekawa M. Calmodulin-binding autoinhibitory domain controls "pH-sensing" in the Na+/H+ exchanger NHE1 through sequence specific interaction. *Biochemist* 1997;**36**:12854–61.
21. Aharonovitz O, Zaun HC, Balla T, York JD, Orlowski J, Grinstein S. Intracellular pH regulation by Na(+)/H(+) exchange requires phosphatidylinositol 4,5-bisphosphate. *J Cell Biol* 2000;**150**:213–24.
22. Meima ME, Webb BA, Witkowska HE, Barber DL. The sodium–hydrogen exchanger NHE1 is an Akt substrate necessary for actin filament reorganization by growth factors. *J Biol Chem* 2009;**284**:26666–75.
23. Serafino A, Moroni N, Psaila R, Zonfrillo M, Andreola F, Wannenes F, et al. Anti-proliferative effect of atrial natriuretic peptide on colorectal cancer cells: evidence for an Akt-mediated cross-talk between NHE1 activity and Wnt/β-catenin signaling. *Biochim Biophys Acta (BBA)—Mol Basis Dis* 2012;**1822**(6):1004–18.
24. Reshkin SJ, Bellizzi A, Albarani V, Guerra L, Tommasino M, Paradiso A, et al. Phosphoinositide 3-kinase is involved in the tumor-specific activation of human breast cancer cell Na(+)/H(+) exchange, motility, and invasion induced by serum deprivation. *J Biol Chem* 2000;**275**:5361–9.
25. Cardone RA, Bellizi A, Busco G, Weinman EJ, Dell'Aquila ME, Casavola V, et al. The NHERF1 PDZ2 domain regulates PKA-RhoA-p38-mediated NHE1 activation and invasión in breast tumor cells. *Mol Biol Cell* 2007;**18**(5):1768–80.
26. Tominaga T, Ishizaki T, Narumita S, Barber DL. P160ROCK meditaes RhoA activation of Na-H exchange. *EMBO J* 1998;**17**(16):4712–22.
27. Khaled AR, Moor AN, Li A, Kim K, Ferris DKMuegge K, et al. Trophic factor withdrawal: p38 mitogen-activated protein kinase activates NHE1 which induces intracellular alkalinization. *Mol Cell Biol* 2001;**21**(22):7545–57.
28. Gillis D, Shrode LD, Krump E, Howard CM, Ruble EA, Tibbles LA, et al. Osmotic stimulation of the Na+/H+ exchanger NHE1: relationship to the activation of three MAPK pathways. *J Membr Biol* 2001;**181**(3):205–14.
29. Bianchini L, L'Allenain G, Poyssegur J. The p42/p44 mitogen-activated protein kinase cacade is determinant in mediating activation of the Na+/H+ exchanger (NHE1 isoform 1) response to growth factors. *J Biol Chem* 1997;**272**(1):271–9.

30. Haworth RS, McCann C, Snabaitis AK, Roberts NA, Avkiran M. Stimulation of the plasma membrane Na+/H+ exchanger NHE1 by sustained intracellular acidosis. Evidence for a novel mechanism mediated by the ERK pathway. *J Biol Chem* 2003;**278**(34):31676–84.
31. Bouaboula M, Bianchini L, McKenzie FR, Pouyssegur J, Casellas P. Cannabinoid receptor CB1 activates the Na+/H+ exchanger NHE1 isoform via Gi-mediated mitogen activated protein kinase signaling transduction pathways. *FEBS Lett* 1999;**449**(1):61–5.
32. Shimoda LA, Fallon M, Pisarcik S, Wang J, Semenza GL. HIF-1 regulates hypoxic induction of NHE1 expression and alkalinization of intracellular pH in pulmonary arterial myocytes. *Am J Phys Lung Cell Mol Phys* 2006;**291**(5):941–9.
33. Azriel-Tamir H, Sharif H, Schwartz B, Hershfinkel M. Extracellular zink triggers ERK-dependent activation of Na+/H+ exchange in colonocytes mediated by the zinc-sensing receptor. *J Biol Chem* 2004;**279**(50):51804–16.
34. Chiang Y, Chou CY, Hsu KF, Huang YF, Shen MR. EGF upregulates Na+/H+ exchanger NHE1 by post-translational regulation that is important for cervical cancer cell invasiveness. *J Cell Physiol* 2008;**214**(3):810–9.
35. Wakabayashi S, Bertrand B, Shigekawa M, et al. Growth factor activation and "H(+)-sensing" of the Na+/H+ exchanger isoform 1 (NHE1). Evidence for an additional mechanism not requiring direct phosphorylation. *J Biol Chem* 1994;**269**(8):5583–8.
36. Cingolani OH, Perez NG, Ennis IL, Alvarez MC, Mosca SM, Schinella GR, et al. In vivo key role of reactive oxygen species and NHE1 activation determining excessive cardiac hypertrophy. *Pflugers Arch—Eur J Physiol* 2011;**462**(5):733.
37. Akram S, Teong HF, Fliegel L, Pervaiz S, Clement MV. Reactive oxygen species-mediated regulation of the Na+-H+ exchanger 1 gene expression connects intracellular redox status with cells' sensitivity to death triggers. *Cell Death Differ* 2006;**13**(4):628.
38. Voekl J, Pasma V, Ahmed MS, Walker B, Szteyn K, Kuhl D, et al. Sgk1-dependent stimulation of cardiac Na+/H+ exchanger Nhe1 by dexamethasone. *Cell Physiol Biochem* 2013;**32**(1):25–38.
39. Bertrand B, Wakabayashi S, Ikeda T, Pouyssegur J, Shigekawa M. The Na+/H+ exchanger isoform 1 (NHE1) is a novel member of the calmodulin-binding proteins. Identification and characterization of calmodulin-binding sites. *J Biol Chem* 1994;**269**(18):13703–9.
40. Sauvage M, Maziere P, Fathallah H, Giraud F. Insulin stimulates NHE1 activity by sequential activation of phosphatidylinositol 3-kinase and protein kinase C in human erythrocytes. *FEBS J* 2000;**267**(4):955–62.
41. Bourguignon LY, Singleton PA, Diedrich F, Stern R, Gilad E. CD44 interaction with Na+/H+ exchanger (NHE1) creates acidic microenvironments leading to hyaluronidase-2 and cathepsin B activation and breast tumor cell invasion. *J Biol Chem* 2004;**279**(26):26991–7007.
42. Snabaitis AK, Cuello F, Avkiran M. Protein kinase B/Akt phosphorylates and inhibits the cardiac Na+/H+ exchanger NHE1. *Circ Res* 2008;**103**:881–90.
43. Kapus A, Grinstein S, Wasan S, Kandasamy R, Orlowski J. Functional characterization of three isoforms of the Na+/H+ exchanger stably expressed in Chinese hamster ovary cells. ATP dependence, osmotic sensitivity, and role in cell proliferation. *J Biol Chem* 1994;**269**(38):23544–52.
44. Gillies RJ, Gatenby RA. Hypoxia and adaptive landscapes in the evolution of carcinogenesis. *Cancer Metastasis Rev* 2007;**26**(2):311–7.
45. Reshkin SJ, Bellizzi A, Caldeira S, Albarani V, Malanchi I, Poignee M, et al. Na+/H+ exchanger-dependent intracellular alkalinization is an early event in malignant transformation and plays an essential role in the development of subsequent transformation-associated phenotypes. *FASEB J* 2000;**14**:2185–97.
46. Gerson DF. Intracellular pH and the mitotic cycle in Physarum and mammalian cells. In: *Cell cycle regulation*. London: Academic Press; 1978. p. 105–31.
47. Sharma M, Astekar M, Soi S, Manjunatha BS, Shetty DC, Radhakrishnan R. pH gradient reversal: an emerging hallmark of cancers. *Recent Pat Anticancer Drug Discov* 2015;**10**(3):244–58.
48. Putnam RW. Effect of insulin on intracellular pH in frog skeletal muscle fibers. *Am J Phys* 1985;**248**:330–6.

49. Moolenaar WH, Tsien RY, van der Saag PT, de Laat SW. Na+/H+ exchange and cytoplasmic pH in the action of growth factors in human fibroblasts. *Nature* 1983;**304**:645–8.
50. Moolenaar WH, Tertoolen LG, de Laat SW. Phorbol ester and diacylglycerol mimic growth factors in raising cytoplasmic pH. *Nature* 1984;**312**:371–4.
51. Burns CP, Rozengurt E. Serum, platelet-derived growth factor, vasopressin and phorbol esters increase intracellular pH in Swiss 3T3 cells. *Biochem Biophys Res Commun* 1983;**116**:931–8.
52. Cassel D, Rothenberg P, Zhuang YX, Deuel TF, Glaser L. Platelet-derived growth factor stimulates Na+/H+ exchange and induces cytoplasmic alkalinization in NR6 cells. *Proc Natl Acad Sci U S A* 1983;**80**:6224–8.
53. Doppler W, Jaggi R, Groner B. Induction of v-mos and activated Ha-ras oncogene expression in quiescent NIH 3T3 cells causes intracellular alkalinisation and cell-cycle progression. *Gene* 1987;**54**:147–53.
54. Rotin D, Steele-Norwood D, Grinstein S, Tannock I. Requirement of the Na+/H+ exchanger for tumor growth. *Cancer Res* 1989;**49**(1):205–11.
55. Erdogan S, FitzHarris G, Tartia AP, Baltz JM. Mechanisms regulating intracellular pH are activated during growth of the mouse oocyte coincident with acquisition of meiotic competence. *Dev Biol* 2005;**286**(1):352–60.
56. Karagiannis J, Young PG. Intracellular pH homeostasis during cell-cycle progression and growth state transition in Schizosaccharomyces pombe. *J Cell Sci* 2001;**114**:2929–41.
57. Racker E. Bioenergetics and the problem of tumor growth. *Am Sci* 1972;**60**:56–63.
58. Schwartz L, Supuran C, Alfarouk OK. The Warburg effect and the hallmarks of Cancer. *Anti Cancer Agents Med Chem* 2017;**17**(2):164–70.
59. Dechant R, Saad S, Ibanez AJ, Peter M. Cytosolic pH regulates cell growth through distinct GTPases, Arf1 and Gtr1, to promote Ras/PKA and TORC1 activity. *Mol Cell* 2014;**55**:409–21.
60. Dechant R, Peter M. Cytosolic pH: a conserved regulator of cell growth? *Mol Cell Oncol* 2014;**1**:4. https://doi.org/10.4161/23723548.2014.969643.
61. Hagag N, Lacal JC, Graber M, Aaronson S, Viola MV. Microinjection of ras p21 induces a rapid rise in intracellular pH. *Mol Cell Biol* 1987;**7**:1984–8.
62. Karagiannis J, Young PG. Intracellular pH homeostasis during cell-cycle progression and growth state transition in Schizosaccharomyces pombe. *J Cell Sci* 2001;**114**:2929–41.
63. Gonzalez B, de Graaf A, Renaud M, Sahm H. Dynamic in vivo31P nuclear magnetic resonance study of Saccharomyces cerevisiae in glucose-limited chemostat culture during the aerobic–anaerobic shift. *Yeast* 2000;**16**(6):483–97.
64. Miller RT, Counillon L, Pages G, Lifton RP, Sardet C, Pouyssegur J. Structure of the 5′-flanking regulatory region and gene for the human growth factor-activatable Na/H exchanger NHE1. *J Biol Chem* 1991;**266**(17):10813–9.
65. Reshkin SJ, Bellizzi A, Albarani V, Guerra L, Tommasino M, Paradiso A, et al. Phosphoinositide 3-kinase is involved in the tumor-specific activation of human breast cancer cell Na+/H+ exchange, motility, and invasion induced by serum deprivation. *J Biol Chem* 2000;**275**(8):5361–9.
66. Reshkin SJ, Bellizzi A, Cardone RA, Tommasino M, Casavola V, Paradiso A. Paclitaxel induces apoptosis via protein kinase A- and p38 mitogen-activated protein-dependent inhibition of the Na+/H+ exchanger (NHE) NHE isoform 1 in human breast cancer cells. *Clin Cancer Res* 2003;**9**(6):2366–73.
67. Amith SR, Wilkinson JM, Baksh S, Fliegel L. The Na^+/H^+ exchanger (NHE1) as a novel co-adjuvant target in paclitaxel therapy of triple-negative breast cancer cells. *Oncotarget* 2015;**6**(2):1262–75.
68. Harguindey S, Arranz JL, Polo Orozco JD, Rauch C, Fais S, Cardone RA, et al. Cariporide and other new and powerful NHE1 inhibitors as potentially selective anticancer drugs—an integral molecular/biochemical/metabolic/clinical approach after one hundred years of cancer research. *J Transl Med* 2013;**11**(1):282.
69. Reshkin SJ, Cardone RA, Harguindey S. Na+-H+ exchanger, pH regulation and cancer. *Recent Pat Anticancer Drug Discov* 2013;**8**:85–99.
70. Li S, Bao P, Li Z, Ouyang HF, Wu C, Qian G. Inhibition of proliferation and apoptosis induced by a Na+/H+ exchanger-1 (NHE1) antisense gene on drug-resistant human small cell lung cancer cells. *Oncol Rep* 2009;**21**(5):1243–9.

71. Rebillard A, Tekpli X, Meurette O, Sergent O, LeMoigne-Muller G, Vernhet L, et al. Cisplatin-induced apoptosis involves membrane fluidification via inhibition of NHE1 in human colon cancer cells. *Cancer Res* 2007;**67**(16):7865–74.
72. Manzella TM, Provost JJ, Wallert MA. Characterizing the role of the Na+-H+ exchanger isoform 1 (NHE1) in cell proliferation and migration in ovarian cancer cells. *FASEB J* 2017;**31**(1 Suppl). 934.17.
73. Kooiker AJ, Provost JJ, Wallert MA. The role of the Na+-H+ exchanger isoform 1 (NHE1) and Calcineurin B homologous protein isoform 2 (CHP2) on cell proliferation and migration in squamous cell carcinoma of the lung. *FASEB J* 2017;**31**(1 Suppl). 934.16.
74. Miraglia E, Viarisio D, Riganti C, Costamagna C, Ghigo D, Bosia A. Na+/H+ exchanger activity is increased in doxorubicin-resistant human colon cancer cells and its modulation modifies the sensitivity of the cells to doxorubicin. *Int J Cancer* 2005;**115**(6):924–9.
75. McLean LA, Roscoe J, Jorgensen NK, Gorin FA, Cala PM. Malignant gliomas display altered pH regulation by NHE1 compared with nontransformed astrocytes. *Am J Phys Cell Physiol* 2000;**278**(4): C676–88.
76. Flinck M, Kramer SH, Schnipper J, Andersen AP, Pedersen SF. The acid-base transport proteins NHE1 and NBCn1 regulate cell cycle progression in human breast cancer cells. *Cell Cycle* 2018;**17**(9):1056–67.
77. Andersen AP, Samsøe-Petersen J, Oernbo EK, Boedtkjer E, Moreira JM, Kveiborg M, et al. The net acid extruders NHE1, NBCn1 and MCT4 promote mammary tumor growth through distinct but overlapping mechanisms. *Int J Cancer* 2018;**42**(12):2529–42.
78. Guan X, Luo L, Begum G, Kohanbash G, Song Q, Rao A, et al. Elevated Na/H exchanger 1 (SLC9A1) emerges as a marker for tumorigenesis and prognosis in gliomas. *J Exp Clin Cancer Res* 2018;**37**(1):255.
79. Guan X, Hasan MN, Begum G, Kohanbash G, Carney KE, Pigott VM, et al. Blockade of Na/H exchanger stimulates glioma tumor immunogenicity and enhances combinatorial TMZ and anti-PD-1 therapy. *Cell Death Dis* 2018;**9**(10):1010.
80. Simchowitz L, Cragoe EJ. Regulation of human neutrophil chemotaxis by intracellular pH. *J Biol Chem* 1986;**261**(14):6492–500.
81. Klein M, Seeger P, Schuricht B, Alper SL, Schwab A. Polarization of Na+/H+ and Cl−/HCO3− exchangers in migrating renal epithelial cells. *J Gen Physiol* 2000;**115**(5):599–608.
82. Denker SP, Barber DL. Cell migration requires both ion translocation and cytoskeletal anchoring by the Na-H exchanger NHE1. *J Cell Biol* 2002;**159**(6):1087–96.
83. Dieterich P, Dreval V, Schwab A. Comprehensive analysis of single cell migration. *Pflugers Arch* 2003;**445**(Suppl):S44.
84. Rosengren S, Henson PM, Worthen GS. Migration-associated volume changes in neutrophils fac ilitate the migratory process in vitro. *Am J Phys Cell Phys* 1994;**267**(6):C1623–32.
85. Ritter M, Schratzberger P, Rossmann H, Wöll E, Seiler K, Seidler U, et al. Effect of inhibitors of Na+/H+-exchange and gastric H+/K+ ATPase on cell volume, intracellular pH and migration of human polymorphonuclear leucocytes. *Br J Pharmacol* 1998;**124**(4):627–38.
86. Lauffenburger DA, Horwitz AF. Cell migration: a physically integrated molecular process. *Cell* 1996;**84**(3):359–69.
87. Denker SP, Huang DC, Orlowski J, Furthmayr H, Barber DL. Direct binding of the Na–H exchanger NHE1 to ERM proteins regulates the cortical cytoskeleton and cell shape independently of H+ translocation. *Mol Cell* 2000;**6**(6):1425–36.
88. Greco MR, Antelmi E, Busco G, Guerra L, Rubino R, Casavola V, et al. Protease activity at invadopodial focal digestive areas is dependent on NHE1-driven acidic pHe. *Oncol Rep* 2014;**31**(2):940–6.
89. Lin Y, Wang J, Jin W, Wang L, Li H, Ma L, et al. NHE1 mediates migration and invasion of HeLa cells via regulating the expression and localization of MT1-MMP. *Cell Biochem Funct* 2012;**30**(1):41–6.
90. Folkman J. Tumour angiogenesis: therapeutic implications. *N Engl J Med* 1971;**285**:1182–6.
91. Mo XG, Chen QW, Li XS, Zheng MM, Ke DZ, Deng W, et al. Suppression of NHE1 by small interfering RNA inhibits HIF-1α-induced angiogenesis in vitro via modulation of calpain activity. *Microvasc Res* 2011;**81**(2):160–8.

92. Gao W, Chang G, Wang J, Jin W, Wang L, Lin Y, et al. Inhibition of K562 leukemia angiogenesis and growth by selective Na+/H+ exchanger inhibitor cariporide through down-regulation of pro-angiogenesis factor VEGF. *Leuk Res* 2011;**35**(11):1506–11.
93. Sills AK, Williams JI, Tyler BM, Epstein DS, Sipos EP, Davis JD, et al. Squalamine inhibits angiogenesis and solid tumor growth in vivo and perturbs embryonic vasculature. *Cancer Res* 1998;**58**(13):2784–92.
94. Zasloff M, Williams J, inventors, Magainin Pharmaceuticals Inc, assignee. *Treatment of carcinomas using squalamine in combination with other anti-cancer agents*. United States patent US 6,147,060,2000. Nov 14.
95. Akhter S, Nath SK, Bowser JM, Tse CM, Williams J, Zasloff M, et al. Squalamine, a novel aminosterol antibiotic, is a specific inhibitor of epithelial brush border Na+/H+ exchanger isoform NHE3. *FASEB J* 1996;**10**(3):514.
96. Bhargava P, Marshall JL, Dahut W, Rizvi N, Trocky N, Hait H, et al. A phase I and pharmacokinetic study of squalamine, a novel antiangiogenic agent, in patients with advanced cancers. *Clin Cancer Res* 2001;**7**(12):3912–9.
97. He B, Deng C, Zhang M, Zou D, Xu M. Reduction of intracellular pH inhibits the expression of VEGF in K562 cells after targeted inhibition of the Na+/H+ exchanger. *Leuk Res* 2007;**31**(4):507–14.
98. Harguindey S, Orive G, Pedraz JL, Paradiso A, Reshkin SJ. The role of pH dynamics and the Na+/H+ antiporter in the etiopathogenesis and treatment of cancer. Two faces of the same coin—one single nature. *Biochim Biophys Acta Rev Cancer* 2005;**1756**(1):1–24.
99. Perez NG, Piaggio MR, Ennis IL, Garciarena CD, Morales C, Escudero EM, et al. Phosphodiesterase 5A inhibition induces Na+/H+ exchanger blockade and protection against myocardial infarction. *Hypertension* 2007;**49**:1095–103.
100. Ganapathy VA, Balkovetz DF, Miyamoto YU, Ganapathy ME, Mahesh VB, Devoe LD, et al. Inhibition of human placental Na+-H+ exchanger by cimetidine. *J Pharmacol Exp Ther* 1986;**239**(1):192–7.
101. Liakopoulos V, Zarogiannis S, Kourti P, Hatzoglou C, Karioti A, Arampatzis S, et al. Effect of cimetidine on the electrophysiologic profile of isolated visceral sheep peritoneum. In: *Advances in peritoneal dialysis. Conference on Peritoneal Dialysis,* Vol. 25; 2009. p. 20–3.
102. Cantiello HF, Lanier SM. Alpha 2-adrenergic receptors and the Na+/H+ exchanger in the intestinal epithelial cell line, HT-29. *J Biol Chem* 1989;**264**(27):16000–7.
103. Guerra L, Di Sole F, Valenti G, Ronco PM, Perlino E, Casavola V, et al. Polarized distribution of Na+/H+ exchanger isoforms in rabbit collecting duct cells. *Kidney Int* 1998;**53**(5):1269–77.
104. Mihaila RG. A minireview on NHE1 inhibitors. A rediscovered hope in oncohematology. *Biomed Pap Med Fac Univ Palacky Olomouc Czech Repub* 2015;**159**(4):519–26.
105. Zasloff M, Adams AP, Beckerman B, Campbell A, Han Z, Luijten E, et al. Squalamine as a broad-spectrum systemic antiviral agent with therapeutic potential. *Proc Natl Acad Sci* 2011;**108**(38):15978–83.
106. Akhter S, Nath SK, Tse CM, Williams J, Zasloff M, Donowitz M. Squalamine, a novel cationic steroid, specifically inhibits the brush-border Na+/H+ exchanger isoform NHE3. *Am J Phys* 1999;**276**(1 Pt 1):C136–44.
107. Pietras RJ, Weinberg OK. Antiangiogenic steroids in human cancer therapy. *Evid Based Complement Alternat Med* 2005;**2**(1):49–57. https://doi.org/10.1093/ecam/neh066.

CHAPTER 6

Voltage gated sodium channels

Introduction

Neurons and excitable tissues in general express voltage gated sodium channels (VGSC), proteins also known as NaV channels. However, tumor cells can also express these proteins. Sodium channels were first described by Hodgkin and Huxley in 1952 and knowledge about structure and physiology of VGSCs are mainly the result of seminal studies by William Catterall.[1]

Sodium channels are glycosylated transmembrane proteins. They form passages in the cell membrane for the selective penetration of sodium into the intracellular space according to an electrochemical gradient. Voltage gated sodium channels (VGSC or NaV channels) refers to the mechanism that triggers these proteins to allow or impede sodium movement across the cellular membrane.

There are nine known VGSCs (NaV1.1 to Nav1.9) that are members of the VGSC superfamily. NaV1.1, 1.2, 1.3, and 1.6 are found in the central nervous system. NaV1.4 is found in striated muscle and NaV1.5 in cardiac muscle.[2] The neonatal splice variant of NaV1.5 (nNaV1.5) represents a particularly important isoform because it has been found in highly invasive tumors in vivo and in vitro.[3]

VGSCs are formed by one large subunit (α) and other smaller subunits (β). The α subunit, the core of the channel, forms the pore for Na^+ trafficking and is fully functional by itself, even without the β subunits[1, 2, 4]; so that when a cell expresses the VGSC α subunits it is able to conduct sodium into the cell. Figs. 1 and 2 show the structure of VGSC.

VGSCs modulate the movement of Na^+ across the cell membrane and the inflow of this electrolyte spikes the action potential in excitable tissues.[6] VGSCs are closed at resting membrane potentials. When they are activated, the channel, which is closed (closed channel), undergoes a fast opening of the pore (open channel) through a conformational change, resulting in Na^+ influx into the cell and further membrane depolarization. As soon as the channel is inactivated (inactivated channel), it becomes unavailable for further activation for a variable time after which it again reaches the closed channel stage and can then be activated again (Fig. 2 and Box 1). The typical characteristics of VGSCs are[7]:

(1) fast gating (time elapsed between opening and closing);
(2) voltage-dependent activation;
(3) rapid inactivation;
(4) selective sodium ion influx.

An Innovative Approach to Understanding and Treating Cancer: Targeting pH
https://doi.org/10.1016/B978-0-12-819059-3.00006-X
© 2020 Elsevier Inc.
All rights reserved.

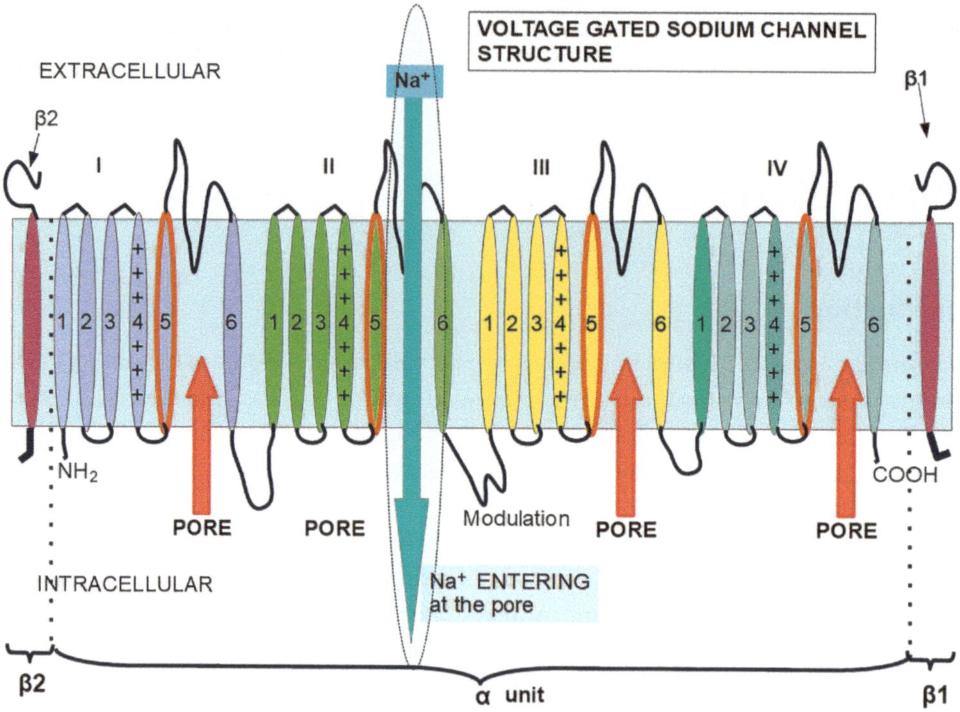

Fig. 1 An idealized drawing of α and β subunits of VGSCs. The pore forming α subunit is formed by four domains (I, II, III, and IV). Each domain is formed by six transmembrane segments. The three dimensional arrangement of the four domains determines the formation of one pore (see Fig. 2) that is modulated by a longer modulatory intracellular portion. Two other intracellular portions (between segment I–II and segment III–IV) are in charge of the inactivation. Voltage sensors are marked with a positive sign (+). Carboxy-terminal and amino-terminal portions are intracellular. Non-pore forming β subunits are not essential for NaV pore functioning. However, these β subunits exert modulatory actions on the α subunit. β subunits form relationships with integrins and intervene in adhesion functions.

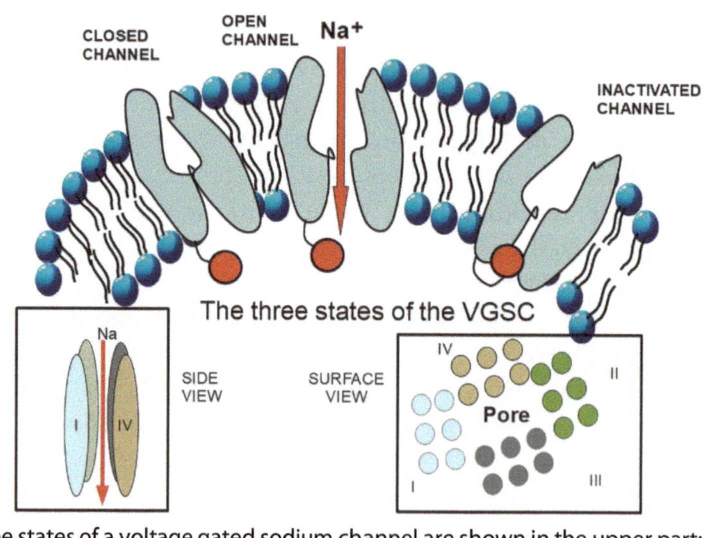

Fig. 2 The three states of a voltage gated sodium channel are shown in the upper part: closed channel, open channel and inactivated channel. The lower left quadrant shows a side view of VGSC and the right quadrant presents a surface view. For a detailed description see Ref. 5.

BOX 1 Sequence of events at the VGSCs.

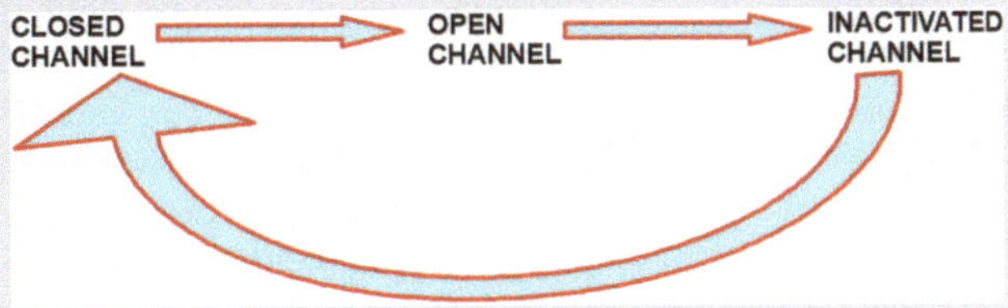

VGSCs and cancer

NaV channels are necessary for the cellular production of action potentials in electrically excitable cells such as nerves and muscle fibers. These cells are rich in NaVs. They can be found in some normal, non-electrically excitable cells such as astrocytes and macrophages but their presence is low or absent in normal breast, lung or prostate cells. In non-excitable normal cells these channels perform the so-called non canonical functions such as secretion, motility and others.[8–10]

It is well known that VGSC expression appears in cancer cells; VGSCs are not expressed in their normal counterparts and play significant roles in disease progression. Many tumors, but not all, overexpress VGSCs. The role played by VGSCs in highly invasive breast and prostate cancer has been extensively studied and the VGSCs are usually located near the NHE1 (sodium hydrogen exchanger 1) in the cell membrane of the advancing motile and invadopodial sectors of the malignant cell. The relationship between them is of utmost importance in the process of migration and invasion (see chapter on migration and invadopodia).

Table 1 shows examples of tumors where VGSCs were found and the role they play.

Based on the evidence shown in Table 1 it seems reasonable that targeting these channels may represent a legitimate way of reducing or blocking the invasive/metastatic process.

Table 1 VGSC functional over-expression in different cancer tissues.

Cancer	Reference	Findings
Human breast cancer	Fraser et al. 2005[11]	VGSC (neonatal isoform of NaV1.5) was significantly upregulated in metastatic cells. VGSC activity increased endocytosis, migration and invasion
NSCLC (non-small cell lung cancer)	Roger et al. 2007[12]	Strongly metastatic cell lines showed functional VGSCs while normal cells lack them. Channel inhibition with Tetrodotoxin (TTX) reduced invasiveness by 50%

Continued

Table 1 VGSC functional over-expression in different cancer tissues—cont'd

Cancer	Reference	Findings
SCLC (small cell lung cancer)	Blandino, et al. 1995[13]	VGSCs were over-expressed in SCLC
Cervical cancer cells	Diaz et al. 2007[14]	$Na_v1.2$, $Na_v1.4$, $Na_v1.6$, and $Na_v1.7$ transcripts were detected in cervical cancer cell specimens
Prostate cancer	Bennett et al. 2004[15]	VGSC expression increased with invasion ability that could be blocked with TTX. Increased VGSC's expression was enough to increase invasive phenotypes
Metastatic Ovarian cancer cells	Gao et al. 2010[16]	Highly metastatic ovarian cells showed significantly elevated expression of Nav1.2, Nav1.4, Nav1.5, and Nav1.7. TTX reduced migration and invasion around 50%
Human colon cancer cells	House, et al. 2010[17]	It was found that *SCN5A,* the gene of the α subunit of VGSC, was a regulator of the invasive phenotype.
T-lymphocytes (Jurkat cells)	Fraser et al. 2004[18]	Jurkat cells expressed VGSC and this protein had an important role in invasiveness
Pancreatic cancer cells	Sato et al. 1994[19]	MIA-PaCa-2 and CAV cells were tested in vitro and in vivo with phenytoin (PHEN). Both cell lines showed growth inhibition in a dose-dependent manner. This might be due to VGSC overexpression according to our criteria (the authors think that this is due to calcium channel blocking)
Mesothelial neoplastic cells	Fulgenzi et al. 2006[20]	These cells expressed VGSCs, particularly NaV1.2, NaV1.6, and NaV1.7. TTX decreased cell motility and migration
Cervical cancer	Hernandez-Plata et al. 2012[21]	The overexpression of NaV1.6 in cervical cancer was strongly associated with migration
Prostate cancer	Aktas et al. 2018[22]	Naringenin[a] decreased prostatic cancer cell movement and metastasis by inhibiting VGSCs.
Breast cancer	Nelson et al. 2015[23]	NaV1.5 α subunit potentiated invasion and migration in metastatic breast cancer. Down-regulation with phenytoin decreased these activities in vivo
Glioma	Shrey et al. 2002[24]	NaV1.6 which is the isoform usually found in central nervous system was almost absent in gliomas. However, all the gliomas overexpressed one or more other isoforms

[a]Naringenin is a bioflavonoid that can be found in certain fruits such as grapefruit, tomatoes, sour orange, etc.

Sodium channel proteins and cancer

In 1995 Grimes et al.[25] investigated the differential electrophysiological characteristics of VGSCs in two different rodent prostate cancer cell lines: the Mat-Ly-Lu cell line which is highly metastatic (more than 90% of metastasis to lung and lymph nodes under experimental conditions) and the AT-2 cell line with a much lower metastatic potential (less than 10% of metastasis). They found fundamental differences in electrophysiological features between these two cell lines, which maintained direct relationship with in vitro invasiveness. Inward sodium currents were detected only in the Mat-Ly-Lu cell line; the Tetrodotoxin (TTX, a powerful inhibitor of VGSCs) inhibition of VGSC significantly reduced the capacity for invasion (mean reduction 33%). On the other hand, TTX showed no effect on invasion of AT-2 cell lines. The TTX-induced reduction of invasion had a direct correlation with the amount of cells expressing VGSC in the culture. No fundamental differences in the potassium channels were found, except for a lower density of potassium channels in the Mat-Ly-Lu cell line. The authors concluded that ion channels may be involved in malignant cell behavior and VGSCs could play a role in the metastatic process. This was the first significant evidence for a role of NaV channels in the malignant phenotype.

The second important evidence came in 1997 from the research by Laniado et al.[26] They investigated the presence of VGSC in human prostate cell lines. As in the Grimes research, they used two different cell lines: one with a low metastatic potential, the LN-Cap cell line which is androgen-dependent and expresses prostate-specific antigen, and the PC-3 line which is more malignant, does not express prostate-specific antigen and shows a high rate of metastatic potential. Like Grimes et al., they found that PC-3, the more malignant cell line, expressed VGSC protein and the TTX inhibition of this channel significantly reduced invasion. LNCap cells did not express VGSC. One of the conclusions reached by the authors was that when cells express functional VGSCs, this confers a selective advantage to cancer cells regarding migration and distant metastasis.

In cell lines of both humans and rodents, not all cells in the highly malignant cell cultures showed the presence of the VGSC protein. For example, in PC-3 cell culture only 10% of cells expressed a functional VGSC protein. This is the reason why the authors considered these cells as a clonal evolution that gave pro-tumoral and pro-invasive advantages. Cellular heterogeneity regarding VGSCs may represent different phenotypes in the same tumor, with those cells expressing the NaV channels being more invasive than the rest.

The correlation between VGSC protein expression and invasiveness in human and rat prostate cancer cells was confirmed by Smith et al.[27] whom compared seven lines of rat prostate carcinoma cells with different metastatic ability, and nine human prostate carcinoma cell lines. In general, invading capacity of the basal membrane and metastatic ability maintained a positive correlation with the percentage of cells expressing VGSCs. However, these authors concluded that NaV channel presence may represent a prerequisite for the invasive phenotype but other requirements must also be met for a full blown invasive phenotype. We may assume that these "other requirements" are probably represented by other membrane proteins like NHE1, membrane carbonic anhydrases, caveolin1, intracellular alkalinization, and extracellular acidification. Fraser et al.[28] determined the key role played by VGSCs in prostate cancer cells in invasion and motility and showed that TTX and phenytoin (PHEN), that are known VGSC blockers, decreased motility and invasiveness while channel openers increased motility.

Furthermore, the increased invasion ability in VGSC-expressing cancer cells is not limited to prostate. The same features were found comparing breast cancer cell lines of different invading abilities: MCF-7 (estrogen receptor positive and low invading capacity), MDA-MB-231 (estrogen negative and high invading capacity), and MDA-MB-468 (estrogen receptor negative and intermediate invading capacity).[29] MDA-MB-231 cells had a fast inward flux of Na^+, which was not found in the less invasive cell lines. The fast inward sodium flux was blocked by TTX. This blockage also decreased invasion in MDA-MB-231 cell lines. TTX had no effects on proliferation in the three lines examined.

The conclusions were that:

(1) cells with high invasion ability had a fast Na^+ influx;
(2) blocking the fast Na^+ influx decreased invasion substantially; and
(3) blocking the fast Na^+ influx had no effects on proliferation.

Antioxidant response in cancer manifested by an increase in the antioxidant system (glutathione, superoxide dismutase, total nitrate, etc.) is an essential feature for the survival of malignant cells. Experimenting in a rat model of induced breast cancer, Baciotglu et al.[30] showed the importance of inhibiting VGSCs in order to decrease the antioxidant response. They observed a survival improvement in rats treated with a VGSC blocker.

The location and relations of VGSCs in malignant cells

An important location of VGSCs in cancer cells is in a cellular region directly involved in migration and invasion: the invadopodia. (See chapter on Invadopodia.)

Invadopodia are actin-rich protrusions of the plasma membrane that are greatly involved in the degradation of the extracellular matrix (ECM) and migration.

Figs. 3 and 4 summarize how invadopodia works and the relationship between VGSC and invadopodia.

According to Brisson et al.[31] NaV 1.5 Na^+ channels allostericaly regulate the NHE1 exchanger, increasing proton extrusion with extracellular matrix acidification that promotes invasion and migration by activating cysteine cathepsins and degrading the extracellular matrix. Inhibiting VGSCs with TTX significantly reduced proteolytic activity at the invadopodia. When cathepsines were inhibited, the TTX blocking of VGSCs did not further decrease the proteolytic activity at the invadopodia.[32]

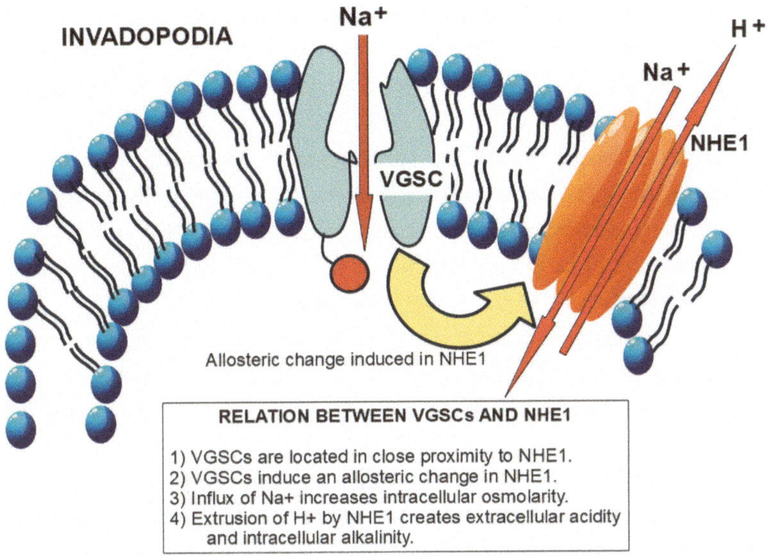

Fig. 3 VGSCs allosterically activate the sodium hydrogen exchanger-1 (NHE1) at higher pH and increase intracellular osmolarity. Proton extrusion through NHE1 acidifies the extracellular matrix.

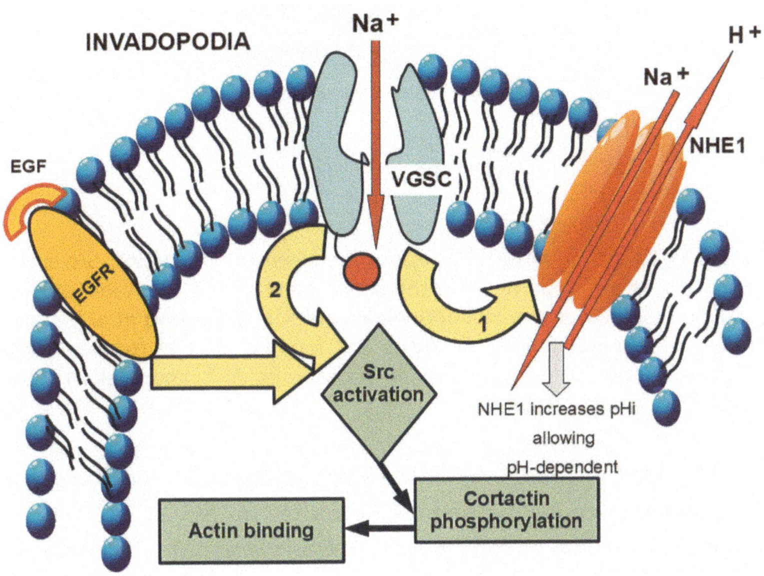

Fig. 4 VGSCs' actions at the invadopodia: (1) allosterical modification of NHE1 that becomes active at a higher pH than normal; (2) activation of Src in cooperation with EGFR, phosphorylating Cortactin and increasing actin binding. It is postulated that there is a feedback loop starting with MMPs products inducing new invadopodia development. For more details on this issue see chapter on invadopodia.

Mader et al. described a second mechanism of invasion promotion through the EGFR-Scr-cortactin pathway.[33] There, Src was activated by growth signals and Src phosphorylated cortactin. Cortactin is involved in the upregulation and secretion of MMP-9 and MMP-2.[34]

There are nine different VGSC α subunits and four different β subunits. The expression of these subunits can vary in the different tumor cells.[4] For example, NaV 1.5 is overexpressed in astrocytoma, breast and colon cancer. NaV 1.7 is found in breast, prostate and non small cell lung cancer (NSCLC) and NaV 1.6 in cervical and prostate cancer. So that the α subunits seem to be tissue specific.

Beside VGSC, the main players in the invadopodia complex, are Src kinase, cortactin and RhoA GTPase. The exact relation among these players is not fully known and needs further research. (For further reading on invadopodia and cortactin see Refs. 35, 36.)

One possible relation between invadopodia-Src-VGSCs is shown in Fig. 5.

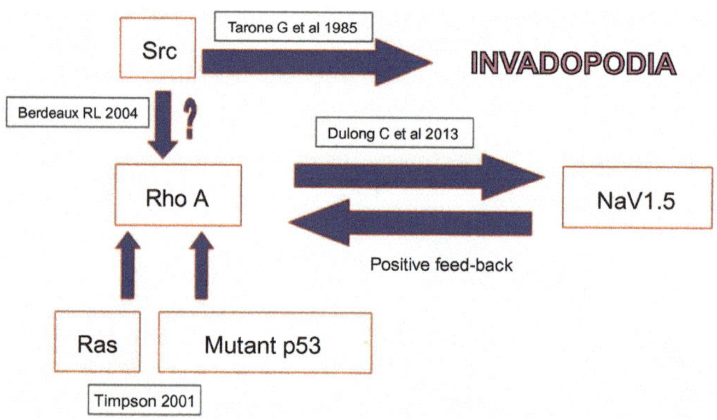

Fig. 5 In 1986, Tarone et al.[37] reported the relation between the oncogenic Src and promotion of invadopodia. Berdeaux et al.[38] reported that the small molecule GTPase RhoA activity is regulated by oncogenic Src and localizes in the invadopodia complex and Durlong et al.[39] showed that Rho-A regulates the expression and activity of NaV1.5 and found a positive feedback between NaV1.5 and RhoA in breast cancer cells. According to Timpson et al.[40] cooperation between mutant p53and oncogenic Ras activates RhoA.

Onganer and Djamgoz[41] suggested that VGSC upregulation enhances the metastatic phenotype by increasing endocytic membrane activity in SCLC.

A conclusion can be drawn from Figs. 4 and 5: a synergistic inhibition of invasion may be achieved by blocking VGSCs and Src simultaneously. This issue will be further discussed in the chapter on pharmacological interventions. Evodiamine, a nutraceutical, is an inhibitor of Src and has anti-migratory and anti-invasive effects.

Andrikopoulus et al.[42] have proved that VGSCs have pro-angiogenic functions by significantly increasing VEGF signaling in endothelial cells. Endothelial cells express NaV1.5 and NaV1.7. TTX block or NaV1.5 RNAi decreased endothelial cell proliferation and tubular differentiation that are essential steps in the angiogenesis process.

Different functions of α subunit and β subunit in relation to cancer

α subunit:

(a) increases migration and invasion;
(b) increases angiogenesis; and
(c) participates in matrix remodeling at the invadopodia level.

β1 subunit:

(a) increases adhesion[43];
(b) promotes invadopodia[44]; and
(c) increases tumor growth and metastasis.[44]

VGSC β subunit and cancer

β subunits are active participants in the cancer process playing roles as signaling molecules, and participating in cell adhesion and cell migration.[45] In prostate cancer, tumors with high metastatic potential had β subunit expression more than fourfold higher than those with low metastatic potential. The β1 subunit, which is under androgenic control, was the more significantly increased.[46]

Association of ion channel regulators

As will be repeatedly explained throughout this book, the association of different inhibitors targeting diverse mechanisms of Na^+ and/or H^+ trafficking proteins produces better results in cancer treatment than just individual attacks. However, an interesting idea was proposed by Gould et al.,[47] i.e., that anticancer results were better when stimulating one channel and inhibiting another. They inhibited Na^+ and K^+ pumps with ouabain while stimulating VGSCs with sub-lethal electric currents in MDA-MB231 breast cancer xenografts. This produced an inward flux of Na^+ that could not leave the cell leading to cellular swelling within seconds and posterior lysis. Other cell lines that did not express VGSCs were unaffected by the procedure. The same experiment was successful in other tumors like prostate, colon and lung carcinomas. Treated animals showed longer survival and slower tumor growth.

Clinical implications and conclusions

The important implications of VGSCs in cancer progression and invasion led Litan and Langhans[48] to postulate that cancer is a channelopathy.

Many malignant tumors overexpress VGSCs, particularly at the level of invadopodia.

NaV1.5 embryonic isoform is the most frequently found in cancer, but other isoforms can also be found.

The participation of VGSCs in cancer is many fold, such as:

(1) Stimulatory/inducer effects on NHE1.
(2) Promoter of invadopodia formation.
(3) Increased Na^+ augments osmotic pressure inside the cell, activating aquaporins that allow water entrance.
(4) Migration is facilitated through 1 and 2.
(5) Interrelation with integrins through the β subunits also increases adhesion mechanisms that intervene in migration.
(6) Neoangiogenesis.

Understanding the pro-tumoral mechanisms of VGSCs opens the possibility to associating drugs in a synergistic manner. For example the triple inhibition of VGSCs, NHE1 and Src should be an effective downregulator of invasion.

There are many available drugs that inhibit VGSCs. Among them, topiramate would be the best choice because it has very low interaction with other drugs and at the same time it is a carbonic anhydrase inhibitor acidifying the cytoplasm. The inhibition of NHE1 with an existing drug like amiloride, although weak at clinically achievable concentrations, should contribute to VGSCs blocking. Finally evodiamine has shown Src inhibiting effects.

The triple association of topiramate, amiloride and evodiamine at usual doses is a non-toxic combination that deserves to be tested in order to achieve a decrease in the migration/invasion process. This subject will be further discussed in the chapter on pharmacological interventions.

An anti-NaV1.5 antibody that showed anticancer properties in cells with high expression of this channel has been developed recently.[49] When tested in vitro and in vivo, this antibody reduced mitotic activity, motility, and invasiveness.

References

1. Catterall WA. Voltage gated sodium channels at 60: structure, function and pathophysiology. *J Physiol* 2012;**590**(11):2577–89.
2. Termin A, Martinborough E, Wilson D. Recent advances in voltage-gated sodium channel blockers: therapeutic potential as drug targets in the CNS. *Annu Rep Med Chem* 2008;**43**:43–60 [Chapter 3].
3. Brackenbury WJ, Chioni AM, Diss JK, Djamgoz MB. The neonatal splice variant of Nav1.5 potentiates in vitro invasive behaviour of MDA-MB-231 human breast cancer cells. *Breast Cancer Res Treat* 2007;**101**(2):149–60.
4. Brackenbury WJ. Voltage gated sodium channels and metastatic disease. *Channels* 2012;**6**(5): 352–61.
5. Ahern CA, Payandeh J, Bosmans F, Chanda B. The hitchhiker's guide to the voltage-gated sodium channel galaxy. *J Gen Physiol* 2016;**147**:1–24.

6. Lee CH, Ruben PC. Interaction between voltage-gated sodium channels and the neurotoxin, tetrodotoxin. *Channels (Austin)* 2008;**2**(6):407–12.
7. Catterall WA. Structure and function of voltage-gated sodium channels at 13 atomic resolution. *Exp Physiol* 2014;**99**:35–51.
8. Black JA, Waxman SG. Noncanonical roles of voltage-gated sodium channels. *Neuron* 2013;**80**: 280–91.
9. Black JA, Liu S, Waxman SG. Sodium channel activity modulates multiple functions in microglia. *Glia* 2009;**57**:1072–81.
10. Kis-Toth K, Hajdu P, Bacskai I, Szilagyi O, Papp F, Szanto A, et al. Voltage-gated sodium channel Nav1.7 maintains the membrane potential and regulates the activation and chemokine-induced migration of a monocyte-derived dendritic cell subset. *J Immunol* 2011;**187**:1273–80.
11. Fraser SP, Diss JKJ, Chioni AM, Mycielska ME, Pan H, Yamaci RF, et al. Voltage-gated sodium channel expression and potentiation of human breast cancer metastasis. *Clin Cancer Res* 2005;**11**: 5381–9.
12. Roger S, Rollin J, Barascu A, Besson P, Raynal P, Iochmann S. Voltage-gated sodium channels potentiate the invasive capacities of human non-small-cell lung cancer cell lines. *Int J Biochem Cell Biol* 2007;**39**(4):774–86.
13. Blandino JKW, Viglione MP, Bradley WA, Oie HK, Kim YI. Voltage-dependent sodium channels in human small-cell lung cancer cells: role in action potentials and inhibition by Lambert-Eaton syndrome IgG. *J Membr Biol* 1995;**143**:153–63.
14. Diaz D, Delgadillo DM, Hernandez Gallegos E, Hinojosa LM, Ortiz CS, Berumen J, et al. Functional expression of voltage-gated sodium channels in primary cultures of human cervical cancer. *J Cell Physiol* 2007;**210**(2):469–78.
15. Bennett ES, Smith BA, Harper JM. Voltage gated sodium channels confer invasive properties on human prostate cancer cells. *Plugers Arch* 2004;**447**:908–14.
16. Gao R, Shen Y, Cai J, Lei M, Wang Z. Expression of voltage-gated sodium channel α subunit in human ovarian cancer. *Oncol Rep* 2010;**23**:1293–9.
17. House CD, Vaske CJ, Schwartz AM, Obias V, Frank B, Luu T, et al. Voltage-gated Na^+ channel SCN5A is a key regulator of a gene transcriptional network that controls colon cancer invasion. *Cancer Res* 2010;**70**:6957.
18. Fraser SP, Diss JKJ, Lloyd LJ, Pani F, Chioni AM, George AJ, et al. T-lymphocyte invasiveness: control by voltage gated Na^+ channel activity. *FEBS Lett* 2004;**569**:191–4.
19. Sato K, Ishizuka J, Cooper CW, Chung DH, Tsuchiya T, Uchida T, et al. Inhibitory effect of calcium channel blockers on growth of pancreatic cancer cells. *Pancreas* 1994;**9**(2):193–202.
20. Fulgenzi G, Graciotti L, Faronato M, Soldovieri MV, Miceli F, Amoroso S, et al. Human neoplastic mesothelial cells express voltage-gated sodium channels involved in cell motility. *Int J Biochem Cell Biol* 2006;**38**(7):1146–59.
21. Hernandez-Plata E, Ortiz CS, Marquina-Castillo B, Medina-Martinez I, Alfaro A, Berumen J, et al. Overexpression of Nav1.6 channels is associated with the invasion capacity of human cervical cancer. *Int J Cancer* 2012;**130**(9):2013–23.
22. Aktas HG, Akgun T. Naringenin inhibits prostate cancer metastasis by blocking voltage-gated sodium channels. *Biomed Pharmacother* 2018;**106**:770–5.
23. Nelson M, Yang M, Millican-Slater R, Brackenbury WJ. Nav1.5 regulates breast tumor growth and metastatic dissemination in vivo. *Oncotarget* 2015;**6**(32):32914–29.
24. Schrey M, Codina C, Kraft R, Beetz C, Kalff R, Wolfl S, et al. Molecular characterization of voltage-gated sodium channels in human gliomas. *Neuroreport* 2002;**13**:2493–8.
25. Grimes JA, Fraser SP, Stephens GJ, Downing JE, Laniado ME, Foster CS, et al. Differential expression of voltage activated Na^+ currents in two prostatic tumour cell lines: contribution to invasiveness in vitro. *FEBS Lett* 1995;**369**:290–4.
26. Laniado ME, Lalani EN, Fraser SP, Grimes JA, Bhangal G, Djamgoz MB, et al. Expression and functional analysis of voltage-activated Na^+ channels in human prostate cancer cell lines and their contribution to invasion in vitro. *Am J Pathol* 1997;**150**(4):1213–21.

27. Smith P, Rhodes NP, Shortland AP, Fraser SP, Djamgoz MB, Ke Y, et al. Sodium channel protein expression enhances the invasiveness of rat and human prostate cancer cells. *FEBS Lett* 1998;**423**: 19–24.
28. Fraser SP, Salvador V, Manning EA, Mizal J, Altun S, Raza M, et al. Contribution of functional voltage-gated Na^+ channel expression to cell behaviors involved in the metastatic cascade in rat prostate cancer: I. Lateral motility. *J Cell Physiol* 2003;**195**(3):479–87.
29. Roger S, Besson P, Le Guennec JY. Involvement of a novel fast inward sodium current in the invasion capacity of a breast cancer cell line. *Biochim Biophys Acta* 2003;**1616**:107–11.
30. Batcioglu K, Uyumlu AB, Satilmis B, Yildirim B, Yucel N, Demirtas H, et al. Oxidative stress in the in vivo DMBA rat model of breast cancer: suppression by a voltage-gated sodium channel inhibitor (RS100642). *Basic Clin Pharmacol Toxicol* 2012;**111**(2):137–41.
31. Brisson L, Driffort V, Benoist L, Poet M, Counillon L, Antelmi E, et al. NaV1.5 Na^+ channels allosterically regulate the NHE-1 exchanger and promote the activity of breast cancer cell invadopodia. *J Cell Sci* 2013;**126**:4835–42.
32. Gillet L, Roger S, Besson P, Lecaille F, Jacques GO, Bougnoux P, et al. Voltage gated sodium channel activity promotes cysteine cathepsin-dependent invasiveness and colony growth of human cancer cells. *J Biol Chem* 2009;**284**:8680–91.
33. Mader CC, Oser M, Magalhaes MA, Bravo-Cordero JJ, Condeelis J, Koleske AJ, et al. An EGFR-Src-Arg-Cortactin pathway mediates functional maturation of invadopodia and breast cancer cell invasion. *Cancer Res* 2011;**71**(5):1730–41.
34. Clark ES, Whigham AS, Yarbrough WG, Weaver AM. Cortactin is an essential regulator of matrix metalloproteinase secretion and extracellular matrix degradation in invadopodia. *Cancer Res* 2007;**67**(9):4227–35.
35. Murphy D, Courtneidge M. The "ins" and "outs" of podosomes and invadopodia: characteristics, formation and function. *Nat Rev Mol Cell Biol* 2011;**12**:413–26.
36. Cosen-Binker LI, Andras K. Cortactin: the gray eminence of the cytoskeleton. *Physiology* 2006;**21**:352–61.
37. Tarone G, Cirillo D, Giancotti FG, Comoglio PM, Marchisio PC. Rous sarcoma virus-transformed fibroblasts adhere primarily at discrete protrusions of the ventral membrane called podosomes. *Exp Cell Res* 1985;**159**(1):141–57.
38. Berdeaux RL, Diaz B, Kim L, Martin S. Active Rho is localized to podosomes induced by oncogenic Src and is required for their assembly and function. *J Cell Biol* 2004;**166**(3):317–23.
39. Dulong C, Fang YJ, Gest C, Zhou MH, Patte-Mensah C, Mensah-Nyagan AG, et al. The small GTPase RhoA regulates the expression and function of the sodium channel Nav1.5 in breast cancer cells. *Int J Oncol* 2014;**44**:539–47.
40. Timpson P, McGhee EJ, Morton JP, von Kriegsheim A, Schwarz JP, Karim SA, et al. Spatial regulation of RhoA activity during pancreatic cancer cell invasion driven by mutant p53. *Cancer Res* 2011;**71**(3):747–57.
41. Onganer PU, Djamgoz MB. Small-cell lung cancer (human): potentiation of endocytic membrane activity by voltage-gated Na(+) channel expression in vitro. *J Membr Biol* 2005;**204**(2):67–75.
42. Andrikopoulos P, Fraser SP, Patterson L, Ahmad Z, Burcu H, Ottaviani D, et al. Angiogenic functions of voltage gated Na^+ channels in human endothelial cells. Modulation of vascular endothelial growth factor (VEGF) signalling. *J Biol Chem* 2011;**286**(19):16846–60.
43. Chioni AM, Brackenbury WJ, Calhoun JD, Isom L, Djamgoz MBA. A novel adhesion molecule in human breast cancer cells: voltage gated Na^+ channel β1 subunit. *Int J Biochem Cell Biol* 2009; **41**(5):1216–27.
44. Nelson M, Millican-Slater R, Forrest LC, Brackenbury WJ. The sodium channel β1 subunit mediates outgrowth of neurite-like processes on breast cancer cells and promotes tumor growth and metastasis. *Int J Cancer* 2014;**135**(10):2338–51.
45. Bouza AA, Isom LL. Voltage-gated sodium channel β subunits and their related diseases. In: Chahine M, editor. *Voltage gated sodium channels: structure, function and channelopaties. Handbook of experimental pharmacology*. vol. 246. Cham: Springer; 2017. p. 423–50.

46. Diss JK, Fraser SP, Walker MM, Patel A, Latchman DS, Djamgoz MB. Beta-subunits of voltage-gated sodium channels in human prostate cancer: quantitative in vitro and in vivo analyses of mRNA expression. *Prostate Cancer Prostatic Dis* 2008;**11**:325–33.
47. Gould III HJ, Norleans J, Ward TD, Reid C, Paul D. Selective lysis of breast carcinomas by simultaneous stimulation of sodium channels and blockade of sodium pumps. *Oncotarget* 2018;**9**(21).
48. Litan A, Langhans SA. Cancer as a channelopaty: ion channels and pumps in tumor development and progression. *Front Cell Neurosci* 2015;**9**:86.
49. Gao R, Cao T, Chen H, Cai J, Lei M, Wang Z. *Nav1.5-E3 antibody inhibits cancer progression. Transl Cancer Res* 2019. Downloaded from:http://tcr.amegroups.com/article/view/26422.

CHAPTER 7

Carbonic anhydrases

Introduction

Malignant tissues have a very interesting characteristic: they modify their own micro-environmental parameters. And they do so with a very high flexibility and adaptation to permanently changing conditions. pH, pO_2, and interstitial fluid pressure have additive influence on tumor growth, proliferation, migration, and invasion, and enhance cell survival of the more fit in a Darwinian selection fashion.[1] Intracellular and extracellular pH depend strongly on ion channels, lactate transporters and enzymes that create extracellular acidity and intracellular alkalinity, inverting the normal pH gradient. Carbonic anhydrases are a group of enzymes with the ability to modify intra and extracellular pH and therefore influence proliferation, migration and invasion, and contribute significantly to the pH gradient inversion.

The family of carbonic anhydrases (CAs), also known as carbonic dehydratases, are a group of metallo-enzymes known since 1933[2] that catalyze in a reversible form the conversion of carbon dioxide and water to bicarbonate and a proton (Fig. 1).

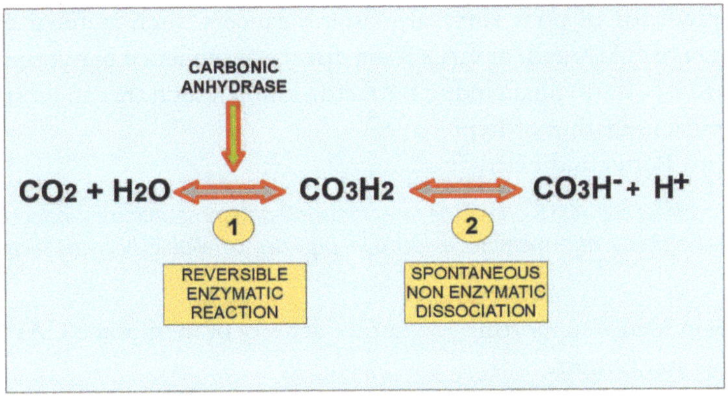

Fig. 1 Carbonic anhydrase catalyzes step 1 in a reversible manner. Step 2, the dissociation of carbonic acid into bicarbonate and a proton, is a spontaneous phenomenon.

An Innovative Approach to Understanding and Treating Cancerr: Targeting pH
https://doi.org/10.1016/B978-0-12-819059-3.00007-1
© 2020 Elsevier Inc.
All rights reserved.

This chemical reaction is essential for the life of prokaryotes and eukaryotes. Hence, they are widely distributed and can be found in mammals, plants and bacteria. These enzymes participate indirectly in ion transportation, and directly in CO_2 exchange and acid-base regulation in the animal kingdom but also in CO_2 fixation and photosynthesis in plants.

In spite of the fact that there are 16 known isoforms of CAs (including some without enzymatic activity),[3] the rest of our description will be limited mainly to membrane isoforms CA-IX and CA-XII due to their importance in cancer and in the pH gradient inversion concept.[4]

Carbonic anhydrases can be:

1. cytosolic such as CA-I, CA-II, CA III and CA-VII;
2. located in cell membranes like CA-IV, CA-IX, CA-XII, and CA-XIV;
3. mitochondrial: CA-V;
4. secreted in saliva: CA-VI.
5. There are also CAs without enzymatic activity.

As we have explained in previous chapters, alkaline intracellular pH (pHi) and acidic extracellular pH (pHe) are metabolic hallmarks of cancer. The acidic tumor microenvironment has a deep impact on cancer development, promoting ECM degradation, migration, invasion and metastasis. Lactic acid was considered the main acidifying factor of the peri-tumor environment, but further research has cast doubts on the real responsibility of lactates in this issue. Nowadays we know that CO_2 and H^+ export are very important sources of ECM acidification. To achieve this, the activity of membrane CAs and NHE1 play capital roles.

CA-IX is a membrane CA; its overexpression induced and regulated by hypoxia, is an independent predictor of poor survival in some cancers, such as renal carcinoma.[5, 6] Over-expression of CAIX and/or CAXII is a direct consequence of hypoxia and its transcription factor HIF-1α (hypoxia inducible factor 1alpha), such that an increased CAIX is considered a reliable marker of hypoxia.[7, 8]

According to Potter and Harris[9]:

> *"Carbonic anhydrase IX (CAIX) is one of the most inducible and most uniformly induced genes and because of its stability and membrane location provides a reliable histochemical marker of hypoxia."*

There is strong evidence supporting pro-tumor activity of membrane CAs IX and XII in different organs and tumors.

However CAIX is also strongly dependent on another transcription factor that "cooperates" with HIF-1α: Sp1[10] (Fig. 2).

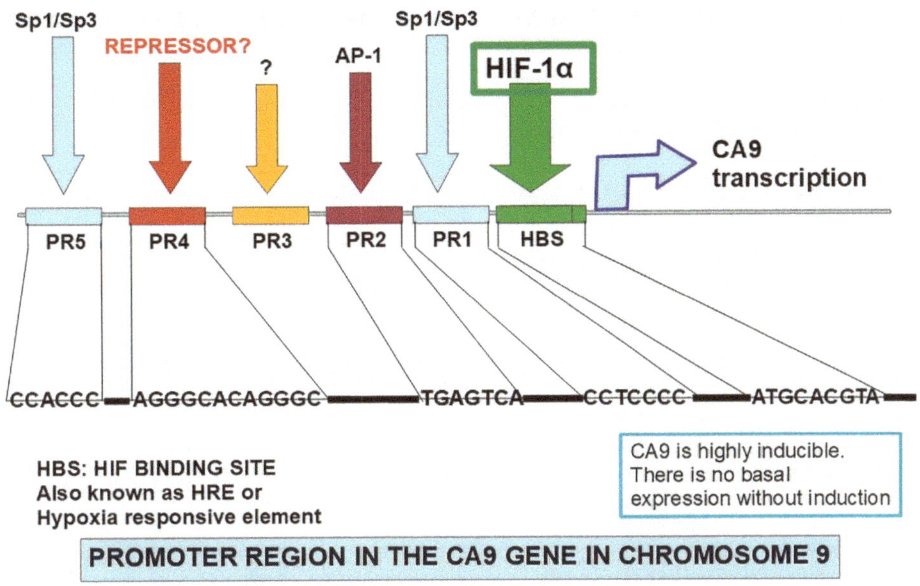

Fig. 2 Promoter region of the CAIX gene that shows this enzyme's transcription control mechanisms. It is under the control of the transcription factor HIF (hypoxia inducible factor), The AP-1 transcription factor and the Sp1 transcription factor that act as co-regulators of HIF. There are two other sites whose binding protein has not been identified: PR4 and PR3.

CAIX is increased in the plasma of patients with renal cell carcinoma, and it has been suggested that it may be a good diagnostic marker for the disease.[11]

Inhibiting membrane carbonic anhydrases became a goal in order to curb growth and development of malignancies.[3, 12] In the experimental environment the inhibition of CAs has shown encouraging results in osteosarcoma,[13] glioblastoma,[14] breast,[15] cervical,[16] prostate,[17, 18] lung,[19, 20] colon, renal[21] cancers, T-cell lymphoma[22] and many other cancers.

Furthermore, the disruption of NHE1 and CAIX showed that NHE1 knockdown inhibited growth of colon cancer cells whether in normoxia or hypoxia while knockdown of CAIX did so mainly in severe hypoxic conditions.[23] In the case of NHE1 and CAIX knockdown CAXII was strongly induced with a potential to compensate for the loss of the two regulators. This finding brings us to a central concept in the treatment of the pH inversion gradient: it is not enough to inhibit one or two mechanisms in order to achieve results. The aim should be to inhibit as many mechanisms as possible.

Inhibition of CAs sensitizes cancer cells to conventional chemotherapies.[24, 25]

The CAIX gene

In the 1990s Silvia Pastorekova and her associates:

1. found that there was a membrane protein with carbonic anhydrase activity[26];
2. and this protein was overexpressed in cancer cells[27];

3. and they finally isolated and sequenced the gene responsible for its codification[28]: the CAIX gene.

The chemical reaction catalyzed by CAs

There is strong evidence supporting pro-tumor activity of membrane CAs IX and XII in different organs and tumors. In addition to the membrane CAIX there is a soluble secreted form of the enzyme that can be found in serum (sCAIX) (Fig. 3).

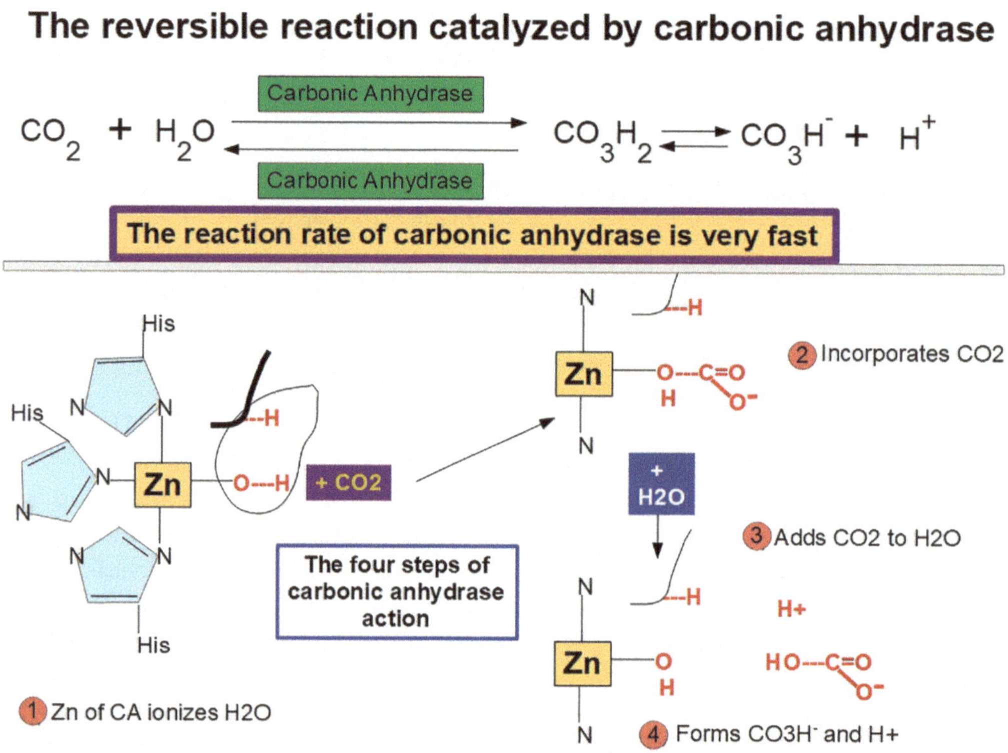

Fig. 3 Mechanism of action of carbonic anhydrase (CA). This enzymes have a Zn nucleus bound to three molecules of histidine. Zn has the ability to ionize a molecule of water (1). Then it accepts a molecule of CO_2 (2) and with a new H_2O molecule produces the final product, which is one molecule of bicarbonate plus a proton (3) and (4). The reaction is fully reversible, but the reverse reaction is much slower.

Structure of membrane carbonic anhydrases

These two CAs (IX and XII) show characteristics that differ from the other members of the CAs family (Fig. 4):

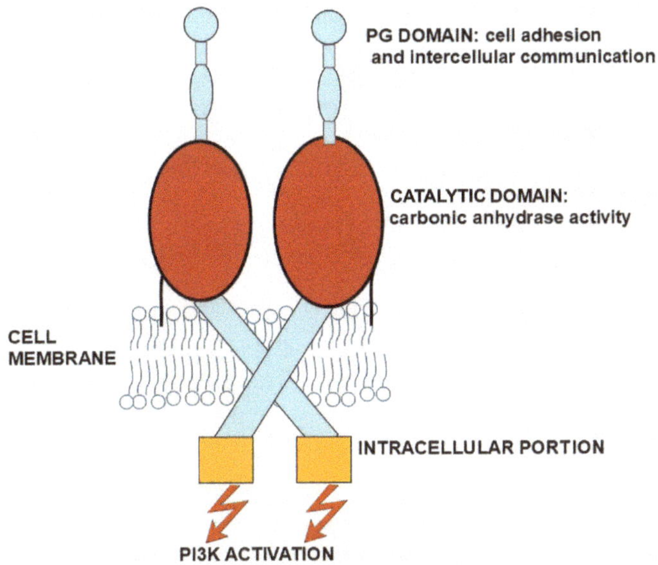

Fig. 4 Carbonic anhydrase IX dimer structure at the cell membrane. N-terminal proteoglycan-like (PG) domain, a CA catalytic domain, a transmembrane segment *(light blue)*, and an intracellular portion. The CA-IX protein is found as a dimer bound by a disulfide bond between the same Cys residue located on two CA catalytic domains. For further details read Alterio et al.[29]

They are transmembrane proteins with an intracytoplasmic domain, a transmembrane domain and an extracellular catalytic domain. The extracellular catalytic domain is very similar to the catalytic domain of the other members of the family. CAIX also has a proteoglycan domain which intervenes in tumorigenesis.[30] In hypoxic cells, CAIX can be located in the cell's nucleus in certain moments of its life and is involved in gene transcription.[31]

Hilvo et al.[32] described these transmembrane CAs as follows:

> *"CA-IX and XII are tumor-related members of the CA family, and their expression is induced by hypoxia. Of these two, CA-IX is the more exceptional because it has been proposed to contribute to cell proliferation, cell adhesion, and malignant cell invasion. It is generally expressed in a very limited number of normal tissues (mainly in the gastrointestinal tract), whereas it is highly expressed in several cancers that arise from CA-IX-negative tissues, such as renal, lung, cervical, ovarian, esophageal, and breast carcinomas"*

CAIX is the most active enzyme for CO_2 hydration.

HIF-1α is the main transcriptional regulator of CAIX, which means that it is a hypoxia responsive protein.[33]

CA mediated mechanism of the pH inversion

Membrane CAs play a central role in the inversion of the pH cancer gradient through the mechanism described in Fig. 5. Along with CAIX and CAXII there is another CA of

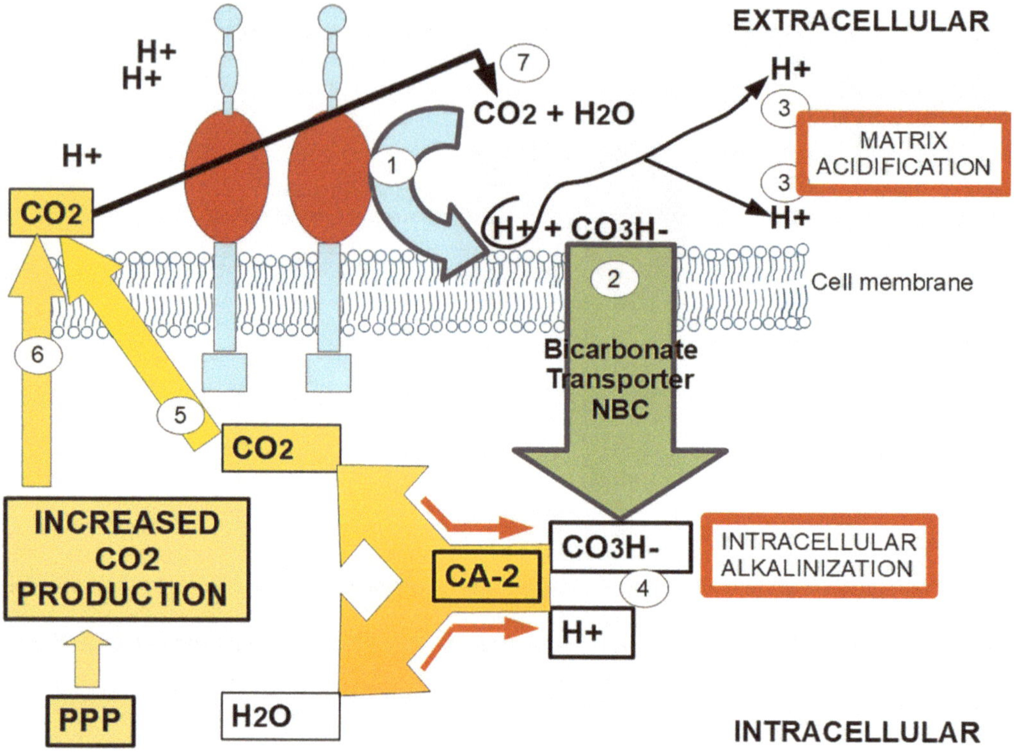

Fig. 5 Mechanism of the inversion of the pH gradient by CAs.[34] (1)There is an excess of CO_2 in the extracellular matrix. CAs facilitate the hydration of CO_2 to produce CO3H2. This carbonic acid dissociates spontaneously. in CO_3H^- and H^+. (2) A bicarbonate transporter, the sodium bicarbonate co-transporter or NBC, imports bicarbonate into the cell contributing to intracellular alkalinization while. (3) The H^+ remains in the matrix contributing to its acidification. (4) The bicarbonate that has been internalized "captures" cytoplasmic protons (H^+) to produce carbonic acid, that dissociates into CO_2 and H_2O through CAII's activity. (5) CO_2 is extruded to the matrix through diffusion and aquaporins. (6) There is also excessive production of CO_2 through the Pentose Phosphate Pathway (PPP) that also diffuses to the matrix. (7) The CO_2 is again processed by the membrane CAs thus repeating the cycle. All the carbonic anhydrases described here are reversible.

interest in cancer: CAII (or CA-2). Fig. 5 shows how CAII participates in the chain of events that leads to the pH paradigm. There is also another mechanism by which CAII has a role in cancer: it increases lactate movement through CAII's association with MCTs (monocarboxylate transporters). CAII facilitates proton transfer between CAII and MCTs increasing transport activity.[35]

The pro-tumor activities of CAs are not limited to CAIX and CAXII as can be seen in the above mentioned example. CAII and CAIV also participate in tumor development. The pro-tumor effects of CAIX and CAXII are not limited to the inversion of the pH gradient. CAIX also plays a role in cell adhesion.[36]

Membrane carbonic anhydrases and cancer

CO_2 is the main oxidative product of cell metabolism and plays an important role in the acid-base balance of the cell.[37, 38] Therefore, the mechanisms involved in handling CO_2 are essential for maintaining cell homeostasis. CAs are responsible for this trafficking.

CAIX was originally found in HeLa cells (as a protein that was initially called MN but shortly afterwards it was identified as a CA), and its gene was identified by the same research team. Four years later Tureci et al. discovered CAXII in a renal cell cancer.[39] These two CAs were associated with malignant cells from the beginning; this association in different tumors will be discussed below.

Breast cancer

Schütze et al.[40]: Higher CAIX expression was associated with negative hormone receptors, positive nodes and shorter disease free survival.

Hussain et al.[41]: Expression of CAIX is an independent predictor of survival, with worse prognosis in the higher expression group of patients.

Chia et al.[42]: 48% of a cohort of women with breast cancer showed CAIX overexpression. This group was associated with worse prognosis, negative hormone receptors and tumor necrosis.

Barnett et al.[43]: found an estrogen responsive enhancer region (ERAR) at 6 Kb distance from the transcription initiation of CAXII. This enhancer is activated by the binding of estrogen to the estrogen receptor alpha (ERα) and it recruits RNA polymerase and steroid receptor coactivators to the promoter region of the gene. This binding starts CAXII gene transcription (Fig. 6).

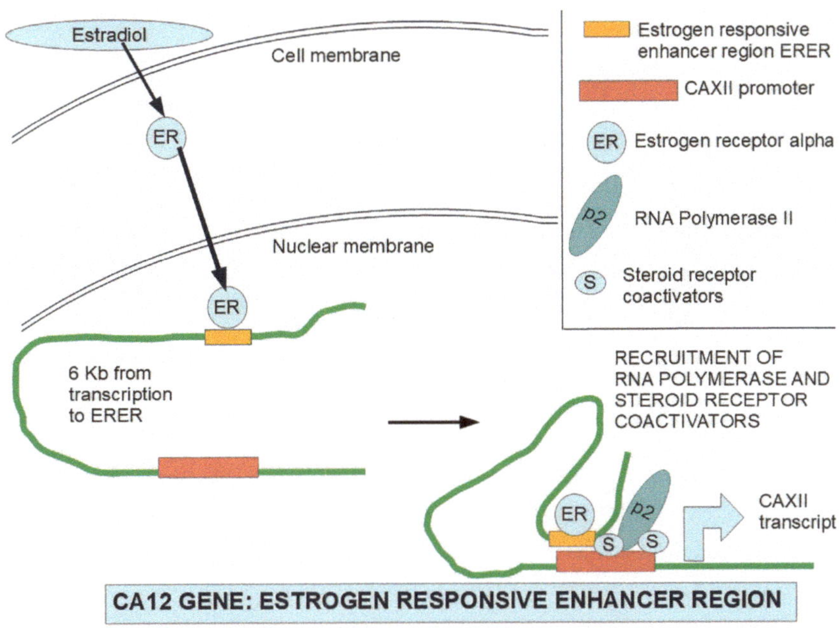

Fig. 6 Estrogenic hormones up-regulate CAXII transcription.

Span et al.[44] in 253 cases of breast cancer did not find correlation between CAIX expression and relapse free survival. However they reported some interesting conclusions:

1. CAIX over-expression was usually found in high grade steroid receptor negative tumors.
2. Patients with a high level of CAIX benefited less from adjuvant therapy.

This last conclusion seems quite logical because a high level of membrane CAs is strongly related to hypoxia, and hypoxia is a key factor in therapeutic failure.

On the other hand, the study by Generali et al.[45] differs fundamentally from that by Span et al. They found that CAIX immunostaining correlated with poor disease free survival. However, both studies concluded that positive CAIX is a marker of poor response to adjuvant therapy.

Li et al.[46] described CAIX activity in MDA-MB-231 breast cancer cells. They found that CAIX increased its activity with hypoxia and reduced pHe, as usually found in cancer.

In patients with circulating cells of metastatic breast cancer, Müller et al.[47] found that 35% had a significant increase of serum CAIX and these patients had a shorter progression free survival (PFS). Lock et al.[48] found that targeting CAIX depleted cancer stem cells in hypoxic niches in breast cancer. The authors considered this enzyme as an important mediator in stemness of cancer cells in hypoxic niches (Table 1).

Table 1 A summary of CAs role in different cancers

Tumor	Reference and findings
B-cell lymphoma	Chen et al.[49]: Lymphomas had a moderately increased CAIX expression which was correlated with pHe in xenograft tumor models
Colorectal cancer	Saarnio et al.[50]: CA IX is expressed in areas with high proliferation. It is coincidental with areas with high Ki-67 expression
	Kummola et al.[51]: CAI, CAII and CAXIII are downregulated in neoplastic tissues. They suggested that these three CAs may have tumor suppressor activity
Pancreatic cancer	Kivela et al.[52]: CAIX was overexpressed in hyperplastic ductal epithelium, but CAXII was poorly expressed
Head and neck squamous cell carcinoma	Beasley et al.[53]: Hypoxia was a strong inducer of CAIX in all the HNSCC cell lines examined. It was highly expressed in the peri-necrotic areas. The immunostaining of CAIX started at a 80 μm distance from vessels
Oral squamous cell carcinoma	Yang et al.[54]: CAIX regulated migration and invasion in oral squamous cell carcinoma
Non small cell lung carcinoma (NSCLC)	Giatromanolaki et al.[55]: 107 cases were examined and 36.5% had an increased expression of CAIX, particularly around necrotic areas. The expression of CAIX was linked to poor outcome in NSCLC due to increased angiogenesis, apoptosis inhibition, and cell-cell adhesion disruption

Table 1 A summary of CAs role in different cancers—cont'd

Tumor	Reference and findings
Small cell lung cancer. (SCLC)	Bryant et al.[56]: mice with SCLC xenograft showed that administering S4, a CAIX inhibitor, alone or in combination with cisplatin, reduced tumor growth. The combination therapy should develop a synergy between the drugs
Renal cancer	Atkins et al.[57]: CAIX is a predictor of outcome where high CAIX expression has a better prognosis
	Genega et al.[58]: CAIX is usually expressed in clear cell renal carcinoma and is associated with the disease's grade
Prostate carcinoma	Fiaschi et al.[59]: CAIX of tumor associated fibroblasts was a driver of epithelial mesenchymal transition
Nasopharyngeal carcinoma	Hui et al.[60]: Co-expression of HIF-1α, CA IX, and VEGF was found in nasopharyngeal carcinoma, and is associated with poor prognosis
Cervical cancer	Olive et al.[61]: CAIX was a reliable marker of ischemia.
	Liao et al.[62]: 21% of patients with early stage cervical cancer showed overexpression of CAIX. It was associated with tumor size and worse overall survival
Uterine cancer	Hynninen et al.[63]: CAIX and CAII were overexpressed in uterine adenocarcinomas. Normal endometrium showed higher CAXII expression
Vulvar cancer	Choschzick et al.[64]: Vulvar cancer showed high expression of CAIX which was associated with worse prognosis and regional metastasis
Gastric cancer	Chen et al.[65]: CAIX expression was usually high in normal gastric mucosa but this was lost in gastric cancers. Gastric cancers that showed persistent CAIX expression had a shorter survival
Mesothelioma	Ramsey et al.[66]: All cases of metastatic renal cell carcinoma and primary mesothelioma showed over-expression of CAIX
Glioblastoma (GBM)	Proescholdt et al.[67]: CAIX Overexpression was a marker of poor survival in GBM. CAIX knockdown improved the results of chemo and radiotherapy in vitro with less invasion and more apoptosis
	Said et al.[68]: CAIX Overexpression in GBM was related to hypoxia
Ovarian tumors	Hynninen et al.[69]: Overexpression of CAIX and CAXII was founding almost all ovarian malignant tumors
Bladder cancer	Klatte et al.[70]: 71% of bladder cancers had an overexpressed CAIX which was associated with poor survival. CAIX was a predictor of recurrence

Continued

Table 1 A summary of CAs role in different cancers—cont'd

Tumor	Reference and findings
Osteosarcoma	Perut et al.[71]: CAIX inhibitor 3, a new compound showed cytotoxicity on osteosarcoma cells under hypoxia and modulated extra and intracellular pH producing cell cycle arrest. In vivo it reduced tumor growth and produced necrosis. Andreucci, et al.[24]: The new CAIX inhibitor SLC-0111 potentiated the cytotoxicity of conventional chemotherapeutics like dacarbazine and temozolomide in melanoma treatment.
Melanoma	Puech et al.[72]: Acetazolamide and umbelliferone's effect on melanoma cells A375 were compared. Both reduced pHi and downregulated CAIX. Umbelliferone was slightly more effective. Both increased dacarbazine efficacy

In hematological malignancies (acute myeloid leukemia, acute lymphoblastic leukemia and chronic myelomonocytic leukemia) CAII was overexpressed frequently.[73] CAII was overexpressed in bone tumors (giant cells)[74] and osteoclasts[75] and CAIX in osteosarcoma cells.[76]

The conclusion is that almost all the malignant tumors can overexpress CAIX or CAXII and in some cases CAII. This overexpression is strongly associated with and even dependent on hypoxic conditions, such that CAIX is an hypoxia marker in certain tumor models.[77]

Clinical trials

The web page of Clinicaltrials.gov has only registered two studies for acetazolamide in cancer to be started sometime in the future:

NCT03467360: Inhibition of Carbonic Anhydrase in Combination With Platinum and Etoposide-based Radiochemotherapy in Patients With Localized Small Cell Lung Cancer.

NCT03011671: Phase I Study of Safety and Tolerability of Acetazolamide with Temozolomide.

Clinical and therapeutic implications

In 1940, Thaddeus Man and David Keilin found that the recently discovered sulfonamides accounted for CA inhibition if H molecules in the sulfonamide function were conserved.[78] These sulfonamides had no antibacterial action but were potent diuretics. After multiple changes in the original molecule, in 1952 acetazolamide, a sulfonamide and powerful inhibitor of CA was introduced in clinical practice as a diuretic. It is a FDA approved drug for the treatment of glaucoma and congestive heart failure. It is increasingly being used for the treatment of endocraneal hypertension of diverse etiologies and as an adjuvant treatment for certain cases of epilepsia. It is not approved for cancer treatment. However, it is known as a powerful pan-inhibitor of CAs with very low toxicity and low cost. No sound clinical trials have been performed as a cancer drug (Fig. 7).

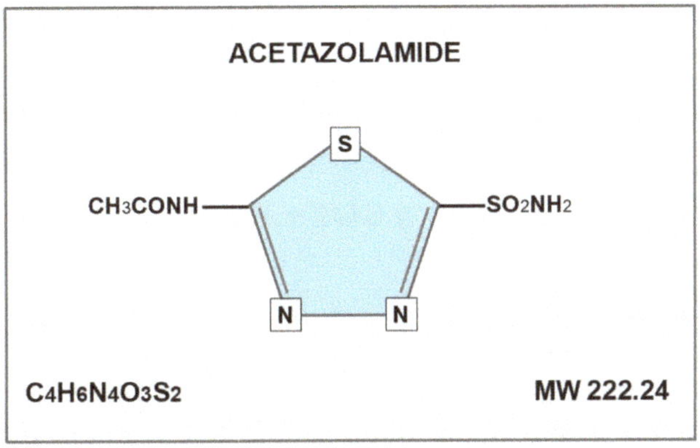

Fig. 7 Structural and synthetic formulas of acetazolamide. MW = molecular weight. Acetazolamide, an inhibitor of the enzyme carbonic anhydrase, is a white to yellowish crystalline, odorless powder, it is weakly acidic, very slightly soluble in water and slightly soluble in alcohol. Adverse effects very frequently include paresthesias; kidney stones are less frequent, and metabolic acidosis, lethargy, and appetite suppression may ensue at high doses. Blood dyscrasias can rarely be seen.

It should not be viewed as a stand alone pharmaceutical in the treatment of cancer, but it should be studied as part of a proton reversal scheme or as a complementary drug for usual mainstream treatments.

Since the anti-cancer effects of CA inhibition were discovered, hundreds of new compounds were tested seeking specific inhibition of CAIX and CAXII. Many of them have shown encouraging results in the laboratory, but none reached the stage of clinical trials. The model molecule is still acetazolamide.

A poorly studied property of acetazolamide is its vasodilator effect. Pickkers et al.[79] found that this drug activated KCa channels increasing blood flow in vivo. To what extent this action increases blood flow to tumors, reducing the hypoxic pro-tumor effects is an issue that remains to be investigated. Acetazolamide is also an inhibitor of aquaporin 1 and it reduces angiogenesis.[80, 81] However, the main anti-tumor activity of acetazolamide remains the CA inhibition and intracellular acidification.

Topiramate

Topiramate is a voltage gated sodium channel blocker approved by FDA for the treatment of epilepsy. It also inhibits intracellular CAII and CAIV[99] and can also decrease intracellular pH on its own.[100] Fig. 8 shows that acetazolamide and topiramate act on different parts of the pH inversion scheme. This should suggest that their blocking effects may be complementary and a synergistic action can be expected. However, this association was never tested in cancer (Fig. 8).

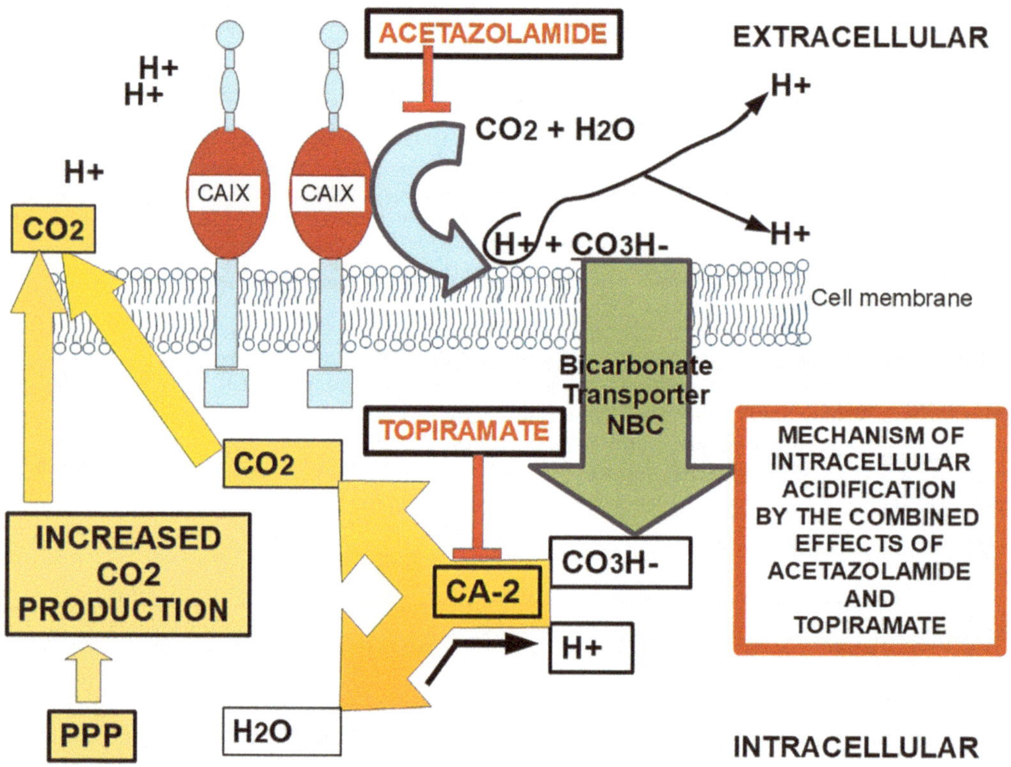

Fig. 8 Mechanism of intracellular acidification by the combined effects of acetazolamide and topiramate.

In the last 20 years, CA inhibition has emerged as a promising approach for pharmacologic intervention in cancer.[101] The evidence in Table 2 shows the reason why the design of CA inhibitors is such a dynamic field today. This ongoing research, has not yet developed a replacement for the safe, time-tested, useful, FDA approved acetazolamide which has not yet been approved for use in cancer.

Table 2 Evidence of anti-tumor activity of acetazolamide (Diamox)

Reference	Findings
Teicher et al.[82]	Acetazolamide used alone in a xerografted fibrosarcoma model produced a tumor growth delay of 2.3 days, and produced additive growth delays with CDDP and Melphalan
Yang et al.[83]	In Lewis lung carcinoma, acetazolamide suppressed metastasis
Parkkila et al.[21]	10 μM acetazolamide inhibited the relative invasion rate of 4 renal carcinoma cell lines between 18% and 74%
Ahlskog et al.[84]	Two acetazolamide derivatives with restricted binding to CA IX and XII, in vivo studies showed the preferential targeting of tumor cells to cause tumor retardation in a SK-RC-52 xenograft model of cancer
Mohktari et al.[85]	Combination of acetazolamide, and sulforaphane, reduced the viability and growth of bronchial carcinoid cell lines
Faes et al.[86]	Targeting the hypoxic tumor compartment by knockdown of CAIX using short hairpin RNA or by chemical inhibition with acetazolamide enhanced the anti-cancer activity of rapamycin
Cianchi et al.[87]	They evaluated the antitumor activity of three CA inhibitors: acetazolamide and two newly synthesized sulfonamides with high affinity for CAIX: TR1 and GA15. The effects of the three compounds on cell proliferation and apoptosis were evaluated in CA IX-positive HeLa and 786-O cells and CA IX-negative 786-O/von Hippel-Lindau (VHL) cells. They significantly reduced cell proliferation and induced apoptosis in HeLa and 786-O cells. Moreover, all three inhibitors decreased intracellular pH and increased ceramide production. The three inhibitors did not show any antitumor activity in CAIX-negative cells and failed to lower pHi or increase intracellular ceramide levels in the negative cells
Kopecka et al.[88]	Human colon cancer cells were explored for the expression of Pgp on the cell surface. CAXII was up-regulated on the surface of chemoresistant cells, similarly to Pgp. In MDR phenotype cells three things were increased: HIF-1α, Pgp and CAXII. CAXII and Pgp physically interacted on the cell surface. CAXII silencing or pharmacological inhibition with acetazolamide decreased the ATPase activity of Pgp by altering the optimal pH at which Pgp worked and promoted chemosensitization to Pgp substrates in MDR cells
Said et al.[89]	Acetazolamide and a derivative SU.D2 displayed inhibitory characteristics to hypoxia induced CAIX expression in four glioblastoma (GBM) cell lines. Parallel experiments using CAIX-siRNA confirmed these results. Application of 150–500 nM of the glycolysis inhibitor Chetomin under similar oxygenation conditions led to a sharply reduced expression of both CAIX protein and mRNA levels, indicating a clear glucose availability involvement for the hypoxic HIF-1α and CAIX expression in GBM cells

Continued

Table 2 Evidence of anti-tumor activity of acetazolamide (Diamox)—cont'd

Reference	Findings
Kasai et al.[90]	Acetazolamide potentiated the activity of Phatalidil- 5 Flurouracil. The potentiation was evaluated in terms of suppression of tumor growth and prolongation of the life-span of tumor-bearing animals
Mokhtari et al.[91]	Acetazolamide increased the anti-tumor potential of HDAC inhibitor in neuroblastoma
Li et al.[92]	Acetazolamide alone could sharply reduce the number of lung metastasis and of primary tumor growth, in a dose-dependent manner. And, it significantly inhibited CA activity in Lewis lung carcinoma. Acetazolamide suppression of tumor growth, the number of metastases and CA activity were not significantly altered when adding sodium bicarbonate. The mechanism of acetazolamide on anti-tumor and its anti-metastasis actions might partly involve either inhibiting the CA activity or reducing AQP1 water channel protein expression
Zhang et al.[93]	In rectal cancer CA-IX was overexpressed in 74.2% of patients, while in rectal adenoma it was only 16.7% and not at all in normal rectal tissue Acetazolamide significantly enhanced the ability of 5-fluorouracil at different concentrations to kill the rectal cancer cells in vitro
Shiverick et al.[94]	A low, therapeutically relevant dose of acetazolamide suppressed the growth of prostate xenograft tumors in mice. These data suggest that selective inhibition of CA enzymes could be a complement to prostate cancer treatment. Treatment with acetazolamide inhibited tumor cell proliferation, induced apoptotic cell death in tumor cells, and appeared to have differential effects on lymphatic vs vascular endothelial cells. Extracellular acidification was only partially blocked by acetazolamide
Islam et al.[95]	In bladder cancer, the combination of acetazolamide with sulforaphane induced dose-dependent suppression of growth and proliferation and induced apoptosis through caspase-3 and PARP activation. Both active p-Akt (Ser473) and p-S6 were significantly down-regulated by this treatment with a concomitant inhibition of Akt kinase activity. They also found down-regulated expression of carbonic anhydrase 9 (CAIX) concomitant with reductions in both E-cadherin, N-cadherin, and vimentin proteins mitigating the epithelial-to-mesenchymal transition (EMT)
Ivanova et al.[96]	Inhibition of CAIX reduced the self-renewal capacity of breast cancer cells, and the combination of doxorubicin and CAIX inhibition is an attractive therapeutic strategy in basal-like and triple-negative breast cancer
Bin et al.[97]	In a colon cancer xenograft model acetazolamide suppressed the xenograft tumor growth by inhibiting the expression of AQP-1 (acquaporin 1)
Noma et al.[98]	Acetazolamide treatment reduced the number of intestinal polyps up to 50% compared to normal controls in a Min mice Apc mutant model, probably by activating Nrf2

But as Shiverick et al. established, while acetazolamide decreases extracellular acidity, and also modifies pHi, it cannot by itself invert the cancer pH paradigm, because there are other proton extruders such as NHE1, VGSCs, V-ATPase proton pumps, and sodium bicarbonate co-transporters. For a successful pharmacologic intervention it will be necessary to inhibit as many proton extruders or bicarbonate importers and lactate transporters as possible.

Celecoxib

Celecoxib is a selective COX2 inhibitor that has been approved by the FDA for the treatment of osteoarthrosis. Weber et al.[102] in 2004 found that, in addition to inhibiting COX2, it is a nanomolar inhibitor of CAI, CAII, CAIV and CAIX.

Conclusions

CAIX and CAXII are essential participants in the pH inversion process found in cancer. The overexpression of CAIX is strongly associated with hypoxia and is an independent marker of bad prognosis. Downregulation of these CA isoenzymes improves standard cancer treatments and is part of the reversion of the abnormal cancer pH paradigm. The ideal CA inhibitor has not been found yet, but in the meantime there are two drugs that can be repurposed for this inhibition: acetazolamide and topiramate. Furthermore, there are molecular fundamentals that hint towards a synergy between them. The search for an isoform-specific inhibitor continues. However, when one isoform is pharmacologically inhibited, another isoform increases its expression and/or activity. Thus, a pan-inhibitor like acetazolamide or some of its newer derivatives should remain valuable tools.

CA inhibition can be advantageously used together with standard conventional treatments. Unfortunately, and in spite of all the evidence gathered favoring CA inhibition in cancer, there are no clinical trials actively going on.

References

1. Matsubara T, DiResta GR, Kakunaga S, Li D, Healey JH. Additive influence of extracellular pH, oxygen tension, and pressure on invasiveness and survival of human osteosarcoma cells. *Front Oncol* 2013;**3**:199.
2. Stadie WC, O'Brien H. The catalytic hydration of carbon dioxide and dehydration of carbonic acid by enzyme isolated from red blood cells. *J Biol Chem* 1933;**103**:521–9.
3. Supuran CT, Scozzafava A, Casini A. Carbonic anhydrase inhibitors. *Med Res Rev* 2003;**23**(2):146–89.
4. Swietach P, Wigfield S, Cobden P, Supuran CT, Harris AL, Vaughan-Jones RD. Tumor-associated carbonic anhydrase 9 spatially coordinates intracellular pH in three-dimensional multicellular growths. *J Biol Chem* 2008;**283**:20473–83.

5. Bui MH, Seligson D, Han KR, Pantuck AJ, Dorey FJ, Huang Y, et al. Carbonic anhydrase IX is an independent predictor of survival in advanced renal clear cell carcinoma: implications for prognosis and therapy. *Clin Cancer Res* 2003;**9**(2):802–11.
6. Švastová E, Hulíková A, Rafajová M, Zat'ovičová M, Gibadulinová A, Casini A, et al. Hypoxia activates the capacity of tumor-associated carbonic anhydrase IX to acidify extracellular pH. *FEBS Lett* 2004;**577**(3):439–45.
7. Loncaster JA, Harris AL, Davidson SE, Logue JP, Hunter RD, Wycoff CC, et al. Carbonic anhydrase (CA IX) expression, a potential new intrinsic marker of hypoxia: correlations with tumor oxygen measurements and prognosis in locally advanced carcinoma of the cervix. *Cancer Res* 2001; **61**(17):6394–9.
8. Kaluz S, Kaluzová M, Liao SY, Lerman M, Stanbridge EJ. Transcriptional control of the tumor-and hypoxia- marker carbonic anhydrase 9: a one transcription factor (HIF-1) show? *Biochim Biophys Acta Rev Cancer* 2009;**1795**(2):162–72.
9. Potter C, Harris AL. Hypoxia inducible carbonic anhydrase IX, marker of tumour: hypoxia, survival pathway and therapy target. *Cell Cycle* 2004;**3**(2):159–62.
10. Kaluz S, Kaluzová M, Stanbridge EJ. Expression of the hypoxia marker carbonic anhydrase IX is critically dependent on SP1 activity. Identification of a novel type of hypoxia-responsive enhancer. *Cancer Res* 2003;**63**(5):917–22.
11. Lucarini L, Magnelli L, Schiavone N, Crisci A, Innocenti A, Puccetti L, et al. Plasmatic carbonic anhydrase IX as a diagnostic marker for clear cell renal cell carcinoma. *J Enzyme Inhib Med Chem* 2018;**33**(1):234–40.
12. Supuran CT. Carbonic anhydrase inhibitors as emerging agents for the treatment and imaging of hypoxic tumors. *Expert Opin Investig Drugs* 2018;**27**(12):963–70.
13. Okuno K, Matsubara T, Nakamura T, Iino T, Kakimoto T, Asanuma K, et al. Carbonic anhydrase IX enhances tumor cell proliferation and tumor progression in osteosarcoma. *OncoTargets Ther* 2018;**11**:6879.
14. Boyd NH, Walker K, Fried J, Hackney JR, McDonald PC, Benavides GA, et al. Addition of carbonic anhydrase 9 inhibitor SLC-0111 to temozolomide treatment delays glioblastoma growth in vivo [published online ahead of print, 2017 Dec 21]. *JCI Insight* 2017;2(24). https://doi.org/10.1172/jci.insight.92928.
15. Zandberga E, Zayakin P, Ābols A, Pūpola D, Trapencieris P, Linē A. Depletion of carbonic anhydrase IX abrogates hypoxia-induced overexpression of stanniocalcin-1 in triple negative breast cancer cells. *Cancer Biol Ther* 2017;**18**(8):596–605.
16. Koyuncu I, Gonel A, Kocyigit A, Temiz E, Durgun M, Supuran CT. Selective inhibition of carbonic anhydrase-IX by sulphonamide derivatives induces pH and reactive oxygen species-mediated apoptosis in cervical cancer HeLa cells. *J Enzyme Inhib Med Chem* 2018;**33**(1):1137–49.
17. Riemann A, Güttler A, Haupt V, Wichmann H, Reime S, Bache M, et al. Inhibition of carbonic anhydrase IX by Ureidosulfonamide inhibitor U104 reduces prostate cancer cell growth, but does not modulate daunorubicin or cisplatin cytotoxicity. *Oncol Res* 2018;**26**(2):191–200.
18. Supuran CT, Scozzafava A. Carbonic anhydrase inhibitors: aromatic sulfonamides and disulfonamides act as efficient tumor growth inhibitors. *J Enzym Inhib* 2000;**15**(6):597–610.
19. Sowa T, Menju T, Chen-Yoshikawa TF, Takahashi K, Nishikawa S, Nakanishi T, et al. Hypoxia-inducible factor 1 promotes chemoresistance of lung cancer by inducing carbonic anhydrase IX expression. *Cancer Med* 2017;**6**(1):288–97.
20. Kon-no H, Ishii G, Nagai K, Yoshida J, Nishimura M, Nara M, et al. Carbonic anhydrase IX expression is associated with tumor progression and a poor prognosis of lung adenocarcinoma. *Lung Cancer* 2006;**54**(3):409–18.
21. Parkkila S, Rajaniemi H, Parkkila AK, Kivelä J, Waheed A, Pastoreková S, et al. Carbonic anhydrase inhibitor suppresses invasion of renal cancer cells in vitro. *Proc Natl Acad Sci U S A* 2000;**97**(5):2220–4.
22. Lounnas N, Rosilio C, Nebout M, Mary D, Griessinger E, Neffati Z, et al. Pharmacological inhibition of carbonic anhydrase XII interferes with cell proliferation and induces cell apoptosis in T-cell lymphomas. *Cancer Lett* 2013;**333**:76–88.
23. Parks SK, Cormerais Y, Durivault J, Pouyssegur J. Genetic disruption of the pHi-regulating proteins Na+/H+ exchanger 1 (SLC9A1) and carbonic anhydrase 9 severely reduces growth of colon cancer cells. *Oncotarget* 2016;**8**(6):10225–37.

24. Andreucci E, Ruzzolini J, Peppicelli S, Bianchini F, Laurenzana A, Carta F, et al. The carbonic anhydrase IX inhibitor SLC-0111 sensitises cancer cells to conventional chemotherapy. *J Enzyme Inhib Med Chem* 2019;**34**(1):117–23.
25. Bryant JL, Gieling RG, Meredith SL, Allen TJ, Walker L, Telfer BA, et al. Novel carbonic anhydrase IX-targeted therapy enhances the anti-tumour effects of cisplatin in small cell lung cancer. *Int J Cancer* 2018;**142**(1):191–201.
26. Pastoreková S, Závadová Z, Kostál M, Babusíková O, Závada J. A novel quasi-viral agent MaTu, is a two component system. *Virology* 1992;**187**(2):620–6.
27. Závada J, Závadová Z, Pastoreková S, Ciampor F, Pastorek J, Zelník V. Expresión of MaTu-MN protein in human tumor cultures and in clinical specimens. *Int J Cancer* 1993;**54**(2):268–74.
28. Opavskýa R, Pastoreková S, Zelnika V, Gibadulinova A, Stanbridgeb EJ, et al. Human *MN/CA9* gene, a novel member of the carbonic anhydrase family: structure and exon to protein domain relationships. *Genomics* 1996;**33**(3):480–7.
29. Alterio V, Hilvo M, Di Fiore A, Supuran CT, Pan P, Parkkila S, et al. Crystal structure of the catalytic domain of the tumor-associated human carbonic anhydrase IX. *Proc Natl Acad Sci U S A* 2009;**106**(38):16233–8.
30. Svastová E, Zilka N, Zat'ovicová M, Gibadulinová A, Ciampor F, Pastorek J, et al. Carbonic anhydrase IX reduces E-cadherin-mediated adhesion of MDCK cells via interaction with beta-catenin. *Exp Cell Res* 2003;**290**:332–45.
31. Buanne P, Renzone G, Monteleone F, Vitale M, Monti SM, Sandomenico A, et al. Characterization of carbonic anhydrase IX interactome reveals proteins assisting its nuclear localization in hypoxic cells. *J Proteome Res* 2013;**12**:282–92.
32. Hilvo M, Baranauskiene L, Salzano AM, Scaloni A, Matulis D, Innocenti A, et al. Biochemical characterization of CA IX, one of the most active carbonic anhydrase isozymes. *J Biol Chem* 2008;**283**:27799–809.
33. Wykoff CC, Beasley NJ, Watson PH, Turner KJ, Pastorek J, Sibtain A, et al. Hypoxia-inducible expression of tumor-associated carbonic anhydrases. *Cancer Res* 2000;**60**:7075–83.
34. Swietach P, Vaughan-Jones RD, Harris AL. Regulation of tumor pH and the role of carbonic anhydrase 9. *Cancer Metastasis Rev* 2007;**26**(2):299–310.
35. Noor SI, Jamali S, Ames S, Langer S, Deitmer JW, Becker HM. A surface proton antenna in carbonic anhydrase II supports lactate transport in cancer cells. *elife* 2018;**7**:.
36. Pastorekova S, Parkkila S, Zavada J. Tumor-associated carbonic anhydrases and their clinical significance. *Adv Clin Chem* 2006;**42**:167–216.
37. Supuran CT. Carbonic anhydrases: novel therapeutic applications for inhibitors and activators. *Nat Rev Drug Discov* 2008;**7**:168–81.
38. Supuran CT. Carbonic anhydrase inhibitors. *Bioorg Med Chem Lett* 2010;**20**:3467–74.
39. Tureci O, Sahin U, Vollmar E, Siemer S, Göttert E, Seitz G, et al. Human carbonic anhydrase XII: cDNA cloning, expression, and chromosomal localization of a carbonic anhydrase gene that is overexpressed in some renal cell cancers. *Proc Natl Acad Sci U S A* 1998;**95**:7608–13.
40. Schütze D, Milde-Langosch K, Witzel I, Rody A, Karn T, Schmidt M, et al. Relevance of cellular and serum carbonic anhydrase IX in primary breast cancer. *J Cancer Res Clin Oncol* 2013;**139**(5):747–54.
41. Hussain SA, Ganesan R, Reynolds G, Gross L, Stevens A, Pastorek J, et al. Hypoxia-regulated carbonic anhydrase IX expression is associated with poor survival in patients with invasive breast cancer. *Br J Cancer* 2007;**96**(1):104.
42. Chia SK, Wyhoff CC, Watson PH, Han C, Leek RD, Pastorek J, et al. Prognostic significance of a novel hypoxia regulated marker, carbonic anhydrase IX in invasive breast carcinoma. *J Cin Oncol* 2001;**19**:3660–8.
43. Barnett DH, Sheng S, Charn TH, Waheed A, Sly WS, YoLin C, et al. Estrogen receptor regulation of carbonic anhydrase XII through a distal enhancer in breast cancer. *Cancer Res* 2008;**68**(9): 3505–15.
44. Span PN, Bussink J, Manders P, Beex LVAM, Sweep CGJ. Carbonic anhydrase-9 expression levels and prognosis in human breast cancer: association with treatment outcome. *Br J Cancer* 2003;**89**:271.
45. Generali D, Fox SB, Berruti A, Brizzi MP, Campo L, Bonardi S, et al. Role of carbonic anhydrase IX expression in prediction of the efficacy and outcome of primary epirubicin/tamoxifen therapy for breast cancer. *Endocr Relat Cancer* 2006;**13**(3):921–30.

46. Li Y, Tu C, Wang H, Silverman DN, Frost SC. Catalysis and pH control by membrane-associated carbonic anhydrase IX in MDA-MB-231 breast cancer cells. *J Biol Chem* 2011;**286**:15789–96.
47. Müller V, Riethdorf S, Rack B, Janni W, Fasching PA, Solomayer E, et al. Prospective evaluation of serum tissue inhibitor of metalloproteinase 1 and carbonic anhydrase IX in correlation to circulating tumor cells in patients with metastatic breast cancer. *Breast Cancer Res* 2011;**13**(4):R71.
48. Lock F, McDonald P, Lou Y, Serrano I, Chafe S, Ostlund C, et al. Targeting carbonic anhydrase IX depletes breast cancer stem cells within the hypoxic niche. *Oncogene* 2013;**32**(44):5210.
49. Chen LQ, Howison CM, Spier C, Stopeck AT, Malm SW, Pagel MD, et al. Assessment of carbonic anhydrase IX expression and extracellular pH in B-cell lymphoma cell line models. *Leuk Lymphoma* 2015;**56**:1432–9.
50. Saarnio J, Parkkila S, Parkkila AK, Haukipuro K, Pastoreková S, Pastorek J, et al. Immunohistochemical study of colorectal tumors for expression of a novel transmembrane carbonic anhydrase, MN/CA IX, with potential value as a marker of cell proliferation. *Am J Pathol* 1998;**153**:279–85.
51. Kummola L, Hämäläinen JM, Kivelä J, Kivelä AJ, Saarnio J, et al. Expression of a novel carbonic anhydrase, CA XIII, in normal and neoplastic colorectal mucosa. *BMC Cancer* 2005;**5**:41.
52. Kivelä AJ, Parkkila S, Saarnio J, Karttunen TJ, Kivelä J, Parkkila AK, et al. Expression of transmembrane carbonic anhydrase isoenzymes IX and XII in normal human pancreas and pancreatic tumours. *Histochem Cell Biol* 2000;**114**:197–204.
53. Beasley NJP, Wykoff CC, Watson PH, Leek R, Turley H, Gatter K, et al. Carbonic anhydrase IX, an endogenous hypoxia marker, expression in head and neck squamous cell carcinoma and its relationship to hypoxia, necrosis, and microvessel density. *Cancer Res* 2001;**61**(13):5262–7.
54. Yang JS, Lin CW, Chuang CY, Su SC, Lin SH, Yang SF. Carbonic anhydrase IX overexpression regulates the migration and progression in oral squamous cell carcinoma. *Tumor Biol* 2015;**36**(12):9517–24.
55. Giatromanolaki A, Koukourakis MI, Sivridis E, Pastorek J, Wykoff CC, Gatter KC, et al. Expression of hypoxia- inducible carbonic anhydrase-9 relates to angiogenic pathways and independently to poor outcome in non-small cell lung cancer. *Cancer Res* 2001;**61**(21):7992–8.
56. Bryant JL, Gieling RG, Meredith SL, Allen TJ, Walker L, Telfer BA, et al. Novel carbonic anhydrase IX-targeted therapy enhances the antitumour effects of cisplatin in small cell lung cancer. *Int J Cancer* 2018;**142**(1):191–201.
57. Atkins M, Regan M, McDermott D, Mier J, Stanbridge E, Youmans A, et al. Carbonic anhydrase IX expression predicts outcome of interleukin 2 therapy for renal cancer. *Clin Cancer Res* 2005;**11**(10):3714–21.
58. Genega EM, Ghebremichael M, Najarian R, Fu Y, Wang Y, Argani P, et al. Carbonic anhydrase IX expression in renal neoplasms: correlation with tumor type and grade. *Am J Clin Pathol* 2010;**134**(6):873–9.
59. Fiaschi T, Giannoni E, Taddei L, Cirri P, Marini A, Pintus G, et al. Carbonic anhydrase IX from cancer-associated fibroblasts drives epithelial-mesenchymal transition in prostate carcinoma cells. *Cell Cycle* 2013;**12**(11):1791–801.
60. Hui EP, Chan AT, Pezzella F, Turley H, To KF, Poon TC, et al. Coexpression of hypoxia-inducible factors 1α and 2α, carbonic anhydrase IX, and vascular endothelial growth factor in nasopharyngeal carcinoma and relationship to survival. *Clin Cancer Res* 2002;**8**(8):2595–604.
61. Olive PL, Aquino-Parsons C, MacPhail SH, Liao SY, Raleigh JA, Lerman MI, et al. Carbonic anhydrase 9 as an endogenous marker for hypoxic cells in cervical cancer. *Cancer Res* 2001;**61**(24):8924–9.
62. Liao SY, Darcy KM, Randall LM, Tian C, Monk BJ, Burger RA, et al. Prognostic relevance of carbonic anhydrase-IX in high-risk, early-stage cervical cancer: a Gynecologic Oncology Group study. *Gynecol Oncol* 2010;**116**(3):452–8.
63. Hynninen P, Parkkila S, Huhtala H, Pastorekova S, Pastorek J, Waheed A, et al. Carbonic anhydrase isozymes II, IX, and XII in uterine tumors. *APMIS* 2012;**120**(2):117–29.
64. Choschzick M, Woelber L, Hess S, Zu Eulenburg C, Schwarz J, Simon R, et al. Overexpression of carbonic anhydrase IX (CAIX) in vulvar cancer is associated with tumor progression and development of locoregional lymph node metastases. *Virchows Arch* 2010;**456**(5):483–90.

65. Chen J, Röcken C, Hoffmann J, Krüger S, Lendeckel U, Rocco A, et al. Expression of carbonic anhydrase 9 at the invasion front of gastric cancers. *Gut* 2005;**54**(7):920–7.
66. Ramsey ML, Yuh BJ, Johnson MT, Yeldani AT, Zynger DL. Carbonic anhydrase IX is expressed in mesothelioma and metastatic clear cell renal cell carcinoma of the lung. *Virchows Arch* 2012;**460**(1):89–93.
67. Proescholdt MA, Merrill MJ, Stoerr EM, Lohmeier A, Pohl F, Brawanski A. Function of carbonic anhydrase IX in glioblastoma multiforme. *Neuro-Oncology* 2012;**14**(11):1357–66.
68. Said HM, Staab A, Hagemann C, Vince GH, Katzer A, Flentje M, et al. Distinct patterns of hypoxic expression of carbonic anhydrase IX (CA IX) in human malignant glioma cell lines. *J Neuro-Oncol* 2007;**81**(1):27–38.
69. Hynninen P, Vaskivuo L, Saarnio J, Haapasalo H, Kivelä J, Pastorekova S, et al. Expression of transmembrane carbonic anhydrases IX and XII in ovarian tumours. *Histopathology* 2006;**49**(6): 594–602.
70. Klatte T, Seligson DB, Rao JY, Yu H, de Martino M, Kawaoka K, et al. Carbonic anhydrase IX in bladder cancer. *Cancer* 2009;**115**(7):1448–58.
71. Perut F, Carta F, Bonuccelli G, Grisendi G, Di Pompo G, Avnet S, et al. Carbonic anhydrase IX inhibition is an effective strategy for osteosarcoma treatment. *Expert Opin Ther Targets* 2015;**19**(12):1593–605.
72. Puech C, Chatard M, Felder-Flesch D, Prevot N, Perek N. Umbelliferone decreases intracellular pH and sensitizes melanoma cell line A375 to Dacarbazin. Comparison with acetazolamide. *Curr Mol Pharmacol* 2018;**11**(2):133–9.
73. Leppilampi M, Koistinen P, Savolainen ER, Hannuksela J, Parkkila AK, Niemelä O, et al. The expression of carbonic anhydrase II in hematological malignancies. *Clin Cancer Res* 2002;**8**(7):2240–5.
74. Toyosawa S, Ogawa Y, Chang CK, Hong SS, Yagi T, Kuwahara H, et al. Histochemistry of tartrate-resistant acid phosphatase and carbonic anhydrase isoenzyme II in osteoclast-like giant cells in bone tumours. *Virchows Arch A* 1991;**418**(3):255–61.
75. Zheng MH, Fan Y, Wysocki S, Wood DJ, Papadimitriou JM. Detection of mRNA for carbonic anhydrase II in human osteoclast-like cells by in situ hybridization. *J Bone Miner Res* 1993; **8**(1):113–8.
76. Kaluzová M, Kaluz S, Lerman MI, Stanbridge EJ. DNA damage is a prerequisite for p53-mediated proteasomal degradation of HIF-1α in hypoxic cells and downregulation of the hypoxia marker carbonic anhydrase IX. *Mol Cell Biol* 2004;**24**(13):5757–66.
77. Shin KH, Diaz-Gonzalez JA, Russell J, Chen Q, Burgman P, Li XF, et al. Detecting changes in tumor hypoxia with carbonic anhydrase IX and pimonidazole. *Cancer Biol Ther* 2007;**6**(1):70–5.
78. Sneader W. *Drug discovery: a history*. John Wiley & Sons; 2005. p.390, ISBN: 9780471899792 [Archived from the original on 2016-12-28].
79. Pickkers P, Hughes AD, Russel FGM, Thien T, Smits P. In vivo evidence for KCa channel opening properties of acetazolamide in the human vasculature. *Br J Pharmacol* 2001;**132**:443–50.
80. Xiang Y, Ma B, Li T, Gao JW, Yu HM, Li XJ. Acetazolamide inhibits aquaporin-1 protein expression and angiogenesis. *Acta Pharmacol Sin* 2004;**25**:812–6.
81. Mu SM, Ma B, Yu HM, Li XJ. Inhibition of AQP1 expression by acetazolamide and its relation to intracellular pH and Ca2+. *Basic Med Sci Clin* 2003;**23**:287–91.
82. Teicher BA, Liu SD, Liu JT, Holden SA, Herman TS. A carbonic anhydrase inhibitor as a potential modulator of cancer therapies. *Anticancer Res* 1993;**13**(5A):1549–56.
83. Yang XI, Ma B, Tao LI. Acetazolamide suppresses tumor metastasis and related protein expression in mice bearing Lewis lung carcinoma. *Acta Pharmacol Sin* 2002;**23**(8):745–51.
84. Ahlskog JK, Dumelin CE, Trüssel S, Mårlind J, Neri D. In vivo targeting of tumor-associated carbonic anhydrases using acetazolamide derivatives. *Bioorg Med Chem Lett* 2009;**19**(16):4851–6.
85. Mokhtari RB, Kumar S, Islam SS, Yazdanpanah M, Adeli K, Cutz E, et al. Combination of carbonic anhydrase inhibitor, acetazolamide, and sulforaphane, reduces the viability and growth of bronchial carcinoid cell lines. *BMC Cancer* 2013;**13**(1):378.
86. Faes S, Planche A, Uldry E, Santoro T, Pythoud C, Stehle JC, et al. Targeting carbonic anhydrase IX improves the anti-cancer efficacy of mTOR inhibitors. *Oncotarget* 2016;**7**(24):36666.

87. Cianchi F, Vinci MC, Supuran CT, Peruzzi B, De Giuli P, Fasolis G, et al. Selective inhibition of carbonic anhydrase IX decreases cell proliferation and induces ceramide-mediated apoptosis in human cancer cells. *J Pharmacol Exp Ther* 2010;**334**(3):710–9.
88. Kopecka J, Campia I, Jacobs A, Frei AP, Ghigo D, Wollscheid B, et al. Carbonic anhydrase XII is a new therapeutic target to overcome chemoresistance in cancer cells. *Oncotarget* 2015;**6**(9):6776.
89. Said HM, Hagemann C, Carta F, Katzer A, Polat B, Staab A, et al. Hypoxia induced CA9 inhibitory targeting by two different sulfonamide derivatives including acetazolamide in human glioblastoma. *Bioorg Med Chem* 2013;**21**(13):3949–57.
90. Kasai H, Tonda K, Hirata M. Potentiation of antitumor activity of 1-phthalidyl 5-fluorouracil by acetazolamide. *Cancer Chemother Pharmacol* 1986;**16**(1):55–7.
91. Mokhtari RB, Baluch N, Tsui MK, Kumar S, Homayouni TS, Aitken K, et al. Acetazolamide potentiates the anti-tumor potential of HDACi, MS-275, in neuroblastoma. *BMC Cancer* 2017;**17**(1):156.
92. Li XJ, Xiang Y, Ma B, Qi XQ. Effects of acetazolamide combined with or without NaHCO3 on suppressing neoplasm growth, metastasis and aquaporin-1 (AQP1) protein expression. *Int J Mol Sci* 2007;**8**(3):229–40.
93. Zhang B, Huang H, Wang Z, Liu X. Relations of carbonic anhydrase-IX with tumorigenesis and metastasis of rectal cancer. *Chin J Gen Surg* 2011;**10**:013.
94. Shiverick K, Selman Y, Mira J, Porvasnik S, Medrano T, McKenna R, et al. Abstract# 1791: Differential effects of carbonic anhydrase inhibition in human prostate cancer cells and the microenvironment in mouse xenograft tumors. In: *AACR annual meeting—April 18–22*; 2009. p. 1791.
95. Islam SS, Mokhtari RB, Akbari P, Hatina J, Yeger H, Farhat WA. Simultaneous targeting of bladder tumor growth, survival, and epithelial-to-mesenchymal transition with a novel therapeutic combination of acetazolamide (AZ) and sulforaphane (SFN). *Target Oncol* 2016;**11**(2):209–27.
96. Ivanova L, Zandberga E, Silina K, Kalnina Z, Abols A, Endzelins E, et al. Prognostic relevance of carbonic anhydrase IX expression is distinct in various subtypes of breast cancer and its silencing suppresses self-renewal capacity of breast cancer cells. *Cancer Chemother Pharmacol* 2015;**75**(2):235–46.
97. Bin K, Shi-Peng Z. Acetazolamide inhibits aquaporin-1 expression and colon cancer xenograft tumor growth. *Hepato-Gastroenterology* 2011;**58**(110–111):1502–6.
98. Noma N, Fujii G, Miyamoto S, Komiya M, Nahanishi R, Shimura M, et al. Impact of acetazolamide, a carbonic anhydrase inhibitor, on the development of intestinal polyps in min mice. *Int J Mol Sci* 2017;**18**(4):851.
99. Shank RP, Gardocki JF, Streeter AJ, Maryanoff BE. An overview of the preclinical aspects of topiramate: pharmacology, pharmacokinetics, and mechanism of action. *Epilepsia* 2000;**41**:3–9.
100. Marathe K, McVicar N, Li A, Bellyou M, Meakin S, Bartha R. Topiramate induces acute intracellular acidification in glioblastoma. *J Neuro-Oncol* 2016;**130**(3):465–72.
101. Tafreshi NK, Lloyd MC, Bui MM, Gillies RJ, Morse DL. Chapter 13: carbonic anhydrase IX as an imaging and therapeutic target for tumors and metastases. *Subcell Biochem* 2014;**75**:221–54.
102. Weber A, Casini A, Heine A, Kuhn D, Supuran CT, Scozzafava A, et al. Unexpected nanomolar inhibition of carbonic anhydrase by COX-2-selective celecoxib: new pharmacological opportunities due to related binding site recognition. *J Med Chem* 2004;**47**(3):550–7.

CHAPTER 8

The vacuolar H^+ ATPase proton pump

Introduction

Regulation of pH, that actually means regulation of proton (H^+) concentration in the different compartments, is an essential requirement in the homeostasis of the biological world.[1] This issue is so critical for cell survival that many overlapping systems of pH control and handling are in charge of this function. Throughout the book we review the most important systems for proton management. Vacuolar H^+ ATPase proton pump (PP) is one of the key players in what we have called the pH-tome, a system that includes buffers, membrane proteins, enzymes, transporters, exchangers, channels, and metabolites that regulate pH (Fig. 1).

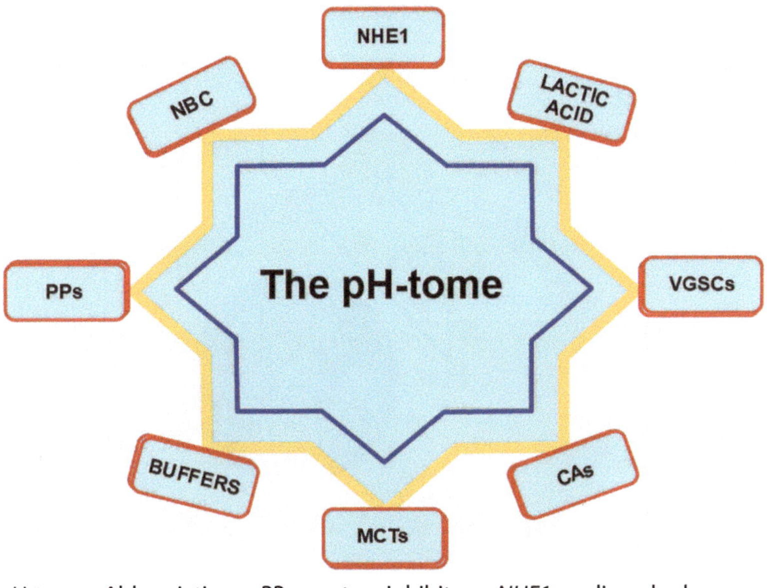

Fig. 1 The pH-tome. Abbreviations: *PPs*, proton inhibitors; *NHE1*, sodium hydrogen exchanger 1; *VGSCs*, voltage gated sodium channels; *CAs*, carbonic anhydrases; *NBC*, sodium bicarbonate cotransporter; *MCTs*, monocarboxylate transporters.

All these apparently independent actors are interconnected by a common denominator: H^+ concentration. Importantly, they are not fully independent of each other. Every day a new crosstalk or signaling or chemical binding is discovered among the members of the pH-tome. Voltage gated sodium channels are able to induce allosteric changes in

An Innovative Approach to Understanding and Treating Cancer: Targeting pH
https://doi.org/10.1016/B978-0-12-819059-3.00008-3
© 2020 Elsevier Inc.
All rights reserved.

NHE1 (sodium/hydrogen exchanger 1); NBCs (sodium/bicarbonate cotransporters) have the ability to increase or decrease cellular buffer content; carbonic anhydrases interact with monocarboxylate transporters, etc. And pH is the conductor of all these active orchestra members, so that they all play the same music.

Proton pumps (i.e., Vacular H^+-ATPase pumps or V-ATPases) seem the more independent only because their relations have not been fully elucidated or investigated.

V-ATPases are a group of membrane enzymes that couple ATP hydrolysis to the transport of H^+ ions across membranes. They are highly conserved multi-enzyme complexes and vital components of all eukaryotes that are located in the membranes of many organelles. They are responsible for low intravacuolar pH, mainly in lysosomes and endosomes. For many years it was thought that the vacuolar H+ ATPases' activity was limited to acidifying intracellular compartments.[2] However, this activity is not restricted to organelle membranes and can also be found in the plasma membrane,[3] where V-ATPases pump protons ($H^{+)}$ to the surrounding extracellular space, thus modifying cytoplasmic and extracellular pH with energy expenditure (ATP hydrolysis) (Fig. 2).

Therefore, the best definition is "ATP-driven proton pumps that function to both acidify intracellular compartments and to transport protons across the plasma membrane".[4]

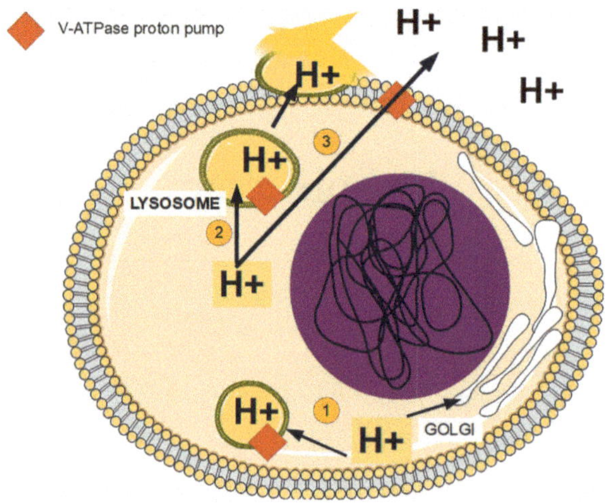

Fig. 2 V-ATPase activity in the cell: (1) acidifying internal organelles like lysosomes and Golgi apparatus (PPs in organelles' membranes; (2) exporting H^+ through vesicles and lysosomes (PPs in lysosomal membrane); (3) extruding H^+ directly (plasmatic membrane PPs). In all the cases, the final result is a decrease of H^+ in the cytoplasm (intracellular alkalinization) and extracellular acidification in 2 and 3. By regulating the pH of intracellular compartments, the V-ATPases indirectly regulate enzyme activity, act on the dissociation of ligands from receptors and promote the transport of substrates across membranes. Organelles have different internal pHs, whose homeostasis is a necessary condition for their normal function. For example lysosomes are acidic while mitochondria are alkaline.

Structure, regulation and function

V-ATPases are complex multimeric proteins with two functionally separate entities[5]:

(1) Domain V1: composed by eight subunits that perform the ATP hydrolysis.

(2) Domain Vo: composed of five subunits that are in charge of H^+ Transport through the membrane (Fig. 3). There are multiple tissue-specific isoforms for both domains.

Regulation of the V-ATPases seems to be carried out by associating or dissociating its two subunits V0 and V1. According to Croos and Müller[6] and Kettner et al.,[7] hydrolysis of

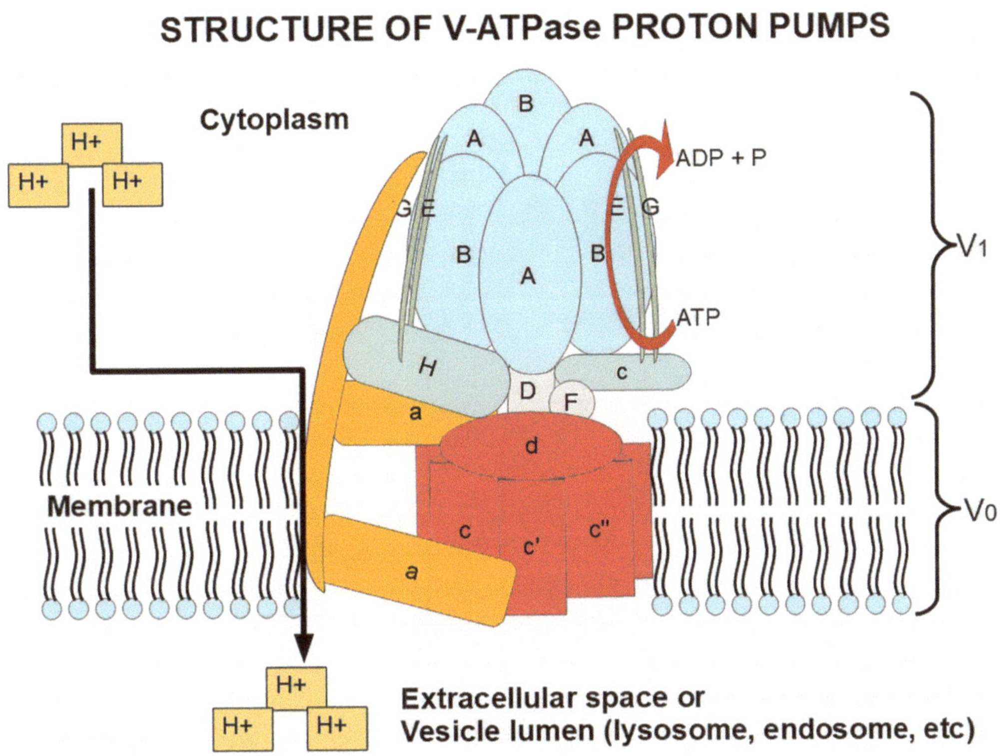

Fig. 3 V-H^+ ATPase at work. Export of H^+ from the cytoplasm to an intracellular vesicle (endosome, lysosome, or secretory vesicle) or to the extracellular space. The two parts of the V-H^+ATPase proton pump molecule can be seen: V0 embedded in the membrane, and V1 intracytoplasmatic. Hydrolysis of ATP takes place in V1 in the cytoplasm.[8–10] The association/dissociation between V0 and V1 regulates the proton pump. ATP hydrolysis generates the energy for a 360 degrees rotation of V0 producing the inflow (to the vesicle) or the outflow to the extracellular space of $H^{+\$}$. The two domains, V_1 and V_O, functionally couple ATP hydrolysis and proton transport. V_O, membrane-bound is the path for proton trafficking. There are drugs with the ability to uncouple the functional relation between the two domains.[11] As an example of proton pump functioning, the V-ATPase in spermatozoa acidifies the acrosome that permits the activation of proteolytic enzymes necessary for penetration into the egg.

one ATP by the V1 domain generates the necessary energy for a 360 degrees rotation of the V0 ring and the "pushing" of 2 to 4 cytosolic H^+ across the membrane against their electrochemical gradient (Fig. 3). The fact that PPs work against an electrochemical gradient is of paramount importance, because the extracellular concentration of H^+ is very high in malignant tissues. Therefore, the H^+ concentration is higher in the extracellular space than inside the cell and energy expenditure is necessary for going against the H^+ gradient and, in this way, can greatly increase the extracellular H^+ concentration. This is shown schematically in Fig. 4.

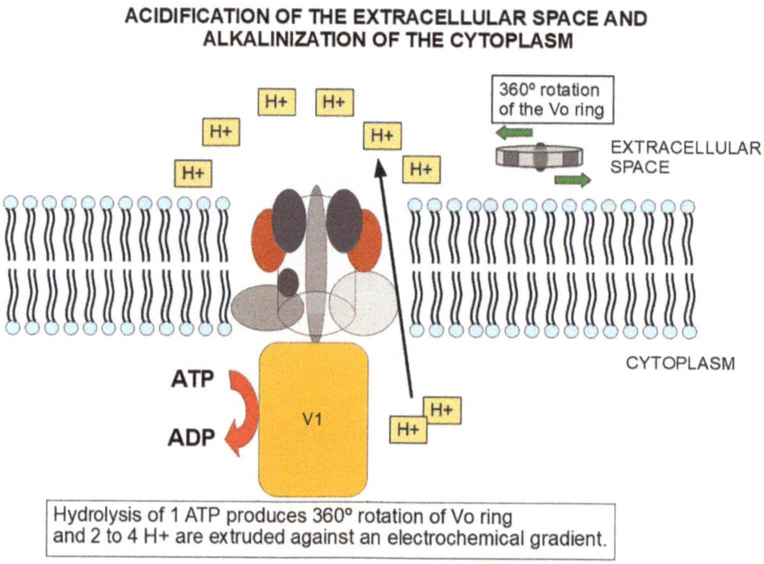

Fig. 4 The energy obtained from ATP hydrolysis is used by the PP to produce a full rotation of the V0 subunit and 2 to 4H+ are exported against an electrochemical gradient.

Function of the V-ATPases

PPs have both canonical and non canonical functions and both are related to cancer.

Canonical functions: transportation of protons across membranes. Besides pumping protons to the extracellular space, PPs are in charge of acidifying lysosomes. This is an important and essential step for protein processing and degradation in endocytosis and autophagy.[12] The acidity generated by PPs in secretory vesicles activates cathepsins which will play a major role in extracellular matrix degradation that requires a pH of around 4.5.[13] In bone metastasis osteoclast activity is essential for bone reabsorption. Osteoclasts have the isoform a3 PP, which are overexpressed in their vesicles, participate actively in bone reabsorption and are associated with cancer progression. For a detailed review see Holiday.[14]

Noncanonical functions: The non canonical functions are those not associated with proton trafficking through membranes. PPs interact with microtubules,[15,16] create the acidic environment for dendritic cells antigen presentation in vesicles[17] and act on mTOR (mammalian target of rapamycin) through lysosomes. V-ATPase proton pumps are necessary for mTOR detection of amino acids present in lysosomes, inducing mTOR migration towards the lysosomal surface.[18] At the same time, mTOR is a regulator of V-ATPase expression through transcription factor EB (TFEB).[19] (TFEB is a master transcription factor for lysosomal generation). The PP regulator protein also increases mTOR activity in osteoclasts augmenting bone metastasis in breast cancer.[20] Rojas et al. described the participation of PPs in endothelial cell migration.[21] Their participation in the migration process has been further confirmed in breast cancer cells[22] and prostate cancer cells.[23]

PPs and cancer

Interest in PPs in cancer started in 1993 when their presence was detected in the cell membranes of many human tumors.[24] Comparing different breast cancer cell lines, Sennoune et al.[25] found that those that were highly metastatic had a significant increase in proton pump expression. Many cancer cells overexpress V-ATPase compared with their normal counterparts. This has been confirmed in hepatoblastoma,[26] pancreatic cancer,[27] oral squamous cell carcinoma,[28] drug resistant non small cell lung cancer,[29] breast cancer,[30–33] and ovarian cancer[34] among others. Furthermore, the over-expression of a yeast plasma membrane V-ATPase transfected into normal mouse fibroblasts increased intracellular pH and induced a tumorigenic phenotype.[35,36]

It has also been found that the inhibition of tumor vacuolar V-ATPase produced a lower intracellular pH and induced apoptosis in cisplatin resistant ovarian cancer cells[37] and increased the cytotoxicity of paclitaxel.[38]

Bafilomycin A1, an experimental and specific inhibitor of PPs significantly decreased the intracellular pH (pHi) of tumor cells that had high PP expression.[39,40] This PP inhibitor also showed clear anti-cancer effects.[41]

Capecci and Forgac[42] have shown that invasion of the highly metastatic MCF10CA1a breast cancer cells but not the parental non metastatic MCF10A cells was inhibited by concanamycin (a powerful V-ATPase inhibitor). The highly metastatic cells expressed increased V-ATPase at the plasma membrane compared with non metastatic cells. Highly invasive cells also over-expressed a subunit of V-ATPase known as a3, which was responsible for the plasma membrane-location of the proton pump and the highly invasive phenotype of the cells. Inhibiting a3 eliminated the cells' ability to invade.

McGuire et al.,[43] wrote "*... that breast tumor cells (by an unknown mechanism) up-regulate the a3 isoform, which results in increased trafficking of V-ATPase complexes to the plasma membrane, where they increase tumor cell invasion*".

Di Cristofori et al.[44] showed that the content of secreted microvesicles in glioblastoma are regulated by V-ATPase. These microvesicles play an important role in cross-talk between tumor cells and non-neoplastic parenchyma and have a role in glioblastoma growth. Importantly, down-regulation of the V-ATPases decreased the activity of secreted microvesicles.

Besides their activity on intracellular and extracellular pH where they play an important role in the pH gradient inversion,[45] V-ATPases, show many other pro-tumoral effects, part of which were mentioned above in the non canonical functions. They also regulate Notch signaling in triple-negative breast cancer,[46] modulate metalloprotease isoforms in pancreatic cancer cells,[47] immunomodulate neutrophils associated with tumors[48] and are essential for cancer invasion.[49] V-ATPases also play an important role as mediators of cancer-related inflammation,[50] angiogenesis,[51] autophagy,[52] pH-mediated drug resistance,[53] and migration.[54] A summary of these studies can be seen in Fig. 5.

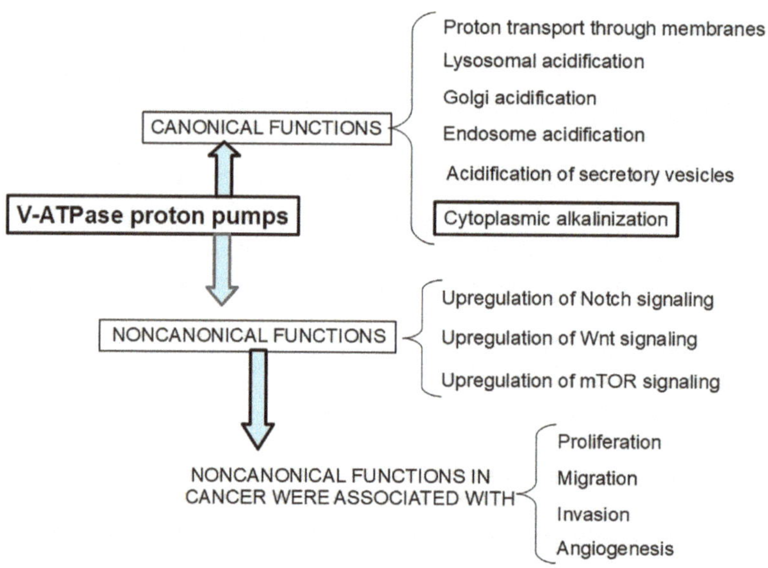

Fig. 5 Functions of PPs in normal cells and cancer.

V-ATPase proton pumps' contributions to the pH paradigm in cancer development

- Extrusion of intracellular protons to the extracellular space, contributing to the creation of an alkaline intracellular media and an acidic extracellular space.
- Acidification of lysosomes essential for activation of cathepsines,
- Promotion of migration through a3 subunit. The mechanism is unknown.
- Regulation of the content of secreted micro vesicles that communicate with other malignant cells (found in glioblastoma).

There is also evidence that chemotherapeutic treatments that damage DNA increase the expression of V-ATPase genes through transcription factors Sp1 and Oct1. This is considered an anti-apoptotic defense and when a V-ATPase inhibitor was used in association with certain chemotherapeutics, enhanced apoptosis in human epidermoid cancer Kb cells was achieved.[55]

The foregoing explains why it is necessary to target PPs.[56,57]

Proton pump inhibition

Since 2004 there is strong evidence that proton pump inhibitors (PPIs) sensitize cancer cells to chemotherapeutic treatments.[58] This increased sensitivity was confirmed for many different PPIs and in many types of tumors.[59] However, there were no highly effective and specific V-ATPase inhibitors in the past. In the early years—1980s—of experimental research only BafilomycinA and concavalin were available. Their high toxicity excluded both from clinical use.

However, since the discovery of proton pump inhibitors such as omeprazole for the treatment of esophageal and gastroduodenal hyperacidity-related diseases the landscape has changed completely. Indeed, PPIs usually used for the treatment of gastroduodenal ulcers, gastroesophageal reflux disease, and Zollinger-Ellison syndrome have shown antitumoral activity. These PPI effects are particularly strong in an acidic environment like the one usually found in tumors.[60]

Besides acidifying the cytoplasm, inhibition of V-ATPases decreased the population of pro-tumoral macrophages associated with the tumor microenvironment,[61] decreased tumor growth, and overcame cisplatin resistance in ovarian cancer cells.[37] Furthermore, PPIs had the ability to induce apoptosis in malignant cells without affecting the normal ones.[62] Canitano et al.[63] found that lansoprazole, and to a lesser extent omeprazole, showed significant antitumor activity in multiple myeloma cells. Similar results were found in melanoma[64] and human B cell lymphoma.[65] Standard chemotherapy cytotoxicity was increased with the concomitant administration of PPIs.[66]

Other unrelated drugs also show PP inhibition activities, like disulfiram.[67,68] It is a V-ATPase inhibitor, which has been used in the treatment of alcoholism and shows interesting anti-tumoral activity. Probably, most of this activity is not related to V-ATPase inhibition.

Finding the appropriate PPI

The different PPIs used in the treatment of non-cancerous gastrointestinal disorders are not uniformly active against cancer. While Lugini et al.[69] proposed lansoprazole as the best choice for targeting cancer, esomeprazol seems to be taking the lead. There have been clinical trials with PPIs as complementary treatments associated with standard chemotherapeutic protocols:

Clinical trial NCT02595372 (https://ClinicalTrials.gov) is based on a preliminary retrospective study that showed that PPI intake in breast cancer patients during chemotherapy significantly improved overall survival. The trial is trying to determine the exact role of PPIs in cancer (using omeprazole).

Clinical trial NCT01748500 ongoing at the University Health Network in Toronto is an interventional study to determine the benefits of pantoprazole in the treatment of castrate-resistant prostate cancer treated with docetaxel.

Clinical trial NCT01069081 is exploring whether adding a PPI (esomeprazole at high doses) to docetaxel and cisplatin chemotherapy improves efficacy and does not affect tolerability in metastatic breast cancer.

Clinical trial NCT01163903 is a dose-finding, phase I study to determine the recommended phase II dose for the combination of doxorubicin and pantoprazole in patients with advanced tumors and no standard treatment options.

The case for esomeprazole (ESO)

- ESO reduced prostaglandin E2 in tissues.[70]
- ESO inhibited growth of cells (MDA-MB-468) in vitro in a dose-dependent manner, through increase of intracellular acidification. No effect on normal breast epithelial cells.[71]
- Taking PPIs chronically increased the risk of pancreatic cancer. This risk increase was even more important with esomeprazole.[72]
- ESO decreased tumor cell survival, adhesion and migration and increased cytotoxicity of cisplatin and 5-FU in human adeno and squamous cell carcinoma of esophagus.[73]
- ESO increased the cytotoxicity of paclitaxel in cervical cancer.[74]
- Intermittent high-dose ESO increased cytotoxicity of chemotherapy in metastatic breast cancer.[75]
- ESO induced apoptosis through increased ROS (reactive oxygen species) production in malignant Barret's esophagus cells.[76]
- ESO increased the sensitivity of tumor cells to chemotherapeutic treatments in lung cancer cells.[77]
- ESO increased the sensitivity of tumor cells to chemotherapeutic treatments in small cell lung cancer cells.[78]
- Intermittent high dose of ESO improved progression-free survival in patients with metastatic breast cancer treated with chemotherapy.[79]

The case for pantoprazole (Panto)

- Panto increased the cytotoxicity of doxorubicin in solid tumors and increased endosomal pH.[80]
- Panto reversed the pH gradient inversion and decreased multidrug resistance in gastric adenocarcinoma, decreasing cell viability in vitro and in vivo. It also decreased the signaling of the V-ATPases-mTOR-HIF-1 pathway.[81]

- Panto decreased the expression of HIF-1α and inhibited tumor growth in human gastric adenocarcinomas cells.[82]
- Panto inhibited invasion by down-regulating Akt/GSK-β/β-catenin signaling in adryamicin resistant gastric cancer cells.[83]
- Panto inhibited protein degradation systems sensitizing the cell to different kinds of stress.[84]
- Synergic effect of Panto and docetaxel inibited epithelial-mesenchymal transition in gastric cancer cells.[85]
- Panto inhibited autophagy increasing the apoptotic effects of chemotherapy.[86]
- Panto increased the chemosensitivity to 5-FU in colon cancer cells.[87]
- Panto induced apoptosis in leukemic cells inhibiting multidrug resistance.[88]
- Panto induced apoptosis and decreased NF-kB signaling in glioblastoma cells.[89]
- Panto increased tumoricidal activity of tumor-associated macrophages (TAMs) in a murine model of T cell lymphoma.[90]
- Panto induced apoptosis in gastric cancer cells by inhibiting Stat3.[91]

The case for lansoprazole (Lan)

- Lansoprazole increased sensitization of human melanoma cells to low doses of paclitaxel. This was confirmed in a xenograft model.[92]
- Lansoprazole induced apoptosis in human breast cancer cells by inhibiting proton extrusion. In xenografts, lansoprazole produced alkalization of lysosomes and increased ROS.[93]
- PPIs have inhibitory action on FASN (Fatty Acid Synthase) but lansoprazole is more effective than other PPIs.[94]
- Lan and carbonic anhydrase inhibitors synergize against melanoma cells.[95]
- Lan decreases migration and invasiveness of osteosarcoma cells.[96]

The case for omeprazole (Ome)

- Ome reduced the viability of Ewing sarcoma cells.[97]
- Rats that were fed with Ome showed decreased formation of aberrant crypts in a murine model of azoxymethane-induced crypt formation. Ome increased p21 expression in colon cancer cell lines and decreased antiapoptotic protein expression.[98]
- Ome decreased proliferation of pancreatic cancer cells.[99]
- Ome showed ligand capacity to aryl hydrocarbon receptor, decreasing cell invasion and metastasis in ER-negative breast cancer.[100]

While the chronic use of PPIs was discouraged by some studies showing an increased risk of pancreatic cancer,[101] other studies did not confirm this higher risk.[102] The issue is unsettled.

However, here we are dealing with a treatment for cancer, which means that the patients already have cancer.

A proof of concept

High dose proton pump inhibitors were used in dogs with naturally occurring cancers and reversal of chemoresistance was achieved as was an improvement in their quality of life.[103]

Conclusions

1. V-ATPases regulate cytoplasmic pH in association with other membrane ion channels.
2. They also regulate pH in endosomes, lysosomes, Golgi apparatus and other organelles.
3. They consume energy.
4. They are overexpressed in many tumors.
5. The plasma membrane PPs are increased in many malignant cancers.
6. V-ATPases participate in migration, invasion and cathepsin activation.
7. They are easily targetable.
8. They play an important role in invasion in breast cancer cells, prostate cancer cells, and in many other malignant processes.
9. V-ATPase inhibition increases anti-tumor effects of chemotherapy.
10. Many PPIs have other anti-tumor effects in addition to reversing the pH inversion.

References

1. Harvey WR. Physiology of V-ATPases. *J Exp Biol* 1992;**172**(1):1–7.
2. Forgac M. Structure, function and regulation of the vacuolar (H+)-ATPases. *FEBS Lett* 1998; **440**(3):258–63.
3. Finbow ME, Harrison MA. The vacuolar H+ ATPase: a universal proton pump of eukaryotes. *Biochem J* 1997;**324**(Pt. 3):697–712.
4. Jefferies KC, Cipriano DJ, Forgac M. Function, structure and regulation of the vacuolar (H+)-ATPases. *Arch Biochem Biophys* 2008;**476**(1):33–42.
5. Toei M, Saum R, Forgac M. Regulation and isoform function of the V-ATPases. *Biochemistry* 2010; **49**(23):4715–23.
6. Cross RL, Müller V. The evolution of A-, F-, and V-type ATP synthases and ATPases: reversals in function and changes in the H+/ATP coupling ratio. *FEBS Lett* 2004;**576**:1–4.
7. Kettner C, Bertl A, Obermeyer G, Slayman C, Bihler H. Electrophysiological analysis of the yeast V-type proton pump: variable coupling ratio and proton shunt. *Biophys J* 2003;**85**:3730–8.
8. Forgac M. Vacuolar ATPases: rotary proton pumps in physiology and pathophysiology. *Nat Rev Mol Cell Biol* 2007;**8**(11):917–29.
9. Koltai T. Cancer: fundamentals behind pH targeting and the double-edged approach. *Onco Targets Ther* 2016;**9**:6343–60.
10. Benlekbir S, Bueler SA, Rubinstein JL. Structure of the vacuolar-type ATPase from *Saccharomyces cerevisiae* at 11-A° resolution. *Nat Struct Mol Biol* 2012;**19**:1356–62.

11. Chan CY, Prudom C, Raines SM, Charkhzarrin S, Melman SD, De Haro LP, et al. Inhibitors of V-ATPase proton transport reveal uncoupling functions of tether linking cytosolic and membrane domains of V0 subunit a (Vph1p). *J Biol Chem* 2012;**287**(13):10236–50.
12. Saftig P, Klumperman J. Lysosome biogenesis and lysosomal membrane proteins: trafficking meets function. *Nat Rev Mol Cell Biol* 2009;**10**:623–35.
13. Pungercar JR, Caglic D, Sajid M, Dolinar M, Vasiljeva O, Pozgan U, et al. Autocatalytic processing of procathepsin B is triggered by proenzyme activity. *FEBS J* 2009;**276**:660–8.
14. Holliday LS. Vacuolar H+-ATPase: an essential multitasking enzyme in physiology and pathophysiology. *New J Sci* 2014;**2014**:1–21.
15. Xu T, Forgac M. Microtubules are involved in glucose-dependent dissociation of the yeast vacuolar [H+]-ATPase in vivo. *J Biol Chem* 2001;**276**:24855–61.
16. Tabke K, Albertmelcher A, Vitavska O, Huss M, Schmitz HP, Wieczorek H. Reversible disassembly of the yeast V-ATPase revisited under in vivo conditions. *Biochem J* 2014;**462**:185–97.
17. Trombetta ES, Ebersold M, Garrett W, Pypaert M, Mellman I. Activation oflysosomal function during dendritic cell maturation. *Science* 2003;**299**:1400–3.
18. Zoncu R, Bar-Peled L, Efeyan A, Wang S, Sancak Y, Sabatini DM. mTORC1 senses lysosomal amino acids through an inside-out mechanism that requires the vacuolar H(+)-ATPase. *Science* 2011; **334**:678–83.
19. Peña-Llopis S, Vega-Rubin- de -Celis S, Schwartz JC, Wolff NC, Tran TA, Zou L, et al. Regulation of TFEB and V-ATPases by mTORC1. *EMBO J* 2011;**30**(16):3242–58.
20. McConnell M, Feng S, Chen W, Zhu G, Shen D, Ponnazhagan S, et al. Osteoclast proton pump regulator Atp6v1c1 enhances breast cancer growth by activating the mTORC1 pathway and bone metastasis by increasing V-ATPase activity. *Oncotarget* 2017;**8**(29):47675.
21. Rojas JD, Sennoune SR, Maiti D, Bakunts K, Reuveni M, Sanka SC, et al. Vacuolar-type H+-ATPases at the plasma membrane regulate pH and cell migration in microvascular endothelial cells. *Am J Physiol Heart Circ Physiol* 2006;**291**(3):1147–57.
22. Cotter K, Capecci J, Sennoune S, Huss M, Maier M, Martinez-Zaguilan R, et al. Activity of plasma membrane V-ATPases is critical for the invasion of MDA-MB231 breast cancer cells. *J Biol Chem* 2015;**290**(6):3680–92.
23. Licon-Munoz Y, Vera M, Fordyce Colleen A, Parra KJ. F-actin reorganization by V-ATPase inhibition in prostate cancer. *Biol Open* 2017;**6**(11):1734–44.
24. Martinez-Zaguilan R, Lynch RM, Martinez GM, Gillies RJ. Vacuolar-type H(+) ATPases are functionally expressed in plasma membranes of human tumor cells. *Am J Physiol* 1993;**265**(4 Pt. 1): 1015–29.
25. Sennoune SR, Bakunts K, Martinez GM, Chua-Tuan JL, Kebir Y, Attaya MN, et al. Vacuolar H+ ATPase in human breast cancer cells with distinct metastatic potential: distribution and functional activity. *Am J Physiol Cell Physiol* 2004;**286**(6):1443–52.
26. Morimura T, Fujita K, Akita M, Nagashima M, Satomi A. The proton pump inhibitor inhibits cell growth and induces apoptosis in human hepatoblastoma. *Pediatr Surg Int* 2008;**24**(10):1087–94.
27. Ohta T, Numata M, Yagishita H, Futagami F, Tsukioka Y, Kitagawa H, et al. Expression of 16 kDa proteolipid of vacuolar-type H (+)-ATPase in human pancreatic cancer. *Br J Cancer* 1996;**73**(12):1511.
28. Garcia-Garcia A, Perez-Sayans Garcia M, Rodriguez MJ, Antunez-Lopez J, Barros-Angueira F, et al. Immunohistochemical localization of C1 subunit of V-ATPase (ATPase C1) in oral squamous cell cancer and normal oral mucosa. *Biotech Histochem* 2012;**87**(2):133–9.
29. Lu Q, Lu S, Huang L, Wang T, Wan Y, Zhou CX, et al. The expression of V-ATPase is associated with drug resistance and pathology of non-small-cell lung cancer. *Diagn Pathol* 2013;**8**(1):145.
30. You H, Jin J, Shu H, Yu B, De Milito A, Lozupone F, et al. Small interfering RNA targeting the subunit ATP6L of proton pump V-ATPase overcomes chemoresistance of breast cancer cells. *Cancer Lett* 2009;**280**(1):110–9.
31. Montcourrier P, Silver I, Farnoud R, Bird I, Rochefort H. Breast cancer cells have a high capacity to acidify extracellular milieu by a dual mechanism. *Clin Exp Metastasis* 1997;**15**(4):382–92.
32. Montcourrier P, Mangeat PH, Valembois C, Salazar G, Sahuquet A, Duperray C, et al. Characterization of very acidic phagosomes in breast cancer cells and their association with invasion. *J Cell Sci* 1994;**107** (9):2381–91.

33. Faronato M, Nguyen VT, Patten DK, Lombardo Y, Steel JH, Patel N, et al. DMXL2 drives epithelial to mesenchymal transition in hormonal therapy resistant breast cancer through notch hyper-activation. *Oncotarget* 2015;**6**(26):22467.
34. Kulshrestha A, Katara GK, Ibrahim S, Pamarthy S, Jaiswal MK, Sachs AG, et al. Vacuolar ATPase 'a2'isoform exhibits distinct cell surface accumulation and modulates matrix metalloproteinase activity in ovarian cancer. *Oncotarget* 2015;**6**(6):3797.
35. Perona R, Serrano R. Increased pH and tumorigenicity of fibroblasts expressing a yeast proton pump. *Nature* 1988;**334**(6181):438–40.
36. Perona R, Portillo F, Giraldez F, Serrano R. Transformation and pH homeostasis of fibroblasts expressing yeast H(+)-ATPase containing site-directed mutations. *Mol Cell Biol* 1990;**10**(8):4110–5.
37. Kulshrestha A, Katara GK, Ginter J, Pamarthy S, Ibrahim SA, Jaiswal MK, et al. Selective inhibition of tumor cell associated vacuolar-ATPase 'a2' isoform overcomes cisplatin resistance in ovarian cancer cells. *Mol Oncol* 2016;**10**(6):789–805.
38. Whitehurst AW, Bodemann BO, Cardenas J, Ferguson D, Girard L, Peyton M, et al. Synthetic lethal screen identification of chemo sensitizer loci in cancer cells. *Nature* 2007;**445**(7137):815–9.
39. Bowman EJ, Siebers A, Altendorf K. Bafilomycins: a class of inhibitors of membrane ATPases from microorganisms, animal cells, and plant cells. *Proc Natl Acad Sci U S A* 1988;**85**:7972–6.
40. McSheehy PM, Troy H, Kelland LR, Judson IR, Leach MO, Griffiths JR. Increased tumour extracellular pH induced by bafilomycin A1 inhibits tumour growth and mitosis in vivo and alters 5-fluorouracil pharmacokinetics. *Eur J Cancer* 2003;**39**:532–40.
41. Ohta T, Arakawa H, Futagami F, Fushida S, Kitagawa H, Kayahara M, et al. Bafilomycin A1 induces apoptosis in the human pancreatic cancer cell line Capan-1. *J Pathol* 1998;**185**:324–30.
42. Capecci J, Forgac M. The function of vacuolar ATPase (V-ATPase) a subunit isoforms in invasiveness of MCF10a and MCF10CA1a human breast cancer cells. *J Biol Chem* 2013;**288**:32731–41.
43. McGuire C, Cotter K, Stransky L, Forgac M. Regulation of V-ATPase assembly and function of V-ATPases in tumor cell invasiveness. *Biochim Biophys Acta* 2016;**1857**(8):1213–8.
44. Di Cristofori A, Ferrero S, Bertolini I, Gaudioso G, Russo MV, Berno V, et al. The vacuolar H+ ATPase is a novel therapeutic target for glioblastoma. *Oncotarget* 2015;**6**(19):17514.
45. Spugnini EP, Sonveaux P, Stock C, Perez-Sayans M, De Milito A, Avnet S, et al. Proton channels and exchangers in cancer. *Biochim Biophys Acta* 2015;**1848**:2715–26.
46. Pamarthy S, Jaiswal MK, Kulshreshtha A, Katara GK, Gilman-Sachs A, Beaman KD. The vacuolar ATPase a2-subunit regulates notch signaling in triple-negative breast cancer cells. *Oncotarget* 2015; **6**(33):34206–20.
47. Chung C, Mader CC, Schmitz JC, Atladottir J, Fitchev P, Cornwell ML, et al. The vacuolar-ATPase (V-ATPase) modulates matrix metalloproteinase (MMP) isoforms in human pancreatic cancer. *Lab Invest* 2011;**91**(5):732–43.
48. Ibrahim SA, Katara GK, Kulshrestha A, Jaiswal MK, Amin MA, Beaman KD. Breast cancer associated a2 isoform vacuolar ATPase immunomodulates neutrophils: potential role in tumor progression. *Oncotarget* 2015;**6**(32):33033–45.
49. Cotter K, Capecci J, Sennoune S, Huss M, Maier M, Martinez-Zaguilan R, Forgac M. The activity of plasma membrane V-ATPases is critical for the invasion of breast cancer cells. *Cancer Res* 2015; **75**(15 Suppl):1437. https://doi.org/10.1158/1538-7445.AM2015-1437.
50. Kwong C, Gilman Sachs A, Beaman K. Tumor associated a2 vacuolar ATPase acts as a key mediator of cancer-related inflammation by inducing pro-tumorigenic properties in monocytes. *J Immunol* 2011; **186**(3):1781–9.
51. Rath S, Liebl J, Fürst R, Vollmar AM, Zahler S. Regulation of endothelial signaling and migration by v-ATPase. *Angiogenesis* 2014;**17**(3):587–601.
52. Mijaljica D, Prescott M, Devenish RJ. V-ATPase engagement in autophagic processes. *Autophagy* 2011;**7**(6):666–8.
53. Martinez-Zaguilan R, Raghunand N, Lynch RM, Bellamy W, Martinez GM, Rojas B, et al. pH and drug resistance. I. Functional expression of plasmalemmal V-type H^+-ATPase in drug-resistant human breast carcinoma cell lines. *Biochem Pharmacol* 1999;**57**(9):1037–46.
54. Feng S, Cai M, Liu P, Wei L, Wang J, Qi J, et al. Atp6v1c1 may regulate filament actin arrangement in breast cancer cells. *PLoS One* 2014;**9**(1):e84833.

55. Torigoe T, Izumi H, Ishiguchi H, Uramoto H, Murakami T, Ise T, et al. Enhanced expression of the human vacuolar H+-ATPase c subunit gene (ATP6L) in response to anticancer agents. *J Biol Chem* 2002; **277**(39):36534–43.
56. Fais S, De Milito A, You H, Qin W. Targeting vacuolar H+-ATPases as a new strategy against cancer. *Cancer Res* 2007;**67**(22):10627–30.
57. Stransky L, Cotter K, Forgac M. The function of V-ATPases in cancer. *Physiol Rev* 2016;**96**:1071–91.
58. Luciani F, Spada M, De Milito A, Molinari A, Rivoltini L, Montinaro A, et al. Effect of proton pump inhibitor pretreatment on resistance of solid tumors to cytotoxic drugs. *J Natl Cancer Inst* 2004; **96**:1702–13.
59. Azzarito T, Venturi G, Cesolini A, Fais S. Lansoprazole induces sensitivity to suboptimal doses of paclitaxel in human melanoma. *Cancer Lett* 2015;**356**:697–703.
60. Larsson H, Mattson H, Sundell G, Carlsson E. Animal pharmacodynamics of omeprazole. A survey of its pharmacological properties in vivo. *Scand J Gastroenterol Suppl* 1985;**108**:23–35.
61. Katara GK, Kulshrestha A, Jaiswal MK, Pamarthy S, Gilman-Sachs A, Beaman KD. Inhibition of vacuolar ATPase subunit in tumor cells delays tumor growth by decreasing the essential macrophage population in the tumor microenvironment. *Oncogene* 2015;**35**(8):1058–65.
62. Yeo M, Kim DK, Kim YB, Oh TY, Lee JE, Cho SW, et al. Selective induction of apoptosis with proton pump inhibitor in gastric cancer cells. *Clin Cancer Res* 2004;**10**(24):8687–96.
63. Canitano A, Iessi E, Spugnini EP, Federici C, Fais S. Proton pump inhibitors induce a caspase-independent antitumor effect against human multiple myeloma. *Cancer Lett* 2016;**376**(2):278–83.
64. De Milito A, Canese R, Marino ML, Borghi M, Iero M, Villa A, et al. pH-dependent antitumor activity of proton pump inhibitors against human melanoma is mediated by inhibition of tumor acidity. *Int J Cancer* 2010;**127**(1):207–19.
65. De Milito A, Iessi E, Logozzi M, Lozupone F, Spada M, Marino ML, et al. Proton pump inhibitors induce apoptosis of human B-cell tumors through a caspase-independent mechanism. *Cancer Res* 2007; **67**(11):5408–17.
66. Wang BY, Zhang J, Wang JL, Sun S, Wang ZH, Wang LP, et al. Intermittent high dose proton pump inhibitor enhances the antitumor effects of chemotherapy in metastatic breast cancer. *J Exp Clin Cancer Res* 2015;**34**(1):85.
67. Johnson RM, Allen C, Melman SD, Waller A, Young SM, Sklar LA, et al. Identification of inhibitors of vacuolar proton-translocating ATPase pumps in yeast by high-throughput screening flow cytometry. *Anal Biochem* 2010;**398**(2):203–11.
68. Parra KJ. Vacuolar ATPase: a model proton pump for antifungal drug discovery. In: Tegos G, Mylonakis E, editors. *Antimicrobial drug discovery: emerging strategies*; 2012. p. 89–100.
69. Lugini L, Federici C, Borghi M, Azzarito T, Marino ML, Cesolini A, et al. Proton pump inhibitors while belonging to the same family of generic drugs show different anti-tumor effect. *J Enzyme Inhib Med Chem* 2016;**31**:538–45.
70. Falk GW, Buttar NS, Foster NR, Ziegler KL, DeMars CJ, Romero Y, et al. A combination of esomeprazole and aspirin reduces tissue concentrations of prostaglandin E2 in patients with Barrett's esophagus. *Gastroenterology* 2012;**143**(4):917–26.
71. Goh W, Sleptsova-Freidrich I, Petrovic N. Use of proton pump inhibitors as adjunct treatment for triple-negative breast cancers. An introductory study. *J Pharm Pharm Sci* 2014;**17**(3):439–46.
72. Lai SW, Sung FC, Lin CL, Liao KF. Use of proton pump inhibitors correlates with increased risk of pancreatic cancer: a case-control study in Taiwan. *Kuwait Med J* 2014;**46**:44–8.
73. Lindner K, Borchardt C, Schöpp M, Bürgers A, Stock C, Hussey DJ, et al. Proton pump inhibitors (PPIs) impact on tumour cell survival, metastatic potential and chemotherapy resistance, and affect expression of resistance-relevant miRNAs in esophageal cancer. *J Exp Clin Cancer Res* 2014; **33**(1):73.
74. Song T, Jeon HK, Hong JE, Choi JJ, Kim TJ, Choi CH, et al. Proton pump inhibition enhances the cytotoxicity of paclitaxel in cervical cancer. *Cancer Res Treat* 2017;**49**(3):595.
75. Wang BY, Zhang J, Wang JL, Sun S, Wang ZH, Wang LP, et al. Intermittent high dose proton pump inhibitor enhances the antitumor effects of chemotherapy in metastatic breast cancer. *J Exp Clin Cancer Res* 2015;**34**(1):85.

76. Hormi-Carver KK, Zhang X, Huo X, Yu C, Zhang HY, Spechler SJ, et al. W1881 in neoplastic Barrett's epithelial cells, esomeprazole induces apoptosis through the production of reactive oxygen species: a potential role for PPIs in the chemotherapy of Barrett's cancer. *Gastroenterology* 2010; **138**(5):S-758.
77. Yilmaztepe OA, Korkmaz Ş, Budak F, Özkaya G, Ulukaya E, Oral H. The effects of esomeprazole combined with classical cancer drugs on lung cancer cells and peripheral blood mononuclear cells. In: Frontiers in immunology. Conference abstract: 15th international congress of immunology (ICI). vol. 898; 2013. https://doi.org/10.3389/conf.fimmu.
78. Oral AY, Oral HB, Sarimahmut M, Cevatemre B, Özkaya G, Korkmaz Ş, et al. Combination of esomeprazole with chemotherapeutics results in more pronounced cytotoxic effect via apoptosis on A549 non-small-cell lung cancer cell line. *Turk J Biol* 2017;**41**(1):231–41.
79. Hu X, Wang B, Sun S, Chiesi A, Wang J, Zhang J, et al. Abstract P6-11-01: intermittent high dose proton pump inhibitor improves progression free survival as compared to standard chemotherapy in the first line treatment of patients with metastatic breast cancer. *Cancer Res* 2012;**72**(24). Abstract P6–11-01.
80. Patel KJ, Lee C, Tan Q, Tannock IF. Use of the proton pump inhibitor pantoprazole to modify the distribution and activity of doxorubicin: a potential strategy to improve the therapy of solid tumors. *Clin Cancer Res* 2013;**19**(24):6766–76.
81. Chen M, Huang SL, Zhang XQ, Zhang B, Zhu H, Yang VW, et al. Reversal effects of pantoprazole on multidrug resistance in human gastric adenocarcinoma cells by down-regulating the V-ATPases/mTOR/HIF-1α/P-gp and MRP1 signaling pathway in vitro and in vivo. *J Cell Biochem* 2012; **113**(7):2474–87.
82. Shen Y, Wu Y, Chen M, Shen W, Huang S, Zhang L, et al. Effects of pantoprazole as a HIF-1α inhibitor on human gastric adenocarcinoma sgc-7901 cells. *Neoplasma* 2012;**59**(2):142–9.
83. Zhang B, Yang Y, Shi X, Liao W, Chen M, Cheng AS, et al. Proton pump inhibitor pantoprazole abrogates adriamycin-resistant gastric cancer cell invasiveness via suppression of Akt/GSK-β/β-catenin signaling and epithelial–mesenchymal transition. *Cancer Lett* 2015;**356**(2):704–12.
84. Cao Y, Chen M, Tang D, Yan H, Ding X, Zhou F, et al. The proton pump inhibitor pantoprazole disrupts protein degradation systems and sensitizes cancer cells to death under various stresses. *Cell Death Dis* 2018;**9**(6):604.
85. Kam S, Joo MK, Koh JS, Lee BJ, Park JJ, Chun HJ, et al. Synergic effect of pantoprazole and docetaxel to inhibit EMT via induction of SHP-1 in gastric cancer cells. *J Gastroenterol Hepatol* 2018;**33**:185.
86. Tan Q, Joshua AM, Wang M, Bristow RG, Wouters BG, Allen CJ, et al. Up-regulation of autophagy is a mechanism of resistance to chemotherapy and can be inhibited by pantoprazole to increase drug sensitivity. *Cancer Chemother Pharmacol* 2017;**79**(5):959–69.
87. Wang X, Liu C, Wang J, Fan Y, Wang Z, Wang Y. Proton pump inhibitors increase the chemosensitivity of patients with advanced colorectal cancer. *Oncotarget* 2017;**8**(35):58801.
88. Liu M, Tang R, Jiang Y. Pantoprazole induces apoptosis of leukemic cells by inhibiting expression of P-glycoprotein/multidrug resistance-associated protein-1 through PI3K/AKT/mTOR signaling. *Indian J Hematol Blood Transfus* 2017;**33**(4):500–8.
89. Geeviman K, Babu D, Babu PP. Pantoprazole induces mitochondrial apoptosis and attenuates NF-κB signaling in glioma cells. *Cell Mol Neurobiol* 2018;**38**(8):1491–504.
90. Vishvakarma NK, Singh SM. Immunopotentiating effect of proton pump inhibitor pantoprazole in a lymphoma-bearing murine host: implication in antitumor activation of tumor-associated macrophages. *Immunol Lett* 2010;**134**(1):83–92.
91. Huang S, Chen M, Ding X, Zhang X, Zou X. Proton pump inhibitor selectively suppresses proliferation and restores the chemosensitivity of gastric cancer cells by inhibiting STAT3 signaling pathway. *Int Immunopharmacol* 2013;**17**(3):585–92.
92. Azzarito T, Venturi G, Cesolini A, Fais S. Lansoprazole induces sensitivity to suboptimal doses of paclitaxel in human melanoma. *Cancer Lett* 2015;**356**(2 Pt. B):697–703.
93. Zhang S, Wang Y, Li SJ. Lansoprazole induces apoptosis of breast cancer cells through inhibition of intracellular proton extrusion. *Biochem Biophys Res Commun* 2014;**448**(4):424–9.
94. Zhang JT. *Targeting FASN for breast cancer treatment by repositioning PPIs.* Indianapolis, United States: Trustees of Indiana University; 2017. Jan 1 Available at *https://apps.dtic.mil/dtic/tr/fulltext/u2/1035028.pdf.*

95. Federici C, Lugini L, Marino ML, Carta F, Iessi E, Azzarito T, et al. Lansoprazole and carbonic anhydrase IX inhibitors sinergize against human melanoma cells. *J Enzyme Inhib Med Chem* 2016;**31**: 119–25.
96. Li X, Yan ML, Yu Q. Identification of candidate drugs for the treatment of metastatic osteosarcoma through a subpathway analysis method. *Oncol Lett* 2017;**13**(6):4378–84.
97. Avnet S, Di Pompo G, Lemma S, Salerno M, Perut F, Bonuccelli G, et al. V-ATPase is a candidate therapeutic target for Ewing sarcoma. *Biochim Biophys Acta* 2013;**1832**(8):1105–16.
98. Patlolla JM, Zhang Y, Li Q, Steele VE, Rao CV. Anticarcinogenic properties of omeprazole against human colon cancer cells and azoxymethane-induced colonic aberrant crypt foci formation in rats. *Int J Oncol* 2012;**40**(1):170–5.
99. Udelnow A, Kreyes A, Ellinger S, Landfester K, Walther P, Klapperstueck T, et al. Omeprazole inhibits proliferation and modulates autophagy in pancreatic cancer cells. *PLoS One* 2011;**6**(5):e20143.
100. Jin UH, Lee SO, Pfent C, Safe S. The aryl hydrocarbon receptor ligand omeprazole inhibits breast cancer cell invasion and metastasis. *BMC Cancer* 2014;**14**:498.
101. Peng YC, Lin CL, Hsu WY, Lu IT, Yeh HZ, Chang CS, et al. Proton pump inhibitor use is associated with risk of pancreatic cancer: a nested case–control study. *Dose-Response* 2018;**16**(4):1559325818803283. https://doi.org/10.1177/1559325818803283.
102. Hicks B, Friis S, Pottegård A. Use of proton pump inhibitors and risk of pancreatic cancer. *Pharmacoepidemiol Drug Saf* 2018;**27**(8):926–30.
103. Spugnini EP, Baldi A, Buglioni S, Carocci F, de Bazzichini GM, Betti G, et al. Lansoprazole as a rescue agent in chemoresistant tumors: a phase I/II study in companion animals with spontaneously occurring tumors. *J Transl Med* 2011;**9**(1):221.

CHAPTER 9

The sodium bicarbonate cotransporter (NBC) family

Introduction

CO_2 inside the cell induces a fall of pHi.[1] Thomas[2] showed that non-CO_2 buffers are much less effective in lowering pHi than CO_2. This finding shows that CO_2 is a powerful intracellular acidifier. The reason for this intracellular acidification has been known since the 1940s: CO_2 is hydrated to carbonic acid, which immediately dissociates into a proton (H^+) and a bicarbonate ion (CO_3H^-).[3]

CO_2 is a highly diffusive gas, but the diffusion of CO_2 from the cell depends mainly on the external concentration of CO_2 (CO_2 gradient).

Malignant cells face here a double problem:

(1) CO_2 production is increased

(2) Extracellular CO_2 is increased also due to the highly hypoxic microenvironment in which they live. Furthermore, the scanty and ineffective blood vessels lead to a decrease of CO_2 clearance.

These elements would create a very dangerous situation for malignant cell survival, unless an effective way of CO_2 handling is developed.

The malignant cells put three basic mechanisms in place to solve the problem:

(a) they generate new blood vessels (angiogenesis);

(b) they activate an effective method to reduce CO_2 in the extracellular space and

(c) they overexpress proteins with the ability to reduce intracellular pH.

These two last steps are strongly interconnected and regulated by a master protein: HIF-1α and complemented by the participation of a transcription factor: Sp1.

How do they reduce the amount of extracellular CO_2?

They achieve this by overexpressing and/or overactivating two membrane carbonic anhydrases (CA): CAIX and CAXII. These CAs located on the cell membrane exert their action on the CO_2 of the extracellular space through hydration. Hydrated CO_2 is the carbonic acid (CO_3H_2) that spontaneously dissociates into H^+ and CO_3H^-. While part of the H^+ remains in the extracellular space (ECS) contributing to its acidity, CO_3H^- is handled in a different way: it is reintroduced into the cell by a family of membrane proteins called sodium/bicarbonate cotransporters (NBC).

The activity of another protein, which is an exchanger and not a transporter, extrudes H^+: the sodium/hydrogen exchanger 1 (NHE1).

An Innovative Approach to Understanding and Treating Cancer: Targeting pH
https://doi.org/10.1016/B978-0-12-819059-3.00009-5
© 2020 Elsevier Inc.
All rights reserved.

These three groups of membrane proteins, NHE1, NBC, and CAs are the pivotal structure of the intracellular alkalinization process.

However, they are not alone in their struggle against intracellular acidification. The cellular buffer system, monocarboxylate transporters (MCTs) and V-ATPase proton pumps complete the team. We have called this association of highly coordinated cellular components the "pH-tome". It not only compensates for the acid overload produced in malignant cells, but by working together it can overshoot the pro-alkaline activity, so that these cells end up with a higher intracellular pHi than their normal counterparts (Fig. 1).

Rapid bicarbonate transport across the plasma membrane is essential for maintaining cellular and whole body pH, and disposing of the metabolic waste CO_2. For this purpose, multicellular organisms have developed a complex system that is based on carbonic anhydrases, sodium hydrogen exchangers and bicarbonate transporters and an efficient circulatory system that removes CO_2. In normal cells, NHE1 and NBC represent the main modifiers of intracellular pH.[4]

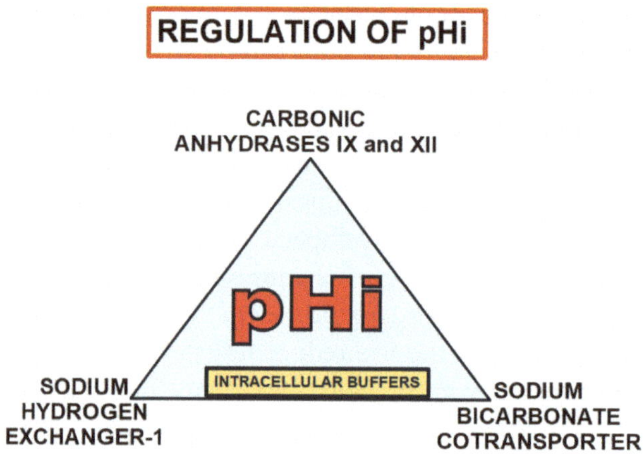

Fig. 1 Regulatory elements of the pHi.

Bicarbonate can be transported by different mechanisms but the two best-known and most important are:

(A) Anion exchangers or Cl^-/CO_3H^- exchangers.

(B) Na^+/bicarbonate cotransporters or NBC.

(A) **Anion exchangers** move bicarbonate into or out of the cell accompanied by chloride, thereby either alkalinizing or acidifying the cell, according to the direction. Anion exchangers not only maintain physiological pH, they are also involved in volume regulation and acid/base secretion.

(B) Na^+/HCO_3^- co-transport proteins (NBC) located in the plasma membrane, mediate the coupled movement of Na^+ and HCO_3^- across plasma membranes in an electrogenic manner. Altered NBC activity caused by genetic changes results in pathophysiologic conditions. The Na/HCO_3 cotransporter is the major HCO_3^- transporter of the renal proximal tubule, and also plays a significant role in Na^+ reabsorption. NBC transporters are important for the regulation of intracellular pH (pHi) in most cells and thereby also regulate blood pH.[5]

Our description will be limited to the NBC family due to its importance in cancer: SLC4A4, SLC4A5, SLC4A7, SLC4A8 and SLC4A10 to be more precise that are importers of Na^+ and CO_3H^-. The most studied isoform in relation to cancer is SLC4A7 (also known as NBCn1).

The Na^+-coupled bicarbonate (NCB) transporter family includes three different groups of transporters[6]: This family is also known as SLC4A.

(1) electrogenic NBCs (E): NBCe1 and NBCe2
(2) electroneutral NBCs (EN): NBCn1 and NBCn2
(3) Na^+-dependent Cl^--HCO_3^- exchanger (NDCBE)

Electrogenic means that their activity produces a shift of membrane potential, while those that are electroneutral lack this ability. The electrogenic change increases the membrane potential to more negative values.

The SLC4A family has a capital and vital function in mammals: it forms an essential part in the chain of CO_2 transport from the systemic capillaries to the pulmonary capillaries.[7]

As the naming system of this family is somewhat confusing, Table 1 shows a list of NBCs with their proteins, functions and possible role in cancer.

SLC4A4 (NBCe1) will be used as an example for describing how the NBCs work to alkalinize pHi. For each Na^+ molecule that they import they also import 2 or 3 $CO3H^-$ molecules. This stoichiometry depends on the cells where they are located: 2:1 in collecting ducts and 3:1 in proximal tubules of the kidney.[10] Importantly, the stoichiometry is phosphorylation dependent.[11,12]

NBC structure and isoforms

NBC is composed of 1035 amino acids (may vary between different isoforms) and it has 10 to 14 transmembrane segments. Usually it is found as a dimer but functionally it can work as a monomer.[13]

Many isoforms have been described but not all of them participate in cancer. SLC4A1 is also related to cancer but by a mechanism not involved in pH modification.[14]

At invadopodia, NBCe1 (SLC4A4) co-localizes with carbonic anhydrase IX and anion exchanger 2 (AE2) and seems to form part of the migration apparatus. There are interactions among these three molecules.[15]

Table 1 The different known isoforms of the NBC family.

Gene	Protein	Imports to the cell	Extrudes	Function	Implicated in cancer
SLC4A4 Electrogenic	NBCe1 Splice variants NBCe1-A NBCe1-B NBCe1-C NBCe1-D NBCe1-E	Na^+ and CO_3H^-		pHi alkalinizer	
SLC4A5 Electrogenic	NBCe2	Na^+ and CO_3H^-		pHi alkalinizer	
SLC4A7 Electroneutral	NBCn1	Na^+ and CO_3H^-		pHi alkalinizer	
SLC4A8 Electroneutral		Na^+ and CO_3H^-		pHi alkalinizer	
SLC4A9 Electroneutral		Na^+ and CO_3H^-		pHi alkalinizer	
SLC4A10 Electroneutral		Na^+ and CO_3H^-		pHi alkalinizer	
SLC4A1	AE1	Cl^-[a]	Na^+ and CO_3H^-	pHi acidifier	
SLC4A2	AE2	Cl^-[a]	Na^+ and CO_3H^-	pHi acidifier	
SLC4A3	AE3	Cl^-[a]	Na^+ and CO_3H^-	pHi acidifier	
SLC4A9	AE4	(?)	(?)	See Peña-Münzenmayer[8]	
SLC4A11	BTR1	Borate			
SLC26 family					
SLC26A3		Cl^-	CO_3H^-	pHi acidifier	
SLC26A4		Cl^-	CO_3H^-	pHi acidifier	
SLC26A7		Cl^-	CO_3H^-	pHi acidifier	
SLC26A9		Cl^-	CO_3H^-	pHi acidifier	
SLC26A6		CO_3H^-	Cl^-	Acid loader/ extruder	
SLC26A11		CO_3H^-	Cl^-	Acid loader/ extruder	
SLC26A8		Unknown	Unknown	Unknown	

[a] The direction of movement of Cl^- depends on the Cl^- gradient. Usually Cl^- is imported. Therefore, AE1, AE2 and AE3 work like Cl^-/CO_3H^- exchangers where Cl^- is imported and CO_3H^- is extruded. The net result is cellular acidification.[9]

NBC structure

NBCs are membrane proteins with 10 to 14 segments spanning the plasmatic membrane (Fig. 2). They are dimeric and cystein bridges creating the union that stabilizes the structure. However, they are fully functional in the monomeric stage.

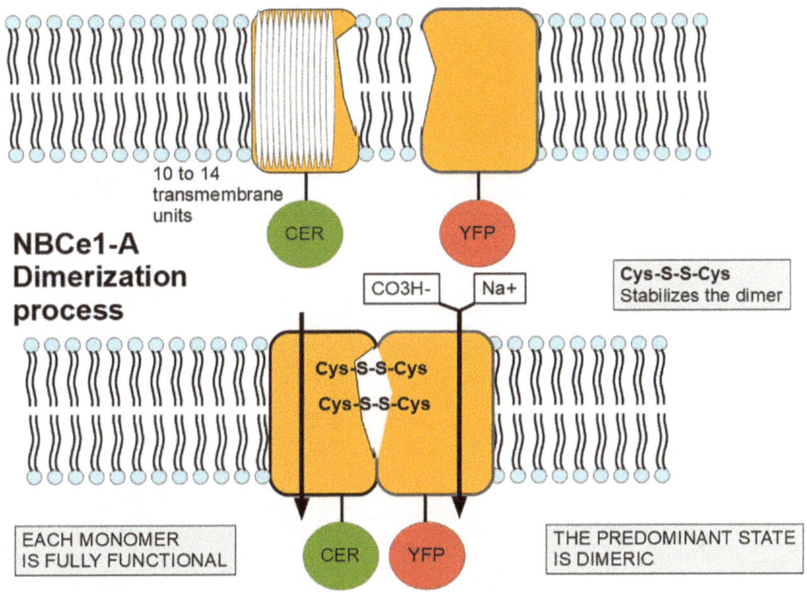

Fig. 2 Structure of NBC isoform NBCe1-A.[16]

NBC modulation in normal tissues

The role of the sodium bicarbonate cotransporter (NBC) family is of utmost importance in acid-base homeostasis because bicarbonate transport mechanisms and sodium/hydrogen exchange are the main regulators of pH in animal cells. Almost all tissues express a member of the NBC family but especially in kidney, stomach, pancreas, intestine, reproductive organs and the central nervous system. The NBC family has been studied extensively in the kidney because this is where it plays a pivotal role in the acid-base equilibrium of the whole organism. Its most important function in the kidney is at the proximal tubule $HCO3^-$ reabsorption where it works in close association with carbonic anhydrases.

The thyroid gland is an interesting example. Stimulation of the gland with thyrotropin (TSH) increases the intracellular pH, which is maintained at a pHi of 7.6 while TSH stimulation is active. This pH is achieved through NBC activity. The mechanism induced by TSH is not the increase of NBC expression, but the blocking of its degradation.[17] Another interesting example of NBC's activity in non-malignant tissues is that of the intervertebral nucleus pulposus. This tissue is always under hypoxic conditions. The extracellular bicarbonate generated by membrane carbonic anhydrases, which is recycled into the cells by NBCs, is vital for an adequate condition of the nucleus pulposus cells. Long term inhibition of carbonic anhydrases compromises cell viability.[18]

Aldosterone increases expression of NBC in heart muscle.[19]

Regulation of NBCs (Fig. 3)

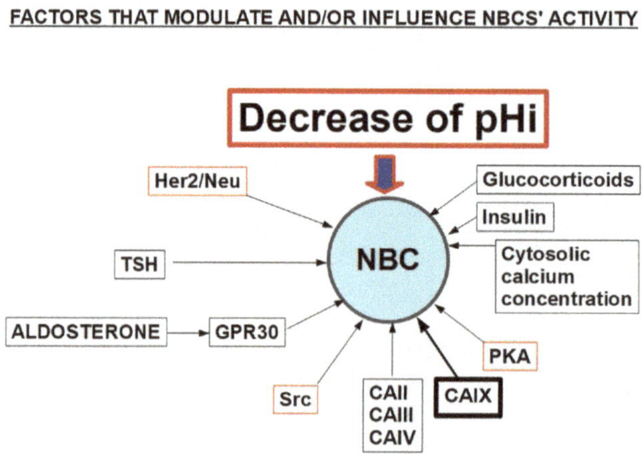

Fig. 3 Regulation of NBC.[20–23] Decrease of pHi is the most powerful of all the stimulants for NBC activity.

NBCs in disease

Under situations of hypoxia as in stroke, cells become acidic. Increased acidity activates NHE1 and NBC in order to compensate this pHi imbalance. This favors reperfusion injury.[24,25]

The best-known diseases related to NBC mutations include distal and proximal renal tubular acidosis, hemolytic anemia, ocular abnormalities, including blindness, and auditory impairment.

Cancer

Tumors are extremely hypoxic with a poorly developed vascular system, so that the circulatory system is hardly available for CO_2 removal. However, tumors develop overexpressed/overactive membrane carbonic anhydrases (CAIX and CAXII) coupled with overexpressed/overactive bicarbonate transport systems as shown in Fig. 4. The result of carbonic anhydrases and NBCs working together is an increase in cytoplasmic pH. The excess of CO2 produced by the malignant cell diffuses to the cell surface where

CAIX and CAXII hydrate the CO2 to carbonic acid. The spontaneous dissociation of carbonic acid generates the bicarbonate that is imported into the cell by NBCs.

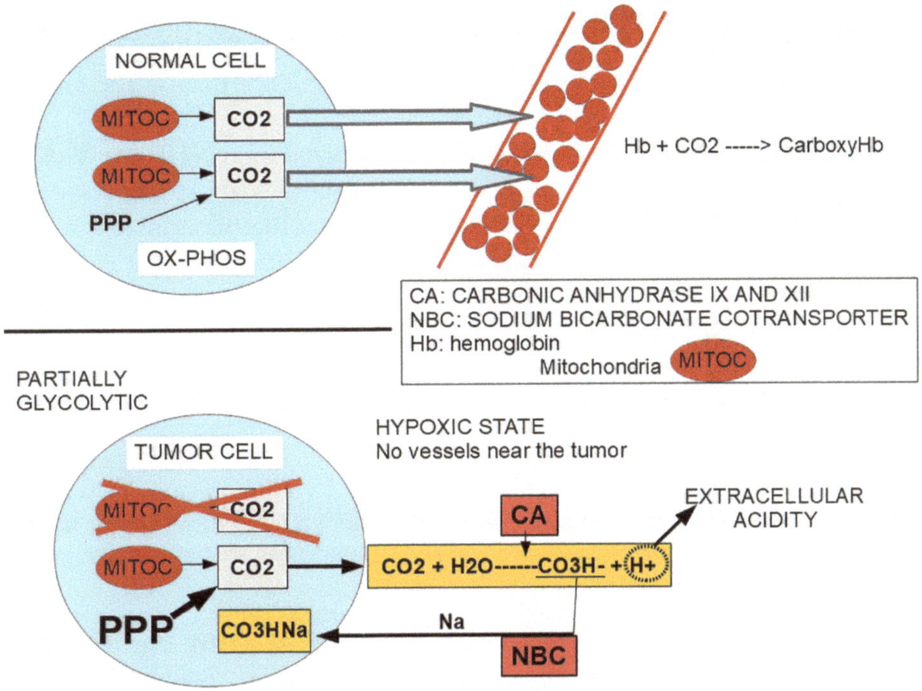

Fig. 4 Normal non-hypoxic cells under oxidative phosphorylation (OXPHOS) metabolism produce CO_2 which diffuses to the blood where hemoglobin in the form of carboxyhemoglobin transports it to the lungs where finally CO_2 is expelled. CO_2 needs to be removed from the body because it is acid (forms CO_3H_2 with water) and it would alter pH homeostasis. Tumor cells on the contrary have a limited OXPHOS activity because metabolism is predominantly glycolytic. But there are no blood vessels nearby, so CO_2 has to be disposed off in a different way. Carbonic anhydrase hydrates CO_2 at the plasma membrane generating CO_3H_2, which is immediately dissociated into CO_3H^- and H^+. While the proton remains in the ECS participating in the acidification process, CO_3H^- is reintroduced into the cell with Na^+ by NBCs participating in producing the intracellular alkalinity. As can be seen, carbonic anhydrases IX and XII work in tandem with NBCs. While CO_2 is a gas, which can move freely out of the cell, bicarbonate cannot do the same and requires a specialized mechanism in order to move through membranes.

NBC and cancer

Boedtkjer et al.[26] compared normal breast tissue with human breast cancer regarding specifically NBC1 (an isoform of NBC), and they found that malignant tissue:

- Had 30% higher expression of NBC1.
- NBC was the predominant form of pHi alkalinization.

- Inhibition of NBC by 200 μM 4,4′ diisothiocyanatostilbene 2,2′ disulfonic acid reached 34 ± 9% in human primary breast carcinomas.
- At intracellular pH (pHi) levels >6.6, CO_2/HCO_3^- dependent mechanisms accounted for >90% of total net acid extrusion.
- Na^+/H^+ exchange activity was prominent only at lower pHi values.
- Steady-state pHi was 0.35 ± 0.06 units lower in the absence than in the presence of CO_2/HCO_3^-.

NBCs were never as popular in studying cancer pathogenesis as NHE1 or V-ATPases, but researchers are gradually realizing that NBC transporters are strong players in the issue, at least in breast cancer, and probably in many other tumors as investigations progress. A small example of the evidence showing NBCs activities in cancer is summarized in Table 2.

Table 2 Evidence of NBCs′ role in cancer.

Reference	Findings
Boedtkjer et al.[26]	As explained above
Long et al.[27]	Association between SNP at 3p24/SLC4A7 and breast cancer risk in Chinese women.(Genome-wide association studies)
Chen et al.[28]	Idem risks of SNP at 3p24/SLC4A7
Gorbatenko et al.[29]	ErbB2 (Her2/Neu) controls NBC1 in breast cancer. The activity of ErbB2 increased the expression of NBC1 but not NHE1. The rate of pH(i) recovery after acid loading increased by 50% upon ErbB2 expression. Knockdown experiments confirmed the involvement of both NHE1 and NBCn1 in acid extrusion. NHE1 inhibition sensitized ErbB2-expressing cells to cisplatin-induced apoptosis. NBCn1 inhibition attenuated cathepsin release. NHE1 and NBCn1 have different functions in breast cancer cells.
Ahmed et al.[30]	Breast cancer risk increased by polymorphism at 3p24/SLC4A7 (Genome-wide association studies)
Sueta et al.[31]	Idem in Japanese population
Lauritzen et al.[32]	ErbB2 induced expression of NBCn1, but not of NHE1 in breast carcinoma cells. However, knockdown of NBC and NHE1experiments showed overexpression of both. NHE1 inhibition increased sensitivity to cisplatin, but NBC inhibition did not change the viability of the cells.
Kumar et al.[33]	Coronary endothelial cells under ischemia are protected from apoptosis by SLC4A7. Antiapoptotic action?
Parks et al.[34]	Participation of SLC4A4 in growth and migration of colon and breast cancer cells was found.
McIntyre et al.[35]	Inhibition of SLC4A4 or SLC4A9 acidified cellular cytoplasm and induced apoptosis in MDA-MB-233 breast cancer cells and U87 glioma cells.
Lee et al.[36]	In human breast carcinogenesis acid extrusion was increased due to NBC activity which increased intracellular pH by approximately 0.3 units compared to normal breast tissue.

NBC and migration

NHE1 and NBC seem to have similar actions regarding pHi increases. NHE1 is an essential figure in the migration progress (see chapter on invadopodia). However, NBC is much less important than NHE1 in migration.[37]

Should NBC be inhibited for cancer treatment?

In order to achieve a maximal intracellular acidification, it may be important to inhibit NBC. However, there are no clinically available NBC inhibitors. The fact that NBC imports bicarbonate, if membrane carbonic anhydrases are disabled, the importance of NBC is relatively reduced. In theory, CA inhibition may partially reduce the need to inhibit NBC. There is evidence of cross-talk between membrane CAs and NBCs: Orlowsky et al.[38] showed that there is binding of CAIX to the extracellular loop 4 of NBCe1 and this binding increases NBC activity. There are also two indirect ways of intercommunication:

(1) pHi

(2) peri-cellular bicarbonate that is generated through CA activity and dissociation of the carbonic acid thus formed.

Lee et al.[39] have further shown the relationship between ErbB2 (Her2/neu) and NBC in mice with ErbB2 induced breast cancer:

(1) NBCn1 was increased 2 fold compared to normal breast tissue;

(2) NBCn1 elevated intracellular pH;

(3) Blocking NBCn1 delayed ErbB2-induced carcinogenesis;

(4) Knocking out NBC1 delayed tumor growth.

Another interesting finding in this report was that, in spite of a twofold increase in NBCn1 protein, the NBCn1 mRNA level decreased by fourfold. This leads us to theorize that the mechanism of action of ErbB2 may be similar to that found in the thyroid with TSH: a decrease in the destruction/inactivation of NCB protein.

Cytoplasmic CAII and CAIII can increase NBC activity in Xenopus.[40]

The inhibition of Src (an important kinase in the formation of invadopodia) decreased NBC activity in normal submandibular glands of rats. This Src regulatory activity was specific for NBCe1 but not for NBCn1.[41] The importance of this Src/NBCe1 relation in cancer has not been determined yet.

Conclusions

1. Bicarbonate transporters cooperate in the alkalinization of the intracellular milieu thus promoting tumor growth.
2. Inhibiting these transporters inhibits tumor growth.
3. There is a functional relationship between carbonic anhydrase IX and NBC.
4. The importance of bicarbonate transporters in cancer was not fully understood until recently.

5. We believe that bicarbonate transporters may act in some situations as a NHE1 proxy. This seems to happen when pHi is above 6.6.
6. The association of NBC with carbonic anhydrase IX increases the alkaline influx into the cell.
7. NBC and NHE1 share many functional similarities, but NBC activity is negligible in migration while NHE1 has an important role.

References

1. Roos A, Boron WF. Intracellular pH. *Physiol Rev* 1981;**61**(2):296–434.
2. Thomas RC. Intracellular pH of snail neurones measured with a new pH-sensitive glass microelectrode. *J Physiol (Lond)* 1974;**238**:159–80.
3. Jacobs MH. Some aspects of cell permeability to weak electrolytes. *Cold Spring Harb Symp Quant Biol* 1940;**8**:30–9.
4. Bevensee MO, Weed RA, Boron WF. Intracellular pH regulation in cultured astrocytes from rat hippocampus: I. Role of HCO_3^-. *J Gen Physiol* 1997;**110**(4):453–65.
5. Boron WF, Fong P, Hediger MA, Boulpaep EL, Romero MF. The electrogenic Na/HCO3 cotransporter. *Wien Klin Wochenschr* 1997;**109**(12 – 13):445–56.
6. Boron WF, Chen L, Parker MD. Modular structure of sodium-coupled bicarbonate transporters. *J Exp Biol* 2009;**212**(11):1697–706.
7. Romero MF, Fulton CM, Boron WF. The SLC4 family of HCO – 3 transporters. *Pflugers Arch—Eur J Physiol* 2004;**447**(5):495–509.
8. Peña-Münzenmayer G, George AT, Shull GE, Melvin JE, Catalán MA. Ae4 (Slc4a9) is an electroneutral monovalent cation-dependent Cl –/HCO3 – exchanger. *J Gen Physiol* 2016;**147**(5):423–36.
9. Alper SL, Darman RB, Chernova MN, Dahlberg CG. The AE gene family of Cl– /HCO3 exchangers. *J Nephrol* 2002;**15**(Suppl. 5):S41–53.
10. Gross E, Hawkins K, Abuladze N, Pushkin A, Cotton CU, Hopfer U, Kurtz I. The stoichiometry of the electrogenic sodium bicarbonate cotransporter NBC1 is cell-type dependent. *J Physiol* 2001;**531**(3):597–603.
11. Gross E, Hawkins K, Pushkin A, Sassani P, Dukkipati R, Abuladze N, et al. Phosphorylation of Ser982 in the sodium bicarbonate cotransporter kNBC1 shifts the HCO3 –: Na + stoichiometry from 3: 1 to 2: 1 in murine proximal tubule cells. *J Physiol* 2001;**537**(3):659–65.
12. Gross E, Kurtz I. Structural determinants and significance of regulation of electrogenic Na +-HCO 3 – cotransporter stoichiometry. *Am J Physiol Ren Physiol* 2002;**283**(5):F876–87.
13. Kao L, Sassani P, Azimov R, Pushkin A, Abuladze N, Peti-Peterdi J, et al. Oligomeric structure and minimal functional unit of the electrogenic sodium bicarbonate cotransporter NBCe1-A. *J Biol Chem* 2008;**283**(39):26782–94.
14. Shen WW, Wu J, Cai L, Liu BY, Gao Y, Chen GQ, et al. Expression of anion exchanger 1 sequestrates p16 in the cytoplasm in gastric and colonic adenocarcinoma. *Neoplasia* 2007;**9**:812–9.
15. Svastova E, Witarski W, Csaderova L, Kosik I, Skvarkova L, Hulikova A, et al. Carbonic anhydrase IX interacts with bicarbonate transporters in lamellipodia and increases cell migration via its catalytic domain. *J Biol Chem* 2012;**287**(5):3392–402.
16. Bernardo AA, Bernardo CM, Espiritu DJ, Arruda JA. The sodium bicarbonate cotransporter: structure, function, and regulation. *Semin Nephrol* 2006;**26**(5):352–60. WB Saunders.
17. Virreira Bermudez M, Jin L, Djerbib S, De Deken X, Miot F, Massart C, et al. Expression, localization and regulation of the sodium bicarbonate cotransporter NBCe1 in the thyroid. *Thyroid* 2019. https://doi.org/10.1089/thy.2017.0576.
18. Silagi ES, Schoepflin ZR, Seifert EL, Merceron C, Schipani E, Shapiro IM, Risbud MV. Bicarbonate recycling by HIF-1–dependent carbonic anhydrase isoforms 9 and 12 is critical in maintaining intracellular pH and viability of nucleus pulposus cells. *J Bone Miner Res* 2018;**33**(2):338–55.

19. Orlowski A, De Giusti VC, Ciancio MC, Espejo MS, Aiello EA. The cardiac electrogenic sodium/bicarbonate cotransporter (NBCe1) is activated by aldosterone through the G protein-coupled receptor 30 (GPR 30). *Channels* 2016;**10**(5):428–34.
20. Gross E, Hawkins K, Pushkin A, Sassani P, Dukkipati R, Abuladze N, et al. Phosphorylation of ser(982) in the sodium bicarbonate cotransporter kNBC1 shifts the HCO(3)(−): Na(+) stoichiometry from 3: 1 to 2: 1 in murine proximal tubule cells. *J Physiol* 2001;**537**(3):659–65.
21. Fantinelli JC, Orlowski A, Aiello EA, Mosca SM. The electrogenic cardiac sodium bicarbonate co-transporter (NBCe1) contributes to the reperfusion injury. *Cardiovasc Pathol* 2014;**23**(4):224–30.
22. Ruiz OS, Wang LJ, Pahlavan P, Arruda JA. Regulation of renal Na-HCO3 cotransporter: III. Presence and modulation by glucocorticoids in primary cultures of the proximal tubule. *Kidney Int* 1995;**47**:1669–76.
23. Ruiz OS, Qiu YY, Cardoso LR, Arruda JA. Regulation of the renal Na-HCO3 cotransporter: IX. Modulation by insulin, epidermal growth factor and carbachol. *Regul Pept* 1998;**77**:155–61.
24. Khandoudi N, Albadine J, Robert P, Krief S, Berrebi-Bertrand I, Martin X, et al. Inhibition of the cardiac electrogenic sodium bicarbonate cotransporter reduces ischemic injury. *Cardiovasc Res* 2001;**52**(3):387–96.
25. Fantinelli JC, Orlowski A, Aiello EA, Mosca SM. The electrogenic cardiac sodium bicarbonate co-transporter (NBCe1) contributes to the reperfusion injury. *Cardiovasc Pathol* 2014;**23**(4):224–30.
26. Boedtkjer E, Moreira J, Mele M, Vahl P, Wielenga VT, Christiansen PM, et al. Contribution of Na+, HCO3−-cotransport to cellular pH control in human breast cancer: a role for the breast cancer susceptibility locus NBCn1 (SLC4A7). *Int J Cancer* 2013;**132**(6):1288–99.
27. Long J, Shu XO, Cai Q, Gao YT, Zheng Y, Li G, et al. Evaluation of breast cancer susceptibility loci in Chinese women. *Cancer Epidemiol Biomarkers Prev* 2010;**19**:2357–65.
28. Chen W, Zhong R, Ming J, Zou L, Zhu B, Lu X, et al. The SLC4A7 variant rs4973768 is associated with breast cancer risk: evidence from a case–control study and a meta-analysis. *Breast Cancer Res Treat* 2012;**136**(3):847–57.
29. Gorbatenko A, Olesen CW, Lauritzen G, Valen E, Pedersen SF. Regulation of the Na+, HCO−3cotransporter NBCn1 (SLC4A7) by a constitutively active ErbB2 receptor in MCF-7 breast cancer cells. *FASEB J* 2014;**28**:350–63.
30. Ahmed S, Thomas G, Ghoussaini M, Healey CS, Humphreys MK, Platte R, et al. Newly discovered breast cancer susceptibility loci on 3p24 and 17q23.2. *Nat Genet* 2009;**41**:585–90.
31. Sueta A, Ito H, Kawase T, Hirose K, Hosono S, Yatabe Y, et al. A genetic risk predictor for breast cancer using a combination of low-penetrance polymorphisms in a Japanese population. *Breast Cancer Res Treat* 2012;**132**:711–21.
32. Lauritzen G, Jensen MB, Boedtkjer E, Dybboe R, Aalkjær C, Nylandsted J, Pedersen SF. NBCn1 and NHE1 expression and activity in ΔNErbB2 receptor-expressing MCF-7 breast cancer cells: contributions to pHi regulation and chemotherapy resistance. *Exp Cell Res* 2010;**316**(15):2538–53.
33. Kumar S, Flacke JP, Kostin S, Appukuttan A, Reusch HP, Ladilov Y. SLC4A7 sodium bicarbonate co-transporter controls mitochondrial apoptosis in ischaemic coronary endothelial cells. *Cardiovasc Res* 2010;**89**(2):392–400.
34. Parks SK, Pouyssegur J. The Na+/HCO3− co-transporter SLC4A4 plays a role in growth and migration of colon and breast cancer cells. *J Cell Physiol* 2015;**230**(8):1954–63.
35. McIntyre A, Hulikova A, Ledaki I, Snell C, Singleton D, Steers G, et al. Disrupting hypoxia-induced bicarbonate transport acidifies tumor cells and suppresses tumor growth. *Cancer Res* 2016. https://doi.org/10.1158/0008-5472.CAN-15-1862. Jan 1:canres-1862.
36. Lee S, Mele M, Vahl P, Christiansen PM, Jensen VE, Boedtkjer E. Na+, HCO3−-cotransport is functionally upregulated during human breast carcinogenesis and required for the inverted pH gradient across the plasma membrane. *Pflugers Arch—Eur J Physiol* 2015;**467**(2):367–77.
37. Schwab A, Rossmann H, Klein M, Dieterich P, Gassner B, Neff C, et al. Functional role of Na+–HCO3− cotransport in migration of transformed renal epithelial cells. *J Physiol* 2005;**568**(2):445–58.
38. Orlowski A, De Giusti VC, Morgan PE, Aiello EA, Álvarez BV. Binding of carbonic anhydrase IX to extracellular loop 4 of the NBCe1 Na+/HCO3− cotransporter enhances NBCe1-mediated HCO3− influx in the rat heart. *Am J Physiol Cell Physiol* 2012;**303**(1):C69–80.

39. Lee S, Axelsen TV, Jessen N, Pedersen SF, Vahl P, Boedtkjer E. Na+,HCO3-cotransporter NBCn1 (Slc4a7) accelerates ErbB2-induced breast cancer development and tumor growth in mice. *Oncogene* 2018;**37**(41):5569–84.
40. Schüler C, Becker HM, McKenna R, Deitmer JW. Transport activity of the sodium bicarbonate cotransporter NBCe1 is enhanced by different isoforms of carbonic anhydrase. *PLoS One* 2011;**6**:e27167.
41. Namkoong E, Shin YH, Bae JS, Choi S, Kim M, Kim N, et al. Role of sodium bicarbonate cotransporters in intracellular pH regulation and their regulatory mechanisms in human submandibular glands. *PLoS One* 2015;**10**(9):e0138368.

CHAPTER 10

pH gradient inversion, aquaporins and cancer

Introduction

All living organisms are highly dependent on the movement of water across cell membranes. Unicellular organisms, prokaryotes and eukaryotes alike, regulate their relationship with the environment through the trafficking of water, ions, nutrients and gases (WINGs). Multicellular organisms, particularly those that are organized in tissues, develop different compartments that require strict regulation of WINGs' movement in order to keep homeostasis.

Water constitutes around 70% of the mass of most living organisms, so the orderly distribution of WINGs, and particularly water, is required to maintain proper fluid and ion balance within different anatomic compartments. Cell membranes are hydrophobic, however, water is known to diffuse through them. This diffusion is a slow process that is insufficient for many physiological needs. Water diffusion without water channels is 2.5 times slower than with osmotic water channels.[1] Water channels, whose existence was suspected but not proved until the 1980s, are inactivated by eliminating osmotic pressure, leaving diffusion as the only mechanism for water movement.[2–3] In this case water movement is slow. This clear difference of water speed across the plasmatic membrane of erythrocytes (with and without osmotic pressure) was the factor that stimulated the search for water channels (Fig. 1). Membranes are basically hydrophobic and heterogeneous,[4] so that water diffusion is a slow process and not uniformly distributed. To overcome this evolutionary limitation, a family of membrane channel proteins evolved for rapid transport of water across biological membranes when an osmotic gradient exists.[5] 14 members of this family have been so far identified in higher mammals and many more in plants. These proteins, called "aquaporins," (AQP) were found in all life forms, including archaea, eubacteria, fungi, and plants. The members of this family are phylogenetically highly conserved.[6–7]

In 2003 the Nobel Prize in Chemistry was awarded to Peter Agre for the discovery of aquaporins (AQPs) in 1987.[8–10] Since the initial discovery considerable knowledge has been gathered on AQPs, but there is still much more to know, therefore the subject of AQPs is currently the object of an ever-growing interest not only in cancer research but in many other fields of medicine.

An Innovative Approach to Understanding and Treating Cancer: Targeting pH
https://doi.org/10.1016/B978-0-12-819059-3.00010-1
© 2020 Elsevier Inc.
All rights reserved.

Regarding cancer, seven aquaporins have been identified in close relationship with invasion, metastasis and angiogenesis. Although the exact mechanism is not fully known, we discuss here on the possible mechanisms for this behavior, including some speculation on the subject.

Basically, AQPs are water (and/or glycerin) channels that are induced by hypertonicity.

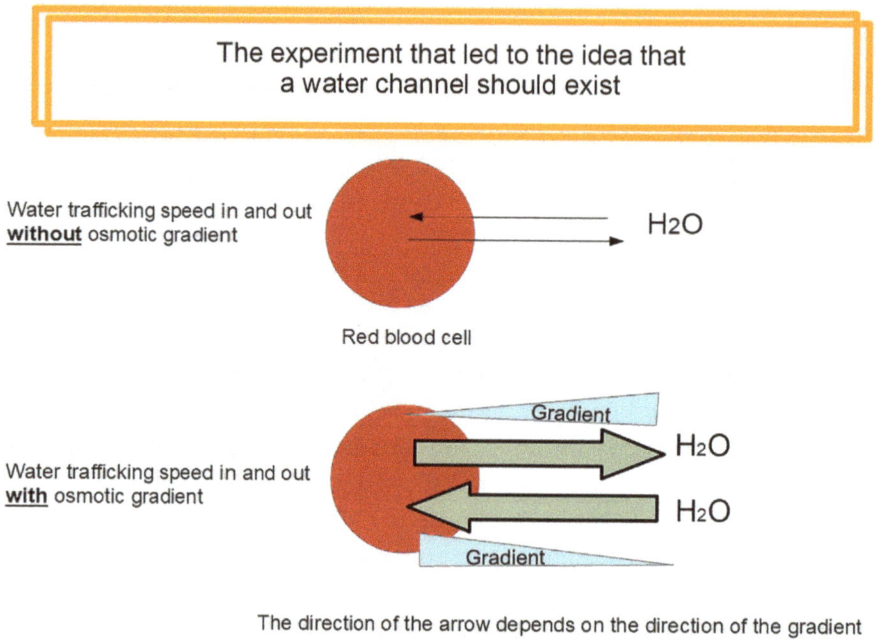

Fig. 1 A hyperosmolar gradient is needed for swift movement of water in the direction of the gradient. This speed is not achieved by diffusion alone.

The different kinds and roles of aquaporins

Aquaporin 1's (AQP1) expression is induced by a hyperosmolar gradient. The promoter region of the AQP1 gene has an osmolar element response. The MAPkinase's cascade is absolutely necessary for the AQP1's gene expression. Hyperosmolarity increases the MAPkinase cascade's phosphorylation, thus the activation of this cascade. If one or more members of this cascade are inhibited the AQP1's expression is strongly attenuated.[11]

AQP-facilitated water permeability is required for swift reabsorption of water in the kidneys and for water fluxes in the brain, eyes, internal ear and lungs, among other tissues. Furthermore, AQPs are strongly related to cell motility and angiogenesis. Aquaglyceroporins are also necessary for glycerol uptake in the liver[12] and they participate in skin hydration and adipocyte metabolism.[13] Some AQPs transport hydrogen peroxide.[14] Others may facilitate the flux of CO_2 and other gases.[15–16] However, this latter issue has been a matter of controversy.[17–18]

From a functional point of view, AQPs can be divided into three groups[19]:

I. Water selective AQPs, the so called classical or orthodox AQPs: AQP0, AQP1, AQP2, AQP4 and AQP5. Only water can go through their pores.

II. Aquaglyceroporins permeable to water and glycerin: AQP3, AQP7, AQP9, AQP10. Glycerin and water are the usual passengers, but other molecules like ammonia can also use these pores. There is a controversy still going on about CO2.

III. Miscellaneous AQPs, or unorthodox AQPs: AQP6, AQP 8, and superaquaporins AQP11 and AQP12, whose permeability has not been fully elucidated. This last group (AQP11 and AQP12) can only be found in animals but not in plants.[20]

Some AQPs are organ specific, this is the case of AQP0 that can only be found in the eye lens.[21] AQP1 is widely distributed. AQP2 is mainly located in the kidney's collecting ducts. AQP3 cooperates in skin moistening and AQP4 is predominantly found in astroglia.

Apart from malignancy, there are different human diseases related to AQPs. For example, mutations in AQP2 cause nephrogenic diabetes insipidus and AQP0 mutations have been associated with cataract formation. However, humans lacking the AQP1 protein are phenotypically normal.[22]

pH and aquaporins

Cytosolic pH regulates aquaporin activity in plants,[23] decreasing its activity when pHi decreases under hypoxic conditions. The opposite has not been experimentally shown, but common sense indicates that an alkaline pH should increase its activity. Aquaporin has segments of its structure that function as pH sensors. These segments have been described in plants, but they are phylogenetically conserved in mammals. Although little is known about the mechanism of pH regulation of AQPs in cancer, we shall raise some hypothesis on the issue.

It seems that mammalian aquaporins, besides osmotic regulation, may also be regulated by external pH, at least this is the case for AQP 3, but not for AQP 0, 1, 2, 4 and 5.[24] However, other AQPs besides AQP3 are pH sensitive. In plants, the pH sensing is also intracellular in *Arabidopsis thaliana*, a small flowering plant in which the lateral emergence of roots is accelerated by the plant's equivalent to AQPs and delayed by its absence.[25] This plant's aquaporin has a phylogenetically conserved pH sensor at the Histidine 193 residue that regulates its functioning. AQPs are stimulated by auxins, the plants equivalent to growth factors. In many aspects the lateral root emergence follows similar steps to the invadopodia in tumors.

Aquaporine's structure

The human AQP1 is a small protein of 269 amino acids of 28 kilodalton that spans the membranes. It has six bilayer spanning domains and permits passive transport of water along an osmotic gradient. Its gene is located on chromosome 7 (7p14.3). Each of these domains is an alpha helix. In membranes, AQPs adopt a homotetramer structure in which each monomer is constituted by the six membrane spanning domains with a centrally oriented pore. The amino and carboxy-terminal ends are cytoplasmic. The C terminal region has three residues that can be phosphorylated. This phosphorylation determines the functional status of the channel[26] (Fig. 2). The six domains are slightly tilted (not seen

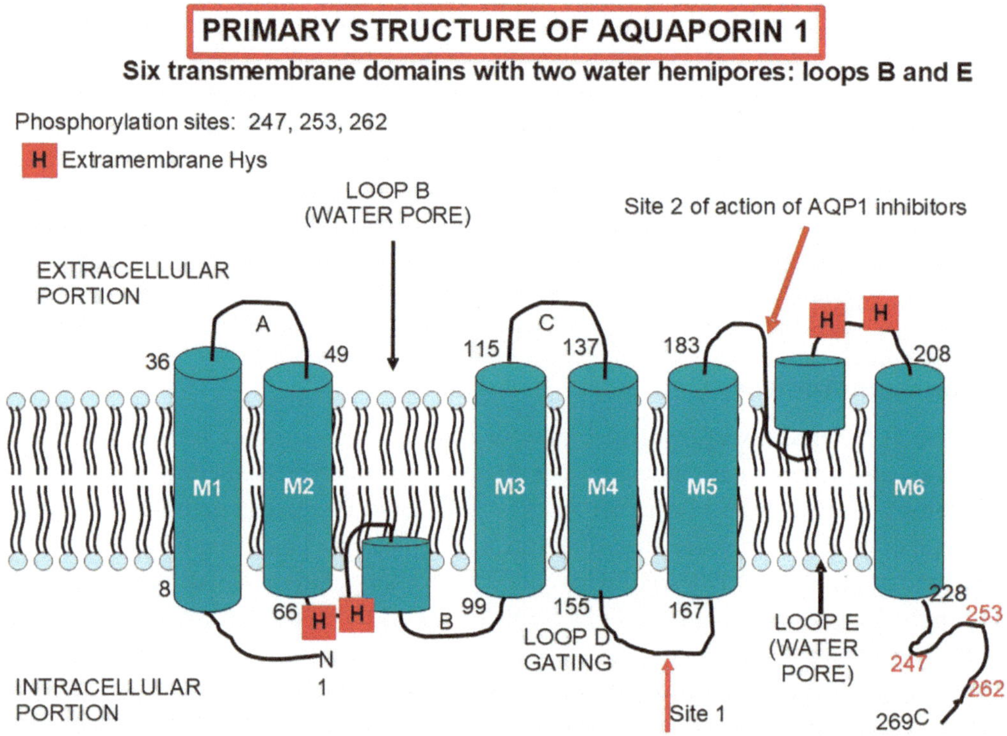

Fig. 2 Primary structure of human AQP1 at the cell membrane. This monomeric structure of 28 Kd with one pore is organized in a homotetrameric structure with a total of 4 pores. Numbers indicate approximately the amino acids position of this protein that has a total of 269. There are two asparagine-proline-alanine motifs (NPA) located in 76-78 and 192-194 positions. The first one is in an intracellular loop while the second one is in an extracellular loop. The phosphorylation of amino acids in the carboxy-terminal end produces conformational changes that play an important role in the gating processes in AQP4.[27] According to Horner et al.,[28] the pore's width in the constriction zone is the equivalent to one molecule of water and 4 to 8 water molecules long. Two intracellular histidine amino acids and two extracellular may be seen in the figure. In theory these 4 Hys may act as intracellular and extracellular pH sensors.

in Fig. 2: primary structure of aquaporins). They resemble an hourglass structure, wider in the external and internal pore and narrower in the middle. The pore has actually two narrow places: the most internal acts as a selective filter allowing the flow of water and excluding other molecules (Figs. 3 and 4).

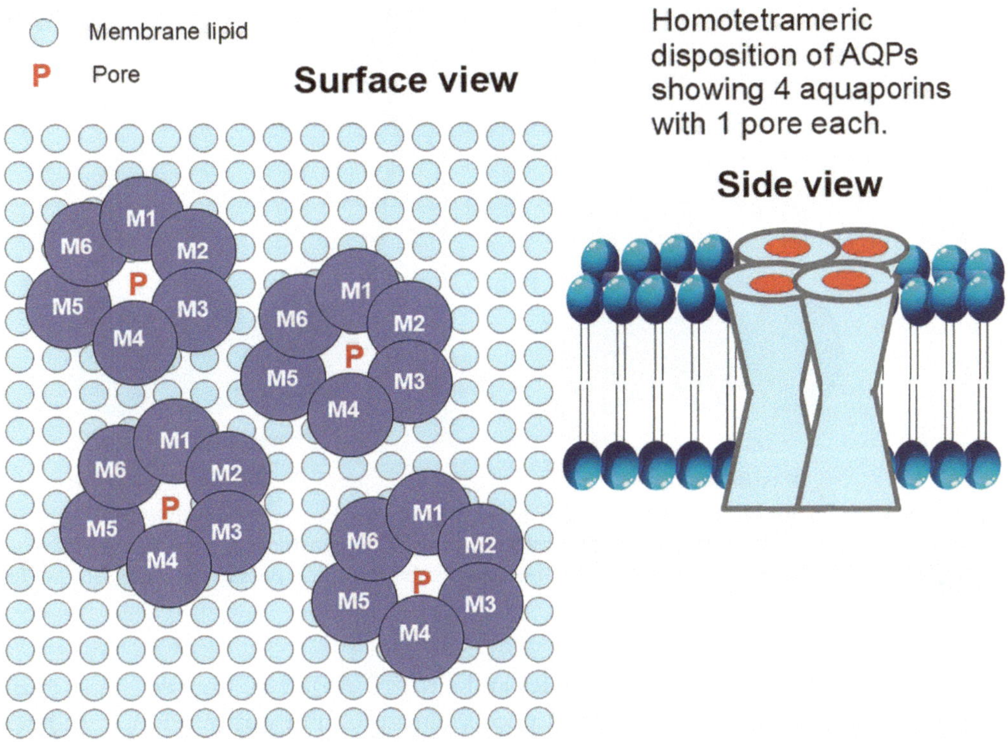

Fig. 3 A view of AQP1 from the cell surface[29] and a side view. In both, the homotetrameric disposition can be seen. In the right upper corner there is a tridimensional view of the homotetrameric structure.

Aquaporin trafficking

AQPs need to reach membranes in order to become functional. This movement from cytoplasm to membrane is the trafficking. Alterations of AQPs' trafficking may cause different diseases. For example, dry eyes in Sjögren's syndrome are produced by abnormal trafficking of AQP5 in the lacrimal glands, where AQP5 remains in the cytoplasm without reaching the cell membrane.[31–32]

When isolated hepatocytes are stimulated with glucagon, AQP8 is mobilized from intracellular vesicles and inserted in the cell membrane.[33] Inhibition of PKA (phosphokinase A) or microtubules with colchicine impedes this trafficking. Therefore AQP8 trafficking is PKA (phosphokinase A) dependent. Trafficking of AQP1 is osmotically

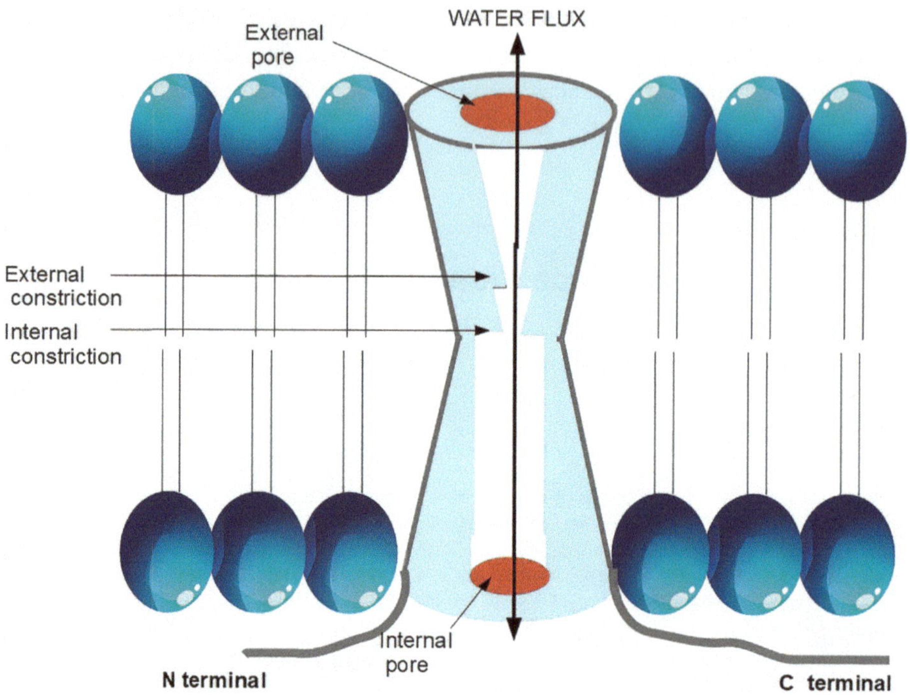

Fig. 4 Details of the hourglass structure. The two constrictions at the pore are visible. The hourglass model was built according to the description by Jung et al.[30] The constriction areas in the hourglass model decrease the size of the pore to 2–3 Å. This size is enough for the passage of water but not for other ions.

dependent and mediated by microtubules and PKC.[34] AQP3 is "moved" to the cell surface by phosphokinase C (PKC) and phospholipase C, but not by PKA.[35] AQP4's trafficking is PKA phosphorylation-dependent.[36] This means that a lactate increase due to glycolytic metabolism may increase, at least theoretically, AQP4's activity through PKA activation. However, this still needs experimental confirmation. AQPs trafficking mechanisms are different for each type of AQP.

Aquaporins regulation

The AQP1 gene promoter segment has:

- 3 consensus sequences for HIF.[37–39]
- 1 sequence for the MAP Kinases.
- 1 sequence for hyperosmolar sensing.
- Consensus sequences for Sp1, AP1 and AP2.
- 1 *E*-box with carbohydrate responsive element (ChRE).[40] In this consensus sequence, glucose-6-phosphate seems to be the up-regulator.[41] The ChRE represents the link

between carbohydrate metabolism and AQP1. When glucose uptake is increased, so does AQP1 expression. Besides osmotic pressure, there are many other factors that modulate AQP's activity and expression.

Cancer cells are usually under strong influence of growth factors, particularly EGF/EGFR signaling through the MAP Kinase pathway increasing the expression of AQP3.[42] PKA and PKC also play a role in AQPs activation. For example PKC regulates AQP4 in glioma, increasing invasive behavior.[43] MiR320 is an AQP1 inhibitor that is frequently reduced in breast cancer.[44] Its reduction increases migration and invasiveness. Loss of Mir320 increases progression of glioma. In this case Mir320 targets AQP4.[45] The same Mir targets different AQPs in different tissues.

Fig. 5 summarizes the different factors that regulate AQPs' activation.

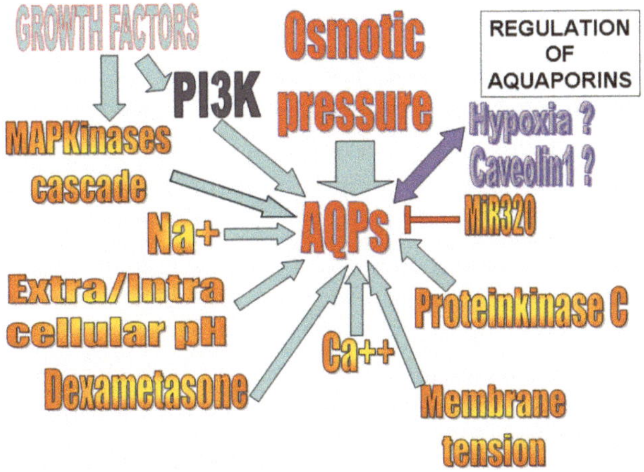

Fig. 5 Aquaporins are mainly regulated by osmotic pressure, but there are many other factors that regulate aquaporins' gating. The MAP Kinase cascades, PI3K pathway signaling, pH (intra and extracellular), divalent calcium, membrane tension, PKC are the best known. Estrogens regulate water movement at the uterus by regulating aquaporins.[46] Hypoxia probably has an effect on AQPs, but it may be exerted through pH changes. Caveolin-1 has been described as a possible AQP modulator. Aquaporin is frequently concentrated in caveolae.[47]

Aquaporins in disease

Aquaporins play a role in different diseases.[48] Their importance in some pathologies is slowly being uncovered.[49] AQP4 is a key player in neuromyelitis optica.[50–52] Sjögren disease is characterized by anti-AQP antibodies.[53,54] It is well established that AQP2 mutation produces renal diabetes insipidus,[55] but AQPs also show alterations in autosomal dominant polycystic kidneys.[56] AQPs are partially responsible for brain edema in cerebro-vascular disease,[57] ischemic stroke,[58] hydrocephalus,[59] cholestasis and liver cirrhosis,[60,61] Meniere's disease,[62,63] heart failure,[64] ulcerative colitis and inflammatory bowel disease in general,[65,66] diseases of the male reproductive system,[67] chronic obstructive pulmonary disease[68], and Alzheimer's disease.[69]

Aquaporins in cancer

Recent research has identified AQPs 1 to 9 (except 6 and 7) to be strongly associated to cancer proliferation, migration, invasion, metastasis and angiogenesis interacting with actins and contributing to the epithelial-mesenchymal transition.[70–75] Certain tumor cells express elevated aquaporin levels and in many cases there seem to be a correlation between tumor grade and AQP expression, while AQP's inhibition has a limiting effect on tumor growth and spread.[76]

Analysis of AQPs pro-cancer activities

Hoque et al.[77] examined 10 different cell lines of non small cell lung cancer (NSCLC) and found AQP1's expression in seven of them. It was not expressed in squamous cell lung carcinoma nor in normal lung tissue, but approximately two thirds of adenocarcinomas and bronchio-alveolar carcinomas overexpressed AQP1. When they forced the expression of AQP1 in NIH-3T3 cells (mouse embryo fibroblast cell line), these cells developed transformed phenotypes, meaning that AQP1 had oncogenic properties. AQP1 was overexpressed in tumor vascular endothelium, and AQP1 knockout mice showed impaired angiogenesis. Saadoun et al.[78] found little expression of AQP1in normal astrocytes, however, astrocytomas usually overexpressed it, particularly at the vascular endothelium. This led to the idea that brain edema, which is so common in brain tumors, is caused by AQP1's activity.

Aquaporins in migration

In 2002, the first evidence of AQPs' role in migration was suggested by Loitto et al.[79] They showed that AQP9 was an essential protein in lamellipodia formation for neutrophil migration. By neutralizing it with the specific antibody, motility of the neutrophils was blocked. Saadoun et al.[80] found that AQP1 had the same ability.

The precise mechanisms involved in AQPs' enhancement of migration are not fully known, but they are probably related with their function as "osmotic guardians" at the critical site where they accumulate: the invadopodia. Papadopoulos et al.[81] suggested that this mechanism was related with osmotic water flow (see Box 1).

AQPs are essential constituents of the migration apparatus because most of the time migrating cells do so in very narrow spaces.[82] This requires the morphological flexibility of extruding and importing water in different parts of the cell (see Figs. 6 and 9 on the osmotic engine model). But it also requires the permanent creation of a new adequate

BOX 1

ACTIN POLYMERIZATION/ DEPOLYMERIZATION	+	INCREASED TRANSMEMBRANE ION FLUXES	=	RAPID OSMOLAR CHANGES

microenvironment, whether degrading existing matrix[83] or following existing tracks.[84] In the first case NHE1 and proton exporters are the major players, creating the acidic environment that activates proteolytic enzymes (see Fig. 7). In the second case the main player is AQP granting cellular flexibility to migrating cells. Actually both mechanisms are required in the long journey from the tumor to surrounding tissues, lymph nodes, circulation and metastasis. In this journey it encounters tracks and narrow spaces to go through. Therefore, the proton extruding system and the aquaporin water trading are complementary parts of the same machinery for migration.

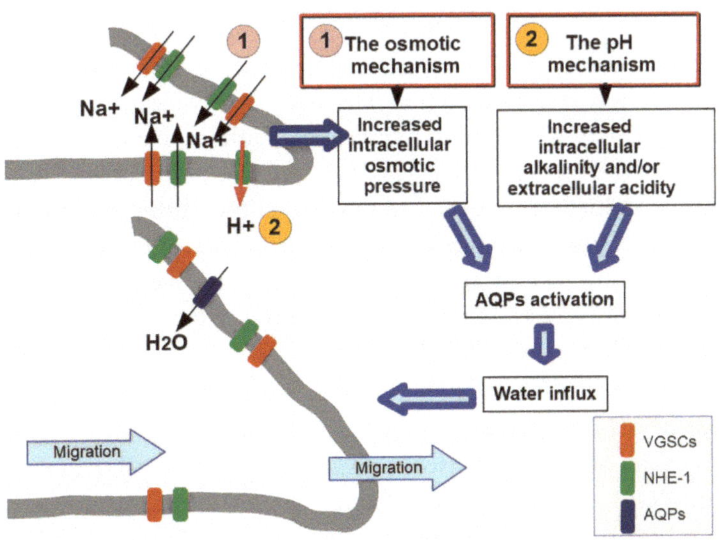

Fig. 6 Osmotic engine model of migration in which AQPs create a net inflow of water at the cell leading edge. (1) Increased intracellular osmolarity due to Na+ influx produced by NHE1 drives water influx by aquaporins. This migration model does not need actin depolymerization-polymerization or myosin contractility.[85] (2) Secondarily a pH induction of AQPs is present in some isoforms, e.g., AQP3.

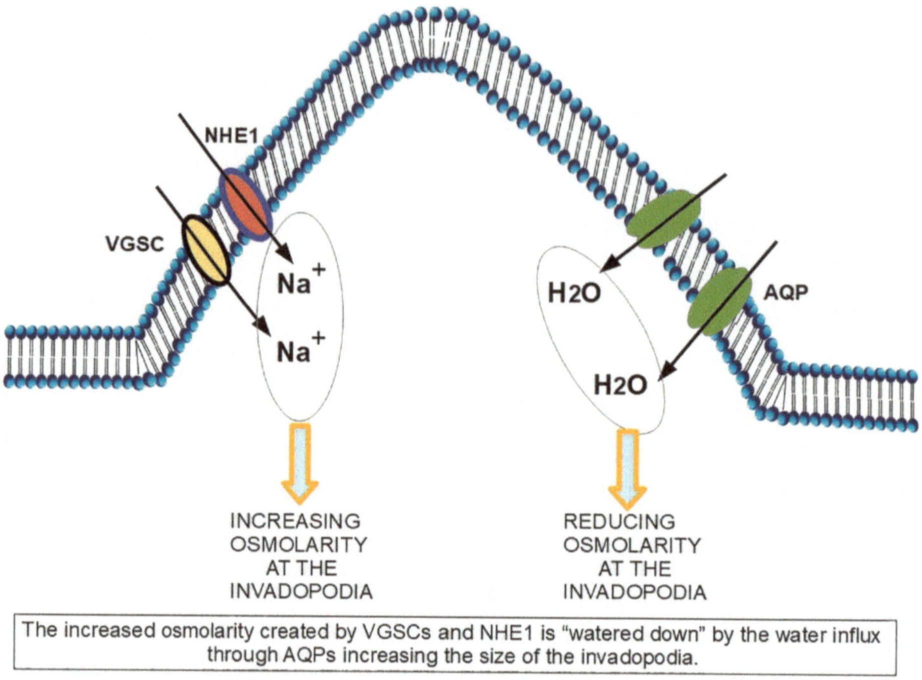

Fig. 7 Canonical function of AQPs at the invadopodia.

The pH theory

AQPs activity is strongly related with pH. This issue is very clear with AQP3 but has not been proven in the case of AQP1. This hypotheses mantains that if the necessary pH conditions are not met, the capacity of AQPs detecting this pH may lead to gate the water influx and migration does not proceed in that direction. If the pH conditions are met, AQPs modulation is only through osmotic signals and migration continues in the intended direction. This would also mean that pH sensitive AQPs, besides acting as "osmotic guardians" also act as "migration direction guardians". This issue is still hypothetical and requires further research.

Reciprocal role of pH and AQPs

The increased activity of ion transporters and channels (mainly NHE1 (sodium hydrogen exchanger 1 and VGSCs (voltage gated sodium channels) at the invadopodia increases the influx of Na^+ and creates a locally hyperosmolar condition that requires AQPs to dilute it through the influx of water. According to Hayashi et al.[86]: "*The ability of cells to transport excess H^+ from intracellular to extracellular space may also require movement of H_2O in the same direction.*" Furthermore, these authors have also reported that AQP1 in gliomas participate in the production of the extracellular acidity. These are probably the basic mechanisms involved, but the deep inversion of the pH gradient at the invadopodia level may be also an activator of AQPs' activity. AQP3 in humans, AQP5 in rats and many plant AQPs are pH sensitive and their activity is blocked when intracellular pH diminishes. The increased transmembrane ion fluxes are tightly related with the over-expression and over-activity of membrane-bound transporters that can contribute to create the pH gradient inversion such as NHE1, NBC (sodium bicarbonate co-transporter), membrane enzymes like CAs (carbonic anhydrases) and ion membrane channels like the VGSCs. Many CA inhibitors also decrease AQP's activities. This possibly means a close association between the two proteins. AQPs are osmotically driven, but we presume that secondarily they may be pH driven with activation on increased intracellular pH or decreased extracellular pH, changes that reach its maximal expression at the leading edge (see chapter on invadopodia).

NHE1, a key player in the pH gradient inversion, is also responsible for cell swelling (through the influx of Na^+) and migration in some kinds of cancer cells and also in human polymorphonuclear leucocytes.[87] Therefore, NHE1 should be considered as one of the many regulators of AQPs, whether through osmotic effects or pH effects. VGSCs would probably have a similar relationship with AQPs.

Aquaporins in invadopodia

In the "organelle" responsible for invasion, the invadopodia, AQPs co-express with the ion channels responsible for the pH inversion. Furthermore, NaV1.5, NHE1, and AQPs locate in caveolin-1 rich portions of the invadopodia membrane and the three of them seem to interact with caveolin-1.[88] The existence of possible pH sensors in AQPs (intra and extracellular) suggests that there is intercommunication between the pH inversion mechanism and AQPs and this signaling could be the pH level, besides the osmolarity. VGSCs, particularly the fetal isoform of Nav1.5, NHE1, caveolin.1, Src, cortactin and AQPs work as a unit in order to stimulate cellular migration and invasion. We call this the "migration complex". As we can see in Figs. 7 and 8, the first step in the invadopodia development process should be the activation of NHE1, thus creating a hyperosmolar intracellular milieu at the invadopodia due to the influx of Na^+. Secondarily, AQPs are activated and the influx of water decreases the hyperosmolar milieu. It is probable that caveolin 1, activated by Src, participates in this process by further stimulating AQP activity. Caveolin 1 also has osmolarity sensors.

Hypoxia is another activator of AQPs by decreasing cholesterol content of the lipid raft.

The hypothesis described in Fig. 8 is based on the following concepts and references

NaV1.5 induces NHE1.[89,90]

NaV1.5 and NHE1 co-locate in caveolin rich lipid rafts [90].

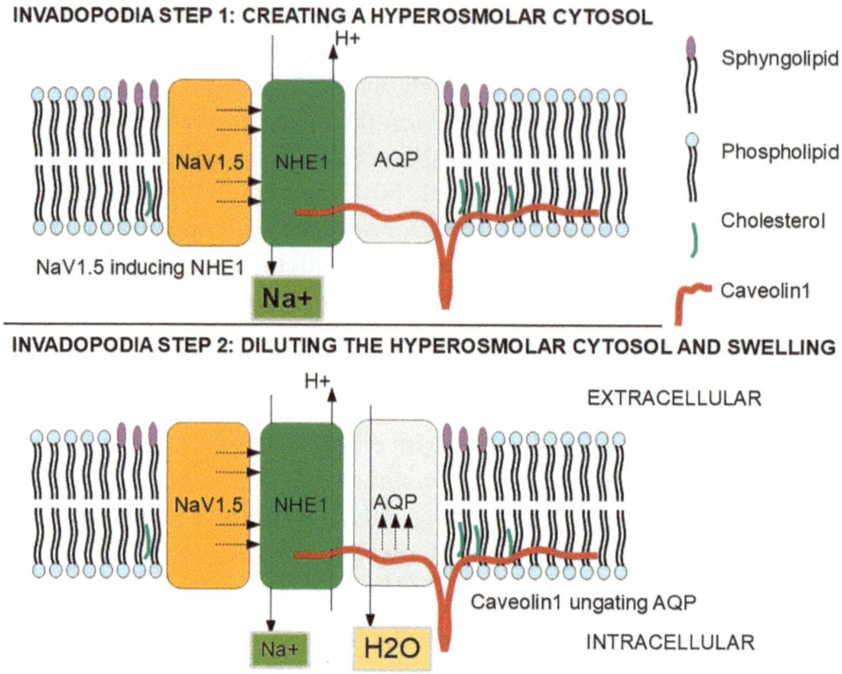

Fig. 8 A hypothesis of how AQPs work at the invadopodia. Interrelations among NaV1.5, NHE1, AQPs and caveolin 1 in a lipid raft. NaV1.5 induces allosterical modifications of NHE1 increasing its sensitivity to pH. Caveolin-1 moderates aquaporin activity.

AQPs and caveolin co-locate in the membrane.[91–93]

AQPs are located in lipid rafts.[94]

Na^+ increases the tolerance to water stress and the expression of AQPs' genes in plants.[95]

Src kinase and Rho/Rock activates caveolin1 by phosphorylation of tyrosine.[96]

AQP1 relocalizes actin protein and activates RhoA and Rac.[97]

Caveolin1 may modulate AQPs.[98]

Point mutations of AQP1 may allow the passage of H^+.[99,100] In this case we presume that AQP1 may participate directly in the pH gradient process.

The ischemic factor: according to Tong et al.,[101] hypoxia decreases the cholesterol content of the cell membrane decreasing its thickness, thus increasing the permeability of water channels.

Fig. 9 illustrates how AQPs and NHE1 could interact in regulating the cycle between invadopodial function and motility involved in the processes of intravasation and extravasation.

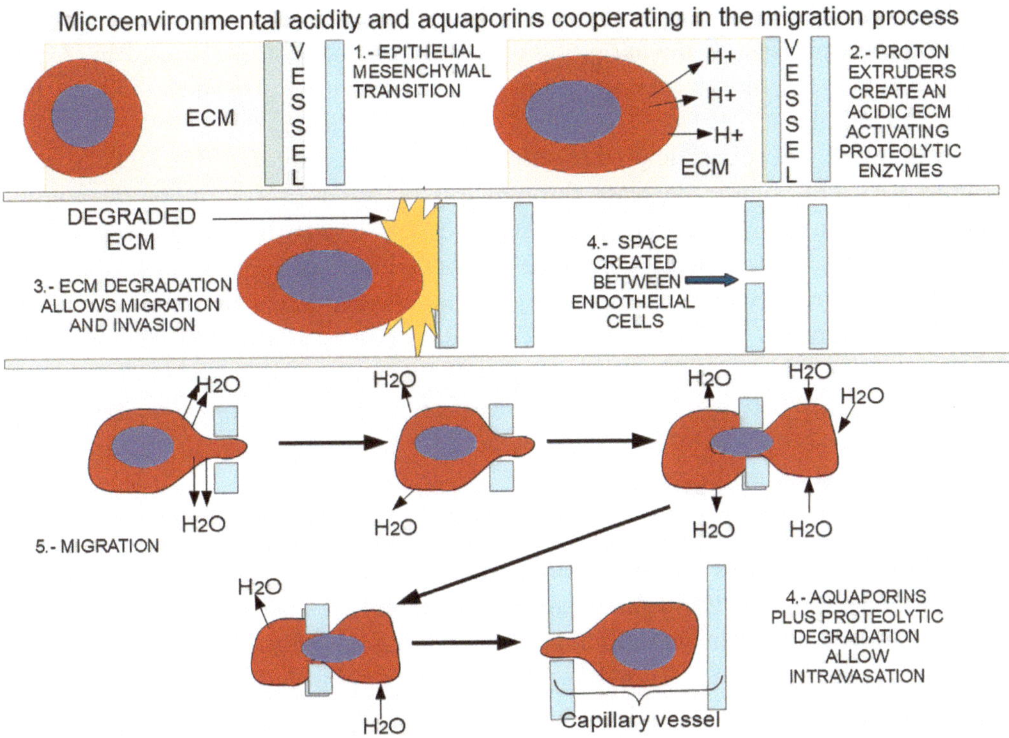

Fig. 9 Proton extruders and aquaporins co-participating in the intravasation and extravasation processes fundamental for metastasis.

AQPs and angiogenesis

AQPs are strong angiogenesis inducers, particularly AQP1.[102,103] This happens not only in malignant tumors, but also in other diseases such as cirrhosis.[104] AQP1 null-mice show difficulties generating new vessels.[105] Silencing AQP1 inhibits angiogenesis in chicken embryo membrane test.[106] In the same line, Kaneko et al.[107] found in human retinal endothelial cells under hypoxic conditions that AQP1 was strongly overexpressed in proliferating vessels. Its inhibition reduced tube formation. These findings were independent of VEGF activity but when both were inhibited, the effect was additive.

AQPs and apoptosis

In the aforementioned experiment of Hoque et al. (AQP1 transfection producing malignant phenotype) one of the transformation characteristics found in fibroblasts transfected with AQP1 was that they became resistant to apoptosis. In normal cells, the inhibition of aquaporins decreased the rate of apoptosis.[108] The mechanism seems related with the necessary cell shrinkage of the apoptotic process, shrinkage that is induced by AQPs. On the contrary, in cancer cells apoptosis is increased with aquaporin inhibition (see Tables 1, 3 and 4).

Furthermore, the apoptotic process shows from a very early stage intracellular acidification (IA) and volume decrease through loss of water (AVD: apoptotic volume decrease). AVD is accomplished through AQPs activity. IA may have many causes: inhibition of NHE1, inhibition of V-ATPase proton pumps, translocation of SHP1 (Src homology 2 bearing tyrosine phosphatase) to the cell membrane, but mainly through K^+ and Cl^- channels.[109] The fact that AVD and IA coexist in the same time frame[110] and also that the inhibition of any of these two players may inhibit apoptosis, suggests a close relationship between them, albeit it has not been experimentally established.

AQPs and lactic acid extrusion

In Arabidopsis plants there is an AQP (NIP2:1) with fully developed abilities to extrude intracellular lactic acid in a similar fashion as MCTs do. A lactate permeable AQP can be found in bacteria.[111] AQP9 in mammals has this ability, as shall be discussed latter. A shorter isoform of AQP9 has been found in the inner mitochondrial wall, which has been proposed to be a lactic acid channel (intracellular lactate shuttle?).[112]

On the other hand, increased lactic acid levels in the brain increases the expression of AQP4. This would be a mechanism involved in the induction of brain edema after a hypoxic episode.[113] At the same time AQPs increase HIF-1 alpha activation.

In keratinocytes caveolin1 is frequently found in the vicinity of AQPs in the plasmatic membrane[114] but the interaction between these two membrane proteins has not been clearly established.[98] It has been suggested that caveolin may act as an "osmometer" modulating AQPs' activity.[115] In rat kidneys AQP2 co-localizes with caveolin-1 and their trafficking is also shared.[116]

AQPs and pH

Anthony et al.[117] and Boassa et al.[118] have shown that AQP1, besides its water channel function, can also work as a non voltage gated ion channel. Zampighi et al.[119] and Ehring et al.[120,121] found the same in AQP0 in the lens. AQP6 in oocytes also seems to be an ion channel.[122] However, there is no evidence that AQPs contribute to the pH gradient inversion, but the ion gating function of these proteins may be a contributor mechanism to that effect. On the other hand, there is evidence that AQPs may be modulated by intracellular and extracellular pH as is the case of AQP5. The AQP5 function is not modified by extracellular pH. But when AQP5 is phosphorylated, the expression and trafficking are increased and extracellular pH activates the functioning of this water channel; and when pH increases the activity of AQP5 is increased too. In cancer, AQP5 is usually phosphorylated.[123] This means that in cancer AQP5 is pH sensitive.

Yasui et al. working with Xenopus oocytes, the cell most frequently used for testing AQPs functioning, showed that AQP0, 3 and 6 had a pH dependent activity.[124,125]

Finally, AQP9 is permeable to lactic acid and weak monocarboxylic acids in general,[126] so that it may have a role in lactic acid extrusion when the monocarboxylate transporter (MCT) capacity is surpassed by excessive lactic acid formation.

Intracellular pH

AQPs are highly conserved molecules throughout phylogenia, so that what happens in plants and trees may be possibly applied also to mammals. In tree roots, AQPs are modulated by cytosolic pH under hypoxic conditions. This protects the roots from flooding of soils in which the excess of water creates hypoxia. The cytosolic acidosis created by hypoxia activates AQP gating thus impeding the inflow of additional water.[127–129] AQP4 is also a pH dependent channel, and in this case the pH sensing depends on the intracellular histidine 95.[130]

Extracellular pH

Zeuthen and Klaerke[131] found that at a low extracellular pH, AQP3 gated for water but not for glycerol in AQP3 expressed in Xenopus oocytes. This external pH gating was not found in AQPs 0, 1, 2, 4, and 5. However a similar pH gating was found in AQP5 in rats.[132] Regarding cancer, this finding means that low pHe predisposes for the uptake of glycerine, which acts as a precursor for lipid synthesis. Zelenina et al.[133] confirmed the regulatory activity of the extracellular pH on AQP3 in lung cells.

AQP1 and CO_2

Aquaporins have as their main function the selective filtering of water and glycerine, but there is experimental evidence that certain non polar gases like CO2 and nitric oxide may also use these channels,[134] although this concept is not universally accepted.[135] In erythrocytes, AQP1 has been shown to be a channel for CO2 flow and is responsible for 60% of the pCO2[16]. The CO2 is hydrated to $CO3H^-$ and H^+. While the H^+ is extruded by NHE1, the remaining CO_3H^- contributes to cellular alkalinization. This CO2 transport mechanism, which was first described by Prasad et al.,[15] also needs the cooperation of intracellular CAs.

Importance and evidence of different AQPs in cancer

The following tables (Tables 1, 3, and 4) describe the accumulated evidence for the role of single AQPs' in cancer, while Table 5 describes those reports of studies on more than one isoform.

Table 2: Anti-tumoral actions of AQP1.

Table 1 AQP1 and cancer

Reference	Cancer	Findings
Zhang et al.[136]	Bladder	Inhibition of AQP1 increased sensitivity to chemotherapeutic drugs in bladder cancer cells.
Oshio et al.[137]	Brain	36 glial tumors were studied. All showed overexpression of AQP1.
El Hindi et al.[138]	Brain	There was a positive correlation between AQP1 expression and angiogenesis in human astrocytomas
Deb et al.[139]	Brain	Biopsy specimens from 30 patients were studied: AQP1 was not increased in meningiomas but it was overexpressed in gliomas and ependymal tumors. There was a positive correlation between AQP1 expression and increased malignancy. Peritumoral overexpression of AQP1 in high grade tumors suggests an explanation for perilesional edema.
Saadoun et al.[78]	Brain	In normal brain, AQP1 was found in the endothelium of microvessels and none was found in the brain parenchyma. In astrocytomas, AQP1 was present in tumor cells. The amount of AQP1 correlated with the grade of malignancy. In highly malignant astrocytomas AQP1 was not only in the membrane but also in the cytoplasm. It was not detected in metastatic cells.
Jiang[97]	Colon	AQP1 increased migration in colon cancer cells through increased water permeability and relocalization of actin protein and activation of RhoA and Rac.
Dorward et al.[140]	Colon	The blockade of AQP1 in colon cancer cells that expressed this AQP, reduced migration and angiogenesis.
Bin et al.[141]	Colon	Acetazolamide suppressed xenografted colon tumor's growth in nude mice through inhibition of AQP1.
Chen et al.[142]	Hemangioblastoma	26 cases of human hemangioblastomas showed overexpression of AQP1 when compared with normal brains. Cystic hemangioblastomas had a higher expression than solid ones.
Mazal et al.[143]	Liver	AQP1 was overexpressed in cholangiocarcinoma but not in hepatocarcinoma.
Guan et al.[144]	Larynx	20 cases of laryngeal tumors and 15 normal cases were examined. A marked overexpression of AQP1 in tumors was found.

Table 1 AQP1 and cancer—cont'd

Reference	Cancer	Findings
Bellezza et al.[145]	Lung adenocarcinoma	Patients who overexpressed AQP1 (46% out of 185 cases) had a shorter disease free survival.
Xiang et al.[146]	Lewis lung carcinoma	Acetazolamide inhibited AQP1 and decreased angiogenesis. AQP1 was mainly localized in endothelial cells.
Li et al.[147]	Lewis lung carcinoma	Acetazolamide reduced growth, metastasis and AQP1 expression. These effects were not further enhanced by combining acetazolamide with sodium bicarbonate.
Simone et al.[148]	Melanoma	Inhibition of AQP1 in a melanoma model of injected cells in mice, decreased lung metastasis, angiogenesis and MMP2.
Nicchia et al.[149]	Melanoma	Inhibition of AQP1 with siRNAs in a mouse model of melanoma decreased angiogenesis and tumor proliferation.
Hu et al.[150]	Melanoma and mice BC cells	Cells expressing AQP1 had a three-fold increase in migration and metastasis when compared with cells without AQP1 expression.
Esteva-Font et al.[151]	Metastatic breast adenocarcinoma model	In AQP1 −/− mice the number of lung metastasis was much lower than in AQP1 +/+ mice. Also the volume and vascular density was much lower in the first group.
Li et al.[152]	Nasopharyngheal cancer	Normal and nasopharyngeal tumor tissue expressed AQP1 but the expression was much higher in tumors.
Chetry et al.[153]	Ovary	Ovarian cancer patients under chemotherapy that expressed AQP0 and 1 had worse prognosis. On the other hand those that expressed AQP5, AQP8, and AQP10 showed a better overall survival
Wang et al.[154]	Ovary	AQP1 was frequently overexpressed in ovarian cancer cells and blocking AQP1 suppressed cell viability, migration and invasion, while promoting apoptosis.
Yang et al.[155]	Ovary	AQP1 was mainly expressed in microvessels. AQP1 was highly increased in tumors as compared to normal counterparts. Intratumoral vessel density correlated with AQP1 expression.
Pan et al.[156]	Prostate	Inhibition of AQP1 attenuated prostate cancer cells migration.
Park et al.[157]	Prostate	AQP1 was a biochemical prognostic factor for recurrence

Continued

Table 1 AQP1 and cancer—cont'd

Reference	Cancer	Findings
Wu et al.[158]	Osteosarcoma	AQP1 knockout downregulated migration, invasion and proliferation of osteosarcoma cells.
Qin et al.[159]	Breast	AQP1 expression in the cytoplasm of breast cancer cells meant a bad prognosis.
Liu et al.[160]	Urothelial cancer	AQP1 had 50% higher expression in urothelial cancer cells as compared with normal urothelium.
Saadoun et al.[161]	Vascular endothelia and tumors in general	Motile cells expressing AQP1 showed prominent membrane ruffles at the leading edge, an area of rapid water fluxes. AQP1 deficient cells showed impaired migration and impaired angiogenesis.

Table 2 Aquaporin 1 has also anti-tumoral effects

Reference	Tumor	Anti-tumoral actions of AQP1
Kao et al.[162]	Mesothelioma	Paradoxically, a two cohort study (both with mesothelioma) showed a direct correlation between prognosis and expression of AQP1. Those patients that expressed AQP1 had a better overall survival than those who did not express it. Since normal mesothelial cells express AQP1, the authors proposed that the better survival was due to a more differentiated type of cells when AQP1 is expressed.
Angelico et al.[163]	Mesothelioma	Better prognosis with high expression of AQP1.

Conclusions on AQP1

(1) It is overexpressed in many solid tumors, including brain, bladder, lung, ovary, melanoma, prostate, breast, hemangioma, larynx, bone and liver.

(2) Its overexpression worsens prognosis, while its down-regulation decreases migration, invasion, metastasis and angiogenesis.

(3) However, in mesothelioma AQP1 over-expression is correlated with a better prognosis. Two independent studies confirmed this finding (Table 2).

Table 3 describes the evidence on AQP3's relation to cancer.

Table 3 AQP3 and cancer

Reference	Cancer	Findings
Huang et al.[164]	Breast	The AQP3 has an estrogen responsive element in the promoter region, that when stimulated by estrogen it expressed AQP3, increasing migration and invasion through epithelial mesenchymal transition and reorganization of actin cytoskeleton.
Cao et al.[165]	Breast (BC)	Treating BC cells with FGF-2 (fibroblast growth factor-2) increased AQP3 expression and induced cell migration. Silencing AQP3 expression inhibited FGF-2 induced cell migration.
Arif et al.[166]	Breast	MDA-MB 231 breast cancer cells reduced migration, invasion and proliferation when AQP3 was downregulated.
Kusayama et al.[167]	Esophagus and oral squamous cancer	Increased expression of AQP3 was found in oral and esophageal squamous cell carcinoma. Inhibition of AQP3 induced apoptosis and decreased cellular adhesion. Inhibition also decreased the Fak signaling pathway and Erk and MAPK phosphorylation.
Wen et al.[168]	Gastric	Helicobacter pylory-driven gastric carcinogenesis was related to overexpression of AQP3 through a pathway: ROS–HIF-1α–AQP3–ROS loop.
Chen et al.[169]	Gastric	AQP3 presence in human gastric carcinoma cells was associated with epithelial-mesenchymal transformation (EMT) and poor prognosis. When cultivated these AQP3 positive cells, proliferation, migration and invasion were increased.
Xu et al.[170]	Gastric	AQP3 increased metalloproteases' activity in tumors through the PI3K/Akt pathway (Akt phosphorylation).
Chen et al.[171]	Gastric	Knockdown of AQP3 in gastric carcinoma cell line induced apoptosis. Glycerol uptake and lipid synthesis were impaired.
Jiang et al.[172]	Gastric	MiR-874, a known inhibitor of gastric cancer proliferation and migration, achieved this inhibition through downregulation of AQP3.
Xia et al.[173]	Experimental lung cancer	AQP3 knockdown reduced angiogenesis, growth and invasion. Also reduced cellular glycerin, mitochondrial ATP production, metalloproteases' activity, Akt activation, and prolonged the life of mice bearing tumor xenografts.
Liu et al.[174]	Pancreas	EGF-EGFR-ERK pathway increased the activity of AQP3 in a dose dependent manner in cultured pancreatic cancer cells and increased migration.
Huang et al.[175]	Pancreas	AQP3 was significantly associated with mTOR signaling in pancreatic adenocarcinoma. A pathway was described in which decreased MiRNA-874 allowed the increased expression of AQP3 and increased mTOR activity. Increasing MiRNA-874 or inhibiting AQP3 decreased proliferation and induced apoptosis.

Continued

Table 3 AQP3 and cancer—cont'd

Reference	Cancer	Findings
Ismail et al.[176]	Prostate	Freezing of prostate cancer cells increased the expression of AQP3. The inhibition of AQP3 with Hg or siRNA made these cells more sensitive to cryo-injury, decreasing cells' viability compared with control cells.
Niu et al.[177]	Thyroid	**Immunohistopathology**: AQP3 was found in 91% of medullary carcinomas and none in non follicular cell tumors. AQP4 was found in 100% of follicular adenomas, 90% of follicular carcinomas, and 85% of papillary carcinomas. It was negative in all medullary carcinomas and undifferentiated carcinomas. **RT-PCR:** showed AQP3 mRNA expression only in medullary carcinomas and AQP4 mRNA expression in follicular cell-derived tumors except for undifferentiated carcinomas.

Conclusions on AQP3

(1) It is overexpressed in many solid tumors, including lung, gastric, prostate, breast, esophagus, thyroid and pancreas.
(2) In a similar way to AQP1, it is associated with migration and invasion.
(3) It is associated with many oncogenic pathways.
(4) Downregulation of AQP3 induces apoptosis and/or decreases proliferation.

Table 4: AQP4 and its association with glioblastoma.

Table 4 AQP4 and glioma

Reference	Cancer	Findings
Badaut et al.[178]	Glioma	Knockdown of AQP4 decreased water movement in astrocytes.
McCoy et al.[179]	Glioma	All biopsies of glioma patients showed overexpression of AQP1 and 4. But when these cells were grown as cell lines they lost completely both aquaporins. Only AQP1 enhanced cell growth and migration. AQP4 increased cell adhesion.
Xiong et al.[180]	Glioma	MiRNA 320a targeted AQP4 and decreased migration and invasion in glioma.
Ding et al.[181]	Glioma	AQP4 was the most important member of the family in the brain, participating in edema after stroke or injuries. It is significantly up-regulated in gliomas through cell invasion.
Ding et al.[182]	Glioma	Inhibition of AQP4 decreased migration and invasion through down-regulation of MMP2 and F-actin polymerization. AQP4 participated in cell adhesion.
Ding et al.[183]	Glioma	Inhibition of AQP4 induced apoptosis in glioblastoma.

Conclusions on AQP4

(1) AQP4 is a key participant in migration and invasion of glioblastoma.

(2) It should be regarded as an important target in glioma treatment.

According to Yang et al.[184] AQP4 over-expression in glioma is a consequence of increased edema due to VEGF activity, with the purpose to reduce edema. This controversial view does not explain the inhibition of migration, invasion and apoptosis achieved with AQP4 knockdown.

Table 5: Combinations of different AQPs.

Table 5 Multiple AQPs and cancer

Reference	Cancer	Findings
Shi et al.[185]	Breast (BC)	BC and normal breast cells expressed AQP1, 3, 4, 5,10, 11,12. AQPs 3, 4 and 5 showed higher expression in BC than in normal counterparts. AQP1 and 4 were expressed in cell membranes and its expression was higher in cancer. AQP4 was expressed also in the cytoplasm and was expressed markedly stronger in normal than in cancer tissues. AQP5 was expressed mainly in cell membranes in carcinoma tissues, but was almost absent in normal breast tissues.
Dai et al.[186]	Breast	AQP7 was a promoter of invasive behavior in breast cancer cells.
Shi et al.[187]	Colon	Silencing AQP5 in colon cancer cell lines, drug resistance diminished through decreased p38MAPK signaling.
Moon et al.[188]	Colon	Expression of AQP1 and 5 appeared early in colon cancerization (dysplasia) and was maintained throughout progression.
Kang et al.[189]	Colon	There was correlation between AQP1, AQP3, and AQP5 expression and lymph node metastasis in patients with colon cancer.
Li et al. [190]	Colon	The silencing of AQP5 increased sensitivity of colon cancer cells to 5-FU through inhibition of the Wnt/B catenin signaling pathway.
Wang et al.[191]	Colon	Silencing AQP5 decreased migration and invasion, and also inhibited the Wnt/β-catenin pathway signaling.
Shen et al.[192]	Gastric	Normal gastric mucosa and gastric cancer expressed AQP1, 3, 4, 5 and 11 and AQP4 was not found in cancer. AQP3 and AQP5 expression were associated with a metastatic stage.
Zhang et al.[193]	Lung	Overexpression of AQP5 in lung cancer indicated a poor prognosis. Lung cancer cells that overexpressed AQP5 also show enhanced proliferation and migration. The MAP kinase pathways were also upregulated.

Continued

Table 5 Multiple AQPs and cancer—cont'd

Reference	Cancer	Findings
Zhang et al.[194]	Lung	Suppressing the expression of AQP5 decreased growth of A549 lung cancer cells in vivo and in vitro.
Chae et al.[195]	Lung	AQP5 promoted invasion in NSCLC (non-small cell lung cancer).
Direito et al.[196]	Pancreas	AQP 3 and AQP5 were overexpressed in pancreatic ductal carcinoma.
Pust et al.[197]	Prostate	AQP5 was moderately to highly increased in 42% of cases of prostate cancer.
De Qing Wu et al.[198]	Colon	AQP8's overexpression significantly decreased proliferation and aggressiveness of certain colorectal carcinoma cell lines by inactivating PI3K/Akt signaling.

Comments on the contents of the tables

The case of mesothelioma (Table 2) and AQP8 (Table 5) shows that it is very difficult to established fixed rules for AQPs since they may have different specific functions in different tissues and malignant tumors.

AQP1 and AQP4 in glioblastoma require further comments. McCoy et al. showed that AQP1 and AQP4 have a consensus site for PKC phosphorylation. Phosphorylation of AQP4 blocks its permeability and decreases glioblastoma cells migration. On the contrary, AQP1 was unaffected by phosphorylation in the consensus site. Both AQPs are highly expressed in glioblastoma. This research also shows that we have to consider AQPs on an individual basis, because they do not have an uniform behavior in spite of belonging to the same protein family.

There is a complex relationship between dexametasone and AQP1 in glioblastoma and probably in other tumors too.[199] Dexametasone, as many other steroids, decreases glioblastoma's proliferation and also shows the same ability in lymphoid malignancies and other tumors, inducing cell arrest in G1.[200–203] On the other hand, dexametasone increases AQP1 expression in a dose-dependent manner and also increases migration. Therefore, dexametasone has a dual action by inhibiting proliferation and increasing migration through AQP1 overexpression. AQP1 knockdown inhibited dexametasone induced migration.[204]

AQP3's over-expression in cancer is strongly related with the increased activation of metalloproteases, including MT1-MMP, MMP2 and MMP9, achieving this activation through phosphorylation of Akt.

Other evidence that suggests the relationship of AQPs with pH and cancer are:

1. The AQP1 gene's promoter area between −2257 and −2251 is the site for the Spi B transcription factor involved in pH regulation of the gene.[205]
2. AQP1's relation with hypoxia: animals under chronic hypoxic conditions up-regulate AQP1 in lung cells. Knocking down AQP1 increases the expression of hypoxia inducible genes. This is the reason why it has been proposed that AQP1 may also work as an O_2 channel or at least as a "facilitator" of O_2 influx into the cell.
3. Mesenchymal stem cells show an autocrine signal that increases migration through AQP1 and CXCR4 overexpression.[206]
4. AQP1 upregulates the β-catenin pathway.[207]
5. AQP1 upregulates FAK (focal adhesion kinase)[208] that participates in migration.

Aquaporins and ion channels seem to work in coordination in a tandem fashion to modulate cellular volume, and move the advancing edge of the invadopodia forward.

AQP inhibitors

Two different mechanisms may be at work in AQP inhibition:

(1) Direct binding of the inhibitor compound interacting with different amino acids of the channel, and so inducing a conformational change that activates gating. This would be the case of inhibitors such as tetraethylammonium, mercurials, gold, silver and other heavy metal compounds. Mercurial inhibitors were between the first investigated AQP inhibitors. Mercurials are very potent AQP1 inhibitors,[209] but they are highly toxic for living cells, thus, they are not useful for clinical use. Au and Ag compounds, as many other heavy metal compounds, inhibit AQP1.[210] Au-phen, a synthetic gold compound, reduces proliferation in rapidly growing cells by targeting AQP3.[211]

(2) Intracellular acidification that acts on the histidine amino acids of AQPs' pH sensors, so activating the conformational change and gating. Acetazolamide and topiramate, besides using the first mechanism, probably also present this second mechanism.

While the first mechanism has been experimentally confirmed, the second one is still in a hypothetical stage.

Other AQP inhibitors are:

Tetraethylammonium (TEA) interacts with a tyrosine residue at loop E 8 and has been used as an experimental AQP1 inhibitor.[212] The inhibition occurs *via* the interaction with a tyrosine residue at the external part of the water pore.[213,214] However, Yang et al.[215] could not confirm TEA's inhibitory actions. Other tetralkylammonium compounds like tetrapropylammonium showed AQP inhibitory actions. In the latter, this inhibitory action was specific for AQP1.[213]

Piroxicam has been proposed as an AQP4 inhibitor based on in silico evaluation.[216]

Topiramate is a FDA approved drug (2012) for the treatment of epilepsy, but it has also been used for off label indications like migraine and obesity. Topiramate is also responsible for cases of metabolic acidosis.[217] Marathe et al.[218] showed that topiramate produces an important intracellular acidification in glioblastoma. The reason for this acidification is probably based in its VGSC inhibitor properties[219] and carbonic anhydrases inhibition.[220,221] Monzani et al.[222] described topiramate as an inhibitor of AQPs 1, 4 and 5. The inhibition of AQP4 by antiepileptics was also confirmed by in silico tests.[223]

At the experimental level, Bing et al.[224] showed that topiramate reduced tumor growth and metastasis in the Lewis lung carcinoma model by inhibiting AQP1 and CAs.

Acetazolamide (AZA) is a powerful non specific pan-inhibitor of carbonic anhydrases and has shown inhibitory properties on AQP1[225–228] and AQP4.[229] AZA binds directly to AQP1, but the exact mechanism of action is not clear. On the contrary, Yang et al.[230] found no inhibitory action of AZA on AQP1. This controversial finding may have been caused by the different type of cells used in the experiments. Furthermore, Zhang et al.[231] found that the diuretic effects of AZA where related to increased trafficking, ubiquitination and destruction of AQP1 at the proteosomal level. Tanimura et al.[232] confirmed AQP4 reversible inhibition by AZA.

Intracellular acidity produced by AZA seems also to contribute to its anti AQP effects. Combining AZA with sodium bicarbonate did not further enhance AZA's anti-metastatic and anti-growth effects.[133] The fact that bicarbonate decreases the extracellular acidity reinforces the idea that AZA's aquaporin inhibitory activity is strongly dependent on pH. It all seems to indicate that AQPs are inhibited by AZA and topiramate only when the acid base balance is altered, as it happens in cancer. It is concluded that topiramate or AZA need an increased intracellular pH and a decreased extracellular pH (pH gradient inversion) in order to exert their inhibitory activities.

Bumetanide is a loop diuretic with inhibitory actions on AQP1 and AQP4. A derivative, 4 amyno pyridine carboxamide showed inhibition of these two AQPs. The site of action seems to be the intracellular part of the channel. Bumetanide derivatives such as AqB007 and AqB011[233] inhibit AQP1 while AqB013[234] inhibits AQP1 and AQP4. AqB013 also inhibits migration, invasion and angiogenesis of cancer colon cells.[235]

Torsemide is another loop diuretic with a pyridine-sulphonylurea structure, which is in clinical use for hypertension and congestive heart failure. It blocks the intracellular pore of AQP1.[236]

Anordio, this antiestrogenic compound has shown inhibitory actions on AQP1 by decreasing the expression of this water channel.[237]

Curcumin downregulates AQP4 in rat's choroidal plexus cells in a dose-dependent manner.[238] A similar effect was observed on AQP3 in ovarian cancer cells.[239] Its very poor absorption and bioavailability makes curcumin inappropriate for clinical use until new pharmaceutically improved forms are developed. Other possible problem is its anti-oxidant effect which may interfere with conventional chemotherapeutic drugs.

Bacopaside I and II are triterpenoides that have shown the ability to block AQP1 activity, reducing colon cancer cell migration,[240] inducing apoptosis and inhibiting angiogenesis in vitro.[241]

Dimethyl sufoxide (DMSO) has been identified as an AQP1 inhibitor.[242]

Copper is an inhibitor of AQP3. This inhibition is potentiated by acidity.[243]

To what extent is aquaporin inhibition clinically possible?

Non-specific inhibition of many different AQPs is highly toxic, particularly for the kidneys, eyes and nervous system as they are very dependant on normal AQP functioning. From a clinical standpoint, only partial and specific AQP inhibition should be useful in cancer treatment if toxicity is to be avoided. Acetazolamide as an AQP4 and topiramate as an AQP1 inhibitors should be the best therapeutic options at the present time. We conclude that if the theory here developed proves correct and AQPs activities are demonstrated to work downstream of the pH inversion paradigm of cancer is correct, the treatment of the cancer-selective pH abnormalities should take care of the AQP management without needing any further drugs besides the two mentioned.

Conclusions

Strong and abundant evidence shows that AQPs are connected with cancer, participating in migration, invasion, metastasis and angiogenesis. In many cases AQPs become vital for cancer cell survival, because its inhibition induces apoptosis. The fact that there are fourteen different kinds of AQPs, with slightly different functions and completely different distributions, complicates the study of these water channels. A few of them, like AQP1 and AQP3 are the most commonly found in cancers, but the other AQPs may appear in certain malignant tissues, like AQP4 in gliomas.

AQPs participate in the pH inversion paradigm in cancer. Even if we cannot determine the exact mechanism, there are a few hypotheses that still need experimental confirmation. According to the new theory here considered, the pH inversion gradient at the invadopodia is an activator of AQPs. This increases water influx and in this way these channels become key participants of the migration process. Probably the main mechanism that leads to AQPs activation is the increased Na^+ inflow produced by NHE1 and VGSCs, particularly at the invadopodia level. The abundant evidence of AQPs participation in cancer progression makes of certain AQPs a legitimate target for anti-cancer treatments.

What is beyond doubts is that there is a chain of events in the migration process in which NHE1 and VGSCs represent the main initial actors. AQPs are part of this chain of events because if they are therapeutically down-regulated, the initially hyperosmolar invadopodia are unable to replenish water and so migration is not possible. What is proposed here is that AQPs' activities are downstream of NHE1, therefore acidifying the cell should create the necessary conditions for AQPs' inhibition. Other hypothesis that fits with the previous one is that cellular acidification down-regulates AQPs probably through AQPs pH sensors. AQP3, 4, and 5 are pH sensitive and AQP9 can function as a lactate transporter. AQP1, however, has not shown pH sensitivity in spite of the fact that its chemical structure should allow it.

Finally, is worth taking into account that many of the treatments addressed against the cancer pH paradigm are also useful for AQP inhibition, like acetazolamide and topiramate.

Based on the evidence gathered in this chapter we may conclude that:

(1) Migration is greatly handicapped with AQPs' inhibition.
(2) Full AQP inhibition is not possible because it has essential housekeeping functions.
(3) VGSCs and NHE1 create the hyperosmolar cytoplasm that sets AQPs in motion.
(4) Osmolarity is the signaling pathway between AQPs and NHE1.
(5) Secondarily, pH may represent another signaling pathway between AQPs and NHE1, but the evidence is insufficient to maintain this theory.

References

1. Paganelli CV, Solomon AK. The rate of exchange of tritiated water across the human red cell membrane. *J Gen Physiol* 1957;**41**(2):259–77.
2. Farmer REL, Macey RI. Perturbation of red cell volume: rectification of osmotic flow. *Biochim Biophys Acta Biomembr* 1970;**196**(1):53–65.
3. Macey RL, Farmer RE. Inhibition of water and solute permeability in human red cells. *Biochim Biophys Acta Biomembr* 1970;**211**(1):104–6.
4. Marrink SJ, Berendsen HJ. Simulation of water transport through a lipid membrane. *J Phys Chem* 1994;**98**(15):4155–68.
5. King LS, Yasui M, Agre P. Aquaporins in health and disease. *Mol Med Today* 2000;**6**(2):60–5.
6. Verkman AS, Anderson MO, Papadopoulos MC. Aquaporins: important but elusive drug targets. *Nat Rev Drug Discov* 2014;**13**(4):259.
7. Ishibashi K, Kondo S, Hara S, Morishita Y. The evolutionary aspects of aquaporin family. *Am J Physiol Regul Integr Comp Physiol* 2011;**300**(3):566–76.
8. Agre P, Saboori AM, Asimos A, Smith BL. Purification and partial characterization of the Mr 30,000 integral membrane protein associated with the erythrocyte Rh(D) antigen. *J Biol Chem* 1987;**262**:17497–503.
9. Preston GM, Agre P. Isolation of the cDNA for erythrocyte integral membrane protein of 28 kilodaltons: member of an ancient channel family. *Proc Natl Acad Sci U S A* 1991;**88**:11110–4.
10. Preston GM, Carroll TP, Guggino WP, Agre P. Appearance of water channels in Xenopus oocytes expressing red cell CHIP28 protein. *Science* 1992;**256**:385–7.
11. Umenishi F, Schrier RW. Hypertonicity-induced aquaporin-1 (AQP1) expression is mediated by the activation of MAPK pathways and hypertonicity-responsive element in the AQP1 gene. *J Biol Chem* 2003;**278**(18):15765–70.

12. Beitz E, Golldack A, Rothert M, von Bülow J. Challenges and achievements in the therapeutic modulation of aquaporin functionality. *Pharmacol Ther* 2015;**155**:22–35.
13. Hara-Chikuma M, Verkman AS. Physiological roles of glycerol-transporting aquaporins: the aquaglyceroporins. *Cell Mol Life Sci* 2006;**63**(12):1386–92.
14. Bienert GP, Møller AL, Kristiansen KA, Schulz A, Møller IM, Schjoerring JK, et al. Specific aquaporins facilitate the diffusion of hydrogen peroxide across membranes. *J Biol Chem* 2007;**282**:1183–92.
15. Prasad GR, Coury LA, Finn F, Zeidel ML. Reconstituted aquaporin 1 water channels transport CO_2 across membranes. *J Biol Chem* 1998;**273**(50):33123–6.
16. Endeward V, Musa-Aziz R, Cooper GJ, Chen LM, Pelletier MF, Virkki LV, et al. Evidence that aquaporin 1 is a major pathway for CO_2 transport across the human erythrocyte membrane. *FASEB J* 2006;**20**(12):1974–81.
17. Hub JS, de Groot BL. Does CO_2 permeate through aquaporin-1? *Biophys J* 2006;**91**(3):842–8.
18. Fang X, Yang B, Matthay MA, Verkman AS. Evidence against aquaporin-1-dependent CO_2 permeability in lung and kidney. *J Physiol* 2002;**542**(1):63–9.
19. Benga G. On the definition, nomenclature and classification of water channel proteins (aquaporins and relatives). *Mol Aspects Med* 2012;**33**(5–6):514–7.
20. Ishibashi K, Tanaka Y, Morishita Y. The role of mammalian superaquaporins inside the cell. *Biochim Biophys Acta* 2014;**1840**:1507–12.
21. Abdel-Sater KA. Physiology of the aquaporins. *Am J Biomed Sci* 2018;**10**(3):167–83. https://doi.org/10.5099/aj180300167.
22. Verkman AS, Mitra AK. Structure and function of aquaporin water channels. *Am J Physiol* 2000;**78**: F13–28.
23. Tournaire Roux C, Stuka M, Javot H, Gout E, Gerbeau P, Luu DT, et al. Cytosolic pH regulates root water transport during anoxic stress through gating of aquaporins. *Nature* 2003;**425**:393–7.
24. Srivastava J, Barber DL, Jacobson MP. Intracellular pH sensors: design principles and functional significance. *Physiology (Bethesda)* 2007;**22**(1):30–9.
25. Péret B, Li G, Zhao J, Band LR, Voß U, Postaire O, et al. Auxin regulates aquaporin function to facilitate lateral root emergence. *Nat Cell Biol* 2012;**14**(10):991.
26. Fujiyoshi Y, Mitsuoka K, De Groot BL, Philippsen A, Grubmüller H, Agre P, et al. Structure and function of water channels. *Curr Opin Struct Biol* 2002;**12**(4):509–15.
27. Kitchen P, Conner AC. Control of the aquaporin-4 channel water permeability by structural dynamics of aromatic/arginine selectivity filter residues. *Biochemistry* 2015;**54**(45):6753–5.
28. Horner A, Zocher F, Preiner J, Ollinger N, Siligan C, Akimov SA, et al. The mobility of single-file water molecules is governed by the number of H-bonds they may form with channel-lining residues. *Sci Adv* 2015;**1**(2):e1400083.
29. Yu J, Yool AJ, Schulten K, Tajkhorshid E. Mechanism of gating and ion conductivity of a possible tetrameric pore in aquaporin-1. *Structure* 2006;**14**:1411–23.
30. Jung JS, Preston GM, Smith BL, Guggino WB, Agre P. Molecular structure of the water channel through aquaporin chip—the hourglass model. *J Biol Chem* 1994;**269**:14648–54.
31. Tsubota K, Hirai S, King LS, Agre P, Ishida N. Defective cellular trafficking of lacrimal gland aquaporin-5 in Sjögren's syndrome. *Lancet* 2001;**357**(9257):688–9.
32. Lee BH, Gauna AE, Perez G, Park YJ, Pauley KM, Kawai T, et al. Autoantibodies against muscarinic type 3 receptor in Sjögren's syndrome inhibit aquaporin 5 trafficking. *PLoS One* 2013;**8**(1):e53113.
33. Gradilone SA, García F, Huebert RC, Tietz PS, Larocca MC, Kierbel A, et al. Glucagon induces the plasma membrane insertion of functional aquaporin-8 water channels in isolated rat hepatocytes. *Hepatology* 2003;**37**(6):1435–41.
34. Conner MT, Conner AC, Brown JE, Bill RM. Membrane trafficking of aquaporin 1 is mediated by protein kinase C via microtubules and regulated by tonicity. *Biochemistry* 2010;**49**(5):821–3.
35. Yasui H, Kubota M, Iguchi K, Usui S, Kiho T, Hirano K. Membrane trafficking of aquaporin 3 induced by epinephrine. *Biochem Biophys Res Commun* 2008;**373**(4):613–7.
36. Kitchen P, Day RE, Taylor LH, Salman MM, Bill RM, Conner MT, et al. Identification and molecular mechanisms of the rapid tonicity-induced relocalization of the aquaporin 4 channel. *J Biol Chem* 2015;**290**:16873–81.

37. Kaneko K, Yagui K, Tanaka A, Yoshihara K, Ishikawa K, et al. Aquaporin 1 is required for hypoxia-inducible angiogenesis in human retinal vascular endothelial cells. *Microvasc Res* 2008;**75**:297–301.
38. Echevarria M, Muñoz-Cabello AM, Sanchez-Silva R, Toledo-Aral JJ, Lopez-Barneo J. Development of cytosolic hypoxia and HIF stabilization are facilitated by aquaporin 1 expression. *J Biol Chem* 2007;**282**:30207–15.
39. Abreu-Rodríguez I, Sánchez Silva R, Martins AP, Soveral G, Toledo-Aral JJ, López-Barneo J, et al. Functional and transcriptional induction of aquaporin-1 gene by hypoxia; analysis of promoter and role of Hif-1α. *PLoS One* 2011;**6**(12):e28385.
40. Umenishi F, Verkman AS. Isolation of the human aquaporin-1 promoter and functional characterization in human erythroleukemia cell lines. *Genomics* 1998;**47**(3):341–9.
41. Foufelle F, Girard J, Ferre P. Glucose regulation of gene expression. *Curr Opin Clin Nutr Metab Care* 1998;**1**:323–8.
42. Cao C, Sun Y, Healey S, Bi Z, Hu G, Wan S, et al. EGFR-mediated expression of aquaporin-3 is involved in human skin fibroblast migration. *Biochem J* 2006;**400**(2):225–34.
43. McCoy ES, Haas BR, Sontheimer H. Water permeability through aquaporin-4 is regulated by protein kinase C and becomes rate-limiting for glioma invasion. *Neuroscience* 2010;**168**(4):971–81.
44. Luo L, Yang R, Zhao S, Chen Y, Hong S, Wang K, et al. Decreased miR-320 expression is associated with breast cancer progression, cell migration, and invasiveness via targeting aquaporin 1. *Acta Biochim Biophys Sin* 2018;**50**(5):473–80.
45. Xiong W, Ran J, Jiang R, Guo P, Shi X, Li H, et al. miRNA-320a inhibits glioma cell invasion and migration by directly targeting aquaporin 4. *Oncol Rep* 2018;**39**(4):1939–47.
46. Jablonski EM, McConnell NA, Hughes Jr. FM, Huet-Hudson YM. Estrogen regulation of aquaporins in the mouse uterus: potential roles in uterine water movement. *Biol Reprod* 2003;**69**(5):1481–7.
47. Schnitzer JE, Oh P. Aquaporin-1 in plasma membrane and caveolae provides mercury-sensitive water channels across lung endothelium. *Am J Physiol Heart Circ Physiol* 1996;**270**(1):416–22.
48. Agre P, King LS, Yasui M, Guggino WB, Ottersen OP, Fujiyoshi Y, et al. Aquaporin water channels–from atomic structure to clinical medicine. *J Physiol* 2002;**542**:3–16.
49. King LS, Yasui M. Aquaporins and disease: lessons from mice to humans. *Trends Endocrinol Metab* 2002;**13**(8):355–60.
50. Verkman AS. Aquaporins: translating bench research to human disease. *J Exp Biol* 2009;**212**(11):1707–15.
51. Jarius S, Paul F, Franciotta D, Waters P, Zipp F, Hohlfeld R, et al. Mechanisms of disease: aquaporin-4 antibodies in neuromyelitis optica. *Nat Rev Neurol* 2008;**4**(4):202.
52. Xia Peng C, Li HY, Wang W, Wang J, Wang L, Gang Xu Q, et al. Retinal segmented layers with strong aquaporin-4 expression suffered more injuries in neuromyelitis optica spectrum disorders compared with optic neuritis with aquaporin-4 antibody seronegativity detected by optical coherence tomography. *Br J Ophthalmol* 2017;**101**(8):1032–7.
53. Tzartos JS, Stergiou C, Daoussis D, Zisimopoulou P, Andonopoulos AP, Zolota V, et al. Antibodies to aquaporins are frequent in patients with primary Sjögren's syndrome. *Rheumatology* 2017;**56**(12):2114–22.
54. Steinfeld S, Cogan E, King LS, Agre P, Kiss R, Delporte C. Abnormal distribution of aquaporin-5 water channel protein in salivary glands from Sjögren's syndrome patients. *Lab Invest* 2001;**81**(2):143.
55. Mulders SM, Bichet DG, Rijss JP, Kamsteeg EJ, Arthus MF, Lonergan M, et al. An aquaporin-2 water channel mutant which causes autosomal dominant nephrogenic diabetes insipidus is retained in the Golgi complex. *J Clin Invest* 1998;**102**:57–66.
56. Devuyst O, Burrow CR, Smith BL, Agre P, Knepper MA, Wilson PD. Expression of aquaporins-1 and-2 during nephrogenesis and in autosomal dominant polycystic kidney disease. *Am J Physiol Ren Physiol* 1996;**271**(1):169–83.
57. Badaut J, Ashwal S, Adami A, Tone B, Recker R, Spagnoli D, et al. Brain water mobility decreases after astrocytic aquaporin-4 inhibition using RNA interference. *J Cereb Blood Flow Metab* 2011;**31**(3):819–31.
58. Vella J, Zammit C, Di Giovanni G, Muscat R, Valentino M. The central role of aquaporins in the pathophysiology of ischemic stroke. *Front Cell Neurosci* 2015;**8**(9):108.

59. Filippidis AS, Kalani MY, Rekate HL. Hydrocephalus and aquaporins: lessons learned from the bench. *Childs Nerv Syst* 2011;**27**(1):27–33.
60. Portincasa P, Palasciano G, Svelto M, Calamita G. Aquaporins in the hepatobiliary tract. Which, where and what they do in health and disease. *Eur J Clin Invest* 2008;**38**(1):1–10.
61. Lehmann GL, Larocca MC, Soria LR, Marinelli RA. Aquaporins: their role in cholestatic liver disease. *World J Gastroenterol: WJG* 2008;**14**(46):7059.
62. Ishiyama G, López IA, Ishiyama A. Aquaporins and Meniere's disease. *Curr Opin Otolaryngol Head Neck Surg* 2006;**14**(5):332–6.
63. Takeda T, Taguchi D. Aquaporins as potential drug targets for Meniere's disease and its related diseases. In: *Aquaporins*. Berlin, Heidelberg: Springer; 2009. p. 171–84.
64. Schrier RW, Cadnapaphornchai MA, Ohara M. Water retention and aquaporins in heart failure, liver disease and pregnancy. *J R Soc Med* 2001;**94**(6):265–9.
65. Hardin JA, Wallace LE, Wong JF, O'loughlin EV, Urbanski SJ, Gall DG, et al. Aquaporin expression is downregulated in a murine model of colitis and in patients with ulcerative colitis, Crohn's disease and infectious colitis. *Cell Tissue Res* 2004;**318**(2):313–23.
66. Ricanek P, Lunde LK, Frye SA, Støen M, Nygård S, Morth JP, et al. Reduced expression of aquaporins in human intestinal mucosa in early stage inflammatory bowel disease. *Clin Exp Gastroenterol* 2015;**8**:49.
67. Hermo L, Smith CE. Thirsty business: cell, region, and membrane specificity of aquaporins in the testis, efferent ducts, and epididymis and factors regulating their expression. *J Androl* 2011;**32**(6):565–75.
68. Calero C, López-Campos JL, Izquierdo LG, Sánchez-Silva R, López-Villalobos JL, Sáenz-Coronilla FJ, et al. Expression of aquaporins in bronchial tissue and lung parenchyma of patients with chronic obstructive pulmonary disease. *Multidiscip Respir Med* 2014;**9**(1):29.
69. Hoshi A, Yamamoto T, Shimizu K, Ugawa Y, Nishizawa M, Takahashi H, et al. Characteristics of aquaporin expression surrounding senile plaques and cerebral amyloid angiopathy in Alzheimer disease. *J Neuropathol Exp Neurol* 2012;**71**(8):750–9.
70. Nico B, Ribatti D. Aquaporins in tumor growth and angiogenesis. *Cancer Lett* 2010;**294**(2):135–8.
71. Liao ZQ, Ye M, Yu PG, Xiao C, Lin FY. Glioma-associated oncogene homolog1 (Gli1)-aquaporin1 pathway promotes glioma cell metastasis. *BMB Rep* 2016;**49**(7):394.
72. Reca A, Szpilbarg N, Damiano AE. The blocking of aquaporin-3 (AQP3) impairs extravillous trophoblast cell migration. *Biochem Biophys Res Commun* 2018;**499**(2):227–32.
73. Wang W, Li Q, Yang T, Li D, Ding F, Sun H, Bai G. RNA interference-mediated silencing of aquaporin (AQP)-5 hinders angiogenesis of colorectal tumor by suppressing the production of vascular endothelial growth factor. *Neoplasma* 2018;**65**(1):55–65.
74. Vacca A, Frigeri A, Ribatti D, Nicchia GP, Nico B, Ria R, et al. Microvessel overexpression of aquaporin 1 parallels bone marrow angiogenesis in patients with active multiple myeloma. *Br J Haematol* 2001;**113**:415–21.
75. Yang WY, Tan ZF, Dong DW, Ding Y, Meng H, Zhao Y, et al. Association of aquaporin-1 with tumor migration, invasion and vasculogenic mimicry in glioblastoma multiforme. *Mol Med Rep* 2018;**17**(2):3206–11.
76. Ribatti D, Ranieri G, Annese T, Nico B. Aquaporins in cancer. *Biochim Biophys Acta Gen Subj* 2014;**1840**(5):1550–3.
77. Hoque MO, Soria JC, Woo J, Lee T, Lee J, Jang SJ, et al. Aquaporin 1 is overexpressed in lung cancer and stimulates NIH-3T3 cell proliferation and anchorage-independent growth. *Am J Pathol* 2006;**168**:1345–53.
78. Saadoun S, Papadopoulos MC, Davies DC, Bell BA, Krishna S. Increased aquaporin 1 water channel expression in human brain tumours. *Br J Cancer* 2002;**87**(6):621.
79. Loitto VM, Forslund T, Sundqvist T, Magnusson KE, Gustafsson M. Neutrophil leukocyte motility requires directed water influx. *J Leukoc Biol* 2002;**71**(2):212–22.
80. Saadoun S, Papadopoulos MC, Hara-Chikuma M, Verkman AS. Impairment of angiogenesis and cell migration by targeted aquaporin-1 gene disruption. *Nature* 2005;**434**(7034):786.
81. Papadopoulos MC, Saadoun S, Verkman AS. Aquaporins and cell migration. *Pflugers Arch—Eur J Physiol* 2008;**456**(4):693–700.

82. Stroka KM, Gu Z, Sun SX, Konstantopoulos K. Bioengineering paradigms for cell migration in confined microenvironments. *Curr Opin Cell Biol* 2014;**30**:41–50.
83. Wolf K, Wu YI, Liu Y, Geiger J, Tam E, Overall C, et al. Multi-step pericellular proteolysis controls the transition from individual to collective cancer cell invasion. *Nat Cell Biol* 2007;**9**(8):893–904.
84. Lämmermann T, Sixt M. Mechanical modes of 'amoeboid' cell migration. *Curr Opin Cell Biol* 2009;**21**(5):636–44.
85. Stroka KM, Jiang H, Chen SH, Tong Z, Wirtz D, Sun SX, et al. Water permeation drives tumor cell migration in confined microenvironments. *Cell* 2014;**157**(3):611–23.
86. Hayashi Y, Edwards NA, Proescholdt MA, Oldfield EH, Merril MJ. Regulation and function of aquaporin-1 in glioma cells. *Neoplasia* 2007;**9**(9):777–87.
87. Ritter M, Schratzberger P, Rossmann H, Woll E, Seiler K, Seidler U, et al. Effect of inhibitors of Na+/H+-exchange and gastric H+/K+ ATPase on cell volume, intracellular pH and migration of human polymorphonuclear leucocytes. *Br J Pharmacol* 1998;**124**:627–38.
88. Brisson L, Driffort V, Benoist L, Poet M, Counillon L, Antelmi E, et al. NaV1.5 Na+ channels allosterically regulate the NHE-1 exchanger and promote the activity of breast cancer cell invadopodia. *J Cell Sci* 2013;**126**(21):4835–42.
89. Brisson L, Gillet L, Calaghan S, Besson P, Le Guennec JY, Roger S, Gore J. Na V 1.5 enhances breast cancer cell invasiveness by increasing NHE1-dependent H+ efflux in caveolae. *Oncogene* 2011;**30**(17):2070.
90. Wannous R, Bon E, Gillet L, Chamouton J, Weber G, Brisson L, et al. Suppression of PPARβ, and DHA treatment, inhibit NaV1. 5 and NHE-1 pro-invasive activities. *Pflugers Arch—Eur J Physiol* 2015;**467**(6):1249–59.
91. Page E, Winterfield J, Goings G, Bastawrous A, Upshaw-Earley J, Doyle D. Water channel proteins in rat cardiac myocyte caveolae: osmolarity-dependent reversible internalization. *Am J Physiol Heart Circ Physiol* 1998;**274**(6):1988–2000.
92. Zheng X, Bollag WB. Aquaporin 3 colocates with phospholipase d2 in caveolin-rich membrane microdomains and is downregulated upon keratinocyte differentiation. *J Invest Dermatol* 2003;**121**(6):1487–95.
93. Au CG, Cooper ST, Lo HP, Compton AG, Yang N, Wintour EM, et al. Expression of aquaporin 1 in human cardiac and skeletal muscle. *J Mol Cell Cardiol* 2004;**36**(5):655–62.
94. Mazzone A, Tietz P, Jefferson J, Pagano R, LaRusso NF. Isolation and characterization of lipid microdomains from apical and basolateral plasma membranes of rat hepatocytes. *Hepatology* 2006;**43**(2):287–96.
95. Li Z, Peng D, Zhang X, Peng Y, Chen M, Ma X, et al. Na+ induces the tolerance to water stress in white clover associated with osmotic adjustment and aquaporins-mediated water transport and balance in root and leaf. *Environ Exp Bot* 2017;**144**:11–24.
96. Joshi B, Strugnell SS, Goetz JG, Kojic LD, Cox ME, Griffith OL, et al. Phosphorylated caveolin-1 regulates Rho/ROCK-dependent focal adhesion dynamics and tumor cell migration and invasion. *Cancer Res* 2008;**68**(20):8210–20.
97. Jiang Y. Aquaporin-1 activity of plasma membrane affects HT20 colon cancer cell migration. *IUBMB Life* 2009;**61**(10):1001–9.
98. Jablonski EM, Hughes FM. The potential role of caveolin-1 in inhibition of aquaporins during the AVD. *Biol Cell* 2006;**98**(1):33–42.
99. Li H, Chen H, Steinbronn C, Wu B, Beitz E, Zeuthen T, Voth GA. Enhancement of proton conductance by mutations of the selectivity filter of aquaporin-1. *J Mol Biol* 2011;**407**(4):607–20.
100. Beitz E, Wu B, Holm LM, Schultz JE, Zeuthen T. Point mutations in the aromatic/arginine region in aquaporin 1 allow passage of urea, glycerol, ammonia, and protons. *Proc Natl Acad Sci U S A* 2006;**103**(2):269–74.
101. Tong J, Briggs MM, McIntosh TJ. Water permeability of aquaporin-4 channel depends on bilayer composition, thickness, and elasticity. *Biophys J* 2012;**103**(9):1899–908.
102. Clapp C, de la Escalera GM. Aquaporin-1: a novel promoter of tumor angiogenesis. *Trends Endocrinol Metab* 2006;**17**(1):1–2.
103. Verkman AS. Aquaporin water channels and endothelial cell function. *J Anat* 2002;**200**(6):617–27.

104. Huebert RC, Vasdev MM, Shergill U, Das A, Huang BQ, Charlton MR, et al. Aquaporin-1 facilitates angiogenic invasion in the pathological neovasculature that accompanies cirrhosis. *Hepatology* 2010;**52**(1):238–48.
105. Verkman AS, Hara-Chikuma M, Papadopoulos MC. Aquaporins—new players in cancer biology. *J Mol Med* 2008;**86**(5):523–9.
106. Camerino GM, Nicchia GP, Dinardo MM, Ribatti D, Svelto M, Frigeri A. In vivo silencing of aquaporin-1 by RNA interference inhibits angiogenesis in the chick embryo chorioallantoic membrane assay. *Cell Mol Biol (Noisy-le-Grand)* 2006;**52**(7):51–6.
107. Kaneko K, Yagui K, Tanaka A, Yoshihara K, Ishikawa K, Takahashi K, et al. Aquaporin 1 is required for hypoxia-inducible angiogenesis in human retinal vascular endothelial cells. *Microvasc Res* 2008;**75**(3):297–301.
108. Jablonski EM, Webb AN, McConnell NA, Riley MC, Hughes Jr. FM. Plasma membrane aquaporin activity can affect the rate of apoptosis but is inhibited after apoptotic volume decrease. *Am J Physiol Cell Physiol* 2004;**286**(4):975–85.
109. Okada Y, Maeno E, Shimizu T, Dezaki K, Wang J, Morishima S. Receptor-mediated control of regulatory volume decrease (RVD) and apoptotic volume decrease (AVD). *J Physiol* 2001;**532**(1):3–16.
110. Thangaraju M, Sharma K, Liu D, Shen S-H, Srikant CB. Interdependent regulation of intracellular acidification and SHP-1 in apoptosis. *Cancer Res* 1999;**59**(7):1649–54.
111. Bienert GP, Desguin B, Chaumont F, Hols P. Channel-mediated lactic acid transport: a novel function for aquaglyceroporins in bacteria. *Biochem J* 2013;**454**(3):559–70.
112. Amiry-Moghaddam M, Lindland H, Zelenin S, Roberg BA, Gundersen BB, Petersen P, et al. Brain mitochondria contain aquaporin water channels: evidence for the expression of a short AQP9 isoform in the inner mitochondrial membrane. *FASEB J* 2005;**19**(11):1459–67.
113. Morishima T, Aoyama M, Iida Y, Yamamoto N, Hirate H, Arima H, et al. Lactic acid increases aquaporin 4 expression on the cell membrane of cultured rat astrocytes. *Neurosci Res* 2008;**61**(1):18–26.
114. Zheng X, Bollag WB. Aquaporin 3 colocates with phospholipase d2 in caveolin-rich membrane microdomains and is downregulated upon keratinocyte differentiation. *J Invest Dermatol* 2003;**121**(6):1487–95.
115. Page E, Winterfield J, Goings G, Bastawrous A, Upshaw-Earley J, Doyle D. Water channel proteins in rat cardiac myocyte caveolae: osmolarity-dependent reversible internalization. *Am J Physiol Heart Circ Physiol* 1998;**274**(6):1988–2000.
116. Aoki T, Suzuki T, Hagiwara H, Kuwahara M, Sasaki S, Takata K, Matsuzaki T. Close association of aquaporin-2 internalization with caveolin-1. *Acta Histochem Cytochem* 2012;**45**(2):139–46.
117. Anthony TL, Brooks HL, Boassa D, Leonov S, Yanochko GM, Regan JW, Yool AJ. Cloned human aquaporin-1 is a cyclic GMP-gated ion channel. *Mol Pharmacol* 2000;**57**(3):576–88.
118. Boassa D, Yool AJ. A fascinating tail: cyclic GMP activation of aquaporin-1 ion channels. *Trends Pharmacol Sci* 2002;**23**:558–62.
119. Zampighi GA, Hall JE, Kreman M. Purified lens junctional protein forms channels in planar lipid films. *Proc Natl Acad Sci U S A* 1985;**82**:8468–72.
120. Ehring GR, Lagos N, Zampighi GA, Hall JE. Phosphorylation modulates the voltage dependence of channels reconstituted from the major intrinsic protein of lens fiber membranes. *J Membr Biol* 1992;**126**:75–88.
121. Ehring GR, Zampighi G, Horwitz J, Bok D, Hall JE. Properties of channels reconstituted from the major intrinsic protein of lens fiber membranes. *J Gen Physiol* 1990;**96**:631–64.
122. Yasui M, Hazama A, Kwon TH, Nielsen S, Guggino WB, Agre P. Rapid gating and anion permeability of an intracellular aquaporin. *Nature* 1999;**402**:184–7.
123. Mósca AF, Rodrigues C, Martins AP, Soveral G. Short-term regulation of aquaporin-5 by combined phosphorylation and pH. *Free Radic Biol Med* 2018;**120**(1):135–6.
124. Németh-Cahalan KL, Hall JE. pH and calcium regulate the water permeability of aquaporin 0. *J Biol Chem* 2000;**275**(10):6777–82.
125. Németh-Cahalan KL, Kalman K, Hall JE. Molecular basis of pH and Ca2+ regulation of aquaporin water permeability. *J Gen Physiol* 2004;**123**(5):573–80.

126. Rothert M, Rönfeldt D, Beitz E. Electrostatic attraction of weak monoacid anions increases probability for protonation and passage through aquaporins. *J Biol Chem* 2017;**292**(22):9358–64.
127. Bellati J, Alleva K, Soto G, Vitali V, Jozefkowicz C, Amodeo G. Intracellular pH sensing is altered by plasma membrane PIP aquaporin co-expression. *Plant Mol Biol* 2010;**74**(1–2):105–18.
128. Amodeo G, Sutka M, Dorr R, Parisi M. Protoplasmic pH modifies water and solute transfer in beta vulgaris root vacuoles. *J Membr Biol* 2002;**187**:175–84.
129. Fischer M, Kaldenhoff R. On the pH regulation of plant aquaporins. *J Biol Chem* 2008;**283**(49):33889–92.
130. Kaptan S, Assentoft M, Schneider HP, Fenton RA, Deitmer JW, MacAulay N, et al. H95 is a pH-dependent gate in aquaporin 4. *Structure* 2015;**23**(12):2309–18.
131. Zeuthen T, Klaerke DA. Transport of water and glycerol in aquaporin 3 is gated by H+. *J Biol Chem* 1999;**274**:21631–6.
132. Rodrigues C, Mósca AF, Martins AP, Nobre T, Prista C, Antunes F, et al. Rat aquaporin-5 is pH-gated induced by phosphorylation and is implicated in oxidative stress. *Int J Mol Sci* 2016;**17**(12):2090.
133. Zelenina M, Bondar AA, Zelenin S, Aperia A. Nickel and extracellular acidification inhibit the water permeability of human aquaporin-3 in lung epithelial cells. *J Biol Chem* 2003;**278**(32):30037–43.
134. Wu B, Beitz E. Aquaporins with selectivity for unconventional permeants. *Cell Mol Life Sci* 2007;**64**:2413–21.
135. Verkman AS. Does aquaporin-1 pass gas? An opposing view. *J Physiol* 2002;**542**(1):31.
136. Zhang X, Chen Y, Dong L, Shi B. Effect of selective inhibition of aquaporin 1 on chemotherapy sensitivity of J82 human bladder cancer cells. *Oncol Lett* 2018;**15**(3):3864–9.
137. Oshio K, Binder DK, Bollen A, Verkman AS, Berger MS, Manley GT. Aquaporin-1 expression in human glial tumors suggests a potential novel therapeutic target for tumor-associated edema. *Acta Neurochir Suppl* 2003;**86**:499–502.
138. El Hindy N, Bankfalvi A, Herring A, Adamzik M, Lambertz N, Zhu Y, et al. Correlation of aquaporin-1 water channel protein expression with tumor angiogenesis in human astrocytoma. *Anticancer Res* 2013;**33**:609–13.
139. Deb P, Pal S, Dutta V, Boruah D, Chandran VM, Bhatoe HS. Correlation of expression pattern of aquaporin-1 in primary central nervous system tumors with tumor type, grade, proliferation, microvessel density, contrast-enhancement and perilesional edema. *J Cancer Res Ther* 2012;**8**(4):571–7.
140. Dorward HS, Du A, Bruhn MA, Wrin J, Pei JV, Evdokiou A, et al. Pharmacological blockade of aquaporin-1 water channel by AqB013 restricts migration and invasiveness of colon cancer cells and prevents endothelial tube formation in vitro. *J Exp Clin Cancer Res* 2016;**35**(1):36.
141. Bin K, Shi-Peng Z. Acetazolamide inhibits aquaporin-1 expression and colon cancer xenograft tumor growth. *Hepatogastroenterology* 2011;**58**(110 – 111):1502–6.
142. Chen Y, Tachibana O, Oda M, Xu R, Hamada JI, Yamashita J, et al. Increased expression of aquaporin 1 in human hemangioblastomas and its correlation with cyst formation. *J Neurooncol* 2006;**80**(3):219–25.
143. Mazal PR, Susani M, Wrba F, Haitel A. Diagnostic significance of aquaporin-1 in liver tumors. *Hum Pathol* 2005;**36**(11):1226–31.
144. Guan B, Zhu D, Dong Z, Yang Z. Expression and distribution of aquaporin 1 in laryngeal carcinoma. *Lin Chung Er Bi Yan Hou Tou Jing Wai Ke Za Zhi* 2007;**21**:269–72.
145. Bellezza G, Vannucci J, Bianconi F, Metro G, Del Sordo R, Andolfi M, et al. Prognostic implication of aquaporin 1 overexpression in resected lung adenocarcinoma. *Interact Cardiovasc Thorac Surg* 2017;**25**(6):856–61.
146. Xiang Y, Ma B, Li T, Gao JW, Yu HM, Li XJ. Acetazolamide inhibits aquaporin-1 protein expression and angiogenesis. *Acta Pharmacol Sin* 2004;**25**:812–6.
147. Li XJ, Xiang Y, Ma B, Qi XQ. Effects of acetazolamide combined with or without $NaHCO_3$ on suppressing neoplasm growth, metastasis and aquaporin-1 (AQP1) protein expression. *Int J Mol Sci* 2007;**8**(3):229–40.
148. Simone L, Gargano CD, Pisani F, Cibelli A, Mola MG, Frigeri A, et al. Aquaporin-1 inhibition reduces metastatic formation in a mouse model of melanoma. *J Cell Mol Med* 2018;**22**(2):904–12.

149. Nicchia GP, Stigliano C, Sparaneo A, Rossi A, Frigeri A, Svelto M. Inhibition of aquaporin-1 dependent angiogenesis impairs tumour growth in a mouse model of melanoma. *J Mol Med* 2013;**91**(5):613–23.
150. Hu J, Verkman AS. Increased migration and metastatic potential of tumor cells expressing aquaporin water channels. *FASEB J* 2006;**20**(11):1892–4.
151. Esteva-Font C, Jin BJ, Verkman AS. Aquaporin-1 gene deletion reduces breast tumor growth and lung metastasis in tumor-producing MMTV-PyVT mice. *FASEB J* 2014;**28**(3):1446–53.
152. Li Q, Zhang B. Expression of aquaporin-1 in nasopharyngeal cancer tissues. *J Otolaryngol Head Neck Surg* 2010;**39**:511–5.
153. Chetry M, Li S, Liu H, Hu X, Zhu X. Prognostic values of aquaporins mRNA expression in human ovarian cancer. *Biosci Rep* 2018;**38**(2):BSR20180108. https://doi.org/10.1042/BSR20180108.
154. Wang Y, Fan Y, Zheng C, Zhang X. Knockdown of AQP1 inhibits growth and invasion of human ovarian cancer cells. *Mol Med Rep* 2017;**16**(4):5499–504.
155. Yang JH, Shi YF, Chen XD, Qi WJ. The influence of aquaporin-1 and microvessel density on ovarian carcinogenesis and ascites formation. *Int J Gynecol Cancer* 2006;**16**(S1):400–5.
156. Guo H, Han J, Hao F, An Y, Xu Y, et al. Ginsenoside Rg3 attenuates cell migration via inhibition of aquaporin 1 expression in PC-3M prostate cancer cells. *Eur J Pharmacol* 2012;**683**(1–3):27–34.
157. Park JY, Yoon GS. Overexpression of aquaporin-1 is a prognostic factor for biochemical recurrence in prostate adenocarcinoma. *Pathol Oncol Res* 2017;**23**(1):189.
158. Wu Z, Li S, Liu J, Shi Y, Wang J, Chen D, et al. RNAi-mediated silencing of AQP1 expression inhibited the proliferation, invasion and tumorigenesis of osteosarcoma cells. *Cancer Biol Ther* 2015;**16**:1332–40.
159. Qin F, Zhang H, Shao Y, Liu X, Yang L, Huang Y, et al. Expression of aquaporin1, a water channel protein, in cytoplasm is negatively correlated with prognosis of breast cancer patients. *Oncotarget* 2016;**7**(7):8143.
160. Liu J, Zhang WY, Ding DG. Expression of aquaporin 1 in bladder uroepithelial cell carcinoma and its relevance to recurrence. *Asian Pac J Cancer Prev* 2015;**16**(9):3973.
161. Saadoun S, Papadopoulos MC, Hara-Chikuma M, Verkman AS. Impairment of angiogenesis and cell migration by targeted aquaporin-1 gene disruption. *Nature* 2005;**434**(7034):786.
162. Kao SC, Armstrong N, Condon B, Griggs K, McCaughan B, Maltby S, et al. Aquaporin 1 is an independent prognostic factor in pleural malignant mesothelioma. *Cancer* 2012;**118**(11):2952–61.
163. Angelico G, Caltabiano R, Loreto C, Ieni A, Tuccari G, Ledda C, et al. Immunohistochemical expression of aquaporin-1 in fluoro-edenite-induced malignant mesothelioma: a preliminary report. *Int J Mol Sci* 2018;**19**(3):685.
164. Huang YT, Zhou J, Shi S, Xu HY, Qu F, Zhang D, et al. Identification of estrogen response element in aquaporin-3 gene that mediates estrogen-induced cell migration and invasion in estrogen receptor-positive breast cancer. *Sci Rep* 2015;**5**:12484.
165. Cao XC, Zhang WR, Cao WF, Liu BW, Zhang F, Zhao HM, et al. Aquaporin3 is required for FGF-2-induced migration of human breast cancers. *PLoS One* 2013;**8**(2):e56735.
166. Arif M, Kitchen P, Conner MT, Hill EJ, Nagel D, Bill RM, et al. Downregulation of aquaporin 3 inhibits cellular proliferation, migration and invasion in the MDA-MB-231 breast cancer cell line. *Oncol Lett* 2018;**16**(1):713–20.
167. Kusayama M, Wada K, Nagata M, Ishimoto S, Takahashi H, Yoneda M, et al. Critical role of aquaporin 3 on growth of human esophageal and oral squamous cell carcinoma. *Cancer Sci* 2011;**102**(6):1128–36.
168. Wen J, Wang Y, Gao C, Zhang G, You Q, Zhang W, et al. Helicobacter pylori infection promotes aquaporin 3 expression via the ROS–HIF-1α–AQP3–ROS loop in stomach mucosa: a potential novel mechanism for cancer pathogenesis. *Oncogene* 2018;**22**:1.
169. Chen J, Wang T, Zhou YC, Gao F, Zhang ZH, Xu H, et al. Aquaporin 3 promotes epithelial-mesenchymal transition in gastric cancer. *J Exp Clin Cancer Res* 2014;**33**(1):38.
170. Xu H, Xu Y, Zhang W, Shen L, Yang L, Xu Z. Aquaporin-3 positively regulates matrix metalloproteinases via PI3K/AKT signal pathway in human gastric carcinoma SGC7901 cells. *J Exp Clin Cancer Res* 2011;**30**(1):86.

171. Chen L, Li Z, Zhang Q, Wei S, Li B, Zhang X, et al. Silencing of AQP3 induces apoptosis of gastric cancer cells via downregulation of glycerol intake and downstream inhibition of lipogenesis and autophagy. *Onco Targets Ther* 2017;**10**:2791.
172. Jiang B, Li Z, Zhang W, Wang H, Zhi X, Feng J, et al. miR-874 inhibits cell proliferation, migration and invasion through targeting aquaporin-3 in gastric cancer. *J Gastroenterol* 2014;**49**(6):1011–25.
173. Xia H, Ma YF, Yu CH, Li YJ, Tang J, Li JB, et al. Aquaporin 3 knockdown suppresses tumour growth and angiogenesis in experimental non-small cell lung cancer. *Exp Physiol* 2014;**99**(7):974–84.
174. Liu W, Wang K, Gong K, Li X, Luo K. Epidermal growth factor enhances MPC-83 pancreatic cancer cell migration through the upregulation of aquaporin 3. *Mol Med Rep* 2012;**6**(3):607–10.
175. Huang X, Huang L, Shao M. Aquaporin 3 facilitates tumor growth in pancreatic cancer by modulating mTOR signaling. *Biochem Biophys Res Commun* 2017;**486**(4):1097–102.
176. Ismail M, Bokaee S, Davies J, Harrington KJ, Pandha H. Inhibition of the aquaporin 3 water channel increases the sensitivity of prostate cancer cells to cryotherapy. *Br J Cancer* 2009;**100**(12):1889.
177. Niu D, Kondo T, Nakazawa T, Kawasaki T, Yamane T, Mochizuki K, et al. Differential expression of aquaporins and its diagnostic utility in thyroid cancer. *PLoS One* 2012;**7**(7):e40770.
178. Badaut J, Ashwal S, Obenaus A. Aquaporins in cerebrovascular disease: a target for treatment of brain edema. *Cerebrovasc Dis* 2011;**31**(6):521–31.
179. McCoy E, Sontheimer H. Expression and function of water channels (aquaporins) in migrating malignant astrocytes. *Glia* 2007;**55**:1034–43.
180. Xiong W, Ran J, Jiang R, Guo P, Shi X, Li H, et al. miRNA-320a inhibits glioma cell invasion and migration by directly targeting aquaporin 4. *Oncol Rep* 2018;**39**(4):1939–47.
181. Ding T, Gu F, Fu L, Ma YJ. Aquaporin-4 in glioma invasion and an analysis of molecular mechanisms. *J Clin Neurosci* 2010;**17**(11):1359–61.
182. Ding T, Ma Y, Li W, Liu X, Ying G, Fu L, Gu F. Role of aquaporin-4 in the regulation of migration and invasion of human glioma cells. *Int J Oncol* 2011;**38**(6):1521–31.
183. Ding T, Zhou Y, Sun K, Jiang W, Li W, Liu X, et al. Knockdown a water channel protein, aquaporin-4, induced glioblastoma cell apoptosis. *PLoS One* 2013;**8**(8):e66751.
184. Yang L, Wang X, Zhen S, Zhang S, Kang D, Lin Z. Aquaporin-4 upregulated expression in glioma tissue is a reaction to glioma-associated edema induced by vascular endothelial growth factor. *Oncol Rep* 2012;**28**:1633–8.
185. Shi Z, Zhang T, Luo L, Zhao H, Cheng J, Xiang J, Zhao C. Aquaporins in human breast cancer: identification and involvement in carcinogenesis of breast cancer. *J Surg Oncol* 2012;**106**(3):267–72.
186. Dai C, Arceo J, Arnold J, Wu J, Dovichi NJ, Sreekumar A, et al. Abstract 3475: novel correlation-based network analysis of breast tumor metabolism identifies the glycerol channel protein aquaporin-7 as a regulator of breast cancer metastasis. *Cancer Res* 2018;**78**(13 Suppl):3475. https://doi.org/10.1158/1538-7445.AM2018-3475.
187. Shi X, Wu S, Yang Y, Tang L, Wang Y, Dong J, et al. AQP5 silencing suppresses p38 MAPK signaling and improves drug resistance in colon cancer cells. *Tumour Biol* 2014;**35**(7):7035.
188. Moon C, Soria JC, Jang SJ, Lee J, Hoque MO, Sibony M, et al. Involvement of aquaporins in colorectal carcinogenesis. *Oncogene* 2003;**22**:6699–703.
189. Kang BW, Kim JG, Lee SJ, Chae YS, Jeong JY, Yoon GS, et al. Expression of aquaporin-1, aquaporin-3, and aquaporin-5 correlates with nodal metastasis in colon cancer. *Oncology* 2015;**88**(6):369–76.
190. Li Q, Yang T, Li D, Ding F, Bai G, Wang W, Sun H. Knockdown of aquaporin5 sensitizes colorectal cancer cells to 5-fluorouracil via inhibition of the Wnt/β-catenin signaling pathway. *Biochem Cell Biol* 2018;**96**:572–9.
191. Wang W, Li Q, Yang T, Li D, Ding F, Sun H, Bai G. Anti-cancer effect of aquaporin 5 silencing in colorectal cancer cells in association with inhibition of Wnt/β-catenin pathway. *Cytotechnology* 2018;**70**(2):615–24.
192. Shen L, Zhu Z, Huang Y, Shu Y, Sun M, Xu H, et al. Expression profile of multiple aquaporins in human gastric carcinoma and its clinical significance. *Biomed Pharmacother* 2010;**64**(5):313–8.
193. Zhang Z, Chen Z, Song Y, Zhang P, Hu J, Bai C. Expression of aquaporin 5 increases proliferation and metastasis potential of lung cancer. *J Pathol* 2010;**221**(2):210–20.

194. Zhang L, Lu J, Zhou H, Du Z, Zhang G. Silencing of aquaporin 5 inhibits the growth of A549 lung cancer cells in vitro and in vivo. *Int J Oncol* 2018;**52**(5):1643–50.
195. Chae YK, Woo J, Kim MJ, Kang SK, Kim MS, Lee J, et al. Expression of aquaporin 5 (AQP5) promotes tumor invasion in human non small cell lung cancer. *PLoS One* 2008;**3**:e2162.
196. Direito I, Paulino J, Vigia E, Brito MA, Soveral G. Differential expression of aquaporin-3 and aquaporin-5 in pancreatic ductal adenocarcinomas. *J Surg Oncol* 2017;**115**(8):980–96.
197. Pust A, Kylies D, Hube-Magg C, Kluth M, Minner S, Koop C, et al. Aquaporin 5 expression is frequent in prostate cancer and shows a dichotomous correlation with tumor phenotype and PSA recurrence. *Hum Pathol* 2015;**48**:102–10.
198. De Qing Wu ZF, Wang KJ, Feng XY, Lv ZJ, Li Y, Jian ZX. AQP8 inhibits colorectal cancer growth and metastasis by down-regulating PI3K/AKT signaling and PCDH7 expression. *Am J Cancer Res* 2018;**8**(2):266.
199. Guan Y, Chen J, Zhan Y, Lu H. Effects of dexamethasone on C6 cell proliferation, migration and invasion through the upregulation of AQP1. *Oncol Lett* 2018;**15**(5):7595–602.
200. Funakoshi Y, Shiono H, Inoue M, Kadota Y, Ohta M, Matsuda H, et al. Glucocorticoids induce G1 cell cycle arrest in human neoplastic thymic epithelial cells. *J Cancer Res Clin Oncol* 2005;**131**:314–22.
201. Cabrelle A, Maschio N, Carraro S, Frezzato F, Binotto G, Gattazzo C, et al. Apoptotic effect of cyclosporin a and dexamethasone in malignant cells of patients with B-chronic lymphocytic leukemia. *J Biol Regul Homeost Agents* 2009;**23**:239–50.
202. Harmon JM, Norman MR, Fowlkes BJ, Thompson EB. Dexamethasone induces irreversible G1 arrest and death of a human lymphoid cell line. *J Cell Physiol* 1979;**98**(2):267–78.
203. Greenberg AK, Hu J, Basu S, Hay J, Reibman J, Yie TA, et al. Glucocorticoids inhibit lung cancer cell growth through both the extracellular signal-related kinase pathway and cell cycle regulators. *Am J Respir Cell Mol Biol* 2002;**27**(3):320–8.
204. Guan Y, Chen J, Zhan Y, Lu H. Effects of dexamethasone on C6 cell proliferation, migration and invasion through the upregulation of AQP1. *Oncol Lett* 2018;**15**(5):7595–602.
205. Zhai Y, Xu H, Shen Q, Schaefer F, Schmitt CP, Chen J, et al. pH-mediated upregulation of AQP1 gene expression through the Spi-B transcription factor. *BMC Mol Biol* 2018;**19**:4.
206. Pelagalli A, Nardelli A, Lucarelli E, Zannetti A, Brunetti A. Autocrine signals increase ovine mesenchymal stem cells migration through aquaporin-1 and CXCR4 overexpression. *J Cell Physiol* 2018;**233**(8):6241–9.
207. Meng F, Rui Y, Xu L, Wan C, Jiang X, Li G. AQP1 enhances migration of bone marrow mesenchymal stem cells through regulation of fak and β-catenin. *Stem Cells Dev* 2014;**23**:66–75.
208. Meng F, Rui Y, Xu L, Wan C, Jiang X, Gang LG. Aqp1 enhances migration of bone marrow mesenchymal stem cells through regulation of FAK and β-catenin. *Stem Cells Dev* 2014;**23**(1):66–75.
209. Haddoub R, Rutzler M, Robin A, Flitsch SL. Design, synthesis and assaying of potential aquaporin inhibitors. In: Beitz E, editor. *Aquaporins, handbook of experimental pharmacology*. vol. 190. Berlin Heidelberg: Springer-Verlag; 2009. p. 385.
210. Niemietz CM, Tyerman SD. New potent inhibitors of aquaporins: silver and gold compounds inhibit aquaporins of plant and human origin. *FEBS Lett* 2002;**531**:443–7.
211. Serna A, Galán-Cobo A, Rodrigues C, Sánchez-Gomar I, Toledo-Aral JJ, Moura TF, et al. Functional inhibition of aquaporin-3 with a gold-based compound induces blockage of cell proliferation. *J Cell Physiol* 2014;**229**(11):1787–801.
212. Brooks HL, Regan JW, Yool AJ. Inhibition of aquaporin-1 water permeability by tetraethylammonium: involvement of the loop E pore region. *Mol Pharmacol* 2000;**57**(5):1021–6.
213. Detmers FJ, De Groot BL, Müller EM, Hinton A, Konings IB, Sze M, et al. Quaternary ammonium compounds as water channel blockers specificity, potency, and site of action. *J Biol Chem* 2006;**281**(20):14207–14.
214. Yool A, Brokl O, Pannabecker T, Dantzler W, Stamer D. TEA block of water flux in AQP1 channels expressed in kidney thin limbs of Henle's loop and a kidney-derived cell line. *BMC Physiol* 2002;**2**:1–8.
215. Yang B, Kim JK, Verkman AS. Comparative efficacy of HgCl2 with candidate aquaporin-1 inhibitors DMSO, gold, TEA+ and acetazolamide. *FEBS Lett* 2006;**580**(28–29):6679–84.

216. Bhattacharya P, Pandey AK, Paul S, Patnaik R. Neuroprotective potential of piroxicam in cerebral ischemia: an in silico evaluation of the hypothesis to explore its therapeutic efficacy by inhibition of aquaporin-4 and acid sensing ion channel1a. *Med Hypotheses* 2012;**79**(3):352–7.
217. Mirza N, Marson AG, Pirmohamed M. Effect of topiramate on acid–base balance: extent, mechanism and effects. *Br J Clin Pharmacol* 2009;**68**(5):655–61.
218. Marathe K, McVicar N, Li A, Bellyou M, Meakin S, Bartha R. Topiramate induces acute cellular acidification in glioblastoma. *J Neurooncol* 2016;**130**(3):465–72.
219. Shank RP, Gardocki JF, Streeter AJ, Maryanoff BE. An overview of the preclinical aspects of topiramate: pharmacology, pharmacokinetics, and mechanism of action. *Epilepsia* 2000;**41**(s1):3–9.
220. Supuran CT. Carbonic anhydrases: novel therapeutic applications for inhibitors and activators. *Nature* 2008;**7**:168.
221. Shank RP, Smith-Swintosky VL, Maryanoff BE. Carbonic anhydrase inhibition. Insight into the chemistry of zonisamide, topiramate, and the sulfamide cognate of topiramate. *J Enzyme Inhib Med Chem* 2008;**23**(2):271–6.
222. Monzani E, Shtil AA, La Porta CA. The water channels, new druggable targets to combat cancer cell survival, invasiveness and metastasis. *Curr Drug Targets* 2007;**8**:1132–7.
223. Huber VJ, Tsujita M, Kwee IL, Nakada T. Inhibition of aquaporin 4 by antiepileptic drugs. *Bioorg Med Chem* 2009;**17**(1):418–24.
224. Bing MA, Yang X, Li T, Yu HM, Li XJ. Inhibitory effect of topiramate on Lewis lung carcinoma metastasis and its relation with AQP1 water channel. *Acta Pharmacol Sin* 2004;**25**(1):54–60.
225. Bin K, Shi-Peng Z. Acetazolamide inhibits aquaporin-1 expression and colon cancer xenograft tumor growth. *Hepatogastroenterology* 2011;**58**(110–111):1502–6.
226. Gao J, Wang X, Chang Y, Zhang J, Song Q, Yu H, Li X. Acetazolamide inhibits osmotic water permeability by interaction with aquaporin-1. *Anal Biochem* 2006;**350**(2):165–70.
227. Xiang Y, Ma B, Li T, Gao JW, Yu HM, Li XJ. Acetazolamide inhibits aquaporin-1 protein expression and angiogenesis. *Acta Pharmacol Sin* 2004;**25**:812–6.
228. Ran X, Wang H, Chen Y, Zeng Z, Zhou Q, Zheng R, et al. Aquaporin-1 expression and angiogenesis in rabbit chronic myocardial ischemia is decreased by acetazolamide. *Heart Vessels* 2010;**25**(3):237–47.
229. Tanimura Y, Hiroaki Y, Fujiyoshi Y. Acetazolamide reversibly inhibits water conduction by aquaporin-4. *J Struct Biol* 2009;**166**(1):16–21.
230. Yang B, Kim JK, Verkman AS. Comparative efficacy of HgCl2 with candidate aquaporin-1 inhibitors DMSO, gold, TEA+ and acetazolamide. *FEBS Lett* 2006;**580**(28–29):6679–84.
231. Zhang J, An Y, Gao J, Han J, Pan X, Pan Y, et al. Aquaporin-1 translocation and degradation mediates the water transportation mechanism of acetazolamide. *PLoS One* 2012;**7**(9):e45976.
232. Tanimura Y, Hiroaki Y, Fujiyoshi Y. Acetazolamide reversibly inhibits water conduction by aquaporin-4. *J Struct Biol* 2009;**166**(1):16–21.
233. Kourghi M, Pei JV, De Ieso ML, Flynn G, Yool AJ. Bumetanide derivatives AqB007 and AqB011 selectively block the aquaporin-1 ion channel conductance and slow cancer cell migration. *Mol Pharmacol* 2015;**115**:1016–8.
234. Migliati E, Meurice N, DuBois P, Fang JS, Somasekharan S, Beckett E, et al. Inhibition of aquaporin-1 and aquaporin-4 water permeability by a derivative of the loop diuretic bumetanide acting at an internal poreoccluding binding site. *Mol Pharmacol* 2009;**76**:105–12.
235. Dorward HS, Du A, Bruhn MA, Wrin J, Pei JV, Evdokiou A, et al. Pharmacological blockade of aquaporin-1 water channel by AqB013 restricts migration and invasiveness of colon cancer cells and prevents endothelial tube formation in vitro. *Nat Rev Cancer* 2016;**35**:36. https://doi.org/10.1186/s13046-016- 0310-6.
236. Naeini VF, Foroutan M, Maddah M, Rémond Y, Baniassadi M. Determinative factors in inhibition of aquaporin by different pharmaceuticals: atomic scale overview by molecular dynamics simulation. *Biochim Biophys Acta Gen Subj* 2018;**1862**(12):2815–23.
237. Bing MA, Xiang Y, Sheng-Mei MU, Tao LI, He-Ming YU, Xue-Jun LI. Effects of acetazolamide and anordiol on osmotic water permeability in AQP1-cRNA injected Xenopus oocyte. *Acta Pharmacol Sin* 2004;**25**:90–7.

238. Nabiuni M, Nazari Z, Safaeinejad Z, Delfan B, Miyan JA. Curcumin downregulates aquaporin-1 expression in cultured rat choroid plexus cells. *J Med Food* 2013;**16**:504–10.
239. Terlikowska KM, Witkowska AM, Zujko ME, Dobrzycka B, Terlikowski SJ. Potential application of curcumin and its analogues in the treatment strategy of patients with primary epithelial ovarian cancer. *Int J Mol Sci* 2014;**15**:21703–22.
240. Pei JV, Kourghi M, De Ieso ML, Campbell EM, Dorward HS, Hardingham JE, et al. Differential inhibition of water and ion channel activities of mammalian aquaporin-1 by two structurally related bacopaside compounds derived from the medicinal plant *Bacopa monnieri*. *Mol Pharmacol* 2016;**90**:496–507.
241. Palethorpe HM, Tomita Y, Smith E, Pei JV, Townsend AR, Price TJ, et al. The aquaporin 1 inhibitor bacopaside II reduces endothelial cell migration and tubulogenesis and induces apoptosis. *Int J Mol Sci* 2018;**19**(3):653.
242. Yamaguchi T, Iwata Y, Miura S, Kawada K. Reinvestigation of drugs and chemicals as aquaporin-1 inhibitors using pressure-induced hemolysis in human erythrocytes. *Biol Pharm Bull* 2012;**35**(11):2088–91.
243. Zelenina M, Tritto S, Bondar AA, Zelenin S, Aperia A. Copper inhibits the water and glycerol permeability of aquaporin-3. *J Biol Chem* 2004;**279**(50):51939–43.

CHAPTER 11

Migration, invasion, invadopodia, and the inversion of the pH gradient

Cancer cell migration, invasion and metastasis

Most of the different cancer types that kill their host do so through growth and invasion, which lead to metastasis. Invasion and metastasis require two basic abilities: motility[1] and degradation of the normal tissue surrounding the tumor.[2] These two abilities permit penetration.[3] Invasion is degradation of surrounding tissues followed by cell migration into them. Metastasis is cell migration to distant organs, which requires the migration into lymphatic or vascular or perineural tissues and from there, to other organs, again, through lymphatic or vascular circulation or via perineural spread.[4] When the cells reach an appropriate tissue to colonize, cell invasion and migration are again the leading processes, allowing the metastatic cell to invade the new organ.[5] The common denominator is always migration and invasion.[6] Different solid tumors use similar or different mechanisms of migration.[7,8] However, tissue proteolysis and migration are always a pre-requisite for invasion and metastasis. Therefore, attacking these processes should be a main issue to fight cancer spreading, independently of the different genotypic characteristics of tumors. The targeting of migration and invasion has turned out to be more complicated than initially thought. Although invasion is the common process leading to metastasis, treating it became a challenge in which there have been no successes so far. The reason for this failure is probably due to the several different "strategies" the invasive tumor cell may adopt. However, the altered pH generated by cancer is present in most of these "migration strategies" and it is so closely associated with migration that by modifying the first we should disrupt the latter.

Migration is cell movement while invasion is cell migration plus penetration into and through structures (basement membrane, extracellular matrix, interstitial stroma or across the vascular membrane). Invasion requires proteolysis of the extracellular matrix (ECM), release from existing adhesions and creation of new adhesions. Cross-talk between invadopodia and proteins in the extracellular matrix has emerged as an important issue in this process.[9]

The first step in migration is a morphological change in the cell to favor a better interaction with surrounding tissues called epithelial-mesenchymal transition (EMT). The EMT that an epithelial cell undergoes to adopt a mesenchymal phenotype is not only a change of shape but it is a complete program that modifies the cell's biochemical functioning and signaling pathways[10,11] (Fig. 1).

https://doi.org/10.1016/B978-0-12-819059-3.00011-3
© 2020 Elsevier Inc.
All rights reserved.

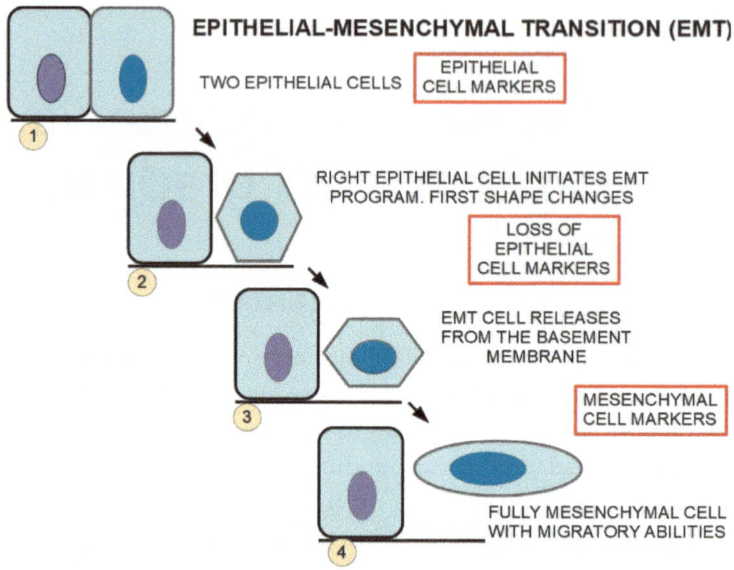

Fig. 1 (1) Two epithelial cells attached to the basal membrane. (2) The right cell undergoing EMT, starts to change its shape and becomes loose from attachments to the other, normal cell. (3) The cell in transition liberates itself from the basal membrane. (4) The cell achieves migratory abilities.

Important structures for achieving migration and invasion in malignant cells are the adhesion receptors (mainly integrins)[12,13] that recognize various proteins of the ECM and the invadopodium (pl. **invadopodia**) also known generically as invadosomes. **Focal adhesions** are areas where transmembrane proteins and extracellular proteins are in contact, represent a connection between cytosolic and ECM proteins[14,15] and probably interact with invadopodia. In fibrosarcoma focal adhesions promote proteolysis and a functional similarity between focal adhesions and invadopodia has been suggested.[16] Futhermore, focal adhesions are rich in NHE1 as are invadopodia,[17,18] and similarly the down-regulation of NHE1 decreases focal adhesion activity.[19]

Migration and lamellipodia/podosomes vs invasion and invadopodia/invadosomes

Invadopodia (plural of invadopodium, a name coined by Chen[20]) are structures similar to lamellipodia[21] of non-invading cells, but with one major difference: while invadopodia invade and destroy surrounding tissues, lamellipodia or podosome do not or do it on a minor scale (deep vs. shallow penetration in the matrix[22]). There are also many other differences described in Fig. 2 but a deeper discussion of this issue goes beyond the objectives of this chapter. Invadopodia are probably the site where the pH gradient inversion is most noticeable, and the place where the pH-cancer association is most remarkable.[23] While the processes of migration and invasion are different, they share a fundamental dependence on the altered pH dynamics discussed in this book.

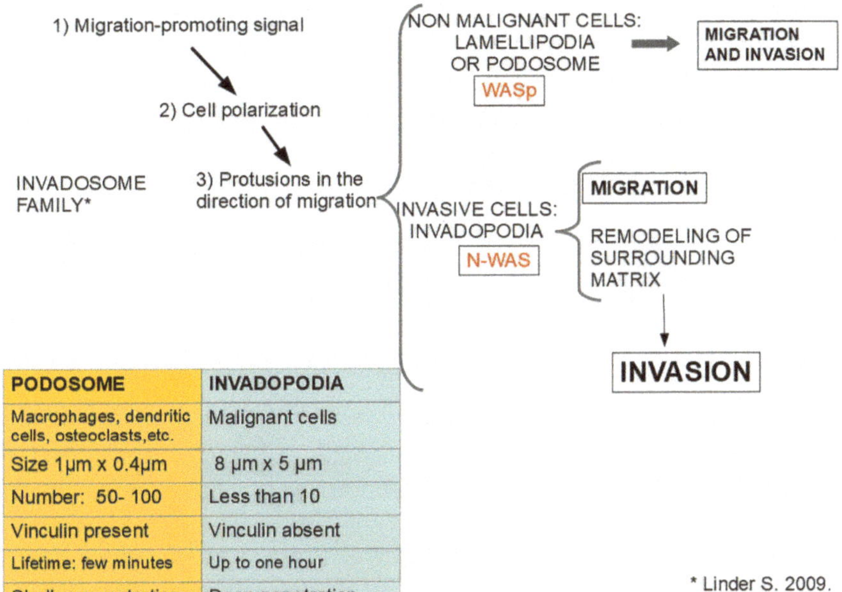

PODOSOME	INVADOPODIA
Macrophages, dendritic cells, osteoclasts,etc.	Malignant cells
Size 1μm x 0.4μm	8 μm x 5 μm
Number: 50- 100	Less than 10
Vinculin present	Vinculin absent
Lifetime: few minutes	Up to one hour
Shallow penetration	Deep penetration

Fig. 2 A simplified diagram differentiating lamellipodia from invadopodia.[24] Migration and invasion are highly integrated processes. The Wiscott-Aldrich syndrome protein (WASP) family members present in lamellipodia, WASp, differ from that found in invadopodia, N-WASP.[25] The important difference between invadosomes in cancer and normal cells is the matrix degrading capacity, which is very high in the first case.[22]

Migrating cells are highly polarized along their axis of movement. Furthermore, polarization is the cell's first response to a migration-promoting signal.[26] Polarization means that the cellular organelles, including the actin cytoskeleton adopt a peculiar distribution pattern and intracellular and extracellular pH gradients are created.[27,28] The polarization pattern of transformed migrating cells is not different from the polarization pattern found in normal migrating cells. The inhibition of NHE1 eliminates the intracellular gradient[4,29] and blocks migration.

The forward ends of mobile cells have thin cytoplasmic dynamic protrusions that contain no organelles. Migration is possible only if polarization is maintained,[30] and that is precisely one of the functions of polarized NHE1 expression from the front to the rear ends. There is no migration without polarization, and there is no polarization without the inversion of the pH gradient (Fig. 3). As Le Clainche and Carlier[31] clearly explained in the case of non-invasive cells: "*To migrate, a cell first extends protrusions such as lamellipodia and filopodia, forms adhesions, and finally retracts its tail. The actin cytoskeleton plays a major role in this process.*"

Even if the more evident features are located in the advancing front of the cell, the activity of the rear end is equally important. Indeed, migration is a highly coordinated process and depends on both the front and rear end. In addition to loosing water in the rear end, the cell must also sever its adhesions to the matrix or other cells. The different pH conditions in the rear end participate in the release from other structures (Fig. 3).

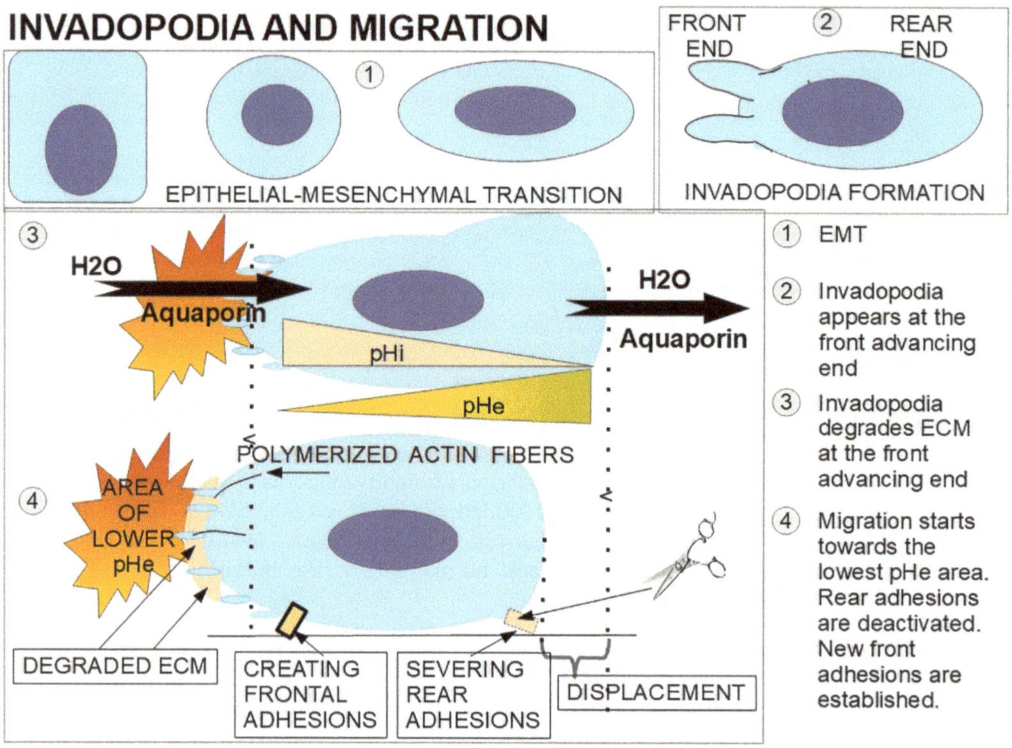

Fig. 3 Panel (1) epithelial-mesenchymal transition. Panel (2) The polarized cell develops invadopodia in the front end. Panel (3) Loss of water at the rear, gain of water at the front and actin fibers pulling the structures move the cell forward. Panel (4) An area of degraded ECM can be seen at the front, between invadopodia protrusions. Polymerized actin is present at the invadopodia. For a detailed description see.[32] The flux of water and ions is essential for the movement.[33] See also Chapter 10 on aquaporins. The creation and disintegration of adhesions between the cell and extracellular matrix is a continuous and dynamic process during migration,[34] These adhesions are pH dependent.[35] The cell is polarized and there is an intracellular and extracellular pH gradient.

Invadopodia are protruding structures that intervene in migration and extracellular matrix degradation[36] and their appearance determines the direction of movement. Their size is considerably larger than those of lamellipodia.[37] The characteristics of the ECM influence the number and activity of invadopodia.[38] This means that invadopodia can "sense" the stiffness of the ECM it must degrade.[39] Invasive cancer cells have many invadopodia (usually from 1 to 10). For the sake of clarity in the figures we show only one or a few invadopodia, which are disproportionately enlarged in relation to the cell.

Four important features of invasion are:

(1) freeing from basal membrane;

(2) digestion of extracellular matrix and;

(3) the establishment of adhesions between the advancing front and surrounding tissues.

(4) Interaction between matrix proteins and invadopodial proteins.

In this site, the advancing front must generate the adequate pH conditions for adhesions. Therefore, the NHE1 is a fundamental factor not only in generating a polarized cellular structure and actin polymerization[40] but also in creating these adhesions[41,42] and adequate conditions for metalloproteases to degrade tissues (Figs. 3 and 4).

Indeed, invadopodia are very rich in NHE1. The importance of this issue can be deduced from the fact that transformed fibroblasts with mutated NHE1 are unable to migrate.[43] Migration is a coordinated cycle of protrusion of the invadopodia at the front and retraction at the rear end. Invadopodia are actin-rich protrusions that seem to be the central machinery for creating an adequate microenvironment and internal milieu leading to migration, invasion and eventual metastasis.[44] Invadopodia are also able to degrade

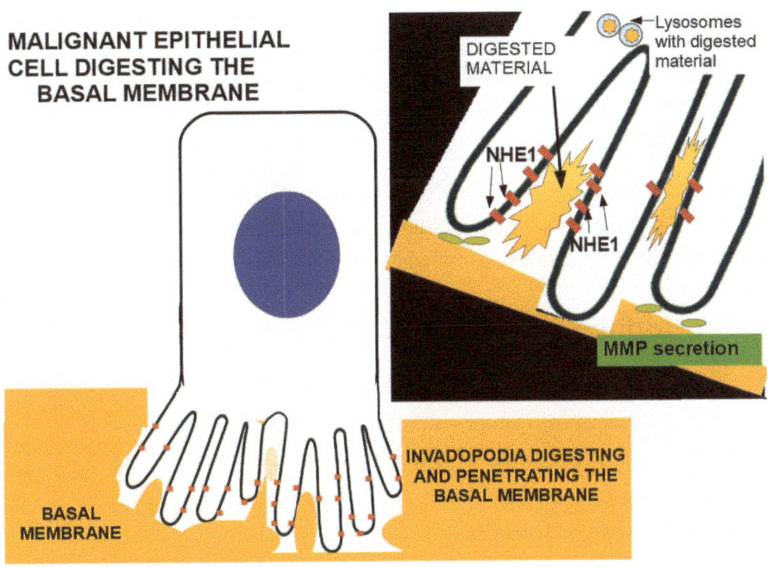

Fig. 4 A malignant epithelial cell is invading the basal membrane. Red dots on the central figure show the abundant presence of NHE1. Digested material through the activity of metalloproteases and cathepsins secreted by the invadopodia or by membrane inserted MT1-MMP can be seen between invadopodia. The proteolysing enzymes are activated by the extracellular acidity generated by NHE1.

BOX 1

INVADOPODIA AT A GLIMPSE

STRUCTURAL ASPECT: dependent on the actin-cortactin interaction. Inhibition of cortactin aborts invadopodia formation.

FUNCTIONAL ASPECT: dependent on metalloproteases, mainly MT1-MMP. Inhibition of MT1-MMP does not impede invadopodia formation, however it cannot degrade the ECM.*

FUNCTIONAL COMPONENTS:**
ACTIN REGULATORS: Cortactin and neural Wiskott-Aldrich syndrome protein (N-WASP)
ADAPTOR PROTEINS: Tyr kinase substrates
PROTEINASES

FUNCTIONAL RELATIONS:***
Cortactin is a regulator of MT1-MMP membrane insertion.

References: *[46]; **[47]; ***[48].

the extracellular matrix in order to create the adequate space for invasion. The degraded matrix is internalized and further processed by lysosomes, and this requires an increased acidity also in them where the V-ATPase proton pump creates the lysosomal acidic milieu. At the same time invadopodia process the actin fibers and polymerize them allowing cellular movements. NHE1 creates an intracellular and extracellular pH gradient that polarizes the cell and determines the direction of migration.[45] Maintaining the polarization is a pre-condition for migration. NHE1 stimulation increases the intracellular gradient while NHE1 inhibition abolishes it.

A summary of the characteristics of invadopodia is described in Box 1.

Voltage gated sodium channels (VGSCs)

NHE1 is not alone in the process of creating the intracellular and extracellular gradients, it requires the participation of VGSCs (voltage gated sodium channels) in its vicinity. Both intervene in the process of creating a highly acidic ECM (extracellular matrix), adequate for the activation of cathepsins and MMPs (matrix metalloproteases) that will degrade the matrix. However, if NHE1 is inhibited, this gradient disappears.[45]

VGSCs and NHE1 serve two different functions in invadopodia:

1. Ion dependent
   - (a) Creating an extracellular pH gradient.
   - (b) Participating in the cellular polarization to create an intracellular pH gradient.

 (c) By increasing extracellular proton load, the created acidity allows a better activation of proteolytic enzymes that degrade the ECM.
 (d) Cell volume regulation (migration entails cell volume increase in the front and cell volume decrease in the rear). Aquaporins have an important role in this process. (See Chapter 10 on aquaporins). The sodium imported by NHE1 and VGSCs create the adequate osmolar environment for AQPs activity.
2. Protein dependent
 (a) VGSCs increase NHE1 activity.
 (b) Adhesion activity of beta subunits of VGSCs.
 (c) VGSCs increase Src activity (that activates cortactin and actin polymerization).
 (d) NHE1 participates in the cytoskeletal modifications necessary for migration.

House et al.[49] found that the VGSC, NaV1.5, controlled a network of genes that included Wnt signaling, cell migration and cell cycle control in colon cancer cells.

VGSC activity promoted MAPK activation inducing transcriptional changes in invasion-related genes.[50] The conclusion is that VGSCs and NHE1 have pro-tumor effects that go beyond their channel function and are essential for migration and invasion. This also means that these channels are important targets for an integral treatment of cancer, as we shall see in the chapters dedicated to pharmacology and therapeutics. There are non-toxic drugs available for the targeting of these channels, such as phenytoin and amiloride derivatives, just to mention a few.

Epidermal growth factor (EGF) has a substantial role in the upregulation of VGSCs.[51] EGF/EGFR signaling increased metastatic potential of prostate cancer PC-3M cells and increased NaV1.7's mRNA expression and protein. In addition to this regulation by EGF, Fraser et al.[52] suggest the possibility of auto-regulation of these channels, in their words "the VGSC is self-sustaining, with its activity promoting further functional channel expression".

Formation of invadopodia

Artym et al.[53] proposed four steps in invadopodia formation:
1. membrane cortactin aggregation at membranes adherent to matrix;
2. MT1-MMP accumulation at the region of cortactin accumulation;
3. matrix degradation at the invadopodia region, and
4. subsequent cortactin dissociation from the area of continued MT1-MMP accumulation associated with foci of degraded matrix. These steps will be seen in the next figures.

The formation of invadopodia follows these well defined stages. Each stage is dependent on the fulfillment of the previous one. The following figures (Figs. 5–9) show the steps of the invadopodia formation induced by growth factors.

Invadopodia formation steps 1, 2 and 3

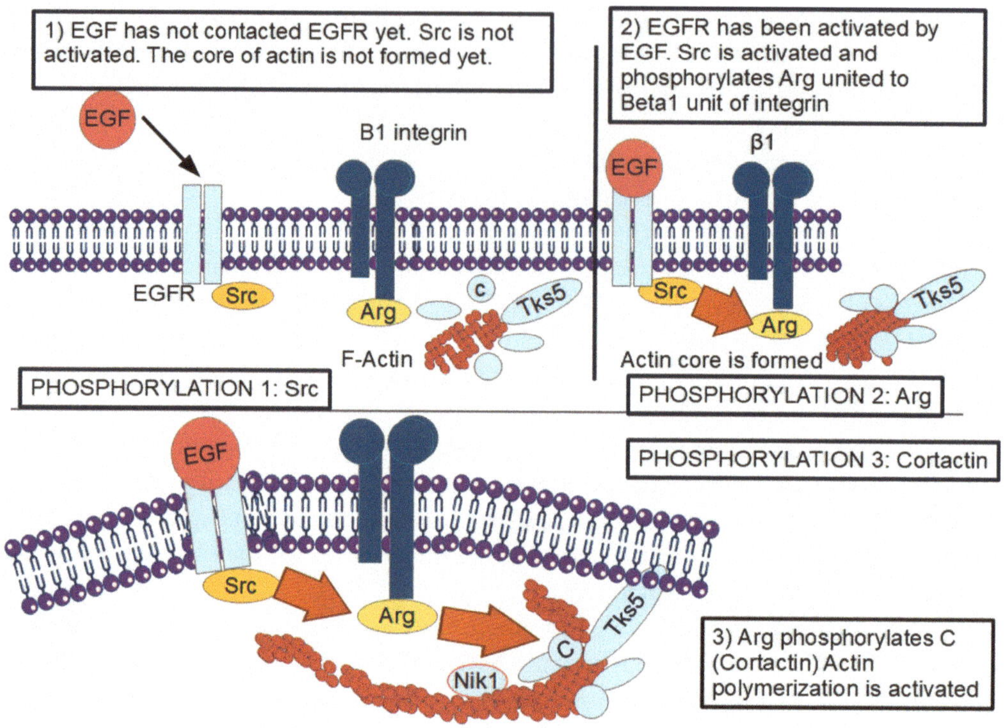

Fig. 5 Growth factors are determinant for the formation of invadopodia. Invadopodia precursor structures are rich in actin regulators, like cortactin, N-WASp, Arp2/3, cofilin, fascin, and others, but are not yet capable of degrading the ECM. Invadopdia formation is the final result of a chain of phosphorylations started by growth factors interacting with their receptors. When EGF binds with EGFR, its intracellular portion recruits and activates the kinase Src. Then Src phosphorylates the protein Arg *(right upper panel)*, which is bound to the Beta 1 unit of an integrin receptor. The actin core is formed (cortactin, cofilin, N-WASp, Arp2/3 and Tks5). *Lower panel*: Arg activates cortactin and the polymerization of actin is initiated. Tks5 anchors the actin to the cell membrane.

Invadopodia formation steps 4 and 5

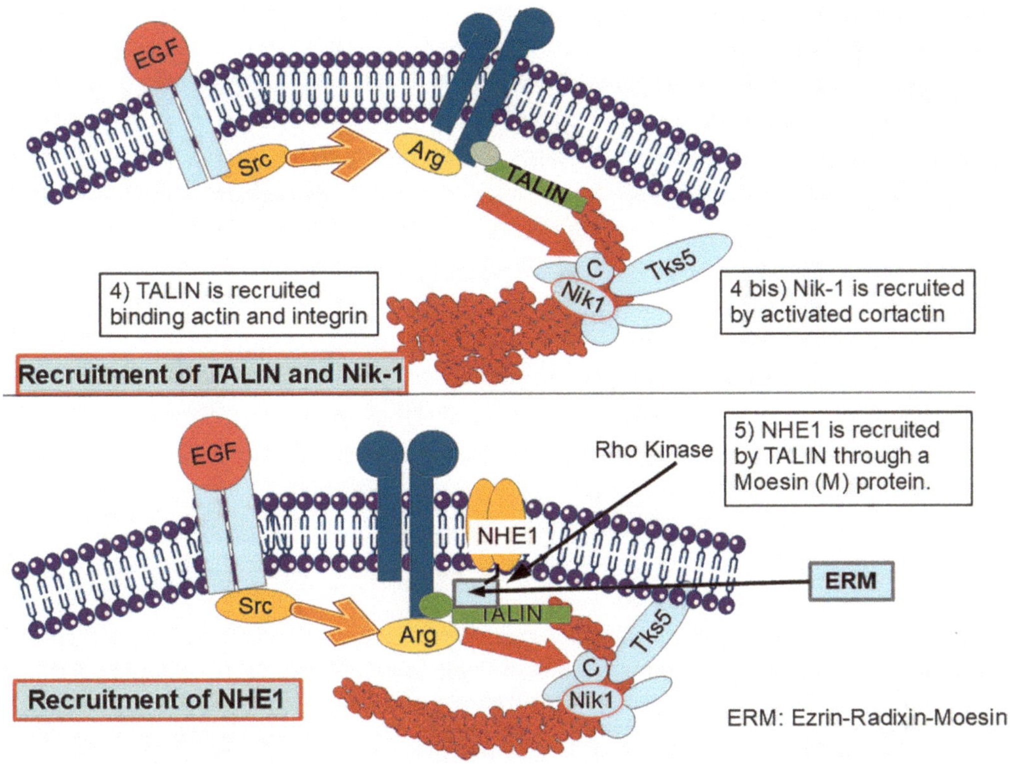

Fig. 6 *Upper panel*: Talin and Nik-1 are recruited. Cortactin represents a key figure in this chain, because as Clark et al.[54] have shown, the secretion of matrix metalloproteinases (MMPs) 2 and 9, and the membrane location of MT1-MMP is correlated with the expression of cortactin. According to these findings, cortactin is not only involved in actin polymerization but also in metalloproteinases expression and trafficking. At this point, there is the recruitment of both NHE1 and VGSC to the invadopodia. The activity of NHE1 is fundamental to the formation (through its binding to the cytoskeleton) and the creation of the highly acidic ECM via its proton secretion activities. NHE1's cytoskeleton anchoring is iindependent from its ion exchange activities.[55] The relationship between NHE1, the ezrin-radixin-moesin protein and EGFR are explained in Fig. 7. The functions of the VGSCs in the invadopodia are illustrated in Figs. 8 and 9.

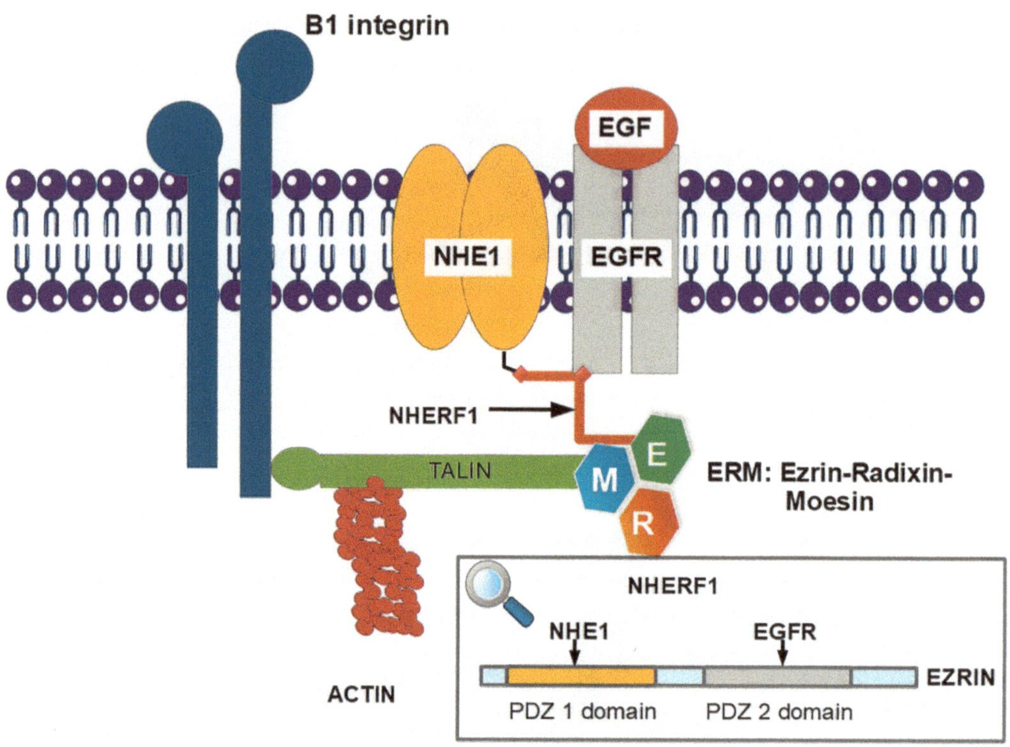

Fig. 7 NHE1 recruited by TALIN through a Moesin[56] will be responsible for H+ extrusion and Na+ import and at the same time, regulates the cytoskeletal anchoring.[57] This will create a low pH in the ECM that is necessary for the activation of proteolytic enzymes that will degrade the ECM. As shown in the next two figures, NHE1 activity is increased by nearby VGSCs. The evagination of the plasma membrane starts to be visible. Extracellular pH reaches its lowest point at the head of this evagination. Ezrin and Moesin over-expression in the cytoplasm were found in highly invasive tumors such as glioblastoma[58] and oral squamous cell carcinoma,[59] where they seem to migrate from the membrane to intracytoplasmic sites. It is highly possible, although hypothetical, that the union between NHE1 and the ERM complex is mediated through NHERF1 protein (NHE regulating factor 1) that works as a multifunctional adapter protein. Cytoplasmic NHERF1 has been found over-expressed in glioblastoma, a very invasive tumor, intervening in migration and invasion.[60] Its inhibition arrested migration. In breast cancer tumor cells that overexpressed NHERF1 had invadopodia in the leading edge.[61] There is evidence of the interaction between NHE1 and NHERF1[62] and between NHE1 and EGFR through NHEFR-1 as shown in the figure. The binding of these proteins is through the two PZD domains of NHERF1.

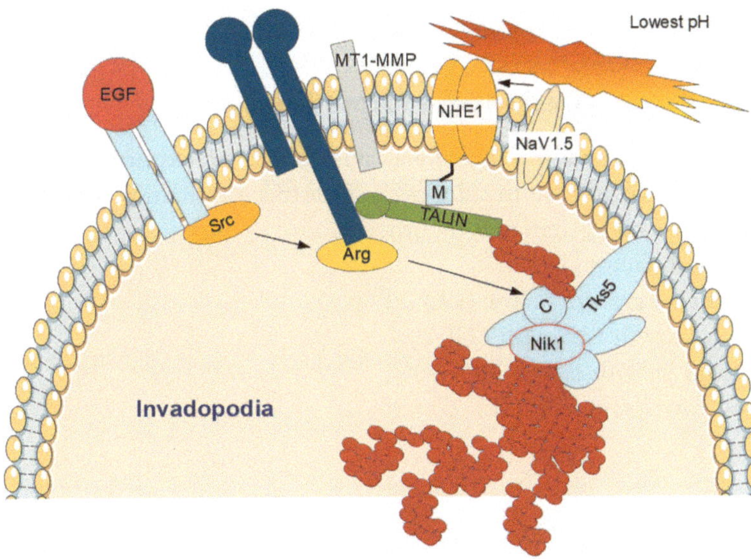

Fig. 8 VGSC (NaV1.5) increases NHE1 activity. In this fully developed invadopodium, the presence of MT1-MMP at the membrane surface means that the ECM degradation is in progress. Aggregation of F-actin and cortactin initiates the accumulation of MT1-MMP at nascent invadopodia.[63] MT1-MMP is synthesized as a latent pro-enzyme and is transported to the plasma membrane. Extracellular acidity is required for activation of these pro-enzymes.

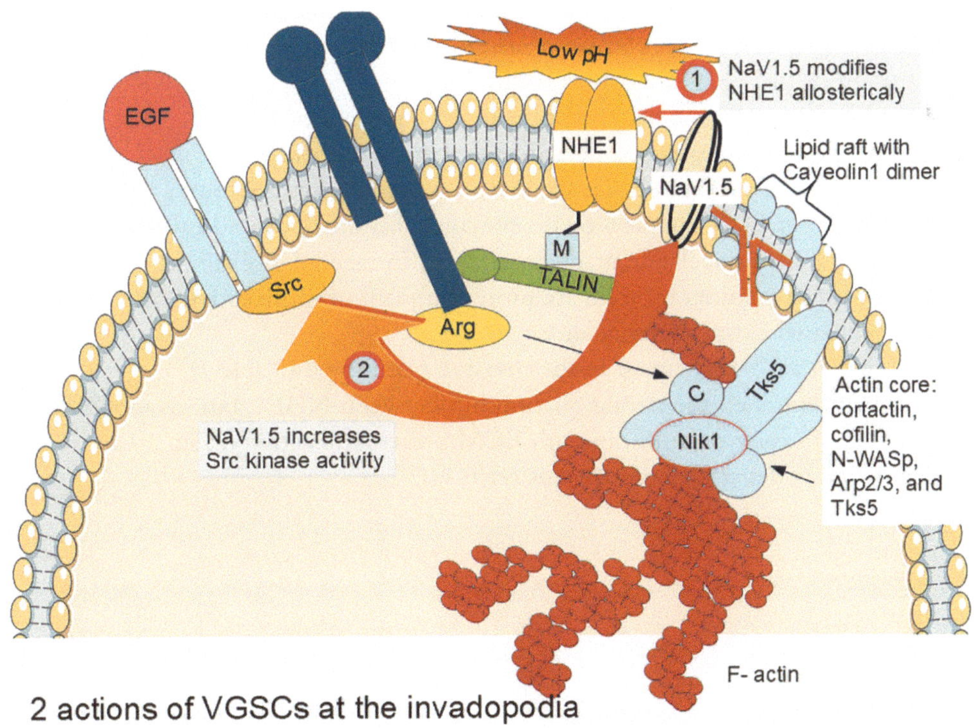

Fig. 9 The two important functions of VGSCs in invadopodia are: (1) acidification of ECM through stimulation of NHE1 and (2) increasing Src kinase activity which, in turn, increases actin polymerization through cortactin.[64,65] Carrithers et al.[66] have found that a splice variant of NaV1.6 participated in the control of invadopodia formation and suggested that this control could be performed through intracellular sodium release (in macrophages and melanoma cells).

Factors that induce invadopodia formation

Inversion of the pH gradient through the activity of proton exporters, mainly NHE1, is essential for the formation of invadopodia. But this mechanism of pH inversion is strongly influenced and/or initiated by other factors:

- Epithelial growth factor (EGF)[67];
- Vascular epithelial growth factor (VEGF)[68];
- Transforming growth factor β (TGF β)[69];
- Platelet derived growth factor (PDGF)[70];
- Hepatocyte growth factor (HGF)[71];
- Hypoxia.[68,72,73]

pH and invasion

Estrella et al.[74] showed that the tumor areas with lower pHe were the most active in invasion and the areas with normal pHe had no invasion. The invasive edges were rich in GLUT1 and NHE1. Oral administration of sodium bicarbonate reduced ECM acidity and resulted in reduced invasion and tumor growth.[75] This finding confirms the acid-mediated tumor migration/invasion hypothesis.[76] This concept is based on the:

(1) increased production of protons in malignant cells;
(2) extrusion of these excess protons to the ECM;
(3) creation of an acidic microenvironment surrounding the tumor;
(4) death of normal surrounding cells due to increased acidity;
(5) development of tumor cells resistance to the acidic environment;
(6) degradation of ECM by proteolytic enzymes that are activated by acidity;
(7) creation of an intracellular gradient with an intracellular alkalinity that is higher where proton extruder proteins are more abundant or more active, correspondingly with increased extracellular acidity.

All these elements co-participate in the invasive phenotype (Table 1).

We have mentioned the strong relationship between NHE1 and invadopodia. And both maintain a strong relationship with ECM proteolytic activity. Fig. 10 and Table 2 summarize the evidence that sustain this relation.

Table 1 Acid-mediated tumor migration/invasion: proof of concept

Reference	Findings
Robey et al. [32]	Bicarbonate decreases metastasis by reducing ECM acidity.
Rofstad et al. [77]	Acidic ECM promotes metastasis.
Ishiguro et al. [78]	Chloride channel 4 exchanger is a proton extruder that increases motility and migration in colon cancer cells.
Roshin et al. [79]	Acidic ECM induces a redistribution of cathepsin B vesicles towards the cell periphery and enhanced secretion of activated cathepsin B.
Webb et al. [80]	Based on a mathematical model the authors show that increased acidity has no important influence on MMP activity but cathepsin B activity is significantly increased.[a]
Calorini et al. [81]	Extracellular acidity is an important factor in tumor invasion and dissemination.
Kulshrestha et al. [82]	Vacuolar ATPase isoform a2 that acidifies ECM is also associated with cortactin at the front edges of invasive ovarian malignant cells. Normal ovarian cells did not express V-ATPase on the surface.
Glunde et al. [83]	Lysosomes participate in carrying proteolytic enzymes that degrade the ECM. Lysosome trafficking, size and quantity is strongly influenced by ECM acidity.
Fan et al. [84]	Overexpression of the gene AGPAT9 decreases invasion and metastasis through inhibition of V-ATPase proton pump and decreasing ECM acidification.

[a]The authors conclusion is based on extracellular pHs that were higher than those usually found in tumors.

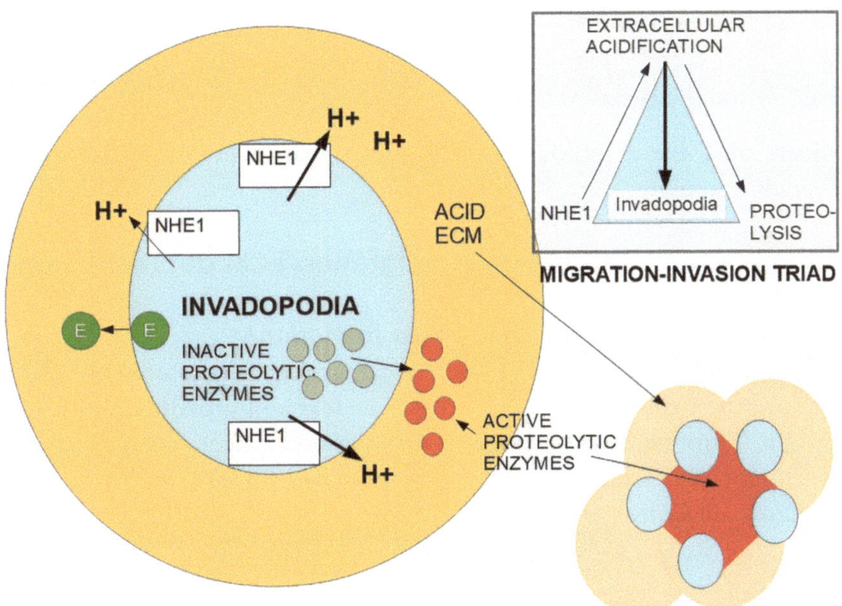

Fig. 10 *Left*: Invadopodia, rich in NHE1 extrude protons that acidify the ECM and alkalinize the intracellular milieu. The inactive proteolytic enzymes in the invadopodia are released into the ECM, where acidity promotes its activation remodeling the peritumoral environment. *E*, Exosomes. *Right upper quadrant*: shows the migration-invasion triad. The *arrow* from ECM acidity to invadopodia represents the reshaping of invadopodia by extracellular acidity. *Right lower quadrant*: shows an area of maximum proteolytic activity and acidity between invadopodia units. Are invadopodia units working coordinately?

Table 2 The NHE1-invadopodia relationship

Reference	Findings
Busco et al. [85]	Invadopodial NHE1 creates the acidic extracellular environment that promotes the degradation of ECM through proteolytic activation. NHE1 is located in the invadopodia in the same site were maximal acidity and proteolysis were achieved in the ECM.
Lucien et al. [86]	Hypoxia is one of the known elements that stimulates invadopodia formation. The authors found that the mechanism behind this stimulation was the activation of the p90 ribosomal S6 kinase, which stimulated invadopodia formation and also induced site-specific phosphorylation of NHE-1 activating the exchanger.
Hoshino et al. [87]	There is a synergistic relation between invadopodia and exosomes. Many of these exosomes contain proteolytic enzymes. Inhibition of invadopodia decreases the amount of exosomes and the inhibition of exosome formation decreases invadopodia formation.
Greco et al. [88]	Invadopodia is the site where protease activity is regulated. Inhibiting NHE1 increases protease secretion but inhibits its activity.
Magahlaes et al. [89]	Cell invasion is regulated by cortactin phosphorylation in a pH-dependent manner.
Beaty et al. [90]	Talin recruits NHE1 to the invadopodia complex through the protein Moesin promoting metastasis.

The NHE1-invadopodia-proteolysis relationship

Conclusions of Table 2: NHE1 plays a pivotal role in invadopodia formation and matrix degrading activity.

Clinical implications: Targeting the migration/ECM degradation/invasion axis

Which are the main players in invadopodia formation?

If we want to avoid or decrease migration and invasion, it is of utmost importance to clearly identify the main participants. Growth factors and growth factor receptors are the initial step. Through their signaling two events take place:

(1) Activation of NHE1 and VGSCs that create a polarized cell, extracellular acidity and intracellular alkalinity.

(2) Activation of Src and a chain of following activations leads to actin polymerization.

This short description allows the identification of the main players:

(a) NHE1;

(b) Growth factors and growth factor receptors;

(c) Src;

(d) VGSCs;

(e) Intracellular alkalinization;
(f) Extracellular acidity;
(g) Polarization of the cell;
(h) Proteolytic enzymes.

Although, all these seven players are amenable to pharmacological interventions, we believe that blocking NHE1 and Src simultaneously should reduce migration and invasion in a significant manner. NHE1 inhibition will decrease the pH gradient inversion at the invadopodia, eliminating intracellular alkalinization, extracellular acidity and cell polarization (factors e, f & g). The normalization of ECM acidity would also impede the activation of the proteolytic enzymes decreasing ECM digestion. By blocking Src the actin polymerization would be substantially decreased hindering migration. A rational treatment for stopping migration and invasion should target growth factors, NHE1 and Src simultaneously. This can be achieved with existing drugs if they are appropriately combined.

Targeting Src

Is a 9-member family of similar kinases that represent a hub in cancer development because integrates tyrosine kinase receptors signaling with STAT transcription factors, integrin adhesion molecules, actins, focal adhesion kinases (FAK) and paxillin. All these elements play an important role in migration and metastasis.[91–93] As Linder said, it is a master switch of invadopodia.[94] The gene that originates this protein, in words of Finn "*is the oldest known oncogene initially identified by Peyton Rous in 1911 as the transforming agent in chicken sarcomas.*"[95] Src protein is a non receptor tyrosine kinase that acts as an intermediary between tyrosine kinase receptors and other effector molecules. The expression of Src increases invadopodia formation and ECM degradation (Box 2).[96]

BOX 2

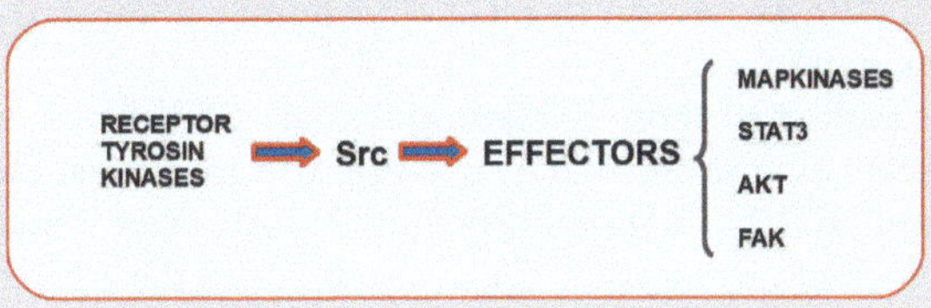

Targeting Src as a monotherapy has not shown good clinical results,[97] therefore it should be part of a broader and integral approach. There are a few nutraceuticals and FDA approved drugs that can be useful for inhibiting Src.

(1) Dasatinib- is an oral small-molecule inhibitor of the BCR-ABL kinase, the SRC family members, c-KIT and PDGFR (PDGF receptor). Approved by the FDA in 2006 for chronic myeloid leukemia, it has been tested in breast cancer.[98]

(2) Imatinib and Nilotinib (ABL kinase inhibitors) have decreased invadopodia precursor formation and cortactin phosphorylation in breast cancer cells.[99]

(3) Bosutinib- belongs to the same family as dasatinib, is also an Src inhibitor.[100]

(4) Evodiamine (EVO) is a quinolone alkaloid compound obtained from a fruit described in traditional Chinese medicine. It has been in use for many centuries for the treatment of headaches, menstrual problems, and other ailments. In the western world, it is known as a controversial weight loss product and is sold as a nutritional supplement (Table 3).

Table 3 Evidence of the anti-migratory effects of evodiamine (EVO)

Reference	Findings
Du et al.[102]	EVO reduced metastasis in MDA-MB-231 breast cancer cells.
Ogasawara et al.[103]	EVO showed anti-invasive and anti-metastatic effects on Lewis lung carcinoma, B16-F10 melanoma and colon 26-L5 carcinoma. The experiments with 26-L5 tumor cell inoculation after incubation with EVO significantly decreased the number of liver and lung metastases.
Ogasawara et al.[104]	After examining 75 natural compounds, searching their inhibitory effect on migration and proliferation of colon cancer cells, EVO was found to be the most potent and selective inhibitor of tumor cell migration. This inhibition was achieved with concentrations 20 times lower than the one needed to inhibit tumor proliferation.
Peng et al.[105]	EVO inhibited the migration of HOME1 and CNE1 nasopharyngeal cancer cells without affecting proliferation; it decreased MMP2 expression and activity without affecting MMP9, and inhibited the translocation of NF-kB p65. The authors attributed these actions to the attenuation of ERK1/2 phosphorylation.
Yuan et al.[106]	EVO modified the expression of many genes related to migration in renal cell carcinoma.
Zhao et al.[107]	EVO inactivated the JAK2/STAT3 pathway, which led to a decrease in MMP3 expression in HCT-116 human colorectal cells and decreased migration.
Takada et al.[108]	The authors propose EVO modulation of NF-kB as the main mechanism for decreasing proliferation, invasion and metastatic behavior.
Wu et al.[109]	When HUVEC is stimulated with TGF-β1 there is increased migration. EVO blocked this increase. EVO also blocked the activation of Smad2, Smad3, ERK1/2, and Akt, and nuclear translocation of Smad4 in HUVEC.
Ge et al.[110]	EVO decreased expression of metalloproteinases and cell adhesion molecules through activation of the expression of peroxisome proliferator-activated receptor γ (PPARγ) and its translocation to the nucleus.
Heo et al.[111]	The antimigratory properties of Evo in LIGHT-induced migration of monocytes was attributed to a decrease in light-induced production of ROS, IL8, IL6, TNF alfa, monocyte chemoattractant protein-1, ICAM-1, CCR1 and CCR2.
Zhou et al.[112]	EVO inhibited migration of osteosarcoma cells.

In recent years, researchers found that **EVO** had anti-cancer effects: cytotoxicity and decreased invasion and migration in vitro in malignant cells while showing minimal damage to normal cells.[101]

The inhibition of Src phosphorylation by EVO[113–115] is further reinforced by berberine.[116,117]

Berberine is an isoquinoline alkaloid found in certain plants, and has been used in China for more than 3000 years. Among the multiple anticancer effects of berberine, it has the ability to inhibit heparanase's degradation of the extracellular matrix.[118,119] Berberine reduces migration in melanoma through inhibition of COX2 and prostaglandins.[120] FAK, MMP2, MMP9, AP1, integrin β1 and NF-kB activation are also downregulated by berberine.[121–123] HIF-1α has been targeted in vitro and in vivo[124] and reverses epithelio-mesenchymal transition.[125]

Reversion of the inverted pH gradient. The drugs involved in this activity include amiloride and its derivatives (NHE1 inhibition), proton pump inhibitors like esomeprazole, pantoprazole, lansoprazole (inhibition of Vacuolar ATPase proton pump), VGSCs inhibitors (phenytoine,[126] carbamazepine, topiramate) are extensively described in the specific chapters. Probably two or more of these drugs should be associated to achieve a significant decrease in migration/invasion.

Fig. 11 is a summary of existing drugs that can be used for blocking invadopodia formation and migration/invasion.

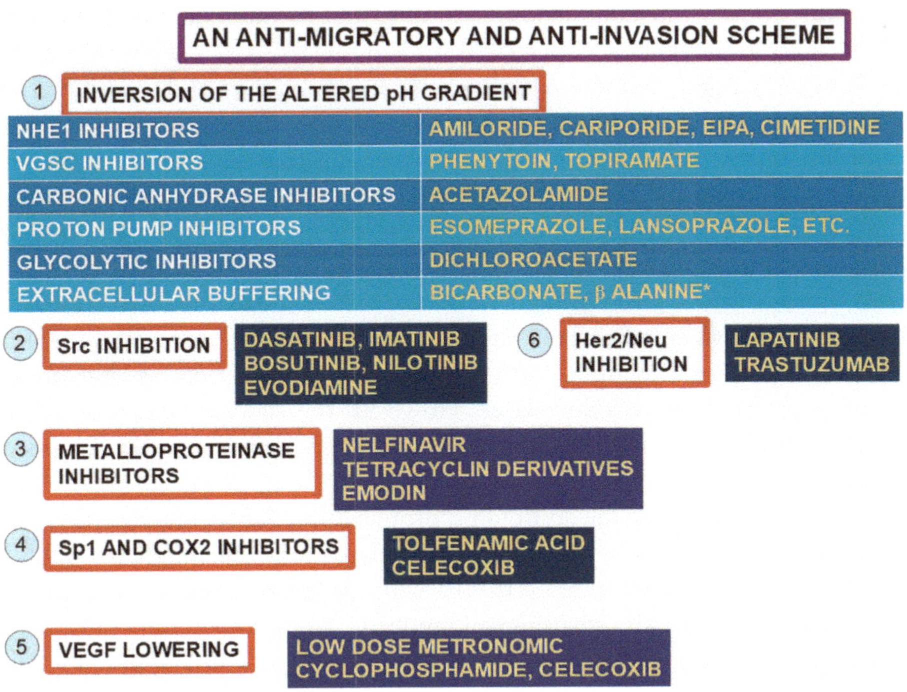

Fig. 11 Chemotherapeutic and repurposed drugs and nutraceuticals with the ability to decrease migration and invadopodia formation. These drugs will be further discussed in Chapters 13–16. *β-alanine reduces extracellular acidity through its metabolite carnosine.[127]

Additional inhibitors of migration and invasion are

- Indol-3-carbinol found in cruciferous vegetables.[128]
- Curcumin[129];
- COX2 inhibitors that may be specific like Celecoxib[130] or non-specific COX1/2 drugs like indometacina[131];
- Daphnetin (dihydroxycoumarin) decreases migration by targeting the PI3K/Akt axis[132];
- Quercetin reduces migration through various mechanisms: down-regulation of MMP-2, MMP-9, VEGF, GLUT1, PKM2, LDH-A, the Akt/mTOR axis and reducing glucose uptake, and lactic acid production. All this actions reduced ECM acidity.[133]
- Auranofin an inhibitor of the thioredoxin system inhibited invasion and migration in a highly malignant breast cancer cell line.[134]
- Alendronate, a biphosphonate used for the treatment and prevention of osteoporosis, has anti-migratory effects through disruption of the cytoskeleton, cofilin downregulation[135] and affecting the mevalonate pathway.[136] The evidence showing the inhibitory effects of alendronate and other biphosphonates is important enough to justify further research.[137–141]

Clinical implications

There is a significant association between over-expression of invadopodia-associated genes and metastasis.[142] Invadopodia, as the main tool for migration and invasion, is a legitimate target to prevent cancer progression and metastasis.

We have chosen glioblastoma multiforme as an example to understand the importance of migration in cancer evolution because it is one of the most invasive types of cancers. Most of the patients with glioblastoma multiforme have a very short survival (15 months in average) and high mortality. Usual treatment is radiation and temozolomide after surgery.[143] Often a few months after treatment, those who responded show a relapse, usually in the vicinity of the original tumor. This is related to difficulties of complete tumor ablation given its highly invasive nature added also to anatomical inaccessible locations. Invadopodia in the glioblastoma cell form the core of its highly infiltrating and invasive behavior.[144] We believe that invadopodia is also the main mechanism used by these cells for relapse. However, the classical treatment of the disease does absolutely nothing to prevent further invasion once the treatment is completed.

Kislin et al.[60] found that the scaffolding protein NHERF1, which is associated to invadopodial development, was markedly expressed on the advancing rim of highly invasive glioblastoma. The depletion of NHERF1 with siRNA blocked migration and invasion and increased sensitivity to temozolomide.[145] They also found morphological changes in actin fibers (Fig. 12).

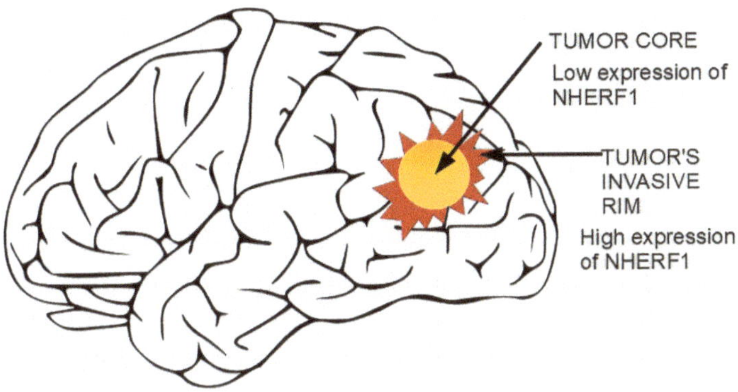

Fig. 12 The high expression of NHERF1 at the invasive rim of glioblastomas is a proof of concept of the importance of the invadopodia complex in relapsing disease.

In this respect, there are therapies that deserve to be tested. One of them should be keeping the pH gradient in normal values. This simple mechanism would retard or even impede the activity of invadopodia. Other anti-invadopodial schemes seem to be successful at the cellular level, e.g., vinorelbine and paclitaxel addressing microtubules and migration.[146] Therefore, an anti-pH approach associated with vinorelbine or paclitaxel after the completion of the mainstay treatment may increase the time to relapse and eventually prolong survival. The necessity of these kind of anti-invadopodial treatments is even more evident if we consider that the temozolomide-radiotherapy treatments increase the invasive abilities of residual glioblastoma cells.

Propranolol, a well known β1 and β2 adrenergic blocker, has shown anti-migratory abilities through the decreased expression of MMP 9 besides other anti-cancer properties.[147–152] Survival time was significantly increased when a new experimental drug VT-122 (chromomodulated combination of propranolol with the COX2 inhibitor etodolac) was used in patients with relapsing glioblastoma receiving low dose metronomic temozolomide.[153]

Therefore, propranolol combined with the anti-pH inversion approach and vinorelbine in a metronomic fashion deserves to be explored as a possible treatment after completion of the standard therapies.

Cimetidine, a histamine receptor H2 antagonist, that is at the same time a cell acidifier through NHE1 inhibition, showed added value in the treatment of glioblastoma because it also reduced migration by *E*-selectin down-regulation and it had the advantage to penetrate the blood-brain barrier.[154] Anti-migratory effects and survival improvement were found when temozolomide was associated with cimetidine in mice bearing human glioblastoma orthotopic xenografts.[155]

Many glioblastomas are resistant to temozolimide. Lena et al.[156] reported that cell acidification with lonidamine and betulinic acid were able to sensitize temozolomide-resistant glioblastoma cells to this alkalyting agent. Furthermore, lonidamine is an inhibitor of migration and invasion.[157] This is a proof of concept that intracellular acidification in glioma acts in more than one way against the tumor.

Conclusions

Migration and invasion are complex processes involving many structures, proteins, and genes and are essential for enabling cancer cells to metastasize.

Invadopodia are the main participant in the invasive process and is absolutely dependent on pH changes inside and outside the cell. Disrupting the pH gradient inversion diminishes migrating abilities to the point of substantially reducing the metastatic risk. This disruption can be achieved by combining existing and FDA-approved drugs (for other purposes) like amiloride, phenytoin, topiramate, esomeprazol, bicarbonate, some nutraceuticals such as quercetin (cellular acidification), carbonic anhydrase inhibitors, like acetazolamide, and Src inhibitors, for example, the FDA-approved dasatinib and the nutraceuticals evodiamine and berberine. There are other drugs that, in spite of, their interesting actions are not being used, like cariporide which is a powerful NHE1 inhibitor. Cariporide was originally tested for preventing the ischemia/reperfusion damage. But it showed few benefits and never entered clinical practice. It was never tested in cancer.

Blocking migration and invasion is not a one drug job as clinical experience has shown. We have to look for the best possible combination of drugs to achieve this major goal in cancer therapy. The drugs described in this chapter are just a few of the many that can be associated with others that reverse the inverted pH gradient.

References

1. Stracke ML, Aznavoorian SA, Beckner ME, Liotta LA, Schiffmann E. Cell motility, a principal requirement for metastasis. In: *Cell motility factors*. Basel: Birkhäuser, 1991. p. 147–62.
2. Mareel M, Leroy A. Clinical, cellular, and molecular aspects of cancer invasion. *Physiol Rev* 2003;**83**:337–76.
3. Björklund M, Koivunen E. Gelatinase-mediated migration and invasion of cancer cells. *Biochim Biophys Acta* 2005;**1755**(1):37–69.
4. Price JE, Carr D, Jones LD, Messer P, Tarin D. Experimental analysis of factors affecting metastatic spread using naturally occurring tumours. *Invasion Metastasis* 1982;**2**(2):77–112.
5. Massagué J, Obenauf AC. Metastatic colonization by circulating tumour cells. *Nature* 2016; **529**(7586):298.
6. Liotta LA, Stetler-Stevenson WG. Tumor invasion and metastasis: an imbalance of positive and negative regulation. *Cancer Res* 1991;**51**(Suppl. 18): 5054s–9s.
7. Friedl P, Wolf K. Tumour-cell invasion and migration: diversity and escape mechanisms. *Nat Rev Cancer* 2003;**3**(5):362.
8. Friedl P, Bröcker EB. The biology of cell locomotion within three-dimensional extracellular matrix. *Cell Mol Life Sci* 2000;**57**:41–64.

9. Di Martino J, Henriet E, Ezzoukhry Z, Goetz JG, Moreau V, Saltel F. The microenvironment controls invadosome plasticity. *J Cell Sci* 2016;**129**(9):1759–68.
10. Kalluri R, Weinberg RA. The basics of epithelial-mesenchymal transition. *J Clin Invest* 2009; **119**(6):1420–8.
11. Kalluri R, Neilson EG. Epithelial-mesenchymal transition and its implications for fibrosis. *J Clin Invest* 2003;**112**(12):1776–84.
12. Zamir E, Geiger B. Molecular complexity and dynamics of cell-matrix adhesions. *J Cell Sci* 2001;**114**:3583–90.
13. Regen CM, Horwitz AF. Dynamics of beta 1 integrin mediated adhesive contacts in motile fibroblasts. *J Cell Biol* 1992;**119**:1347–59.
14. Zaidel-Bar R, Ballestrem C, Kam Z, Geiger B. Early molecular events in the assembly of matrix adhesions at the leading edge of migrating cells. *J Cell Sci* 2003;**116**:4605–13.
15. Wolfenson H, Henis YI, Geiger B, Bershadsky AD. The heel and toe of the cell,s foot: a multifaceted approach for understanding the structure and dynamics of focal adhesions. *Cell Motil Cytoskeleton* 2009;**66**:1017–29.
16. Wang Y, McNiven MA. Invasive matrix degradation at focal adhesions occurs via protease recruitment by a FAK–p130Cas complex. *J Cell Biol* 2012;**196**(3):375–85.
17. Grinstein S, Woodside M, Waddell TK, Downey GP, Orlowski J, Pouyssegur J, et al. Focal localization of the NHE-1 isoform of the Na+/H+ antiport: assessment of effects on intracellular pH. *EMBO J* 1993;**12**:5209–18.
18. Plopper GE, McNamee HP, Dike LE, Bojanowski K, Ingber DE. Convergence of integrin and growth factor receptor signaling pathways within the focal adhesion complex. *Mol Biol Cell* 1995;**6**:1349–65.
19. Tominaga T, Ishizaki T, Narumiya S, Barber DL. p160ROCK mediates RhoA activation of Na–H exchange. *EMBO J* 1998;**17**(16):4712–22.
20. Chen WT. Proteolytic activity of specialized surface protrusions formed at rosette contact sites of transformed cells. *J Exp Zool* 1989;**251**(2):167–85.
21. Small JV, Stradal T, Vignal E, Rottner K. The lamellipodium: where motility begins. *Trends Cell Biol* 2002;**12**(3):112–20.
22. Linder S. The matrix corroded: podosomes and invadopodia in extracellular matrix degradation. *Trends Cell Biol* 2007;**17**:107–17.
23. Brisson L, Reshkin SJ, Gore J, Roger S. pH regulators in invadosomal functioning: proton delivery for matrix tasting. *Eur J Cell Biol* 2012;**91**(11–12):847–60.
24. Linder S. Invadosomes at a glance. *J Cell Sci* 2009;**122**(17):3009–13.
25. DesMarais V, Yamaguchi H, Oser M, Soon L, Mouneimne G, Sarmiento C, et al. N-WASP and cortactin are involved in invadopodium-dependent chemotaxis to EGF in breast tumor cells. *Cell Motil Cytoskeleton* 2009;**66**(6):303–16.
26. Ridley AJ, Schwartz MA, Burridge K, Firtel RA, Ginsberg MH, Borisy G, et al. Cell migration: integrating signals from front to back. *Science* 2003;**302**(5651):1704–9.
27. Martin C, Pedersen SF, Schwab A, Stock C. Intracellular pH gradients in migrating cells. *Am J Phys Cell Phys* 2011;**300**:490–5.
28. Wang W, Goswami S, Sahai E, Wyckoff JB, Segall JE, Condeelis JS. Tumor cells caught in the act of invading: their strategy for enhanced cell motility. *Trends Cell Biol* 2005;**15**:138–45.
29. Stock C, Mueller M, Kraehling H, Mally S, Noel J, Eder C, et al. pH nanoenvironment at the surface of single melanoma cells. *Cell Physiol Biochem* 2007;**20**:679–86.
30. Schwab A, Fabian A, Hanley PJ, Stock C. Role of ion channels and transporters in cell migration. *Physiol Rev* 2012;**92**:1865–913.
31. Le Clainche C, Carlier MF. Regulation of actin assembly associated with protrusion and adhesion in cell migration. *Physiol Rev* 2008;**88**:489–513.
32. Yamaguchi H, Condeelis J. Regulation of the actin cytoskeleton in cancer cell migration and invasion. *Biochim Biophys Acta Mol Cell Res* 2007;**1773**(5):642–52.
33. Schwab A, Nechyporuk-Zloy V, Fabian A, Stock C. Cells move when ions and water flow. *Pflugers Arch—Eur J Physiol* 2007;**453**(4):421–32.

34. Webb DJ, Parsons JT, Horwitz AF. Adhesion assembly, disassembly and turnover in migrating cells—over and over and over again. *Nat Cell Biol* 2002;**4**:E97–100.
35. Stock C, Schwab A. Role of the Na/H exchanger NHE1 in cell migration. *Acta Physiol (Oxford)* 2006;**187**:149–57.
36. Hoshino D, Branch KM, Weaver AM. Signaling inputs to invadopodia and podosomes. *J Cell Sci* 2013;**126**:2979–89.
37. Enderling H, Alexander NR, Clark ES, Branch KM, Estrada L, Crooke C, et al. Dependence of invadopodia function on collagen fiber spacing and cross-linking: computational modeling and experimental evidence. *Biophys J* 2008;**95**:2203–18.
38. Parekh A, Ruppender NS, Branch KM, Sewell-Loftin MK, Lin J, Boyer PD, et al. Sensing and modulation of invadopodia across a wide range of rigidities. *Biophys J* 2011;**100**:573–82.
39. Destaing O, Block MR, Planus E, Albiges-Rizo C. Invadosome regulation by adhesion signaling. *Curr Opin Cell Biol* 2011;**23**(5):597–606.
40. Denker SP, Barber DL. Ion transport proteins anchor and regulate the cytoskeleton. *Curr Opin Cell Biol* 2002;**14**(2):214–20.
41. Krähling H, Mally S, Eble JA, Noel J, Schwab A, Stock C. The glycocalyx maintains a cell surface pH nanoenvironment crucial for integrin-mediated migration of human melanoma cells. *Pflugers Arch* 2009;**458**:1069–83.
42. Ludwig FT, Schwab A, Stock C. The Na+/H+-exchanger (NHE1) generates pH nanodomains at focal adhesions. *J Cell Physiol* 2013;**228**:1351–8.
43. Denker SP, Barber DL. Cell migration requires both ion translocation and cytoskeletal anchoring by the Na-H exchanger NHE1. *J Cell Biol* 2002;**159**:1087–96.
44. Stock C, Pedersen SF. Roles of pH and the Na+/H+ exchanger NHE1 in cancer: from cell biology and animal models to an emerging translational perspective? *Semin Cancer Biol* 2017;**43**:5–16.
45. Stock C, Mueller M, Kraehling H, Mally S, Noel J, Eder C, et al. pH nanoenvironment at the surface of single melanoma cells. *Cell Physiol Biochem* 2007;**20**:679–86.
46. Artym VV, Zhang Y, Seiller-Moiseiwitsch F, Yamada KM, Mueller SC. Dynamic interactions of cortactin and membrane type 1 matrix metalloproteinase at invadopodia defining the stages of invadopodia formation and function. *Cancer Res* 2006;**66**(6):3034–43.
47. Murphy DA, Courtneidge SA. The 'ins' and 'outs' of podosomes and invadopodia characteristics, formation and function. *Nat Rev Mol Cell Biol* 2011;**12**(7):413.
48. Clark ES, Whigham AS, Yarbrough WG, Weaver AM. Cortactin is an essential regulator of matrix metalloproteinase secretion and extracellular matrix degradation in invadopodia. *Cancer Res* 2007;**67**(9):4227–35.
49. House CD, Vaske CJ, Schwartz AM, et al. Voltage-gated Na+ channel SCN5A is a key regulator of a gene transcriptional network that controls colon cancer invasion. *Cancer Res* 2010;**70**:6957–67.
50. House CD, Wang BD, Ceniccola K, Obias V, Frank B, Luu T, et al. Voltage-gated Na+ channel activity increases colon cancer transcriptional activity and invasion via persistent MAPK signaling. *Sci Rep* 2015;**5**:11541.
51. Uysal-Onganer P, Djamgoz MB. Epidermal growth factor potentiates in vitro metastatic behaviour of human prostate cancer PC-3M cells: involvement of voltage-gated sodium channel. *Mol Cancer* 2007;**6**:76.
52. Fraser SP, Ozerlat-Gunduz I, Brackenbury WJ, et al. Regulation of voltage-gated sodium channel expression in cancer: hormones, growth factors and auto-regulation. *Philos Trans R Soc B* 2014; **369**(1638):20130105.
53. Artym VV, Zhang Y, Seillier-Moiseiwitsch F, et al. Dynamic interactions of cortactin and membrane type 1 matrix metalloproteinase at invadopodia: defining the stages of invadopodia formation and function. *Cancer Res* 2006;**66**:3034–43.
54. Clark ES, Whigham AS, Yarbrough WG, Weaver AM. Cortactin is an essential regulator of matrix metalloproteinase secretion and extracellular matrix degradation in invadopodia. *Cancer Res* 2007;**67**(9): 4227–35.
55. Denker SP, Barber D. Cell migration requires both ion translocation and cytoskeletal anchoring by the Na-H exchanger NHE1. *J Cell Biol* 2002;**159**(6):1087–96.

56. Beaty BT, Wang Y, Bravo-Cordero JJ. Talin regulates moesin-NHE-1 recruitment to invadopodia and promotes mammary tumor metastasis. *J Cell Biol* 2014;**205**:737. jcb-201312046.
57. Denker SP, Huang DC, Orlowski J, Furthmayr H, Barber DL. Direct binding of the Na-H exchanger NHE1 to ERM proteins regulates the cortical cytoskeleton and cell shape independently of H+ translocation. *Mol Cell* 2000;**6**(6):1425–36.
58. Iwadate Y, Sakaida T, Hiwasa T, Nagai Y, Ishikura H, Takiguchi M, et al. Molecular classification and survival prediction in human gliomas based on proteome analysis. *Cancer Res* 2004;**64**(7): 2496–501.
59. Kobayashi H, Sagara J, Masumoto J, Kurita H, Kurashina K, Taniguchi S. Shifts in cellular localization of moesin in normal oral epithelium, oral epithelial dysplasia, verrucous carcinoma and oral squamous cell carcinoma. *J Oral Pathol Med* 2003;**32**(6):344–9.
60. Kislin KL, McDonough WS, Eschbacher JM, Armstrong BA, Berens ME. NHERF-1: modulator of glioblastoma cell migration and invasion. *Neoplasia* 2009;**11**(4):377–IN7.
61. Cardone RA, Bellizzi A, Busco G, Weinman EJ, Dell' Aquila ME, Casavola V, et al. The NHERF1 PDZ2 domain regulates PKA-RhoA-p38 –mediated NHE1 activation and invasion in breast tumor cells. *Mol Biol Cell* 2017;**18**:1768–80.
62. Cardone RA, Greco MR, Zeeberg K, Zaccagnino A, Saccomano M, Bellizzi A, et al. A novel NHE1-centered signaling cassette drives epidermal growth factor receptor–dependent pancreatic tumor metastasis and is a target for combination therapy. *Neoplasia* 2015;**17**(2):155–66.
63. Poincloux R, Lizárraga F, Chavrier P. Matrix invasion by tumour cells: a focus on MT1-MMP trafficking to invadopodia. *J Cell Sci* 2009;**122**(17):3015–24.
64. Brisson L, Driffort V, Benoist L, Poet M, Counillon L, Antelmi E, et al. Nav1.5 Na+ channels allosterically regulate the NHE-1 exchanger and promote the activity of breast cancer cell invadopodia. *J Cell Sci* 2013;**126**:4835–42.
65. Brisson L, Gillet L, Calaghan S, Besson P, Le Guennec JY, Roger S, et al. Na(V)1.5 enhances breast cancer cell invasiveness by increasing NHE1-dependent H(+) efflux in caveolae. *Oncogene* 2011;**30**:2070–6.
66. Carrithers MD, Chatterjee G, Carrithers LM, Offoha R, Iheagwara U, Rahner C, et al. Regulation of podosome formation in macrophages by a splice variant of the sodium channel SCN8A. *J Biol Chem* 2009;**284**:8114–26.
67. Mader CC, Oser M, Magalhaes MA, Bravo-Cordero JJ, Condeelis J, Koleske AJ, et al. An EGFR-Src-Arg-cortactin pathway mediates functional maturation of invadopodia and breast cancer cell invasion. *Cancer Res* 2011;**71**:1730–41.
68. Lucas JT, Salimath BP, Slomiany MG, Rosenzweig SA. Regulation of invasive behavior by vascular endothelial growth factor is HEF1- dependent. *Oncogene* 2010;**29**:4449–59.
69. Mandal S, Johnson KR, Wheelock MJ. TGF-beta induces formation of F-actin cores and matrix degradation in human breast cancer cells via distinct signaling pathways. *Exp Cell Res* 2008;**314**:3478–93.
70. Quintavalle M, Elia L, Condorelli G, Courtneidge SA. MicroRNA control of podosome formation in vascular smooth muscle cells in vivo and in vitro. *J Cell Biol* 2010;**189**:13–22.
71. Rajadurai CV, Havrylov S, Zaoui K, Vaillancourt R, Stuible M, Naujokas M, et al. Met receptor tyrosine kinase signals through a cortactin-Gab1 scaffold complex, to mediate invadopodia. *J Cell Sci* 2012;**125**:2940–53.
72. Mukhopadhyay D, Tsiokas L, Zhou XM, Foster D, Brugge JS, Sukhatme VP. Hypoxic induction of human vascular endothelial growth factor expression through c-Src activation. *Nature* 1995; **375**:577–81.
73. Hoffmann C, Mao X, Brown-Clay J, Moreau F, Al Absi A, Wurzer H, et al. Hypoxia promotes breast cancer cell invasion through HIF-1α-mediated up-regulation of the invadopodial actin bundling protein CSRP2. *Sci Rep* 2018;**8**(1):10191.
74. Estrella V, Chen T, Lloyd M, Wojtkowiak J, Cornnell HH, Ibrahim-Hashim A, et al. Acidity generated by the tumor microenvironment drives local invasion. *Cancer Res* 2013. https://doi.org/10.1158/0008-5472.CAN-12-2796.
75. Robey IF, Baggett BK, Kirkpatrick ND, Roe DJ, Dosescu J, Sloane BF, et al. Bicarbonate increases tumor pH and inhibits spontaneous metastases. *Cancer Res* 2009;**69**:2260–8.

76. Gatenby RA, Gawlinski ET, Gmitro AF, Kaylor B, Gillies RJ. Acid mediated tumor invasion: a multidisciplinary study. *Cancer Res* 2006;**66**:5216–23.
77. Rofstad EK, Mathiesen B, Kindem K, Galappathi K. Acidic extracellular pH promotes experimental metastasis of human melanoma cells in athymic nude mice. *Cancer Res* 2006;**66**:6699–707.
78. Ishiguro T, Avila H, Lin SY, Nakamura T, Yamamoto M, Boyd DD. Gene trapping identifies chloride channel 4 as a novel inducer of colon cancer cell migration, invasion and metastases. *Br J Cancer* 2010;**102**(4):774.
79. Rozhin J, Sameni M, Ziegler G, Sloane BF. Pericellular pH affects distribution and secretion of cathepsin B in malignant cells. *Cancer Res* 1994;**54**(24):6517–25.
80. Webb SD, Sherratt JA, Fish RG. Alterations in proteolytic activity at low pH and its association with invasion: a theoretical model. *Clin Exp Metastasis* 1999;**17**:397–407.
81. Calorini L, Peppicelli S, Bianchini F. Extracellular acidity as favouring factor of tumor progression and metastatic dissemination. *Exp Oncol* 2012;**34**(2):79–84.
82. Kulshrestha A, Katara GK, Ibrahim S, Pamarthy S, Jaiswal MK, Sachs AG, et al. Vacuolar ATPase "a2" isoform exhibits distinct cell surface accumulation and modulates matrix metalloproteinase activity in ovarian cancer. *Oncotarget* 2015;**6**(6):3797–810.
83. Glunde K, Guggino SE, Solaiyappan M, Pathak AP, Ichikawa Y, et al. Extracellular acidification alters lysosomal trafficking in human breast cancer cells. *Neoplasia* 2003;**5**(6):533–45.
84. Fan S, Wang Y, Wu Z, Zhang Z, Lu J, Li M, et al. AGPAT9 suppresses cell growth, invasion and metastasis by counteracting acidic tumor microenvironment through KLF4/LASS2/V-ATPase signaling pathway in breast cancer. *Oncotarget* 2015;**6**(21):18406–17.
85. Busco G, Cardone RA, Greco MR, Bellizzi A, Colella M, Antelmi E, et al. NHE1 promotes invadopodial ECM proteolysis through acidification of the peri-invadopodial space. *FASEB J* 2010; **24**(10):3903–15.
86. Lucien F, Brochu-Gaudreau K, Arsenault D, Harper K, Dubois CM. Hypoxia-induced invadopodia formation involves activation of NHE-1 by the p90 ribosomal S6 kinase (p90RSK). *PLoS One* 2011; **6**(12):e28851.
87. Hoshino D, Kirkbride KC, Costello K, Clark ES, Sinha S, Grega-Larson N, et al. Exosome secretion is enhanced by invadopodia and drives invasive behavior. *Cell Rep* 2013;**5**(5):1159–68.
88. Greco MR, Antelmi E, Busco G, Guerra L, Rubino R, Casavola V, Reshkin SJ, et al. Protease activity at invadopodial focal digestive areas is dependent on NHE1-driven acidic pHe. *Oncol Rep* 2014; **31**(2):940–6.
89. Magalhaes MA, Larson DR, Mader CC, Bravo-Cordero JJ, Gil-Henn H, Oser M, et al. Cortactin phosphorylation regulates cell invasion through a pH-dependent pathway. *J Cell Biol* 2011;**195**(5):903–20.
90. Beaty BT, Wang Y, Bravo-Cordero JJ, Sharma VP, Miskolci V, Hodgson L, et al. Talin regulates moesin–NHE-1 recruitment to invadopodia and promotes mammary tumor metastasis. *J Cell Biol* 2014; **205**(5):737–51.
91. Kim LC, Song LJ, Haura EB. Src kinases as therapeutic targets for cancer. *Nat Rev Clin Oncol* 2009;**6**:587–95.
92. Playford MP, Schaller MD. The interplay between Src and integrins in normal and tumor biology. *Oncogene* 2004;**23**(48):7928–46.
93. Yeatman TJ. A renaissance for SRC. *Nat Rev Cancer* 2004;**4**:470–80.
94. Linder S, Kopp P. Podosomes at a glance. *J Cell Sci* 2005;**118**:2079–82.
95. Finn RS. Targeting Src in breast cancer. *Ann Oncol* 2008;**19**(8):1379–86.
96. Artym VV, Zhang Y, Seillier-Moiseiwitsch F, Yamada KM, Mueller SC. Dynamic interactions of cortactin and membrane type 1 matrix metalloproteinase at invadopodia: defining the stages of invadopodia formation and function. *Cancer Res* 2006;**66**(6):3034–43.
97. Zhang S, Yu D. Targeting Src family kinases in anti-cancer therapies: turning promise into triumph. *Trends Pharmacol Sci* 2012;**33**(3):122–8.
98. Mayer EL, Baurain JF, Sparano J, Strauss L, Campone M, Fumoleau P, et al. A phase 2 trial of dasatinib in patients with advanced HER2-positive and/or hormone receptor-positive breast cancer. *Clin Cancer Res* 2011;**17**(21):6897–904.

99. Meirson T, Genna A, Lukic N, Makhnii T, Alter J, Sharma VP, et al. Targeting invadopodia mediated breast cancer metastasis by using ABL kinase inhibitors. *Oncotarget* 2018;**9**(31):22158.
100. Vultur A, Buettner R, Kowolik C, Liang W, Smith D, Boschelli F, et al. SKI-606 (bosutinib), a novel Src kinase inhibitor, suppresses migration and invasion of human breast cancer cells. *Mol Cancer Ther* 2008;**7**(5):1185–94.
101. Koltai T. Re-purposing evodiamine as an anti-cancer drug: effects on migration and apoptosis. *Open Access J Oncol Med* 2018;**1**(3). https://doi.org/10.32474/oajom.2018.01.000111.
102. Du J, Wang XF, Zhou QM, Zhang TL, Lu YY, Zhang H, et al. Evodiamine induces apoptosis and inhibits metastasis in MDA-MB-231 human breast cancer cells in vitro and in vivo. *Oncol Rep* 2013; **30**(2):685–94.
103. Ogasawara M, Matsunaga T, Takahashi S, Saiki I, Suzuki H. Anti-invasive and metastatic activities of evodiamine. *Biol Pharm Bull* 2002;**25**:1491–3.
104. Ogasawara M, Matsubara T, Suzuki H. Screening of natural compounds for inhibitory activity on colon cancer cell migration. *Biol Pharm Bull* 2001;**24**:720–3.
105. Peng X, Zhang Q, Zeng Y, Li J, Wang L, Ai P. Evodiamine inhibits the migration and invasion of nasopharyngeal carcinoma cells in vitro via repressing MMP-2 expression. *Cancer Chemother Pharmacol* 2015;**76**(6):1173–84.
106. Yuan XL, Zhang P, Liu XM, Du YM, Hou XD, Cheng S, et al. Cytological assessments and transcriptome profiling demonstrate that evodiamine inhibits growth and induces apoptosis in a renal carcinoma cell line. *Sci Rep* 2017;**7**(1):12572.
107. Zhao LC, Li J, Liao K, Luo N, Shi QQ, Feng ZQ, Chen DL. Evodiamine induces apoptosis and inhibits migration of HCT-116 human colorectal cancer cells. *Int J Mol Sci* 2015; **16**(11):27411–21.
108. Takada Y, Kobayashi Y, Aggarwal B. Evodiamine abolishes constitutive and inducible NF-kB activation by inhibiting IkBα kinase activation, thereby suppressing NF-kB regulated antiapoptotic and metastatic gene expression, up-regulating apoptosis and inhibiting invasion. *J Biol Chem* 2005;**280**(17):17203–12.
109. Wu QQ, Xiao Y, Jiang XH, Yuan Y, Yang Z, Chang W, et al. Evodiamine attenuates TGF-β 1-induced fibroblast activation and endothelial to mesenchymal transition. *Mol Cell Biochem* 2017;**430**(1–2):81–90.
110. Ge X, Chen S, Liu M, Liang T, Liu C. Evodiamine attenuates PDGF-BB-induced migration of rat vascular smooth muscle cells through activating PPARγ. *Int J Mol Sci* 2015;**16**(12):28180–93.
111. Heo SK, Yun HJ, Yi HS, Noh EK, Park SD. Evodiamine and rutaecarpine inhibit migration by LIGHT via suppression of NADPH oxidase activation. *J Cell Biochem* 2009;**107**(1):123–33.
112. Zhou Y, Hu J. Evodiamine induces apoptosis, G2/M cell cycle arrest, and inhibition of cell migration and invasion in human osteosarcoma cells via Raf/MEK/ERK signalling pathway. *Med Sci Monit* 2018;**24**:5874.
113. Liao W, Li B, Li L, Yan L, Wang Z, An X, et al. Erk1/2, CDK8, Src and Ck1e mediate *Evodia rutaecarpa* induced hepatotoxicity in Mice. *Chin Med* 2015;**6**(02):97.
114. Verlicchi A, Karachaliou N, Lazzari C, Servat CC, Capitan AG, Servat JC, et al. P3. 01-038 STAT3 and Src-YAP1 inhibition results in greater necitumumab sensitivity in lung squamous cell carcinoma. *J Thorac Oncol* 2017;**12**(1):S1143.
115. Yang J, Cai X, Lu W, Hu C, Xu X, Yu Q, et al. Evodiamine inhibits STAT3 signaling by inducing phosphatase shatterproof 1 in hepatocellular carcinoma cells. *Cancer Lett* 2013;**328**(2):243–51.
116. Wu CM, Li TM, Tan TW, Fong YC, Tang CH. Berberine reduces the metastasis of chondrosarcoma by modulating the alpha v beta 3 integrin and the PKC delta, c-Src, and AP-1 signaling pathways. *Evid Based Complement Alternat Med* 2013;**2013**:423164.
117. Wang XN, Han X, Xu LN, Yin LH, Xu YW, Qi Y, et al. Enhancement of apoptosis of human hepatocellular carcinoma SMMC-7721 cells through synergy of berberine and evodiamine. *Phytomedicine* 2008;**15**(12):1062–8.
118. Yan L, Yan K, Kun W, Xu L, Ma Q, Tang Y, et al. Berberine inhibits the migration and invasion of T24 bladder cancer cells via reducing the expression of heparanase. *Tumour Biol* 2013;**34**:215–21.

119. Peng PL, Hsieh YS, Wang CJ, Hsu JL, Chou FP. Inhibitory effect of berberine on the invasion of human lung cancer cells via decreased productions of urokinase-plasminogen activator and matrix metalloproteinase-2. *Toxicol Appl Pharmacol* 2006;**214**(1):8–15.
120. Singh T, Vaid M, Katiyar N, Sharma S, Katiyar SK. Berberine, an isoquinoline alkaloid, inhibits melanoma cancer cell migration by reducing the expressions of cyclooxygenase-2, prostaglandin E 2 and prostaglandin E 2 receptors. *Carcinogenesis* 2010;**32**(1):86–92.
121. Ho YT, Yang JS, Li TC, Lin JJ, Lin JG, Lai KC, et al. Berberine suppresses in vitro migration and invasion of human SCC-4 tongue squamous cancer cells through the inhibitions of FAK, IKK, NF-κB, u-PA and MMP-2 and-9. *Cancer Lett* 2009;**279**(2):155–62.
122. Kim S, Choi JH, Kim JB, Nam SJ, Yang JH, Kim JH, et al. Berberine suppresses TNF-α-induced MMP-9 and cell invasion through inhibition of AP-1 activity in MDA-MB-231 human breast cancer cells. *Molecules* 2008;**13**(12):2975–85.
123. Park JJ, Seo SM, Kim EJ, Lee YJ, Ko YG, Ha J, et al. Berberine inhibits human colon cancer cell migration via AMP-activated protein kinase-mediated downregulation of integrin β1 signaling. *Biochem Biophys Res Commun* 2012;**426**(4):461–7.
124. Yang X, Yang B, Cai J, Zhang C, Zhang Q, Xu L, et al. Berberine enhances radiosensitivity of esophageal squamous cancer by targeting HIF-1α in vitro and in vivo. *Cancer Biol Ther* 2013;**14**(11):1068–73.
125. Chu SC, Yu CC, Hsu LS, Chen KS, Su MY, Chen PN. Berberine reverses epithelial-to-mesenchymal transition and inhibits metastasis and tumor-induced angiogenesis in human cervical cancer cells. *Mol Pharmacol* 2014;**86**:609–23. mol-114.
126. Yang M, Kozminski DJ, Wold LA, Modak R, Calhoun JD, Isom LL, et al. Therapeutic potential for phenytoin: targeting Na v 1.5 sodium channels to reduce migration and invasion in metastatic breast cancer. *Breast Cancer Res Treat* 2012;**134**(2):603–15.
127. Vaughan RA, Gannon NP, Garcia-Smith R, Licon-Munoz Y, Barberena MA, Bisoffi M, et al. β-alanine suppresses malignant breast epithelial cell aggressiveness through alterations in metabolism and cellular acidity in vitro. *Mol Cancer* 2014;**13**(1):14.
128. Meng Q, Goldberg ID, Rosen EM, Fan S. Inhibitory effects of indole-3-carbinol on invasion and migration in human breast cancer cells. *Breast Cancer Res Treat* 2000;**63**(2):147–52.
129. Yang CL, Liu YY, Ma YG, Xue YX, Liu DG, Ren Y, et al. Curcumin blocks small cell lung cancer cells migration, invasion, angiogenesis, cell cycle and neoplasia through Janus kinase-STAT3 signalling pathway. *PLoS One* 2012;**7**(5):e37960.
130. Reed S, Li H, Li C, Lin J. Celecoxib inhibits STAT3 phosphorylation and suppresses cell migration and colony forming ability in rhabdomyosarcoma cells. *Biochem Biophys Res Commun* 2011;**407**(3):450–5.
131. Rozic JG, Chakraborty C, Lala PK. Cyclooxygenase inhibitors retard murine mammary tumor progression by reducing tumor cell migration, invasiveness and angiogenesis. *Int J Cancer* 2001; **93**(4):497–506.
132. He Z, Dong W, Yao K, Qin C, Duan B. Daphnetin inhibits proliferation and glycolysis in colorectal cancer cells by regulating the PI3K/Akt signaling pathway. *RSC Adv* 2018;**8**(60):34483–90.
133. Jia L, Huang S, Yin X, Zan Y, Guo Y, Han L. Quercetin suppresses the mobility of breast cancer by suppressing glycolysis through Akt-mTOR pathway mediated autophagy induction. *Life Sci* 2018; **208**:123–30.
134. Bhatia M, McGrath KL, Di Trapani G, Charoentong P, Shah F, King MM, et al. The thioredoxin system in breast cancer cell invasion and migration. *Redox Biol* 2016;**8**:68–78.
135. Virtanen SS, Ishizu T, Sandholm JA, Löyttyniemi E, Kalervo Väänänen H, Tuomela JM, et al. Alendronate-induced disruption of actin cytoskeleton and inhibition of migration/invasion are associated with cofilin downregulation in PC-3 prostate cancer cells. *Oncotarget* 2018;**9**(66):32593–608.
136. Virtanen SS, Kalervo Väänänen H, Härkönen PL, Lakkakorpi PT. Alendronate inhibits invasion of PC-3 prostate cancer cells by affecting the mevalonate pathway. *Cancer Res* 2002;**62**(9):2708–14.
137. Sawada K, Morishige KI, Tahara M, Kawagishi R, Ikebuchi Y, Tasaka K, et al. Alendronate inhibits lysophosphatidic acid-induced migration of human ovarian cancer cells by attenuating the activation of rho. *Cancer Res* 2002;**62**(21):6015–20.
138. Molinuevo MS, Bruzzone L, Cortizo AM. (2007). Alendronate induces anti-migratory effects and inhibition of neutral phosphatases in UMR106 osteosarcoma cells. *Eur J Pharmacol* 2007; **562**(1–2):28–33.

139. Heikkilä P, Teronen O, Moilanen M, Konttinen YT, Hanemaaijer R, Laitinen M, et al. Bisphosphonates inhibit stromelysin-1 (MMP-3), matrix metalloelastase (MMP-12), collagenase-3 (MMP-13) and enamelysin (MMP-20), but not urokinase-type plasminogen activator, and diminish invasion and migration of human malignant and endothelial cell lines. *Anti-Cancer Drugs* 2002;**13**(3):245–54.
140. Jung J, Park JS, Righesso L, Pabst AM, Al-Nawas B, Kwon YD, et al. Effects of an oral bisphosphonate and three intravenous bisphosphonates on several cell types in vitro. *Clin Oral Investig* 2018; **22**(7):2527–34.
141. Rogers M, Kalra S, Moukharskaya J, Chakraborty K, Niyazi M, Krishnan K, et al. Synergistic growth inhibition of PC3 prostate cancer cells with low-dose combinations of simvastatin and alendronate. *Anticancer Res* 2015;**35**(4):1851–9.
142. Meirson T, Genn-H G. Targeting invadopodia for blocking breast cancer metastasis. *Drug Resist Updat* 2018;**39**:1–7.
143. Stupp R, Masson WP, van den Bent MJ, Weller M, Fisher B, Taphoorn MJB, et al. Radiotherapy plus concomitant and adjuvant temozolomide for glioblastoma. *N Engl J Med* 2005;**352**(10):987–96.
144. Stylli S, Kaye A, Lock P. Invadopodia: at the cutting edge of tumor invasion. *J Clin Neurosci* 2008; **15**(7):725–37.
145. Kislin KL, McDonough WS, Eschbacher JM, Armstrong BA, Berens ME. NHERF-1: modulator of glioblastoma cell migration and invasion. *Neoplasia* 2009;**11**(4):377–IN7.
146. Whitehead CA, Nguyen HPT, Morokoff AP, Luwor RB, Paradiso L, Kaye AH, et al. Inhibition of radiation and temozolomide-induced invadopodia activity in glioma cells using FDA-approved drugs. *Transl Oncol* 2018;**11**(6):1406–18.
147. Lamy S, Lachambre MP, Lord-Dufour S, Beliveau R. Propranolol suppresses angiogenesis in vitro: inhibition of proliferation, migration, and differentiation of endothelial cells. *Vasc Pharmacol* 2010;**53**(5–6):200–8.
148. Masur K, Niggemann B, Zanker KS, Entschladen F. Norepinephrine-induced migration of SW 480 colon carcinoma cells is inhibited by β-blockers. *Cancer Res* 2001;**61**(7):2866–9.
149. İşeri ÖD, Sahin FI, Terzi YK, Yurtcu E, Erdem SR, Sariaglioglu F. Beta-adrenoreceptor antagonists reduce cancer cell proliferation, invasion, and migration. *Pharm Biol* 2014;**52**(11):1374–81.
150. Annabi B, Lachambre MP, Plouffe K, Moumdjian R, Beliveau R. Propranolol adrenergic blockade inhibits human brain endothelial cells tubulogenesis and matrix metalloproteinase-9 secretion. *Pharmacol Res* 2009;**60**(5):438–45.
151. Guo K, Ma Q, Wang L, Hu H, Li J, Zhang D, et al. Norepinephrine-induced invasion by pancreatic cancer cells is inhibited by propranolol. *Oncol Rep* 2009;**22**(4):825–30.
152. De Giorgi V, Grazzini M, Benemei S, Marchionni N, Botteri E, Pennacchioli E, et al. Propranolol for off-label treatment of patients with melanoma: results from a cohort study. *JAMA Oncol* 2018;**4**(2): e172908.
153. Shahid T, Bhattacharyya GS, Malhotra H, Govindbabu KS, Parikh KM, Ranade A, et al. Continuous low dose temozolomide with or without VT-122 in glioblastoma. *J Clin Oncol* 2015;**33**(15 Suppl): e13010.
154. Lefranc F, Yeaton P, Brotchi J, Kiss R. Cimetidine, an unexpected anti-tumor agent, and its potential for the treatment of glioblastoma. *Int J Oncol* 2006;**28**(5):1021–30.
155. Lefranc F, James S, Camby I, Gaussin JF, Darro F, Brotchi J, et al. Combined cimetidine and temozolomide, compared with temozolomide alone: significant increase in nude mice bearing U373 human glioblastoma multiforme orthotopic xenografts. *J Neurosurg* 2005;**102**(4):706–14.
156. Lena A, Rechichi M, Salvetti A, Bartoli B, Vecchio D, Scarcelli V, et al. Drugs targeting the mitochondrial pore act as citotoxic and cytostatic agents in temozolomide-resistant glioma cells. *J Transl Med* 2009;**7**(1):13.
157. Santos FM, Goncalves VM, Pinheiro S, Vieira AF, Paredes J, Schmitt FC, et al. Differential sensitivities to lactate transport inhibitors of breast cancer cell lines. *Endocr Relat Cancer* 2014;**21**(1):1–40.

CHAPTER 12

The Specificity protein 1 (Sp1) transcription factor

Introduction

Specificity protein 1 (Sp1) is a general transcription factor of the Krüpper-like factor (KLF) family. This family has 25 different members that work as transcription factors.[1,2]

From the Sp subfamily, we are mainly interested in Sp1, Sp2, and Sp3 that seem to share activities on many genes related to cancer and to the creation of the inverted pH gradient, namely HIF-1α, (hypoxia inducible factor 1 alpha), CAIX (carbonic anhydrase 9), NHE1 (sodium/ hydrogen exchanger 1), NaV1.5 (isoform of voltage gated sodium channel), V-ATPase proton pumps and CD147 (basigin). The description will be focused on Sp1, the best known of the group. It is not clear if Sp3 and Sp4 share all the characteristics of Sp1.

The members of this subfamily bind the DNA on sites known as Sp1 sites. Family members have different translational properties and can regulate each other. Sp1 was first found as a promoter for the transcription of the SV40 major Immediate Early gene in HeLa cells.[3] Sp1 is a transcription factor that contributes to the expression of housekeeping genes but it also intervenes in the regulation of other genes, many of them involved in cancerization and cancer evolution. It may activate or repress genes in response to physiologic or pathologic stimuli. Spl is a sequence-specific, DNA-binding protein. It enhances transcription by RNA polymerase II 10-to 50-fold in genes whose promoter regions contain at least one properly positioned GC box.[4] Sp1's mechanism of action is the recruitment of the transcription machinery to the promoter region (Fig. 1).

Pro-tumoral transcription factors, like Sp1 are able to induce the overexpression of essential genes for proliferation, migration, invasion and metastasis.[5]

Regarding the pH-tome Sp1 is one of the promoters of HIF-1α. Then HIF-1α as a promoter and Sp1 as an enhancer or co-promoter regulate the expression of at least five important players of the pH-tome: NHE1, CAIX, basigin (a chaperone of monocarboxylate transporters), a protein that forms part of proton pumps, and NaV1.5 voltage gated sodium channel. Therefore, Sp1's role in the pH-tome and its own pro-tumoral behavior make Sp1 a "protein of interest". One logical way to reduce

An Innovative Approach to Understanding and Treating Cancer: Targeting pH
https://doi.org/10.1016/B978-0-12-819059-3.00012-5
© 2020 Elsevier Inc.
All rights reserved.

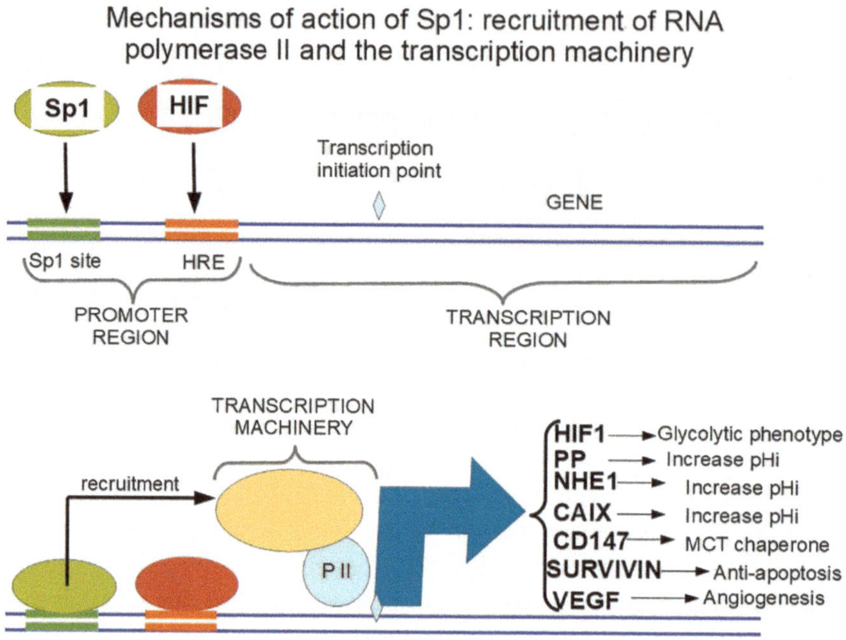

Fig. 1 Sp1 binds to the Sp1 site and recruits the transcription machinery initiating the transcription process. HIF1 is the major transcription factor and both, Sp1 and HIF are essential for efficient transcription. *PP*, proton pump.

the expression of CAIX, CAXII, NHE1, and HIF-1α is to reduce their induction by promoters and enhancers. The induction of many of the above mentioned proteins is regulated by HIF-1α that acts as the main transcription factor for them. But other transcription factors are also necessary to achieve accelerated transcription. The promoter region of gene CA9 that encodes the protein CAIX has a HRE (Hypoxia Responsive Element) where HIF-1α binds to transactivate CAIX mRNA transcription. But HIF-1α requires co-activators. At least two binding sites for Sp1 have been identified in the CAIX gene.[6] Sp1 binding to one of these co-activator sites, and PR1 are essential for efficient transcription.[7]

Fig. 2 shows the linear structure of the Sp1 protein with its domains and binding sites.

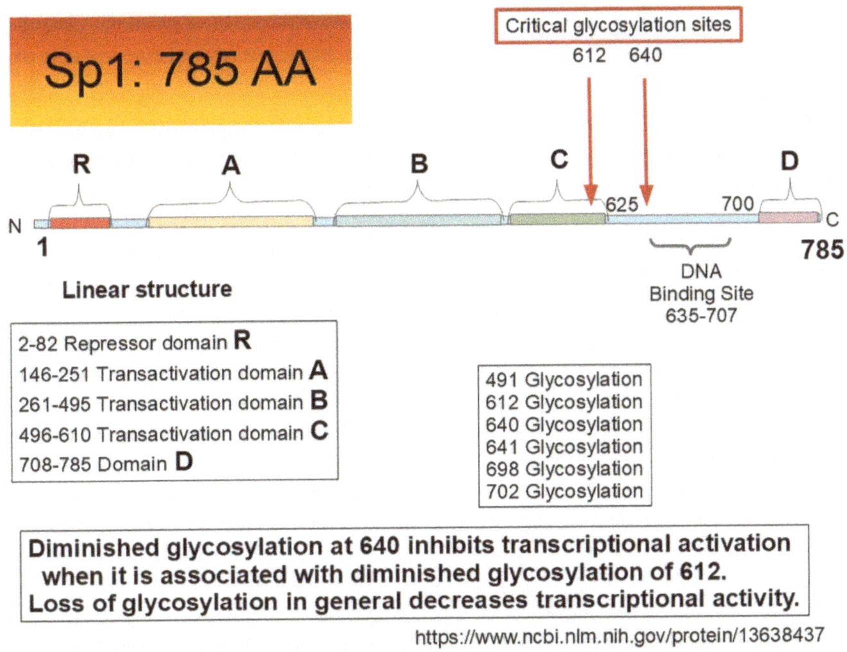

Fig. 2 Linear representation of Sp1transcription factor's structure.[8] It is a protein of 785 amino acids (AA) with 5 domains, 3 of which act in the transactivation process of transcription. The DNA binding site is between AA 635 and 707. It also has 6 glycosylated sites. Glycosylation of sites 640 and 612 are critical for the activity of Sp1 and is one of the posttranscriptional mechanisms of modulation. Loss of glycosylation decreases transcriptional activity. Phosphorylation of Sp1 is another major mechanism of posttranslational modulation.[9]

Sp1 transcription factor is a pro-tumoral protein

Sp1 transcription factor is a promoter of VEGF transcription after Ras activation,[10,11] and a promoter of human monocyte chemoattractant protein-1.[12] There is synergy between Sp1 and the estrogen receptor.[13] Sp1 interacts with transcription factors that regulate cell cycle and DNA synthesis like cyclines, CDKs and DNA repair and protein synthesis genes.[14] Included in its pro-tumoral activities is the transactivation of COX2 in gliomas[15] and induces interleukin 10 expression in T regulatory cells and macrophages,[16] increases TGF-β1,[17] just to mention a few pro-tumoral activities. The suppression of Sp1 induces apoptosis in cancer cells.[18,19] Banerjee et al.[20] found that down-regulation of Sp1 with triptolide produced pancreatic cancer cell death. The pro-tumor activity and/or the advantages of its inhibition has been confirmed in many different malignant tumors such as gastric,[21–23] prostate,[24] lung,[25,26] ovary,[27] breast,[28–32] colon,[33–34] bladder,[35] fibrosarcoma,[36] liver,[37] myeloid leukemia,[38] multiple myeloma,[39] pancreas, kidney,[40] and mesothelioma. Survivin is a downstream target of Sp1 stimulation and is a key figure in many cancers such as Ewing sarcoma. FoxO3a is also a downstream target of Sp1 and has an important role in colon cancer.[41] Increased expression of Sp1 is a sign of poor prognosis.[42–44]

It would go beyond the scope of this chapter to describe all the pro-tumor activities of Sp1 which have been extensively reviewed by Beishline et al.[45] Instead, we shall focus on the possible mechanisms for down-regulating/inhibiting Sp1 as an activator of CAIX, NHE1, VGSCs, proton pumps, basigin, and HIF, because the Sp1 suppressors may be well positioned to collaborate with the pH inversion treatment. (Fig. 3).

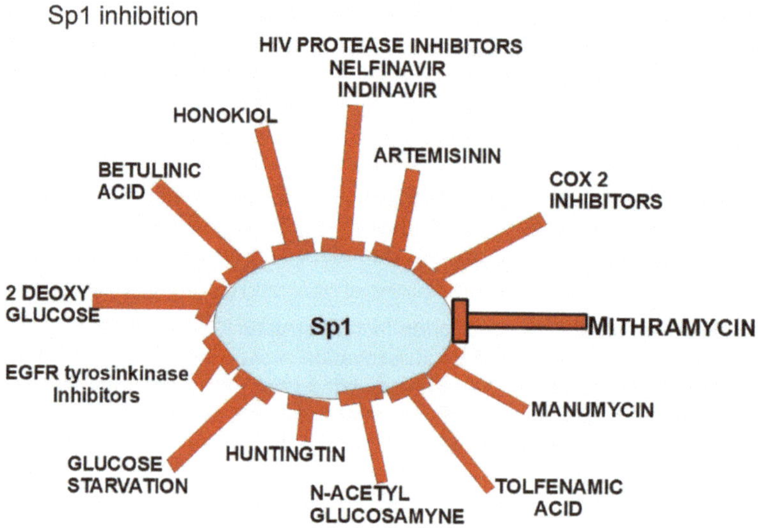

Fig. 3 Drugs and compounds that inhibit Sp1.

Inhibiting Sp1

There are many Sp1 inhibitors with very low or no toxicity at all. Of the Sp1 inhibitors shown in Fig. 3, there are a few which have no toxic effects like betulinic acid and its derivatives[46] or others with low toxicity that have long been used clinically, like HIV protease inhibitors,[47,48] tolfenamic acid[49] and cox-2 inhibitors such as celecoxib.[50,51] Mithramycin, with intermediate toxicity is the most efficient.[52] Its mechanism of action consists in interfering with Sp1 binding to DNA, by cross-binding CG rich segments of DNA found in Sp1 sites.[53] Under experimental conditions this antibiotic showed an important decrease in CAIX gene expression (around 4 fold).[54]

Fig. 4 shows the sites of action of different Sp1 inhibitors.

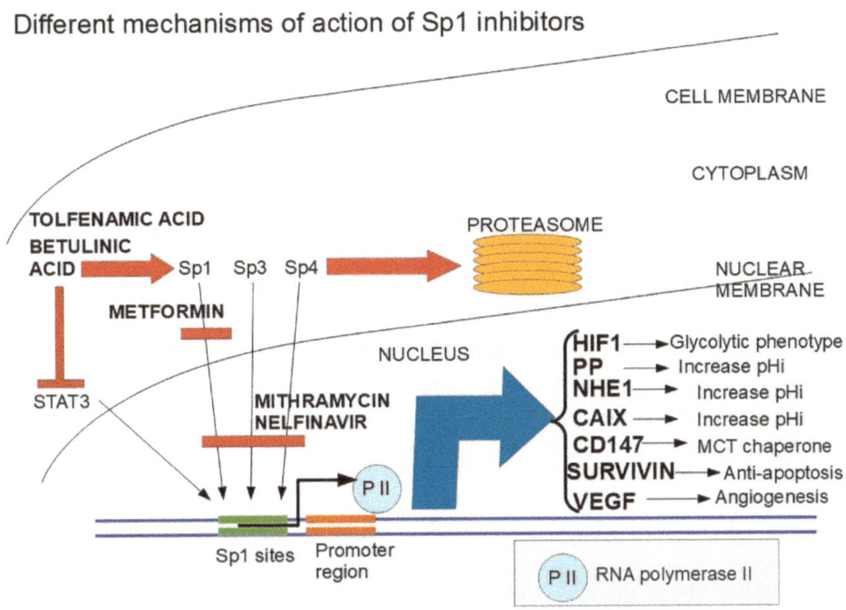

Fig. 4 Mechanism of action of Sp1 inhibitors. COX2 inhibitors, tolfenamic acid and betulinic acid increase the destruction of Sp transcription factors. Metformin inhibits the trafficking of Sp1 from the cytoplasm to the nucleus. Nelfinavir (and other HIV protease inhibitors) and mithramycin block Sp transcription factors binding to the enhancer region of the gene. Mithramycin also decreases Sp1 expression[55] (not shown in the figure).

Betulinic acid and tolfenamic acid have a different mechanism of action on Sp1: they increase its proteasomal destruction. Therefore, synergistic activity between mithramycin and betulinic acid may be expected. This has been found in pancreatic cancer.[56]

Betulinic acid is a natural terpenoid with multiple anticancer effects, and it does not affect normal cells. Betulinic acid penetrates the blood-brain-barrier and is highly effective in temozolide-resistant glioblastoma cells that usually show high expression of the Sp1 transcription factor. Betulinic acid inhibited Sp1 expression in these cells and induced apoptosis.[57] Betulinic acid decreased Sp1 expression in lung cancer cells inhibiting growth.[58]

Tolfenamic acid (TA) showed the strongest down-regulation of Sp-1, Sp-3 and Sp-4 among NSAIDS, as well as synergy with mithramycin in vitro and in vivo.[59] TA produced growth inhibition and antiangiogenesis in an in vivo model of pancreatic cancer but the doses required for the anti-tumor activity were higher than those clinically achievable. However, when mithramycin and tolfenamic acid were used in association, the doses required for anti-tumor activity were significantly lower for both drugs and toxicity was nil.[60] The abundant evidence on tolfenamic acid as a Sp1 down-regulator and its anticancer effects[61] is summarized in Table 1.

Table 1 Evidence for tolfenamic acid (TA) as an anticancer drug and Sp1 down-regulator.

Ref.	Findings
62	TA inhibited Sp1 activity in ovarian cancer cells and decreased growth, expression of c-Met and survivin.
63	TA inhibited lung cancer cell proliferation in vivo and decreased c-Met expression through Sp proteins inhibition.
64	TA enhanced sensitivity to radiation in pancreatic cancer by down-regulating Sp1 and survivin.
65	TA associated with curcumin decreased colon cancer cells growth through modulation of Sp1 and ROS (reactive oxygen species).
66	TA inhibited colon cancer cells growth through Sp1 degradation.
67	TA exerted preventive action on cervical cancer development by down-regulating Sp1, in vitro and in vivo.
68	In 6 canine cell lines of 3 different cancers (osteosarcoma, mammary carcinoma, and melanoma). TA downregulated Sp1 and 4 and all the downstream targets of Sp1 were also inhibited, decreasing cell survival in a dose dependent manner. Apoptosis was increased.
69	TA induced apoptosis and inhibited proliferation of neuroblastoma cells.
70	TA inhibited rhabdomyosarcoma cell growth through Sp1 inhibition.
71	TA inhibited ErbB2 expression in breast cancer cells and decreased growth.
72	TA inhibited growth and induced apoptosis in head and neck cancer.
73	A cisplatin and TA association had an increased inhibitory effect on ovarian cancer cell proliferation, Sp1 and surviving.
74	Anaplastic thyroid cancer cells treated with TA showed inhibited growth and apoptosis. ROS was generated. It also induced NAG1 (NSAIDs activated gene 1) that has anti-tumorigenic activity.
75	TA downregulated β-cathenin in colon cancer cells.
76	ER (endoplasmic reticulum) stress was an important contributory factor to TA's antitumoral actions in colon cancer cells.
77	TA induced apoptosis through Sp1 inhibition in androgen independent prostate cancer cells.
78	TA inhibited prostate cancer cell growth in vitro and in vivo.
79	TA showed anti-cancer effects in 4 different leukemia cell lines: decreased Sp1 and survivin, induced apoptosis in a time- and dose-dependent manner. TA down-regulated CDC2, Cyclin D3 and pRb arresting growth.
80	In medulloblastoma cell lines, TA decreased the expression of Sp1 and survivin in vivo and in vitro. In vivo TA caused a 40% decrease in tumor weight and volume.
81	In an NMBA induced murine tumor model TA reduced the development of esophageal cancer by 33%. In the rats treated with TA that developed tumors, the size and volume of the tumors were significantly lower than in the untreated group.

Table 1 **Conclusions:** TA is an effective Sp1 inhibitor through proteasomal degradation. This inhibition decreased the expression of Met and survivin. TA decreased expression of ErbB2 in breast cancer. In esophageal cancer, TA showed chemopreventive effects.

TA has also other important actions in oncology: it reduces NF-kB signaling[82] and reduces prostaglandin E2 synthesis, thus acting as a COX2 inhibitor. Mefenamic acid, a close relative of TA, uncouples oxidative phosphorylation in mitochondria and reduces ATP production.[83,84] Masuchi et al.[85] found that the diphenylamine structure of mefenamic acid was the reason for the uncoupling of oxidative phosphorylation. TA has also a diphenylamine structure, therefore the same effect can be expected. TA is an inhibitor of calcium influx into the cell.[86]

Mithramycin has been tested in the treatment of cancer patients with chronic myeloid leukemia[87], testicular cancer,[88] anaplastic glioma,[89] Ewing's sarcoma,[90] and some other malignancies,[91] but it has never been accepted as a standard cancer treatment due to its toxicity, and since 2000 it has been discontinued. However, a National Cancer Institute (USA)-sponsored study on mithramycin for mesothelioma, lung and esophageal cancer is underway (NCT01624090).[92]

Tolfenamic acid plus mithramycin may represent a rational association with pH inversion drugs, since the combination of the two drugs uses a lower dosage of mithramycin thereby decreasing its toxic effects.

Nelfinavir an HIV protease inhibitor has shown outstanding features against cancer (described in detail in[93]). Box 1 is a summary of nelvinavir's anticancer activities.

BOX 1 Nelfinavir's anticancer effects

Down-regulation of Akt and HIF-lα.
Decreased Sp1 phosphorylation and decreased binding to the VEGF promoter.
Increased tumor oxygenation.[94,95]
Inducer of endoplasmic reticulum stress-unfolded protein response pathway.
Activation of eEF2K leading to the phosphorylation of eEF2.
Decreased expression of MMP-9 and MMP-2.
Down-regulation of CDK-2, VEGF, bFGF, NF-kB, STAT-3, HIF-1α, IGF, EGFR, survivin, BCRP, androgen receptor, proteasome, fatty acid synthase (FAS).
Decreased cellular ATP concentration.
Up-regulation of TRAIL receptor DR5, and Bax.
Increased radio sensitivity, and increased autophagy.

However, with the exception of Kaposi's sarcoma, Nelfinavir has never been tested in well planned clinical trials. With its much lower toxicity, nelfinavir could be a good alternative to mithramycin.

Metformin down-regulates Sp1, Sp3 and Sp4[96] and it probably acts through AMPK activation and intracellular acidification that reduces Sp1 translocation from the cytoplasm to the nucleus.

Penfluridol, is an anti-psychotic drug which, by down-regulating Sp1 decreases the expression of integrins β1 and β3, overexpressed in breast cancer.[97]

Quercetin, a natural flavonoid was identified as a Sp1 down-regulator in mesothelioma.[98]

Celastrol is a natural triterpenoid that in addition to down-regulating Sp1[99] has many other anti-cancer effects.[100]

Arsenic trioxide has many anti-tumoral effects including down-regulation of Sp1.[101]

Why Sp1 down-regulation has a role in the targeting of the pH-tome?

(1) NHE1, NHE2 and NHE3 genes have one or more Sp1 sites in their promoter regions.[102,103] NHE-1, in particular, has 4 Sp1 sites in the promoter region.[104]

(2) The VGSCs comprise a family of 10 genes (genes *SCN1A to SCN11A*). Nav1.3, a product of gene Scn3a, has Sp1 sites in its promoter area.[105] Transcriptional regulation of Nav1.5, the most important VGSC regarding cancer, has not been fully clarified[106] but in humans, immediately upstream of the transcription initiation nucleotide, there are three GC boxes (Sp1 sites) that could serve for Sp1 binding.[107] Nav1.8 contains Sp1 sites.[108]

(3) Vacuolar H^+ ATPase proton pump also has many Sp1 sites in the promoter segment of unit B2.[109]

(4) CAIX has two Sp1 sites in the promoter region.

(5) Sp1 regulates the expression of CD147/basigin (the monocarboxylate transporter chaperone) as found in lung cancer[110] and hepatocarcinoma.[111]

(6) Sp1 is part of cancer cells' mechanisms for adapting to hypoxia.[112]

Based on this evidence, we can presume that Sp1 down-regulation may affect the entire proton extruder system, as well as, decreasing HIF-1α, the master regulator of the pH-tome genes. A logical, however speculative conclusion is that Sp1 is the second most important transcription factor associated with the pH gradient inversion (HIF is the first). The association of pH- reversion drugs with Sp1 inhibitors is an attractive concept that deserves further investigation.

Association of drugs

The association of nelfinavir, metformin and tolfenamic acid seems a rational approach for down-regulating Sp1 with low or no toxicity. The fact that these three drugs act on different steps of Sp1 promotion allows the presumption of a synergistic down-regulation of Sp1. All three drugs have other significant anti-cancer effects that may be important in the general treatment landscape.

Nelfinavir is an inhibitor of eIF2α through phosphorylation. This inhibition blocks translational initiation.[113] Although, It remains to be seen whether nelfinavir actually reduces Sp1 transcription under clinical conditions, but it most likely does decrease the transcription rate of many HIF-1α/Sp-1 dependent pro-tumor genes.

A fourth compound should also be considered in this scheme: betulinic acid. The reasons for its eventual inclusion are:

(1) It is a Sp1 and Sp3 activity down-regulator.
(2) It is an inducer of apoptosis in malignant cells.[114]
(3) In vitro studies showed that betulinic acid enhanced apoptosis and reduced proliferation in malignant cells without affecting normal cells.[115]
(4) In vivo, it decreased angiogenesis, proliferation and invasion through decreased HIF-1α and STAT3 activity.[116]
(5) Toxicity is very low.

Sp1 also has anti-tumor functions as shown in Fig. 5 but the pro-tumoral activity outweighs the anti-tumor effects.

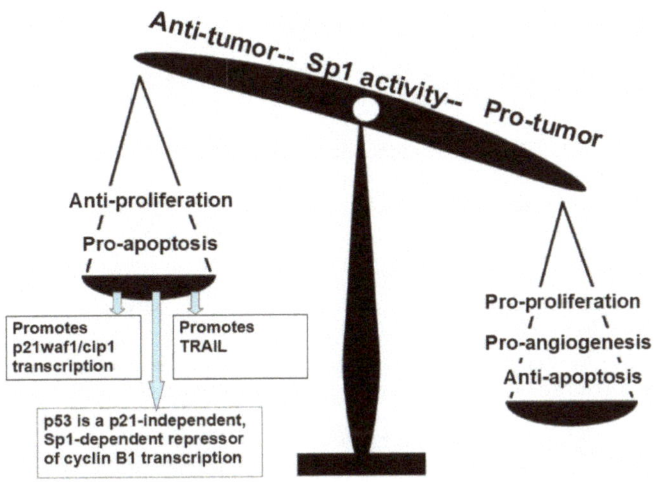

Fig. 5 Pro-tumor and anti-tumor effects of Sp1: the pro-tumor activities outweigh the anti-tumor effects.[117–121]

Conclusions

Many compounds that show cancer prevention activities, such as aniinflammatory drugs (celecoxib, sulindac, aspirin), polyphenols (resveratrol, quercetin, curcumin), metformin, betulinic acid, celastrol, etc., share one common characteristic that strongly contributes to their positive effects: they induce down-regulation of the Sp transcription factors.

Sp1 transcription factor is a modulator/promoter/activator of the expression of genes associated with the pH-tome: genes of NHE1, NaV1.5, V-ATPase proton pumps and CAIX. Furthermore, it is also associated with the hypoxia response and HIF-1α. Down-regulation of Sp1 in the experimental setting has shown benefits in cancer treatment. There are many compounds that can decrease the expression/activity of Sp1. Although the most effective one, mithramycin, is toxic, there are others completely devoid of toxicity. Among them, tolfenamic acid, celecoxib and betulinic acid should be seriously considered as good complements for the treatment of the pH gradient inversion. Tolfenamic acid in particular deserves special attention due to the abundant evidence on its anticancer properties and lack of toxicity. Its well known dosages and limited adverse effects are also factors that can reduce the length and costs of clinical trials. Unfortunately, while this drug has been used for many years as a pain reliever for migraine, it is no longer available on the market for human utilization. However, it can be found for veterinary use.

References

1. Suske G, Bruford E, Philipsen S. Mammalian SP/KLF transcription factors: bring in the family. *Genomics* 2005;**85**:551–6.
2. Black AR, Black JD, Azizkahn Clifford J. Sp1 and krüppel-like factor family of transcription factors in cell growth regulation and cancer. *J Cell Physiol* 2001;**188**(2):143–60.
3. Dynan WS, Tjian R. The promoter-specific transcription factor Sp1 binds to upstream sequences in the SV40 early promoter. *Cell* 1983;**35**:79–87.
4. Kadonaga JT, Jones KA, Tjian R. Promoter-specific activation of RNA polymerase II transcription by Sp1. *Trends Biochem Sci* 1986;**11**(1):20–3.
5. Rauf A, Ali S, Khan MT. The expanding role of Sp1 in pancreatic cancer: tumorigenic and clinical. In: Nagaraju GP, Bramhachari PV, editors. *Role of transcription factors in gastrointestinal malignancies.* Springer; 2018. p. 389.
6. Kaluz S, Kaluzová M, Liao S-Y, Lerman M, Stanbridge EJ. Transcriptional control of the tumor- and hypoxia-marker *carbonic anhydrase 9*: a one transcription factor (HIF-1) show? *Biochim Biophys Acta* 2009;**1795**(2):162–72.
7. Kaluz S, Kaluzova M, Stanbridge EJ. Expression of the Hipoxia marker carbonic anhydrase IX is critically dependant on Sp1 activity. identification of a novel type of hypoxia-responsive enhancement. *Cancer Res* 2003;**63**:917–22.
8. https://www.ncbi.nlm.nih.gov/protein/13638437.
9. Tan NY, Khachigian LM. Sp1 phosphorylation and its regulation of gene transcription. *Mol Cell Biol* 2009;**29**(10):2483–8.
10. Pal S, Claffey KP, Cohen HT, Mukhopadhyay D. Activation of Sp1-mediated vascular permeability factor/vascular endothelial growth factor transcription requires specific interaction with protein kinase C zeta. *Biol Chem* 1998;**273**:26277–80.
11. Milanini J, Vinals F, Pouysségur J, Pagès G. p42/p44 MAP kinase module plays a key role in the transcriptional regulation of the vascular endothelial growth factor gene in fibroblasts. *J Biol Chem* 1998;**273**:18165–72.
12. Ueda A, Okuda K, Ohno S, Shirai A, Igarashi T, Matsunaga K, et al. NF-kappa B and Sp1 regulate transcription of the human monocyte chemoattractant protein-1 gene. *J Immunol* 1994;**153**(5): 2052–63.
13. Porter WB, Saville B, Hoivik D, Safe S. Functional synergy between the transcription factor Sp1 and the estrogen receptor. *Mol Endocrinol* 1997;**11**(11):1569–80.
14. Karlseder J, Rotheneder, Wintersberger E. Interaction of Sp1 with the growth- and cell cycle-regulated transcription factor E2F. *Mol Cell Biol* 1996;**16**:1659–67.
15. Xu K, Shu HK. EGFR activation results in enhanced cyclooxygenase-2 expression through p38 mitogen-activated protein kinase–dependent activation of the SP1/SP3 transcription factors in human gliomas. *Cancer Res* 2007;**67**(13):6121–9.

16. Ma W, Lim W, Gee K, Aucoin S, Nandan D, Kozlowski M, et al. The p38 mitogen-activated kinase pathway regulates the human interleukin-10 promoter via the activation of Sp1 transcription factor in lipopolysaccharide-stimulated human macrophages. *J Biol Chem* 2001;**276**(17):13664–74.
17. Zhang HW, Wang EW, Li LX, Yi SH, Li LC, Xu FL, et al. A regulatory loop involving miR-29c and Sp1 elevates the TGF-beta1 mediated epithelial-to-mesenchymal transition in lung cancer. *Oncotarget* 2016;**7**:85905–16.
18. Zhao Y, Zhang W, Guo Z, Ma F, Wu Y, Bai Y, et al. Inhibition of the transcription factor Sp1 suppresses colon cancer stem cell growth and induces apoptosis in vitro and in nude mouse xenografts. *Oncol Rep* 2013;**30**(4):1782–92.
19. Chae JI, Jeon YJ, Shim JH. Downregulation of Sp1 is involved in honokiol-induced cell cycle arrest and apoptosis in human malignant pleural mesothelioma cells. *Oncol Rep* 2013;**29**(6):2318–24.
20. Banerjee S, Sangwan V, McGinn O, Chugh R, Dudeja V, Vickers SM, et al. Triptolide-induced cell death in pancreatic cancer is mediated by O-GlcNAc modification of transcription factor Sp1. *J Biol Chem* 2013;**288**(47):33927–38.
21. Wang L, Wei D, Huang S, Peng Z, Le X, Wu TT, et al. Transcription factor Sp1 expression is a significant predictor of survival in human gastric cancer. *Clin Cancer Res* 2003;**9**(17):6371–80.
22. Qiu T, Zhou X, Wang J, Du Y, Xu J, Huang Z, et al. MiR-145, miR-133a and miR-133b inhibit proliferation, migration, invasion and cell cycle progression via targeting transcription factor Sp1 in gastric cancer. *FEBS Lett* 2014;**588**(7):1168–77.
23. Zhou C, Ji J, Cai Q, Shi M, Chen X, Yu Y, et al. MTA2 promotes gastric cancer cells invasion and is transcriptionally regulated by Sp1. *Mol Cancer* 2013;**12**(1):102.
24. Willoughby JA, Sundar SN, Cheung M, Tin AS, Modiano J, Firestone GL. Artemisinin blocks prostate cancer growth and cell cycle progression by disrupting Sp1 interactions with the cyclin-dependent kinase-4 (CDK4) promoter and inhibiting CDK4 gene expression. *J Biol Chem* 2009;**284**(4):2203–13.
25. Hsu TI, Wang MC, Chen SY, Huang ST, Yeh YM, Su WC, et al. Betulinic acid decreases specificity protein 1 (Sp1) level via increasing the sumoylation of Sp1 to inhibit lung cancer growth. *Mol Pharmacol* 2012;**82**(6):1115–28.
26. Hsu TI, Wang MC, Chen SY, Yeh YM, Su WC, Chang WC, et al. Sp1 expression regulates lung tumor progression. *Oncogene* 2012;**31**:3973–88.
27. Previdi S, Malek A, Albertini V, Riva C, Capella C, Broggini M, et al. Inhibition of Sp1-dependent transcription and antitumor activity of the new aureolic acid analogues mithramycin SDK and SK in human ovarian cancer xenografts. *Gynecol Oncol* 2010;**118**(2):182–8.
28. Zannetti A, Del Vecchio S, Romanelli A, Scala S, Saviano M, Stoppelli MP, et al. Inhibition of Sp1 activity by a decoy PNA–DNA chimera prevents urokinase receptor expression and migration of breast cancer cells. *Biochem Pharmacol* 2005;**70**(9):1277–87.
29. Cram EJ, Liu BD, Bjeldanes LF, Firestone GL. Indole-3-carbinol inhibits CDK6 expression in human MCF-7 breast cancer cells by disrupting Sp1 transcription factor interactions with a composite element in the CDK6 gene promoter. *J Biol Chem* 2001;**276**(25):22332–40.
30. Higashimura Y, Nakajima Y, Yamaji R, Harada N, Shibasaki F, Nakano Y, et al. Up-regulation of glyceraldehyde-3-phosphate dehydrogenase gene expression by HIF-1 activity depending on Sp1 in hypoxic breast cancer cells. *Arch Biochem Biophys* 2011;**509**(1):1–8.
31. Yue L, Li L, Liu F, Hu N, Zhang W, Bai X, et al. The oncoprotein HBXIP activates transcriptional coregulatory protein LMO4 via Sp1 to promote proliferation of breast cancer cells. *Carcinogenesis* 2013;**34**:927–35. bgs399.
32. Kim K, Chadalapaka G, Lee SO, Yamada D, Sastre-Garau X, Defossez PA, et al. Identification of oncogenic microRNA-17-92/ZBTB4/specificity protein axis in breast cancer. *Oncogene* 2012;**31**:1034–44.
33. Waby JS, Chirakkal H, Yu C, Griffiths GJ, Benson RS, Bingle CD, et al. Sp1 acetylation is associated with loss of DNA binding at promoters associated with cell cycle arrest and cell death in a colon cell line. *Mol Cancer* 2010;**9**(1):275.
34. Wang F, Ma YL, Zhang P, Shen TY, Shi CZ, Yang YZ, et al. SP1 mediates the link between methylation of the tumour suppressor miR-149 and outcome in colorectal cancer. *J Pathol* 2013;**229**:12–24.
35. Jutooru I, Chadalapaka G, Sreevalsan S, Lei P, Barhoumi R, Burghardt R, et al. Arsenic trioxide downregulates specificity protein (Sp) transcription factors and inhibits bladder cancer cell and tumor growth. *Exp Cell Res* 2010;**316**(13):2174–88.

36. Lou Z, O'Reilly S, Liang H, Maher VM, Sleight SD, JJ MC. Down-regulation of overexpressed sp1 protein in human fibrosarcoma cell lines inhibits tumor formation. *Cancer Res* 2005;**65**(3): 1007–17.
37. Wang J, Liu X, Ni P, Gu Z, Fan Q. SP1 is required for basal activation and chromatin accessibility of CD151 promoter in liver cancer cells. *Biochem Biophys Res Commun* 2010;**393**(2):291–6.
38. Liu S, Wu LC, Pang J, Santhanam R, Schwind S, Wu YZ, et al. Sp1/NFκB/HDAC/miR-29b regulatory network in KIT-driven myeloid leukemia. *Cancer Cell* 2010;**17**(4):333–47.
39. Fulciniti M, Amin S, Nanjappa P, Rodig S, Prabhala R, Li C, et al. Significant biological role of sp1 transactivation in multiple myeloma. *Clin Cancer Res* 2011;**17**(20):6500–9.
40. Hedrick E, Cheng Y, Jin UH, Kim K, Safe S. Specificity protein (Sp) transcription factors Sp1, Sp3 and Sp4 are non-oncogene addiction genes in cancer cells. *Oncotarget* 2016;**7**(16):22245–56.
41. Yu Y, Peng K, Li H, Zhuang R, Wang Y, Li W, et al. SP1 upregulated FoxO3a promotes tumor progression in colorectal cancer. *Oncol Rep* 2018;**39**(5):2235–42.
42. Maurer GD, Leupold JH, Schewe DM, Biller T, Kates RE, Hornung HM, et al. Analysis of specific transcriptional regulators as early predictors of independent prognostic relevance in resected colorectal cancer. *Clin Cancer Res* 2007;**13**:1123–32.
43. Yao JC, Wang L, Wei D, Gong W, Hassan M, Wu TT, et al. Association between expression of transcription factor Sp1 and increased vascular endothelial growth factor expression, advanced stage, and poor survival in patients with resected gastric cancer. *Clin Cancer Res* 2004;**10**:4109–17.
44. Bedolla RG, Gong J, Prihoda TJ, Yeh IT, Thompson IM, Ghosh R, et al. Predictive value of Sp1/Sp3/FLIP signature for prostate cancer recurrence. *PLoS One* 2012;**7**:e44917.
45. Beishline K, Azizkhan-Clifford J. Sp1 and the "hallmarks of cancer" *FEBS J* 2015;**282**:224–58.
46. Chintharlapalli S, Papineni S, Ramaiah SK, Safe S. Betulinic acid inhibits prostate cancer growth through inhibition of specificity protein transcription factors. *Cancer Res* 2007;**67**(6).
47. Pore N. The HIV protease inhibitor nelfinavir downregulates HIF-1A and VEGF expression and increases tumor oxygenation. *Cancer Res* 2006;**66**(8):226–7.
48. Barillari G, Iovane A, Bacigalupo I, Labbaye C, Chiozzini C, Sernicola L, et al. The HIV protease inhibitor indinavir down-regulates the expression of the pro-angiogenic MT1-MMP by human endothelial cells. *Angiogenesis* 2014;**17**(4):831–8.
49. Papineni S, Chintharlapalli S, Abdelrahim M, Lee SO, Burghardt R, Abudayyeh A, et al. Tolfenamic acid inhibits esophageal cancer through repression of specificity proteins and c-Met. *Carcinogenesis* 2009;**30**(7):1193–201.
50. Abdelrahim M, Safe S. Cyclooxygenase-2 inhibitors decrease vascular endothelial growth factor expression in colon cancer cells by enhanced degradation of Sp1 and Sp4 proteins. *Mol Pharmacol* 2005;**68**(2):317–29.
51. Wei D, Wang L, He Y, Xiong HQ, Abbruzzese JL, Xie K. Celecoxib inhibits vascular endothelial growth factor expression in and reduces angiogenesis and metastasis of human pancreatic cancer via suppression of Sp1 transcription factor activity. *Cancer Res* 2004;**64**(6):2030–8.
52. Yuan P, Wang L, Wei D, Zhang J, Jia Z, Li Q, et al. Therapeutic inhibition of Sp1 expression in growing tumors by mithramycin A correlates directly with potent antiangiogenic effects on human pancreatic cancer. *Cancer* 2007;**110**(12):2682–90.
53. Majee S, Chakrabarti A. Membrane interaction of an antitumor antibiotic, mithramycin, with anionic phospholipid vesicles. *Biochem Pharmacol* 1999;**57**(9):981–7.
54. Rao M, Atay SM, Shukla V, Hong Y, Upham T, Ripley RT, et al. Mithramycin depletes specificity protein 1 and activates p53 to mediate senescence and apoptosis of malignant pleural mesothelioma cells. *Cancer Res* 2016;**22**(5):1197–210.
55. Shelake S, Sankpal UT, Bowman WP, Wise M, Ray A, Basha R. Targeting specificity protein 1 transcription factor and survivin using tolfenamic acid for inhibiting Ewing sarcoma cell growth. *Investig New Drugs* 2017;**35**(2):158–65.
56. Gao Y, Jia Z, Kong X, Li Q, Chang DZ, Wei D, et al. Combining betulinic acid and mithramycin a effectively suppresses pancreatic cancer by inhibiting proliferation, invasion, and angiogenesis. *Cancer Res* 2011;**71**:5182–93.
57. Lo WL, Huang YN, Hsu TI, Chuang JY. The effect of betulinic acid on temozolomide-resistant glioblastoma cells. *Neuro-Oncology* 2017;**19**:61.

58. Hsu TI, Wang MC, Chen SY, Huang ST, Yeh YM, Su WC, et al. Betulinic acid decreases specificity protein 1 (Sp1) level via increasing the sumoylation of sp1 to inhibit lung cancer growth. *Mol Pharmacol* 2012;**82**:1115–28.
59. Abdelrahim M, Baker CH, Abbruzzese JL, Safe S. Tolfenamic acid and pancreatic cancer growth, angiogenesis, and Sp protein degradation. *J Natl Cancer Inst* 2006;**98**(12):855–68.
60. Jia Z, Gao Y, Wang L, Li Q, Zhang J, Le X, et al. Combined treatment of pancreatic cancer with mithramycin A and tolfenamic acid promotes Sp1 degradation and synergistic antitumor activity. *Cancer Res* 2010;**70**(3):1111–9.
61. Basha R, Baker CH, Sankpal UT, Ahmad S, Safe S, Abbruzzese JL, et al. Therapeutic applications of NSAIDS in cancer: special emphasis on tolfenamic acid. *Front Biosci (Schol Ed)* 2011;**3**:797–805.
62. Basha R, Ingersoll SB, Sankpal UT, Ahmad S, Baker CH, Edwards JR, et al. Tolfenamic acid inhibits ovarian cancer cell growth and decreases the expression of c-Met and survivin through suppressing specificity protein transcription factors. *Gynecol Oncol* 2011;**122**(1):163–70.
63. Colon J, Basha MR, Madero-Visbal R, Konduri S, Baker CH, Herrera LJ, et al. Tolfenamic acid decreases c-Met expression through Sp proteins degradation and inhibits lung cancer cells growth and tumor formation in orthotopic mice. *Investig New Drugs* 2011;**29**(1):41–51.
64. Konduri S, Colon J, Baker CH, Safe S, Abbruzzese JL, Abudayyeh A, et al. Tolfenamic acid enhances pancreatic cancer cell and tumor response to radiation therapy by inhibiting survivin protein expression. *Mol Cancer Ther* 2009;**8**(3):533–42.
65. Sankpal UT, Nagaraju GP, Gottipolu SR, Hurtado M, Jordan CG, Simecka JW, et al. Combination of tolfenamic acid and curcumin induces colon cancer cell growth inhibition through modulating specific transcription factors and reactive oxygen species. *Oncotarget* 2016;**7**(3):3186.
66. Pathi S, Li X, Safe S. Tolfenamic acid inhibits colon cancer cell and tumor growth and induces degradation of specificity protein (Sp) transcription factors. *Mol Carcinog* 2014;**53**(S1):E53–61.
67. Shim JH, Shin JA, Jung JY, Choi KH, Choi ES, Cho NP, et al. Chemopreventive effect of tolfenamic acid on KB human cervical cancer cells and tumor xenograft by downregulating specificity protein 1. *Eur J Cancer Prev* 2011;**20**(2):102–11.
68. Wilson H, Chadalapaka G, Jutooru I, Sheppard S, Pfent C, Safe S. Effect of tolfenamic acid on canine cancer cell proliferation, specificity protein (Sp) transcription factors, and S p-regulated proteins in canine osteosarcoma, mammary carcinoma, and melanoma cells. *J Vet Intern Med* 2012;**26**(4):977–86.
69. Eslin D, Sankpal UT, Lee C, Sutphin RM, Maliakal P, Currier E, et al. Tolfenamic acid inhibits neuroblastoma cell proliferation and induces apoptosis: a novel therapeutic agent for neuroblastoma. *Mol Carcinog* 2013;**52**(5):377–86.
70. Chadalapaka G, Jutooru I, Sreevalsan S, Pathi S, Kim K, Chen C, et al. Inhibition of rhabdomyosarcoma cell and tumor growth by targeting specificity protein (Sp) transcription factors. *Int J Cancer* 2013;**132**(4):795–806.
71. Liu X, Abdelrahim M, Abudayyeh A, Lei P, Safe S. The nonsteroidal anti-inflammatory drug tolfenamic acid inhibits BT474 and SKBR3 breast cancer cell and tumor growth by repressing erbB2 expression. *Mol Cancer Ther* 2009;**1535**:7163.
72. Kang SU, Shin YS, Hwang HS, Baek SJ, Lee SH, Kim CH. Tolfenamic acid induces apoptosis and growth inhibition in head and neck cancer: involvement of NAG-1 expression. *PLoS One* 2012;**7**(4):e34988.
73. Sankpal UT, Ingersoll SB, Ahmad S, Holloway RW, Bhat VB, Simecka JW, et al. Association of Sp1 and survivin in epithelial ovarian cancer: Sp1 inhibitor and cisplatin, a novel combination for inhibiting epithelial ovarian cancer cell proliferation. *Tumor Biol* 2016;**37**(10):14259–69.
74. Chang JW, Kang SU, Choi JW, Shin YS, Baek SJ, Lee SH, et al. Tolfenamic acid induces apoptosis and growth inhibition in anaplastic thyroid cancer: involvement of nonsteroidal anti-inflammatory drug-activated gene-1 expression and intracellular reactive oxygen species generation. *Free Radic Biol Med* 2014;**67**:115–30.
75. Ha T, Lou Z, Baek SJ, Lee SH. Tolfenamic acid downregulates β-catenin in colon cancer. *Int Immunopharmacol* 2016;**35**:287–93.
76. Zhang X, Lee SH, Min KW, MF ME, Jeong J, Li Q, et al. The involvement of endoplasmic reticulum stress in the suppression of colorectal tumorigenesis by tolfenamic acid. *Cancer Prev Res* 2013; https://doi.org/10.1158/1940-6207.CAPR-13-0220.

77. Choi ES, Shim JH, Jung JY, Kim HJ, Choi KH, Shin JA, et al. Apoptotic effect of tolfenamic acid in androgen receptor-independent prostate cancer cell and xenograft tumor through specificity protein 1. *Cancer Sci* 2011;**102**(4):742–8.
78. Sankpal UT, Abdelrahim M, Connelly SF, Lee CM, Madero-Visbal R, Colon J, et al. Small molecule tolfenamic acid inhibits PC-3 cell proliferation and invasion in vitro, and tumor growth in orthotopic mouse model for prostate cancer. *Prostate* 2012;**72**(15):1648–58.
79. Sutphin RM, Connelly SF, Lee CM, Sankpal UT, Eslin D, Khan M, et al. Anti-leukemic response of a NSAID, tolfenamic acid. *Target Oncol* 2014;**9**(2):135–44.
80. Eslin D, Lee C, Sankpal UT, Maliakal P, Sutphin RM, Abraham L, et al. Anticancer activity of tolfenamic acid in medulloblastoma: a preclinical study. *Tumor Biol* 2013;**34**(5):2781–9.
81. Maliakal P, Abdelrahim M, Sankpal UT, Maliakal C, Baker CH, Safe S, et al. Chemopreventive effects of tolfenamic acid against esophageal tumorigenesis in rats. *Investig New Drugs* 2012;**30**(3):853–61.
82. Shao HJ, Lou Z, Jeong JB, Kim KJ, Lee J, Lee SH. Tolfenamic acid suppresses inflammatory stimuli-mediated activation of NF-κB signaling. *Biomol Ther* 2015;**23**(1):39.
83. Uyemura SA, Santos AC, Mingatto FE, Jordani MC, Curti C. Diclofenac sodium and mefenamic acid: potent inducers of the membrane permeability transition in renal cortex mitochondria. *Arch Biochem Biophys* 1997;**342**(2):231–5.
84. Mingatto FE, Santos AC, Uyemura SA, Jordani MC, Curti C. In vitro interaction of nonsteroidal anti-inflammatory drugs on oxidative phosphorylation of rat kidney mitochondria: respiration and ATP synthesis. *Arch Biochem Biophys* 1996;**334**(2):303–8.
85. Masubuchi Y, Yamada S, Horie T. Diphenylamine as an important structure of nonsteroidal anti-inflammatory drugs to uncouple mitochondrial oxidative phosphorylation. *Biochem Pharmacol* 1999;**58**:861–5.
86. Kankaanranta H, Moilanen E. Flufenamic and tolfenamic acids inhibit calcium influx in human polymorphonuclear leukocytes. *Mol Pharmacol* 1995;**47**(5):1006–13.
87. Dutcher JP, Coletti D, Paietta E, Wiernik PH. A pilot study of alpha-interferon and plicamycin for accelerated phase of chronic myeloid leukemia. *Leuk Res* 1997;**21**(5):375–80.
88. Kennedy BJ, Torkelson JL. Long-term follow-up of stage III testicular carcinoma treated with mithramycin (plicamycin). *Med Pediatr Oncol* 1995;**24**(5):327–8.
89. Walker MD, Alexander Jr E, Hunt WE, Leventhal CM, Mahaley Jr MS, Mealey J, et al. Evaluation of mithramycin in the treatment of anaplastic gliomas. *J Neurosurg* 1976;**44**(6):655–67.
90. Kofman S, Perlia CP, Economou SG. Mithramycin in the treatment of metastatic Ewing's sarcoma. *Cancer* 1973;**31**(4):889–93.
91. Ansfield FJ. Clinical studies with mithramycin. *Oncology* 1969;**23**(4):283–8.
92. https://clinicaltrials.gov/ct2/show/NCT01624090.
93. Koltai T. Nelfinavir and other protease inhibitors in cancer: mechanisms involved in anticancer activity. *F1000Res* 2015;**4**:9.
94. Pore N. The HIV protease inhibitor nelfinavir downregulates HIF-1α and VEGF expression and increases tumor oxygenation. *Cancer Res* 2006;**66**(8):226–7.
95. Barillari G, Iovane A, Bacigalupo I, Labayye C, Chiozzini C, Sernicola N, et al. The HIV protease inhibitor indinavir down-regulates the expression of the pro-angiogenic MT1-MMP by human endothelial cells. *Angiogenesis* 2014;**17**(4):831–8.
96. Nair V, Pathi S, Jutooru I, Sreevalsan S, Basha R, Abdelrahim M, et al. Metformin inhibits pancreatic cancer cell and tumor growth and downregulates Sp transcription factors. *Carcinogenesis* 2013;**34**(12):2870–9.
97. Hedrick E, Li X, Safe S. Penfluridol represses integrin expression in breast cancer through induction of reactive oxygen species and downregulation of Sp transcription factors. *Mol Cancer Ther* 2017;**16**(1):205–16.
98. Chae JI, Cho JH, Lee K, Choi NJ, Seo KS, Kim SB, et al. Role of transcription factor Sp1 in the quercetin-mediated inhibitory effect on human malignant pleural mesothelioma. *Int J Mol Med* 2012;**30**(4):835–41.
99. Chadalapaka G, Jutooru I, Safe S. Celastrol decreases specificity proteins (Sp) and fibroblast growth factor receptor-3 (FGFR3) in bladder cancer cells. *Carcinogenesis* 2012;**33**:886–94.
100. Zhu H, Liu X-W, Cai T-Y, Cao J, Tu C-X, Lu W, et al. Celastrol acts as a potent antimetastatic agent targeting integrin and inhibiting cell-extracellular matrix adhesion, in part via the p38 mitogen-activated protein kinase pathway. *J Pharmacol Exp Ther* 2010;**334**(2):489–99.

101. Jutooru I, Chadalapaka G, Sreevalsan S, Lei P, Barhoumi R, Burghardt R, et al. Arsenic trioxide downregulates specificity protein (Sp) transcription factors and inhibits bladder cancer cell and tumor growth. *Exp Cell Res* 2010;**316**:2174–88.
102. Malakooti J, Dahdal RY, Dudeja PK, Layden TJ, Ramaswamy K. The human Na+/H+ exchanger *NHE2* gene: genomic organization and promoter characterization. *Am J Physiol Gastrointest Liver Physiol* 2001;**280**(4):763–73.
103. Malakooti J, Memark VC, Dudeja PK, Ramaswami K. Molecular cloning and functional analysis of the human Na+/H+exchanger NHE3 promoter. *Am J Physiol Gastrointest Liver Physiol* 2002; **282**(3):491–500.
104. Miller RT, Counillon L, Pages G, Lifton RP, Sardet C, Pouyssegur J. Structure of the 5′-flanking regulatory region and gene for the human growth factor-activatable Na/H exchanger NHE-1. *J Biol Chem* 1991;**266**:10813–9.
105. Deng GF, Qin JM, Sun XS, Kuang ZY, Su T, Zhao QH, et al. Promoter analysis of mouse Scn3a gene and regulation of the promoter activity by GC box and CpG methylation. *J Mol Neurosci* 2011; **44**(2):115–21.
106. Rook MB, Evers MM, Vos MA, Bierhuizen MF. Biology of cardiac sodium channel Nav1. 5 expression. *Cardiovasc Res* 2011;**93**(1):12–23.
107. Loussouarn G, Sternberg D, Nicole S, Marionneau C, Le Bouffant F, Toumaniantz G, et al. Physiological and pathophysiological insights of Nav1. 4 and Nav1. 5 comparison. *Front Pharmacol* 2016;**6**:314.
108. Puhl III HL, Ikeda SR. Identification of the sensory neuron specific regulatory region for the mouse gene encoding the voltage-gated sodium channel NaV1. 8. *J Neurochem* 2008;**106**(3):1209–24.
109. Lee BS, Krits I, Crane-Zelkovic MK, Gluck SL. A novel transcription factor regulates expression of the vacuolar H+-ATPase B2 subunit through AP-2 sites during monocytic differentiation. *J Biol Chem* 1997;**272**(1):174–81.
110. Kong LM, Liao CG, Fei F, Guo X, Xing JL, Chen ZN. Transcription factor Sp1 regulates expression of cancer-associated molecule CD147 in human lung cancer. *Cancer Sci* 2010;**101**:1463–70.
111. Kong LM, Yao L, Lu N, Dong YL, Zhang J, Wang YQ, et al. Interaction of KLF6 and Sp1 regulates basigin-2 expression mediated proliferation, invasion and metastasis in hepatocellular carcinoma. *Oncotarget* 2016;**7**:27975–87.
112. Koizume S, Miyagi Y. Diverse mechanisms of Sp1 dependent transcriptional regulation potentially involved in the adaptive response of cancer cells to oxygen-deficient conditions. *Cancer* 2016;**8**(1):2.
113. De Gassart A, Martinon F. Translating the anticancer properties of eEF2K. *Cell Cycle* 2017; **16**(4):299–300.
114. Pisha E, Chai H, Lee IS, Chagwedera TE, Farnsworth NR, Cordell GA, et al. Discovery of betulinic acid as a selective inhibitor of human melanoma that functions by induction of apoptosis. *Nat Med* 1995;**1**:1046–51.
115. Zuco V, Supino R, Righeltis SC, Cleris L, Marchesi E, et al. Selective cytotoxicity of betulinic acid on tumor cell lines, but not on normal cells. *Cancer Lett* 2002;**175**:17.
116. Shin J, Lee HJ, Jung DB, Jung JH, Lee HJ, Lee EO, et al. Suppression of STAT3 and HIF-1 alpha mediates anti-angiogenic activity of betulinic acid in hypoxic PC-3 prostate cancer cells. *PLoS One* 2011;**6**(6):e21492.
117. Xiao H, Hasegawa T, Isobe K. p300 collaborates with Sp1 and Sp3 in p21 waf1/cip1 promoter activation induced by histone deacetylase inhibitor. *J Biol Chem* 2000;**275**:1371. https://doi.org/10.1074/jbc.275.2.1371.
118. Innocente SA, Lee JM. p53 is a NF-Y- and p21-independent, Sp1-dependent repressor of cyclin B1 transcription. *FEBS Lett* 2005;**579**(5):1001–7.
119. Kavurma MM, Khachigian LM. Sp1 inhibits proliferation and induces apoptosis in vascular smooth muscle cells by repressing p21WAF1/Cip1 transcription and cyclin D1-Cdk4-p21WAF1/Cip1 complex formation. *J Biol Chem* 2003;**278**:32537.
120. Xu J, Zhou JY, Wei WZ, Philipsen S, Wu GS. Sp1-mediated TRAIL induction in chemosensitization. *Cancer Res* 2008;**68**(16):6718–26.
121. Li L, Gao P, Li Y, Shen Y, Xie J, Sun D, et al. JMJD2A-dependent silencing of Sp1 in advanced breast cancer promotes metastasis by downregulation of DIRAS3. *Breast Cancer Res Treat* 2014; **147**(3):487–500.

PART II

Therapeutics

CHAPTER 13

The inverted pH gradient in cancer: Pharmacological interventions. Part I

Introduction

The abnormal pH gradient in cancer is defined by a higher pHi and a lower pHe than in normal tissues. Therefore, modifying the pH paradigm in cancer means bringing pH back to the same levels of normal tissues: a lower intracellular pH (pHi) and a relatively higher extracellular pH (pHe).

Does this have any practical importance?

As shown in the first part of this book, there is strong evidence to the effect that normalizing pH or achieving a very low intracellular pH, decreases tumor growth, invasion, migration and metastasis. Therefore, manipulating the extracellular and intracellular pH of tumors has shown considerable potential in cancer treatment.[1–3]

The research by Robey et al.[4] is a good example to help understand what a simple pharmacological intervention may accomplish. They administered oral NaHCO3 (sodium bicarbonate) to mice harboring a metastatic model of breast cancer. The treatment increased the extracellular pH but had no major effects on the intracellular pH. The biologic effect consisted in a significant reduction of spontaneous metastasis and lymph node involvement. The number of circulating malignant cells was not decreased by the bicarbonate treatment. This led the authors to the conclusion that the extracellular alkalinization had not impeded intravasation of tumor cells but decreased extravasation and colonization. This conclusion was further supported by the fact that the intrasplenic injection of bicarbonate reduced hepatic metastases. They also showed that these findings can differ according to tumor type. While it worked with prostate cancer cells, bicarbonate did not reduce metastases in B16 melanoma. Two conclusions can be drawn from these experiments:

(1) modifying the pHe of the niche may reduce the chances for metastatic colonization;
(2) niche requirements may be tumor-specific.

Another publication by Robey et al.[5] showed a longer survival rate in mice with tumors treated with chronic administration of bicarbonate. The conclusion in this case is that pH modifications may alter the tumor's biology and progression.

The extracellular acidosis of tumors facilitates invasion (acid-mediated invasion model) so that the reduction of extracellular acidity should inhibit malignant tumor

An Innovative Approach to Understanding and Treating Cancer: Targeting pH
https://doi.org/10.1016/B978-0-12-819059-3.00013-7
© 2020 Elsevier Inc.
All rights reserved.

invasion. In the experiments by Robey et al. the excess of extracellular H^+ (protons) combines with bicarbonate and generates water and CO2.[6] In malignant tumors there is already an excess of CO2 production. The buffering of the extracellular matrix with 2-imidazole-1-3-ethoxycarbonylpropionic acid (IEPA) showed similar results: reduced metastasis in a mouse model of prostate cancer.[7]

Gullino et al.[8] in 1967 found that the interstitial fluid of Walker carcinomas had a pH of 6.99 and a pCO2 of 79 mmHg, almost twice the pCO2 plasma level. In another publication in 1965 by Gullino et al.,[9] he wrote:

> *"In 5 different types of rat tumors, the pH of the interstitial fluid ranged from 7.193 to 6.948, the CO_2 tension from 59 to 84 mm Hg, the dissolved CO_2 from 2.0 to 2.6 mM per liter, and the concentration of bicarbonate from 18.6 to 24.3 mM per liter. Compared with the plasma of blood afferent to the tumor, the pH of the tumor interstitial fluid was 0.2 to 0.4 units lower, the pCO_2 16 to 39 mm Hg higher, the dissolved CO_2 about 1 mM per liter higher, and the concentration of bicarbonate 4 to 6 mM per liter lower. The same differences were observed when the tumor interstitial fluid was compared with the interstitial fluid of the normal subcutaneous area, though the differences were slightly less when the comparison was made with the plasma of the efferent blood of the tumor or with the lymph of the thoracic duct of the host. A change in the acid-base status of the interstitial fluid of the tumor was produced either by an increase of the CO_2 concentration in the air inhaled by the host or by ingestion of sodium bicarbonate or ammonium chloride. When a depression of pH was produced in the blood plasma, a depression roughly equal was found in the interstitial fluid of the tumor. As the pH of this fluid is normally quite low, exceptionally low pH values (6.6–6.7) could be induced exclusively in the tumor. The CO_2 tension increased in the tumor to levels twice as high as that in plasma or in the other body fluids of the host, and values of more than 200 mm Hg were reached in the interstitial fluid of the tumor. The amount of dissolved CO_2 changed proportionally while the concentration of the bicarbonate showed less drastic oscillations."*

In this pioneering research of more than half a century ago, it was clearly established that there is an increase of CO2 at the level of tumor's extracellular space (ECS), with increased acidity and lower bicarbonate levels.

Administering IEPA or bicarbonate introduced a full modification of the ECS acid-base balance and decreased extracellular acidity. This is probably the simplest pharmacological intervention, which should have effect mainly on the pH of the ECS. It retards migration, invasion and decreases metastasis, but has little effect on proliferation. It is not enough to increase extracellular pH (pHe) to decrease proliferation, the intracellular pH (pHi) needs to be modified making it less alkaline. Xiao-Fang et al.[10] induced apoptosis by decreasing intracellular pH. The first conclusion from these experiments is that by modifying a tumors pHe and pHi in an appropriate way, a clear anti-tumoral result is achievable. There is strong evidence that pHi is the main player in the inversion of the pH gradient in tumors.[11,12]

Therefore, the oncologist's task should be to address both pHe and pHi: alkalinizing the first and acidifying the second.

The objectives of the pH-centered treatment of cancer are:

(1) To decrease proliferation by reducing intracellular alkalinity (through intracellular acidification).
(2) To decrease migration, invasion, metastasis and immunologic escape blocking by increasing extracellular pH (through extracellular alkalinization).

The experiments described above show that addressing the pHe, although effective for metastatic reduction is insufficient for decreasing proliferation. To achieve more important and lasting results, both pHs need to be modified simultaneously. Furthermore, the reduction of the intracellular pH in general increases extracellular pH but the inverse situation does not happen because extracellular alkalinization with bicarbonate had no effect on intracellular alkalinity in the experiments by Robey et al.

The importance of addressing the pH paradigm in cancer

All the previous chapters described the intimate relationship between cancer and the pH gradient inversion in detail. It is not necessary to repeat those concepts here, but a graphic summary will help understand how this unbalanced or dysfunctional pH conspires against the patient and the standard treatments, thus favoring the tumor's progression and evolution (Fig. 1).

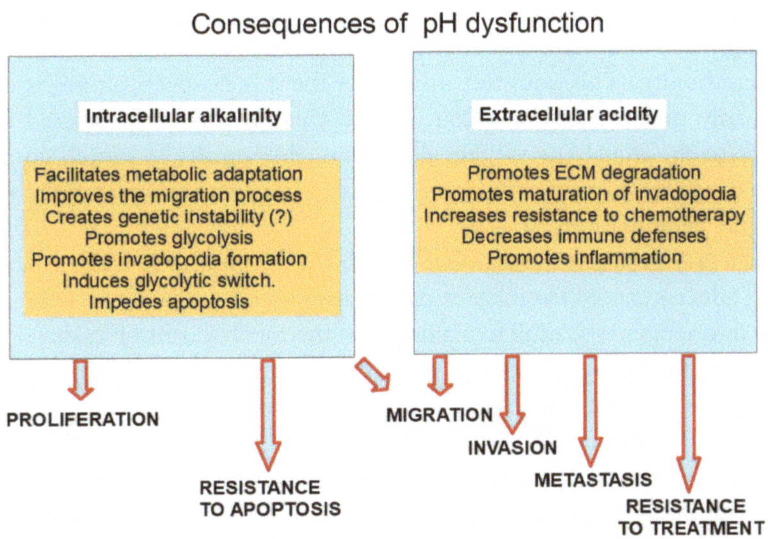

Fig. 1 The consequences of the pH inversion gradient are essential for cancer development. This figure is based on references.[13–18]

Historical perspective

A possible role for alkalinity in the development of malignancies was already suggested back in 1906.[19] Then, Warburg in the 1920s discovered the acidity of tumors. However, he did not suspect that in spite of this acidity, tumor cells were strongly alkaline in the inside. During the 50s, 60s and early 70s, a considerable number of clinical studies associated acid-base changes with either the onset of malignancy or tumor regressions.[20] Many decades had to elapse before the cancer research community began understanding the real consequences of Otto Warburg's initial discoveries on the deregulation of pH in cancer.[21,22]

Now, basic researchers have arrived at the conclusion that pH alterations in cancer have a capital importance in this disease's origin, development and progression.[23–29] This new approach to the cancer problem is based on the fact that abnormal pH is a universal finding in all tumors. The main abnormality, the inversion of the pH gradient, which was considered a mere consequence of the metabolic changes is now considered the cause and not the consequence.[30,31] However, what is so evident to the world of basic research, seems not to be so clear to those who treat patients with cancer. The average oncologist is not aware of this issue.

What comes first glycolysis or pH inversion?

Another unique hallmark[32] of cancer cells is their shift to glycolytic metabolism over oxidative phosphorylation, even under aerobic conditions. This was first described by Otto Warburg[33] and is known as the Warburg Effect. For many years, aerobic glycolysis was considered the sole cause of the alterations found in pH. Over the past 30 years, this notion lost credibility. This does not mean that there is no deep interrelation between the two. It only means that the initiator of all this process is the altered intracellular pH that leads to the metabolic switch. After that, the metabolic switch cooperates and makes the extracellular space even more acidic. However, the metabolic switch does not initiate pH deregulation.

During the last few years the number of publications trying to explain how and why the Warburg effect is caused has increased dramatically. However, apart from considering its early role in oncogenesis, a full explanation of the mechanisms of its development and how it relates or even leads to cancer is still unresolved.[34] Early experiments showed that glycolytic metabolism first appears very early in the oncogene-driven transformation of normal cells and that its development is dependent on the initial cytoplasmic alkalinization.[11,35,36] A rapid shift in cell metabolism follows alkalinization (Fig. 2). The interest in the relationships between pH and cancer dates from the 1920s[37] and the dependence of the rates of glycolysis on pH was already described more than four decades ago.[38,39] However, at that time, emphasis was mainly on a physiological and not an oncological perspective.[40]

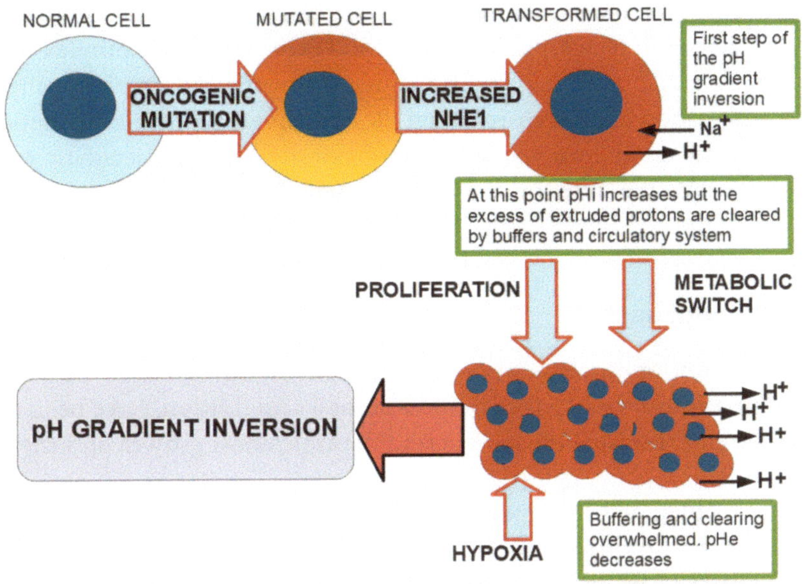

Fig. 2 How the pH gradient inversion develops in the early stages of the malignant tumor.

Which is more important pHe or pHi?

Both pH abnormalities have to be dealt with for effective cancer treatment. This means that the question posed in the title has no major sense. However, there are four issues in the pH paradigm that suggest that the principal player is pHi:

(1) It is extremely difficult for a cell, whether normal or malignant to enter the mitotic cycle with a pHi under 7.2.[41]

(2) pHe acidification is a consequence of increased intracellular proton production and extrusion[42] that overwhelm proton clearance and buffering.

(3) Alkalinizing the ECS decreases migration and invasion but does not alter proliferation.

(4) Acidifying the intracellular milieu decreases proliferation and induces apoptosis in malignant cells.[43–45]

Presently, there is also enough evidence to accept that both pHi and pHe are highly important in driving the increased dependence on glycolysis and decreased dependence on OXPHOS as the tumor progresses.[46,47] It is the alkaline pHi in cancer cells which is the primordial driver of this metabolic shift and this change is one of the "cornerstones" in the altered metabolism underlying neoplastic transformation and progression.[11] Quach et al.[48] reached the conclusion that the Warburg Effect may be completely explained through the elevation of pHi in cancer cells.

On the other hand, lower extracellular pH shows profound effects on tumor cell gene expression, including genes involved in glycolysis.[49] The relationships between pH and the glycolytic chain has been discussed in the chapter on metabolism and has also been thoroughly reviewed in recent publications.[50,51] Therefore, these complex dynamics of the role of pH in metabolism starts with the first oncogene-driven pHi change and it is this very early alteration of pH dynamics and the consequent metabolic disruptions which set the stage for the conditions necessary for tumor growth and metastatic progression, further highlighting the role of the pH gradient inversion in carcinogenesis and metastasis.[11,52,53]

Therapeutic approach

From the standpoint of cancer treatment, recent attempts to control and reverse intracellular alkalinization and extracellular/intratumoral acidification, have been considered a viable and original therapeutic approach to control tumor growth and the metastatic process.[54]

The first step observed in the apoptotic process is cytoplasmic acidification.[55] The first step in malignant transformation is cytoplasmic alkalinization.[11] From the birth of a cancer (malignization) to its death (apoptosis), the first finding is always a modified pHi. Blocking initial cytoplasmic alkalinization by inhibiting NHE1 impedes transformation. On the other hand, cytoplasmic alkalinization is a serious hurdle for apoptosis. The starting process in the pH gradient inversion starts at the pHi, therefore, the therapeutic effort should be directed there. In spite of deciding that pHi acidification should be the primary goal of treatment, pHe alkalinization must not be overlooked.

Cellular acidification

Inhibiting H^+ efflux produced cell death in invasive primary tumor cell lines and this was related to a secondary mitochondrial depolarization.[56] Proton pump V-ATPase inhibitors (PPIs) have been repeatedly proposed in preclinical and clinical studies, but were not always successful when associated with chemotherapy,[57–59] or in overcoming resistance to certain chemotherapeutic drugs.[60] The utilization of alkaline diets using bicarbonate or other buffers has also been tested in attempts to decrease the interstitial acidification of malignant tumors, but did not prove effective in bedside oncology, because of toxicity, lack of patient compliance, or lack of effects. However, it is important to remember that tumor extracellular acidification is not the cause but the consequence of a pathological induction of intracellular alkalinity and a result of proton (H^+) extrusion driven by the up-regulation of one or more membrane-bound proton transporters. This is the reason why we support a more etiological therapeutic approach. Rather than struggling directly against extracellular acidification of malignant tumors, intracellular pH is the main target.

The fundamentals of the pH-centered treatment of cancer are shown in Fig. 3.

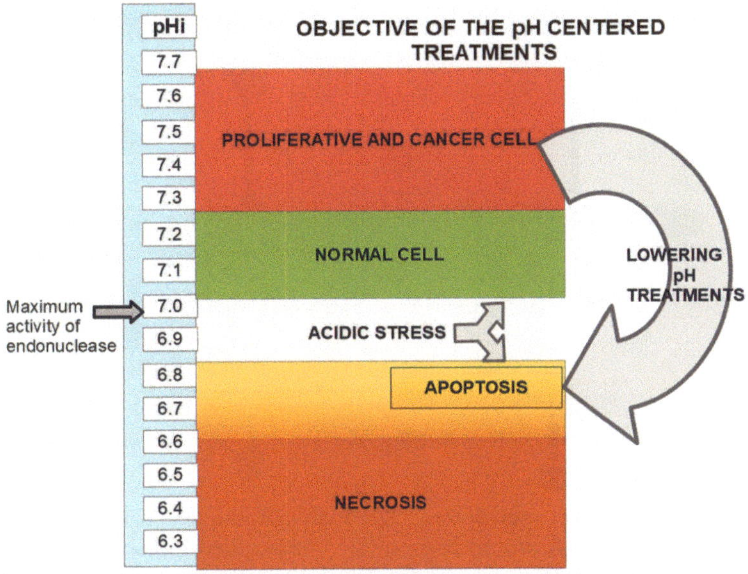

Fig. 3 The main objective of the pH centered treatments is to lower intracellular pH (pHi) that drives the malignant cells into apoptosis. Decreasing the intracellular pH produces acidic stress. In this phase the process is reversible. However if pHi is further decreased apoptosis or necrosis will finally ensue. The new therapeutic strategies are directed to achieving intracellular acidification of all malignant cells without affecting normal cells. This intracellular acidification must achieve a pHi low enough to induce a chain reaction that leads to apoptosis. The maximum activity of a fundamental enzyme in the apoptotic process such as the endonuclease DNAase I occurs around a pHi of 7.0.[61]

Importantly, the final aim will always be to induce a selective intracellular hyper-acidification of cancer cells by blocking acid extrusion through the utilization of pharmacological dosages of a cohort of proton transport inhibitors (cocktail of proton extruders inhibitors) and/or other cell acidifiers with known anticancer properties. The targets for treatment should be those directly involved in the creation of intracellular alkalosis. These targets and their mechanism of action have been identified.

Targets: These are NHE1 (sodium hydrogen exchanger 1), VGSCs (voltage gated sodium channels), CA (membrane carbonic anhydrases), MTCs 1 and 4 (monocarboxylate transporters) and those that create the metabolic conditions for an increased production of protons: glycolytic metabolism. Sp1 transcription factor is an indirect target that reduces the expression of NHE1, VGSCs, CAIX, and proton pumps (Fig. 4).

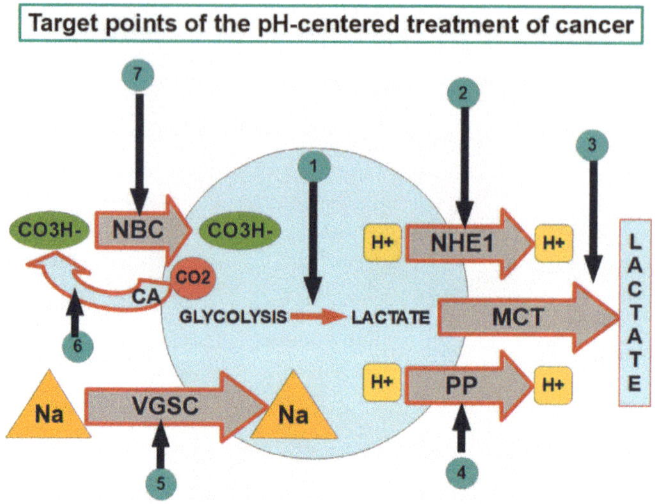

Fig. 4 Targeting (1) glycolytic metabolism; (2) sodium/hydrogen exchanger 1 (NHE1); (3) monocarboxylate transporters; (4) proton pumps; (5) voltage gated sodium channels; (6) carbonic anhydrases; (7) sodium/bicarbonate cotransporter. This is the rational basis for building the cocktail of proton extrusion inhibitors.

Proton transport inhibitors and proton pump inhibitors will secondarily decrease the extracellular acidification of malignant tumors known to stimulate local invasion and the metastatic process. It is clear that therapeutic intracellular acidification is fundamental in any attempt to induce selective apoptosis of malignant cells. Thus, the drugs that have the ability to modify the pH gradient inversion will be analyzed in this and the following chapters.

Mechanism of action: As stated above, the mechanism that creates intracellular alkalosis is mainly proton extrusion.

The following classification of drugs according to their main mechanism of action will demonstrate the therapeutic choices (Table 1).

The transcription factor Sp1 that acts as co-promoter for HIF-1α, NHE-1 and carbonic anhydrase IX, should also be considered a targetable pro-tumoral factor.

Table 1 Drugs that produce intracellular acidification and/or extracellular alkalinization

Cell acidifiers	
Inhibitors of ion trafficking	
CHANNEL/EXCHANGER BLOCKERS ------------------------------------	
NHE1	AMILORIDE CARIPORIDE EIPA TROGLITAZONE CIMETIDINE
VGSCs	PHENITOYN TOPIRAMATE FLUNARIZIN CARBAMACEPINE RANOLAZINE RILUZOLE
PROTON PUMP	LANSOPRAZOLE ESOMEPRAZOLE PANTOPRAZOLE OMEPRAZOLE
TRANSPORTER BLOCKERS ------	--------------------------------
MCT INHIBITORS	QUERCETIN STATINS
BASIGIN INHIB.	NOT AVAILABLE FOR CLINICAL USE
NBC BLOCKERS	NOT AVAILABLE FOR CLINICAL USE
ENZYME BLOCKERS -------------	--------------------------------
CA INHIBITORS	ACETAZOLAMIDE
METABOLIC MODIFIERS ------------	------------------
GLYCOLYTIC INHIBITORS	DICHLOROACETATE (DCA) 2-DEOXY-D-GLUCOSE (2DG) ALFA LIPOIC ACID (ALA) LONIDAMINE
OXPHOS INHIBITORS -----------	--------------------------------
	METFORMIN PHENFORMIN
DRUGS WITH UNPRECISED MECHANISM	
	SALINOMYCIN DISULFIRAM 3-BROMOPYRUVATE (3BP)
EXTRACELLULAR ALKALINIZERS	SODIUM BICARBONATE

Attacking NHE1

The type 1 Na^+/H^+ exchanger (NHE1) catalyzes the electroneutral exchange of H^+ and Na^+, and it is ubiquitously expressed at the plasma membrane of all mammalian cells.[62] It is considered a "house-keeping" intracellular H^+ regulator, protecting normal cells from intracellular acidification.[63] NHE1 is quiescent at physiological pHi (pHi $\geq$7.2), but its activity increases rapidly upon intracellular acidification.[64] Its activity is tightly regulated and its sensitivity to pHi increases in situations that can be found in the context of cancer, such as activation by growth factors, hormones, different mitogens and environmental carcinogens.[65–67] It is also involved in cancer cell motility and matrix degradation (see chapter on invadopodia). It is well established that there is an increased expression and/or activity of either one or more pH regulators in most, or all, cancer cells of different origins.[68,69] Among the known pH regulators, NHE1 has drawn a great deal of attention because it is overexpressed and/or over-activated in many types of cancer, from breast to ovarian cancer cells.[70–72] Such hyperactivity is directly associated with cancer cell survival, migration, invasion and metastatic progression.[73] For all these reasons, during the last few years NHE1 has become a very important target in selective cancer therapeutics.[74–76] Most recently, inhibition of NHE1 and cancer proton reversal have been correctly regarded as the latest concept in cancer treatment.[77–80]

Anticancer potential of NHE inhibitors

Decreasing NHE1 expression or inhibiting its activity leads to acidification of the intracellular space, inhibition of glycolysis, tumor cell growth arrest and selective apoptosis.[81] It seems that cancer cells are highly dependent on NHE1 for their survival. Cariporide, a selective and potent NHE1 inhibitor, reduces proliferation and induces apoptosis by decreasing intracellular pH in cholangiocarcinoma cells[82] (for a detailed description on cariporide see the next chapter). Since intracellular acidification driven by inhibition of NHE1 also inhibits the expression of vascular endothelial growth factor (VEGF),[83] we can understand why cariporide decreases angiogenesis and the growth of leukemia cells by inhibiting the Na^+/H^+ exchanger.[84] NHE1 inhibition has also proven useful in overcoming metastatic progression and multiple drug resistance (MDR).[85] Besides, in combination with inhibitors of the Na^+-dependent Ca^{2+} transporter, cariporide induces a non-apoptotic death in glioma cells.[86] This drug also acts synergistically with erlotinib in reducing growth and invasion in pancreatic ductal adenocarcinoma.[87] What is most surprising is that, in spite of all these promising results, cariporide has not raised any interest to facilitate its preclinical and/or translational research and application in the field of oncology, especially since its international patent literally states that "*there is a surprising prolongation of life of cancer in the elderly to an extent which has to date been achievable by no other group of drugs or by any natural products*". It was also added that "*this is a unique effect of NHE inhibitors like cariporide* (CARIPORIDE PATENT WO2004007480, SANOFI-AVENTIS, 2005)".

Perhaps even more importantly and promising, the 5-aryl-4-(4-(5-methyl-1*H*-imidazol-4-yl) piperididn-1-yl) pyrimidine analog ("Compound 9t") was reported to be 500-fold more potent against NHE1 than cariporide and to have a greater selectivity for NHE1 over NHE2 (1400-fold).[88] Additionally, Compound 9t is orally bioavailable, has low side effects in mice and shows a significantly improved safety profile over other NHE1 inhibitors (BRISTOL-MYERS SQUIB PATENT WO 0127107 A2, PCT/US00/27, 2001,US 6887870 B1; EP 1224183 B1). Surprisingly too, this drug has never been tested as an anticancer drug to date and, somehow, has been lost—missing in inaction—due to a lack of further interest concerning its possible utilization in preclinical and/or clinical oncology in spite of its most promising antitumoral characteristics and selective anticancer potential.[89]

There is another potent and selective NHE1 inhibitor, the aminophenoxazine derivative Phx-3, that has been shown to trigger apoptosis in a variety of cancer cell lines. In animal models, Phx-3 effectively, and without noticeable toxicity, reversed tumor growth after subcutaneous injection of adult leukemia cells.[90,91] In any case, all these new and selective NHE1 inhibitors show a great potential to become powerful anticancer agents in preclinical trials and, eventually, in cancer patients. Current research along this line in different malignancies shows that several proton transporters (PTs) and their inhibitors (PTIs) are very selective as diagnostic, predictive and potential anticancer agents.[92–94]

While decreasing intracellular pH induces a cell death program, apoptosis, para-apoptotic or necrotic cell death in malignant cells, the elevation of cellular pH by different methods protects cancer cells by preventing them from entering the apoptotic cascade. This feature even suggests a universality of phenomena concerning the role of pH in cell fate.[95] On the low pHi therapeutic-apoptotic side, striking results in leukemia cells were initially reported using the amiloride derivative and potent NHE1 inhibitor 5-(*N*,*N*-hexamethylene) amiloride (HMA). This compound decreases the pHi below the cellular survival threshold leading to selective apoptosis in a variety of human leukemic cells.[96] An increasing number of later studies has led to the conclusion that inducing a low pHi-mediated apoptosis can be a cancer-specific therapeutic strategy for all cancer cells and tissues.[97–101] Most recently, selective cytotoxicity of hexametylene amiloride was found in breast cancer cells independently of their molecular profile, proliferative status or species of origin, suggesting a necrotic cell death mechanism common to all mammary tumor subtypes: the lysosomal death pathway.[102] In this same sense, it was suggested that the use of ciprofloxacin acting as a "lysosomal poison" associated with anti-proton therapy may produce increased cytotoxicity.[103]

Moreover, acting not only on NHE1 but on other PTs at the same time, leads to a significant control of malignant growth and tumor angiogenesis, also showing an important role in overcoming MDR.[104,105] Marches et al.,[106] showed the intimate link between cancer biochemistry, molecular biology and cancer-related immunity by demonstrating that the anti-IgM-mediated induction of cell death in human B lymphoma cells is dependent on NHE1 inhibition and subsequent intracellular acidification. This publication combined three different fields of oncology research: biochemistry, molecular biology and cancer immunity under one broader embracing unit. The concerted utilization of proton transport inhibitors as a primary cancer treatment, as well as an adjuvant measure in overcoming MDR, has been advanced in several publications approached from different perspectives, ranging from basic physics to bedside therapeutics.[107–109] A successful, etiological and radical treatment of cancer would thus be mediated by the activation of an acidic pHi-apoptotic chain reaction cascade ending in cancer cell death. Finally, preliminary trials in a clinical setting using different proton export inhibitors, either on their own or associated with chemotherapy, have been recently published in basic studies, animal models[110,111] and in cancer patients.[112–114] However, some studies have also shown a negative effect of the combination of certain chemotherapy protocols with proton pump inhibitors.[115]

Reversing the altered pH gradient: Rational association of inhibitors

Malignant cells have shown an incredible versatility in resisting all kinds of treatments. Many practicing oncologist will have experienced this problem frequently. Sooner or later, advanced tumors become impossible to tame. There is no reason to believe that this is not going to happen with pH oriented treatments. We may inhibit NHE1, but proton extrusion may go on with the activity of other extruders such as NBC and MCTs. This concept leads us to the core of the pH-centered treatment: more than one type of proton extruder needs to be inhibited. Most of the available drugs partially inhibit the channel, transporter or enzyme on which they act. This means that a complete block of proton extrusion is not possible with low toxicity. Even in the theoretical case that it would be possible to fully inhibit proton extrusion, this would result in high toxicity for normal cells, which also use these systems. After all, NHE1, carbonic anhydrases, NBC and VGSCs are housekeeping proteins and necessary for the normal functioning of non malignant cells. Therefore, we have to accept the partial reduction of the proton export system as a valid therapeutic goal without a complete shut down. The partial inhibition of many of the proton exporters is a more realistic therapeutic objective than the full inhibition of one or more exporters. To achieve this partial inhibition of many exporters it is necessary to associate many different drugs addressing to each of them. The rational treatment of the inverted pH gradient is not possible by targeting only one or perhaps two of the responsible proteins. It requires targeting the whole system as suggested in Fig. 5, and even other hallmarks of cancer like angiogenesis. The reduction of hypoxia would also be an important contributor to the scheme presented here, because the transporters, channels and enzymes whose overexpression forms part of the pH gradient inversion are HIFα-dependent.

Fig. 5 shows a theoretical and ideal model of how the inhibition of different proton exporters would work through drug association.

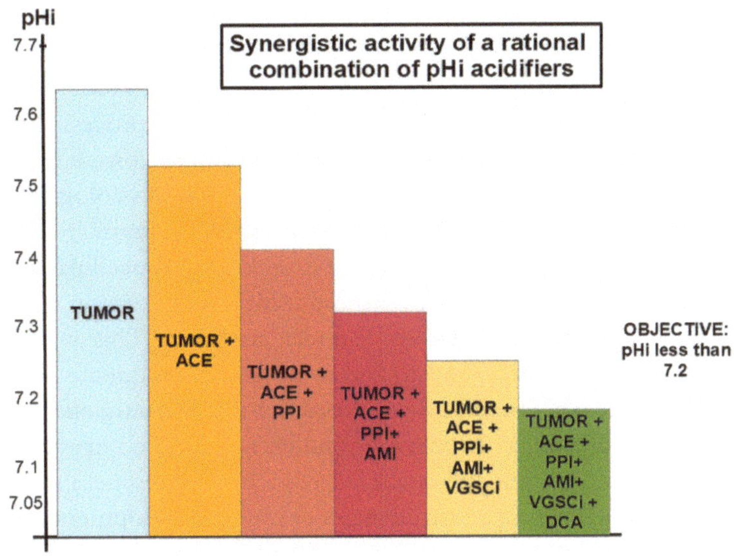

Fig. 5 An idealized view of how the association of drugs may modify the tumor's intracellular pH. Even if any one drug cannot achieve a real inversion of the pH gradient, a rational association of several can. *ACE*, acetazolamide; *PPI*, proton pump inhibitors; *AMI*, amiloride; *VGSCi*, voltage gated channel inhibitors; *DCA*, dichloroacetate. The minimal objective is to lower pHi to <7.2.

Each of the cell acidifiers shown in Fig. 5 acts on a different proton extruder. Acetazolamide is a non-specific carbonic anhydrase inhibitor. Proton pump inhibitors like omeprazole, pantoprazole, lansoprazole, esomeprazole block proton extrusion through V-ATPase proton pump inhibition. Amiloride is a weak NHE1 inhibitor. Voltage gated sodium channels are inhibited by anti-convulsants like phenytoin, carbamacepine, flunarizine, topiramate, etc. Dichloroacetate inhibits the glycolytic pathway reducing the production of lactic acid. This last drug goes somewhat against the logical expectations, because the reduction of lactic acid production should increase intracellular alkalinity. However, DCA increases intracellular acidity (see next chapters).

Conclusions

1. Historically, classical medicine recognized systemic acid-base balance as the fundamental parameter for defining the general concept of homeostasis. More than a century later this concept can now be extended to the cellular level where understanding the abnormal pH dynamics occurring in cancer has become an entire field on its own.
2. The inverted pH gradient of cancer explained in this book allows a greater understanding of the many faces and stages of malignant disease, from cell transformation to local growth, dissemination and the metastatic process, as a single, integrated process driven by acid-base abnormalities.

3. Regarding cancer etiopathogenesis and treatment, the primary therapeutic objective of the inverted pH gradient approach ("the pH-centered paradigm") is to manipulate the dysregulated pH abnormalities of cancer cells and tissues in order to reverse tumor growth, control local invasion and deactivate the metastatic process. These abnormal cellular pH-dynamics of cancer and its consequences can be considered a chronic failure in maintaining cellular acid-base homeostasis within physiological parameters.

4. The main therapeutic aim of the pH-centric paradigm in translational and bedside oncology is the induction of a pro-apoptotic selective intracellular acidification of cancer cells using different membrane-bound inhibitors of proton transport, as well as other non-proton transport derived cellular acidifiers. This would reverse the inverted pH gradient of malignant cells and tissues and consequently increase tumoral interstitial pH. At the same time, this would inhibit the Warburg effect, act as an anti-glycolytic measure, alkalinize the intratumoral extracellular compartment, thus controlling cancer growth and the metastatic process. This represents a rational approach to cancer therapeutics encompassing all stages of cancer development at the same time and it has the potential of being exploited in the treatment of many, if not all, malignant solid tumors and leukemias.

5. Since cancer cells use different proton pumps and transporters to protect themselves from an intracellular accumulation of hydrogen ions, it is probable that an integrated and concerted utilization of known proton extrusion inhibitors in pharmacological dosages will be necessary in the treatment of human malignancies. This is due to the fact that there are multiple pH regulators at the membrane of cancer cells that can be co-expressed at the same time, even in the most hostile conditions, and also that the inhibition of one transporter will be compensated by the over-activation of others.

6. One of the main aims of this contribution is to promote the new pH-centered approach in the therapeutic utilization of proton transport inhibitors and other non proton transport-derived intracellular acidifiers as well as drugs affecting ion channel physiopathology in human cancer.

References

1. Harguindey S, Cragoe Jr EJ. The Na+/H+ antiporter in oncology in the light of the spontaneous regression of cancer and cell metabolism. *Med Hypotheses* 1992;**39**:229–37.
2. McCarty MF, Whitaker J. Manipulating tumor acidification as a cancer treatment strategy. *Altern Med Rev* 2010;**15**:264–72.
3. Gerweck LE, Seetharaman K. Cellular pH gradient in tumor versus normal tissue: potential exploitation for the treatment of cancer. *Cancer Res* 1996;**56**:1194–8.
4. Robey IF, Baggett BK, Kirkpatrick ND, Roe DJ, Dosescu J, Sloane BF, et al. Bicarbonate increases tumor pH and inhibits spontaneous metastases. *Cancer Res* 2009;**69**(6):2260–8.
5. Robey IF, Martin NK. Bicarbonate and dichloroacetate: evaluating pH altering therapies in a mouse model for metastatic breast cancer. *BMC Cancer* 2011;**11**(1):235.

6. Silva AS, Yunes JA, Gillies RJ, Gatenby RA. The potential role of systemic buffers in reducing intratumoral extracellular pH and acid-mediated invasion. *Cancer Res* 2009;**69**(6):2677–84.
7. Ibrahim Hashim A, Cornnell HH, Coelho Ribeiro Mde L, Abrahams D, Cunningham J, Lloyd M, et al. Reduction of metastasis using a non-volatile buffer. *Clin Exp Metastasis* 2011;**28**:841–9.
8. Gullino PM, Grantham FH, Courtney AH. Utilization of oxygen by transplanted tumors in vivo. *Cancer Res* 1967;**27**(6 Pt. 1):1020–30.
9. Gullino PM, Grantham FH, Smith SH, Haggerty AC. Modifications of the acid-base status of the internal milieu of tumors. *J Natl Cancer Inst* 1965;**34**(6):857–69.
10. Xiao-Fang C, Zheng C-L, Akiyama S-I, Tomoda A. 2-Aminophenoxazine-3-one and 2-amino-4, 4α-dihydro-4α, 7-dimethyl-3H-phenoxazine-3-one cause cellular apoptosis by reducing higher intracellular pH in cancer cells. *Proc Jpn Acad Ser B Phys Biol Sci* 2011;**87**:199.
11. Reshkin SJ, Bellizzi A, Caldeira S, Albarani V, Malanchi I, Poignee M, et al. Na+/H+ exchanger-dependent intracellular alkalinization is an early event in malignant transformation and plays an essential role in the development of subsequent transformation-associated phenotypes. *FASEB J* 2000;**14**:2185–97.
12. Reshkin SJ, Greco MR, Cardone RA. Role of pHi, and proton transporters in oncogene-driven neoplastic transformation. *Philos Trans R Soc Lond Ser B Biol Sci* 2014;**369**:20130100.
13. Mahoney BP, Raghunand N, Baggett B, Gillies RJ. Tumor acidity, ion trapping and chemotherapeutics: I. Acid pH affects the distribution of chemotherapeutic agents in vitro. *Biochem Pharmacol* 2003;**66**:1207–18.
14. Pope BJ, Zierler-Gould KM, Kühne R, Weeds AG, Ball LJ. Solution structure of human cofilin: actin binding, pH sensitivity, and relationship to actin-depolymerizing factor. *J Biol Chem* 2004;**279**:4840–8.
15. Huber V, Camisaschi C, Berza A, Ferro S, Lugini L, Triulzi T. Cancer acidity: an ultimate frontier of tumor immune escape and a novel target of immunomodulation. *Semin Cancer Biol* 2017;**43**:74–89.
16. Webb BA, Chimenti M, Jacobson MP, Barber DL. Dysregulated pH: a perfect storm for cancer progression. *Nat Rev Cancer* 2011;**11**(9):671.
17. Stock C, Schwab A. Protons make tumor cells move like clockwork. *Pflugers Arch* 2009;**458**:981–92.
18. Reynolds TY, Rockwell S, Glazer PM. Genetic instability induced by the tumor microenvironment. *Cancer Res* 1996;**56**:5754–7.
19. Moore B, Wilson FP. A clinical method of haemalkimetry, with application to determination of the reactivity of the inorganic salts of the serum in malignant disease and other conditions. *Biochem J* 1906; **1**(6–7):297–327.
20. Harguindey S. Hydrogen ion dynamics and cancer: an appraisal. *Med Pediatr Oncol* 1982;**10**(3):217–36.
21. Harguindey S, Arranz JL, Polo Orozco JD, Rauch C, Fais S, Cardone RA, et al. Cariporide and other new and powerful NHE1 inhibitors as potentially selective anticancer drugs—an integral molecular/biochemical/metabolic/clinical approach after one hundred years of cancer research. *J Transl Med* 2013;**11**:282.
22. Harguindey S, Rehkin SJ. The new pH-centric anticancer paradigm in oncology and medicine. *Semin Cancer Biol* 2017;**3**:1–180.
23. Huber V, De Milito A, Harguindey S, Reshkin SJ, Wahl ML, Rauch C, et al. Proton dynamics in cancer. *J Transl Med* 2010;**8**:57.
24. Neri D, Supuran CT. Interfering with pH regulation in tumours as a therapeutic strategy. *Nat Rev Drug Discov* 2011;**10**:767–77.
25. Baldini N, Sonveaux P, De Milito A, Supuran CT, Otto AM, Stock CM, et al. Metabolism and microenvironment in cancer plasticity. Meeting report of the 2nd meeting of the International Society of Cancer Metabolism. *Cancer Metab* 2016;**4**:1.
26. Spugnini EP, Sonveaux P, Stock C, Perez-Sayans M, De Milito A, Avnet S, et al. Proton channels and exchangers in cancer. *Biochim Biophys Acta* 2015;**1848**:2715–26.
27. Harguindey S. Integrating fields of cancer research through pivotal mechanisms and synthetic final pathways: a unifying and creative overview. *Med Hypotheses* 2002;**58**:444–52.
28. Swietach P, Vaughan-Jones RD, Harris AL, Hulikova A. The chemistry, physiology and pathology of pH in cancer. *Philos Trans R Soc Lond Ser B Biol Sci* 2014;**369**:20130099.
29. Harguindey S. The unitary strategy of cancer cells: the hydrogen ion as a multidimensional unitary factor and carbohydrate metabolism (second of three parts). *Ann Med Int* 1994;**11**(6):299–303.

30. Harguindey S, Orive G, Luis Pedraz J, Paradiso A, Reshkin SJ. The role of pH dynamics and the Na+/H+ antiporter in the etiopathogenesis and treatment of cancer. Two faces of the same coin–one single nature. *Biochim Biophys Acta* 2005;**1756**:1–24.
31. Reshkin SJ, Cardone RA, Harguindey S. Na+-H+ exchanger, pH regulation and cancer. *Recent Pat Anticancer Drug Discov* 2013;**8**:85–99.
32. Hanahan D, Weinberg RA. Hallmarks of cancer: the next generation. *Cell* 2011;**144**(5):646–74.
33. Warburg O. On respiratory impairment in cancer cells. *Science* 1956;**124**:269–70.
34. Liberti MV, Locasale JW. The Warburg effect: how does it benefit cancer cells? *Trends Biochem Sci* 2016;**41**(3):211–8.
35. Doppler W, Jaggi R, Groner B. Induction of v-mos and activated Ha-ras oncogene expression in quiescent NIH 3T3 cells causes intracellular alkalinisation and cell-cycle progression. *Gene* 1987;**54**:147–53.
36. Hagag N, Lacal JC, Graber M, Aaronson S, Viola MV. Microinjection of ras p21 induces a rapid rise in intracellular pH. *Mol Cell Biol* 1987;**7**:1984–8.
37. Goldfeder A. Über die in bösartigen Geschwülsten vorkommendenpH-Werte. *Z Krebsforsch* 1929;**29**(1–2):134–46.
38. Halperin ML, Connors HP, Relman AS, Karnovsky ML. Factors that control the effect of pH on glycolysis in leukocytes. *J Biol Chem* 1969;**244**(2):364–90.
39. Burr MJ. The relationship between pH and aerobic glycolysis in human and canine erythrocytes. *Comp Biochem Physiol B* 1972;**4**:687–94.
40. Relman AS. Metabolic consequences of acid-base disorders. *Kidney Int* 1972;**1**:347–59.
41. Pouysségur J, Franchi A, L'Allemain G, Paris S. Cytoplasmic pH, a key determinant of growth factor induced DNA synthesis in quiescent fibroblasts. *FEBS Lett* 1985;**190**:115–9.
42. Harguindey S, Arranz JL, Wahl ML, Orive G, Reshkin SJ. Proton transport inhibitors as potentially selective anticancer drugs. *Anticancer Res* 2009;**29**(6):2127–36.
43. Chiche J, Ilc K, Laferrière J, Trottier E, Dayan F, Mazure NM, et al. Hypoxia-inducible carbonic anhydrase IX and XII promote tumor cell growth by counteracting acidosis through the regulation of the intracellular pH. *Cancer Res* 2009;**69**(1):358–68.
44. Lagadic-Gossmann D, Huc L, Lecureur V. Alterations of intracellular pH homeostasis in apoptosis: origins and roles. *Cell Death Differ* 2004;**11**(9):953.
45. Rich IN, Worthington-White D, Garden OA, Musk P. Apoptosis ofleukemic cells accompanies reduction in intracellular pH after targeted inhibition of the Na+/H+ exchanger. *Blood* 2000;**95**(4):1427–34.
46. Porporato PE, Dhup S, Dadhich RK, Copetti T, Sonveaux P. Anticancer targets in the glycolytic metabolism of tumors: a comprehensive review. *Front Pharmacol* 2011;**2**:49.
47. Calderon-Montano JM, Burgos-Moron E, Perez-Guerrero C, Salvador J, Robles A, Lopez-Lazaro M. Role of the intracellular pH in the metabolic switch between oxidative phosphorylation and aerobic glycolysis-relevance to cancer. *WebmedCentral* 2011;**2**.
48. Quach CHT, Jung K-H, Lee JH, Park JW, Moon SH, Cho YS, et al. Mild alkalization acutely triggers the Warburg effect by enhancing hexokinase activity via voltage-dependent anion channel binding. *PLoS One* 2016;**11**.
49. Chen J, Merl D, Peterson CW, Wu J, Liu PY, Yin H, et al. Lactic acidosis triggers starvation response with paradoxical induction of TXNIP through MondoA. *PLoS Genet* 2010;**6**.
50. Alfarouk KO, Verduzco D, Rauch C, Muddathir AK, Adil HH, Elhassan GO, et al. Glycolysis, tumor metabolism, cancer growth and dissemination. A new pH-based etiopathogenic perspective and therapeutic approach to an old cancer question. *Oncoscience* 2014;**1**:777–802.
51. Payen VL, Porporato PE, Baselet B, Sonveaux P. Metabolic changes associated with tumor metastasis, part 1: tumor pH, glycolysis and the pentose phosphate pathway. *Cell Mol Life Sci* 2015;**73**:1–16.
52. Hardonnière K, Huc L, Sergent O, Holme JA, Lagadic-Gossman. Environmental carcinogenesis and pH homeostasis: not only a matter of dysregulated metabolism. *Semin Cancer Biol* 2017;**3**:9–65.
53. Cardone R, Casavola V, Reshkin SJ. The role of disturbed pH dynamics and the Na+/H+ exchanger in metastasis. *Nat Rev Cancer* 2005;**5**(19):786–95.
54. Harguindey S, Stanciu D, Devesa J, Alfarouk K, Cardone RA, Polo Orozco JD, et al. Cellular acidification as a new approach to cancer treatment and to the understanding and therapeutics of neurodegenerative diseases. *Semin Cancer Biol* 2017;**43**:157–79.

55. Gottlieb RA, Nordberg J, Skowronski E, Babior BM. Apoptosis induced in Jurkat cells by several agents is preceded by intracellular acidification. *Proc Natl Acad Sci* 1996;**93**(2):654–8.
56. Grillo-Hill BK, Choi C, Jimenez-Vidal M, Barber DL. Increased H(+) efflux is sufficient to induce dysplasia and necessary for viability with oncogene expression. *elife* 2015;**4**. https://doi.org/10.7554/eLife.03270.
57. Fais S. Evidence-based support for the use of proton pump inhibitors in cancer therapy. *J Transl Med* 2015;**13**:368.
58. Fais S. A nonmainstream approach against cancer. *J Enzyme Inhib Med Chem* 2016;**31**:1–8.
59. Sun J, Ilich AI, Kim CA, Chu MP, Wong GG, Ghosh S, et al. Concomitant administration of proton pump inhibitors and capecitabine is associated with increased recurrence risk in early stage colorectal cancer patients. *Clin Colorectal Cancer* 2016;**15**:257–63.
60. Yu M, Lee C, Wang M, Tannock IF. Influence of the proton pump inhibitor lansoprazole on distribution and activity of doxorubicin in solid tumors. *Cancer Sci* 2015;**106**:1438–47.
61. Heon JP. Effects of intracellular pH on apoptosis in HL-60 human leukemia cells. *Yonsei Med J* 1995;**36**:473–9.
62. Orlowski J, Grinstein S. Na+/H+ exchangers of mammalian cells. *J Biol Chem* 1997;**272**:22373–6.
63. Counillon L, Pouyssegur J. The expanding family of eucaryotic Na(+)/H(+) exchangers. *J Biol Chem* 2000;**275**:1–4.
64. Lacroix J, Poet M, Maehrel C, Counillon L. A mechanism for the activation of the Na/H exchanger NHE-1 by cytoplasmic acidification and mitogens. *EMBO Rep* 2004;**5**:91–6.
65. Paris S, Pouyssegur J. Growth factors activate the Na+/H+ antiporter in quiescent fibroblasts by increasing its affinity for intracellular H+. *J Biol Chem* 1984;**259**:10989–94.
66. Hardonniere K, Saunier E, Lemarie A, Fernier M, Gallais I, Helies-Toussaint C, et al. The environmental carcinogen benzo[a]pyrene induces a Warburg-like metabolic reprogramming dependent on NHE1 and associated with cell survival. *Sci Rep* 2016;**6**:30776.
67. Pedraz-Cuesta E, Fredsted J, Jensen HH, Bornebusch A, Nejsum LN, Kragelund BB, et al. Prolactin signaling stimulates invasion via Na(+)/H(+) exchanger NHE1 in T47D human breast Cancer cells. *Mol Endocrinol* 2016;**30**:693–708.
68. Aredia F, Scovassi AI. Multiple effects of intracellular pH modulation in cancer cells. *Cancer Cell Microenviron* 2014;**1**:1. https://doi.org/10.14800/ccm.14136.
69. Sanhueza C, Araos J, Naranjo L, Villalobos R, Westermeier F, Salomon C, et al. Modulation of intracellular pH in human ovarian cancer. *Curr Mol Med* 2016;**16**:23–32.
70. Schammim R, Amith SR, Fong S, Baksh S, Fliegel L. Na (+)/H (+)exchange in the tumour microenvironment: does NHE1 drive breast cancer carcinogenesis? *Int J Dev Biol* 2015;**59**:367–77.
71. Amith SR, Wilkinson JM, Fliegel L. Assessing Na+/H+ exchange and cell effector functionality in metastatic breast cancer. *Biochim Open* 2016;**2**:16–23.
72. Sanhueza C, Araos J, Naranjo L, Toledo F, Beltran AR, Ramirez MA, et al. Sodium/proton exchanger isoform 1 regulates intracellular pH and cell proliferation in human ovarian cancer. *Biochim Biophys Acta* 2016;**1863**:81–91.
73. Brisson L, Reshkin SJ, Gore J, Roger S. pH regulators in invadosomal functioning: proton delivery for matrix tasting. *Eur J Cell Biol* 2012;**91**:847–60.
74. Acevedo-Olvera LF, Diaz-Garcia H, Parra-Barrera A, Caceres-Perez AA, Gutierrez-Iglesias G, Rangel-Corona R, et al. Inhibition of the Na+/H+ antiporter induces cell death in TF-1 erythroleukemia cells stimulated by the stem cell factor. *Cytokine* 2015;**75**:142–50.
75. Sharma M, Astekar M, Soi S, Manjunatha BS, Shetty DC, Radhakrishnan R. pH gradient reversal: an emerging hallmark of cancers. *Recent Pat Anticancer Drug Discov* 2015;**10**:244–58.
76. Stock C, Ludwig FT, Schwab A. Is the multifunctional Na(+)/H(+) exchanger isoform 1 a potential therapeutic target in cancer? *Curr Med Chem* 2012;**19**:647–60.
77. Izumi H, Torigoe T, Ishiguchi H, Uramoto H, Yoshida Y, Tanabe M, et al. Cellular pH regulators: potentially promising molecular targets for cancer chemotherapy. *Cancer Treat Rev* 2003;**29**:541–9.
78. Provost JJ, Wallert MA. Inside out: targeting NHE1 as an intracellular and extracellular regulator of cancer progression. *Chem Biol Drug Des* 2013;**81**:85–101.
79. Yue Loo S, Ker Xing Chang M, Sui Huay Chua C, Prem Kumar A, Pervaiz S, Veronique Clement M. NHE-1: a promising target for novel anti-cancer therapeutics. *Curr Pharm Des* 2012;**18**:1372–82.

80. Mihaila RG. A minireview on NHE1 inhibitors: a rediscovered hope in oncohematology. *Biomed Pap Med Fac Univ Palacky Olomouc Czech Repub* 2015;**159**:519–26.
81. Barriere H, Poujeol C, Tauc M, Blasi JM, Counillon L, Poujeol P. CFTR modulates programmed cell death by decreasing intracellular pH in Chinese hamster lung fibroblasts. *Am J Phys Cell Phys* 2001;**281**: C810–24.
82. Di Sario A, Bendia E, Omenetti A, De Minicis S, Marzioni M, Kleemann HW, et al. Selective inhibition of ion transport mechanisms regulating intracellular pH reduces proliferation and induces apoptosis in cholangiocarcinoma cells. *Dig Liver Dis* 2007;**39**:60–9.
83. He B, Deng C, Zhang M, Zou D, Xu M. Reduction of intracellular pH inhibits the expression of VEGF in K562 cells after targeted inhibition of the Na+/H+ exchanger. *Leuk Res* 2007;**31**:507–14.
84. Gao W, Chang G, Wang J, Jin W, Wang L, Lin Y, et al. Inhibition of K562 leukemia angiogenesis and growth by selective Na+/H+ exchanger inhibitor cariporide through down-regulation of pro-angiogenesis factor VEGF. *Leuk Res* 2011;**35**:1506–11.
85. Reshkin SJ, Bellizzi A, Cardone RA, Tommasino M, Casavola V, Paradiso A. Paclitaxel induces apoptosis via protein kinase A- and p38 mitogen-activated protein-dependent inhibition of the Na+/H+ exchanger (NHE) NHE isoform 1 in human breast cancer cells. *Clin Cancer Res* 2003;**9**:2366–73.
86. Harley W, Floyd C, Dunn T, Zhang XD, Chen TY, Hegde M, et al. Dual inhibition of sodium-mediated proton and calcium efflux triggers non-apoptotic cell death in malignant gliomas. *Brain Res* 2010;**1363**:159–69.
87. Cardone RA, Greco MR, Zeeberg K, Zaccagnino A, Saccomano M, Bellizzi A, et al. A novel NHE1-centered signaling cassette drives epidermal growth factor receptor-dependent pancreatic tumor metastasis and is a target for combination therapy. *Neoplasia* 2015;**17**:155–66.
88. Atwal KS, O'Neil SV, Ahmad S, Doweyko L, Kirby M, Dorso CR, et al. Synthesis and biological activity of 5-aryl-4-(4-(5-methyl-1H-imidazol-4-yl)piperidin-1-yl)pyrimidine analogs as potent, highly selective, and orally bioavailable NHE-1 inhibitors. *Bioorg Med Chem Lett* 2006;**16**:4796–9.
89. Harguindey S, Koltai T, Reshkin SJ. Curing cancer? Further along the new pH-centric road and paradigm. *Oncoscience* 2018;**5**(5–6):132–3.
90. Nagata H, Che XF, Miyazawa K, Tomoda A, Konishi M, Ubukata H, et al. Rapid decrease of intracellular pH associated with inhibition of Na+/H+ exchanger precedes apoptotic events in the MNK45 and MNK74 gastric cancer cell lines treated with 2-aminophenoxazine-3-one. *Oncol Rep* 2011;**25**:341–6.
91. Che XF, Zheng CL, Akiyama S, Tomoda A. 2-Aminophenoxazine-3-one and 2-amino-4,4alpha-dihydro-4alpha,7-dimethyl-3H-phenoxazine-3-one cause cellular apoptosis by reducing higher intracellular pH in cancer cells. *Proc Jpn Acad Ser B Phys Biol Sci* 2011;**87**:199–213.
92. Tavares-Valente D, Baltazar F, Moreira R, Queiros O. Cancer cell bioenergetics and pH regulation influence breast cancer cell resistance to paclitaxel and doxorubicin. *J Bioenerg Biomembr* 2013;**45**:467–75.
93. Marchiq I, Pouyssegur J. Hypoxia, cancer metabolism and the therapeutic benefit of targeting lactate/H(+) symporters. *J Mol Med (Berl)* 2016;**94**:155–71.
94. Lastraioli E, Iorio J, Arcangeli A. Ion channel expression as promising cancer biomarker. *Biochim Biophys Acta* 2015;**1848**:2685–702.
95. Harguindey S, Pedraz JL, Canero RG, Katin M. Edelfosine, apoptosis, MDR and Na+/H+ exchanger: induction mechanisms and treatment implications. *Apoptosis* 2000;**5**:87–9.
96. Rich IN, Worthington-White D, Garden OA, Musk P. Apoptosis of leukemic cells accompanies reduction in intracellular pH after targeted inhibition of the Na(+)/H(+) exchanger. *Blood* 2000;**95**:1427–34.
97. Harguindey S, Orive G, Cacabelos R, Hevia EM, de Otazu RD, Arranz JL, et al. An integral approach to the etiopathogenesis of human neurodegenerative diseases (HNDDs) and cancer. Possible therapeutic consequences within the frame of the trophic factor withdrawal syndrome (TFWS). *Neuropsychiatr Dis Treat* 2008;**4**:1073–84.
98. Amith SR, Fliegel L. Regulation of the Na+/H+ exchanger (NHE1) in breast cancer metastasis. *Cancer Res* 2013;**73**:1259–64.
99. Torigoe T, Izumi H, Ise T, Murakami T, Uramoto H, Ishiguchi H, et al. Vacuolar H(+)-ATPase: functional mechanisms and potential as a target for cancer chemotherapy. *Anti-Cancer Drugs* 2002;**13**:237–43.

100. Pedersen SF, Stock C. Ion channels and transporters in cancer: pathophysiology, regulation, and clinical potential. *Cancer Res* 2013;**73**:1658–61.
101. Oosterwijk E, Gillies RJ. Targeting ion transport in cancer. *Philos Trans R Soc Lond Ser B Biol Sci* 2014;**369**:20130107.
102. Rowson-Hodel AR, Berg AL, Wald JH, Hatakeyama J, VanderVorst K, Curiel DA, et al. Hexamethylene amiloride engages a novel reactive oxygen species- and lysosome-dependent programmed necrotic mechanism to selectively target breast cancer cells. *Cancer Lett* 2016. https://doi.org/10.1016/j.canlet.2016.02.042.
103. Koltai T. Triple-edged therapy targeting intracellular alkalosis and extracellular acidosis in cancer. *Semin Cancer Biol* 2017;**43**:139–46.
104. Provost JJ, Rastedt D, Canine J, Ngyuen T, Haak A, Kutz C, et al. Urokinase plasminogen activator receptor induced non-small cell lung cancer invasion and metastasis requires NHE1 transporter expression and transport activity. *Cell Oncol* 2012;**35**:95–110.
105. Amith SR, Wilkinson JM, Baksh S, Fliegel L. The Na(+)/H(+) exchanger (NHE1) as a novel co-adjuvant target in paclitaxel therapy of triple-negative breast cancer cells. *Oncotarget* 2015;**6**:1262–75.
106. Marches R, Vitetta ES, Uhr JW. A role for intracellular pH in membrane IgM-mediated cell death of human B lymphomas. *Proc Natl Acad Sci U S A* 2001;**98**:3434–9.
107. Daniel C, Bell C, Burton C, Harguindey S, Reshkin SJ, Rauch C. The role of proton dynamics in the development and maintenance of multidrug resistance in cancer. *Biochim Biophys Acta* 2013;**1832**:606–17.
108. Rauch C, Blanchard A, Wood E, Dillon E, Miriam L, Wahl HS. Cell membranes, cytosolic pH and drug transport in cancer and MDR: physics, biochemistry and molecular biology. In: Meszaros A, Balogh G, editors. *Multiple drug resistance*. Nova Publishers; 2009. p. 1–24.
109. Harguindey S, Gonzalez Molinillo J, Chinchilla D, Reshkin S, Tomoda A. Further along a clinical protocol using a cocktail of PTIs in human cancer. In: ISPDC Abstract Book, 2nd ISPD Meeting, Nice, France, 18-19; 2011.
110. Walsh M, Fais S, Spugnini EP, Harguindey S, Abu Izneid T, Scacco L, et al. Proton pump inhibitors for the treatment of cancer in companion animals. *J Exp Clin Cancer Res* 2015;**34**:93.
111. Harguindey S, Orozco JP, Cuenca M, Fernández MC, Arranz JL. New and powerful NHE1 inhibitors as potential anticancer drugs in bedside oncology: a prospective program of preclinical studies in cats and dogs with spontaneous malignant tumors. In: Frontiers in pharmacology [Conference abstract: 4th annual meeting of the International Society of Proton Dynamics in Cancer]; 2014.
112. Papagerakis S, Bellile E, Peterson LA, Pliakas M, Balaskas K, Selman S, et al. Proton pump inhibitors and histamine 2 blockers are associated with improved overall survival in patients with head and neck squamous carcinoma. *Cancer Prev Res (Phila)* 2014;**7**:1258–69.
113. Wang BY, Zhang J, Wang JL, Sun S, Wang ZH, Wang LP, et al. Intermittent high dose proton pump inhibitor enhances the antitumor effects of chemotherapy in metastatic breast cancer. *J Exp Clin Cancer Res* 2015;**34**:85.
114. Perut F, Carta F, Bonuccelli G, Grisendi G, Di Pompo G, Avnet S, et al. Carbonic anhydrase IX inhibition is an effective strategy for osteosarcoma treatment. *Expert Opin Ther Targets* 2015;**19**:1–13.
115. Chu MP, Hecht JR, Slamon D, Wainberg ZA, Bang YJ, Hoff PM, et al. Association of proton pump inhibitors and capecitabine efficacy in advanced gastroesophageal cancer: secondary analysis of the TRIO-013/LOGiC randomized clinical trial. *JAMA Oncol* 2016;.

CHAPTER 14

Pharmacological interventions part II

NHE1 inhibitors

Introduction

The NHE antiporter system is a family of proteins whose promoter is HIF-1α and the Sp1 (specificity protein 1) transcription factor acts as an enhancer or co-promoter. The main therapeutic objective is to reverse the inverted pH gradient and the most important step in this process is to inhibit NHE1. Since NHE1 initiates the pH gradient inversion from the very beginning of malignant transformation, its inhibition should be a primary goal. However, before considering the inhibition of this antiporter, it is necessary to have some notion about the rest of the family of NHE antiporters and its relation with cancer. This is summarized in Table 1.

Table 1 Description of the NHE family of antiporters

Gene	Protein	Synonim	Possible role in cancer
SLC9A1	NHE1	Sodium/hydrogen exchanger 1	This protein plays a central role in regulating pH homeostasis, cell migration and cell volume. This protein is also involved in tumor growth
SLC9A2	NHE2	Sodium/hydrogen exchanger 2	Similar activities as NHE1?
SLC9A3R1	NHE3	Sodium/hydrogen exchanger 3 Regulator 1	NHE3 phosphorylation via PKCη marks the polarity and orientation of directionally migrating cells. The association of plasminogen with dipeptidyl peptidase IV and NHE3 regulates invasion of human 1-LN prostate tumor cells
SLC9A5	NHE5	Sodium/hydrogen exchanger 5	No relation found with cancer
SLC9A6	NHE6	Sodium/hydrogen exchanger 6	Hypoxia induced mobilization of NHE-6 to the plasma membrane triggers, endosome hyperacidification and chemoresistance

Continued

An Innovative Approach to Understanding and Treating Cancer: Targeting pH
https://doi.org/10.1016/B978-0-12-819059-3.00014-9
© 2020 Elsevier Inc.
All rights reserved.

Table 1 Description of the NHE family of antiporters—cont'd

Gene	Protein	Synonim	Possible role in cancer
SLC9A7	NHE7	Sodium/hydrogen exchanger 7	The encoded protein is primarily localized in the trans-Golgi network and is involved in maintaining pH homeostasis in organelles along the secretory and endocytic pathways. This protein can enhance cell growth of certain breast tumors
SLC9A8	NHE8	Sodium/hydrogen exchanger 8	Not related to cancer
SLC9A9	NHE9	Sodium/hydrogen exchanger 9	Involved in endosome recycling
SL9C1	NHE10	Sodium/hydrogen exchanger 10	Specific for spermatozoa
SLC10	–	NRAMP	Was associated with recurrence of bladder cancer. This has not been fully confirmed[1]

The initial development of NHE1-specific inhibitors took place in cardiology where the hyperactivation of NHE1 in ischemia-reperfusion episodes disrupted the intracellular ion balance, leading to cardiac dysfunction and damage. Therefore, the excessive activity of proton exchangers in the heart had to be decreased in order to reduce ischemic damage. Borchgreving and Jynge[2] found that amiloride provided some protection to the ischemic heart.

Amiloride

Since 1981, when it was approved by the FDA, the potassium-sparing diuretic amiloride has been used for the treatment of high blood pressure, congestive heart failure and cirrhosis related-edema/ascitis. Moolenaar et al. demonstrated that it is a NHE1 blocker, at the same time that NHE1 was discovered in 1981.[3,4] But besides its activity on high blood pressure, other effects brought attention to this drug: anti-tumor and anti-metastatic effects.[5,6] In 1983 Sparks et al.[7] described the effects of amiloride on cancer cells in vivo: "*The effects of amiloride, a reported inhibitor of serum-stimulated sodium influx, were tested on tumor growth, tumor cell proliferation, and intracellular element content of cancer cells in vivo. We have shown previously that cancer cells have high intranuclear levels of sodium compared to those of their normal counterpart cells and have postulated that such a high level of sodium may be involved in the cancer state. We now report that amiloride, when given in a series of injections, inhibited both H6 hepatoma and DMA/J mammary adenocarcinoma growth in vivo in a dose-dependent fashion and that 3 injections of amiloride at a dose of 1.0 μg/g body weight into mice bearing H6 hepatomas resulted in a significant decrease in the intranuclear content of sodium but not the content of magnesium, phosphorus, sulfur, chlorine, or potassium as measured by electron probe X-ray microanalysis in the H6 hepatoma cells. Amiloride at dosages as low as 1.0 μg/g body weight per injection also inhibited tumor cell proliferation as measured by the tritated thymidine autoradiography labeling index. Amiloride caused no changes in the mean profile diameters of metaphase or interphase H6 hepatoma or DMA/J mammary adenocarcinoma cells, suggesting that the action of amiloride on tumor growth was not due to cell volume changes. These data show that amiloride both inhibited tumor growth and decreased the proliferation of the tumor cells in the H6 hepatomas which was correlated with a decreased intranuclear sodium content* (Fig. 1)."

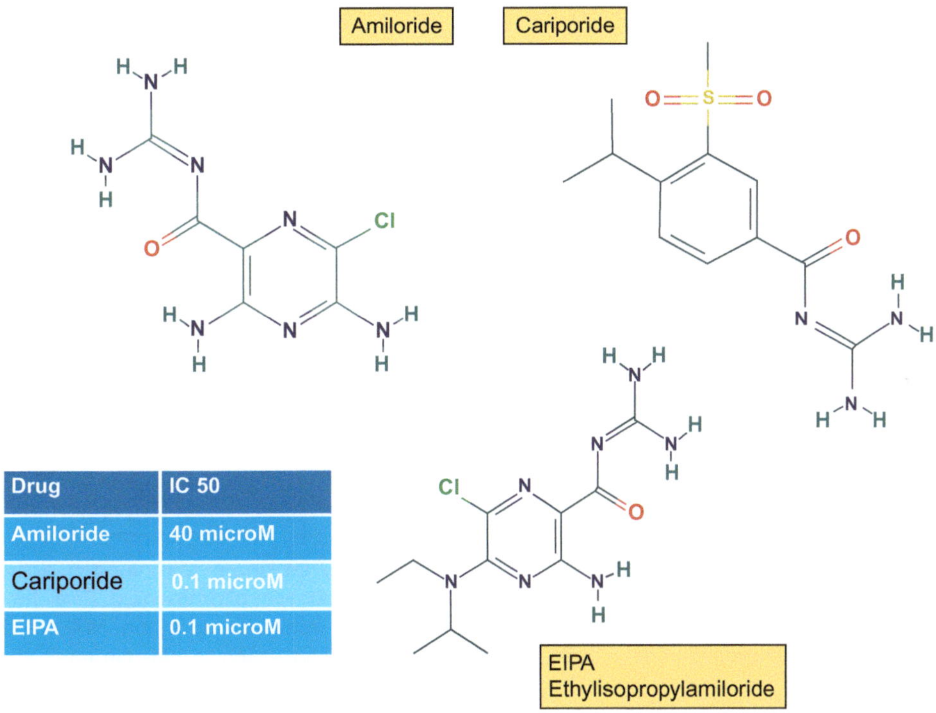

Drug	IC 50
Amiloride	40 microM
Cariporide	0.1 microM
EIPA	0.1 microM

Fig. 1 Amiloride and its more recent derivatives: cariporide and EIPA. Their chemical structure and relative potency are shown. Amiloride is the weakest and less specific inhibitor, while cariporide and ethylisopropylamiloride (EIPA) are far more potent and specific NHE1 inhibitors.[8]

The anti-tumor actions of amiloride are mainly based on two effects:

1. NHE inhibition in vitro and in vivo.[9]
2. uPA (urinary type plasminogen activator) inhibition in vitro and in vivo.[10]

3. The are other actions that affect tumor development but are less evident, such as inhibition of angiogenesis and inhibition of ornithine decarboxylase, an enzyme necessary for DNA synthesis.

Effects 1 and 2 are interconnected, because NHE1 creates extracellular acidity, which activates cathepsins, MMPs and other proteolytic enzymes that are necessary for plasminogen activation. At the same time, plasminogen activates NHE1 establishing a positive feedback. Amiloride blocks this feedback mechanism through inhibition of NHE and impedes plasminogen activation (Fig. 2).

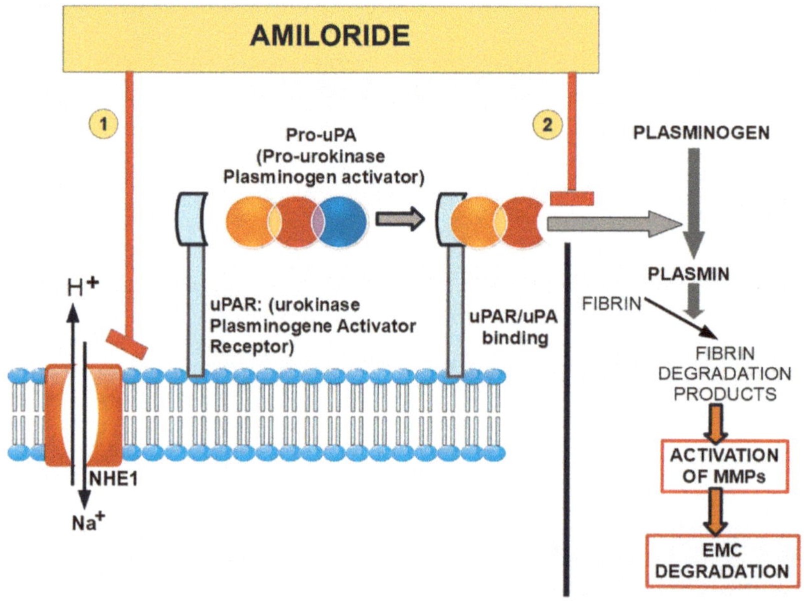

Fig. 2 Mechanisms of anti-tumor effects of amiloride (uPA, urokinase plasminogen activator; uPAR, urokinase plasminogen receptor; u plasminogen, urokinase plasminogen). The binding of uPA to its receptor permits plasminogen activation. (1) Amiloride blocking NHE1 activity. (2) Amiloride blocking plasminogen activation. Plasminogen's conversion to plasmin intervenes in the activation of metalloproteases (MMPs) and extracellular matrix (EMC) degradation. *(This figure is based on Refs. 11–20).*

Several other drugs have been developed and investigated in more recent years in terms of their increased selectivity and potency towards NHE1 inhibition. Cariporide is perhaps the most important of all (Table 2).[21]

Table 2 Evidence of the anti-cancer effects of amiloride

Reference	Findings
Koo et al. 1992[22]	Amiloride inhibited the in vitro growth of human colon cancer cells
Tatsuta et al. 1993[23]	Amiloride prolonged administration of 5.0 mg kg^{-1}, significantly decreased the incidence of gastric cancer in mice that had been transplanted with gastric cancer cells
Wang et al. 1995[24]	Amiloride modulated urokinase gene expression at both transcription and post-transcription levels in human colon cancer cells
Pannocchia et al. 1996[25]	Multidrug resistance (MDR) was reversed by EIPA in a colon cancer cell line (LoVo) with resistance to doxorubicin
Evans et al. 1998[26]	Amiloride suppressed metastasis in a rat mammary cancer model, the effect being dose- and time-dependent due to inhibition of uPA
Evans et al., 2000[27]	Amiloride significantly suppressed the invasive capacity of unaltered human breast cancer cells in Matrigel invasion chambers. Cell proliferation was also suppressed but to a lesser degree
Oka et al. 2003[28]	Amiloride inhibited urothelial cancer invasion by inhibiting uPA
Hedge et al. 2004[29]	At high concentrations (500 μM) amiloride impaired calcium efflux leading to elevations of intracellular calcium, and thus initiating a morphologically necrotic, predominantly caspase-independent, cell death process in glioma
Kim et al. 2005[30]	Amiloride enhanced TRAIL induced apoptosis in prostate cancer cells, promoted the dephosphorylation of Akt, PI3K and PDK-1. In vitro assays showed that amiloride inhibited phosphorylation of kinases and phosphatases by competing with ATP
Chang et al. 2011[31]	Amiloride modulated alternative splicing of oncogenic molecules, decreasing the pro-oncogenic variants
Yang et al. 2011[32]	Amiloride decreased invasiveness and proliferation in a human hepatoma cell line through NHE1 inhibition
Chang et al. 2011[33]	Amiloride resensitized chronic myeloid leukemic cells with Bcr-Abl mutation to imatinib treatment through modulation of alternative splicing of Bcl-x, HIPK3 and BCR/ABL, APAF-1, CRK and survivin
Guan et al. 2014[34]	NHE1 was highly expressed in esophageal adenocarcinoma tissues (66 of 101 cases), but not in normal esophageal squamous cell epithelium (1 of 26 cases). Knockdown of NHE1 expression using NHE1 shRNA or amiloride inhibition suppressed viability and induced apoptosis in esophageal cancer cells
Zheng et al. 2015[35]	Amiloride sensitized pancreatic carcinoma cells to erlotinib in vitro. The mechanism seemed to be inhibition of the PI3K/AKT pathway
Tang et al. 2015[36]	Amiloride significantly inhibited cell proliferation in irradiated glioblastoma cells, so that it modulated cell radiosensitivity involving the Akt phosphorylation and the alternative splicing of APAF1
Rojas et al. 2017[37]	Amiloride-induced apoptosis was observed in a broad panel of multiple myeloma cell lines and in a xenograft mouse model. Amiloride also had a synergistic effect when combined with dexamethasone, melphalan, lenalidomide, and pomalidomide

Continued

Table 2 Evidence of the anti-cancer effects of amiloride—cont'd

Reference	Findings
Zhang et al. 2018[38]	Amiloride inhibited proliferation and migration of lung cancer stem cells and promoted apoptosis. It also decreased the expression of the proteins of the PI3K/Akt/mTOR pathway
Poon et al. 2018[39]	Amiloride and doxorubicine had synergistic effects on canine osteosarcoma cells

Conclusions: Amiloride has a dual anti-tumoral activity: inhibition of NHE1 and uPA. Through these mechanisms it inhibits proliferation, migration, invasion, and metastasis.
After analyzing the evidence shown in this table, three other anti-cancer mechanisms of amiloride become clear: (1) alternative splicing of genes involved in the cancer process; (2) inhibition of phosphorylation of kinases of the PI3K pathway; and (3) reversal of MDR. Regarding this last item, the report by Pannocchia et al.[25] suggests the intimate association between MDR and membranes, and in particular with pH.
Of all the NHE inhibitors, amiloride is the only FDA approved drug and in clinical use as a diuretic, but not as an anticancer drug.

Cancer, amiloride and exosomes

Cancer cells communicate with other cancer cells and with non-cancer cells promoting malignancy. One method they use for intercellular communication are the exosomes, consisting of membrane vesicles that carry proteins and RNA to distant places delivering an oncogenic message[40] (some call it the postman of cancer). Exosomes are part of the endocytic pathway. They are vesicles that form within the cell and are released when trafficked to the plasma membrane.[41] The activity of NHE1 is essential for the release of exosomes. By inhibiting NHE1, dimethyl amiloride (DMA) also decreases the number of exosomes.[42,43] Exosomes also play an important role in metastatic niche pre-conditioning, and DMA has shown inhibiting effects in this case as well.[44] Exosome formation and trafficking are calcium dependent. However calcium channel blockers like verapamil are unable to block exosome formation and release. In this respect, inhibiting NHE1 can be considered a "kill the messenger" tool.

Drawbacks of amiloride

- Cobo et al.[45] found that amiloride can retard the apoptotic process by decreasing endonuclease mediated degradation of nuclear chromatin after using conventional chemotherapy or radiotherapy.
- Pilat et al.[46] did not find any objective benefit from amiloride on the metastatic properties of a prostate cancer in the Dunning rat model.

Cariporide (HOE 642), eniporide and other derivatives

Like amiloride, cariporide was born as an investigational drug for ischemic heart disease in 1995.[47] Further research did not confirm effective protection of myocardium under ischaemia/reperfusion (GUARDIAN study[48]). In 2002, Wong et al.[49] in a study initially addressed to show the cytostatic activity of cell acidifying compounds, found that cariporide was a selective cytostatic agent and that it was not toxic to normal cells. In 2007 this was further confirmed in cholangiocarcinoma cells.[50] Stüwe et al.[51] found that migration had a direct relation with NHE1 activity and not with the pHe in melanoma cells, because when NHE1 was inhibited there was no migration no matter what changes occurred in pHe. But when NHE1 activity was restored migration was reinitiated, and at the same time a pHe gradient was established with proton concentration that was higher at the leading migrating edge and lower at the rear end. This study clearly showed the importance of NHE1 in cancer progression and the ability of cariporide to thwart the process. Later on, there were further confirmations of the anti-migratory and anti-proliferative effects of cariporide in tongue squamous cell carcinoma,[52] HeLa cells,[53] leukemia cells,[54] mesothelioma cells,[55] pancreatic ductal adenocarcinomas cells,[56] and head and neck squamous cell carcinoma.[57] So far, cariporide and eniporide have been the only NHE1 inhibitory compounds that have undergone clinical trials. Unfortunately, those trials were not in the field of cancer but in the cardiological setting and for ischemic-reperfusion injury. Cariporide never entered medical practice after the poor results obtained in the treatment of ischemic heart disease, and it also showed a slightly higher risk for cerebrovascular accidents. Many researchers think that cariporide deserves further studies in oncology because:

- It is a powerful NHE1 inhibitor.
- It decreases migration, invasion and metastasis.
- It inhibits angiogenesis and the growth of leukemia cells by inhibiting NHE
- It has the ability to down-regulate Akt.
- It is useful in overcoming not only metastatic progression but also multiple drug resistance (MDR)
- Cariporide modifies the cancer pH paradigm.
- Its pharmacological dosages are well known.
- Adverse events are proportionally lower than with any other chemotherapeutic drug.
- Human tolerance is good.
- Cariporide has been tested with 11,000 patients in the GUARDIAN protocol and its pharmacodynamics are also well known.

There are many cariporide derivatives[58] such as eniporide and others. They have also been tested in the cardiological setting but never in the treatment of cancer. None of them have been approved for human use.

Compound 9t

The 5-aryl-4-(4-(5-methyl-1H-imidazol-4-yl)piperididn-1-yl) pyrimidine analog (Compound 9t) is an experimental drug that showed the strongest and more selective NHE inhibition so far identified with any anti-NHE1 drug, besides showing scarce toxicity to experimental animals.[59] These authors also reported that Compound 9t is

500-fold more potent against NHE1 than cariporide. It was never tested for cancer and its development has been abandoned for unknown reasons.

Troglitazone and other PPARγ agonists

Troglitazone is a PPARγ agonist that decreases the activity of NHE1 by mechanisms independent from PPARγ agonism[60] in the case of rosiglitazone.[61] Peroxisome proliferator response elements were identified in the promoter region of the NHE1 gene. Those breast cancer cells that expressed high levels of PPARγ when exposed to an agonist, decreased NHE1 transcription.

However, troglitazone was removed from the market because of possible hepatic toxicity. At present we are unable to know if rosiglitazone can replace troglitazone as an NHE1 inhibitor, but in general we may assume that PPARγ agonists could decrease NHE1 expression. This needs further experimental confirmation.

We can conclude that the manipulation of intracellular pH by NHE1 inhibitors is an effective mode of controlling many features of the malignant phenotype.[62]

Voltage gated sodium channel (VGSC) inhibitors

Some ion channels are significantly involved in both the regulation of pHi/pHe in cancer cells and in the acquisition of its proliferative and pro-invasive abilities. Ion channels regulate several cell processes, such as proliferation, resistance to apoptosis, intercellular adhesion, motility and extracellular matrix invasion. They also participate in tumor progression through non-excitable functions.[63–66] Consequently, an altered physiology of ion channels has been proposed as a new hallmark of cancer and as a potential target for selective therapeutics.[67] Some authors consider cancer a "channelopathy."

Voltage-gated sodium channels (Na_V), carrying inward charges upon gating, are known to be responsible for membrane depolarization and critical for "excitable" cells such as neurons, skeletal and cardiac muscle cells, as well as for initiating and propagating action potentials.[68] They were initially identified as being characteristic of excitable cells. However, they are also found to be unexpectedly expressed in cancer cells, especially in those of epithelial origins (prostate, breast, lung, colon, cervix and ovary), while they are not expressed in the same non-cancerous tissues. This has led to considering them among the new hallmarks of cancer. Their function is associated with cancer progression,[69,70] metastases and patient death.[71] In highly aggressive human breast cancer cells, the activity of the $Na_V1.5$ isoform was not associated with the triggering of action potentials but it enhanced extracellular matrix (ECM) degradation and cancer cell invasiveness by increasing the activity of the extracellular cysteine cathepsins B and S,[72] which have maximal activity in acidic conditions. While the mechanical details by which Na_V promotes cancer cell invasiveness are not fully characterized, it has been demonstrated that invasiveness of malignant cells takes place with persistent $Na_V1.5$ activity. This occurs through the interaction of $Na_V1.5$ with NHE1, which by extruding H^+ and so acidifying the pericellular microenvironment then activates cathepsins.[73] Indeed, $Na_V1.5$ activity is responsible for the allosteric modulation of NHE1, rendering it more active at pHi values between 6.4 and 7. It was suggested that the sodium current was responsible for this regulation. Furthermore, NHE1 and $Na_V1.5$ proteins can be co-immunoprecipitated and may physically interact in invadopodia.[74] In summary, the activity of Na_V channels in cancer cells has been identified as an important factor favoring cancer migration through a persistent sodium current that would tend to depolarize the membrane potential of malignant cells,[75] while favoring the reversed proton gradient in cancer cells (Fig. 3).

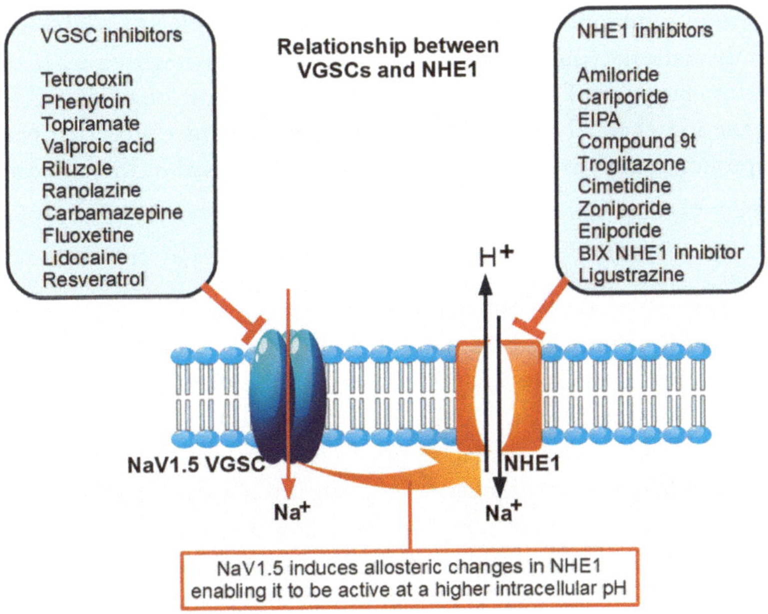

Fig. 3 Relationship between NHE1 and VGSCs.

From a therapeutic standpoint, the use of VGSC-inhibiting drugs has been proposed in order to reduce cancer progression, and increase survival in patients with cancer.[76] Indeed, inhibitors of Na_V that are clinically used for the treatment of other pathologies, such as ranolazine (treatment of chronic angina) or phenytoin (an anticonvulsant), are powerful pharmacological tools that have been shown to prevent metastatic colonization of organs in immunodepressed mice. These studies indicate that Na_V channels inhibitors, already approved for other clinical uses such as anti-arrhythmics, anticonvulsants and anesthetics, as well as other compounds, can be repurposed for anti-invasive cancer treatment and even for the prevention of metastatic development. The list of these drugs follows below.

Tetrodotoxin (TTX)

TTX is the model molecule for the study of VGSC inhibitors and their relative potential. It is also the first molecule discovered with such characteristics. Many biological toxins like those found in scorpions and sea anemones develop their toxicity by modifying VGSCs properties.[77] This toxicity can be achieved by inactivation of VGSC as produced by TTX or, on the contrary, by persistent activation of the channel (veratridine, acinitine and many others).

TTX is a powerful biological neurotoxin found in fishes of the Tetraodontiformes order and certain symbiotic bacteria. TTX irreversibly binds to VGSCs and blocks its activity, mainly in the nervous system. In the animal kingdom it is used as a toxin for defensive or predatory purposes. TTX binds to the extracellular portion of VGSC, disabling the function of the ion channel thus resulting in very poisonous effects that produce death through respiratory paralysis.[78] Clearly, TTX cannot be used in the clinical setting due to its high toxicity.

Phenytoin (diphenylhydantoin)

Phenytoin (PHEN), synthesized by the German chemist Heinrich Blitz in 1908, was introduced into clinical practice in 1937 when its antiepileptic activity was discovered by Putnam and Houston Merrit.[79] PHEN was approved by the FDA in 1953 for the treatment of generalized and partial seizures. It is an anticonvulsant that has been identified as a sodium channel blocker.[80,81] It was initially thought that it might induce lymphoma, and other hematological malignancies in patients under chronic treatment.[82] Fortunately, this carcinogenic effect of PHEN was not confirmed by large epidemiological studies[83] and nowadays it is considered a safe and FDA approved drug. However PHEN diminishes cell-mediated immunity,[84] a detail that should be kept in mind in cancer treatment.

In 1990, Vernillo et al.[85] found that PHEN inhibited bone resorption in rat osteosarcoma cells through significant reduction of collagenase and gelatinase activities. However, Dyce et al.[86] did not find PHEN's gelatinase inhibitory activity in B16 melanoma cells in vitro. This may be evidence of a tissue specific activity, which has not been investigated any further. Dyce et al. did not find important anti-metastatic activity either, in a melanoma tail injection model in mice. But when the data of this publication is examined in detail, it seems that the anti-metastatic activity is not that low since the mean pulmonary colonies were 4.6 ± 3.1 with PHEN vs. 10.2 ± 9.9 without PHEN.

Yang et al.[87] found that NaV 1.5 was over-expressed in breast cancer cells (MDA-MB-231) with high metastatic potential. In these experiments, PHEN was able to reduce migration and invasion at clinically achievable concentrations in MDA-MB-231 cells and showed no effects on MCF-7 cells with low metastatic potential.

PHEN blocks Na^+ channels and shows high affinity for VGSCs in the inactivated state of the channel.[88] Compared with verapamil, lidocaine and carbamazepine, PHEN had intermediate potency between verapamil and lidocaine, with verapamil showing the strongest inhibition and carbamazepine the weakest. Abdul et al.[89] studied the effect of four anticonvulsants (PHEN, carbamazepine, valproate and ethosuxinide) on the secretion of PSA (prostate specific antigen) and interleukin-6 in human prostate cancer cell lines. PHEN and carbamazepine inhibited the secretion of both. Finally, Fadiel et al.[90] found that PHEN is a strong estrogen receptor α antagonist at clinically achievable concentrations and at the same time is a weak agonist. Evidence of PHEN's anti-cancer effects are described in Table 3.

Table 3 Evidence of PHEN's anticancer effects

Reference	Findings
Tittle et al. 1992[91]	PHEN decreased growth in six murine tumor cell lines of lymphoid origin
Lang et al. 1993[92]	PHEN, carbamazepine and lamotrigine reduced glutamate release in rat neuroblastoma cells
Kawamura et al. 1996[93]	PHEN increased vinblastine's cytotoxicity
Lobert et al. 1999[94]	PHEN had inhibitory effects on microtubule assembly and had additive effects when associated with vinblastine
Sikes et al. 2003[95]	New VGSC blockers based on PHEN, showed increased inhibition of prostate cancer growth
Anderson et al. 2003[96]	New VGSC blockers based on the PHEN binding site study were developed. These new VGSC blockers showed potent inhibition of prostate cancer cells growth in an androgen independent PC3 line
Lyu et al. 2008[97]	DPTH-N10 a phenytoin derivative drug, showed strong anti-angiogenic activity. It inhibited HUVEC proliferation and capillary-like tube formation
Lee et al. 2010[98]	DPTH-N10 also inhibits proliferation of the COLO 205 colon cancer cell line
Yang et al. 2012[75]	PHEN inhibited the migration and invasion of MDA-MB-231 cells, with no effect on MCF-7 cells
Brackenbury et al. 2015[99]	$Na_v1.5$ expression was retained on MDA-MB-231 cells in orthotopic xenografts in immune-deficient mice. Three weeks of PHEN treatment in mice significantly reduced tumor growth, the density of Ki67-positive cycling cells in the primary tumors, local invasion, the density of MMP9-positive cells as well as metastasis to the liver, lungs and spleen

Continued

Table 3 Evidence of PHEN's anticancer effects—cont'd

Reference	Findings
Nelson et al. 2015[100]	PHEN at clinically achievable concentrations (at the same doses used for the treatment of epilepsy) reduced breast cancer growth, invasion and metastasis in a xenograft model
Mohammed et al. 2016[101]	PHEN significantly decreased invasion of breast cancer cells partly due to a reduction in the phosphorylation of ERK1/2
Takada et al. 2016[102]	In a large database mining study, chronic use of PHEN was inversely associated with diagnoses of colorectal, lung, gastric, breast and hematological cancers

There is an undesired side effect of PHEN that may represent a drawback for its use in cancer: immune depression.[103–105] PHEN also seems to be an inhibitor of necroptosis,[106] the importance of this action in cancer treatment remains elusive. These two issues deserve further research. Finally, we must mention that PHEN interacts with many other pharmaceuticals, especially those employed in chemotherapy.[107] These reasons and the increased risk of bone fractures[108] led to the exclusion of PHEN from the protocol of pH inversion treatment that will be discussed in Chapter 18.

In summary, the main activities developed by PHEN in relation to cancer are:

VGSC blocking, microtubule polymerization blocking, immunosuppression, calcium channel blocking and enhancement of vinblastine cytotoxicity.

Carbamazepine (CAR)

Carbamazepine is an antiepileptic, sodium channel blocker, pro-autophagic agent and histone deacetylase inhibitor, that has been in use since 1962 for the treatment of seizures, neuropathic pain, schizophrenia, bipolar disorders and trigeminal neuralgia. It was discovered in 1953 by Walter Schindler and approved by FDA shortly after. Its major drawback, as in the case of PHEN, is that it interferes with the metabolism of other drugs.[109–113]

CAR has shown interesting anti-metastatic potential in the experimental setting.[114] It has also shown that it may help prevent prostate cancer.[115]

CAR induces Her2 protein degradation through the proteosome without modifying its production.[116] This activity seems related to histone deacetylase inhibition rather than to VGSC blocking.[117] Growth inhibition in ER-positive breast cancer cell lines seems secondary to a histone deacetylase inhibitory effect.[118]

In summary, the anti-cancer mechanisms of carbamazepine are:

(1) VGSC blocker as anti-metastatic.
(2) Histone deacetylase inhibition.
(3) Her2 degradation by proteasome.

Valproic acid (VAL)

VAL is an anticonvulsant drug that exerts multiple actions related to anti-cancer effects: calcium channel blocker, VGSC blocker, inhibitor of histone deacetylase, it also potentiates the inhibitory activity of GABA, decreases angiogenesis, interferes with MAP kinases pathways and the β catenin-Wnt pathway.[119] VAL is being tested in various clinical trials in leukemias and solid tumors.[120] Most of the anti-tumor activities of VAL seem to be related to the inhibition of histone deacetylase rather than to VGSC blocking.

Topiramate (TOP)

The FDA approved topiramate in 2012 as an anti-epileptic drug. TOP is a voltage gated sodium channel blocker and GABA enhancer. It has five pharmacological effects that are related to its anti-cancer activity:

(1) Non-specific inhibition of carbonic anhydrases (CA), mainly CAI, CAII[121] and CAXII.[122]
(2) Inhibition of VGSCs.
(3) Intracellular acidification due to points 1 and 2.
(4) Inhibition of glutamate receptors.[123] Glutamate receptors are also expressed and functional in some cancers, most notably melanoma and triple negative breast cancer cells.
(5) Inhibition of aquaporins (see Chapter 10 on aquaporins).[124]

The main interest in TOP lies on its ability to induce a rapid intracellular pH decrease,[125] an effect that is probably dependent on VGSCs and CA inhibition.[126] It is also a down-regulator of angiogenesis,[127] and has shown anti-cancer effects on Lewis lung carcinoma.[128]

An important issue to remember is that TOP may interact with metformin, increasing the risk of lactic acidosis.[129]

The five effects described above make TOP a subject of interest for oncological research and cancer therapeutics.

Flunarizine (FLU)

Flunarizine is a calcium channel blocker and a VGSC inhibitor[130–132] that increases oxygen delivery to hypoxic regions in tumors. This characteristic explains why FLU is a sensitizer to radiotherapy.[133] The increased oxygenation of tumors is probably achieved by two mechanisms:

(a) *Vasodilation*: in 1984 Kaelin et al. injected a single dose of the calcium blocker flunarizine in mice with breast adenocarcinomas. After the injection the blood flow to the tumor and metastatic sites increased markedly[134] increasing oxygen availability.[135]

(b) *Decreased blood viscosity*: FLU decreased the viscosity of erythrocytes in suspension under hypoxia, acidic conditions and high levels of lactate.[136]

In 1985 Sezzi et al. reported that the calcium channel blocker flunarizine inhibited the migration of melanoma cells and macrophage-like cancer cells.[137]

FLU is also a VGSC blocker, and as such, it inhibited the growth of different lines of prostate cancer cells with a high VGSC expression.[138]

These properties of FLU, namely, its ability to increase the oxygenation of tumors,[139] also reducing migration and growth, inhibiting VGSC, and enhancing chemotherapeutic activity,[140] transforms it into a drug of great interest in oncology (Table 4).

Table 4 Further evidence of FLU's anticancer activities

Reference	Findings
Sezzi et al. 1984[141]	FLU inhibited the growth rate and decreased the survival of melanoma B16 cells
Silvestrini et al. 1993[142]	FLU modulatedd doxorubicin resistance in colon adenocarcinoma cells
Fink-Puches et al. 1993[143]	FLU inhibited migration in melanoma cells
Castellino et al. 1995[144]	FLU increased the activity of melphalan against rhabdomyosarcoma cells
Gornati et al. 1997[145]	FLU increased the activity of melphalan against drug resistant ovarian cancer cells. It did not enhance cisplatin's activity
Conrad et al. 2010[146]	FLU induced caspase 10-dependent apoptosis in Jurkat leukemia cells. While FLU was cytotoxic for hematological malignancies, it was not so for breast and colon carcinoma cells. FLU also increased ROS in Jurkat cells
Astin et al. 2014[147]	FLU decreased lymphangiogenesis and lymphatic metastasis
Schmeel et al. 2015[148]	FLU showed significant apoptotic effect on myeloma and lymphoma cells
Zheng et al. 2018[149]	FLU induced N-Ras degradation
Allen et al. 2018[150]	Cinnarizine, a close relative of FLU, inhibited mTORC1

Interactions between anticonvulsivants and chemotherapeutic drugs

Carbamazepine, phenytoin, and valproic acid frequently interact with standard chemotherapeutic drugs[151] because they share common metabolic pathways through cytochrome P450 isoenzymes. Carbamazepine and phenobarbital are enzyme-inducing antiepileptics decreasing the concentration of chemotherapeutics, while valproic acid is an inhibitor of the P450 system.[152] Topiramate is one of the antiepileptic compounds with fewer drug interactions although it does show enzyme-inducing effects.[153]

Ranolazine

Ranolazine is an anti-arrhythmic drug indicated for the treatment of chronic angina that was approved by the FDA in 2006. Frequent side effects are dizziness, constipation, headache and nausea.

Ranolazine inhibits the late inward sodium current in heart muscle, so that it works as a sodium channel inhibitor. Ranolazine is metabolized by the CYP3A enzyme.

Driffort et al.[154] demonstrated that ranolazine inhibition of NaV1.5 reduced breast cancer cell invasiveness in vivo and in vitro using the highly invasive MDA-MB-231 breast cancer cell line. This drug also efficiently decreased the activity of the embryonic/neonatal isoform of NaV1.5 (the active isoform usually found in human breast cancer cells). Ranolazine did not modify cell viability. It also decreased the pro-invasive morphology of MDA-MB-231 breast cancer cells. They showed that in cancer cells injected through the tail vein of nude mice, ranolazine at non-toxic doses achieved a significant reduction in metastatic colonization.

Polyphenols

Natural polyphenols[155–157] such as resveratrol,[158,159] and genistein[160,161] have been shown to inhibit VGSC.

Riluzole (RIL)

RIL is an FDA approved drug that has been developed for the treatment of lateral amyotrophic sclerosis. It is a VGSC inhibitor.[162] It has been found to inhibit the proliferation of melanoma cells and induce apoptosis. However, in this case, the mechanism is through the inhibition of glutamate receptors.[163]

Local anesthetics

Local anesthetics are VGSC blockers that suppress pain sensation by inhibiting nociceptive neurons. Lidocaine, one of the most frequently used local anesthetics is an inhibitor of EGFR signaling.[164] It exerts anti-migratory activities through inhibition of CXCR4 signaling and decreased cytoskeletal remodeling.[165] Piegeler et al. found that amid-linked anesthetics inhibited Src activity, thus decreasing cell mobility.[166] They also increase cisplatin's anti-metastatic activity.[167] When lidocaine was infused during a surgical procedure in a mouse model of cancer, the amount of lung metastasis was significantly decreased.[168] Lidocaine decreased human breast cancer cell migration through inhibition of cytoskeletal remodeling.[169] Lidocaine in IV infusion during surgery decreased the number of pulmonary metastasis in a mouse model of breast cancer. These effects were also found with sevorane as an anesthetic, however, when ketamine was used, the number of metastasis increased.[168] Although, all these actions seem independent of VGSC blocking, lidocaine and other amid-linked local anesthetics remain an interesting issue for further research, particularly in the perioperative setting.

Proton pump inhibitors (PPIs)

Vacuolar H^+ATPase is a highly conserved enzyme and a vital component of all eukaryotes.[170] Its location in membranes allows the transport of protons (H^+) through the membrane with energy expenditure. This movement of protons produces important pH changes. For example, proton pumps in lysosomal membranes extrude protons from the cytoplasm into the lysosome. In this way the cytoplasm becomes more alkaline and the lysosome more acidic. Vacuolar H^+ATPase can also be found in the cell membrane. In this case, it pumps H^+ from the cytoplasm into the extracellular substance, with the same cytoplasmic alkalinizing effect.

PPIs are usually used for the treatment of gastroduodenal ulcers, gastroesophagic reflux disease, and Zollinger-Ellison syndrome, but they have also shown anti-tumor activity. PPI's effect is particularly effective in acidic environments as found in tumors.

There is abundant evidence of the anti-tumoral activity of PPIs[171–181] and this activity is mediated by decreased intracellular pH and increased extracellular pH.[182] However, some publications have raised doubts about using PPIs in cancer patients,[183] or using PPIs for very long periods.[184]

The basic issue here is that the patient already has cancer, so when PPIs are used in this situation the risk/benefit ratio of a new cancer favors the use of PPIs as part of a cocktail of proton extruder inhibitors. PPIs do not exert a chemopreventive tool against cancer.[185] And precisely, the idea is not to use them in cancer prevention but in cancer treatment. Furthermore, the "unpleasant" preconditioning of eventual niches for metastasis with PPIs and bicarbonate[186] may reduce the metastatic risk through what has been called "niche engineering."[187]

PPIs' reversion of multidrug resistance has also become another issue that highlights the importance of these drugs in cancer treatment.[188,189]

From the long list of PPIs in clinical use, three of them have been selected for clinical practice, based on certain advantages in the cancer area: lansoprazole, esomeprazole and pantoprazole. (See chapter on treatment.) All of them should be used at much higher doses than those usually prescribed for the treatment of non malignant diseases.

Enrico Pierluigi Spugnini and Stefano Fais discuss a more clinically oriented approach to PPIs in Chapter 21.

Conclusions

The main channel/exchanger ion trafficking inhibitor drugs have been reviewed in this chapter. The goal of all these pharmaceuticals is to target the inverted pH gradient. None of them are stand alone drugs against cancer. They work adequately as "reversal" therapy of the pH abnormality when they are part of a cocktail of pH modifiers that will be fully described in this book. Even if the main culprit of the pH alterations is NHE1, only inhibiting it will not achieve all the desired results in most cases, because malignant cells and tissues manage to survive through other proton exporting mechanisms. Treatment will require dealing with all or almost all of these mechanisms. This concept leads to the creation of an integrated cocktail of pH modifiers.

References

1. Sourake TDM, Shi MJ, Meng XY. The relationship between tumor recurrence and polymorphism of hGPX1 and NRAMP1 in superficial bladder cancer patients: a meta-analysis. *Pan Afr Med J* 2017;**27**:270.
2. Borchgrevink PC, Jynge P. Prolonged diuretic administration and myocardial tolerance to ischaemia. *Pharmacol Toxicol* 1987;**61**(4):254–9.
3. Moolenaar WH, Boonstra J, van der Saag PT, de Laat SW. *Sodium/ proton exchange in mouse neuroblastoma cells. J Biol Chem* 1981;**256**(24):12883–7. 7309738.
4. Moolenaar WH, Mummery CL, Van der Saag PT, De Laat SW. Rapid ionic events and the initiation of growth in serum-stimulated neuroblastoma cells. *Cell* 1981;**23**(3):789–98.
5. Matthews H, Ranson M, Kelso MJ. Anti-tumour/metastasis effects of the potassium-sparing diuretic amiloride: an orally active anti-cancer drug waiting for its call-of-duty? *Int J Cancer* 2011; **129**(9):2051–61.
6. Harguindey S, Orive G, Pedraz JL, Samaniego JM. Curación aparente de un caso de carcinoma de ovario metastático despues de tratamiento crónico con inhibidores del antiportador Na-H^+. *Oncología* 2002;**25**(10):472–6.
7. Sparks RL, Pool TB, Smith NK, Cameron IL. Effects of amiloride on tumor growth and intracellular element content of tumor cells in vivo. *Cancer Res* 1983;**43**(1):73–7.
8. Delvaux M, Bastie MJ, Chentoufi J, Cragoe Jr EJ, Vaysee N, Ribet A. *Amiloride and analogues inhibit Na(+)-H^+ exchange and cell proliferation in AR42J pancreatic cell line. Am J Phys Gastrointest Liver* 1990; **259**(5):842–9. 2173418.
9. Masereel B, Pochet L, Laeckmann D. An overview of inhibitors of Na(+)/H(+) exchanger. *Eur J Med Chem* 2003;**38**(6):547–54.
10. Ding Y, Zhang H, Zhou Z, Zhong M, Chen Q, Wang X, et al. u-PA inhibitor amiloride suppresses peritoneal metastasis in gastric cancer. *World J Surg Oncol* 2012;**10**(1):270.
11. Lokman NA, Ween MP, Oehler MK, Ricciardelli C. The role of annexin A2 in tumorigenesis and cancer progression. *Cancer Microenviron* 2011;**4**(2):199–208.
12. Mali AV, Joshi AA, Hedge MV, Kadam SS. Enterolactone suppresses proliferation, migration and metastasis of MDA-MB-231 breast cancer cells through inhibition of uPA induced plasmin activation and MMPs-mediated ECM remodeling. *Asian Pac J Cancer Prev* 2017;**18**(4):905.
13. Vassalli JD, Belin D. Amiloride selectively inhibits the urokinase-type plasminogen activator. *FEBS Lett* 1987;**214**(1):187–91.
14. Jankun J, Skrzypczak-Jankun E. Molecular basis of specific inhibition of urokinase plasminogen activator by amiloride. *Cancer Biochem Biophys* 1999;**17**(1–2):109–23.
15. Ray B, Bhatti R, Gadarowski J, Bell N, Nasruddin S. Inhibitory effect of amiloride on the urokinase plasminogen activators in prostatic cancer. *Tumor Biol* 1998;**19**(1):60.
16. Danø K, Andreasen PA, Grøndahl-Hansen J, Kristensen P, Nielsen LS, Skriver L. Plasminogen activators, tissue degradation, and cancer. *Adv Cancer Res* 1985;**44**:139–226. Academic Press.
17. Danø K, Berendt N, Høyer-Hansen G, Johnsen M, Lunden LR, Ploug M, et al. Plasminogen activation and cancer. *Thromb Haemost* 2005;**93**(04):676–81.
18. Garty HA, Benos DJ. Characteristics and regulatory mechanisms of the amiloride-blockable Na^+ channel. *Physiol Rev* 1988;**68**(2):309–73.
19. Maidorn RP, Cragoe Jr EJ, Tannock IF. Therapeutic potential of analogues of amiloride: inhibition of the regulation of intracellular pH as a possible mechanism of tumour selective therapy. *Br J Cancer* 1993;**67**(2):297.

20. Harris CA, Fliegel LA. Amiloride and the Na(+)/H(+) exchanger protein: mechanism and significance of the inhibition of the Na(+)/H(+) exchanger. *Int J Mol Med* 1999;**3**(3):315–36.
21. Harguindey S, Arranz JL, Polo Orozco JD, et al. Cariporide and other new and powerful NHE1 inhibitors as potentially selective anticancer drugs- an integral molecular/biochemical/metabolic/clinical approach after one hundred years of cancer research. *J Transl Med* 2013;**11**:282.
22. Koo JY, Parekh D, Townsend CM, Saydjari R, Evers BM, Farre A, et al. Amiloride inhibits the growth of human colon cancer cells in vitro. *Surg Oncol* 1992;**1**(6):385–9.
23. Tatsuta M, Iishi H, Baba M, Uehara H, Nakaizumi A, Taniguchi H. Inhibition by amiloride of gastric carcinogenesis induced by N-methyl-N′-nitro-N-nitrosoguanidine in Wistar rats. *Br J Cancer* 1993; **67**(5):1011.
24. Wang Y, Dang J, Liang X, Doe WF. Amiloride modulates urokinase gene expression at both transcription and post-transcription levels in human colon cancer cells. *Clin Exp Metastasis* 1995;**13**(3):196–202.
25. Pannocchia A, Revelli S, Tamponi G, Giorgianni A, Todde R, Bosia A, et al. Reversal of doxorubicin resistance by the amiloride analogue EIPA in multidrug resistant human colon carcinoma cells. *Cell Biochem Funct* 1996;**14**(1):11–8.
26. Evans DM, Sloan-Stakleff K, Arvan M, Guyton DP. Time and dose dependency of the suppression of pulmonary metastases of rat mammary cancer by amiloride. *Clin Exp Metastasis* 1998;**16**(4):353–7.
27. Evans DM, Sloan-Stakleff K. Suppression of the invasive capacity of human breast cancer cells by inhibition of urokinase plasminogen activator via amiloride and B428. *Am Surg* 2000;**66**(5):460–4.
28. Oka N, Okumura Y, Kanayama HO, Izaki H, Okamoto M, Kido H, et al. Amiloride and urinary trypsin inhibitor inhibit urothelial cancer invasion. *Eur Urol* 2003;**44**(6):737–41.
29. Hegde M, Roscoe J, Cala P, Gorin F. Amiloride kills malignant glioma cells independent of its inhibition of the sodium-hydrogen exchanger. *J Pharmacol Exp Ther* 2004;**310**(1):67–74.
30. Kim KM, Lee YJ. Amiloride augments TRAIL-induced apoptotic death by inhibiting phosphorylation of kinases and phosphatases associated with the PI3K-Akt pathway. *Oncogene* 2005;**24**(3):355.
31. Chang JG, Yang DM, Chang WH, Chow LP, Chan WL, Lin HH, et al. Small molecule amiloride modulates oncogenic RNA alternative splicing to devitalize human cancer cells. *PLoS One* 2011;**6**(6).
32. Yang X, Wang D, Dong W, et al. Suppression of Na^+/H^+ exchanger 1 by RNA interference or amiloride inhibits human hepatoma cell line SMMC-7221 cell invasion. *Med Oncol* 2011; **28**(1):385–90.
33. Chang WH, Liu TC, Yang WK, Lee CC, Lin YH, Chen TY, et al. Amiloride modulates alternative splicing in leukemic cells and resensitizes Bcr-Abl T315I mutant cells to imatinib. *Cancer Res* 2011; **71**(2):383–92.
34. Guan B, Hoque A, Xu X. Amiloride and guggulsterone suppression of esophageal cancer cell growth in vitro and in nude mouse xenografts. *Front Biol* 2014;**9**(1):75–81.
35. Zheng YT, Yang HY, Li T, Zhao B, Shao TF, Xiang XQ, et al. Amiloride sensitizes human pancreatic cancer cells to erlotinib in vitro through inhibition of thrPI3K/AKT signaling pathway. *Acta Pharmacol Sin* 2015;**36**(5):614–26.
36. Tang JY, Chang HW, Chang JG. Modulating roles of amiloride in irradiation-induced antiproliferative effects in glioblastoma multiforme cells involving Akt phosphorylation and the alternative splicing of apoptotic genes. *DNA Cell Biol* 2013;**32**(9):504–10.
37. Rojas EA, Corchete LA, San-Segundo L, Martinez-Blanch JF, Codoner FM, Paino T, et al. Amiloride, an old diuretic drug, is a potential therapeutic agent for multiple myeloma. *Clin Cancer Res* 2017; **23**(21):6602–15.
38. Zhang H, Peng C, Huang H, Lai Y, Hu C, Li F, et al. Effects of amiloride on physiological activity of stem cells of human lung cancer and possible mechanism. *Biochem Biophys Res Commun* 2018; **504**(1):1–5.
39. Poon AC, Inkol JM, Luu AK, Mutsaers AJ. Effects of the potassium-sparing diuretic amiloride on chemotherapy response in canine osteosarcoma cells. *J Vet Intern Med* 2018. https://doi.org/10.1111/jvim.15382.
40. Marleau AM, Chen CS, Joyce JA, Tullis RH. Exosome removal as a therapeutic adjuvant in cancer. *J Transl Med* 2012;**10**(1):134.

41. McCready J, Sims JD, Chan D, Jay DG. Secretion of extracellular hsp90α via exosomes increases cancer cell motility: a role for plasminogen activation. *BMC Cancer* 2010;**10**(1):294.
42. Savina A, Furlán M, Vidal M, Colombo MI. Exosome release is regulated by a calcium-dependent mechanism in K562 cells. *J Biol Chem* 2003;**278**(22):20083–90. JBC Papers in Press. Published on March 14, 2003 as Manuscript M301642200.
43. Chalmin F, Ladoire S, Mignot G, Vincent J, Bruchard M, Remy-Martin JP, et al. Membrane-associated Hsp72 from tumor-derived exosomes mediates STAT3-dependent immunosuppressive function of mouse and human myeloid-derived suppressor cells. *J Clin Invest* 2010;**120**(2):457–71.
44. Zhang L, Zhang S, Yao J, Lowery FJ, Zhang Q, Huang WC, et al. Microenvironment-induced PTEN loss by exosomal microRNA primes brain metastasis outgrowth. *Nature* 2015;**527**(7576):100.
45. Cobo JM, Garcia-Cañero R, Valdez JG, Barrasso AM, Sailer BL, Crissman HA. Attenuation of apoptotic DNA fragmentation by amiloride. *J Cell Physiol* 1998;**175**(1):59–67.
46. Pilat MJ, Lehr JE, Quigley MM, Pienta KJ. The effect of amiloride on the metastatic properties of prostate cancer in the dunning rat model. *Oncol Rep* 1998;**5**(4):889–981.
47. Scholz W, Albus U, Counillon L, Gögelein H, Lang HJ, Linz W, et al. Protective effects of HOE642, a selective sodium-hydrogen exchange subtype 1 inhibitor, on cardiac ischaemia and reperfusion. *Cardiovasc Res* 1995;**29**(2):260–8.
48. Boyce SW, Bartels C, Bolli R, Chaitman B, Chen JC, Chi E, et al. Impact of sodium-hydrogen exchange inhibition by cariporide on death or myocardial infarction in high-risk CABG surgery patients: results of the CABG surgery cohort of the GUARDIAN study. *J Thorac Cardiovasc Surg* 2003;**126**(2):420–7.
49. Wong P, Kleemann HW, Tannock IF. Cytostatic potential of novel agents that inhibit the regulation of intracellular pH. *Br J Cancer* 2002;**87**(2):238–45.
50. Di Sario A, Bendia E, Omenetti A, De Minicis S, Marzioni M, Kleemann HW, et al. Selective inhibition of ion transport mechanisms regulating intracellular pH reduces proliferation and induces apoptosis in cholangiocarcinoma cells. *Dig Liver Dis* 2007;**39**(1):60–9.
51. Stüwe L, Müller M, Fabian A, Waning J, Mally S, Noel J, et al. pH dependence of melanoma cell migration: protons extruded by NHE-1dominate protons of the bulk solution. *J Physiol* 2007; **585**(Pt. 2):351–60.
52. Lv C, Yang X, Yu B, Ma Q, Liu B, Liu Y. Blocking the Na^+/H^+ exchanger 1 with cariporide (HOE642) reduces the hypoxia-induced invasion of human tonge squamous cell carcinoma. *Int J Oral Maxillofac Surg* 2012;**41**(10):1206–10.
53. Lin Y, Wang J, Jin W, Wang L, Li H, Ma L, et al. NHE1 mediates migration and invasion of HeLa cells via regulating the expression and localization of MT1-MMP. *Cell Biochem Funct* 2012;**30**(1):41–6.
54. Gao W, Chang G, Wang J, Jin W, Wang L, Lin Y, et al. Inhibition of K562 leukemia angiogenesis and growth by selective Na^+/H^+ exchanger inhibitor cariporide through down-regulation of pro-angiogenesis factor VEGF. *Leuk Res* 2011;**35**(11):1506–11.
55. Lee YJ, Bae JH, Kim SA, Kim SH, Woo KM, Nam HS, et al. Cariporide enhances the DNA damage and apoptosis in acid tolerable malignant mesothelioma H-2452 cells. *Mol Cell* 2017;**40**(8):567–76.
56. Cardone RA, Greco MR, Zeeberg K, Zaccagnino A, Saccomano M, Bellizzi A, et al. A novel NHE-1 centered signaling cassette drives epidermal growth factor-dependent pancreatic tumor metastasis and is a target for combination therapy. *Neoplasia* 2015;**17**(2):155–66.
57. Kaminota T, Yano H, Skiota K, Nomura N, Yaguchi H, Kirino Y, et al. Elevated Na^+/H^+ exchanger-1 expression enhances the metastatic collective migration of head and neck squamous cell carcinoma cells. *Biochem Biophys Res Commun* 2017;**486**(1):101–7.
58. Baumgarth M, Beier N, Gericke R. (2-Methyl-5-(methylsulfonyl) benzoyl) guanidine Na^+/H^+ antiporter inhibitors. *J Med Chem* 1997;**40**(13):2017–34.
59. Atwal KS, O'Neil SV, Ahmad S, Doweyko L, Kirby M, Dorso CR, et al. Synthesis and biological activity of 5-aryl-4-(4-(5-methyl-1Himidazol-4-yl)piperidin-1-yl)pyrimidine analogs as potent, highly selective, and orally bioavailable NHE-1 inhibitors. *Bioorg Med Chem Lett* 2006;**16**:4796–9.
60. Turturro F, Friday E, Fowler R, Surie D, Welbourne T. Troglitazone acts on cellular pH and DNA synthesis through a peroxisome proliferator-activated receptor γ-independent mechanism in breast cancer-derived cell lines. *Clin Cancer Res* 2004;**10**(20):7022–30.

61. Kumar AP, Quake AL, Chang MK, Zhou T, Lim KS, Singh R, et al. Repression of NHE1 expression by PPARγ activation is a potential new approach for specific inhibition of the growth of tumor cells in vitro and in vivo. *Cancer Res* 2009;**69**(22):8636–44.
62. Aredia F, Scovassi AI. Manipulation of Intracellular pH in cancer cells by NHE1 Inhibitors. *Protein Pept Lett* 2016;**23**. https://doi.org/10.2174/0929866523666161013125536;23.
63. Obara M, Szeliga M, Albrecht J. Regulation of pH in the mammalian central nervous system under normal and pathological conditions: facts and hypotheses. *Neurochem Int* 2008;**52**:905–19.
64. Schwab A, Stock C. Ion channels and transporters in tumour cell migration and invasion. *Philos Trans R Soc Lond Ser B Biol Sci* 2014;**369**.
65. Litan A, Langhans SA. Cancer as a channelopathy: ion channels and pumps in tumor development and progression. *Front Cell Neurosci* 2015;**9**:86.
66. Roger S, Gillet L, Le Guennec JY, Besson P. Voltage-gated sodium channels and cancer: is excitability their primary role? *Front Pharmacol* 2015;**6**:152.
67. Koltai T. Voltage-gated sodium channel as a target for metastatic risk reduction with re-purposed drugs. *F1000Res* 2015;**4**.
68. Catterall WA. Voltage-gated sodium channels at 60: structure, function and pathophysiology. *J Physiol* 2012;**590**:2577–89.
69. Brackenbury WJ. Voltage-gated sodium channels and metastatic disease. *Channels (Austin)* 2012;**6**:352–61.
70. Roger S, Potier M, Vandier C, Besson P, Le Guennec JY. Voltage-gated sodium channels: new targets in cancer therapy? *Curr Pharm Des* 2006;**12**:3681–95.
71. Fraser SP, Diss JK, Chioni AM, Mycielska ME, Pan H, Yamaci RF, et al. Voltage-gated sodium channel expression and potentiation of human breast cancer metastasis. *Clin Cancer Res* 2005;**11**:5381–9.
72. Gillet L, Roger S, Besson P, Lecaille F, Gore J, Bougnoux P, et al. Voltage-gated sodium channel activity promotes cysteine cathepsin-dependent invasiveness and colony growth of human cancer cells. *J Biol Chem* 2009;**284**:8680–91.
73. Brisson L, Gillet L, Calaghan S, Besson P, Le Guennec J, Roger S, et al. NaV1.5 enhances breast cancer cell invasiveness by increasing NHE1-dependent H(+); efflux in caveolae. *Oncogene* 2011;**30**:2070–6.
74. Brisson L, Driffort V, Benoist L, Poet M, Counillon L, Antelmi E, et al. NaV1.5 Na(+) channels allosterically regulate the NHE-1 exchanger and promote the activity of breast cancer cell invadopodia. *J Cell Sci* 2013;**126**:4835–42.
75. Yang M, Brackenbury WJ. Membrane potential and cancer progression. *Front Physiol* 2013;**4**:185.
76. Fairhurst C, Watt I, Martin F, Bland M, Brackenbury WJ. Exposure to sodium channel-inhibiting drugs and cancer survival: protocol for a cohort study using the QResearch primary care database. *BMJ Open* 2014;**4**.
77. Denac H, Mevissen M, Scholtysik G. Structure, function and pharmacology of voltage gated sodium channels. *Naunyn Schmiedeberg's Arch Pharmacol* 2000;**362**:453–79.
78. Lee CH, Ruben PC. Interaction between voltage-gated sodium channels and the neurotoxin, tetrodotoxin. *Channels (Austin)* 2008;**2**(6):407–12.
79. Sehgal VK, Kumar R. Phenytoin beyond epilepsy. *Int J Med Dent Sci* 2018;**4**(2):743–4.
80. Matsuki N, Quandt FN, Ten Eick RE, Yeh JZ. Characterization of the block of sodium channels by phenytoinin mouse neuroblastoma cells. *J Pharmacol Exp Ther* 1984;**228**(2):523–30.
81. Willow M, Gonoi T, Catterall WA. Voltage clamp analysis of the inhibitory actions of diphenylhydantoin and carbamazepine on voltage-sensitive sodium channels in neuroblastoma cells. *Mol Pharmacol* 1985;**27**(5):549–58.
82. Schwinghammer TL, Howrie DL. Phenytoin-induced lymphadenopathy. *Drug Intell Clin Pharm* 1983;**17**(6):460–2.
83. Olsen JH, Boyce JD, Jenssen JP, Fraumeni Jr JF. Cancer among epileptic patients exposed to anticonvulsivant drugs. *J Natl Cancer Inst* 1989;**81**(10):803–8.
84. Okamoto Y, Shimizu K, Tamura K. Effects of phenytoin on cell-mediated immunity. *Cancer Immunol Immunother* 1988;**26**(2):176–9.
85. Vernillo AT, Ramamurthy NS, Lee HM, Rifkin BR. The effect of phenytoin on collagenase and gelatinase activities in UMR 106-01 rat osteoblastic osteosarcoma cells. *Matrix* 1990;**10**(1):27–32.

86. Dyce M, Sharif SF, Whalen GF. Search for anti-metastatic therapy: effects of phenytoin on B16 melanoma metastasis. *J Surg Oncol* 1992;**49**(2):107–12.
87. Yang M, Kozminski DJ, Wold LA, Modak R, Calhoun JD, Isom LL, et al. Therapeutic potential for phenytoin: targeting Nav 1.5 sodium channels to reduce migration and invasion in metastatic breast cancer. *Breast Cancer Res Treat* 2012;**134**:603–15.
88. Ragsdale DS, Scheuer T, Catterall WA. Frequency and voltage dependent inhibition of type IIA Na^+ channels, expressed in a mammalian cell line by local anesthetic, antiarrhytmic and anticonvulsivant drugs. *Mol Pharmacol* 1991;**40**(5):756–65.
89. Abdul M, Hoosein N. Inhibition by anticonvulsants of prostate-specific antigen and interleukin-6 secretion by human prostate cancer cells. *Anticancer Res* 2001;**21**(3B):2045–8.
90. Fadiel A, Song J, Tivon D, Hamza A, Cardozo T, Naftolin F. Phenytoin is an estrogen receptor α-selective modulator that interacts with helix 12. *Reprod Sci* 2015;**22**(2):146–55.
91. Tittle TV, Schaumann BA, Rainey JE, Taylor K. Segregation of the growth slowing effects of valproic acid from phenytoin and carbamazepine on lymphoid tumor cells. *Life Sci* 1992;**50**(14):79–83.
92. Lang DG, Wang CM, Cooper BR. Lamotrigine, phenytoin and carbamazepine interactions on the sodium current present in N4TG1 mouse neuroblastoma cells. *J Pharmacol Exp Ther* 1993;**266**(2):829–35.
93. Kawamura KI, Grabowski D, Weizer K, Bukowski R, Ganapathi R. Modulation of vinblastine cytotoxicity by dilantin (phenytoin) or the protein phosphatase inhibitor okadaic acid involves the potentiation of anti-mitotic effects and induction of apoptosis in human tumour cells. *Br J Cancer* 1996;**73**(2):183–8.
94. Lobert S, Ingram JW, Correia JJ. Additivity of dilantin and vinblastine inhibitory effects on microtubule assembly. *Cancer Res* 1999;**59**(19):4816–22.
95. Sikes RA, Walls AM, Brennen WN, Anderson JD, Choudhury-Mukherjee I, Schenck HA, et al. Therapeutic approaches targeting prostate cancer progression using novel voltage-gated ion channel blockers. *Clin Prostate Cancer* 2003;**2**(3):181–7.
96. Anderson JD, Hansen TP, Lenkowski PW, Walls AM, Choudhury IM, Schenck HA, et al. Voltage-gated sodium channel blockers as cytostatic inhibitors of the androgen-independent prostate cancer cell line PC-3. *Mol Cancer Ther* 2003;**2**(11):1149–54.
97. Liu Y, Wu J, Ho PY, Chen LC, Chen CT, Liang YC, et al. Anti-angiogenic action of 5,5-diphenyl-2-thiohydantoin-N10 (DPTH-N10). *Cancer Lett* 2008;**271**(2):294–305.
98. Lee TS, Chen LC, Liu Y, Wu J, Liang YC, Lee WS. 5,5-Diphenyl-2-thiohydantoin-N10 (DPTH-N10) suppresses proliferation of cultured colon cancer cell line COLO-205 by inhibiting DNA synthesis and activating apoptosis. *Naunyn Schmiedeberg's Arch Pharmacol* 2010;**382**(1):43–50.
99. Brackenbury WJ, Nelson M, Yang M, Dowle AA, Thomas JR. Abstract P6-03-10: targeting voltage-gated sodium channels with the antiepileptic drug phenytoin inhibits breast tumor growth and metastasis. *Cancer Res* 2015;**75**(9 Suppl). https://doi.org/10.1158/1538-7445.SABCS14-P6-03-10.
100. Nelson M, Yang M, Dowle AA, Thomas JR, Brackenbury WJ. The sodium channel-blocking antiepileptic drug phenytoin inhibits breast tumour growth and metastasis. *Mol Cancer* 2015;**14**(1):13.
101. Mohammed FH, Khajah MA, Yang M, Brackenbury WJ, Luqmani YA. Blockade of voltage-gated sodium channels inhibits invasion of endocrine-resistant breast cancer cells. *Int J Oncol* 2016;**48**(1):73–83.
102. Takada M, Fujimoto M, Motomura H, Hosomi K. Inverse association between sodium channel-blocking antiepileptic drug use and cancer: data mining of spontaneous reporting and claims databases. *Int J Med Sci* 2016;**13**(1):48.
103. Sorrell TC, Forbes IJ. Depression of immune competence by phenytoin and carbamazepine. Studies in vivo and in vitro. *Clin Exp Immunol* 1975;**20**(2):273–85.
104. Fontana A, Grob PJ, Sauter R, Joller H. IgA deficiency, epilepsy, and hydantoin medication. *Lancet* 1976;**2**(7979):228–31.
105. Marcoli M, Gatti G, Ippoliti G, Lombardi M, Crema A, Zocchi MT, et al. Effect of chronic anticonvulsant monotherapy on lymphocyte subpopulations in adult epileptic patients. *Hum Exp Toxicol* 1985;**4**(2):147–57.

106. Mässenhausen A, Tonnus W, Himmerkus N, Parmentier S, Saleh D, Rodriquez D, et al. Phenytoin inhibits necroptosis. *Cell Death Dis* 2018;**9**(3):359.
107. Fink KL, Deo AK. Antiepileptic drugs and chemotherapy: potential interactions and impact on treatment of patients with cancer. In: *Handbook of brain tumor chemotherapy, molecular therapeutics, and immunotherapy*. 2nd ed. Academic Press; 2018. p. 45–58.
108. Rizal DS, Oktafia D, Hasmono D. *The effects of phenytoin, carbamazepine, and valproate on bone change. Dama Int J Res (DIJR)* 2018;**3**(2):40–6. Available at:www.damaacademia.com.
109. Stafstrom CE, Nohria V, Loganbill H, Nahouraii R, Boustany RM, DeLong GR. Erythromycin-induced carbamazepine toxicity: a continuing problem. *Arch Pediatr Adolesc Med* 1995;**149**(1):99–101.
110. Dailey JW, Reith ME, Yan QS, Li MY, Jobe PC. Carbamazepine increases extracellular serotonin concentration: lack of antagonism by tetrodotoxin or zero Ca^{2+}. *Eur J Pharmacol* 1997; **328**(2–3):153–62.
111. Spina E, Pisani F, Perucca E. Clinically significant pharmacokinetic drug interactions with carbamazepine. *Clin Pharmacokinet* 1996;**31**(3):198–214.
112. Benoist GE, van der Doelen MJ, ter Heine R, van Erp NP, Mehra N. A clinically relevant decrease in abiraterone exposure associated with carbamazepine use in a patient with castration-resistant metastatic prostate cancer. *Br J Clin Pharmacol* 2018;**84**(5):1064–7.
113. Chu E, DeVita Jr VT. *Physicians' cancer chemotherapy drug manual 2019*. Jones & Bartlett Learning; 2018.
114. Teichmann M, Kretschy N, Kopf S, Jarukamjorn K, Atanasov AG, Viola K, et al. Inhibition of tumour spheroid-induced prometastatic intravasation gates in the lymph endothelial cell barrier by carbamazepine: drug testing in a 3D model. *Arch Toxicol* 2014;**88**(3):691–9.
115. Stettner M, Krämer G, Strauss A, Kvitkina T, Ohle S, Kieseier BC, et al. Long-term antiepileptic treatment with histone deacetylase inhibitors may reduce the risk of prostate cancer. *Eur J Cancer Prev* 2012;**21**(1):55–64.
116. Meng Q, Chen X, Sun L, Zhao C, Sui G, Cai L. Carbamazepine promotes Her-2 protein degradation in breast cancer cells by modulating HDAC6 activity and acetylation of Hsp90. *Mol Cell Biochem* 2011;**348**(1–2):165–71.
117. Beutler AS, Li SD, Nicol R, Walsh MJ. Carbamazepine is an inhibitor of histone deacetylases. *Life Sci* 2005;**76**(26):3107–15.
118. Meng QW, Zhao CH, Xi YH, Cai L, Sun LC, Sui GJ. Inhibitory effect of carbamazepine on proliferation of estrogen-dependent breast cancer cells. *Ai Zheng* 2006;**25**(8):967–73.
119. Kostrouchová M, Kostrouch Z, Kostrouchová M. Valproic acid, a molecular lead to multiple regulatory pathways. *Folia Biol (Praha)* 2007;**53**:37–49.
120. Michaelis M, Doerr HW, Cinatl Jr J. Valproic acid as anti-cancer drug. *Curr Pharm Des* 2007; **13**(33):3378–93.
121. Shank RP, Smith-Swintosky VL, Maryanoff BE. Carbonic anhydrase inhibition. Insight into the characteristics of zonisamide, topiramate, and the sulfamide cognate of topiramate. *J Enzyme Inhib Med Chem* 2008;**23**(2):271–6.
122. Supuran CT. Carbonic anhydrases: novel therapeutic applications for inhibitors and activators. *Nat Rev Drug Discov* 2008;**7**(2):168.
123. Raffa RB, Finno KE, Tallarida CS, Rawls SM. Topiramate-antagonism of L-glutamate-induced paroxysms in planarians. *Eur J Pharmacol* 2010;**649**(1):150–3.
124. Monzani E, Shtil AA, La Porta CA. The water channels, new druggable targets to combat cancer cell survival, invasiveness and metastasis. *Curr Drug Targets* 2007;**8**(10):1132–7.
125. Marathe K, McVicar N, Li A, Bellyou M, Meakin S, Bartha R. Topiramate induces acute intracellular acidification in glioblastoma. *J Neuro-Oncol* 2016;**130**(3):465–72.
126. Shank RP, Gardocki JF, Streeter AJ, Maryanoff BE. An overview of the preclinical aspects of topiramate: pharmacology, pharmacokinetics, and mechanism of action. *Epilepsia* 2000;**41**(s1):3–9.
127. Ma B, Pan Y, Song Q, Tie L, Zhang Y, Xiao Y, et al. The effect of topiramate on tumor-related angiogenesis and on the serum proteome of mice bearing Lewis lung carcinoma. *Eur J Pharmacol* 2011;**663**(1–3):9–16.

128. Bing MA, Yang X, Li T, Yu HM, Li XJ. Inhibitory effect of topiramate on lewis lung carcinoma metastasis and its relation with AQP1 water channel. *Acta Pharmacol Sin* 2004;**25**(1):54–60.
129. Toplak H, Hamann A, Moore T, Masson E, Gorska M, Vercruysse F, et al. Efficacy and safety of topiramate in combination with metformin in the treatment of obese subjects with type 2 diabetes: a randomized, double-blind, placebo-controlled study. *Int J Obes* 2007;**31**:138–46.
130. Kiskin NI, Chizhmakov IV, Tsyndrenko AY, Krishtal OA, Tegtmeier F. R56865 and flunarizine as Na(+)-channel blockers in isolated Purkinje neurons of rat cerebellum. *Neuroscience* 1993;**54**(3):575–85.
131. Ye Q, Yan LY, Xue LJ, Wang Q, Zhou ZK, Xiao H, et al. Flunarizine blocks voltage-gated Na(+) and Ca(2+) currents in cultured rat cortical neurons: a possible locus of action in the prevention of migraine. *Neurosci Lett* 2011;**487**(3):394–9.
132. Fischer W, Kittner H, Regenthal R, De Sarro G. Anticonvulsant profile of flunarizine and relation to Na(+) channel blocking effects. *Basic Clin Pharmacol Toxicol* 2004;**94**(2):79–88.
133. Wood PJ, Hirst DG. Cinnarizine and flunarizine as radiation sensitisers in two murine tumours. *Br J Cancer* 1988;**58**(6):742.
134. Kaelin WG, Shrivastav S, Jirtle RL. Blood flow to primary tumors and lymph node metastases in SMT-2A tumor-bearing rats following intravenous flunarizine. *Cancer Res* 1984;**44**(3):896–9.
135. Vaupel P, Menke H. Blood flow, vascular resistance and oxygen availability in malignant tumours upon intravenous flunarizine. In: *Oxygen transport to tissue IX*. Boston, MA: Springer; 1987. p. 393–8.
136. Kavanagh BD, Coffey BE, Needham D, Hochmuth RM, Dewhirst MW. The effect of flunarizine on erythrocyte suspension viscosity under conditions of extreme hypoxia, low pH, and lactate treatment. *Br J Cancer* 1993;**67**(4):734.
137. Sezzi ML, De Luca G, Materazzi M, Bellelli L. Effects of a calcium-antagonist flunarizine on cancer cell movement and phagocytosis. *Anticancer Res* 1985;**5**(3):265–71.
138. Abdul M, Hoosein N. Voltage-gated sodium ion channels in prostate cancer: expression and activity. *Anticancer Res* 2002;**22**(3):1727–30.
139. Dewhirst MW, Ong ET, Madwed D, Klitzman B, Secomb T, Brizel D, et al. Effects of the calcium channel blocker flunarizine on the hemodynamics and oxygenation of tumor microvasculature. *Radiat Res* 1992;**132**(1):61–8.
140. Bellelli A, Camboni C, de Luca G, Materazzi M, Mattioni M, Sezzi ML, et al. In vitro and in vivo enhancement of vincristine antitumor activity on B16 melanoma cells by calcium antagonist flunarizine. *Oncology* 1987;**44**(1):17–23.
141. Sezzi ML, Zupi G, De GL, Materazzi M, Bellelli L. Effects of a calcium-antagonist (flunarizine) on the in vitro growth of B16 mouse melanoma cells. *Anticancer Res* 1984;**4**(4–5):229–34.
142. Silvestrini R, Zaffaroni N, Costa A, Orlando L, Villa R, Hendriks HR. Flunarizine as a modulator of doxorubicin resistance in human colon-adenocarcinoma cells. *Int J Cancer* 1993;**55**(4):636–9.
143. Fink-Puches R, Helige C, Kerl H, Smolle J, Tritthart HA. Inhibition of melanoma cell directional migration in vitro via different cellular targets. *Exp Dermatol* 1993;**2**(1):17–24.
144. Castellino SM, Friedman HS, Elion GB, Ong ET, Marcelli SL, Page R, et al. Flunarizine enhancement of melphalan activity against drug-sensitive/resistant rhabdomyosarcoma. *Br J Cancer* 1995;**71**(6):1181.
145. Gornati D, Zaffaroni N, Villa R, De CM, Silvestrini R. (1997). Modulation of melphalan and cisplatin cytotoxicity in human ovarian cancer cells resistant to alkylating drugs. *Anti-Cancer Drugs* 1997;**8**(5):509–16.
146. Conrad DM, Furlong SJ, Doucette CD, West KA, Hoskin DW. The Ca^{2+} channel blocker flunarizine induces caspase-10-dependent apoptosis in Jurkat T-leukemia cells. *Apoptosis* 2010;**15**(5):597–607.
147. Astin JW, Jamieson SM, Eng TC, Flores MV, Misa JP, Chien A, et al. An *in vivo* antilymphatic screen in zebrafish identifies novel inhibitors of mammalian lymphangiogenesis and lymphatic-mediated metastasis. *Mol Cancer Ther* 2014;**13**(10):2450–62.
148. Schmeel LC, Schmeel FC, Kim Y, Blaum-Feder S, Endo T, Schmidt-Wolf IG. Flunarizine exhibits in vitro efficacy against lymphoma and multiple myeloma cells. *Anticancer Res* 2015;**35**(3):1369–76.
149. Zheng ZY, Li J, Li F, Zhu Y, Cui K, Wong ST, et al. Induction of N-Ras degradation by flunarizine-mediated autophagy. *Sci Rep* 2018;**8**(1).

150. Allen SA, Tomilov A, Cortopassi GA. Small molecules bind human mTOR protein and inhibit mTORC1 specifically. *Biochem Pharmacol* 2018;**155**:298–304.
151. Yap KY, Chui WK, Chan A. Drug interactions between chemotherapeutic regimens and antiepileptics. *Clin Ther* 2008;**30**(8):1385–407.
152. Vecht CJ, Wagner GL, Wilms EB. Interactions between antiepileptic and chemotherapeutic drugs. *Lancet Neurol* 2003;**2**(7):404–9.
153. Glauser TA. Topiramate. In: Pellock JA, Dodson WE, BFD B, editors. *Pediatric epilepsy: diagnosis and therapy*. New York: Demos Medical Publishing; 2001. p. 475–88.
154. Driffort V, Gillet L, Bon E, Maionneau-Lambot S, Quillier T, Joulin V, et al. Ranolazine inhibits NaV1.5-mediated breast cancer cell invasiveness and lung colonization. *Mol Cancer* 2014;**13**:264.
155. Djamgoz M, Onkal R. Persistent current blockers of voltage-gated sodium channels: a clinical opportunity for controlling metastatic disease. *Recent Pat Anticancer Drug Discov* 2013;**8**:66–84.
156. Ingolfsson HI, Thakur P, Herold KF, Hobart EA, Ramsey NB, Periole X, et al. Phytochemicals perturb membranes and promiscuously alter protein function. *ACS Chem Biol* 2014;**9**:1788–98.
157. Wallace CHR, Baczko I, Jones L, Fercho M, Light PE. Inhibition of cardiac voltage-gated sodium channels by grape polyphenols. *Br J Pharmacol* 2006;**149**:657–65.
158. Weixel KM, Marciszyn A, Alzamora R, Li H, Fisher O, Edinger RS, et al. Resveratrol inhibits the epithelial sodium channel via phosphoinositides and AMP-activated protein kinase in kidney collecting duct cells. *PLoS One* 2013;**8**(19).
159. Fraser SP, Peters A, Fleming-Jones S, Mukhey D, Djamgoz MB. Resveratrol: inhibitory effects on metastatic cell behaviors and voltage-gated Na channel activity in rat prostate cancer in vitro. *Nutr Cancer* 2014;**66**(6):1047–58.
160. Jia Z, Jia Y, Liu B, Zhao Z, Jia Q, Liang H, et al. Genistein inhibits voltage gated sodium currents in SCG neurons through protein tyrosine kinase-dependent and kinase-independent mechanisms. *Pflugers Arch* 2008;**456**(5):857–66.
161. Liu L, Yang T, Simon SA. The protein tyrosine kinase inhibitor, genistein, decreases excitability of nociceptive neurons. *Pain* 2004;**112**(1–2):131–41.
162. Song JH, Huang CS, Nagata K, Yeh JZ, Narahashi T. Differential action of riluzole on tetrodotoxin-sensitive and tetrodotoxin-resistant sodium channels. *J Pharmacol Exp Ther* 1997;**282**(2):707–14.
163. Wall BA, Wangari-Talbot J, Shin SS, Schiff D, Sierra J, Yu LJ, et al. Disruption of GRM1-mediated signalling using riluzole results in DNA damage in melanoma cells. *Pigment Cell Melanoma Res* 2014;**27**(2):263–74.
164. Sakaguchi M, Kuroda Y, Hirose M. The antiproliferative effect of lidocaine on human tongue cancer cells with inhibition of the activity of epidermal growth factor receptor. *Anesth Analg* 2006;**102**(4):1103–7.
165. D'Agostino G, Saporito A, Cecchinato V, Silvestri Y, Borgeat A, Anselmi L, et al. Lidocaine inhibits cytoskeletal remodelling and human breast cancer cell migration. *Br J Anaesth* 2018;**121**(4):962–8.
166. Piegeler T, Votta-Velis G, Liu G, Place AT, Schwartz DE, Beck-Schimmer B, et al. Antimetastatic potential of amide-linked local anaesthetics: inhibition of lung adenocarcinoma cell migration and inflammatory Src signaling independent of sodium channel blockade. *Anesthesiology* 2012;**117**(3):548–59.
167. Freeman J, Crowley PD, Foley AG, Gallagher HC, Iwasaki M, Ma D, et al. Effect of perioperative lidocaine and cisplatin on metastasis in a murine model of breast cancer surgery. *Anticancer Res* 2018;**38**(10):5599–606.
168. Johnson M, Crowley P, Foley A, Xue C, Gallagher H, Buggy D. Does perioperative iv lidocaine infusion during tumour resection surgery reduce metastatic disease in the 4T1 mouse model of breast cancer? *Br J Anaesth* 2018;**120**(1):e1–2.
169. D'Agostino G, Saporito A, Cecchinato V, Silvestri Y, Borgeat A, Anselmi L, et al. Lidocaine inhibits cytoskeletal remodelling and human breast cancer cell migration. *Br J Anaesth* 2018;**121**(4):962–8.
170. Finbow ME, Harrison MA. The vacuolar H^+-ATPase: a universal proton pump of eukaryotes. *Biochem J* 1997;**324**(Pt. 3):697–712.
171. De Milito A, Iessi E, Logozzi M, Lozupone F, Spada M, Marino ML, et al. Proton pump inhibitors induce apoptosis of human B-cell tumors through a caspase-independent mechanism involving reactive oxygen species. *Cancer Res* 2007;**67**(11):5408–17.

172. Spugnini EP, Citro G, Fais S. Proton pump inhibitors as anti vacuolar-ATPases drugs: a novel anti-cancer strategy. *J Exp Clin Cancer Res* 2010;**29**(1):44.
173. Fais S. Proton pump inhibitor-induced tumour cell death by inhibition of a detoxification mechanism. *J Intern Med* 2010;**267**(5):515–25.
174. Luciani F, Spada M, De Milito A, Molinari A, Rivoltini L, Montinaro A, et al. Effect of proton pump inhibitor pretreatment on resistance of solid tumors to cytotoxic drugs. *J Natl Cancer Inst* 2004; **96**(22):1702–13.
175. Fais S. Evidence-based support for the use of proton pump inhibitors in cancer therapy. *J Transl Med* 2015;**13**(1):368.
176. Wang BY, Zhang J, Wang JL, Sun S, Wang ZH, Wang LP, et al. Intermittent high dose proton pump inhibitor enhances the antitumor effects of chemotherapy in metastatic breast cancer. *J Exp Clin Cancer Res* 2015;**34**(1):85.
177. Lindner K, Borchardt C, Schöpp M, Bürgers A, Stock C, Hussey DJ, et al. Proton pump inhibitors (PPIs) impact on tumour cell survival, metastatic potential and chemotherapy resistance, and affect expression of resistance-relevant miRNAs in esophageal cancer. *J Exp Clin Cancer Res* 2014; **33**(1):73.
178. Bellone M, Calcinotto A, Filipazzi P, De Milito A, Fais S, Rivoltini L. The acidity of the tumor micro-environment is a mechanism of immune escape that can be overcome by proton pump inhibitors. *Oncoimmunology* 2013;**2**(1).
179. Morimura T, Fujita K, Akita M, Nagashima M, Satomi A. The proton pump inhibitor inhibits cell growth and induces apoptosis in human hepatoblastoma. *Pediatr Surg Int* 2008;**24**(10):1087–94.
180. Whitton B, Okamoto H, Packham G, Crabb SJ. Vacuolar ATPase as a potential therapeutic target and mediator of treatment resistance in cancer. *Cancer Med* 2018. https://doi.org/10.1002/cam4.1594.
181. Spugnini E, Fais S. Proton pump inhibition and cancer therapeutics: a specific tumor targeting or it is a phenomenon secondary to a systemic buffering? *Semin Cancer Biol* 2017;**43**:111–8.
182. De Milito A, Canese R, Marino ML, Borghi M, Iero M, Villa A, et al. pH-dependent antitumor activity of proton pump inhibitors against human melanoma is mediated by inhibition of tumor acidity. *Int J Cancer* 2010;**127**(1):207–19.
183. Tvingsholm SA, Dehlendorff C, Østerlind K, Friis S, Jäättelä M. Proton pump inhibitor use and cancer mortality. *Int J Cancer* 2018;**143**(6):1315–26.
184. Kobayashi J, Uchida H, Ito J. Long-term proton pump inhibitor use after *Helicobacter pylori* eradication may create a gastric environment for N-nitrosamine formation and gastric cancer development. *Gut* 2018. https://doi.org/10.1136/gutjnl-2018-316592.
185. Hálfdánarson ÓÖ, Fall K, Ogmundsdottir MH, Lund SH, Steingrímsson E, Ogmundsdottir HM, et al. Proton pump inhibitor use and risk of breast cancer, prostate cancer, and malignant melanoma: an Icelandic population-based case-control study. *Pharmacoepidemiol Drug Saf* 2018. https://doi.org/10.1002/pds.4702.
186. Fais S, Venturi G, Gatenby B. Microenvironmental acidosis in carcinogenesis and metastases: new strategies in prevention and therapy. *Cancer Metastasis Rev* 2014;**33**(4):1095–108.
187. Yang KR, Mooney SM, Zarif JC, Coffey DS, Taichman RS, Pienta KJ. Niche inheritance: a cooperative pathway to enhance cancer cell fitness through ecosystem engineering. *J Cell Biochem* 2014;**115**(9):1478–85.
188. Chen M, Lu J, Wei W, Lv Y, Zhang X, Yao Y, et al. Effects of proton pump inhibitors on reversing multidrug resistance via downregulating V-ATPases/PI3K/Akt/mTOR/HIF-1α signaling pathway through TSC1/2 complex and Rheb in human gastric adenocarcinoma cells in vitro and in vivo. *Onco Targets Ther* 2018;**11**:6705.
189. Luciani F, Spada M, De Milito A, Molinari A, Rivoltini L, Montinaro A, et al. Effect of proton pump inhibitor pretreatment on resistance of solid tumors to cytotoxic drugs. *J Natl Cancer Inst* 2004;**96**:1702–13.

CHAPTER 15

Pharmacological interventions part III

Transporter blockers

MCT inhibitors

Lonidamine (LON)

Lonidamine ([1-(2,4-dichlorobenzyl)-1-H-indazol-3-carboxylic acid], LON), first investigated in 1979 as an anti-spermogenic drug,[1] has a limited effect on cancer as a stand-alone compound, but when associated with other anti-cancer treatments its effects are remarkable.

Lonidamine is not only a transporter blocker but is also a metabolic modifier, with some effects similar to 3-bromopyruvate (3BP), in that it is a metabolic acidifier of the intracellular milieu and it depletes cellular ATP.[2] While its exact mechanism of action is not fully understood, it shows clear anti-cancer effects and decreases both aerobic glycolysis and oxidative phosphorylation simultaneously to markedly diminish ATP production. LON affects the energy metabolism of normal and malignant cells in different ways: while it reduces oxygen consumption in both normal and neoplastic cells, it increases the aerobic glycolysis of normal cells and inhibits it in malignant cells.[3] The cellular site of action of LON seems to be at the mitochondrial level: when cells are exposed to LON the mitochondria become swollen and there is loss of their matrix,[1, 4].

The mechanisms postulated for the anti-cancer effects are:

(1) Inhibition of monocarboxylate transporters, MCT1, MCT2 and MCT4 and inhibition of the mitochondrial pyruvate carrier.[5]

(2) Inhibition of lactate efflux and cellular acidification[6]: LON is able to decrease intracellular pH by between 0.25 and 0.45 units according to dose. The depletion of ATP is approximately 50% while inorganic phosphate increases 280% compared to control. The inorganic phosphate also probably plays a role in intracellular acidification.

(3) Cellular acidification through interference with lysosomal acidification[7] and lactate accumulation.

(4) Energy depletion.[8]

(5) Decreased protein synthesis, which is probably secondary to energy depletion.[9]

(6) Depressed oxygen consumption of all the mitochondrial complexes (pan-mitochondrial depressor).[10]

(7) Hexokinase inhibition.[11]

(8) Elevation of intracellular calcium.[12]

An Innovative Approach to Understanding and Treating Cancer: Targeting pH
https://doi.org/10.1016/B978-0-12-819059-3.00015-0
© 2020 Elsevier Inc.
All rights reserved.

(9) Suppression of the pentose phosphate pathway resulting in inhibition of antioxidant generation (NADPH and glutathione).[13]

Based on the existing evidence it can be assumed that the main anti-tumor activity of LON is due to two major effects:

(1) cytoplasmic accumulation of lactic acid produced by inhibition of mitochondrial respiration and MCTs leading to deep intracellular acidification[14] and

(2) marked energy depletion.

We cannot exclude the possibility that most of the other anti-cancer effects are a consequence of cellular acidification. Ben-Horin et al.[15] showed that in breast cancer cells perfused with LON, pHi levels decreased to around 6.5 as lactate accumulated inside, extracellular lactate decreased, and intracellular ATP levels fell to 66% of the initial values. While the association of LON with 2 deoxy-glucose did not increase intracellular acidification, it did decrease intracellular accumulation of lactate and increased ATP depletion. The conclusions arising from these experiments are that LON does not significantly inhibit glycolysis at clinically achievable concentrations and is a strong inhibitor of lactate extrusion. Furthermore, those authors also found that "*...modifications of extracellular pH altered the effects of lonidamine on pH and ATP levels; pH declined more quickly and ATP was more depleted in an acidic environment than in a basic environment.*"

Ben-Yoseph et al.,[6] further confirmed these findings and reported that the intracellular acidification and energy depletion were selective for malignant cells without major changes in non-malignant cells.

Clinical experience with LON associated with diacepam in recurrent glioblastoma showed stabilization of the disease in 50% of cases with a median time-to-progression of 8 weeks. There were no complete or partial responses.[16]

In a phase II clinical study with LON as a stand-alone drug in advanced heavily treated breast cancer, 11% of the patients showed a partial response, 11% a minor response and 7% had stable disease for more than 3 months. No major adverse events were reported with a dose of 600 mg a day. The only side effect was myalgia, which lasted for only 2 weeks.[17] In another clinical trial by the National Cancer Institute of Canada Clinical Trials Group in patients with metastatic breast cancer, only 17% responded to LON alone, but adverse events were serious enough in 20% to interrupt the treatment.[18] When LON was used alone on patients with heavily treated metastatic breast cancer, 60% improved (partial remission in 16% and stable disease in 44%).[19] In a clinical trial with patients with inoperable NSCLC (non small cell lung cancer) receiving chemotherapy, the addition of LON significantly improved the results without adverse effects (43% vs. 24% of major responses).[20] In spite of the promising perspectives of LON in cancer treatment, research seems to have been abandoned. LON may have a place in a pH-centered treatment and in reversion of multidrug resistance (MDR). It has never been tested with amiloride that inhibits NHE1 or with carbonic anhydrase inhibitors. On a speculative basis, the association of these drugs may achieve enough cellular acidification to induce apoptosis.

Fig. 1 shows LON's mechanisms of cell acidification.

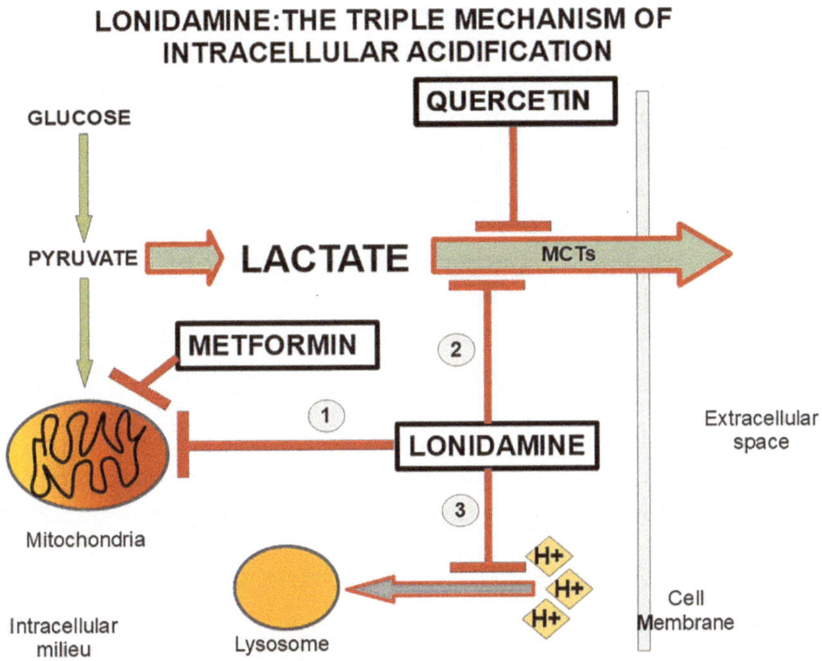

Fig. 1 LON (1) inhibits mitochondrial respiration, promoting lactic acid increase. (2) Lactic acid cannot be extruded because LON also inhibits MCTs. (3) Finally, LON blocks the influx of protons to lysosomes and endosomes reducing the clearance of protons from the cytoplasm. Presumably if proton extrusion from the cell is further blocked with amiloride and proton pump inhibitors, a major hyperacidification could be achieved. Quercetin also inhibits MCTs.

Lonidamine (LON) and multidrug resistance (MDR)

MDR was described by Bielder and Reiehm in 1970,[21] after treating ovarian cancer cells with increasing concentrations of actinomycin D. After a short interval these cells became resistant not only to actinomycin D, but also to many other cytotoxic drugs not used in the initial experiment, such as doxorubicin, daunomycin, vincristine, etoposide, colchicine and others. This initial report showed that the use of cytotoxic drugs had the ability to decrease or fully impede the effects of other not related cytotoxic drugs and resulting in resistance to most of the other drugs. This major problem of modern oncology is a complex one in which there is overexpression of drug extruders of the ATP binding cassette family and protective proteins like clusterin, activation of certain pathways, energy imbalance, etc. and pH gradient inversion plays an important role. LON has shown the ability to reverse MDR in many experimental and clinical settings perhaps due to its cell acidification and energy downregulation.

Evidence of LON's anti-MDR effects

In breast cancer cell lines (MCF-7)

(1) Del Bufalo et al.[22] showed that LON could reverse MDR in different malignant cells: "*...exposure to 50 microg/ml of lonidamine induces apoptosis in adriamycin and nitrosourea-resistant cells (MCF-7 ADR(r) human breast cancer cell line, and LB9 glioblastoma multiform cell line).*" This reversion was independent of p53 status and what is more surprising is that LON did not induce apoptosis in the same cell line that was not drug resistant.

(2) Citro et al.[23] tested the effects of LON in two breast cancer cell lines: MCF-7 WT (wild type) and MCF-7 ADR being the last one resistant to adryamicin. The addition of LON to adryamicin in both cell lines markedly increased lethality but the increased adryamycin cytotoxicity was more significant for the adryamicin resistant cells than for their sensitive counterparts. In this report, the authors attribute the results to three characteristics of LON: (1) LON's interactions with the cell membrane that impede P-gp activity; (2) LON's interference with mitochondrial bound hexokinase that decreases glycolysis; and (3) energy decrease that interferes with the drug extrusion process. Here, we have a different interpretation: LON does not decrease glycolysis at clinically achievable concentrations and, therefore, it is the lower pHi achieved through intracellular lactic acid accumulation that is reversing the MDR.

(3) However Vivi et al.[24] obtained different results. They compared the effects of LON in two lines of breast cancer cells: adryamicin sensitive and adryamicin resistant. They found that LON caused intracellular acidification and ATP depletion in both cell lines but pHi and ATP decreased less in the resistant cells than in the wild type. LON, as expected, induced intracellular lactate accumulation in both. The difference with the previous reports was that intracellular lactate levels increased more in WT than in the adryamicin resistant cells. There is a reason that justifies this controversial finding: LON enhanced glucose uptake in the WT cells, but not in the resistant cells. The excess of glucose produced more lactate in the WT cells. This experiment also proves that LON is not a strong glycolytic inhibitor.

(4) Zupi et al.[25] Compared the effects of LON on two lines of MCF-7 breast cancer cells: sensitive and resistant to adriamycin. Resistant cells showed a 70% reduction of P-gp phosphorylation after 24 h and increased intracellular adriamycin accumulation.

Other cell lines

(5) Li et al.[26] showed that LON induced apoptosis in doxorubicin-resistant human hepatocarcinoma cells. As in the report of Del Bufalo et al. they found that the non-resistant cells were less sensitive to LON.

(6) An experiment by Fanciulli et al.[27] shed more light on the issue. They used two different colon carcinoma cells LoVoDx and LoVoDx10. The two lines are doxorubicin (DOX) resistant and express P-gp (P-170), but the first is less resistant (extrudes less DOX) than the second which extrudes more DOX. When these two cell lines were treated with LON, the extrusion of DOX decreased dramatically in both (100% reduction in LoVoDx and 85% in LoVoDx10) but the LoVoDx10 cells were significantly more sensitive to LON than LoVoDx. The authors attributed this difference to ATP depletion considering that "more resistance requires more ATP for P-gp activity." We have a different interpretation on the issue: P-gp is pH dependent. The more resistant cell line was "more" glycolytic than the less resistant line, as shown by the authors, therefore, LON accumulated more lactate in the LoVoDx10 and reduced more the intracellular pH than in LoVoDx.

(7) Silvestrini et al.[28] found that LON potentiated cisplatin's effects on ovarian cancer cells, sensitive and resistant to cisplatin, on both types of cells, and the increased cytotoxicity was stronger in resistant cells.

(8) In the clinical setting Tomirotti et al.[29] achieved reversal of resistance to adriamycin associated with cyclophosphamide in 32% of patients with metastatic breast cancer by adding LON to the treatment.

Quercetin (QUER)

QUER is a natural flavonoid that can be found in many plants, fruits and vegetables. It is usually sold over the counter as a dietary supplement. A normal diet contains approximately 30–50 mg of QUER. Although it is considered to be merely a nutraceutical, QUER is a compound with clearly defined pharmacological actions. One of them is the inhibition of monocarboxylate transporters (MCTs) (Fig. 1).

As with LON, the importance of inhibiting the lactate extrusion system by MCTs, in order to achieve cell acidification, stems from two facts:

(1) when MCTs transport lactate there is a co-transport of protons and

(2) intracellular acidity strongly increases transporter activity.[30]

QUER has the ability to increase intracellular lactate 3- to 4-fold and it produces a decrease of intracellular pH to 6.9.[31] QUER seems to be a more efficient cell acidifier when compared with LON.

QUER also targets other pro-cancer proteins: anion exchange protein 1 (AE1)[32] and multidrug resistance protein.[33] Synthetic methylated derivatives of QUER downregulated P-glycoprotein and breast cancer resistance protein BCRP.[34] QUER showed pro-apoptotic activities mediated by the PI3K/Akt pathway.[35, 36]

QUER's MCT inhibitory activity makes it the ideal drug to associate with metformin that increases intracellular lactate. Increased lactate with decreased efflux is the perfect storm against the cancer cell. This association had synergistic anti-tumoral effects in prostate cancer cells through downregulation of the VGEF/PI3K/Akt pathway.[37] Further information on QUER's effects in cancer is summarized in Table 1.

Table 1 Anti-cancer effects of QUER.

Reference	Findings
Jia et al. 2018[38]	QUER decreased: **(1)** the mobility of breast cancer cells **(2)** the expression of MMP2, MMP9 and VEGF **(3)** the glucose uptake **(4)** glycolysis: decreasing lactic acid production and **(5)** Induced autophagy by downregulating the PI3K/mTOR pathway
Gong et al. 2018[39]	QUER inhibited DNA double-strand break repair and enhanced radiosensitivity in ovarian cancer cells. It produced endoplasmic reticulum stress
Lei et al. 2018[40]	QUER increased the anti-tumoral activity of irinotecan in gastric cancer cells both in vitro and in vivo
Ward et al. 2018[41]	QUER inhibited anti-apoptotic proteins and cell survival in prostate cancer cells
Li et al. 2018[42]	QUER had anti-metastatic effects on gastric cancer cells. Decreased uPA/uPAR action
Teekaraman et al. 2018[43]	QUER inhibited the growth of metastatic human ovarian cancer cells.
Li S, et al. 2018[44]	QUER enhanced the cytotoxic effects of doxorubicin in human breast cancer cells
Wang et al. 2018[45]	QUER inhibited the growth of breast cancer stem cells
Ma et al. 2018[46]	QUER induced apoptosis in oral cancer cells
Li et al. 2018[47]	QUER inhibited growth of breast cancer stem cells
Alvarez et al. 2018[48]	QUER induced apoptosis in leukemia cells
Cheng et al. 2018[49]	QUER induced apoptosis in melanoma cells
Shu et al. 2018[50]	QUER down-regulated c-Met and reversed resistance to doxorubicin in prostate cancer cells

Conclusions: QUER kills cancer cells without affecting normal cells. Cellular acidification produced by QUER is probably related to this apoptotic ability because the non-cancer cells were not affected. Normal cells are not glycolytic and their lactic acid production is negligible, therefore, QUER does not acidify them and this is the reason for their survival. This suggests that QUER's selective cancer cell killing is a consequence of inhibition of the cancer cells increased MCT expression and activity.

Enzyme inhibitors

Carbonic anhydrase inhibitors (CAI)

Carbonic anhydrases (CA) are a large family of zinc metallo-enzymes that reversibly catalyze a very important biochemical step in metabolism:

$$CO_2 + H_2O \overset{[CA]}{\leftrightarrow} CO_3H^- + H^+$$

These enzymes catalyze this specific chemical reaction and intervene in a great variety of biological processes. The conversion of CO_2 to bicarbonate and a proton is the mechanism by which CAs handle the cellular CO_2 overload that is extruded to extracellular space.

Two members of this family of carbonic anhydrases are important in cancer: the high activity hypoxia-inducible membrane associated CAIX and CAXII because they are the only tumor-associated CA isoenzymes. They are highly expressed in many malignant tumors, but particularly in renal cell carcinoma while they are not found in normal kidneys. CAIX and CAXII (but particularly CAIX) are responsible for intracellular alkalinization and extracellular acidification promoting the survival of malignant cells and their growth.[51] In cancer cells, CAIX and CAXII expression in cooperation with NHE1 is an essential step against cytoplasmic acidification. Furthermore, CAIX expression was able to replace NHE1 as a cell alkalinizer, when the later was experimentally down-regulated. When both, CAIX and CAXII were down-regulated in a mouse xenograft model, tumor growth rate was decreased.[51]

There are many reviews on CA activity,[52–55] and it is discussed in detail in Chapter 7 in the first part of this book. Here, we will focus on acetazolamide (AZD) as the model CA inhibitor. AZD has been used for over fifty years as a FDA approved drug for the treatment of glaucoma and as a diuretic.

In spite of the fact that many new CA inhibitors have been discovered in recent years, AZD associated with NHE1 inhibitors such as amiloride is the drug of choice for reversing the pH gradient.

Evidence of anti-cancer activity with AZD is summarized in Table 2.

Table 2 Relationship between AZD and cancer.

Reference	Findings
Teicher et al. 1993[56]	AZD alone delayed tumor growth in vitro and added further growth delay when a chemotherapeutic drug was associated. When local hyperthermia was applied to the tumor, there was a further growth delay
Parkkila et al. 2000[57]	AZD inhibited invasion of renal cancer cells in vitro
Xiang et al. 2004[58]	AZD inhibited aquaporin 1 and angiogenesis
Cianchi et al. 2010[59]	AZD decreased cell proliferation and induced ceramide-mediated apoptosis in human cancer cells
Bin et al. 2011[60]	AZD inhibited renal cancer xenograft growth. It also inhibited aquaporin 1
Dubois et al. 2011[61]	AZD increased the effects of radiotherapy
Faes et al. 2016[62]	AZD increased the anti-cancer effects of mTOR inhibitors because mTOR inhibitors target only non hypoxic areas of the tumor, while AZD has a predominant action in highly hypoxic parts
Vaeteewoottacharn et al. 2016[63]	AZD increased the effects of anti-angiogenic treatments with bevacizumab in cholangiocarcinoma cells

Continued

Table 2 Relationship between AZD and cancer—cont'd

Reference	Findings
Karakus et al. 2018[64]	AZD affected cell survival, migration and colony formation in vitro in a dose dependent manner in colon cancer cells
Noma et al. 2018[65]	AZD inhibited colonic polyp formation in mice through suppression of IL6 and activation of NRF2
Mokhtary et al. 2018[66]	The association of AZD with sulforaphane reduced proliferation and viability of bronchial carcinoid cells in culture
Islam et al. 2018[67]	The association of AZD with sulforaphane inhibited the epithelial-mesenchymal transition and dose dependent proliferation decreased in bladder cancer cells

Conclusions: AZD has anti-tumor effects, not only at the therapeutic level but also as a preventive. It enhances the cytotoxic activity of chemotherapeutics, radiotherapy and anti-angiogenic treatments.

Drugs with non fully identified mechanisms

3-Bromopyruvate (3BP)

A certain success in treating advanced cancers with 3BP has already been demonstrated in the clinical setting.[68, 69] Multiple reports indicate that the small molecule "energy blocker" 3BP transport into cancer cells is mediated by H^+-coupled or Na^+-coupled MCTs.[70] MCT1, in particular, is essential to its cytotoxic action.[71, 72] The tumoral extracellular acidic pH increases 3BP uptake. As a result, the inverted pH gradient enhances its selective cytotoxicity in tumor cells. Indeed, this typical acid-base abnormality of the malignant tumor microenvironment could also explain the relatively few secondary effects of 3BP described in in vivo studies. In addition, it has also been shown that 3BP inhibits glycolysis and mitochondrial respiration inducing cell death in leukemia cells while also acting as a chemosensitizing agent.[73] Actually, the exact mechanism of action of 3BP is not fully known. While it seems that the main activity is related to inhibition of glyceraldehyde 3 phosphate dehydrogenase[74] (GAPDH), other authors found Hexokinase II to be the main target.[75] 3BP is also an alkylating agent[76] that "alkylates" proteins and, in this context, sixty two cellular proteins were identified as possible targets of 3BP's alkylating activity.[77] This last report confirmed GAPDH as the main target but found many other "secondary targets." 3BP is a general and not a specific alkylating agent. For example, 3BP completely inhibits serine hydroxymethyltransferase I which is an essential enzyme for the glycolytic phenotype[78] (Fig. 2). Thus 3BP is a multitarget molecule in which the more evident action is glycolytic and mitochondrial inhibition, but there are many other actions that have not been fully investigated.

The important anti-cancer activity of 3BP deserves further analysis to determine if this compound will be able to enter clinical practice in the near future. With the exception of one patient who received a treatment with 3BP on a compassionate basis, there is no real experience in humans. Mice treated with 3BP tolerated 8 mg/kg without any sign of toxicity and reached tumoricidal level; 16 mg/kg was hepatotoxic.[81]

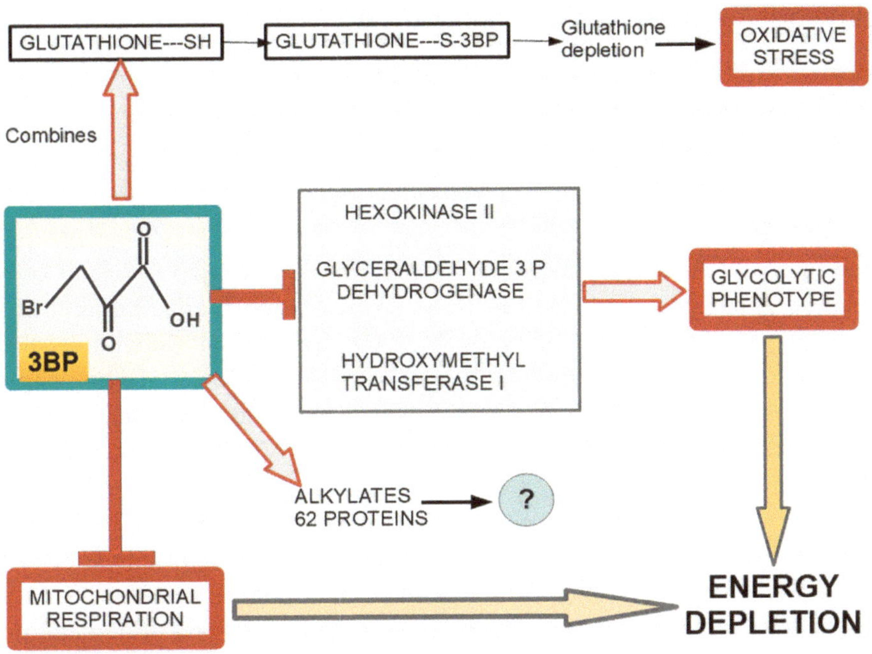

Fig. 2 Known and unknown actions of 3BP. Glutathione depletion. Energy depletion.[79, 80] The implications of the 62 proteins alkylated by 3BP are not known yet.

In a US patent for 3BP in cancer, the patent holders suggest a dose in the range of 100 mg to 200 mg per kg.[82]

3BP has been tested experimentally against all kind of tumors with good results.

Glioblastoma, probably the most difficult to treat, and in which all the classical treatments have had disappointing results, was no exception.

A few examples of these studies in *glioblastoma* are summarized in Table 3.

Table 3 3BP and its activity in glioblastoma.

Reference	Findings
Xu et al. 2016[83]	3BP targeted CD133 positive stem cells of the tumor and inhibited proliferation
Ishiguro et al. 2018[84]	The association of valproate and 3BP increased cytotoxicity in glioblastoma
Ehrke et al. 2015[85]	3BP showed cytotoxic effects on primary rat astrocytes through glutathione depletion and inhibition of glycolysis
Chiasserini et al. 2017[86]	3BP decreased glycolytic enzymes and activated the p38 pathway in glioblastoma cells
Davidescu et al. 2015[87]	Lonidamine and 3BP induced apoptosis in glioma cells but via different pathways. ATP decreased 1 h after treatment with 3BP and 40 h after lonidamide. There was a higher autophagic response with 3BP
El-Sayed 2018[88]	3BP does not cross the blood-brain barrier

Conclusions: 3BP is a useful drug against glioblastoma in the laboratory, but it is very difficult to translate this to the clinical setting: it does not cross the blood-brain barrier and has cytotoxic effects on normal glial cells. Therefore it must be delivered with nanoparticules that enhance its trafficking through the barrier or perhaps through some form of trans-nasal administration.

Miscelaneous mechanisms of acidification

There is a large family of cellular acidifiers indirectly related to PTI (proton transport inhibition) that have shown strong anticancer properties and, here, are being considered as a group for the first time.

Salinomycin (SAL)

Salinomycin is a K^+ ionophore antibiotic isolated from *Streptomyces albus* and known for more than 30 years. It has been used in veterinary medicine as an antimicrobial against Gram negative bacteria and coccidioides. In 2009, Gupta et al.[89] identified SAL in a screening among 16,000 compounds, as having a strong anti-cancer activity, particularly against cancer stem cells. Salinomycin kills cancer stem cells in different types of human cancers by interfering with ABC drug transporters, the Wnt/β-catenin signaling pathway, mitochondrial function and other cancer stem cell (CSC) pathways.[90, 91] By comparing the chemical structures and cellular effects of this drug with those of valinomycin (K^+ ionophore) and nigericin (K^+/H^+ exchanger), these authors concluded that salinomycin mediates K^+/H^+ exchange across the inner mitochondrial membrane. Clinically, this drug has been able to induce partial regressions in heavily pre-treated and therapy-resistant cancer patients.[92] The antitumor mechanisms described for salinomycin's activity include the efflux of intracellular K^+ and a reduction of pHi, an increase in intracellular Ca^{2+} and a down-regulation of MDR. Such antitumor mechanisms are connected to its strong affinity to potassium cations. Salinomycin promotes the outflow of K^+ from the mitochondria and the cytoplasm, and similarly to other ionophores mediates a H^+ / K^+ exchange across lipid membranes. This in turn leads to an increase in intracellular Ca^{2+} and a down-regulation of MDR.[93] The inhibition of mitochondrial Na^+/Ca^{2+}-exchangers (NCX) by CGP-37157 was shown to inhibit the mitochondrial accumulation of Ca^{2+} induced by salinomycin, while maintaining its antineoplastic efficacy.[94] These data not only indicate a route to reduce or prevent salinomycin-induced neuropathy, but also rule out mitochondrial Ca^{2+} accumulation as a main potential anti-cancer mechanism. The antineoplastic effect of this and other ionophores is believed to be mainly due to their ability to directly acidify the intracellular environment since DNA synthesis stops as cell pHi decreases.[95] This confirms a direct relationship between pH and DNA synthesis, a feature that has been known for decades.[96] Finally, it has been recently shown that the cytotoxic effect and inhibition of authopagy associated with salinomycin was dramatically enhanced in acidic conditions. Cellular acidification increasing autophagic elimination of mitochondria (mitophagy), is a characteristic that has also been proposed to underlie the pathogenesis of several neurodegenerative diseases.[97, 98] Therefore, acidity around tumors may represent the fuel supporting the transfer of cations across biological membranes via K^+/H^+ exchange.

Salinomycin also has other anti-tumoral actions: (1) increasing the cytotoxicity of doxorubicin and etoposide[99]; (2) increasing death of ALDH positive lung cancer stem cells; (3) reducing migration and metastasis[100]; (4) increasing stem cell death by sequestering iron into the lysosomes[101]; (5) increasing apoptosis in prostate cancer cells by accumulation of ROS (reactive oxygen species)[102]; (6) inhibiting P-glycoprotein (P-gp)[103]; (7) sensitizing cancer cells to radiation[104]; (8) inducing apoptosis through endoplasmic reticulum stress[105]; (9) up-regulating p21 and down-regulating survivin[106]; (10) inhibiting the Akt/NF-kB axis[107]; (11) synergizing with dichloroacetate cytotoxicity in colon cancer cells.[108] Many of these actions are related to cellular acidification. The important issues to highlight about salinomycin are the strong evidence of its effectiveness in killing cancer stem cells[109] and circulating cancer cells.[110] There is no clinical experience with this drug in humans because the therapeutic concentration range is very narrow. Neurotoxicity is the usual problem with higher doses.

Niclosamide

Niclosamide was discovered in 1958 and was used for the treatment of human tapeworm infestations. It is ineffective against roundworms. It was approved by the FDA but it is not available in the United States. It is a weak lipophilic acid that inhibits glucose metabolism in the worm, including glucose uptake and oxidative phosphorylation.[111] Niclosamide is basically a proton carrier and as such, it targets acidic endosomes, generating antiviral activity via cytoplasmic acidification in the case of the pH-dependent human rhinoviruses and influenza virus.[112]

Niclosamide has been shown to exert an antiproliferative effect in a broad spectrum of cancers, including acute myeloid leukemia cells and solid tumors, like colon cancer, breast cancer and prostate cancer.[113] Besides the fact that niclosamide is known to hit many different intracellular signaling pathways, structure-activity analyses revealed a possible role in cytoplasmic pH control. The pathways under niclosamide's inhibitory influence are: β-catenin,[114] Notch, mTORC1,[115, 116] and STAT3.[117] Experiments in both breast cancer cells and cell-free systems demonstrated that this drug pumps protons from lysosomes to the cytosol against their concentration gradient, therefore lowering cytoplasmic pH.[118] The report by Fonseca et al.[118] that showed decreased mTORC1 signaling is particularly interesting as the reduction of this signaling by niclosamide is through neither inhibition of the PI3K/Akt pathway nor inhibition of the phosphorylation of the mTORC1 complex. The downregulation is exclusively due to intracellular acidification.

Adding niclosamide to a saccharine palladium complex of terpyridine increased apoptosis in breast cancer stem cells.[119]

Table 4 is a summary of further anti-cancer effects of niclosamide.

Table 4 Anti-cancer activities of niclosamide.

Reference	Findings
Wang et al. 2013[120]	Niclosamide was identified as a breast cancer stem cell inhibitor through drug screening
Londoño-Joshi et al. 2014[121]	Niclosamide eliminated cancer stem cells in basal type breast cancer by inhibiting the Wnt/bet catenine signaling
Ye et al. 2014[122]	Niclosamide induced apoptosis and reduced metastasis in a breast cancer model
Liu et al. 2015[123]	Niclosamide reduced migration and invasion in prostate cancer
Liu et al. 2016[124]	Niclosamide inhibited androgen receptor in castration resistant prostate cancer and synergized with abiraterone
Yu et al. 2018[125]	Niclosamide showed synergic activity with sorafenib in renal cell carcinoma
Pan et al. 2018[126]	A Phase I clinical study of abiraterone, prednisone and niclosamide determined that the recommended dose of niclosamide was 1600 mg three times a day (oral administration). No toxicity was found with this schedule
Han et al. 2018[127]	Niclosamide stopped proliferation in head and neck squamous cell carcinoma
Zuo et al. 2018[128]	Niclosamide resensitized cisplatin resistant lung cancer cells
Burock et al. 2018[129]	A phase I study of niclosamide in five patients with metastatic colon cancer showed that a daily dose of 2000 mg was safe and without toxicities. The patient that reached the higher drug concentration showed stable disease
Jin et al. 2010[130]	Niclosamide was effective in targeting myelogenous leukemia stem cells. It generated ROS and inhibited the NF-kB pathway

The NIKOLO trial, in phase I/II aiming to establish the benefits of niclosamide in cancer treatment is ongoing.

Although it is not commercially available in the United States, it is on the market in many other countries as an anti-helminthic drug.

A recent clinical trial (NCT02532114) found that niclosamide's maximum tolerated dose of 500 mg three times a day was not enough to achieve concentrations in the cancer therapeutic range.[131] This study had more than one bias: the population tested was too small, only five patients; the investigators tested only two dose schedules 500 and 1000 mg per os t.i.d. The 1000 mg schedule was considered toxic. No intermediate dose was tested such as 750 mg t.i.d., which may be in the cancer therapeutic range and at the same time may not trigger adverse events. Other on going clinical trials are: NCT02687009, NCT03123978, NCT02519582, NCT02807805.

Disulfiram

The thiocarbamate drug disulfiram, normally used to block the processing of alcohol in the body, is used in the treatment of alcoholism. It is a weak VGSC blocker. However, its anti-cancer activities do not seem to be related to voltage gated channel inhibition but rather to proteasome inhibition.[132] It has been shown that it targets cancer stem cells in triple-negative breast cancer cells.[133] It has the ability to block the P-gp extrusion pump and to inhibit the transcription factor nuclear factor-kappaB (NF-kB). In this way, it sensitizes tumors to chemotherapy, reducing angiogenesis and inhibiting tumor growth in vitro.[134] The exact relationship of disulfiram to pHi is not known. However, an acidic tumor environment enhanced the cytotoxicity of disulfiram/Cu^{2+} complexes in two different tumor lines: MCF-7 breast carcinoma and HT-29 colon carcinoma cells.[135] Disulfiram showed apoptotic effects on glioblastoma cells,[136] prostate cancer cells,[137, 138] breast cancer cells and xenografts,[139] urinary bladder cancer[140] and pancreatic cancer cells,[141] and metastatic colorectal cancer cells[142] among others.

Lipophilic statins (simvastatin, atorvastatin)

Statins show antiproliferative activity against cancer cells. The addition of statins to classical chemotherapeutic treatments increased the cytotoxicity in statin-sensitive tumors.[143] Lipophilic statins in general inhibit MCTs,[144, 145] which induces intracellular acidification.

Population-based studies showed a decrease in cancer risk in general,[146, 147] and in prostate and colorectal cancer (reviewed by Demierre et al.[148]), however these results are controversial.[149]

Diclofenac

Diclofenac has a monocarboxylic acid structure and has been reported to inhibit MCTs.[150] MCTs have a maximum transportation capacity that when surpassed by monocarboxylate molecules to be transported, leads to lactic acid accumulation in the cell. It seems that diclofenac adds to the burden of MCTs and the transporter reaches its threshold with a lower amount of lactic acid. It is also a COX1 and COX2 inhibitor. COX2 inhibition decreases angiogenesis. However, COX2 inhibition and MCTs inhibition seem to be independent antitumoral actions. A significant intracellular accumulation of lactate in cells treated with diclofenac preceded the observed effects on gene expression, suggesting a direct inhibitory effect of diclofenac on lactate efflux.

Other antitumoral actions of diclofenac are related to:

1. attenuation of Wnt/β-catenin signaling in colon cancer[151];
2. modulation of VEGF levels and arginase activity[152];
3. down-regulation of the PI3K/Akt/PTEN pathway and activation of apoptosis[153];
4. down-regulation of telomerase activity inducing apoptosis.[154]

For a review on the antitumoral actions on diclofenac read Pantziarka et al.[155]

Unfortunately, there is also some bad news about diclofenac: it may increase the risk of multidrug resistance (MDR) by increasing the activity of multidrug resistance protein 2 (MRP2).[156]

Perillyl alcohol (POH)

Monoterpene perillyl alcohol (POH) is isolated from the essential oils of several plants. It shows anti-cancer effects presumably related to its Na^+/K^+-ATPase inhibitory properties as well as to its activity as an intracellular acidifier, similar to cardiac glycosides. The exact mechanism of action in cancer has not been elucidated, however it is probably related to TGF-β and the Ras/ERK pathway.[157] POH has been used in the treatment of several malignant tumors, including gliomas. It also induces growth arrest and apoptosis of leukemia[158] and breast cancer cells.[159] There are occasional but impressive reports of complete remissions of recurrent glioblastoma treated with POH via intranasal administration.[160] The intranasal route is supposed to avoid the blood-brain barrier and reach the nervous system through the olfactory nerve.[161] Also, POH administrated via inhalation, concomitantly with oral temozolomide, has been reported to halt the progression of recurrent glioblastoma.[162] Many new compounds for the adequate delivery of perillyl alcohol have been developed recently.

Urocanic acid

This is an intermediate product in the metabolism of histidine. The final metabolite is glutamic acid. When an effective amount of urocanic acid is administered, this drug is able to acidify the cytosol causing inhibition of proliferation of both transformed and non-transformed cells. The use of urocanic acid as an enhancer of other therapeutically active agents has also been reported.[163] Urocanic acid has the ability to transport protons into the cell (this property is called protodynamic) In melanoma cells it decreases intracellular pH by 0.5 and induces apoptosis.[164] The problem is that it provides glutamic acid to the cancer cell, which may enhance the glutaminergic pathway.

Silymarin

This mixture of standardized natural flavonoids has not been described as a cellular acidifier, but it has three actions that should be kept in mind:

(1) silymarin reduces the activity of NHE1[165];
(2) it reduces HIF-1α activation[166]; and
(3) it also reduces glycolytic activity in cancer cells.[167]

Sylimarin has shown many other anti-cancer effects that go beyond the scope of this chapter, but it is worth mentioning that there is strong pre-clinical evidence of the benefits of sylimarin in clear cell renal carcinoma,[168] melanoma,[169] and hepatocarcinoma.[170]

Ciprofloxacin (cipro)

Ciprofloxacin is a "lysosomal poison" that acts as a detergent on lysosomal membranes.[171] The lysosomal content is highly acidic, and the release of its content has an acidifying effect on the cytoplasm.[172] Although Cipro's ability to acidify cytoplasm has not been experimentally confirmed, there is indirect evidence that it may have such an effect. Cipro induces apoptosis in triple negative breast cancer cells,[173] and melanoma cells[174] among many other types of cancer cells. Another mechanism of cell acidification by Cipro may be related to mitochondrial damage, because it inhibits topoisomerase II impeding mitochondrial DNA replication.[175] Glioblastoma is one of the most difficult tumors to treat and one of the main problems is delivering medication through the blood-brain barrier. Cipro can cross this barrier and has cytotoxic effects on glioblastoma cells, inducing apoptosis.[176]

5-Nonyloxytryptamine (5-NOT) and related intracellular pH acidifiers

5-NOT is an agonist of HT (hydroxytryptamine) receptors that produces intracellular acidification and apoptosis and/or necrosis in malignant cells. HT agonists also have other anti-cancer effects like down-regulation of the Akt/mTORC1 pathway[177] and depletion of cellular ATP levels.[178] The anticancer activity of 5-NOT seems to be tissue-specific because while it reduced the viability of glioma cells it did not affect uveal melanoma cells nor normal cells.

Extracellular alkalinization

The experiments by Robbey et al.[179] with sodium bicarbonate (SB) were described in Chapter 13. The administration of SB to animals harboring tumors, decreased invasion and metastasis. This was a good proof of concept of SB's potential to buffer tumor's extracellular acidity and hence block acidity-related tumor effects: migration, invasion and metastasis.

In spite of the fact that acidifying the cytoplasm of cancer cells is the best way to decrease extracellular acidity, there is no reason for refusing extra help from buffering the extracellular side.

SB increases peri-tumoral pH, and in mice with breast tumor xenografts, SB improved the efficacy of doxorubicin.[180] This is a logical and expected finding, because extracellular acidity decreases the access of weak bases to the cell.[181] The host's immunological defenses are also disabled by extracellular acidity.

While SB has shown some benefits in cancer over the short term[182, 183], SB has not been included in our treatment approach (see Chapter 17) for two reasons:

(1) inhibiting NHE1, membrane CAs, and proton pumps substantially modifies intracellular and extracellular pH simultaneously by decreasing the proton overload of the extracellular space (ECS) and

(2) it is difficult to maintain SB administration over the long term due to the hydro-electrolytic imbalance it produces (metabolic alkalosis and hypernatremia[184]).

Therefore, the usefulness of SB should be kept in mind as a possible candidate for reversing the pH gradient inversion, but only for a very short period (not more than five days) and as part of a more integral approach, such as described in the Chapter 18.

Conclusions

This and the two preceding chapters reviewed the potential pharmacological resources for reversing the pH gradient. The main accent was placed on those drugs that are available for the practicing oncologist and can be found on the market. Products with promise, albeit not on the market yet, were also considered.

The treatment of extracellular acidity was not considered a priority, because an efficient intracellular acidification also modifies the extracellular conditions.

There are a few drugs whose importance have been overlooked by mainstream practice. This is the case of quercetin, a drug which health authorities consider a food supplement and not a drug. Probably, this misconception is the result of quercetin's natural origin, normal presence in many foods, and insufficient evaluation of the existing evidence that shows the importance of this drug in cancer treatment. To this we must add, the absolute lack of toxicity. Being the only available inhibitor of MCTs at bedside, this drug deserves further research.

Sylimarin and its main compound silibin is another drug that has been overlooked by oncological research, in spite of the fact that it inhibits two of the culprits of the pH alterations: NHE1 and HIF-1α. Quercetin and sylimarin share the sin of being over-the-counter substances.

The reversal of the pH alterations in cancer requires a metronomic methodology: permanent, daily and uninterrupted reversal of the altered pH. This means that the drugs selected for treatment must also promote patient compliance: they should be non-toxic, well tolerated, and be available at reasonable cost. Furthermore, any drug involved in pH based treatment should be compatible with standard protocols. Based on these premises, a treatment scheme was developed that will be fully discussed in Chapter 17. A metastasis prevention schedule was also developed on the same basic concepts (Chapter 18). Three metabolic modifiers (metformin, dichloroacetate and 2-deoxy-D-glucose) and an immunomodulator (cimetidine) will be discussed in the Chapter 16. These drugs require a deeper consideration due to their possible importance and all the existing controversies that had to be discussed. So, this chapter should be considered a prelude to the next one.

References

1. Cioli V, Bellocci B, Putzolu S, Malomi W, Demartino C. Anti-spermogenic activity of lonidamine (AF-1890) in rabbit. *Ultramicroscopy* 1980;**5**(3):418.
2. Nath K, Nelson DS, Heitjan DF, Leeper DB, Zhou R, Glickson JD. Lonidamine induces intracellular tumor acidification and ATP depletion in breast, prostate and ovarian cancer xenografts and potentiates response to doxorubicin. *NMR Biomed* 2015;**28**(3):281–90.
3. Floridi A, Paggi MG, Marcante ML, Silvestrini B, Caputo A, de Martino C. Lonidamine, a selective inhibitor of aerobic glycolysis of murine tumor cells. *J Natl Cancer Inst* 1981;**66**(3):497–9.
4. De Martinet CD, Battelli T, Paggi MG, Nista A, Marcante ML, D'Atri S, et al. Effects of lonidamine on murine and human tumor cells in vitro. *Oncology* 1984;**41**:15–29.
5. Nancolas B, Guo L, Zhou R, Nath K, Nelson DS, Leeper DB, et al. The anti-tumor agent lonidamine is a potent inhibitor of the mitochondrial pyruvate carrier and plasma membrane monocarboxylate transporters. *Biochem J* 2016;**473**(7):929–36.
6. Ben-Yoseph O, Lyons JC, Song CW, Ross BD. Mechanism of action of lonidamine in the 9L brain tumor model involves inhibition of lactate efflux and intracellular acidification. *J Neuro-Oncol* 1998; **36**(2):149–57.
7. Dell'Antone P, Piergallini L. The antineoplastic drug lonidamine interferes with the acidification mechanism of cell organelles. *Biochim Biophys Acta, Mol Cell Res* 1997;**1358**(1):46–52.
8. Floridi A, D'Atri S, Menichini R, Marcante ML, Nista A, Silvestrini B, et al. The effect of the association of gossypol and lonidamine on the energy metabolism of Ehrlich ascites tumor cells. *Exp Mol Pathol* 1983;**38**(3):322–35.
9. Floridi A, Delpino A, Nista A, Feriozzi R, Marcante ML, Silvestrini B, et al. Effect of lonidamine on protein synthesis in neoplastic cells. *Exp Mol Pathol* 1985;**42**(3):293–305.
10. Macchioni L, Davidescu M, Roberti R, Corazzi L. The energy blockers 3-bromopyruvate and lonidamine: effects on bioenergetics of brain mitochondria. *J Bioenerg Biomembr* 2014;**46**(5):389–94.
11. Floridi A, Paggi MG, D'Atri S, De Martino C, Marcante ML, Silvestrini B, et al. Effect of lonidamine on the energy metabolism of Ehrlich ascites tumor cells. *Cancer Res* 1981;**41**(11):4661–6.
12. Castiglione S, Kennedy KA, Floridi A, Fiskum G. Non-ionophoretic elevation of intracellular Ca^{2+} by lonidamine. *Biochem Pharmacol* 1993;**46**(2):330–2.
13. Guo L, Shestov AA, Worth AJ, Nath K, Nelson DS, Leeper DB, et al. Inhibition of mitochondrial complex II by the anyicancer agent lonidamine. *J Biol Chem* 2016;**291**(1):42–57.
14. Ben-Horin H, Tassini M, Vivi A, Navon G, Kaplan O. Mechanism of action of the antineoplastic drug lonidamine: 31P and 13C nuclear magnetic resonance studies. *Cancer Res* 1995;**55**(13):2814–21.
15. Ben-Horin H, Tassini M, Vivi A, Navon G, Kaplan O. Mechanism of action of the antineoplastic drug lonidamine: 31P and 13 C nuclear magnetic resonance studies. *Cancer Res* 1995;**55**(13):2814–21.
16. Oudard S, Carpentier A, Banu E, Fauchon F, Celerier D, Poupon MF, et al. Phase II study of lonidamine and diazepam in the treatment of recurrent glioblastoma multiforme. *J Neuro-Oncol* 2003; **63**(1):81–6.
17. Mansi JL, De Graeff A, Newell DR, Glaholm J, Button D, Leach MO, et al. A phase II clinical and pharmacokinetic study of lonidamine in patients with advanced breast cancer. *Br J Cancer* 1991; **64**(3):593.
18. Band PR, Maroun J, Pritchard K, Stewart D, Coppin CM, Wilson K, et al. Phase II study of lonidamine in patients with metastatic breast cancer: a National Cancer Institute of Canada Clinical Trials Group Study. *Cancer Treat Rep* 1986;**70**(11):1305–10.
19. Pronzato P, Amoroso D, Bertelli G, Conte PF, Cusimano MP, Ciottoli GB, et al. Phase II study of lonidamine in metastatic breast cancer. *Br J Cancer* 1989;**59**(2):251.
20. Ianniello GP, Cataldis GD, Comella P, Scarpati MD, Maiorino A, Brancaccio L, et al. Cisplatin, epirubicin, and vindesine with or without lonidamine in the treatment of inoperable nonsmall cell lung carcinoma: a multicenter randomized clinical trial. *Cancer* 1996;**78**(1):63–9.

21. Bielder JL, Reiehm H. Cellular resistance to Chinese hamster cells in vitro: cross resistance, radio-autographic studies and cytogenetic studies. *Cancer Res* 1970;**30**:1174–84.
22. Del Bufalo D, Biroccio A, Soddu S, Laudonio N, D'angelo C, Sacchi A, et al. Lonidamine induces apoptosis in drug-resistant cells independently of the p53 gene. *J Clin Invest* 1996;**98**(5):1165–73.
23. Citro G, Cucco C, Verdina A, Zupi G. Reversal of adriamycin resistance by lonidamine in a human breast cancer cell line. *Br J Cancer* 1991;**64**(3):534–6.
24. Vivi A, Tassini M, Ben-Horin H, Navon G, Kaplan O. Comparison of action of the anti-neoplastic drug lonidamine on drug-sensitive and drug-resistant human breast cancer cells: 31 P and 13 C nuclear magnetic resonance studies. *Breast Cancer Res Treat* 1997;**43**(1):15–25.
25. Zupi G, Molinari A, D'Agnano I, Meschini S, Del DB, Matarrese P, et al. Adriamycin resistance modulation induced by lonidamine in human breast cancer cells. *Anticancer Res* 1995;**15**(6B): 2469–77.
26. Li YC, Fung KP, Kwok TT, Lee CY, Suen YK, Kong SK. Mitochondrial targeting drug lonidamine triggered apoptosis in doxorubicin-resistant HepG2 cells. *Life Sci* 2002;**71**(23):2729–40.
27. Fanciulli M, Bruno T, Giovannelli A, Gentile FP, Di Padova M, Rubiu O, et al. Energy metabolism of human LoVo colon carcinoma cells: correlation to drug resistance and influence of lonidamine. *Clin Cancer Res* 2000;**6**(4):1590–7.
28. Silvestrini R, Zaffaroni N, Villa R, Orlandi L, Costa A. Enhancement of cisplatin activity by lonidamine in human ovarian cancer cells. *Int J Cancer* 1992;**52**(5):813–7.
29. Tomirotti M, Bernardo G, Epifani C, Biasioli R, Franchi R, Mensi F, et al. Recovery of response to adriamycin and cyclophosphamide by lonidamine in previously treated metastatic breast-cancer patients. *Int J Oncol* 1993;**3**(2):213–7.
30. Bröer S, Rahman B, Pellegri G, Pellerin L, Martin JL, Verleysdonk S, et al. Comparison of lactate transport in astroglial cells and monocarboxylate transporter 1 (MCT 1) expressing *Xenopus laevis* oocytes expression of two different monocarboxylate transporters in astroglial cells and neurons. *J Biol Chem* 1997;**272**(48):30096–102.
31. Volk C, Kempski B, Kempski OS. Inhibition of lactate export by quercetin acidifies rat glial cells in vitro. *Neurosci Lett* 1997;**223**(2):121–4.
32. Halestrap AP, Price NT. The proton-linked monocarboxylate transporter (mct) family: structure, function and regulation. *Biochem J* 1999;**343**(2):281–99.
33. Chen C, Zhou J, Ji C. Quercetin: a potential drug to reverse multidrug resistance. *Life Sci* 2010;**87**:333–8.
34. Yuan J, Wong IL, Jiang T, Wang SW, Liu T, Wen BJ, et al. Synthesis of methylated quercetin derivatives and their reversal activities on P-gp-and BCRP-mediated multidrug resistance tumour cells. *Eur J Med Chem* 2012;**54**:413–22.
35. Gulati N, Laudet B, Zohrabian VM, Murali RA, Jhanwar-Uniyal ME. The antiproliferative effect of Quercetin in cancer cells is mediated via inhibition of the PI3K-Akt/PKB pathway. *Anticancer Res* 2006;**26**(2A):1177–81.
36. Shah KN, Patel SS. Phosphatidylinositide 3-kinase inhibition: a new potential target for the treatment of polycystic ovarian syndrome. *Pharm Biol* 2016;**54**(6):975–83.
37. Sun S, Gong F, Liu P, Miao Q. Metformin combined with quercetin synergistically repressed prostate cancer cells via inhibition of VEGF/PI3K/Akt signaling pathway. *Gene* 2018;**664**:50–7.
38. Jia L, Huang S, Yin X, Zan Y, Guo Y, Han L. Quercetin suppresses the mobility of breast cancer by suppressing glycolysis through Akt-mTOR pathway mediated autophagy induction. *Life Sci* 2018; **208**:123–30.
39. Gong C, Yang Z, Zhang L, Wang Y, Gong W, Liu Y. Quercetin suppresses DNA double-strand break repair and enhances the radiosensitivity of human ovarian cancer cells via p53-dependent endoplasmic reticulum stress pathway. *Onco Targets Ther* 2018;**11**:17.
40. Lei CS, Hou YC, Pai MH, Lin MT, Yeh SL. Effects of quercetin combined with anticancer drugs on metastasis-associated factors of gastric cancer cells: in vitro and in vivo studies. *J Nutr Biochem* 2018;**51**:105–13.
41. Ward AB, Mir H, Kapur N, Gales DN, Carriere PP, Singh S. Quercetin inhibits prostate cancer by attenuating cell survival and inhibiting anti-apoptotic pathways. *World J Surg Oncol* 2018;**16**(1):108.

42. Li H, Chen C. Quercetin has antimetastatic effects on gastric cancer cells via the interruption of uPA/uPAR function by modulating NF-κb, PKC-δ, ERK1/2, and AMPKα. *Integr Cancer Ther* 2018; **17**(2):511–23.
43. Teekaraman D, Elayapillai SP, Viswanathan MP, Jagadeesan A. Quercetin inhibits human metastatic ovarian cancer cell growth by modulating intrinsic apoptotic pathway in PA-1 cell line. *Chem Biol Interact* 2019;**300**:91–100.
44. Li S, Yuan S, Zhao Q, Wang B, Wang X, Li K. Quercetin enhances chemotherapeutic effect of doxorubicin against human breast cancer cells while reducing toxic side effects of it. *Biomed Pharmacother* 2018;**100**:441–7.
45. Wang R, Yang L, Li S, Ye D, Yang L, Liu Q, et al. Quercetin inhibits breast cancer stem cells via downregulation of aldehyde dehydrogenase 1A1 (ALDH1A1), chemokine receptor type 4 (CXCR4), mucin 1 (MUC1), and epithelial cell adhesion molecule (EpCAM). *Med Sci Monit* 2018;**24**:412.
46. Ma YS, Yao CN, Liu HC, Yu FS, Lin JJ, Lu KW, et al. Quercetin induced apoptosis of human oral cancer SAS cells through mitochondria and endoplasmic reticulum mediated signaling pathways. *Oncol Lett* 2018;**15**(6):9663–72.
47. Li X, Zhou N, Wang J, Liu Z, Wang X, Zhang Q, et al. Quercetin suppresses breast cancer stem cells (CD44+/CD24−) by inhibiting the PI3K/Akt/mTOR-signaling pathway. *Life Sci* 2018;**196**:56–62.
48. Alvarez MC, Maso V, Torello CO, Ferro KP, Saad ST. The polyphenol quercetin induces cell death in leukemia by targeting epigenetic regulators of pro-apoptotic genes. *Clin Epigenetics* 2018;**10**(1):139.
49. Cheng Q, Song L. Antiproliferation and apoptosis-inducing effects of Quercetin in human melanoma CHL-1 cells. *J Pharm Biomed Sci* 2018;**8**(2).
50. Shu Y, Xie B, Liang Z, Chen J. Quercetin reverses the doxorubicin resistance of prostate cancer cells by downregulating the expression of c-met. *Oncol Lett* 2018;**15**(2):2252–8.
51. Chiche J, Ilc K, Laferriere J, Trottier E, Dayan F, Mazure MN, et al. Hypoxia-inducible carbonic anhydrase IX and XII promote tumor cell growth by counteracting acidosis through the regulation of the intracellular pH. *Cancer Res* 2009;**69**(1):358.
52. McDonald PC, Winum J-Y, Supuran CT, Dedhar S. Recent developments in targeting carbonic anhydrase IX for cancer therapeutics. *Oncotarget* 2012;**3**:84–97.
53. Swietach P, Vaughan-Jones RD, Harris AL. Regulation of tumor pH and the role of carbonic anhydrase 9. *Cancer Metastasis Rev* 2007;**26**(2):299–310.
54. Ozensoy Guler O, Capasso C, Supuran CT. A magnificent enzyme superfamily: carbonic anhydrases, their purification and characterization. *J Enzyme Inhib Med Chem* 2016;**31**(5):689–94.
55. Supuran CT. Carbonic anhydrases and metabolism. *Metabolites* 2018;**8**(2):25. https://doi.org/10.3390/metabo8020025.
56. Teicher BA, Liu SD, Liu JT, Holden SA, Herman TS. A carbonic anhydrase inhibitor as a potential modulator of cancer therapies. *Anticancer Res* 1993;**13**(5A):1549–56.
57. Parkkila S, Rajaniemi H, Parkkila AK, Kivelä J, Waheed A, Pastoreková S, et al. Carbonic anhydrase inhibitor suppresses invasion of renal cancer cells in vitro. *Proc Natl Acad Sci* 2000;**97**(5):2220–4.
58. Xiang Y, Ma B, Li T, Gao JW, Yu HM, Li XJ. Acetazolamide inhibits aquaporin-1 protein expression and angiogenesis. *Acta Pharmacol Sin* 2004;**25**:812–6.
59. Cianchi F, Vinci MC, Supuran CT, Peruzzi B, De Giuli P, Fasolis G, et al. Selective inhibition of carbonic anhydrase IX decreases cell proliferation and induces ceramide-mediated apoptosis in human cancer cells. *J Pharmacol Exp Ther* 2010;**334**:710–9.
60. Bin K, Shi-Peng Z. Acetazolamide inhibits aquaporin-1 expression and colon cancer xenograft tumor growth. *Hepato-Gastroenterology* 2011;**58**(110–111):1502–6.
61. Dubois L, Peeters S, Lieuwes NG, Geusens N, Thiry A, Wigfield S, et al. Specific inhibition of carbonic anhydrase IX activity enhances the in vivo therapeutic effect of tumor irradiation. *Radiother Oncol* 2011;**99**:424–31.
62. Faes S, Planche A, Uldry E, Santoro T, Pythoud C, Stehle JC, et al. Targeting carbonic anhydrase IX improves the anti-cancer efficacy of mTOR inhibitors. *Oncotarget* 2016;**7**(24):36666.
63. Vaeteewoottacharn K, Kariya R, Dana P, Fujikawa S, Matsuda K, Ohkuma K, et al. Inhibition of carbonic anhydrase potentiates bevacizumab treatment in cholangiocarcinoma. *Tumour Biol* 2016; **37**(7):9023–35.

64. Karakuş F, Eyol E, Yılmaz K, Ünüvar S. In vitro effect of carbonic anhydrase inhibitor acetazolamide on cell viability, migration and colony formation of colorectal cancer cells. *Biologia* 2018;**73**(6):621–8.
65. Noma N, Fujii G, Miyamoto S, Komiya M, Nakanishi R, Shimura M, et al. Impact of acetazolamide, a carbonic anhydrase. *Infirm Can* 2018;**18**:208.
66. Mokhtari RB, Kumar S, Islam SS, Yazdanpanah M, Adeli K, Cutz E, et al. Combination of carbonic anhydrase inhibitor, acetazolamide, and sulforaphane, reduces the viability and growth of bronchial carcinoid cell lines. *BMC Cancer* 2013;**13**(1):378.
67. Islam SS, Mokhtari RB, Akbari P, Hatina J, Yeger H, Farhat WA. Simultaneous targeting of bladder tumor growth, survival, and epithelial-to-mesenchymal transition with a novel therapeutic combination of acetazolamide (AZ) and sulforaphane (SFN). *Target Oncol* 2016;**11**(2):209–27.
68. Ko YH, Verhoeven HA, Lee MJ, Corbin DJ, Vogl TJ, Pedersen PL. A translational study "case report" on the small molecule "energy blocker" 3-bromopyruvate (3BP) as a potent anticancer agent: from bench side to bedside. *J Bioenerg Biomembr* 2012;**44**:163–70.
69. Ko YH, Smith BL, Wang Y, Pomper MG, Rini DA, Torbenson MS, et al. Advanced cancers: eradication in all cases using 3-bromopyruvate therapy to deplete ATP. *Biochem Biophys Res Commun* 2004;**324**:269–75.
70. Ganapathy-Kanniappan S, Kunjithapatham R, Geschwind JF. Anticancer efficacy of the metabolic blocker 3-bromopyruvate: specific molecular targeting. *Anticancer Res* 2013;**33**(1):13–20.
71. Sprowl-Tanio S, Habowski AN, Pate KT, McQuade MM, Wang K, Edwards RA, et al. Lactate/pyruvate transporter MCT-1 is a direct Wnt target that confers sensitivity to 3-bromopyruvate in colon cancer. *Cancer & Metab* 2016;**4**(1):20.
72. Azevedo-Silva J, Queiros O, Ribeiro A, Baltazar F, Young KH, Pedersen PL, et al. The cytotoxicity of 3-bromopyruvate in breast cancer cells depends on extracellular pH. *Biochem J* 2015;**467**:247–58.
73. Calvino E, Estan MC, Sanchez-Martin C, Brea R, de Blas E, Boyano-Adanez Mdel C, et al. Regulation of death induction and chemosensitizing action of 3-bromopyruvate in myeloid leukemia cells: energy depletion, oxidative stress, and protein kinase activity modulation. *J Pharmacol Exp Ther* 2014;**348**:324–35.
74. Ganapathy-Kanniappan S, Geschwind JF, Kunjithapatham R, Buijs M, Vossen JA, Tchernyshyov I, et al. Glyceraldehyde-3-phosphate dehydrogenase (GAPDH) is pyruvylated during 3-bromopyruvate mediated cancer cell death. *Anticancer Res* 2009;**29**(12):4909–18.
75. Chen Z, Zhang H, Lu W, Huang P. Role of mitochondria-associated hexokinase II in cancer cell death induced by 3-bromopyruvate. *Biochim Biophys Acta* 2009;**1787**(5):553–60.
76. Cardaci S, Desideri E, Ciriolo MR. Targeting aerobic glycolysis: 3-bromopyruvate as a promisinganticancer drug. *J Bioenerg Biomembr* 2012;**44**(1):17–29.
77. Darabedian N, Chen TC, Molina H, Pratt MR, Schönthal AH. Bioorthogonal profiling of a cancer cell proteome identifies a large set of 3-bromopyruvate targets beyond glycolysis. *ACS Chem Biol* 2018; **13**(11):3054–8.
78. Paiardini A, Tramonti A, Schirch D, Guiducci G, di Salvo ML, Fiascarelli A, et al. Differential 3-bromopyruvate inhibition of cytosolic and mitochondrial human serine hydroxymethyltransferase isoforms, key enzymes in cancer metabolic reprogramming. *Biochim Biophys Acta, Proteins Proteomics* 2016;**1864**(11):1506–17.
79. Pan Q, Sun Y, Jin Q, Li Q, Wang Q, Liu H, Zhao S. Hepatotoxicity and nephrotoxicity of 3-bromopyruvate in mice. *Acta Cir Bras* 2016;**31**(11):724–9.
80. Sadowska-Bartosz I, Szewczyk R, Jaremko L, Jaremko M, Bartosz G. Anticancer agent 3-bromopyruvic acid forms a conjugate with glutathione. *Pharmacol Rep* 2016;**68**(2):502–5.
81. Pereira Da Silva AP, El-Bacha T, Kyaw N, Santos Dos RS, Da-Silva WS, Almeida FC, et al. Inhibition of energy-producing pathways of HepG2 cells by 3-bromopyruvate. *Biochem J* 2009;**417**:717–26.
82. Geschwind J-F, Vali M. *Methods and compositions for administration of 3-halopyruvate and related compounds for the treatment of cancer*. U.S. Patent Application No. US 15/284,014, 2017. https://patents.google.com/patent/US20170224641A1/en.
83. Xu DQ, Tan XY, Zhang BW, Wu T, Liu P, Sun SJ, Cao YG. 3-Bromopyruvate inhibits cell proliferation and induces apoptosis in CD133+ population in human glioma. *Tumor Biol* 2016; **37**(3):3543–8.

84. Ishiguro Y, Kobayashi M, Ideno M, Narumi K, Furugen A, Iseki K. Valproate sensitizes human glioblastoma cells to 3-bromopyruvate-induced cytotoxicity. *Int J Pharm* 2018;**551**(1–2):97–102.
85. Ehrke E, Arend C, Dringen R. 3-bromopyruvate inhibits glycolysis, depletes cellular glutathione, and compromises the viability of cultured primary rat astrocytes. *J Neurosci Res* 2015;**93**(7):1138–46.
86. Chiasserini D, Davidescu M, Orvietani PL, Susta F, Macchioni L, Petricciuolo M, et al. 3-Bromopyruvate treatment induces alterations of metabolic and stress-related pathways in glioblastoma cells. *J Proteome* 2017;**152**:329–38.
87. Davidescu M, Macchioni L, Scaramozzino G, Marchetti MC, Migliorati G, Vitale R, et al. The energy blockers bromopyruvate and lonidamine lead GL15 glioblastoma cells to death by different p53-dependent routes. *Sci Rep* 2015;**5**:14343.
88. El Sayed SM. Enhancing anticancer effects, decreasing risks and solving practical problems facing 3-bromopyruvate in clinical oncology: 10 years of research experience. *Int J Nanomedicine* 2018;**13**:4699.
89. Gupta PB, Onder TT, Jiang G, Tao K, Kuperwasser C, Weinberg RA, et al. Identification of selective inhibitors of cancer stem cells by high-throughput screening. *Cell* 2009;**138**(4):645–59.
90. Managò A, Leanza L, Carraretto L, Sassi N, Grancara S, Quintana-Cabrera R, et al. Early effects of the antineoplastic agent salinomycin on mitochondrial function. *Cell Death Dis* 2015;**6**:e1930.
91. Antoszczak M, Huczynski A. Anticancer activity of polyether ionophore-salinomycin. *Anti Cancer Agents Med Chem* 2015;**15**:575–91.
92. Naujokat C, Steinhart R. Salinomycin as a drug for targeting human cancer stem cells. *J Biomed Biotechnol* 2012;**2012**:950658. https://doi.org/10.1155/2012/950658.
93. Boehmerle W, Endres M. Salinomycin induces calpain and cytochrome c-mediated neuronal cell death. *Cell Death Dis* 2011;**2**:e168.
94. Boehmerle W, Muenzfeld H, Springer A, Huehnchen P, Endres M. Specific targeting of neurotoxic side effects and pharmacological profile of the novel cancer stem cell drug salinomycin in mice. *J Mol Med (Berl)* 2014;**92**:889–900.
95. Huczynski A. Salinomycin: a new cancer drug candidate. *Chem Biol Drug Des* 2012;**79**:235–8.
96. Rubin H, Fodge D. Interrelationships of glycolysis, sugar transport and the initiation of DNA synthesis in chick embryo cells. In: Clarkson B, Basega R, editors. *Control of proliferation in animal cells*. vol. 1. New York: Cold Spring Harbor Laboratory; 1974. p. 801–16.
97. Pellegrini P, Dyczynski M, Sbrana FV, Karlgren M, Buoncervello M, Hagg-Olofsson M, et al. Tumor acidosis enhances cytotoxic effects and autophagy inhibition by salinomycin on cancer cell lines and cancer stem cells. *Oncotarget* 2016. https://doi.org/10.18632/oncotarget.9601.
98. Berezhnov AV, Soutar MP, Fedotova EI, Frolova MS, Plun-Favreau H, Zinchenko VP, et al. Intracellular pH modulates autophagy and mitophagy. *J Biol Chem* 2016;**291**:8701–8.
99. Kim JH, Chae M, Kim WK, Kim YJ, Kang HS, Kim HS, Yoon S. Salinomycin sensitizes cancer cells to the effects of doxorubicin and etoposide treatment by increasing DNA damage and reducing p21 protein. *Br J Pharmacol* 2011;**162**(3):773–84.
100. Kopp F, Hermawan A, Oak PS, Herrmann A, Wagner E, Roidl A. Salinomycin treatment reduces metastatic tumor burden by hampering cancer cell migration. *Mol Cancer* 2014;**13**(1):1.
101. Mai TT, Hamaï A, Hienzsch A, Cañeque T, Müller S, Wicinski J, et al. Salinomycin kills cancer stem cells by sequestering iron in lysosomes. *Nat Chem* 2017;**9**(10):1025.
102. Kim KY, Yu SN, Lee SY, Chun SS, Choi YL, Park YM, et al. Salinomycin-induced apoptosis of human prostate cancer cells due to accumulated reactive oxygen species and mitochondrial membrane depolarization. *Biochem Biophys Res Commun* 2011;**413**(1):80–6.
103. Riccioni R, Dupuis ML, Bernabei M, Petrucci E, Pasquini L, Mariani G, et al. The cancer stem cell selective inhibitor salinomycin is a p-glycoprotein inhibitor. *Blood Cell Mol Dis* 2010;**45**(1):86–92.
104. Kim WK, Kim JH, Yoon K, Kim S, Ro J, Kang HS, Yoon S. Salinomycin, a p-glycoprotein inhibitor, sensitizes radiation-treated cancer cells by increasing DNA damage and inducing G2 arrest. *Investig New Drugs* 2012;**30**(4):1311–8.
105. Li T, Su L, Zhong N, Hao X, Zhong D, Singhal S, Liu X. Salinomycin induces cell death with autophagy through activation of endoplasmic reticulum stress in human cancer cells. *Autophagy* 2013;**9**(7):1057–68.

106. Al Dhaheri Y, Attoub S, Arafat K, AbuQamar S, Eid A, Al Faresi N, Iratni R. Salinomycin induces apoptosis and senescence in breast cancer: upregulation of p21, downregulation of survivin and histone H3 and H4 hyperacetylation. *Biochim Biophys Acta Gen Subj* 2013;**1830**(4):3121–35.
107. Parajuli B, Lee HG, Kwon SH, Cha SD, Shin SJ, Lee GH, et al. Salinomycin inhibits Akt/NF-κB and induces apoptosis in cisplatin resistant ovarian cancer cells. *Cancer Epidemiol* 2013;**37**(4):512–7.
108. Skeberdytė A, Sarapinienė I, Aleksander-Krasko J, Stankevičius V, Sužiedėlis K, Jarmalaitė S. Dichloroacetate and salinomycin exert a synergistic cytotoxic effect in colorectal cancer cell lines. *Sci Rep* 2018;**8**(1):17744.
109. Liu L, Wang Q, Mao J, Qin T, Sun Y, Yang J, et al. Salinomycin suppresses cancer cell stemness and attenuates TGF-β-induced epithelial-mesenchymal transition of renal cell carcinoma cells. *Chem Biol Interact* 2018;**296**:145–53.
110. Pizon M, Schott D, Pachmann U, Pachmann K, Szalus N. Abstract A067: salinomycin kills not only circulating epithelial tumor cells (CETCs) but also circulating cancer stem cells (cCSC) and may potentially prevent progression in patients with breast cancer. *Mol Cancer Ther* 2018. https://doi.org/10.1158/1535-7163.TARG-17-A067. Abstract A067.
111. Weinbach NC, Garbus J. Mechanism of action of reagents that uncouple oxidative phosphorylation. *Nature* 1969;**221**(5185):1016–8.
112. Jurgeit A, McDowell R, Stephan M, Meldrum E, Schwendener R, Greber UF. Niclosamadide is a proton carrier and targets bacidic endosomes with broad antiviral effects. *PLoS Pathog* 2012; **8**(10):e1002976.
113. Pan JX, Ding K, Wang CY. Niclosamide, an old antihelminthic agent, demonstrates antitumor activity by blocking multiple signaling pathways of cancer stem cells. *Chin J Cancer* 2012;**31**:178–84.
114. Arend RC, Londono-Joshi AI, Samant RS, Li Y, Conner M, Hidalgo B, et al. Inhibition of Wnt/β-catenin pathway by niclosamide: a therapeutic target for ovarian cancer. *Gynecol Oncol* 2014; **134**(1):112–20.
115. Li Y, Li PK, Roberts MJ, Arend RC, Samant RS, Buchsbaum DJ. Multi-targeted therapy of cancer by niclosamide: a new application for an old drug. *Cancer Lett* 2014;**349**:8–14.
116. Lu W, Lin C, Roberts MJ, Waud WR, Piazza GA, Li Y. Niclosamide suppresses cancer cell growth by inducing Wnt co-receptor LRP6 degradation and inhibiting the Wnt/β-catenin pathway. *PLoS One* 2011;**6**(12):e29290.
117. You S, Li R, Park D, Xie M, Sica GL, Cao Y, Xiao ZQ, Deng X. Disruption of STAT3 by niclosamide reverses radioresistance of human lung cancer. *Mol Cancer Ther* 2014;**13**(3):606–16.
118. Fonseca BD, Diering GH, Bidinosti MA, Dalal K, Alain T, Balgi AD, et al. Structure-activity analysis of niclosamide reveals potential role for cytoplasmic pH in control of mammalian target of rapamycin complex 1 (mTORC1) signaling. *J Biol Chem* 2012;**287**:17530–45.
119. Karakas D, Cevatemre B, Aztopal N, Ari F, Yilmaz VT. UlukuyabE. Addition of niclosamide to palladium (II) saccharinate complex of terpyridine results in enhanced cytotoxic activity inducingbapoptosis on cancer ítem cells of breast cancer. *Bioorg Med Chem* 2015;**23**(17):5580–6.
120. Wang YC, Chao TK, Chang CC, Yo YT, Yu MH, Lai HC. Drug screening identifies niclosamide as an inhibitor of breast cancer stem-like cells. *PLoS One* 2013;**8**(9):e74538.
121. Londoño-Joshi AI, Arend RC, Aristizabal L, Lu W, Samant RS, Metge BJ, et al. Effect of niclosamide on basal-like breast cancers. *Mol Cancer Ther* 2014;**13**(4):800–11. molcanther-0555.
122. Ye T, Xiong Y, Yan Y, Xia Y, Song X, Liu L, et al. The anthelmintic drug niclosamide induces apoptosis, impairs metastasis and reduces immunosuppressive cells in breast cancer model. *PLoS One* 2014;**9**(1):e85887.
123. Liu C, Lou W, Armstrong C, Zhu Y, Evans CP, Gao AC. Niclosamide suppresses cell migration and invasion in enzalutamide resistant prostate cancer cells via Stat3-AR axis inhibition. *Prostate* 2015; **75**(13):1341–53.
124. Liu C, Armstrong C, Zhu Y, Lou W, Gao AC. Niclosamide enhances abiraterone treatment via inhibition of androgen receptor variants in castration resistant prostate cancer. *Oncotarget* 2016; **7**(22):32210.
125. Yu X, Liu F, Zeng L, He F, Zhang R, Yan S, et al. Niclosamide exhibits potent anticancer activity and synergizes with sorafenib in human renal cell cancer cells. *Cell Physiol Biochem* 2018; **47**(3):957–71.

126. Pan CX, Lara P, Evans CP, Parikh M, Dall'era M, Liu C, Robles D, Gao A. Niclosamide in combination with abiraterone and prednisone in men with castration-resistant prostate cancer (CRPC): initial results from a phase Ib/II trial. *J Clin Oncol* 2018;**36**(6 Suppl):192.
127. Han Z, Li Q, Wang Y, Wang L, Wang Y, Guo C. Niclosamide induces cell cycle arrest in G1 phase in head and neck squamous cell carcinoma through let-7d/CDC34 axis. *Front Pharmacol* 2018;**9**:1544.
128. Zuo Y, Yang D, Yu Y, Xiang M, Li H, Yang J, et al. Niclosamide enhances the cytotoxic effect of cisplatin in cisplatin-resistant human lung cancer cells via suppression of lung resistance-related protein and c-myc. *Mol Med Rep* 2018;**17**(3):3497–502.
129. Burock S, Daum S, Tröger H, Kim TD, Krüger S, Rieke DT, et al. Niclosamide a new chemotherapy agent? Pharmacokinetics of the potential anticancer drug in a patient cohort of the NIKOLO trial. *J Clin Oncol* 2018;**36**(15 Suppl):e85887.
130. Jin Y, Lu Z, Ding K, Li J, Du X, Chen C, et al. Antineoplastic mechanisms of niclosamide in acute myelogenous leukemia stem cells: inactivation of the NF-kappaB pathway and generation of reactive oxygen species. *Cancer Res* 2010;**70**(6):2516–27.
131. Schweizer MT, Haugk K, McKiernan JS, Gulati R, Cheng HH, Maes JL, et al. A phase I study of niclosamide in combination with enzalutamide in men with castration-resistant prostate cancer. *PLoS One* 2018;**13**(6):e0198389. https://doi.org/10.1371/journal.pone.0198389.
132. Cvek B, Dvorak Z. The value of proteasome inhibition in cancer: can the old drug, disulfiram, have a bright new future as a novel proteasome inhibitor? *Drug Discov Today* 2008;**13**(15–16):716–22.
133. Liu P, Kumar IS, Kannapan V, Tawari PE, Tang JZ, Juang W, Armesilla AL, et al. Disulfiram targets cancer stem-like cells and reverses resistance and cross-resistance in acquired paclitaxel-resistant triple-negative breast cancer cells. *Br J Cancer* 2013;**109**:1876–85.
134. Brar SS, Grigg C, Wilson KS, Holder Jr WD, Dreau D, Austin C, et al. Disulfiram inhibits activating transcription factor/cyclic AMP-responsive element binding protein and human melanoma growth in a metal-dependent manner in vitro, in mice and in a patient with metastatic disease. *Mol Cancer Ther* 2004;**3**:1049–60.
135. Navratilova J, Hankeova T, Benes P, Smarda J. Acidic pH of tumor microenvironment enhances cytotoxicity of the disulfiram/Cu2+ complex to breast and colon cancer cells. *Chemotherapy* 2013;**59**:112–20.
136. Liu P, Brown S, Goktug T, Channathodiyil P, Kannappan V, Hugnot JP, et al. Cytotoxic effect of disulfiram/copper on human glioblastoma cell lines and ALDH-positive cancer-stem-like cells. *Br J Cancer* 2012;**107**(9):1488.
137. Lin J, Haffner MC, Zhang Y, Lee BH, Brennen WN, Britton J, et al. Disulfiram is a DNA demethylating agent and inhibits prostate cancer cell growth. *Prostate* 2011;**71**(4):333–43.
138. Majera D, Skrott Z, Bouchal J, Bartkova J, Simkova D, Gachechiladze M, et al. Targeting genotoxic and proteotoxic stress-response pathways in human prostate cancer by clinically available PARP inhibitors, vorinostat and disulfiram. *Prostate* 2019;**79**(4):352–62.
139. Chen D, Cui QC, Yang H, Dou QP. Disulfiram, a clinically used anti-alcoholism drug and copper-binding agent, induces apoptotic cell death in breast cancer cultures and xenografts via inhibition of the proteasome activity. *Cancer Res* 2006;**66**(21):10425–33.
140. Irving CC, Tice AJ, Murphy WM. Inhibition of Nn-butyl-N-(4-hydroxybutyl) nitrosamine-induced urinary bladder cancer in rats by administration of disulfiram in the diet. *Cancer Res* 1979;**39**(8):3040–3.
141. Leiting J, Yang L, Hernandez MC, Truty MJ. Medicine cabinet oncology: disulfiram and copper gluconate show similar efficacy in pancreatic cancer cell lines and patient-derived xenograft tissue. *J Am Coll Surg* 2018;**227**(4):e217.
142. Stenvang J, Keinicke H, Nielsen SL, Jandu H, Bartek J, Brünner N. Repurposing disulfiram as a potential novel treatment of drug-resistant metastatic colorectal cancer. In: *Proceedings of the AACR-NCI-EORTC International conference: molecular targets and cancer therapeutics; 2017 Oct 26-30; Philadelphia, PA. Philadelphia (PA): AACR; Molecular cancer therapeutics*. vol. 17. 2018. (1 Suppl):Abstract nr A143.
143. Martirosyan A, Clendening JW, Goard CA, Penn LZ. Lovastatin induces apoptosis of ovarian cancer cells and synergizes with doxorubicin: potential therapeutic relevance. *BMC Cancer* 2010;**10**:103.
144. Bakar NS, Kamali F, Brown CD. Effect of statins on functional expression of membrane transporters in L6 rat skeletal muscle cells. *J Biomed Clin Sci* 2017;**1**(1):17–26.

145. Bakar NSB, Kamali F, Brown CD. Effect of statins on functional expression of membrane transporters in L6 rat skeletal muscle cells. *J Biomed Clin Sci* 2017;**1**(1):17–26.
146. Farwell WR, Scranton RE, Lawler EV, Lew RA, Brophy MT, Fiore LD, et al. The association between statins and cancer incidence in a veterans population. *J Natl Cancer Inst* 2008;**100**(2):134–9.
147. Karp I, Behlouli H, LeLorier J, Pilote L. Statins and cancer risk. *Am J Med* 2008;**121**(4):302–9.
148. Demierre MF, Higgins PD, Gruber SB, Hawk E, Lippman SM. Statins and cancer prevention. *Nat Rev Cancer* 2005;**5**(12):930.
149. Kuoppala J, Lamminpää A, Pukkala E. Statins and cancer: a systematic review and meta-analysis. *Eur J Cancer* 2008;**44**(15):2122–32.
150. Gottfried E, Lang SA, Renner K, Bosserhoff A, Gronwald W, Rehli M, et al. New aspects of an old drug—diclofenac targets MYC and glucose metabolism in tumor cells. *PLoS One* 2013;**8**:e66987.
151. Cho M, Gwak J, Park S, Won J, Kim DE, Yea SS, et al. Diclofenac attenuates Wnt/β-catenin signaling in colon cancer cells by activation of NF-kB. *FEBS Lett* 2005;**579**(20):4213–8.
152. Mayorek N, Naftali-Shani N, Grunewald M. Diclofenac inhibits tumor growth in a murine model of pancreatic cancer by modulation of VEGF levels and arginase activity. *PLoS One* 2010;**5**(9):e12715.
153. Rana C, Piplani H, Vaish V, Nehru B, Sanyai SN. Downregulation of the PI3-K/Akt/PTEN pathway and activation of mitochondrial intrinsic apoptosis by diclofenac and curcumin in colon cancer. *Mol Cell Biochem* 2015;**402**(1–2):225–41.
154. Rana C, Piplani H, Vaish V, Nehru B, Sanyal SN. Downregulation of telomerase activity by diclofenac and curcumin is associated with cell cycle arrest and induction of apoptosis in cancer. *Tumor Biol* 2015;**36**(8):5999–6010.
155. Pantziarka P, Sukhatme V, Bouche G, Meheus L, Sukhatme VP. Repurposing drugs in oncology (ReDo)-diclofenac as an anti-cancer agent. *Ecancermedicalscience* 2016;**10**:610. https://doi.org/10.3332/ecancer.2016.610.
156. Lagas JS, van der Kruijssen CM, van de Wetering K, Beijnen JH, Schinkel AH. Transport of diclofenac by breast cancer resistance protein (ABCG2) and stimulation of multidrug resistance protein 2 (ABCC2)-mediated drug transport by diclofenac and benzbromarone. *Drug Metab Dispos* 2009;**37**(1):129–36.
157. Chaudhary SC, Alam MS, Siddiqui MS, Athar M. Perillyl alcohol attenuates Ras-ERK signaling to inhibit murine skin inflammatiom and tumorigenesis. *Chem Biol Interact* 2009;**179**(2–3):145–53.
158. Clark SS, Perman SM, Sahin MB, Jenkins GJ, Elegbede JA. Antileukemia activity of perillyl alcohol (POH): uncoupling apoptosis from G0/G1 arrest suggests that the primary effect of POH on Bcr/Abl-transformed cells is to induce growth arrest. *Leukemia* 2002;**16**:213–22.
159. Yuri T, Danbara N, Tsujita-Kyutoku M, Kiyozuka Y, Senzaki H, Shikata N, et al. Perillyl alcohol inhibits human breast cancer cell growth in vitro and in vivo. *Breast Cancer Res Treat* 2004;**84**(3):251–60.
160. Fonseca C, Teixeira R, Ramina R, Kovaleski G, Silva J, Nagel J, et al. Case of advanced recurrent glioblastoma successfully treated with monoterpene perillyl alcohol by intranasal administration. *J Cancer Ther* 2011;**2**:16–21.
161. Chen TC, Shonthal AH, Hofman FH. AbstractbB77: perillyl alcohol derivatives as anti-cancer agents. *Clin Cancer Res* 2018;**24**(1 Suppl):B77. AACR international Conference May 4-6 Sao Paulo, Brazil.
162. Da Fonseca CO, Soares IP, Clemençon DS, Rochlin S, Cardeman L, Quirico-Santos T. Perillyl alcohol inhalation concomitant with oral temozolomide halts progression of recurrent inoperable glioblastoma: a case report. *J Histol Histopathol* 2015; https://doi.org/10.7243/2055-091X-2-12.
163. Lasse Leino JL. *Anti-proliferative pharmaceutical composition.* US Patent 8338475 B2, 2012.
164. Laihia JK, Kallio JP, Taimen P, Kujari H, Kahari VM, Leino L. Protodynamic intracellular acidification by cis-urocanic acid. *J Investig Dermatol* 2010;**130**(10):2431–9.
165. Trappoliere M, Caligiuri A, Schmid M, Bertolani C, Failli P, Vizzutti F, et al. Silybin, a component of sylimarin, exerts anti-inflammatory and anti-fibrogenic effects on human hepatic stellate cells. *J Hepatol* 2009;**50**(6):1102–11.
166. Jung HJ, Park JW, Lee JS, Lee SR, Jang BC, Suh SI, et al. Silibinin inhibits expression of HIF-1α through suppression of protein translation in prostate cancer cells. *Biochem Biophys Res Commun* 2009;**390**(1):71–6.

167. Shukla SK, Dasgupta A, Mehla K, Gunda V, Vernucci E, Souchek J, et al. Silibinin-mediated metabolic reprogramming attenuates pancreatic cancer-induced cachexia and tumor growth. *Oncotarget* 2015;**6**(38):41146.
168. Chang HR, Chen PN, Yang SF, Sun YS, Wu SW, Hung TW, et al. Silibinin inhibits the invasion and migration of renal carcinoma 786-O cells in vitro, inhibits the growth of xenografts in vivo and enhances chemosensitivity to 5-fluorouracil and paclitaxel. *Mol Carcinog* 2011;**50**(10):811–23.
169. Vaid M, Singh T, Prasad R, Katiyar SK. Silymarin inhibits melanoma cell growth both in vitro and in vivo by targeting cell cycle regulators, angiogenic biomarkers and induction of apoptosis. *Mol Carcinog* 2015;**54**(11):1328–39.
170. Cui W, Gu F, Hu KQ. Effects and mechanisms of silibinin on human hepatocellular carcinoma xenografts in nude mice. *World J Gastroenterol* 2009;**15**(16):1943–50.
171. Ouedraogo G, Morliere P, Maziere C, Maziere JC, Santus R. Alteration of the endocytotic pathway by photosensitization with fluoroquinolones. *Photochem Photobiol* 2000;**72**:458–63.
172. Balashova N, Dhingra A, Boesze-Battaglia K, Lally ET. Aggregatibacter actinomycetemcomitans leukotoxin induces cytosol acidification in LFA-1 expressing immune cells. *Mol Oral Microbiol* 2016; **31**(1):106–14.
173. Beberok A, Wrzesniok D, Rok J, Rzepka Z, Respondek M, Buszman E. Ciprofloxacin triggers the apoptosis of human triple-negative breast cancer MDA-MB-231 cells via p53/Bax/Bcl-2 signaling pathway. *Int J Oncol* 2018;**52**(5):1727–37.
174. Beberok A, Wrzesniok D, Minecka A, Rock J, Delijewsky M, Rzepka Z, et al. Ciprofloxacin-mediated induction of S-phase cell cycle arrest and apoptosis in COLO829 melanoma cells. *Pharmacol Rep* 2018;**70**(1):6–13.
175. Hangas A, Aasumets K, Kekalainen NJ, Paloheina M, Pohjoismaki JL, Gerhold JM, et al. Ciprofloxacin impairs mitochondrial DNA replication initiation through inhibition of Topoisomerase 2. *Nucleic Acids Res* 2018;**46**(18):9625–36.
176. Zandi A, Zanjani TM, Ziai SA, Poul YK, Hoseini MH. Evaluation of the cytotoxic effects of ciprofloxacin on human glioblastoma A-172 cell line. *Middle East J Cancer* 2017;**8**(3):119–26.
177. Jiney J, Tavares CDJ, Ebelt ND, Lodi A, Edupuganti R, Xie X, et al. Serotonin analogues as inhibitors of breast cancer cell growth. *ACS Med Chem Lett* 2017;**8**(10):1072–6.
178. Anon, n.d., Patent US20140296278A1.
179. Robey IF, Baggett BK, Kirkpatrick ND, Roe DJ, Dosescu J, Sloane BF, et al. Bicarbonate increases tumor pH and inhibits spontaneous metastases. *Cancer Res* 2009;**69**:2260–8.
180. Raghunand N, He X, van Sluis R, Mahoney B, Baggett B, Taylor CW, et al. Enhancement of chemotherapy by manipulation of tumour pH. *Br J Cancer* 1999;**80**:1005–11.
181. Silva AS, Yunes JA, Gillies RJ, Gatenby RA. The potential role of systemic buffers in reducing intratumoral extracellular pH and acid-mediated invasion. *Cancer Res* 2009;**69**:2677–84.
182. Estrella V, Chen T, Lloyd M, Wojtkowiak J, Cornnell HH, et al. Acidity generated by the tumor microenvironment drives local invasion. *Cancer Res* 2013;**73**:1524–35.
183. Ibrahim-Hashim A, Cornnell HH, Abrahams D, Lloyd M, Bui M, Gillies RJ, et al. Systemic buffers inhibit carcinogenesis in TRAMP mice. *J Urol* 2012;**188**:624–31.
184. Fitzgibbons LJ, Snoey ER. Severe metabolic alkalosis due to baking soda ingestion: case reports of two patients with unsuspected antacid overdose. *J Emerg Med* 1999;**17**:57–61.

CHAPTER 16

Pharmacological interventions part IV: Metabolic modifiers

2-Deoxy-D-glucose (2DG)

Glucose enters the cell through the activity of transporters: Gluts (glucose transporters). And once inside, it is phosphorylated to glucose-6-phosphate through the action of an enzyme: hexokinase. Gluts located inside the cells are mobilized to the cell membrane after insulin binds the insulin receptor. When the insulin stimulus is over, Gluts are recycled into the intracellular milieu. Glucose transporters are actually a family of proteins that have specificity for glucose transport (Fig. 1).

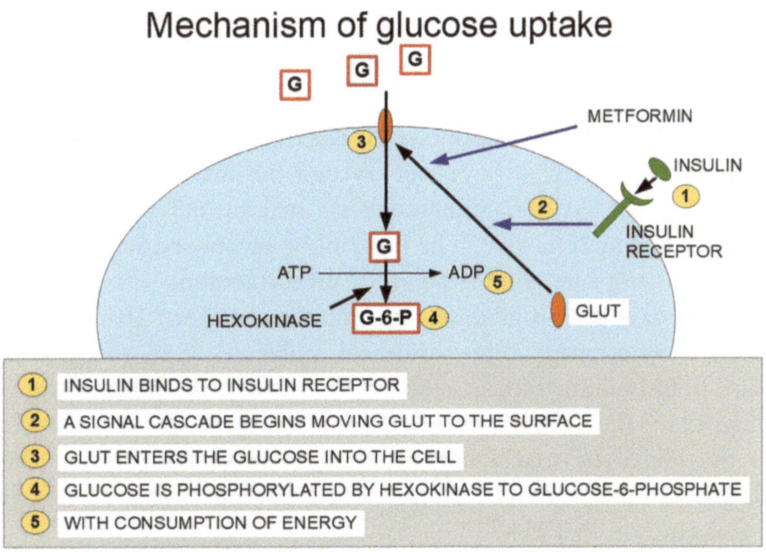

Fig. 1 Upon entering the cytoplasm, glucose is phosphorylated by an hexokinase to glucose-6-phosphate, with energy expenditure. Before this can happen Gluts must traffick to the cell surface induced by insulin binding to its receptor.[1–4]

2-Deoxyglucose (2DG) is a molecule that shares similarities with glucose. 2DG is actually a glucose molecule where the 2-hydroxyl group has been replaced by hydrogen (Box 1).

An Innovative Approach to Understanding and Treating Cancer: Targeting pH
https://doi.org/10.1016/B978-0-12-819059-3.00016-2
© 2020 Elsevier Inc.
All rights reserved.

BOX 1

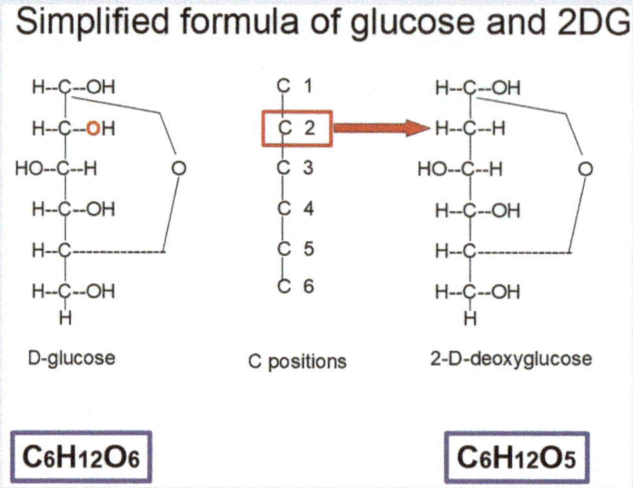

2DG is taken up by Gluts, and transported inside the cell in a similar way as it does with glucose. Once inside the cell, it is phosphorylated by hexokinase to 2-deoxyglucose-6-phosphate (2DG-6-P). At this point there is a striking difference between glucose-6-phosphate (G-6-P) and 2DG-6-P. The first can be glycolysed while the latter cannot, and at the same time impedes glucose downstream metabolism.

Malignant cells have a high glucose consumption, between 10- and 17-fold higher than their normal counterparts. They use these molecules as an energy source and for biosynthesis of other building blocks necessary for growth and proliferation. These cells also have a higher uptake of 2DG. As 2DG cannot undergo glycolysis, and at the same time blocks the glycolytic machinery impeding glucose metabolism, this molecule hampers cell growth and it may be useful for the treatment of cancer as proposed by Woodward and Hudson in 1954.[5]

In their pioneering study they found that:

1. both aerobic and anaerobic glycolysis were inhibited/decreased by 2DG;
2. this inhibition was of a competitive nature (between glucose and 2DG) for the glycolytic system, because when the concentration of glucose was increased the inhibition was reversed;
3. inhibition/decrease of fructolysis by 2DG was 50-fold superior to inhibition/decrease of glycolysis;

4. the level of inhibition/decrease of glycolysis varied among different tumors and sometimes among different samples of the same tumor lineage;
5. when the molar ratio between glucose and 2DG was 1:1 glycolysis inhibition was between 72% and 91%, but when this ratio went to 1:0.5 the inhibition dropped to 50%;
6. in tumors, aerobic glycolysis was less sensitive to 2DG inhibition than anaerobic glycolysis;
7. they suggested that as hexokinase was the first enzyme of the glycolytic pathway, it was possible that 2DG produced its effects through hexokinase inhibition.

Ely et al.[6] found that when 2DG was administered to mice there was a significant growth decrease in Krebs ascites carcinoma and xenografted malignancies. Previous research by Cramer and Woodward showed that 2DG acted as a glucose anti-metabolite in the fermentation and growth of yeast[7] and rat tumor tissues.[8] By 1957, 2DG was established as a possible anti-cancer candidate. Since then, it became the most widely investigated compound in clinical and experimental settings to target tumor metabolism and bioenergetics. Wick et al.[9] working on eviscerated animals confirmed that:

(1) 2DG was responsive to insulin;
(2) insulin increased its intracellular transfer;
(3) 2DG decreased the intracellular transfer of glucose in spite of insulin stimulation;

For these reasons 2DG was considered a metabolic blocker for glucose metabolism.

Wick's experiments were addressed to find 2DG's site of action and they showed that:

(a) 2DG was not oxidized in the cell;
(b) 2DG produced a drop in the cellular metabolic rate;
(c) 2DG markedly reduced glucose oxidation;
(d) the inhibition of glucose oxidation took place before the acetate stage, because when acetate was administered to the cell, 2DG did not decrease acetate's oxidation.

These results localized the block's site in the glycolytic pathway prior to oxidation. They went a step forward and found that 2DG-6-P acted as a competitive inhibitor of phosphogluco- isomerase. The conclusion was that inhibition of phosphogluco-isomerase was 2DG's main site of action, and that a secondary block also occured at the level of hexokinase due to accumulation of G-6-P that inhibits this kinase. (Fig. 2).

Up to this point we have three possible sites of action of 2DG: phosphogluco-isomerase, hexokinase and the intracellular trafficking of glucose.[10, 11] The evidence for inhibition of phosphoglucoisomerase is strong, but inhibition of hexokinase may also be a part of the scheme as can be seen in Fig. 2.

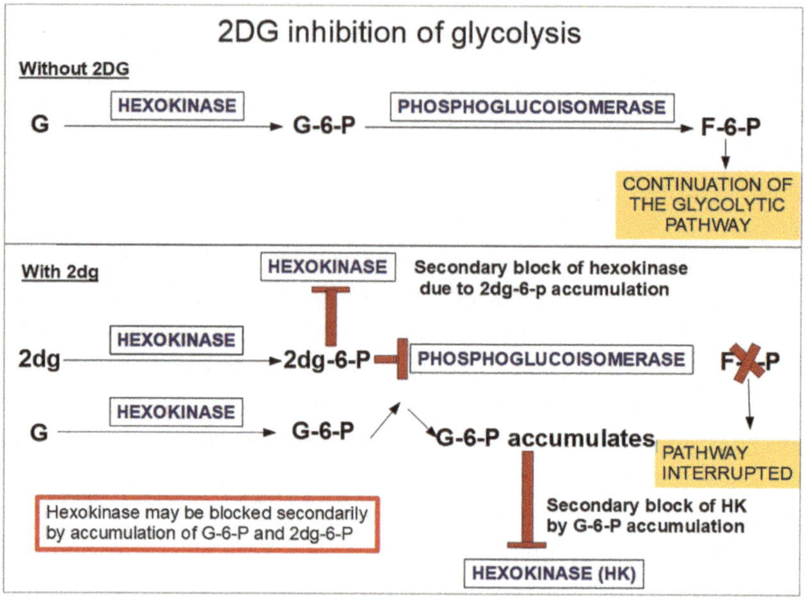

Fig. 2 Glycolysis inhibition by 2DG according to Wick et al.[9] The inhibition of hexokinase by accumulation of 2DG-6-P is not in the original work by Wick et al. but it is quite possible that it is due to the usual inhibition of an enzyme when the product accumulates. In this case the accumulation of G-6-P and 2DG-6-P by inhibition of phosphoglucoisomerase secondarily inhibits hexokinase.

Less evidence exists to sustain an inhibition of intracellular glucose trafficking. Barban and Schulze[12] did not find any impediment for the phosphorylation of glucose when cells were in a highly concentrated 2DG media, so that hexokinase was not the primary target.

Lowering glucose utilization is a way to limit energy production much like starvation or caloric restriction. The actions of 2DG go beyond lowering glucose metabolism by inhibiting phosphohexoisomerase (phosphogluco-isomerase) because extreme low glucose utilization produces:

- decreased protein glycosylation[13];
- which in turn diminishes GRP78 expression, a glucose regulated chaperon molecule that is engaged in preventing endoplasmic reticulum stress (ERS)[14];
- triggering ERS.[15]

Evidence supporting the idea of 2DG induced ERS is its synergistic activity with nelfinavir, another ER stressor.

Under glucose starvation, such as produced by 2DG, the transcription factor Sp1 is deglycosylated and degraded at the proteasome. This implies a reduction in the transcription process of many proteins.[16] Another mechanism of anti-tumoral action was recently described: reduction of NADPH pool, exposing the cell to oxidative stress.[17]

2-Deoxyglucose and cell growth in vitro and in vivo

In vitro

Barban and Schulze[12] showed that 2DG inhibits the growth of human normal cell cultures by inhibiting the metabolism and utilization of glucose. They found that equimolecular amounts of glucose and 2DG inhibited the growth of the two human cell lines tested. Lactic acid production was decreased due to lack of glucose utilization and consequently intracellular ATP also decreased.[18] But these effects were not permanent, because as soon as 2DG was removed from the culture medium, growth resumed. If the cells were kept in the 2DG medium for more than 3 days, growth did not resume and cell death increased. If instead of glucose, the 2DG medium contained fructose, the cells were even more sensitive to 2DG.

2DG also decreased growth of malignant cells in different kinds of tumors in vitro, such as rat fibrosarcoma,[19] human breast cancer,[20] breast cancer stem cells,[21] rat breast cancer,[22] triple negative breast cancer[23], human alveolar rhabdomyosarcoma,[24] prostate,[25] colon,[26] pancreas,[27] and ovarian cancers, mesothelioma, glioblastoma, head and neck squamous cell carcinoma, hepatocarcinoma,[28, 29] non-small cell lung cancer,[30] gastric carcinoma,[31] and neuroblastoma.[32] In some cases, 2DG produced a slight slowdown in cell proliferation and growth, whereas in other cases, it produced complete growth arrest, associated or not with apoptosis. The best cytotoxic results were observed with mesothelioma cells (MSTO-211H) and the worst with HepG2 hepatocarcinoma cells.

In vivo

Rats with subcutaneously injected sarcomas were treated with 2DG. Sarcoma cells showed an average decrease of cellular ATP of 50% while no significant modifications were found in brain ATP content. This means that at the experimental level, 2DG can be used to decrease tumor energy supply without affecting brain metabolism.[33] 2DG produced an average decrease of 0.4 of intracellular pH. Griffiths et al.[34] reported significant cellular acidification of Walker tumors after 2DG, with no concomitant decrease of ATP. Decreased intracellular ATP and/or intracellular acidification can both represent interesting targets for cancer treatment probably linked to lack of nutrients for biosynthesis and decreased DNA repair.[35] At the present, we have no clear explanation for 2DG-induced intracellular acidification although it is clear why it increases extracellular pH: decreased production of lactic acid and decreased pentose phosphate pathway utilization lowering CO_2 production (Fig. 3).

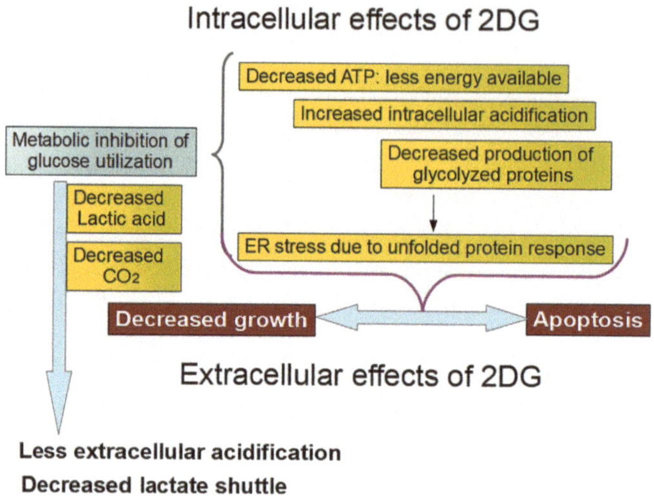

Fig. 3 Intra and extracellular effects of 2DG.

The mechanism of intracellular acidification

2DG decreases intracellular pH.[36] This may seem paradoxical, because 2DG decreases the production of lactic acid. But it is not so. Acidity created by 2DG is achieved by a mechanism that has no relation to lactic acid. The basic pharmacological activity of 2DG is the inhibition of the glycolytic pathway upstream of pyruvate or lactic acid due to inhibition of glucose uptake and metabolism. Therefore, the cell is literally starved. Since glucose is the main source of cellular energy, this starvation leads to the depletion of the ATP pool and no new ATP is formed. Lack of ATP inhibits NHE1 functioning, thus the cell loses the activity of its main proton extrusion mechanism. The final result is intracellular acidification.[37] The interesting issue is that NHE1 requires neither ATP nor energy and its inhibition by 2DG is reversible. However NHE1 is ATP-dependent,[38] although the exact mechanism of this dependency has not been fully clarified.[39] NHE1 has four intracellular loops, the first one (loop I, amino acids 516–590) is an ATP sensor. This segment is probably related to NHE1 inhibition through ATP depletion. Intracellular loop III has an auto-inhibitory function. It has been suggested that ATP depletion is detected by loop I and activates the auto-inhibitory loop III.[40] Therefore, intracellular pH is lowered through inhibition of proton extrusion and not lactic acid accumulation.

The problem for the clinical use of 2DG is that this happens in cancer cells and normal cells alike, although the latter have more ATP reserves than the cancer cells and are slightly less susceptible to NHE1 inhibition. Another problem found with 2DG in glioblastoma was that it increased migration and invasion.[41] In triple negative breast cancer cell lines, the association of metformin and 2DG increased the detachment of viable cancer cells.[42]

Clinical experience

2DG has only mild anti-cancer effects and nowadays it is being investigated as a sensitizer to chemotherapy and radiotherapy. Although, 2DG is one of the most investigated compounds in vitro and in vivo, clinical experience is very limited because 2DG as a stand-alone drug did not show significant results. However, when 2DG was used in association with other agents the results were different. In a phase I dose-escalating study 45 mg/kg was recommended for phase II studies.[43] In this phase I clinical trial seven patients had advanced castrate-resistant prostate cancer with no response to chemotherapy. The administration of 2DG showed no decrease in PSA levels and three patients continued with the trial because they had a stable disease. In another phase I dose-escalating study of 2DG combined with docetaxel in patients with advanced tumors ($n=32$) 66% showed progressive disease and 32% stable disease with only one case (3%) with partial response.[44] A phase I/II trial (NCT00633087) for the treatment of advanced cancer and hormone refractory prostate cancer was conducted at Rutgers, the State University of New Jersey. Results are not available because the study was closed early due to slow accrual.[45] The scarcity of 2DG clinical trials is surprising because it has shown promising anti-cancer effects in the pre-clinical setting.

2DG and multidrug resistance (MDR)

As we briefly described in the previous chapter MDR, is pH and energy dependent. 2DG fulfills these two basic requisites for reversing MDR: lowering pHi and decreasing cellular energy. Evidence supporting this principle can be found in the studies by:

- Kaplan et al.[46] found that 2DG's cytotoxicity was significantly higher in MDR cells than in their parental sensitive cells.
- Lemontt et al.[47] reported a decrease in drug extrusion in MDR melanoma cells treated with rotenone and 2DG. Rotenone is a mitochondrial "poison" that increases glycolysis and decreases energy production.
- Bell et al.[48] showed that 2 DG preferentially killed MDR cells in a KB human carcinoma line when compared with parental-non resistant cells.
- Xue et al.[49] associated metformin with 2DG and found an enhanced reversal of MDR in MCF-7 breast cancer cells with resistance to doxorubicin. Both drugs are cytoplasmic acidifiers and energy depletors.

Other uses and effects of 2DG

2DG has been used as a vehicle to deliver cytotoxic particles to malignant cells. Actually, a derivative of 2DG, 2amino-2DG, was used for this purpose, it was conjugated with cobalt-ferrite magnetic nanoparticles. This compound showed high uptake in malignant breast cancer cells and induced apoptosis even in low concentrations in a dose-dependent manner.[50] 2DG decreases tumor angiogenesis through downregulation of VEGF (vascular endothelial growth factor).[51]

Associations with 2DG

Supposedly 2DG may delay tumor growth but does not produce apoptosis at clinically achievable and tolerable concentrations. This has not been fully proved. That is why 2DG has been associated with other drugs in order to increase cytotoxic power. Basically it has been associated with chemotherapeutics like doxorubicin, anti-autophagy agents and metformin. There is evidence showing that metformin acts in synergy with 2DG to achieve cytotoxicity in malignant cells. This association inhibited both parts of glucose metabolism: glycolysis (2DG) and mitochondrial respiration (metformin) and produced a severe depletion of ATP inducing p53 dependent apoptosis in prostate cancer cells.[52, 53] Cheong et al.[54] found that this association worked mainly due to the added energy limitation they produced. There is also evidence that other "mitochondrial poisons" besides metformin show synergy with 2DG.[55] We must add, that there are probably two other not investigated reasons for the success of this association:

1. An alkaline intracellular milieu is a necessary condition for biosynthesis and mitosis. Metformin, through increased production of lactic acid and 2DG are powerful acidifiers of the intracellular environment, so that both cooperate in this issue.[56]
2. 2DG increases AMPK activity and metformin is also a powerful AMPK activator.[57]

In summary, 2DG and metformin can be considered as the two prongs of an energy restriction effect. However, both induce autophagia through AMPK activation in normoxia. Under hypoxia ATP depletion may have an inhibitory effect on autophagy. Actually ATP depletion and ERS (endoplasmic reticulum stress) may be responsible for autophagy.

The metformin-2DG association has the ability to induce apoptosis in esophageal squamous cell carcinoma,[58] ovarian cancer cells,[59] glioblastoma,[60] colon cancer[26] and has reversed multidrug resistance in breast cancer cells resistant to doxorubicin.[61]

One of the consequences of energy restriction produced by 2DG is autophagy as a defensive or adaptive mechanism to metabolic challenges. Autophagy may lead to survival or to apoptosis depending on many circumstances that are not yet fully known.[62, 63]

Decreasing autophagy pharmacologically has increased 2DG cytotoxicity in glioblastoma cells.[64] Both, metformin and 2DG could lead to autophagy and specific drugs like chloroquine or hydroxychloroquine[65] should be associated to address this issue.

In summary:

(1) Protein processing is carried out in the ER.
(2) Tumor cells have a greater need to produce proteins in an accelerated way.
(3) This high metabolic demand means that there is usually an overload of misfolded and unfolded proteins driving ER to the verge of stress.
(4) 2DG may increase this stress[66] by decreasing glucose utilization and production of glycosylated proteins necessary for the unfolded protein response (UPR).
(5) Metformin also increases the stress by reducing ATP production at the mitochondrial level.

These events lead us to the hypothetical idea that:

(6) Adding ER stressors as celecoxib or nelfinavir[67] would increase the level of ER stress in malignant cells.

(7) The natural response to excessive stress would be: autophagy or apoptosis.

(8) If at the same time, autophagy is inhibited by chloroquine or hydroxycloroquine, the process would end in apoptosis.[68]

2DG and chemotherapeutics

2DG increases the cytotoxic activity of chemotherapeutics like paclitaxel, adriamycin[69] and cisplatin,[70] but antagonizes erlotinib.[71] Hernlund et al.[72] tested 2DG with 18 standard and experimental chemotherapeutic drugs and found that 5FU and etoposide were not potentiated by 2DG, but on the other hand cisplatin and ellipticine were. The molecular mechanism of 2DG increasing the apoptotic effect of cisplatin is related to the ERS. This ERS produces an UPR (unfolded protein response) that decreases the transcription of DNA repair genes and at the same time increases the transcription of pro-apoptotic genes.[73]

Why is 2DG a poor anti-cancer drug?

Clinical experience showed that 2DG is not a stand-alone drug in spite of all its anti-cancer effects. One of the possibilities, raised by Kobayashi et al.[74] is that cancer cells can "escape" the energy restriction by β-oxidation of lipids. As a proof of concept they found that when lipolysis was inhibited with atglistatin, 2DG's cytotoxicity was increased. Another possible escape mechanism is through autophagy. This is based on the finding of autophagy markers in patients treated with 2DG and is logical because 2DG activates AMPK that down-regulates mTOR activity (mTOR is an inhibitor of the autophagy cascade). This is the reason why autophagy inhibitors increase the apoptotic effects of 2DG (Fig. 4).

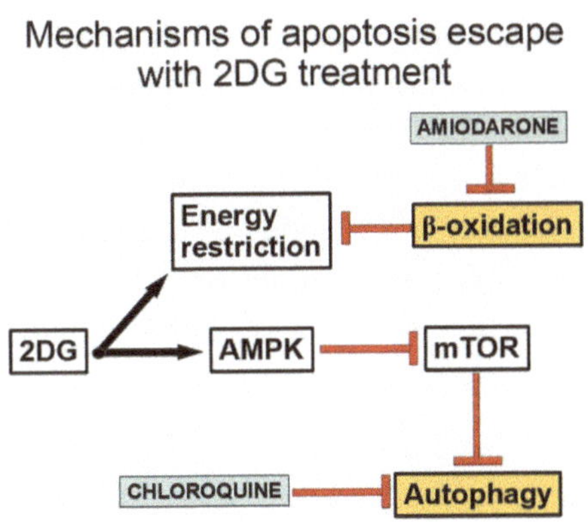

Fig. 4 The two known mechanisms that make 2DG a low apoptotic drug. It also shows that blocking autophagy with chloroquine or some other drug, and the inhibition of β-oxidation of lipids with a drug such as amiodarone[75] may increase the apoptotic effects of 2DG.

Discussion

2DG is not a stand-alone drug. Combination of many pharmacological agents has been the standard protocol used for cancer therapy. These treatments are often associated with an increase in toxicity. Therefore, there is a need for developing new, effective, and less toxic approaches. There are drugs like nelfinavir, chloroquine or metformin with anti-cancer activity and low toxicity that have been tested for other diseases and would not need repeating phase I clinical trials.

On the other side of this path there is a drug like 2DG that has shown interesting anti-cancer activities, in addition to very low toxicity. But this anti-cancer activity was not important enough to enter oncological practice. The idea behind the mechanism of action of 2DG is however interesting enough to explore further for two reasons:

1. this pharmaceutical is the most powerful inhibitor of glucose metabolism which has low human toxicity and
2. malignant cells high dependency on aerobic glycolysis.

Combinatorial schemes would be necessary to make 2DG a useful therapy. The more obvious one is the association of 2DG with other chemotherapeutics. The association of 2DG with metformin (an inhibitor of mitochondrial complex I and down-regulator of oxidative phosphorylation), and/or chloroquine (an inhibitor of autophagy) plus nelfinavir (an ER stressor and Akt down-regulator) could be one possible way to increase 2DG's cytotoxicity.

Conclusions

The basic condition for 2DG's role in cancer is that the tumor should uptake glucose (and therefore 2DG too) at a high rate. This is easily tested with 18F-2DG-positron emission tomography. Mesothelioma cells have been particularly sensitive to 2DG inhibition in vitro. The results of standard treatments for this cancer with surgery, chemo- and radiotherapy have been particularly poor. So that 2DG deserves to be investigated as a complementary drug in these protocols.

Phase I studies have been completed and dose requirements and adverse events are well known. 2DG alone or, as part of a four drug scheme as described above, deserves Phase II studies as adjuvant to chemotherapeutic protocols. 2DG has a role in cancer treatment as part of a scheme as described or as a sensitizer for chemo or radiotherapy. It may be specifically considered in a scheme of intracellular acidification with other acidifiers like amiloride, proton pump inhibitors and carbonic anhydrase blockers. Finally, brief administration in the first week of chemotherapy, alone or associated with metformin would sensitize tumor cells and enhance apoptosis.

Metformin

The incidence of cancer is higher in diabetic populations. This is true for certain tumors, but not for all. The consensus statement of experts assembled by the American Diabetes Association and the American Cancer Society in 2010[76] found that:

> *"Diabetes (primarily type 2) is associated with an increased risk of some cancers (liver, pancreas, endometrium, colon/rectum, breast, and bladder). Diabetes is associated with a reduced risk of prostate cancer. For some other cancer sites, there appears to be no association or the evidence is inconclusive".*

This consensus statement also established that:

> *"Although still limited, early evidence suggests that metformin is associated with a lower risk of cancer and that exogenous insulin is associated with an increased cancer risk. Further research is needed to clarify these issues and evaluate if insulin glargine is more strongly associated with cancer risk compared with other insulins".*

Insulin resistance and hyperinsulinemia might promote carcinogenesis through the insulin receptor (IR) or by increasing the levels of insulin-like growth factors (IGF).[77]

Metformin (Met) is the most commonly prescribed therapy in patients with type 2 diabetes.[78] It has been in use for more than 60 years to treat type 2 diabetes. And, with the exception of lactic acidosis,[79] which rarely occurs, it is a non-toxic and safe drug. Such a prolonged experience means that Met's pharmacodynamic is well known and manageable although its mechanisms of action in cancer are incompletely known.[80] However, the effects are well established: Met reduces hepatic glucose output, increases glucose uptake by muscles (Fig. 5), and reduces the levels of circulating glucose and insulin in patients with insulin resistance and hyperinsulinemia.[81]

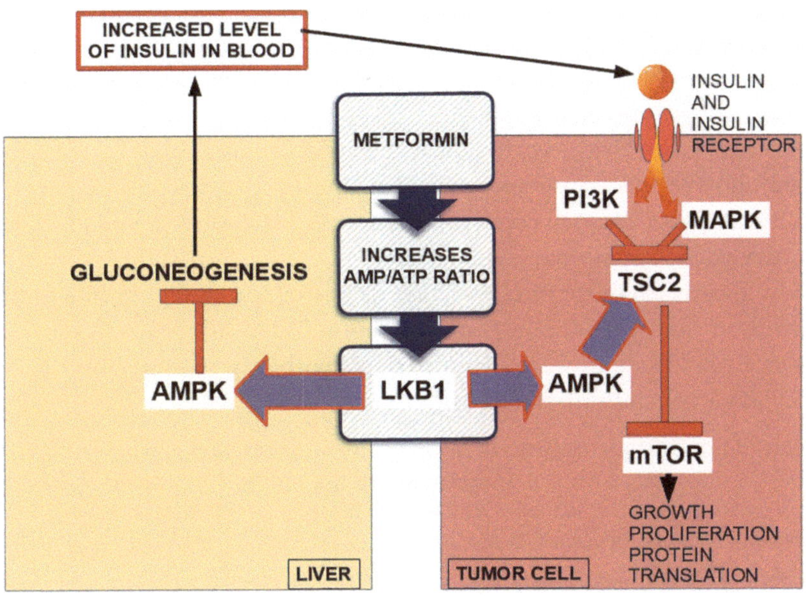

Fig. 5 Mechanism of action of metformin. Metformin down-regulates hepatic gluconeogenesis by activating the LKB1/AMPK pathway. In tumor cells the mechanism is similar: Activated AMPK stimulates TSC2 (tuberous sclerosis complex 2) that inhibits mTOR. The PI3K/Akt and the mapkinases pathways inhibit TSC2. The proposed explanation for metformin's anti-tumoral effects is through its activating action on the LKB1/AMPK pathway. Metformin is a mitochondrial inhibitor of complex I decreasing ATP production. Decreased ATP increases the AMP/ATP ratio which in turn is an activator of AMPK. Through reduced gluconeogenesis the level of insulin is lower and the stimulation on insulin receptors is decreased.

To understand Met's activity in cancer it is necessary to distinguish two different kinds of actions:

(a) those that depend essentially from its insulin lowering abilities and

(b) those exerted directly on the tumor.

The confusion born from the blending of these two different activities confounded many investigations over the past 15 years.

Metformin as an anti-cancer drug

The idea of metformin as an anti-cancer drug actually dates back to 2005:

(1) In 2005, Evans et al.[82] published a report that called attention to the decrease in cancer risk in diabetic patients receiving metformin as part of their treatment. Previous publications showed the importance of the LKB1-AMPK pathway in anti-oncogenesis. There was evidence showing that this pathway was up-regulated by metformin, thus, the question that naturally arouse was whether metformin had anti-cancer effects The publication by Evans et al., had a great impact on the scientific community. It was based on a case-control study of resident population of Scotland (11,876 patients studied), that showed that diabetic patients treated with metformin had a lower relative risk (RR) of having cancer (RR=0.912) and those with a long standing treatment showed even a lower risk (RR=0.782) as compared to diabetics not treated with metformin. In summary, metformin reduced cancer risk in diabetics and this reduction was dose and treatment-duration dependent.

(2) In that same year, 2005, two publications by Anisimov et al.,[83, 84] that were less noticed, showed that transgenic mice Her2/neu with a tendency to develop spontaneous mammary tumors, receiving metformin had a lower incidence and size of mammary adenocarcinomas.

The study by Evans et al. burst into several epidemiological studies, trying to find a place for metformin in cancer treatment (summarized in Table 1).

(3) In 2006, Evans' findings were confirmed by another large population study (10,309 cases).[94] The conclusions of this study were:

> *"Patients with type 2 diabetes exposed to sulfonylureas and exogenous insulin had a significantly increased risk of cancer-related mortality compared with patients exposed to metformin. It is uncertain whether this increased risk is related to a deleterious effect of sulfonylurea and insulin or a protective effect of metformin or due to some unmeasured effect related to both choice of therapy and cancer risk".*

The largest population study (62,809 diabetic patients) conducted by Currie et al.[95] showed that the lowest risk for cancer was found in the patients on Met treatment. It was higher for those on Met and sulfonylureas and the highest risk was in the group

Table 1 Population studies showing Met's beneficial effects in cancer patients.

Reference	Findings
Jiralerspong et al.[85]	Diabetic women with breast cancer who received adjuvant therapy and metformin showed better evolution. But Jiralerspong et al. went a step further and also analyzed patients receiving insulin and found that those women who were on insulin and metformin showed similar results to those on metformin alone. Patients on insulin without metformin had significantly worse results
Libby et al.[86]	New metformin users showed less risk of developing cancer
Decensi et al.[87]	31% reduction of relative risk of cancer in metformin users
Landman et al.[88]	Lower mortality due to cancer in diabetic population on metformin
Murtola et al.[89]	Lower prostate cancer incidence in patients receiving anti-diabetic medication of any kind, metformin and insulin included
Gagnon et al.[90]	Better prognosis in diabetic patients with lung cancer receiving metformin
Rieken et al.[91]	Bladder cancer: diabetics without metformin showed higher risk of recurrence and progression
Spratt et al.[92],	Met use in diabetic patients improved prostate-specific antigen-recurrence-free survival, distant metastases-free survival, prostate cancer-specific mortality and overall survival, compared to diabetic patients not using metformin
Kumar et al.[93]	Met was associated with better survival in ovarian cancer

receiving insulin. When patients received Met and insulin the risk was lower than those who were on insulin alone. The risk modification was not a generalized one: insulin increased the risk for colorectal and pancreatic cancer but not for breast and prostate cancer. Lee et al.,[96] studied a population of 800,000 Taiwanese persons without diabetes at the beginning of a 7-year follow up period. They divided the population into four cohorts: (1) people who developed diabetes type 2 (DM2) and received only Met; (2) people who developed DM2 and were treated with oral medication without Met; (3) people who developed DM2 but received no oral medications; and (4) people who had no DM2 and received no medication of any kind. They found that Met reduced the total incidence of liver, colorectal and pancreatic cancer compared with those diabetics who did not receive any medication or were treated with other oral compounds. Incidence of cancer in the Met treated group was similar or slightly superior to non-diabetics. This shows that Met "brings down" the incidence of cancer in diabetics, but this downturn does not go below the incidence found in non-diabetics.

At this point, there is one question we are unable to answer: is it insulin that increases the risk, or metformin that decreases it? Based on the work of

Currie et al., it seems logical to think that insulin increases the risk and metformin decreases it. This leads to another unanswered question: is the metformin-receiving population the baseline and the statistic is showing an increased risk for those treated with insulin? If the answer is affirmative, actually, we are in the presence of increased RR for "insuliners" with no real RR decrement for "metforminers" compared to non-diabetic population.

All these population based studies should not be extrapolated to the general population. Including the original data from Evans et al. and all the rest mentioned above are the result of comparing one diabetic population receiving metformin with another diabetic population receiving a different treatment, which quite frequently was insulin. Insulin is a growth factor and therefore can be considered a pro-tumoral hormone. This is only the first step of the controversy. The second step arises from the fact that there are epidemiological studies that did not confirm the initial findings of Evans et al. Table 2 is a summary of this conflicting evidence.

Table 2 Epidemiologic evidence against metformin in cancer.

Reference	Findings
Becker et al.[97]	Met did not reduce risk of esophageal cancer
Becker et al.[98]	Met did not reduce risk of endometrial cancer
Ko et al.[99]	Met was not associated with a decreased risk of endometrial cancer
Bodmer et al.[100]	Met did not reduce risk of lung cancer
Bodmer et al.[101]	Met did not reduce risk of colorectal cancer
Kaushic et al.[102]	No benefits with Met in prostate cancer outcome after radical prostatectomy
Niraula et al.[103]	No benefits with Met in prostate cancer treated with docetaxel
Lega et al.[104]	Met did not improve overall survival in breast cancer
Hwang et al.[105]	Met did not improve overall survival in pancreatic cancer
Lega et al.[106]	No association was found between Met and breast cancer stage at initial diagnosis
Smiechowski et al.[107]	Met was not associated with decreased risk of lung cancer
Azoulay et al.[108]	Met did not reduce the risk of prostate cancer
Bayraktar et al.[109]	Met had no impact on the outcome of patients with triple negative breast cancer

It is important to mention an epidemiologic work by Grealdine et al.[110] conducted in Belgium with a population study of 1.5 million patients-year. They concluded that patients with diabetes had a higher prevalence of cancer, but the interesting point of these conclusions was that decrease in risk was associated not only with metformin but also with the use of other anti-diabetic drugs.

In all the cases mentioned in Table 2, the population under scrutiny was large enough to be significant. The question will not be resolved until a study is conducted with two cohorts of non-diabetics, one receiving Met and the other not.

It is also absolutely necessary to discover the molecular mechanisms of metformin in normal and cancer cells if we want to understand the different findings.

Molecular mechanisms of metformin

Controversial results were also found in this issue. Foretz et al.,[111] suggested that LKB1 may possibly be a target of Met, not only in diabetes treatment but also in cancer. With this starting point, many studies at basic level were conducted to confirm metformin's anti-tumor action in order to understand its possible mechanism. Zakikhani et al.[112] confirmed the concepts outlined above and found that metformin had an anti-tumor action in tissue cultures of mammary carcinoma and also confirmed the activation of the LKB1-AMPK pathway and consequent inhibition of mTOR. In the past few years, many studies confirming metformin's anti-tumor action in the experimental setting have been brought to light. However, there are publications that show that in certain cell types, metformin develops a pro-angiogenic action.[113, 114] Under hypoxic conditions metformin has produced pro-angiogenic actions in umbilical cord endothelium,[115] and without hypoxemia, in certain prostate carcinomas[116], muscle[117] and human glial cells. Contrary to the concept of metformin as an agent in cancer prevention, the LKB1-AMPK pathway promotes cell survival in glucose-deprived non-tumor cell cultures.[118] There are also controversies regarding the association of metformin with chemotherapeutic drugs like cisplatin. Janjetovic et al.[119] found decreased cytotoxicity by cisplatin in vitro when it was associated with metformin in glioma, retinoblastoma, fibrosarcoma and leukemia cells. Rattan et al.[120] on the contrary, observed increased cytotoxicity when cisplatin was associated with metformin. Kim et al.[121] found that inhibiting activated AMPK increased cisplatin cytotoxicity by hyperinduction of p53.

Therefore, at present it is not clear how metformin acts in association with chemotherapeutic drugs like cisplatin. However, the association of metformin with doxorubicin,[122] or paclitaxel,[123] increased cytotoxicity.

Table 3 lists the effects of Met at the molecular level that show anti-cancer activity.

The mechanisms proposed for Met's anti-cancer activity can be summarized in the following way:

(1) Activation of the LKB1/AMPK pathway that inhibits mTOR and Akt.

(2) Suppression of mitochondrial production of metabolic intermediaries.

Hypoxia and glucose deprivation conditions are usually found in the microenvironment of highly proliferating tumors. Both conditions increase AMPK activation[139] independently of Met. Therefore, the reasonable doubt is: if the LKB1/AMPK pathway is already up-regulated, what can Met add by up-regulating this pathway?

Table 3 Molecular evidence favoring metformin in cancer.

Reference	Finding
Dowling et al.[124]	Metformin inhibited translation initiation of mRNA under mTOR control in breast cancer cell cultures
Hirsch et al.[125]	Low doses of Met "attacked" cancer stem cells in four different breast cancer cell lines and the combination of Met and doxorubicin killed non-stem cells and stem cells alike in cell culture and in a mouse xenograft model. The combination also prevented relapse in a synergistic manner
Alimova et al.[126]	In tissue culture, Met showed activity against many breast cancer cell lines and reduced expression of cycline D1, E2F1 and decreased activity of MAPK, Akt and mTOR
Goodwin et al.[127]	Met lowered insulin levels in non-diabetic women in early stage breast cancer
Liu et al.[128]	In tissue culture and xenograft models of triple negative breast cancer lines, MET showed growth inhibition, induced apoptosis, decreased EGFR, cyclins D1 and E and partition of PARP
Iliopoulos et al.[129]	In cancer xenograft models, Met decreased the necessary dose of chemotherapeutic drugs to prolong the remission stage
Rocha et al.[123]	In mice xenografted with cancer cells, Met increased AMPK activation and increased cytotoxicity of paclitaxel
Ben Sahra et al.[130]	Met decreased the level of cyclin D1 in vitro and in vivo and decreased growth of prostate cancer cell lines
Buzzai et al.[131]	Met was selectively active against P53 deficient tumor cells decreasing growth
Zhuang et al.[132]	Met arrested growth of breast cancer cell lines activating AMPK, and decreasing cycline D1
Zakihkani et al.[133]	Rapamycin and Met showed differences in activity over Akt. While rapamycin activated Akt by a feed back loop, Met inhibited Akt
Huang et al.[134]	Established the importance of the LKB1-AMPK pathway in tumorigenesis suppression
Gotlieb et al.[135]	Showed Met's anti-tumor activity in ovarian cancer
Dong et al.[136]	Metformin increased effects of chemotherapy on endometrial cancer cells by down-regulating the expression of glyoxalase I (cisplatin and paclitaxel were tested)
Younis et al.[137]	In head and neck squamous cell carcinoma Met inhibited mTOR
Griss et al.[138]	Met suppressed mitochondrial-dependent metabolic intermediates required for cell growth

The in vitro experiments with Met were mainly performed in conditions quite different from that effectively found in a tumor with its environment. Undoubtedly, in cell cultures Met up-regulates the LKB1/AMPK pathway, but in a solid tumor this job is already done by hypoxia and lack of nutrients. Therefore, the anti-tumoral activity of Met is not based on this pathway. Fig. 5 shows the effects of metformin on glucose metabolism and the LKB1-AMPK pathway.

Table 4 summarizes the concepts why the LKB1/AMPK pathway is not related to Met's anti-tumoral activity.

Table 4 Molecular evidence against the LKB1/AMPK pathway as the anti-tumoral action of metformin.

Reference	Findings
Phoenix et al.[113]	Pro-angiogenic activity of metformin was described in ERα negative MDA-MB-435 breast cancer cells
Nagata et al.[115]	Endothelial cells of human umbilical cord (HUVEC): under hypoxia there was activation of AMPK and angiogenesis playing a role in cell migration and tubular structure formation. In normoxia, AMPK activation was not required for angiogenesis
Lee et al.[116]	In DU 145 prostate cancer cells AMPK activation is essential for HIF expression. Blocking AMPK decreased expression of HIF-1 and VEGF
Ouchi et al.[117]	In muscle tissue, activation of AMPK produced expression of VEGF and angiogenesis. AMPK activation was induced through hypoxia and AICAR (similar activity as metformin). AICAR also up-regulated p38 MAPK
Neurath et al.[140]	In human glia hypoxia activated AMPK-alha2 fraction and increased production of VEGFR increasing angiogenesis. Inhibition of AMPK activation blocked overexpression of VEGF and HIF1
Zwetsloot et al.[141]	AMPK regulated expression of VEGF and muscular capilarization
Li et al.[142]	AMPK induced differentiation of endothelial progenitor cells. AICAR increased angiogenesis and compound C decreased it. (Compound C is an AMPK inhibitor)
Park et al.[143]	Activation of AMPK increased growth and cell survival in human prostate cancer cells
Jeon et al.[144]	AMPK regulated NADPH production under cell stress favoring tumor cell survival
Damelin et al.[145]	Metformin "protected" cancer cells through the creation of a reductive state by NADPH when treated with cisplatin

Starting with clinical epidemiology and following with the studies of tumor tissue cultures, a clear anti-tumor action was described for metformin. But its use in preventive clinical settings and treatment of human cancer has not shown undisputable results up to now. The anti-tumor effect of metformin, which seems so evident in the tissue culture setting, does not fully translate into similar effects in the epidemiological and prevention settings. There is then a conundrum: why does a drug that seems so beneficial in the laboratory, not seem to be equally beneficial in the prevention of cancer risk in the clinical setting?

One possible answer is that metformin's anti-tumoral activity is tissue-specific. To find an answer, different tissues will be examined. There are two types of tissues that show particular susceptibility to metformin: endometrium and thyroid.

Endometrial cancer

Metformin does not reduce the risk of endometrial cancer as shown in Table 2, but many of these tumors are found in patients with insulin resistance and the insulin-insulin receptor pathway is overexpressed. However, there is in vitro and clinical evidence of metformin's benefits in this cancer as shown in Table 5.

Table 5 Anti-cancer activities of metformin in endometrial cancer

Reference	Metformin action
Zhang et al.[146]	Metformin reduced the secretion of IGF-1 in endometrial cancer cell lines and also the expression of IGF-1R and deactivated downstream signaling involving the PI-3K/Akt/mTOR pathway inhibiting endometrial carcinoma (EC) growth
Zhan et al.[147]	Metformin inhibited estrogen-mediated proliferation of human endometrial cancer cells
Tabrizi et al.[148]	Metformin was an effective anti-estrogenic agent in control of abnormal endometrial proliferative disorders. (Cases analyzed were not cancer) metformin could induce endometrial atrophy in 21 out of 22 patients)
Schuler et al.[149]	Metformin reduced tumor proliferation in 65% of pre-operative obese patients studied with endometrial carcinoma. Inhibition of the mTOR pathway was found
Laskov et al.[150]	Eleven newly diagnosed, untreated, non-diabetic patients with EC received metformin starting at the diagnostic biopsy until surgery. There was reduction of ki-67 and pS6 expression. In the non-treated control group expression was unchanged
Takahashi et al.[151]	Metformin-treated cells exhibited significantly lower viability and proliferation and significantly more cell cycle arrest than control cells. These cells also exhibited significantly more apoptosis. Met treatment induced autophagy. Inhibition of autophagy, suppressed the anti-proliferative effects of metformin on endometrial cancer cells
Mitsuhashi et al.[152]	Preoperative metformin in EC treatment decreased DNA synthesis in sera and significantly reduced the Ki-67 (44.2%) and topoisomerase IIα, S6 and ERK1/2.
Mu et al.[153]	Evidence suggests that insulin resistance plays a central role in endometrial cancer development
Cantrell et al.[154]	Metformin inhibited growth of endometrial cancer cells in a dose-dependent manner and resulted in arrest and induction of apoptosis
Tan et al.[155]	In vitro invasion in ECC-1 cells was significantly attenuated by sera from women with polycystic ovary syndrome after 6 months of metformin treatment compared to matched controls
Sarfstein et al.[156]	Metformin induced apoptosis, inhibition of proliferation, and migration of USC cell lines through interaction with the IGF pathway

Thyroid cancer

Positive effects of metformin in thyroid cancer were described.[157] Patients with differentiated thyroid cancer were divided into three groups: (a) non-diabetics, (b) diabetics on metformin, and (c) diabetics without metformin. Complete response (CR) to treatment was higher in the diabetics on Met when compared with diabetics without Met. However it was the same as the CR found in non-diabetics. Progression free survival (PFS) was similar between non-diabetics and diabetics on Met. Both groups had a significantly higher PFS than diabetics without Met. The article does not mention which was the treatment besides the Met issue.

In a Taiwanese diabetes II population[158] the hazard ratio was 0.68 for metformin users. This favorable action has been detected in papillary cancer in vitro and in vivo.[159] Chen et al.[160] demonstrated growth-inhibitory effects on differentiated thyroid cells and undifferentiated thyroid carcinoma cells, including thyroid carcinoma stem cells. Klubo-Gwiezdzinska et al.[161] showed anti-tumor activity of metformin in medullary thyroid cancer. Metformin also produces a slight reduction in TSH levels without affecting thyroid hormone levels and decreases iodine uptake.[162] This should be considered in case of using radioiodine. We found no controversies regarding the promise of Met as an adjuvant therapy in thyroid cancer.

Colorectal cancer

Metformin at low dose does not reduce colorectal cancer risk.[163] Metformin at high doses reduces colorectal cancer mortality. Spillane et al.[164] confirmed the lack of risk reduction with low dose Met, but with high doses they found a decrease in mortality.

Breast cancer

Controversial effects of metformin on breast cancer were found. Aksoy et al.[165] described benefits attributed to metformin intake in diabetes: lower incidence of advanced histological grades, lower incidence of triple negatives, higher incidence of positivity to estrogen and progesterone receptors, increased disease-free survival but similar overall survival. Lega et al. (Table 2) did not find increased survival with duration of metformin treatment nor association between Met and tumor stage at diagnosis. Tang et al.[166] found no modifications in breast cancer incidence with Met.

On the other hand Dowling et al.[167] in a study conducted on non-diabetic women with untreated breast cancer administered metformin for 2 weeks after diagnostic biopsy until surgery. The surgical specimens obtained showed decreased levels of IR (insulin receptor) expression and Akt phosphorylation in tumors. The authors remarked that reduced Akt and ERK1/2 phosphorylation, and decreased insulin and IR levels, suggest that the reduction of the insulin-insulin receptor pathway is important in the clinical setting. Metformin decreased aromatase's action.[168]

Metformin has demonstrated anti-tumoral activity in thyroid cancer and endometrial cancer. In other type of cancers (prostate, lung, larynx, melanoma), the results are controversial. The evidence indicates that there is certain tumor or tissue specificity for Met's anti-tumoral effects.

Metformin and the LKB1/AMPK axis

The anti-tumoral activity of metformin was explained through two different mechanisms:

(a) *Insulin dependent*: metformin decreases hyperglycemic conditions in diabetics and therefore decreases the level of insulin and the activation of insulin receptors (IR).

(b) *Insulin independent*: Met induces the activation of the LKB1/AMPK axis.

The insulin-dependent activity is probably the reason for certain benefits regarding cancer in diabetic populations. This mechanism seems to "protect" against the pro-tumoral signaling of IR. And this protection seems to be context-dependent: it is not found in all the tumors, but only in certain cancers (e.g., endometrial and thyroid cancer). This explains some of the controversies found in population-based studies.

The LKB1/AMPK axis was originally described as a tumor-suppressor pathway through the inhibition of mTOR. However, evidence started to accumulate regarding this pathway as an oncogenic one. Actually, it is a dual pathway: pro- and anti-oncogenic and contextual dependent.[169, 170] Figs. 6 and 7.

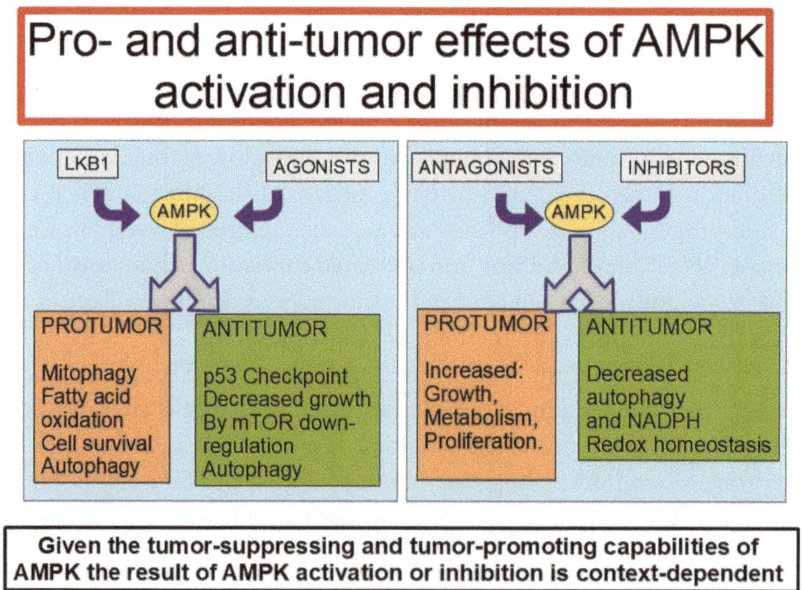

Fig. 6 Inhibition and activation of AMPK has pro- and anti-oncogenic effects.

Finally, there is an explanation for the controversies we found with metformin. This situation makes metformin an unreliable drug for cancer treatment if we expect any effect other than mitochondrial Complex I inhibition. This last effect is always present, no matter the context (Fig. 7).

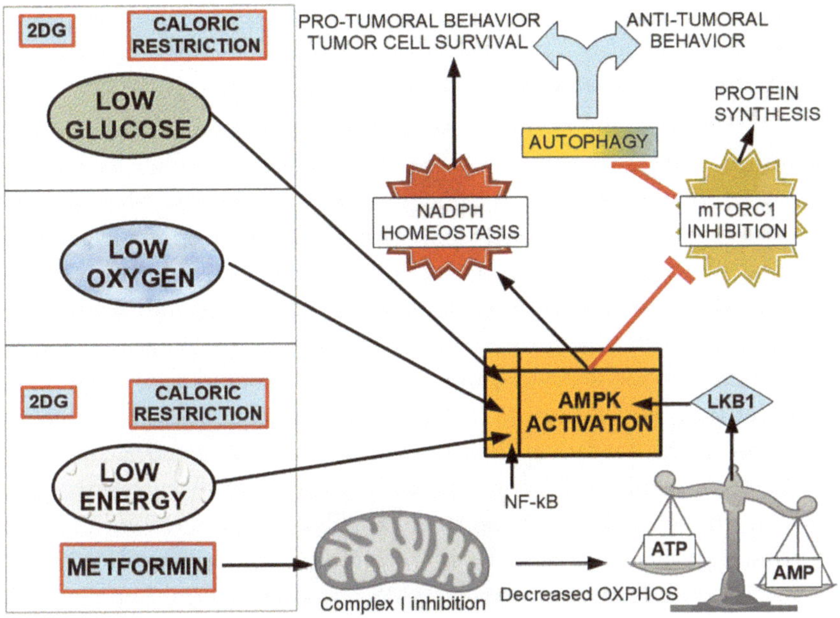

Fig. 7 AMPK activation under different environmental conditions. The dual activity of AMPK is represented by mTORC1 inhibition (anti-oncogenic) and tumor cell survival (pro-oncogenic).

Context-dependent activity of metformin

- Buzzai showed that metformin selectively impairs p53-deficient tumor cell growth.
- In the context of energetic stress and hypoxia, AMPK is activated and protects cells from apoptosis.[171] This last report showed that when hepatic and pancreatic cancer cells where exposed to glucose deprivation, most of the hepatic cells died within 48 h while pancreatic cells survived well beyond that time span. The survival of the pancreatic cancer cells correlated with the cellular expression of AMPK.
- Under energy stress conditions AMPK regulates NADPH promoting tumor cell survival.
- Activated AMPK increased prostate cancer cell growth and survival.
- AMPK seems to play an important role early in the malignant transformation and knockdown of AMPK suppressed tumor growth.[172]
- When metformin was administered to p53 mutated or knockdown cancer cells, apoptosis was the result. But when the cell had a wild type p53 there was no apoptosis

but autophagy instead. The authors considered that Met had a toxic effect on p53 mutated cells.

Liang and Mills[173] proposed an explanation to these contextual effects of AMPK activation: "It is possible that modest activation of AMPK engages cell protective mechanisms resulting in oncogene-like activities, whereas increased magnitude or duration of stress could induce growth arrest or cell death exhibiting a tumor suppressor function."

Non-contextual effects of metformin

Inhibition of mitochondrial Complex I by Met does not depend on the context. This happens in normal and tumor cells.[174, 175] Actually, many researchers consider the inhibition of mitochondrial Complex I as the primary site of action of Met, and the AMPK activation as a consequence of decreased energy balance due to a lower ATP production when oxidative phosphorylation is inhibited.[175–177] There is evidence showing that metformin does not directly act on the LKB1/AMPK axis, but indirectly by increasing AMP.[178–180]

In the presence of glucose, Met inhibited cellular proliferation, but under glucose deprivation Met induced cell death.[181] The direct consequence of this inhibition is a down-regulation of the oxidative phosphorylation with increased lactic acid production.[182] If the accumulation of lactic acid saturates the lactate extrusion mechanisms, namely, monocarboxylate transporters, the intracellular pH decreases. This would also happen if other proton extrusion systems are inhibited and a high dose of metformin is simultaneously administered. However, ROS production would be diminished with metformin,[183] and this would decrease the cytotoxicity of some chemotherapeutic drugs.

Met's ability to decrease intracellular pH has been confirmed by the publication of Melnik et al.[184] They also found that Met blocked the signaling by Wnt/β-catenin in a specific way: only in cancer cells. Interestingly, the mechanism behind the blocking was cellular acidification. And it could also be achieved with many other Complex I inhibitors. The mechanism according to Melnik et al. would be:

(1) Metformin inhibits mitochondrial Complex I.
(2) This produces an intracellular acidification due to lactate accumulation.
(3) Acidification triggers the unfolded protein response (UPR).
(4) UPR induces the transcriptional repressor DDIT3.
(5) DDIT3 blocks Wnt signaling.
(6) Intracellular acidification in general inhibits Wnt signaling.

The fact that Met only acted on cancer cells and not in normal cells, has a very simple explanation: cancer cells through the glycolytic pathway have a very increased production of lactic acid which is further increased by Met. Normal cells metabolizing through the oxidative pathway produce 60-fold less lactic acid. The extrusion of lactate by MCTs has a limit, that when surpassed, increases intracellular lactate that acidifies the cell. The findings by Melnik et al. explain some other findings that had no clear explanation. For example:

(1) the synergistic activity of nelfinavir and metformin on cancer cells[185, 186]: both induce UPR;

(2) the synergistic activity of simvastatin and rosuvastatin with metformin[187, 188]: statins inhibit MCT4 (monocarboxylate transporter 4),[189] therefore, while Met increases lactic acid, statins block its extrusion increasing intracellular acidity;

(3) the synergistic activity when the three (nelfinavir, statins and Met) are associated[190];

(4) statins also inhibit mitochondrial Complex I in muscle[191];

(5) the higher growth inhibition activity of phenformin versus metformin[26]: phenformin produces more lactic acid than metformin. We assume that phenformin can easier saturate MCTs transport capacity. This also explains the higher clinical toxicity of phenformin (Box 2).

BOX 2 Metformin's actions.

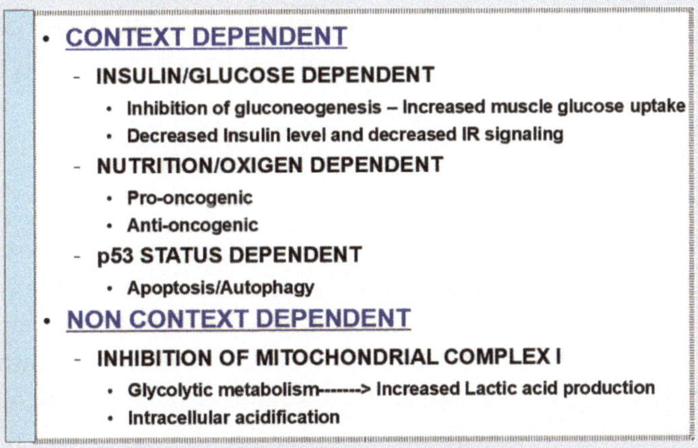

Conclusions

Metformin can acidify the intracellular milieu only if the dose used is high and/or other drugs are used with it, e.g.:

(a) monocarboxylate transporter inhibitors that impede disposing of the lactic acid;

(b) proton extruder inhibitors;

(c) dichloroacetate that when associated with metformin induces a metabolic stress;

(d) 2-deoxy-glucose that would create an energy crisis due to decreased ATP;

(e) statins that down-regulate MCT4; and

(f) nelfinavir that increases UPR synergistically.

Metformin cannot be considered a stand alone treatment for cancer. The anti-cancer effects of metformin in non-diabetics are related to intracellular acidification and not the up-regulation of the LKB-1/AMPK axis. The anti-cancer effects of metformin in diabetics may be related to the down-regulation of the insulin-insulin receptor signaling (which is a pro-proliferative signaling). There is no evidence that this cancer prevention mechanism can be present in non-diabetics.

- Population-based studies are controversial and the findings are limited to diabetic patients.
- Population-based studies cannot be extrapolated to non-diabetic populations.
- Many cellular studies were performed using metformin concentrations that are difficult to achieve in the clinical setting without a serious risk of lactic acidosis.
- Cellular studies have shown that metformin's activity is contextual and therefore its therapeutic value is limited to certain tumors whose exact characteristics have not been clearly determined.
- All the effects of metformin stem from intracellular acidification which is its main cytotoxic mechanism.
- Metformin is cytotoxic for cancer cells when MCTs are simultaneously inhibited.
- Lonidomine can advantageously replace metformin when treating cancer in non-diabetic patients.

Dichloroacetate

Cancer cells differ from normal cells in many ways, including metabolism. Cancer metabolism is bound to cancer's development and progression in such an intricate way, that we may assume that there is no malignancy without the appropriate metabolic changes. Among the many metabolic alterations in cancer, glycolytic metabolism is one of the hallmarks of cancer as opposed to oxidative metabolism which is characteristic of normal cells. Cancer "needs" a glycolytic phenotype in order to avoid cell death.[192] Apoptosis "needs" an oxidative phenotype.[193] Cells that depend on the glycolytic pathways are resistant to cell death.[194–196]

Tumor progression is a Darwinian type of natural selection where the cells with a highly glycolytic activity and repressed mitochondrial function are selected for tumor development.[197] The reversion of the glycolytic phenotype decreases proliferation and increases apoptosis of malignant cells. It is not so difficult to reduce lactic acid production by inducing the cell towards oxidative phosphorylation: this would be the supposed function of dichloroacetate (DCA).

DCA has been used for the treatment of acute lactic acidosis in children and adults,[198, 199] and in cases of congenital conditions in children[200] since the 1970s.[201] In 2007 the group headed by Dr. Michelakis[202] called attention to DCA's effects in cancer due to its ability to inhibit pyruvate dehydrogenase kinase (PDK), thus restoring oxidative phosphorylation. They also found that DCA induced apoptosis and inhibited tumor growth.

However, due to a lack of FDA approval, DCA has not entered the mainstream of oncology practice.

DCA is a small orally available molecule, that is quickly and almost completely absorbed by the digestive system.[203, 204] It is metabolized by GSTZ1 (a glutathione transferase isoform).[205] When DCA was administered to 16 healthy individuals, it lowered plasma glucose, lactate and alanine concentrations. Plasma levels linearly followed those of the administered dose up to 30 mg/kg. 35 mg/kg was considered the most effective dose regarding lactic acid, that fell to 75% below baseline concentrations within 2h of infusion.[206]

The experimental administration of a single dose of DCA may be misleading because as Gonzalez-Leon et al.[207] have shown in rats with repeated administration of DCA, this drug has a surprising feature: it inhibits its own metabolism. This means that the chronic administration of DCA differs from single doses in its plasma concentration, toxicology and metabolism. In cultured normal hepatocytes DCA must be added daily to maintain physiological effects which means that it should be used in a metronomic fashion.[208] The first dose is more rapidly cleared from plasma than the subsequent ones. This is probably due to GSTZ1 inhibition by DCA.

Mechanism of action

The main mechanism of action of DCA is the inhibition of pyruvate dehydrogenase kinase (PDK) and its isoforms. This inhibition increases the flux of pyruvate into the mitochondria, promoting glucose oxidation instead of glycolysis.[209]

Pyruvate dehydrogenase kinase (PDK) is a kinase that inactivates the pyruvate dehydrogenase enzyme complex through phosphorylation. By downregulating the activity of this complex, PDK decreases the oxidation of pyruvate in mitochondria and increases the conversion of pyruvate to lactate in the cytosol.

Actually by inhibiting PDK, DCA "forces" the cancer cell to abandon its preferred metabolic process (aerobic glycolysis), allowing the PDH complex to drive pyruvic acid towards the mitochondrial tricarboxylic cycle.

In cancer cells with glycolytic metabolism, pyruvate dehydrogenase (PDH) is inhibited through phosphorylation performed by pyruvate dehydrogenase kinase (PDK).

Under normal conditions the pyruvate goes into the mitochondria as acetyl-coenzyme A and proceeds towards the Krebs cycle; in cancer cells it remains in the cytoplasm and is converted into lactic acid that finally is extruded from the cell through the activity of monocarboxylate transporters (MCTs). Otto Warburg showed that cancer cells use this mechanism even when oxygen levels are improved (aerobic glycolysis or Warburg effect). Malignant cells are highly dependent on aerobic glycolysis. The main mechanism of DCA's anti-cancer activity is the inhibition of mitochondrial pyruvate dehydrogenase kinase, allowing the activity of pyruvate dehydrogenase and the switch back from the glycolytic metabolism. Fig. 8. This PDK inhibition is not ATP-competitive[210] and leads to caspase-mediated apoptosis.[211]

Another possible mechanism involved in anti-tumor activity is the modification of the intracellular acidity, that paradoxically decreases.[212] And the paradoxical issue should be underlined because DCA produces a decrease in lactic acid production, which would mean that an increase of intracellular pH (pHi) would be expectable. This does not happen, on the contrary, pHi decreases (at least temporarily). This means that there must be other, as yet not fully known mechanisms, playing a role. Anemone et al.[213] found an increase of extracellular pH in vivo in breast tumor bearing mice after treatment with DCA, but after 15 days there was resistance to DCA treatment and extracellular pH and lactic acid levels returned to the original levels.

The decrease of pHi produced by a simple dose of DCA is approximately one third of the pHi decrease produced by a simple injection of cariporide[214] in a glioblastoma model.

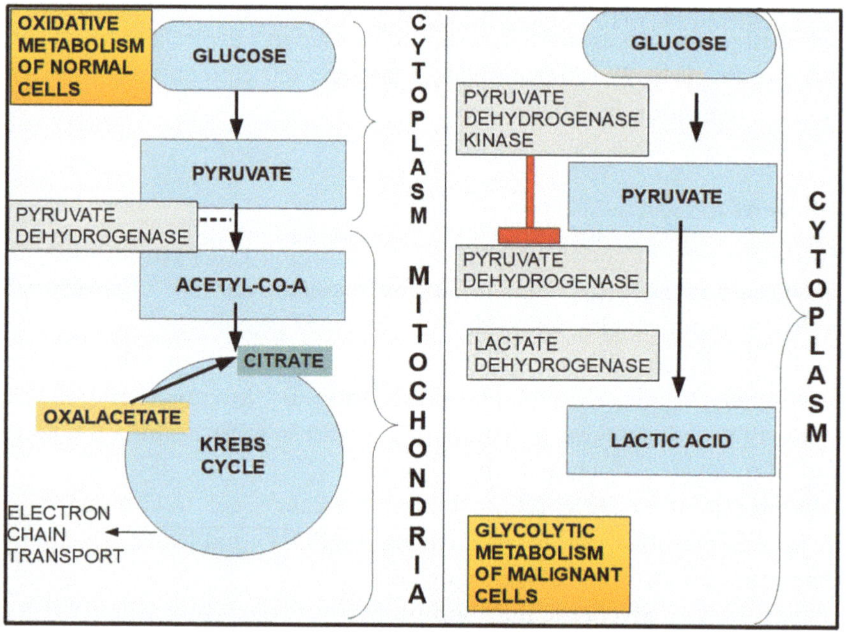

Fig. 8 Glucose metabolism in normal cells (oxidative metabolism) on the left side, and glucose metabolism in the cancer cell (glycolytic metabolism) on the right side. In the first case, the chemical end products are CO_2 and H_2O. In the second case the end product is lactic acid, that is extruded from the cell through MCTs. The activity of PDH is the cornerstone that defines which way the metabolism will go. If PDH is working normally, the cell will adopt an oxidative metabolism, otherwise it will take a glycolytic pathway. The second important fact is that malignant cells show a 2- to 20-fold higher glucose intake which implies on average a 60-fold increase in lactate production. When all this lactate is extruded, it has an impact on extracellular acidity. However, lactate does not seem to be the main cause of extracellular acidity. The mitochondrial PDH irreversibly decarboxylates pyruvate to acetyl coenzyme A, linking glycolysis to the tricarboxylic acid cycle. This is a critical step in carbohydrate metabolism where a decision point is reached: glycolytic or oxidative pathway. Inhibition of PDH by PDK is shown on the right.

DCA also increases the expression of COX2 and the latter increases resistance of tumors to DCA. Therefore a COX2 inhibitor should be associated with DCA treatment.[215]

Other actions of DCA are inhibition of lipogenesis and cholesterol generation.[216, 217]

DCA also shows anti-HIF-1α activity. Sutendra et al.[218] found that DCA suppresses HIF-1α activity and angiogenesis through the inhibition of PDK-II.

DCA produces T regulatory cells (Tregs) differentiation through mechanisms not related to PDK inhibition.[219] Tomsig et al.[220] found that DCA inhibited the cell uptake of lactate through a competitive mechanism. This means that DCA may interrupt the lactate shuttle. This last action has not been confirmed yet in the oncology setting. The mechanism involved seems unrelated to PDK inhibition.

Experimental evidence of DCA activity in cancer

See Tables 6–11.

Table 6 DCA and breast cancer.

Reference	Findings
Sun et al. 2010[221]	The growth of several human breast cancer cell lines was found to be inhibited by DCA in vitro. In vivo there was a 58% reduction in lung metastasis
Gang et al. 2012[222]	Extracellular lactate was reduced by 50% at 1 mM and 5 mM DCA. And cell viability was reduced between 15% and 70% according to malignant cell line. Normal breast cells only suffered a 7% decrease in viability
Wang et al. 2017[223]	DCA increased sensitization to doxorubicin by inhibiting autophagy
Harting et al. 2017[224]	DCA reduced proliferation but did not cause apoptosis in canine breast cancer cells
Woo et al. 2016[225]	DCA synergized with tamoxifen in breast cancer cell lines by down-regulating EGFR
Haugrud et al. 2014[226]	Metformin-induced oxidative damage was increased by DCA. Their combination had synergistic effects
De Preter et al. 2016[227]	DCA decreased the Pentose Phosphate Pathway, the glycolytic pathway and reduced proliferation in different breast cancer and cervical cancer cell lines
Sun et al. 2011[228]	The combination of DCA and arsenic trioxide synergistically inhibited breast cancer cell proliferation and induced apoptosis
Robey et al. 2011[229]	In a MDA-MB231 metastatic breast cancer model in tumor-bearing mice, the combination of sodium bicarbonate and DCA did not show better results than bicarbonate alone
Cheng et al. 2012[55]	The association of DCA with 2 deoxyglucose had synergistic effects on breast cancer cell apoptosis

Continued

Table 6 DCA and breast cancer—cont'd

Reference	Findings
Lefort et al. 2014[230]	DCA and allopurinol was applied in three different breast cell lines (MDA-MB-231 and MCF-7 and normal control cell line, MCF-10A). Treatments of breast cancer cells with DCA and allopurinol resulted in larger changes in metabolites found in extracellular medium than intracellular pools. DCA had no metabolic effects on normal breast cells. Apoptosis was low in the three cell lines. Treatment with low doses of DCA did not lead to high level of apoptosis in MDA-MB-231 cells. Allopurinol decreased the metabolic effects of DCA
Sun et al. 2009[231]	DCA treatment of breast cancer cell lines reversed the glycolytic phenotype
Xintaropoulou et al. 2015[232]	Comparing different glycolytic metabolic inhibitors and blockers DCA showed high effectiveness against MCF-7 cells and less effective against MDA-MB-231 cells
Blackburn et al. 2010[233]	In a murine mammary adenocarcinoma model DCA reduced growth in 28% and less tumor formation. Increased tumor-infiltrating lymphocytes were found in the treated group. There was one case of complete regression of the tumor

Table 7 DCA and prostate cancer.

Reference	Findings
Harting et al. 2017[234]	DCA decreased proliferation, lactic acid production, survivin expression and increased miR-375 in prostate cancer and transitional carcinoma cells
Cao et al. 2008[235]	DCA alone had significant cytotoxic effects and was associated with G1 cell cycle arrest and increased apoptosis. The combination of DCA with irradiation sensitized cells to the radiation's cytotoxic effects
Higgins et al. 2009[236]	Certain prostate cancer cell lines were susceptible to DCA while others were not. For example PC3 and DU145 were unresponsive to DCA suggesting that their glycolytic phenotype was due to mitochondrial dysfunction rather than inhibition of PDH
Olszewski et al. 2010[237]	In prostate cancer cells DCA increased in vitro cytotoxicity of carboplatin and satraplatin, but not of cisplatin and oxaliplatin
Chaudhary et al. 2016[238]	Afro-Americans (AA) responded poorly to prostate cancer treatment. DCA inhibited cell proliferation in AA and Caucasians. DCA increased taxol cell death in Caucasians and sensitized AA cells to doxorubicin
Zhou et al. 2015[239]	In vitro DCA restored paclitaxel sensitivity in prostate cancer cells by increasing citrate accumulation

Table 8 DCA and colon cancer.

Reference	Findings
Madhok et al. 2010[240]	DCA did not reduce growth of non-cancerous cells but *caused* significant decrease in cancer cell proliferation
Lin et al. 2014[241]	DCA induced autophagy in human colon cancer cells with ROS production and mTOR inhibition, and reduced lactate excretion
Delaney et al. 2015[242]	DCA did not increase apoptosis in colon cancer cells, despite decreasing levels of anti-apoptotic protein Mcl-1 but correlated with a decrease in proliferation
Kahn et al. 2016[243]	Case report: in a 57-year-old female with stage IV colon cancer oral DCA treatment resulted in tumor stabilization
Tong et al. 2011[244]	DCA and 5-FU showed synergistic anti-cancer effects on colon cancer cells
Lin et al. 2014[241]	DCA induced autophagy in colon cancer cells and increased the production of ROS. It also inhibited mTOR

Table 9 DCA and melanoma.

Reference	Findings
Khan et al. 2017[245]	A case report of a 32-year-old male with advanced BRAF positive metastatic melanoma receiving only DCA (500 mg three times a day) achieved stable disease and shrinkage of tumor mass for more than 4 years
Abildgaard et al. 2014[246]	DCA reduced glycolytic activity and intracellular ATP levels, and inhibited cellular growth. Co-treatment of BRAFV600E-mutant melanoma cells with DCA and vemurafenib induced a greater reduction in intracellular ATP levels and cellular growth than either compound alone. DCA was effective against cells resistant to vemurafenib
Zheng et al. 2013[247]	DCA increased anti-tumoral effects of capecitabine
Populo et al. 2015[248]	Melanomas overexpress PDKs which makes them a candidate for DCA treatment. DCA produced modifications of melanoma metabolism: decreasing glucose uptake and lactic acid production, down-regulating proliferation
Franco-Molina et al. 2012[249]	Murine melanoma cancer cells were the most susceptible cell lines to DCA treatment in vitro. DCA induced apoptosis, inhibited invasion and angiogenesis. In vivo it reduced the volume and weight of tumors
Chaube et al. 2015[250]	Blocking lactate generation with DCA or oxamates in metformin-treated melanoma cells decreased cell proliferation and tumor progression in mice. Inhibition of complex I alone did not induce apoptosis, whereas inhibiting complex I and lactate generation caused metabolic catastrophe-induced apoptosis

Table 10 DCA and glioblastoma-retinoblastoma.

Reference	Findings
Michelakis et al. 2010[251]	Authors tested whether DCA can reverse cancer-specific metabolic and mitochondrial remodeling in glioblastoma. Freshly isolated glioblastomas from 49 patients showed mitochondrial hyperpolarization, which was rapidly reversed by DCA. In a separate experiment with five patients who had glioblastoma, they treated each patient with oral DCA for up to 15 months. DCA depolarized mitochondria, increased mitochondrial reactive oxygen species, and induced apoptosis in GBM cells, as well as in putative GBM stem cells, both in vitro and in vivo. DCA therapy also inhibited hypoxia-inducible factor-1α, promoted p53 activation, and suppressed angiogenesis both in vivo and in vitro
Duan et al. 2013[252]	DCA inhibited cell proliferation, induced apoptosis, and arrested cells in the S phase. In vivo anti-tumor effect results indicated that DCA inhibited the growth of gliomas in brain tumor-bearing rats and mice. DCA induced ROS production and decreased the mitochondrial membrane potential. Anti-angiogenic effects were also found
Kumar et al. 2013[253]	DCA enhanced results of bevacizumab treatment in a xerotransplanted mouse model of glioblastoma
Kolesnik et al. 2014[254]	Hypoxia enhanced the cytotoxic effects of DCA in glioblastoma cells
Vella et al. 2012[255]	DCA had anti-cancer effects on NB (neuroblastoma) tumor cells which were selectively directed to very malignant NB cells, whereas the more differentiated/less malignant NB cells were refractory to DCA
Sradhanjali et al. 2017[256]	DCA reduced retinoblastoma cell growth and potentiated carboplatin activity
Park et al. 2013[257]	C6 glioma was transplanted into rat brains. DCA modified PDH flux in glioma but not in normal brains in vivo
Fedorchuk et al. 2016[258]	Rats with transplanted C6 glioma cells showed that the DCA administration scheme could cause ambiguous effects: growth or inhibition of growth. Prolonged daily administration showed the best anti-tumor effects. Under hypoxia, anti-cancer efficacy of DCA was significantly increased
Wicks et al. 2014[259]	Local delivery of DCA on experimental brain tumors showed significantly increased survival in rats (oral administration)
Ward 2017[260]	Metformin potentiated DCA-induced superoxide production in glioblastoma VM-M3 cells increasing cytotoxicity

Table 11 DCA and lung cancer.

Reference	Findings
Lu et al. 2018[261]	NSCLC: DCA increased the apoptotic effect of chemotherapeutics by down-regulating autophagy in vivo and in vitro
Yang et al. 2016[262]	NSCLC: DCA showed synergy with EGFR tyrosin kinase inhibitors in NSCLC cell lines

Table 11 DCA and lung cancer—cont'd

Reference	Findings
Allen et al. 2015[263]	NSCLC: DCA interrupted the Warburg effect and decreased proliferation. It also increased the cytotoxicity of radiotherapy
Sun et al. 2017[264]	DCA re-sensitized human lung cancer cells resistant to paclitaxel. There was a synergistic effect between the drugs
Kolesnikov et al. 2015[265]	Lewis lung carcinoma: DCA slightly decreased the proliferation of primary tumor (30% less) and markedly decreased the number and volume of metastases
Fiebiger et al. 2011[266]	Lung carcinoid cells: DCA sensitized lung carcinoid cells (resistant to chemotherapeutics) to carboplatin
Shun et al. 2013[267]	DCA produced apoptosis in A549 lung adenocarcinoma cell line and this effect was synergistic with cisplatin

DCA and some interesting associations

DCA and metformin

The simultaneous use of DCA that decreases lactate production, and metformin that increases lactate production had interesting effects: they synergistically suppressed ovarian cancer cell growth and increased apoptosis,[268] in vivo and in vitro. It is not easy to explain these findings at the molecular level. Searching for an answer to this paradox, Ward et al. found that in glioblastoma cells, complex I inhibition increased DCA cytotoxicity, whether produced by metformin or other complex I inhibitors like rotenone, through increased oxidative stress.[269] This last finding seemed to explain DCA's cytotoxic activity: it is achieved through an increase of oxidative metabolism that increases oxidative stress, which is enhanced with complex I inhibition. This leads us to Fig. 3 that fully shows DCA activity, although there are controversies on this issue.

DCA and COX2 inhibitors

We have noted above that DCA induces COX2 expression, so that DCA should be associated with a COX2 inhibitor to achieve better results.

DCA and lipoic acid

Alpha lipoic acid (ALA) is a cofactor of PDH[270] and inhibits PDK.[271] An increased activity of PDH (as produced by inhibition of PDK) should require increased lipoic acid. Alpha lipoic acid decreases lactate and pyruvate levels in patients with type II diabetes,[272] but at the same time, ALA is a strong anti-oxidant and anti-apoptotic compound.[273] This last issue may create doubts about the utility of associating ALA with DCA. On the other hand, there are publications showing anti-tumoral effects of ALA,[274] and ROS generation through increased oxidative metabolism.[275–277] Feuerecker et al.[278] found that ALA was more effective against proliferation, decreasing lactate production and inducing apoptosis than DCA.

We think, even lacking all the necessary experimental evidence, that for cancer treatment DCA should be associated with lipoic acid and a COX2 inhibitor such as celecoxib.

DCA and 2D-deoxy glucose (2DG)

2DG markedly increased DCA's anti-cancer effects in Lewis lung carcinoma. DCA alone showed a reduction of 60% of metastasis in this model and a decrease of 90% of metastasis volume. DCA and 2DG did not affect the growth of the primary tumor, but the association decreased its volume.[279]

DCA and sulindac

This association increases tumor cell death through increased ROS production.[280]

Mitaplatin

Is a fusion compound formed by the association of two molecules of DCA and one of a platinum compound allowing simultaneous targeting of DNA and mitochondria. The cytotoxicity of mitaplatin in a variety of cancer cell lines equals that of cisplatin with the advantage of very low or no toxicity for normal cells.[281, 282]

A mechanism to "make DCA work" in cancer

Cancer cells are highly dependent on the glycolytic pathway. DCA impedes the use of this pathway by inhibiting PDK. If the cancer cell's dependence on the glycolytic pathway is increased by pharmacological interventions, blocking this pathway would lead to metabolic stress (metabolic catastrophe) whose final outcome should be apoptosis.

This is the rationale behind associating DCA with enhancers of glycolytic dependence.

There is evidence showing that the association of DCA with metformin (a glycolytic enhancer) increases apoptosis.[283] Fig. 9. In melanoma, blocking the RARβ, also increases glycolytic dependence, and when a selective inhibitor of RARβ is associated with DCA a significant apoptosis is achieved.[284]

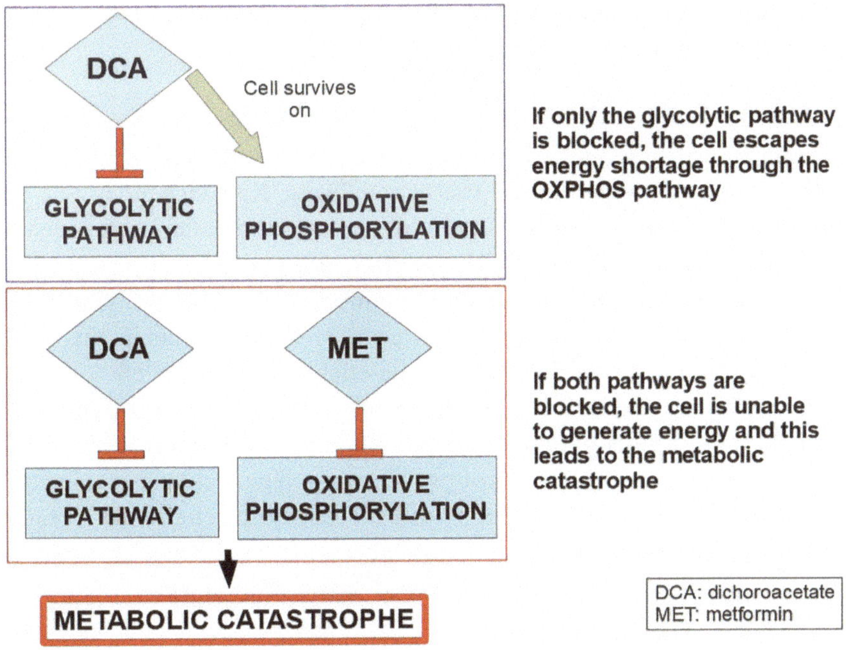

Fig. 9 The metabolic catastrophe is the consequence of the simultaneous inhibition of oxidative phosphorylation and glycolysis, reducing the chances for energy production. However, this is only a theoretical model, because metformin decreases oxidative metabolism but does not fully inhibit it. This means that there is a non-negligible OXPHOS going on, in spite of metformin. Using very high doses of metformin would reduce the chances of escaping the metabolic catastrophe but at the same time it would deeply affect normal cells.

Side effects

The most serious published side effect is reversible peripheral neuropathy.[285] No renal, cardiac or bone marrow toxicity were found.[200] The incidence of polyneuropathy in children with mitochondrial disease receiving 12.5 mg twice a day is 10%, and in adults it reaches 86%. The polymorphism of the glutathione transferase zeta 1 gene in humans may enhance DCA toxicity.[286] Cell death due to DCA treatment predominates in cancer cells, but non-cancer cells may also be affected, although to a less extent.[287]

Clinical cases

DCA is a compound that has been tested in cancer patients with good results.

Some have been already mentioned, but as a summary: Kahn et al. in metastatic lymphoma: 1 case; Michelakis et al. in glioma: 5 cases; Strum et al. in non-Hodgkin lymphoma: 1 case; Ishiguro et al.[288] in cholangiocarcinoma; 1 case; Lemmo et al.[289]

described a case of prolonged survival (a year and a half) after DCA treatment of leptomeningeal carcinomatosis originated in a NSCLC.

Not previously mentioned and with poor results:

NCT01029925 (clinical trials USA): metastatic breast cancer and lung cancer phase II with 7 cases. Due to these serious adverse events, the study was terminated ahead of time.[290] The dose used in this clinical trial was 6.25 mg/Kg twice a day.

NCT01386632: DCA in combination with cisplatin and radiation for squamous cell carcinoma of the head and neck. 25 patients received placebo and 25 received DCA. Both cohorts received radio and chemotherapy concomitantly. Two-year progression-free survival was similar in both groups. Two patients in the DCA group experienced delirium, but no other serious side effects were observed.

NCT01111097: a phase I study to evaluate safety of DCA in 15 patients with recurrent malignant brain tumors (glioblastoma and brain metastasis of other tumors).[291]

Patients were separated into two groups according to genotype: low metabolizers of DCA and quick metabolizers of DCA with different dose schedules for each group. Starting dose 8 mg/kg twice daily that was increased to 12.5 mg/kg twice daily for the quick metabolizers. No patients suffered toxicities, in spite of the relatively high dose.

The result of the trial showed "clinical and radiological evidence of disease stabilization through the first 4 weeks of DCA administration in all 8 evaluable patients."

NCT00566410: is a Canadian phase I trial of DCA in 23 patients with solid tumors. No results have yet been published. They established the initial dose at 12.5 mg/kg/day.

There were other three phase I trials with DCA for cancer treatment, but none have posted results.

Against DCA

Hesche et al. evaluated DCA in pediatric tumor cells and a non-malignant cell line and they found that DCA only moderately decreased cancer cell growth. DCA also decreased the cytotoxicity of cisplatin and doxorubicin. It did not activate caspases and decreased caspase activation by cisplatin.[292] Shahrzad et al.[293] found that DCA had a cytoprotective effect, decreasing apoptosis, in certain cases of colon carcinoma under hypoxic conditions and increased the growth of xenografts of SW480 tumors. In vivo it may decrease radiosensitivity and increase tumor hypoxia.[294] Neuroblastoma progression and proliferation was found in xerotranplanted nude mice treated with DCA.[295]

DCA was tested against three glioblastoma cell lines in a low millimolar range. In all the cases lactic acid production decreased but there was no modification in proliferation or apoptosis rates.[296]

Washington and Quintyne[287] found that not all cancer cell lines are susceptible to DCA cytotoxicity at lower concentrations while at higher concentrations and prolonged exposure they may all undergo apoptosis. On the other hand not all normal cell lines are resistant to DCA cytotoxicity.

DCA responders and DCA non-responders

Even if it is slightly farfetched, it seems that the glycolytic phenotype in cancer may be due to two different conditions, according to metabolic response to DCA:

(1) DCA responders: the glycolytic phenotype is due to up-regulation of PDK.

(2) DCA non-responders: the glycolytic phenotype is due to mitochondrial dysfunction.

This situation is particularly evident in prostate cancer with lines of cells that respond to DCA treatment and show no mitochondrial abnormalities and other lines that do not respond to DCA treatment and exhibit mitochondrial dysfunction. Something similar happens with highly metastatic prostate cancer cells that respond to DCA and low metastatic potential cells that do not respond to DCA.[297]

There is strong evidence indicating that the association of metformin and DCA has enhanced anti-tumoral effects and the association permits lower doses of each component than when they are used separately.

Some important issues regarding future clinical trials:

(1) Human pharmacodynamic features of DCA should be adequately considered.

(2) No stand alone DCA trials are worthwhile.

(3) The polymorphism of the glutathione transferase zeta 1 gene should be tested in the patients included in the study.

(4) The triple metformin-celecoxib-DCA association combined with a standard chemotherapeutic drug would be advisable.

(5) Initial dose of DCA should be increased in further administrations and the total daily dose should not exceed 12.5 mg/kg (in two 6.25 mg/kg doses).

(6) The triple association should be given in a metronomic fashion.

(7) The standard chemotherapeutic drug should not have neurologic side effects that might increase those of DCA.

(8) Vitamin B1 and alfa lipoic acid supplements are advisable.

Conclusions

DCA decreases tumor growth by inhibiting PDK and glycolysis and increasing ROS production. DCA decreases metastasis in number and volume. DCA activity is particularly evident in highly hypoxic tumors.

DCA is not a stand alone drug and should be used in association with conventional chemotherapeutic compounds or others like metformin, omeprazol, bicarbonate, etc.

It has synergic effects with most of standard chemotherapeutic drugs.

DCA does not induce high levels of apoptosis, thus the need for its association with other chemotherapeutics. It should not be used in association with allopurinol.

In order to achieve important metabolic changes in cancer cells, it is necessary to use DCA in a prolonged manner, at high doses, and on a daily basis. It is not devoid of toxicity, particularly when it is used in adults for a long time. It is useful for reversing chemoresistance to chemotherapeutic drugs in highly hypoxic tumors.

Experiments performed in rats or dogs cannot be translated to humans due to important pharmacodynamic differences. Due to its important absorption after oral administration, it is not necessary to use the intravenous route.

Cancer cells in which the glutaminergic pathway is up-regulated will not respond to DCA treatment.

DCA alone or in association with other compounds has never been tested in phase II clinical trials.

Cimetidine in cancer: A different kind of drug

Introduction

Cimetidine (CIM) the prototype of the histamine H2 receptor antagonist was developed for the treatment of gastroduodenal ulcer based on its ability to inhibit gastric acid secretion. However, CIM has shown limited but evident anti-cancer activities in certain tumors. Histamine type-2 receptor antagonists have been extensively used in the past for the treatment of peptic ulcer and gastritis. Unfortunately, it is not available on many markets because CIM has been replaced by the newer proton pump inhibitors. Since 1979 there is growing evidence that CIM may improve the survival of gastrointestinal cancer patients, when used as an adjuvant therapy. It has also been proposed as adjuvant treatment in other types of cancers like melanoma, glioblastoma, renal cell carcinoma, and ovarian cancer. The mechanism of action of CIM in tumor biology is well known. However, as CIM has been fully replaced by the newer hydrogen pump inhibitors, the research on this drug came to a standstill.

CIM is a substituted imidazole with a specific antagonistic effect on histamine H2 receptors. Its main mechanism of action is on parietal cell histamine H2 receptors and through binding these receptors, it inhibits gastric acid secretion.

In 1979, CIM was proposed for the first time as an anti-cancer agent by Armitage and Sidner.[298] They followed the spontaneous remission of two patients with metastatic carcinoma after coincidental use of the drug. Since then, there have been several clinical trials assessing CIM in a range of malignancies with varied and inconclusive results. CIM has been shown to improve the survival of patients with colorectal cancer, melanoma, and renal cell cancer.[299–307] CIM has also been proposed as a potential tool for the treatment of glioblastoma[308] and ovarian cancer.[309]

Mechanisms of cimetidine's anti-tumoral action

Different mechanisms of anti-tumoral action for CIM have been identified:

I. *Immunologic mechanisms*

(1) In 1987, Gifford and Tilberg[310] showed that CIM was able to increase IL-2 production after lymphocyte activation. IL-2 is the most potent growth factor and activator of T lymphocytes and induces the release of mediators such as IFN-γ. According to the authors, these data further suggest that histamine type-2 receptor antagonists may have immunorestorative potential in clinical immunotherapy of IL-2 deficient states. In another report, these authors[311] also showed that CIM reduced cyclosporine's inhibition of interleukin-2 production. The two mentioned publications are the first evidence of CIM's immunomodulatory activity. Thus, CIM increases the ability of IL-2 to activate a number of cells capable of exerting a cytotoxic effect on cancer cells, by increasing its production in T lymphocytes.[312]

(2) Takahashi et al.[313] found that CIM increased the production of interleukin 18 (IL-18) in human monocytes and dendritic cells.

(3) CIM effectively inhibited suppressor T-cell function and thus enhanced the immune response against cancer.[314–316]

(4) CIM increased tumor infiltrating lymphocytes (TIL) and natural killer cells.[317, 318] TIL correlated with a trend towards improved survival at 3 years.[319]

Perioperative CIM administration improved systemic immune response and tumor infiltrating lymphocytes in patients with colorectal cancer. The levels of CD3 and CD4 T-lymphocytes were increased significantly in both low and high dose CIM groups 10 days after surgery. The number of CD19 B cells was also raised by CIM. However, no significant changes were observed in the CD8, CD4/CD8 values.[320]

(5) CIM enhanced the host's anti-tumor cell-mediated immunity by improving the suppressed dendritic cell function in patients with advanced cancer.[321] Fig. 10 left panel.

II. *Modification of cancer cell adhesion*

(6) CIM inhibited cancer cell adhesion to endothelial cells and prevented metastasis by blocking *E*-selectin expression.[322–325] Expression of E-selectin on endothelial cells has been shown to be essential in models of transendothelial migration of colon carcinoma cells.[326]

(7) It was also found that CIM inhibited neural cell adhesion molecule (NCAM) expression and induced apoptosis in salivary gland tumor cells.[327]

The hematogenous metastatic cascade is a complex mechanism, with many molecular details still unknown. A key player in the cascade is the complex interaction of circulating cancer cells with the endothelial cells lining blood vessels. The adhesion cascade is initiated via cell tethering and rolling through selectin mediated interactions. Circulating cancer cells displaying selectin ligands (such as glycoproteins like sialy Lewis X or A) are potentially involved in a series of events that eventually lead to metastasis.

The attachment of circulating cancer cells to vascular endothelium is considered an important initial step in hematogenous metastasis. In order for tumor cells to metastasize, they need to interact with endothelial cells. This requires interaction between cell adhesion molecules, particularly *E*-selectin (on the endothelial surface) and its ligand molecules sialyl Lewis antigens (on the tumor surface).[322, 328] This is precisely the point where CIM impedes the interaction.[329]

Fig. 10 right panel shows the interaction between E-selectin and sialyl Lewis antigen. Fig. 11 shows the mechanism how the circulating tumor cell adheres to the endothelium.

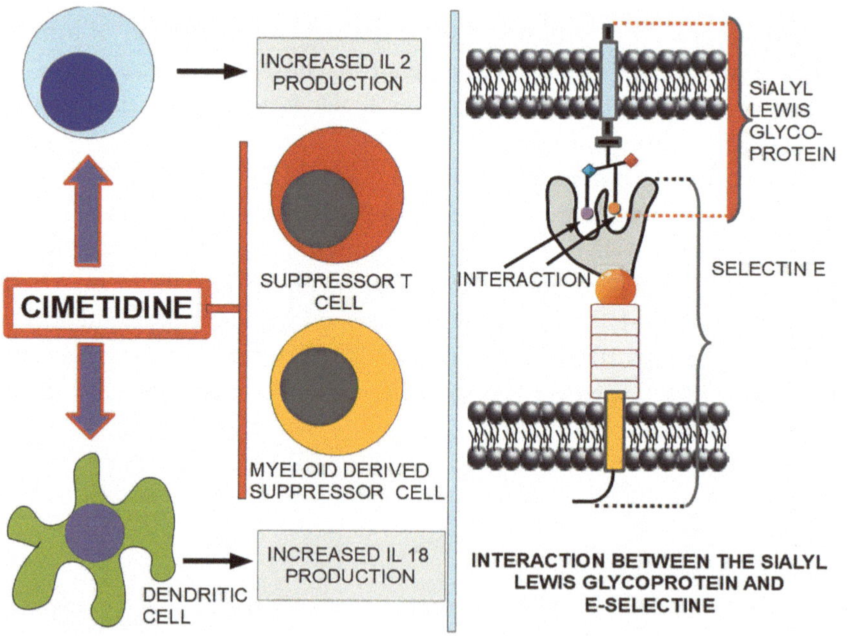

Fig. 10 Left panel: immunological effects of cimetidine. Right panel: Sialyl Lewis antigen in the circulating tumor cell membrane and Selectin-E in the endothelial cell membrane. This interaction allows the circulating cell to adhere to the endothelium.

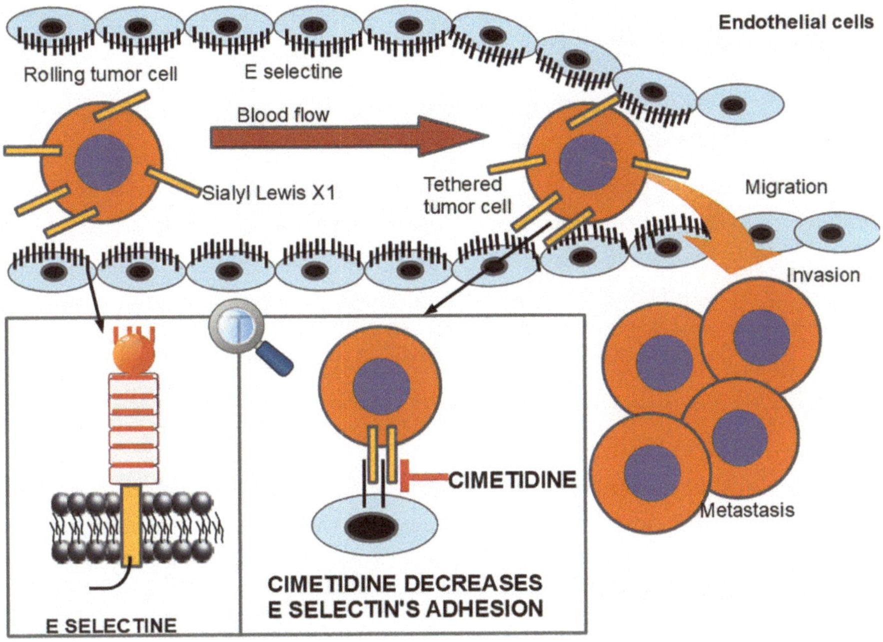

Fig. 11 E-selectin mediates the adhesion of tumor cells to endothelial cells by binding to E-selectin ligands expressed by neutrophils, monocytes, eosinophils, memory-effector T-like lymphocytes, NT killer cells or cancer cells. This interaction is associated with metastatic dissemination. E-selectin participates in the conversion of rolling cells to firm adhesion to endothelial cells. Blocking the sialyl lewis X glycopotein (overexpressed in tumor cells) with E-selectin (expressed or overexpressed in endothelial cells) interaction, CIM decreases tethering of tumor cells and decreases migration and capacity to metastasize.

(Sialyl Lewis glycoprotein is also known with other names such as Sialyl Lewis antigen, CA19-9, carbohydrate antigen 19-9 and is a tumor marker for the progression of pancreatic cancer. It may be increased in other gastrointestinal malignancies such as colorectal cancer and hepatocarcinoma).

III. *Modification of cell signaling and angiogenesis*

(8) CIM inhibited epidermal growth factor induced cell proliferation and migration in hepatocellular carcinoma cell lines by decreasing the concentration of intracellular cAMP levels.[330]

(9) CIM had anti-angiogenic effects by reducing VEGF levels in tumor tissues (Box 3).[331, 332]

However, all the above mentioned anti-cancer effects, as important as they are, are not the reason why we are including CIM in this chapter. The real reason for considering CIM in the treatment of the pH inversion lays in a less-explored ability: inhibition of NHE1.

CIM's potency is similar to amiloride in NHE inhibition.[333–339] The dose of amiloride has un upper limit mainly determined by hyperkalemia. Up to 15 mg a day, if

BOX 3 Summary of CIM's anti-cancer mechanisms.

- IMMUNOMODULATION
- INHIBITION OF CANCER CELL ADHESION TO ENDOTHELIUM
- DECREASED EGF INDUCED PROLIFERATION
- INHIBITORY EFFECT ON PDGF
- REDUCTION OF VEGF AND ANGIOGENESIS
- REVERSAL OF HISTAMINE INHIBITION OF PRODUCTION OF INTERFERON INDUCED PROTEIN (ip10)

associated with another diuretic such as furosemide or hydrochlorotiazide is perfectly tolerable. However, at this dose it does not achieve all the necessary inhibition of NHE1. This is the point where cimetidine could play a role. It would not only add to NHE1 inhibition, but would cooperate in anti-angiogenesis, metastasis prevention and immunomodulation.

CIM's inhibition of NHE1 has not only been overlooked by oncologists but also by basic researchers and although further study is necessary to determine its synergy with amiloride, it is a very useful drug in anti-angiogenic schemes.

For further reading on cimetidine's anti-cancer effects read Pantziarka et al.[340]

References

1. Cushman SW. Cell biology of glucose transport: a key to understanding peripheral glucose utilization, In: Banting lecture. 62nd scientific sessions of the American Diabetes Association; June 14–18, 2002; San Francisco, CA. Diabetes, vol. 51. 2002. Suppl. 2.
2. Cushman SW, Wardzala LJ. Potential mechanism of insulin action on glucose transport in the isolated rat adipose cell. Apparent translocation of intracellular transport systems to the plasma membrane. *J Biol Chem* 1980;**255**:4758–62.
3. Hundal HS, Ramlal T, Reyes R, Leiter LA, Klip A. Cellular mechanism of metformin action involves glucose transporter translocation from an intracellular pool to the plasma membrane in L6 muscle cells. *Endocrinology* 1992;**131**(3):1165–73.
4. Klip A, Leiter LA. Cellular mechanism of action of metformin. *Diabetes Care* 1990;**13**(6):696–704.
5. Woodward GE, Hudson MT. The effect of 2-deoxy-D-glucose on glycolysis and respiration of tumor and normal tissues. *Cancer Res* 1954;**14**(8):599–605.
6. Ely JO. 2-Deoxy-D-glucose as an inhibitor of cancerous growth in animals. *J Franklin Inst* 1954;**268**:157–60.
7. Cramer FB, Woodward GE. 2-Desoxy-D-glucose as an antagonist of glucose in yeast fermentation. *J Franklin Inst* 1952;**253**:354–60.
8. Woodward GE, Cramer FB. 2-Desoxy-D-glucose as an inhibitor of anaerobic glycolysis in tumor tissue. *J Franklin Inst* 1952;**254**:259–60.
9. Wick AN, Drury DR, Nakada HI, Wolfe JB. Localization of the primary metabolic block produced by 2-deoxyglucose. *J Biol Chem* 1957;**224**:963–9.

10. Kipnis DM, Cori CF. Studies of tissue permeability. VI. The penetration and phosphorylation of 2-deoxyglucose in the diafragm of diabetic rats. *J Biol Chem* 1960;**235**(11):3070–6.
11. Kipnis DM, Cori CF. Studies of tissue permeability. V. The penetration and phosphorylation of 2-deoxyglucose in the rat diaphragm. *J Biol Chem* 1959;**234**:171–7.
12. Barban S, Schulze HO. The effects of 2-deoxyglucose on the growth and metabolism of cultured human cells. *J Biol Chem* 1961;**236**(7):1887–91.
13. Ahadova A, Gebert J, von Knebel Doeberitz M, Kopitz J, Kloor M. Dose-dependent effect of 2-deoxy-D-glucose on glycoprotein mannosylation in cancer cells. *IUBMB Life* 2015;**67**(3):218–26.
14. Little E, Ramakrishnan M, Roy B, Gazit G, Lee AS. The glucoseregulated proteins (GRP78 and GRP94): functions, gene regulation, and applications. *Crit Rev Eukaryot Gene Expr* 1994;**4**:1–18.
15. Ishino K, Kudo M, Peng WX, Kure S, Kawahara K, Teduka K, et al. 2-Deoxy-D-glucose increases GFAT1 phosphorylation resulting in endoplasmic reticulum-related apoptosis via disruption of protein N-glycosylation in pancreatic cancer cells. *Biochem Biophys Res Commun* 2018;**501**(3): 668–73.
16. Han I-O, Kudlow JE. Reduced O glycosylation of Sp1 is associated with increased proteasome susceptibility. *Mol Cell Biol* 1997;**17**(5):2550–8.
17. Zhang SQ, Yung KL, Chung SK, Chung SM. Aldo-keto reductases-mediated cytotoxicity of 2-deoxyglucose: a novel anticancer mechanism. *Cancer Sci* 2018;**109**(6):1970.
18. Kang HT, Hwang ES. 2-Deoxyglucose: an anticancer and antiviral therapeutic, but not any more a low glucose mimetic. *Life Sci* 2008;**78**:1392–9.
19. Kern KA, Norton JA. Inhibition of established rat fibrosarcoma growth by the glucose antagonist 2-deoxy-D-glucose. *Surgery* 1987;**102**:380–5.
20. Kaplan O, Navon G, Lyon RC, Faustino PJ, Straka EJ, Cohen JS. Effects of 2-deoxyglucose on drug-sensitive and drug-resistant human breast cancer cells: toxicity and magnetic resonance spectroscopy studies of metabolism. *Cancer Res* 1990;**50**(3):544–51.
21. Ciavardelli D, Rossi C, Barcaroli D, Volpe S, Consalvo A, Zucchelli M, et al. Breast cancer stem cells rely on fermentative glycolysis and are sensitive to 2-deoxyglucose treatment. *Cell Death Dis* 2014;**5**(7):e1336.
22. Zhu Z, Jiang W, McGinley JN, Thompson HJ. 2-Deoxyglucose as an energy restriction mimetic agent: effects on mammary carcinogenesis and on mammary tumor cell growth in vitro. *Cancer Res* 2005;**65**(15):7023–30.
23. Triggle C, Majeed Y, Samuel SM, Ding HM. Antiproliferative effects of metformin in triple negative MDA-MB 231 breast cancer cells exposed to glucose-starved and 2-deoxyglucose. *FASEB J* 2018; **32**(1):9650 Rockville Pike, Bethesda, MD 20814-3998 USA: Federation Amer Soc Exp Biol.
24. Ramírez-Peinado S, Alcázar-Limones F, Lagares-Tena L, El Mjiyad N, Caro-Maldonado A, Tirado OM, et al. 2-Deoxyglucose induces Noxa-dependent apoptosis in alveolar rhabdomyosarcoma. *Cancer Res* 2011;**71**(21):6796–806.
25. Ahmad IM, Abdalla MY, Aykin-Burns N, Simons AL, Oberley LW, Domann FE, et al. 2-Deoxyglucose combined with wild-type p53 overexpression enhances cytotoxicity in human prostate cancer cells via oxidative stress. *Free Radic Biol Med* 2008;**44**(5):826–34.
26. Lea MA, Chacko J, Bolikal S, et al. Addition of 2-deoxyglucose enhances growth inhibition but reverses acidification in colon cancer cells treated with phenformin. *Anticancer Res* 2011;**31**:421–6.
27. Coleman MC, Asbury CR, Daniels D, Du J, Aykin-Burns N, Smith BJ, et al. 2-Deoxy-D-glucose causes cytotoxicity, oxidative stress, and radiosensitization in pancreatic cancer. *Free Radic Biol Med* 2008;**44**(3):322–31.
28. Wang Z, Zhang L, Zhang D, Sun R, Wang Q, Liu X. Glycolysis inhibitor 2-deoxy-D-glucose suppresses carcinogen- induced rat hepatocarcinogenesis by restricting cancer cell metabolism. *Mol Med Rep* 2015;**11**(3):1917–24.
29. Zhang XD, Deslandes E, Villedieu M, Poulain L, Duval M, Gauduchon P, et al. Effect of 2-deoxy-D-glucose on various malignant cell lines in vitro. *Anticancer Res* 2006;**26**:3561–6.
30. Sun L, Liu X, Fu H, Zhou W, Zhong D. 2-Deoxyglucose suppresses ERK phosphorylation in LKB1 and ras wild-type non-small cell lung cancer cells. *PLoS One* 2016;**11**(12):e0168793.

31. Xu Y, Wang Q, Zhang L, Zheng M. 2-Deoxy-D-glucose enhances TRAIL-induced apoptosis in human gastric cancer cells through downregulating JNK-mediated cytoprotective autophagy. *Cancer Chemother Pharmacol* 2018;**81**(3):555–64.
32. Huang CC, Wang SY, Lin LL, Wang PW, Chen TY, Hsu WM, et al. Glycolytic inhibitor 2-deoxyglucose simultaneously targets cancer and endothelial cells to suppress neuroblastoma growth in mice. *Dis Model Mech* 2015;**8**(10):1247–54.
33. Karczmar GS, Arbeit JM, Toy BJ, Speder A, Weiner MW. Selective depletion of tumor ATP by 2-deoxyglucose and insulin, detected by magnetic resonance spectroscopy. *Cancer Res* 1992;**52**:71–6.
34. Griffiths JR, Cady E, Edwards RHT, McCready VR, Wilkie DR, Wiltshaw E. 3′P NMR studies of a human tumour in situ. *Lancet* 1983;**1**:1435–6.
35. Jain VK, Kalia VK, Sharma R, Maharajan V, Menon M. Effects of 2-deoxyglucose on glycolysis, proliferation kinetics, and radiation response of human cancer cells. *Int J Radiat Oncol Biol Phys* 1985;**11**:943–50.
36. Poole DT. Intracellular pH of the Ehrlich ascites tumor cell as it is affected by sugars and sugar derivatives. *J Biol Chem* 1967;**242**(16):3731–6.
37. Cassel D, Katz M, Rotman M. Depletion of cellular ATP inhibits Na^+/H+ antiport in cultured human cells. Modulation of the regulatory effect of intracellular protons on the antiporter activity. *J Biol Chem* 1986;**261**(12):5460–6.
38. Demaurex N, Romanek RR, Orlowski J, Grinstein S. ATP dependence of Na^+/H^+ exchange: nucleotide specificity and assessment of the role of phospholipids. *J Gen Physiol* 1997;**109**(2):117–28.
39. Aharonovitz O, Demaurex N, Woodside M, Grinstein S. ATP dependence is not an intrinsic property of Na^+/H^+ exchanger NHE1: requirement for an ancillary factor. *Am J Physiol Cell Physiol* 1999; **276**(6):1303–11.
40. Ikeda T, Schmitt B, Pouyssegur J, Wakabayashi S, Shigekawa M. Identification of cytoplasmic subdomains that control pH-sensing of the Na^+/H^+ exchanger (NHE1): pH-maintenance, ATP-sensitive, and flexible loop domains. *J Biochem* 1997;**121**(2):295–303.
41. Shukla A, Gupta P, Singh R, Mishra DP. Glycolytic inhibitor 2-deoxy-D-glucose activates migration and invasion in glioblastoma cells through modulation of the miR-7-5p/TFF3 signaling pathway. *Biochem Biophys Res Commun* 2018;**499**(4):829–35.
42. Bizjak M, Malavašič P, Dolinar K, Pohar J, Pirkmajer S, Pavlin M. Combined treatment with metformin and 2-deoxy glucose induces detachment of viable MDA-MB-231 breast cancer cells in vitro. *Sci Rep* 2017;**7**(1):1761.
43. Stein M, Lin H, Jeyamohan C, Dvorzhinski D, Gounder M, Bray K, et al. Targeting tumor metabolism with 2-deoxyglucose in patients with castrate-resistant prostate cancer and advanced malignancies. *Prostate* 2010;**70**(13):1388–94.
44. Raez LE, Papadopoulos K, Rikard AD, Chiorean EG, DiPaola RS, Stein MN, et al. A phase I dose-escalation trial of 2-deoxy-D-glucose alone or combined with docetaxel in patients with advanced solid tumors. *Cancer Chemother Pharmacol* 2013;**71**(2):523–30.
45. https://clinicaltrials.gov/ct2/show/NCT00633087.
46. Kaplan O, Jaroszewski JW, Clarke R, Fairchild CR, Schoenlein P, Goldemberg S, et al. The multidrug resistance phenotype: 31P nuclear magnetic resonance characterization and 2-deoxyglucose toxicity. *Cancer Res* 1991;**51**(6):1638–44.
47. Lemontt JF, Azzaria M, Gros P. Increased mdr gene expression and decreased drug accumulation in multidrug-resistant human melanoma cells. *Cancer Res* 1988;**48**(22):6348–53.
48. Bell SE, Quin DM, Kellet GL, Warr JR. 2-Deoxy-D-glucose preferentially kills multidrug-resistant human KB carcinoma cell lines by apoptosis. *Br J Cancer* 1998;**78**(11):1464.
49. Xue C, Wang C, Sun Y, Meng Q, Liu Z, Huo X, et al. Targeting P-glycoprotein function, p53 and energy metabolism: combination of metformin and 2-deoxyglucose reverses the multidrug resistance of MCF-7/Dox cells. *Oncotarget* 2017;**8**(5):8622.
50. Aşık E, Aslan TN, Volkan M, Güray NT. 2-Amino-2-deoxy-glucose conjugated cobalt ferrite magnetic nanoparticle (2DG-MNP) as a targeting agent for breast cancer cells. *Environ Toxicol Pharmacol* 2016;**41**:272–8.
51. Kunhiraman H, Edatt L, Thekkeveedu S, Poyyakkara A, Raveendran V, Kiran MS, et al. 2-Deoxy glucose modulates expression and biological activity of VEGF in a SIRT-1 dependent mechanism. *J Cell Biochem* 2017;**118**(2):252–62.

52. Sahra IB, Laurent K, Giuliano S, Larbret F, Ponzio G, Gounon P, et al. Targeting cancer cell metabolism: the combination of metformin and 2-deoxyglucose induces p53-dependent apoptosis in prostate cancer cells. *Cancer Res* 2010;**70**(6):2465–75.
53. Ben Sahra I, Tanti JF, Bost F. The combination of metformin and 2 deoxyglucose inhibits autophagy and induces AMPK-dependent apoptosis in prostate cancer cells. *Autophagy* 2010;**6**(5):670–1.
54. Cheong JH, Park ES, Liang J, Dennison JB, Tsavachidou D, Nguyen-Charles C, et al. Dual inhibition of tumor energy pathway by 2-deoxyglucose and metformin is effective against a broad spectrum of preclinical cancer models. *Mol Cancer Ther* 2011;**10**(12):2350–62.
55. Cheng G, Zielonka J, Dranka BP, McAllister D, Mackinnon AC, Joseph J, et al. Mitochondria-targeted drugs synergize with 2-deoxyglucose to trigger breast cancer cell death. *Cancer Res* 2012;**72**(10):2634–44.
56. Koltai T. Triple-edged therapy targeting intracellular alkalosis and extracellular acidosis in cancer. *Semin Cancer Biol* 2017;**43**:139–46.
57. Xi H, Barredo JC, Merchan JR, Lampidis TJ. Endoplasmic reticulum stress induced by 2-deoxyglucose but not glucose starvation activates AMPK through CaMKKβ leading to autophagy. *Biochem Pharmacol* 2013;**85**(10):1463–77.
58. Shafaee A, Islamian JP, Zarei D, Mohammadi M, Nejati-Koshki K, Farajollahi A, et al. Induction of apoptosis by a combination of 2-deoxyglucose and metformin in esophageal squamous cell carcinoma by targeting cancer cell metabolism. *Iran J Med Sci* 2018. Available at: http://ijms.sums.ac.ir/index.php/IJMS/article/view/4050.
59. Zhu J, Zheng Y, Zhang H, Sun H. Targeting cancer cell metabolism: the combination of metformin and 2-deoxyglucose regulates apoptosis in ovarian cancer cells via p38 MAPK/JNK signaling pathway. *Am J Transl Res* 2016;**8**(11):4812.
60. Kim EH, Lee JH, Oh Y, Koh I, Shim JK, Park J, et al. Inhibition of glioblastoma tumorspheres by combined treatment with 2-deoxyglucose and metformin. *Neuro Oncol* 2016;**19**(2):197–207.
61. Xue C, Wang C, Sun Y, Meng Q, Liu Z, Huo X, et al. Targeting P-glycoprotein function, p53 and energy metabolism: combination of metformin and 2-deoxyglucose reverses the multidrug resistance of MCF-7/Dox cells to doxorubicin. *Oncotarget* 2017;**8**(5):8622.
62. Shanmugam M, McBrayer SK, Qian J, Raikoff K, Avram MJ, Singhal S, et al. Targeting glucose consumption and autophagy in myeloma with the novel nucleoside analogue 8-aminoadenosine. *J Biol Chem* 2009;**284**(39):26816–30.
63. Pandey S, Chandravati C. Autophagy in cervical cancer: an emerging therapeutic target. *Asian Pac J Cancer Prev* 2012;**13**(10):4867–71.
64. Wu H, Zhu H, Liu DX, Niu TK, Ren X, Patel R, et al. Silencing of elongation factor-2 kinase potentiates the effect of 2-deoxy-D-glucose against human glioma cells through blunting of autophagy. *Cancer Res* 2009;**69**(6):2453–60.
65. Yang ZJ, Chee CE, Huang S, Sinicrope FA. The role of autophagy in cancer: therapeutic implications. *Mol Cancer Ther* 2011;**10**(9):1533–41.
66. León-Annicchiarico CL, Ramírez-Peinado S, Domínguez-Villanueva D, Gonsberg A, Lampidis TJ, Muñoz-Pinedo C. ATF4 mediates necrosis induced by glucose deprivation and apoptosis induced by 2-deoxyglucose in the same cells. *FEBS J* 2015;**282**(18):3647–58.
67. Koltai T. Nelfinavir and other protease inhibitors in cancer: mechanisms involved in anticancer activity. *F1000Res* 2015;**4**:9.
68. Suh DH, Kim MK, Kim HS, Chung HH, Song YS. Unfolded protein response to autophagy as a promising druggable target for anticancer therapy. *Ann N Y Acad Sci* 2012;**1271**(1):20–32.
69. Maschek G, Savaraj N, Priebe W, Braunschweiger P, Hamilton K, Tidmarsh GF, et al. 2-Deoxy-D-glucose increases the efficacy of adriamycin and paclitaxel in human osteosarcoma and non-small cell lung cancers in vivo. *Cancer Res* 2004;**64**(1):31–4.
70. Jalota A, Kumar M, Das BC, Yadav AK, Chosdol K, Sinha S. Synergistic increase in efficacy of a combination of 2-deoxy-D-glucose and cisplatin in normoxia and hypoxia: switch from autophagy to apoptosis. *Tumour Biol* 2016;**37**(9):12347–58.
71. Sobhakumari A, Orcutt KP, Love-Homan L, Kowalski CE, Parsons AD, Knudson CM, Simons AL. 2-Deoxy-D-glucose suppresses the in vivo antitumor efficacy of erlotinib in head and neck squamous cell carcinoma cells. *Oncol Res* 2016;**24**(1):55–64.

72. Hernlund E, Ihrlund LS, Khan O, Ates YO, Linder S, Panaretakis T, et al. Potentiation of chemotherapeutic drugs by energy metabolism inhibitors 2-deoxyglucose and etomoxir. *Int J Cancer* 2008;**123**(2):476–83.
73. Gaddameedhi S, Chatterjee S. Association between the unfolded protein response, induced by 2-deoxyglucose, and hypersensitivity to cisplatin: a mechanistic study employing molecular genomics. *J Cancer Res Ther* 2009;**5**:61–6.
74. Kobayashi H, Nishimura H, Matsumoto K, Yoshida M. Identification of the determinants of 2-deoxyglucose sensitivity in cancer cells by shRNA library screening. *Biochem Biophys Res Commun* 2015;**467**(1):121–7.
75. Fromenty BE, Fisch CE, Labbe GI, Degott CL, Deschamps DO, Berson AL, et al. Amiodarone inhibits the mitochondrial beta-oxidation of fatty acids and produces microvesicular steatosis of the liver in mice. *J Pharmacol Exp Ther* 1990;**255**(3):1371–6.
76. Giovannucci E, Harlan DM, Archer MC, Bergenstal RM, Gapstur SM, Habel LA, et al. Diabetes and cancer: a consensus report. *Diabetes Care* 2010;**33**:1674–85.
77. Jalving M, Gietema JA, Lefrandt JD, de Jong S, Reyners AK, Gans RO, de Vries EG. Metformin: taking away the candy for cancer? *Eur J Cancer* 2010;**46**:2369–80.
78. Nathan DM, Buse JB, Davidson MB, Ferrannini E, Holman RR, Sherwin R, Zinman B. Medical management of hyperglycemia in type 2 diabetes: a consensus algorithm for the initiation and adjustment of therapy: a consensus statement of the American Diabetes Association and the European Association for the Study of Diabetes. *Diabetes Care* 2009;**32**(1):193–203.
79. Bodmer M, Meier C, Krähenbühl S, Jick SS, Meier CR. Metformin, sulfonylureas or other antidiabetic drugs and the risk of lactic acidosis or hypoglycemia: a nested case-control analysis. *Diabetes Care* 2008;**31**:2086–91.
80. Goodwin PJ, Stambolic V, Lemieux J, Chen BE, Parulekar WR, Gelmon KA, et al. Evaluation of metformin in early breast cancer: a modification of the traditional paradigm for clinical testing of anti-cancer agents. *Breast Cancer Res Treat* 2011;**126**(1):215–20.
81. Shaw RJ, Lamia KA, Vasquez D, Koo SH, Bardeesy N, DePinho RA, et al. The kinase LKB1 mediates glucose homeostasis in liver and therapeutic effects of metformin. *Science* 2005;**310**(5754): 1642–6.
82. Evans JMM, Donelly LA, Emslie_Smith AM, Alessi AR, Morris D. Metformin and reduced risk of cancer in diabetic patients. *BMJ* 2005;**330**(7503):1304–5.
83. Anisimov VN, Berstein LM, Egormin PA, et al. Effect of metformin on life span and on the development of spontaneous mammary tumors in HER-2/neu transgenic mice. *Exp Gerontol* 2005; **40**(8–9):685–93.
84. Anisimov VN, Egormin PA, Bershtein LM, Zabezhinskii MA, Piskunova TS, Popovich IG, et al. Metformin decelerates aging and development of mammary tumors in HER-2/neu transgenic mice. *Bull Exp Biol Med* 2005;**139**(6):721–3.
85. Jiralerspong S, Palla SL, Giordano SH, Meric-Bernstam F, Liedtke C, Barnett CM, et al. Metformin and pathologic complete responses to neoadjuvant chemotherapy in diabetic patients with breast cancer. *J Clin Oncol* 2009;**27**:3297–302.
86. Libby G, Donnelly LA, Donnan PT, Alessi DR, Morris AD, Evans JM. New users of metformin are at low risk of incident cancer: a cohort study among people with type 2 diabetes. *Diabetes Care* 2009;**32**:1620–5.
87. Decensi A, Puntoni M, Goodwin P, Cazzaniga M, Gennari A, Bonanni B, et al. Metformin and cancer risk in diabetic patients: a systematic review and meta-analysis. *Cancer Prev Res (Phila)* 2010;**3**:1451–61.
88. Landman GW, Kleefstra N, van Hateren KJ, Groenier KH, Gans RO, Bilo HJ. Metformin associated with lower cancer mortality in type 2 diabetes: ZODIAC-16. *Diabetes Care* 2010;**33**:322–6.
89. Murtola TJ, Tammela TL, Lahtela J, Auvinen A. Antidiabetic medication and prostate cancer risk: a population-based case-control study. *Am J Epidemiol* 2008;**168**:925–31.
90. Gagnon B, Roseman M, Kasymjanova G, MacDonald N, Kreisman H, Small D. Protective effect of metformin in lung cancer patients. In: ASCO annual meeting. Abstract; 2009. p. e22063.
91. Rieken M, Xylinas E, Kluth L, Crivelli JJ, Chrystal J, Faison T, et al. Association of diabetes mellitus and metformin use with oncological outcomes of patients with non-muscle-invasive bladder cancer. *BJU Int* 2013;**112**(8):1105–12.

92. Spratt DE, Zhang C, Zumsteg ZS, Pei X, Zhang Z, Zelefsky MJ. Metformin and prostate cancer: reduced development of castration-resistant disease and prostate cancer mortality. *Eur Urol* 2013; **63**(4):709–16.
93. Kumar S, Meuter A, Thapa P, Langstraat C, Giri S, Chien J, et al. Metformin intake is associated with better survival in ovarian cancer: a case-control study. *Cancer* 2013;**119**(3):555–62.
94. Bowker SL, Majumdar SR, Veugelers P, Johnson JA. Increased cancer-related mortality for patients with type 2 diabetes who use sulfonylureas or insulin. *Diabetes Care* 2006;**29**(2):254–8.
95. Currie CJ, Poole CD, Gale EA. The influence of glucose-lowering therapies on cancer risk in type 2 diabetes. *Diabetologia* 2009;**52**(9):1766–77.
96. Lee MS, Hsu CC, Wahlqvist ML, Tsai HN, Chang YH, Huang YC. Type 2 diabetes increases and metformin reduces total, colorectal, liver and pancreatic cancer incidences in Taiwanese: a representative population prospective cohort study of 800,000 individuals. *BMC Cancer* 2011;**11**(1):20.
97. Becker C, Meier CR, Jick SS, Bodmer M. Case-control analysis on metformin and cancer of the esophagus. *Cancer Causes Control* 2013;**24**(10):1763–70.
98. Becker C, Jick SS, Meier CR, Bodmer M. Metformin and the risk of endometrial cancer: a case-control analysis. *Gynecol Oncol* 2013;**129**(3):565–9.
99. Ko EM, Stürmer T, Hong JL, Castillo WC, Bae-Jump V, Funk MJ. Metformin and the risk of endometrial cancer: a population-based cohort study. *Gynecol Oncol* 2015;**136**(2):341–7.
100. Bodmer M, Becker C, Jick SS, Meier CR. Metformin does not alter the risk of lung cancer: a case-control analysis. *Lung Cancer* 2012;**78**(2):133–7.
101. Bodmer M, Becker C, Meier C, Jick SS, Meier CR. Use of metformin is not associated with a decreased risk of colorectal cancer: a case-control analysis. *Cancer Epidemiol Biomarkers Prev* 2012;**21**(2):280–6.
102. Kaushik D, Karnes RJ, Eisenberg MS, Rangel LJ, Carlson LE, Bergstralh EJ. Effect of metformin on prostate cancer outcomes after radical prostatectomy. *Urol Oncol* 2014;**32**(1):43.
103. Niraula S, Pond G, de Wit R, Eisenberger M, Tannock IF, Joshua AM. Influence of concurrent medications on outcomes of men with prostate cancer included in the TAX 327 study. *Can Urol Assoc J* 2013;**7**(1–2):E74–81. https://doi.org/10.5489/cuaj.267.
104. Lega IC, Austin PC, Gruneir A, Goodwin PJ, Rochon PA, Lipscombe LL. Association between metformin therapy and mortality after breast cancer: a population-based study. *Diabetes Care* 2013; **36**(10):3018–26.
105. Hwang AL, Haynes K, Hwang WT, Yang YX. Metformin and survival in pancreatic cancer: a retrospective cohort study. *Pancreas* 2013;**42**(7):1054–9.
106. Lega IC, Fung K, Austin PC, Lipscombe LL. Metformin and breast cancer stage at diagnosis: a population-based study. *Curr Oncol* 2017;**24**(2):e85.
107. Smiechowski BB, Azoulay L, Yin H, Pollak MN, Suissa S. The use of metformin and the incidence of lung cancer in patients with type 2 diabetes. *Diabetes Care* 2013;**36**(1):124–9.
108. Azoulay L, Dell'Aniello S, Gagnon B, Pollak M, Suissa S. Metformin and the incidence of prostate cancer in patients with type 2 diabetes. *Cancer Epidemiol Biomarkers Prev* 2011;**20**(2):337–44.
109. Bayraktar S, Hernadez-Aya LF, Lei X, Meric-Bernstam F, Litton JK, Hsu L, et al. Effect of metformin on survival outcomes in diabetic patients with triple receptor-negative breast cancer. *Cancer* 2012; **118**(5):1202–11.
110. Grealdine N, Marc A, Carla T, Chantal M, Stefaan B, Welcome W, Frank B. Relation between diabetes, metformin treatment and the occurrence of malignancies in a Belgium primary care setting. *Diabetes Res Clin Pract* 2012;**97**(2):331–6.
111. Foretz M, Guigas B, Viollet B. From cancer to diabetes treatment: the tumor suppressor LKB1 as a new pharmacological target. *Med Sci (Paris)* 2006;**22**(4):348–50. Available at: https://doi.org/10.1051/medsci/2006224348.
112. Zakikhani M, Dowling R, Fantus IG, Sonenberg N, Pollak M. Metformin is an AMP kinase-dependent growth inhibitor for breast cancer cells. *Cancer Res* 2006;**66**(21):10269–73.
113. Phoenix KN, Vumbaca F, Claffey KP. Therapeutic metformin/AMPK activation promotes the angiogenic phenotype in the ERα negative MDA-MB-435 breast cancer model. *Breast Cancer Res Treat* 2009;**113**(1):101–11.
114. Stambolic V, Woodgett JR, Fantus IG, Pritchard KI, Goodwin PJ. Utility of metformin in breast cancer treatment, is neoangiogenesis a risk factor? *Breast Cancer Res Treat* 2009;**114**:387–9.

115. Nagata D, Mogi M, Walsh K. AMP activated protein kinase (AMPK) signaling in endothelial cells is essential for angiogenesis in response to hypoxic stress. *J Biol Chem* 2003;**278**:31000–6.
116. Lee M, Hwang JT, Lee HJ, Jung SN, Kang I, Chi SG, et al. AMP-activated protein kinase activity is critical for hypoxia-inducible factor-1 transcriptional activity and its target gene expression under hypoxic conditions in DU145 cells. *J Biol Chem* 2003;**278**:39653–61.
117. Ouchi N, Shibata R, Walsh K. AMP-activated protein kinase signaling stimulates VEGF expression and angiogenesis in skeletal muscle. *Circ Res* 2005;**96**:838–46.
118. Jones RG, Plas DR, Kubek S, Buzzai M, Mu J, Xu Y, et al. AMP-activated protein kinase induces a p53 dependent metabolic checkpoint. *Mol Cell* 2005;**18**(3):283–93.
119. Janjetovic K, Vucicevic L, Misirkic M, Vilimanovich U, Tovilovic G, Zogovic N, et al. Metformin reduces cisplatin-mediated apoptotic death of cancer cells through AMPK-independent activation of Akt. *Eur J Pharmacol* 2011;**651**(1–3):41–50.
120. Rattan R, Graham RP, Maguire JL, Giri S, Shridhar V. Metformin suppresses ovarian cancer growth and metastasis with enhancement of cisplatin cytotoxicity in vivo. *Neoplasia* 2011;**13**:483–91.
121. Kim HS, Hwang JT, Yun H, Chi SG, Lee SJ, Kang I, et al. Inhibition of AMP activated protein kinase sensitizes cancer cells to cisplatin induced apoptosis via hyperinduction of p53. *J Biol Chem* 2008;**283**:3731–42.
122. Hirsch HA, Iliopoulos D, Tschilis PN, Struhl K. Metformin selectively targets cancer stem cells, and act together with chemotherapy to block tumor growth and prolong remission. *Cancer Res* 2009; **69**(19):7507–11.
123. Rocha GZ, Dias MM, Ropelle ER, Osorio-Costa F, Rossato FA, Vercesi AE, et al. Metformin amplifies chemotherapy induced AMPK activation and antitumoral growth. *Clin Cancer Res* 2011;**10**:2243.
124. Dowling RJ, Zakikhani M, Fantus IG, Pollak M, Sonenberg N. Metformin inhibits mammalian target of rapamycin–dependent translation initiation in breast cancer cells. *Cancer Res* 2007;**67**(22): 10804–12.
125. Hirsch HA, Iliopoulos D, Tsichlis PN, Struhl K. Metformin selectively targets cancer stem cells, and acts together with chemotherapy to block tumor growth and prolong remission. *Cancer Res* 2009; **69**(19):7507–11.
126. Alimova IN, Liu B, Edgerton SM, Dillon T, Lind SE, Thor AD. Metformin inhibits breast cancer cell growth, colony formation and induces cell cycle arrest in vitro. *Cell Cycle* 2009;**8**(6):909–15.
127. Goodwin PJPK, Ennis M, Clemons MJ, Graham MA, Fantus IG. Metformin lowers fasting insulin levels in non-diabetic hyperisulinemic early stage breast cancer (BC) patients: results of a phase II study. *Breast Cancer Res Treat* 2006;**100**:109.
128. Liu B, Fan Z, Edgerton SM, Deng XS, Alimova IN, Lind SE, et al. Metformin induces unique biological and molecular responses in triple negative breast cancer cells. *Cell Cycle* 2009;**8**(13):2031–40.
129. Iliopoulos D, Hirsch HA, Struhl K. Metformin decreases the dose of chemotherapy for prolonging tumor remission in mouse xenografts involving multiple cancer cell types. *Cancer Res* 2011; **71**(9):3196–201.
130. Ben Sahra I, Laurent K, Loubat A, Giorgetti-Peraldi S, Colosetti P, Auberger P, et al. The antidiabetic drug metformin exerts an antitumoral effect *in vitro* and *in vivo* through a decrease of cyclin D1 level. *Oncogene* 2008;**27**:3576–86.
131. Buzzai M, Jones RG, Amaravadi RK, Lum JJ, DeBerardinis RJ, Zhao F, et al. Systemic treatment with the antidiabetic drug metformin selectively impairs p53-deficient tumor cell growth. *Cancer Res* 2007;**67**:6745–52.
132. Zhuang Y, Miskimins KW. Cell cycle arrest in metformin treated breast cancer cells involves activation of AMPK, downregulatiom of cyclin D1 and requires p27 or p21. *J Mol Signal* 2008;**3**:18.
133. Zakikhani M, Blouin MJ, Piura E, Pollak M. Metformin and rapamycin have distinct effects on the Akt pathway and proliferation in breast cancer cells. *Breast Cancer Res Treat* 2010;**123**(1):271–9.
134. Huang X, Wullschleger S, Shapiro N, McGuire VA, Sakamoto K, Woods YL, et al. Important role of the LKB1-AMPK pathway in suppressing tumorigenesis in PTEN deficient mice. *Biochem J* 2008;**412**:211–21.
135. Gotlieb WH, Saumet J, Beauchamp MC, Gu J, Lau S, Pollak MN, Bruchim I. In vitro metformin antineoplastic activity in epithelial ovarian cancer. *Gynecol Oncol* 2008;**110**:246–50.

136. Dong L, Zhou Q, Zhang Z, Zhu Y, Duan T, Feng Y. Metformin sensitizes endometrial cancer cells to chemotherapy by repressing glyoxalase I expression. *J Obstet Gynaecol Res* 2012;**38**(8):1077–85.
137. Younis RH, Schneider A. Metformin inhibits mTOR pathway in HNSCC cells through REDD1 upregulation, In: Poster 225 presented in Miami (IADR/AADR/CADR 87th general session and exhibition April 1–4; 2009.
138. Griss T, Vincent EE, Egnatchik R, Chen J, Ma EH, Faubert B, et al. Metformin antagonizes cancer cell proliferation by suppressing mitochondrial-dependent biosynthesis. *PLoS Biol* 2015;**13**(12):e1002309.
139. Laderoute KR, Amin K, Calaoagan JM, et al. 5′-AMP-activated protein kinase (AMPK) is induced by low-oxygen and glucose deprivation conditions found in solid-tumor microenvironments. *Mol Cell Biol* 2006;**26**(14):5336–47.
140. Neurath KM, Keough MP, Mikkelsen T, Claffey KP. AMP-dependent protein kinase alpha 2 isoform promotes hypoxia-induced VEGF expression in human glioblastoma. *Glia* 2006;**53**(7):733–43.
141. Zwetsloot KA, Westerkamp LM, Holmes BF, Gavin TP. AMPK regulates basal skeletal muscle capillarization and VEGF expression, but is not necessary for the angiogenic response to exercise. *J Physiol* 2008;**586**(Pt. 24):6021–35.
142. Li X, Han Y, Pang W, Li C, Xie X, Shyy J, Yi Z. AMP-activated protein kinase promotes the differentiation of endothelial progenitor cells. *Arterioscler Thromb Vasc Biol* 2008;**28**:1789–95.
143. Park HU, Suy S, Danner M, Dailey V, et al. AMP-activated protein kinase promotes human prostate cancer cell growth and survival. *Mol Cancer Ther* 2009;**8**:733–41.
144. Jeon M, Chandel NS, Hay N. AMPK regulates NADPH homeostasis to promote tumour cell survival during energy stress. *Nature* 2012;**485**:661–5.
145. Damelin LH, Jivan R, Veale RB, Rousseau AL, Mavri-Damelin D. Metformin induces an intracellular reductive state that protects oesophageal squamous cell carcinoma cells against cisplatin but not copper-bis (thiosemicarbazones). *BMC Cancer* 2014;**14**(1):314.
146. Zhang Y, Li MX, Wang H, Zeng Z, Li XM. Metformin down-regulates endometrial carcinoma cell secretion of IGF-1 and expression of IGF-1R. *Asian Pac J Cancer Prev* 2015;**16**(1):221–5.
147. Zhang J, Zhang B2, Yin Z, Chen F, Liu T, Xu H, et al. Effects of metformin on the estrogen-induced proliferation and the expression of ER in human endometrial cancer cells. *Zhonghua Fu Chan Ke Za Zhi* 2014;**49**(12):932–7.
148. Tabrizi AD, Melli MS, Foroughi M, Ghojazadeh M, Bidadi S. Antiproliferative effect of metformin on the endometrium—a clinical trial. *Asian Pac J Cancer Prev* 2014;**15**(23):10067–70.
149. Schuler KM, Rambally BS, DiFurio MJ, et al. Antiproliferative and metabolic effects of metformin in a preoperative window clinical trial for endometrial cancer. *Cancer Med* 2015;**4**(2):161–73.
150. Laskov I, Drudi L, Beauchamp MC, Yasmeen A, Ferenczy A, Pollak M, et al. Anti-diabetic doses of metformin decrease proliferation markers in tumors of patients with endometrial cancer. *Gynecol Oncol* 2014;**134**(3):607–14.
151. Takahashi A, Kimura F, Yamanaka A, Takebayashi A, Kita N, Takahashi K, et al. Metformin impairs growth of endometrial cancer cells via cell cycle arrest and concomitant autophagy and apoptosis. *Cancer Cell Int* 2014;**14**:53.
152. Mitsuhashi A, Kiyokawa T, Sato Y, Shozu M. Effects of metformin on endometrial cancer cell growth in vivo: a preoperative prospective trial. *Cancer* 2014;**120**(19):2986–95.
153. Mu N, Zhu Y, Wang Y, Zhang H, Xue F. Insulin resistance: a significant risk factor of endometrial cancer. *Gynecol Oncol* 2012;**125**(3):751–7.
154. Cantrell LA, Zhou C, Mendivil A, Malloy KM, Gehrig PA, Bae-Jump VL. Metformin is a potent inhibitor of endometrial cancer cell proliferation—implications for a novel treatment strategy. *Gynecol Oncol* 2010;**116**(1):92–8.
155. Tan BK, Adya R, Chen J, Lehnert H, Sant Cassia LJ, Randeva HS. Metformin treatment exerts antiinvasive and antimetastatic effects in human endometrial carcinoma cells. *J Clin Endocrinol Metab* 2011;**96**(3):808–16.
156. Sarfstein R, Friedman Y, Attias-Geva Z, Fishman A, Bruchim I, Werner H. Metformin downregulates the insulin/IGF-I signaling pathway and inhibits different uterine serous carcinoma (USC) cells proliferation and migration in p53-dependent or -independent manners. *PLoS One* 2013;**8**(4):e61537.
157. Klubo-Gwiezdzinska J, Costello Jr. J, Patel A, Bauer A, Jensen K, Mete M, et al. Treatment with metformin is associated with higher remission rate in diabetic patients with thyroid cancer. *J Clin Endocrinol Metabol* 2013;**98**(8):3269–79.

158. Tseng C-H. Metformin reduces thyroid cancer risk in Taiwanese patients with type 2 diabetes. *PLoS One* 2014;**9**(10):e109852.
159. Cho SW, Yi KH, Han SK, Sun HJ, Kim YA, Oh BC, et al. Therapeutic potential of metformin in papillary thyroid cancer in vitro and in vivo. *Mol Cell Endocrinol* 2014;**393**(1–2):24–9.
160. Chen G, Xu S, Renko K, Derwahl M. Metformin inhibits growth of thyroid carcinoma cells, suppresses self-renewal of derived cancer stem cells, and potentiates the effect of chemotherapeutic agents. *J Clin Endocrinol Metabol* 2012;**97**:510–20.
161. Klubo-Gwiezdzinska J, Jensen K, Costello J, Patel A, Hoperia V, Bauer A, et al. Metformin inhibits growth and decreases resistance to anoikis in medullary thyroid cancer cells. *Endocr Relat Cancer* 2012;**19**:447–56.
162. Pappa T, Alevizaki M. Metformin and thyroid: an update. *Eur Thyroid J* 2013;**2**(1):22–8.
163. Cossor FI, Adams-Campbell LL, Chlebowski RT, Gunter MJ, Johnson K, Martell RE, et al. Diabetes, metformin use, and colorectal cancer survival in postmenopausal women. *Cancer Epidemiol* 2013; **37**(5):742–9.
164. Spillane S, Bennett K, Sharp L, Barron TI. A cohort study of metformin exposure and survival in patients with stage I–III colorectal cancer. *Cancer Epidemiol Biomarkers Prev* 2013;**22**(8):1364–73.
165. Aksoy S, Sendur MA, Altundag K. Demographic and clinico-pathological characteristics in patients with invasive breast cancer receiving metformin. *Med Oncol* 2013;**30**(2):590.
166. Tang GH, Satkunam M, Pond GR, Steinberg GR, Blandino G, Schünemann HJ, Muti P. Association of metformin with breast cancer incidence and mortality in patients with type II diabetes: a GRADE-assessed systematic review and meta-analysis. *Cancer Epidemiol Biomarkers Prev* 2018; **27**(6):627–35.
167. Dowling RJ, Niraula S, Chang MC, Done SJ, Ennis M, McCready DR, et al. Changes in insulin receptor signaling underlie neoadjuvant metformin administration in breast cancer: a prospective window of opportunity neoadjuvant study. *Breast Cancer Res* 2015;**17**(1):540.
168. La Marca A, Morgante G, Palumbo M, Cianci A, Petraglia F, De Leo V. Insulin-lowering treatment reduces aromatase activity in response to follicle-stimulating hormone in women with polycystic ovary syndrome. *Fertil Steril* 2002;**78**(6):1234–9.
169. Faubert B, Vincent EE, Poffenberger MC, Jones RG. The AMP-activated protein kinase (AMPK) and cancer: many faces of a metabolic regulator. *Cancer Lett* 2015;**356**(2):165–70.
170. Vincent EE, Coelho PP, Blagih J, Griss T, Viollet B, Jones RG. Differential effects of AMPK agonists on cell growth and metabolism. *Oncogene* 2015;**34**(28):3627.
171. Kato K, Ogura T, Kishimoto A, Minegishi Y, Nakajima N, Miyazaki M, et al. Critical roles of AMP-activated protein kinase in constitutive tolerance of cancer cells to nutrient deprivation and tumor formation. *Oncogene* 2002;**21**(39):6082.
172. Jang T, Calaoagan JM, Kwon E, Samuelsson S, Recht L, Laderoute KR. 5′-AMP-activated protein kinase activity is elevated early during primary brain tumor development in the rat. *Int J Cancer* 2011; **128**(9):2230–9.
173. Liang J, Mills GB. AMPK: a contextual oncogene or tumor suppressor? *Cancer Res* 2013; **73**(10):2929–35.
174. Brunmair B, Staniek K, Gras F, Scharf N, Althaym A, Clara R, et al. Thiazolidinediones, like metformin, inhibit respiratory complex I: a common mechanism contributing to their antidiabetic actions? *Diabetes* 2004;**53**(4):1052–9.
175. Doran E, Halestrap AP. Evidence that metformin exerts its anti-diabetic effects through inhibition of complex 1 of the mitochondrial respiratory chain. *Biochem J* 2000;**348**(3):607–14.
176. Viollet B, Guigas B, Garcia NS, Leclerc J, Foretz M, Andreelli F. Cellular and molecular mechanisms of metformin: an overview. *Clin Sci* 2012;**122**(6):253–70.
177. El-Mir MY, Nogueira V, Fontaine E, Averet N, Rigoulet M, Leverve X. Dimethylbiguanide inhibits cell respiration via an indirect effect targeted on the respiratory chain complex I. *J Biol Chem* 2000;**275**:223.
178. Hardie DG. Neither LKB1 nor AMPK are the direct targets of metformin. *Gastroenterology* 2006;**131**:973. author reply 974–5.
179. Hawley SA, Gadalla AE, Olsen GS, Hardie DG. The antidiabetic drug metformin activates the AMP-activated protein kinase cascade via an adenine nucleotide-independent mechanism. *Diabetes* 2002;**51**(8):2420–5.

180. Stephenne X, Foretz M, Taleux N, van der Zon G, Sokal E, Hue L, et al. Metformin activates AMP-activated protein kinase in primary human hepatocytes by decreasing cellular energy status. *Diabetologia* 2011;**54**(12):3101–10.
181. Wheaton WW, Weinberg SE, Hamanaka RB, Soberanes S, Sullivan LB, Anso E, et al. Metformin inhibits mitochondrial complex I of cancer cells to reduce tumorigenesis. *Elife* 2014;**3**:e02242.
182. Piel S, Ehinger JK, Elmer E, Hansson MJ. Metformin induces lactate production in peripheral blood mononuclear cells and platelets through specific mitochondrial complex I inhibition. *Acta Physiol* 2015;**213**(1):171–80.
183. Batandier C, Guigas B, Detaille D, El-Mir MY, Fontaine E, Rigoulet M, et al. The ROS production induced by a reverse-electron flux at respiratory-chain complex 1 is hampered by metformin. *J Bioenerg Biomembr* 2006;**38**(1):33–42.
184. Melnik S, Dvornikov D, Müller-Decker K, Depner S, Stannek P, Meister M, et al. Cancer cell specific inhibition of Wnt/β-catenin signaling by forced intracellular acidification. *Cell Discov* 2018;**4**(1):37.
185. Xia C, Chen R, Chen J, Qi Q, Pan Y, Du L, et al. Combining metformin and nelfinavir exhibits synergistic effects against the growth of human cervical cancer cells and xenograft in nude mice. *Sci Rep* 2017;**7**:43373.
186. Hart T, Dider S, Han W, Xu H, Zhao Z, Xie L. Toward repurposing metformin as a precision anti-cancer therapy using structural systems pharmacology. *Sci Rep* 2016;**6**:20441.
187. Babcook MA, Shukla S, Fu P, Vazquez EJ, Puchowicz MA, Molter JP, et al. Synergistic simvastatin and metformin combination chemotherapy for osseous metastatic castration-resistant prostate cancer. *Mol Cancer Ther* 2014;**13**:https://doi.org/10.1158/1535-7163.MCT-14-0451.
188. Pennanen P, Syvälä H, Bläuer M, Savinainen K, Ylikomi T, Tammela TL, et al. The effects of metformin and simvastatin on the growth of LNCaP and RWPE-1 prostate epithelial cell lines. *Eur J Pharmacol* 2016;**788**:160–7.
189. Kobayashi M, Otsuka Y, Itagaki S, Hirano T, Iseki K. Inhibitory effects of statins on human monocarboxylate transporter 4. *Int J Pharm* 2006;**317**(1):19–25.
190. Vali S, Usmani S, Sultana Z, Kumar A, Abbasi T, Vidva R, Cellworks Research India Pvt Ltd, Cellworks Group Inc. *Combination of nelfinavir, metformin and rosuvastatin for treating cancer caused by aberrations in pten/tp53.* United States Patent Application US 15/309,144, 2017.
191. Sirvent P, Bordenave S, Vermaelen M, Roels B, Vassort G, Mercier J, et al. Simvastatin induces impairment in skeletal muscle while heart is protected. *Biochem Biophys Res Commun* 2005;**338**(3):1426–34.
192. Danial NN, Gramm CF, Scorrano L, Zhang CY, Krauss S, Ranger AM, et al. BAD and glucokinase reside in a mitochondrial complex that integrates glycolysis and apoptosis. *Nature* 2003;**424**(6951):952.
193. Tomiyama A, Serizawa S, Tachibana K, Sakurada K, Samejima H, Kuchino Y, et al. Critical role for mitochondrial oxidative phosphorylation in the activation of tumor suppressors Bax and Bak. *J Natl Cancer Inst* 2006;**98**(20):1462–73.
194. Dey R, Moraes CT. Lack of oxidative phosphorylation and low mitochondrial membrane potential decrease susceptibility to apoptosis and do not modulate the protective effect of Bcl-xL in osteosarcoma cells. *J Biol Chem* 2000;**275**(10):7087–94.
195. Park SY, Chang I, Kim JY, Kang SW, Park SH, Singh K, Lee MS. Resistance of mitochondrial DNA-depleted cells against cell death: role of mitochondrial superoxide dismutase. *J Biol Chem* 2004;**279**(9):7512–20.
196. Harris MH, Vander Heiden MG, Kron SJ, Thompson CB. Role of oxidative phosphorylation in Bax toxicity. *Mol Cell Biol* 2000;**20**(10):3590–6.
197. Sánchez-Aragó M, Chamorro M, Cuezva JM. Selection of cancer cells with repressed mitochondria triggers colon cancer progression. *Carcinogenesis* 2010;**31**(4):567–76.
198. Stacpoole PW, Harman EM, Curry SH, Baumgartner TG, Misbin RI. Treatment of lactic acidosis with dichloroacetate. *N Engl J Med* 1983;**309**(7):390–6.
199. Stacpoole PW, Lorenz AC, Thomas RG, Harman EM. Dichloroacetate in the treatment of lactic acidosis. *Ann Intern Med* 1988;**108**(1):58–63.
200. Stacpoole PW, Kerr DS, Barnes C, Bunch ST, Carney PR, Fennell EM, et al. Controlled clinical trial of dichloroacetate for treatment of congenital lactic acidosis in children. *Pediatrics* 2006;**117**(5):1519–31.
201. Irsigler K, Brändle J, Kaspar L, Kritz H, Lageder H, Regal H. Treament of biguanide-induced lactic acidosis with dichloroacetate. 3 case histories. *Arzneimittelforschung* 1979;**29**(3):555–9.

202. Bonnet S, Archer SL, Allalunis-Turner J, Haromy A, Beaulieu C, Thompson R, et al. A mitochondria-K^+ channel axis is suppressed in cancer and its normalization promotes apoptosis and inhibits cancer growth. *Cancer Cell* 2007;**11**(1):37–51.
203. Li T, Schultz I, Keys DA, Campbell JL, Fisher JW. Quantitative evaluation of dichloroacetic acid kinetics in human—a physiologically based pharmacokinetic modeling investigation. *Toxicology* 2008;**245**(1–2):35–48.
204. Stacpoole PW. The pharmacology of dichloroacetate. *Metabolism* 1989;**38**:1124–44.
205. Shroads AL, Guo X, Dixit V, Liu HP, James MO, Stacpoole PW. Age-dependent kinetics and metabolism of dichloroacetate: possible relevance to toxicity. *J Pharmacol Exp Ther* 2008;**324**(3):1163–71.
206. Wells PG, Moore GW, Rabin D, Wilkinson GR, Oates JA, Stacpoole PW. Metabolic effects and pharmacokinetics of intravenously administered dichloroacetate in humans. *Diabetologia* 1980;**19**(2): 109–13.
207. Gonzalez-Leon A, Schultz IR, Xu G, Bull RJ. Pharmacokinetics and metabolism of dichloroacetate in the F344 rat after prior administration in drinking water. *Toxicol Appl Pharmacol* 1997;**146**(2):189–95.
208. Lecellier JH, Daujat-Chavanieu M, Villalba M. The PDK1 inhibitor dichloroacetate controls cholesterol homeostasis through the ERK5/MEF2 pathway. Cellular metabolism regulates anti-oxidant response through ERK5-MEF2 pathway. In: Khan AUH, editor. *Human health and pathology*. Université Montpellier; 2017. p. 167. Available at: https://tel.archives-ouvertes.fr/tel-01684254/document#page=177.
209. Michelakis ED, Webster L, Mackey JR. Dichloroacetate (DCA) as a potential metabolic-targeting therapy for cancer. *Br J Cancer* 2008;**99**(7):989.
210. Whitehouse S, Cooper RH, Randle PJ. Mechanism of activation of pyruvate dehydrogenase by dichloroacetate and other halogenated carboxylic acids. *Biochem J* 1974;**141**(3):761–4.
211. Kankotia S, Stacpoole PW. Dichloroacetate and cancer: new home for an orphan drug? *Biochim Biophys Acta Rev Cancer* 2014;**1846**(2):617–29.
212. Albatany M, Li A, Meakin S, Bartha R. Dichloroacetate induced intracellular acidification in glioblastoma: in vivo detection using AACID-CEST MRI at 9.4 Tesla. *J Neurooncol* 2018;**136**(2):255–62.
213. Anemone A, Consolino L, Conti L, Reineri F, Cavallo F, Aime S, Longo DL. In vivo evaluation of tumour acidosis for assessing the early metabolic response and onset of resistance to dichloroacetate by using magnetic resonance pH imaging. *Int J Oncol* 2017;**51**(2):498–506.
214. Albatany M, Li A, Meakin S, Bartha R. In vivo detection of acute intracellular acidification in glioblastoma multiforme following a single dose of cariporide. *Int J Clin Oncol* 2018;**10**:1–8.
215. Li B, Li X, Xiong H, Zhou P, Ni Z, Yang T, et al. Inhibition of COX2 enhances the chemosensitivity of dichloroacetate in cervical cancer cells. *Oncotarget* 2017;**8**(31).
216. Ribes G, Valette G, Loubatières-Mariani MM. Metabolic effects of sodium dichloroacetate in normal and diabetic dogs. *Diabetes* 1979;**28**(9):852–7.
217. Misbin RI. Effects of dichloroacetate on lipid metabolism in isolated rat liver cells. *Diabetes* 1979; **28**(4):265–71.
218. Sutendra G, Dromparis P, Kinnaird A, Stenson TH, Haromy A, Parker JM, et al. Mitochondrial activation by inhibition of PDKII suppresses HIF1a signaling and angiogenesis in cancer. *Oncogene* 2013; **32**(13):1638.
219. Makita N, Ishiguro J, Suzuki K, Nara F. Dichloroacetate induces regulatory T-cell differentiation and suppresses Th17-cell differentiation by pyruvate dehydrogenase kinase-independent mechanism. *J Pharm Pharmacol* 2017;**69**(1):43–51.
220. Tomsig JL, Gruenstein E, Dimlich RV. Inhibition of lactate-induced swelling by dichloroacetate in human astrocytoma cells. *Brain Res* 1991;**568**(1–2):92–100.
221. Sun RC, Fadia M, Dahlstrom JE, Parish CR, Board PG, Blackburn AC. Reversal of the glycolytic phenotype by dichloroacetate inhibits metastatic breast cancer cell growth in vitro and in vivo. *Breast Cancer Res Treat* 2010;**120**(1):253–60.
222. Gang BP, Dilda PJ, Hogg PJ, Blackburn AC. Dichloroacetate reverses the Warburg effect, inhibiting growth and sensitizing breast cancer cells towards apoptosis, In: Proceedings of the 103rd annual meeting of the American Association for Cancer Research; 2012; Chicago, IL. Philadelphia (PA): AACR; 2012. Cancer Res 72(8 Suppl):Abstract nr 3228.
223. Wang M, Liao C, Hu Y, Pan Q, Jiang J. Sensitization of breast cancer cells to paclitaxel by dichloroacetate through inhibiting autophagy. *Biochem Biophys Res Commun* 2017;**489**(2):103–8.

224. Harting TP, Stubbendorff M, Hammer SC, Schadzek P, Ngezahayo A, Escobar HM, Nolte I. Dichloroacetate affects proliferation but not apoptosis in canine mammary cell lines. *PLoS One* 2017;**12**(6):e0178744.
225. Wo SH, Seo S-K, Park Y, Kim E-K, Seong M-K, Kim H-A, Park I-C. Dichloroacetate potentiates tamoxifen-induced cell death in breast cancer cells via downregulation of the epidermal growth factor receptor. *Oncotarget* 2016;**7**(37):59809–19.
226. Haugrud AB, Zhuang Y, Coppock JD, Miskimins WK. Dichloroacetate enhances apoptotic cell death via oxidative damage and attenuates lactate production in metformin-treated breast cancer cells. *Breast Cancer Res Treat* 2014;**147**(3):539–50.
227. De Preter G, Neveu MA, Danhier P, Brisson L, Payen VL, Porporato PE, Gallez B. Inhibition of the pentose phosphate pathway by dichloroacetate unravels a missing link between aerobic glycolysis and cancer cell proliferation. *Oncotarget* 2016;**7**(3):2910–20.
228. Sun RC, Board PG, Blackburn AC. Targeting metabolism with arsenic trioxide and dichloroacetate in breast cancer cells. *Mol Cancer* 2011;**10**(1):142.
229. Robey IF, Martin NK. Bicarbonate and dichloroacetate: evaluating pH altering therapies in a mouse model for metastatic breast cancer. *BMC Cancer* 2011;**11**(1):235.
230. Lefort N, Brown A, Lloyd V, Ouellette R, Touaibia M, Culf AS, Cuperlovic-Culf M. 1H NMR metabolomics analysis of the effect of dichloroacetate and allopurinol on breast cancers. *J Pharm Biomed Anal* 2014;**93**:77–85.
231. Sun RC, Fadia M, Dahlstrom JE, Parish CR, Board PG, Blackburn AC. Reversal of the glycolytic phenotype by dichloroacetate inhibits metastatic breast cancer cell growth in vitro and in vivo. *Breast Cancer Res Treat* 2009;**120**:253–60.
232. Xintaropoulou C, Ward C, Wise A, Marston H, Turnbull A, Langdon SP. A comparative analysis of inhibitors of the glycolysis pathway in breast and ovarian cancer cell line models. *Oncotarget* 2015; **6**(28):25677.
233. Blackburn AC, Rooke M, Sun RC, Fadia M, Dahlstrom JE, Board PG. Reversal of the glycolytic phenotype with dichloroacetate in a mouse mammary adenocarcinoma model. *Pathology* 2010;**42**:S61–2.
234. Harting T, Stubbendorff M, Willenbrock S, Wagner S, Schadzek P, Ngezahayo A, et al. The effect of dichloroacetate in canine prostate adenocarcinomas and transitional cell carcinomas in vitro. *Int J Oncol* 2016;**49**(6):2341–50.
235. Cao W, Yacoub S, Shiverick KT, Namiki K, Sakai Y, Porvasnik S, et al. Dichloroacetate (DCA) sensitizes both wild-type and over expressing Bcl-2 prostate cancer cells in vitro to radiation. *Prostate* 2008;**68**(11):1223–31.
236. Higgins LH, Withers HG, Garbens A, Love HD, Magnoni L, Hayward SW, et al. Hypoxia and the metabolic phenotype of prostate cancer cells. *Biochim Biophys Acta Bioenerg* 2009;**1787**(12): 1433–43.
237. Olszewski U, Poulsen TT, Ulsperger E, Poulsen HS, Geissler K, Hamilton G. In vitro cytotoxicity of combinations of dichloroacetate with anticancer platinum compounds. *Clin Pharmacol* 2010;**2**:177.
238. Chaudhary AK, Bhat TA, Kumar S, Kumar A, Kumar R, Underwood W, et al. Mitochondrial dysfunction-mediated apoptosis resistance associates with defective heat shock protein response in African-American men with prostate cancer. *Br J Cancer* 2016;**114**(10):1090.
239. Zhou X, Chen R, Yu Z, Li R, Li J, Zhao X, et al. Dichloroacetate restores drug sensitivity in paclitaxel-resistant cells by inducing citric acid accumulation. *Mol Cancer* 2015;**14**(1):63.
240. Madhok BM, Yeluri S, Perry SL, Hughes TA, Jayne DG. Dichloroacetate induces apoptosis and cell-cycle arrest in colorectal cancer cells. *Br J Cancer* 2010;**102**(12):1746.
241. Lin G, Hill DK, Andrejeva G, Boult JK, Troy H, Fong AL, et al. Dichloroacetate induces autophagy in colorectal cancer cells and tumours. *Br J Cancer* 2014;**111**(2):375.
242. Delaney LM, Ho N, Morrison J, Farias NR, Mosser DD, Coomber BL. Dichloroacetate affects proliferation but not survival of human colorectal cancer cells. *Apoptosis* 2015;**20**(1):63–74.
243. Khan A, Andrews D, Blackburn AC. Long-term stabilization of stage 4 colon cancer using sodium dichloroacetate therapy. *World J Clin Cases* 2016;**4**(10):336.
244. Tong J, Xie G, He J, Li J, Pan F, Liang H. Synergistic antitumor effect of dichloroacetate in combination with 5-fluorouracil in colorectal cancer. *J Biomed Biotechnol* 2011;**2011**: https://doi.org/10.1155/2011/740564 740564.

245. Khan A, Andrews D, Shainhouse J, Blackburn AC. Long-term stabilization of metastatic melanoma with sodium dichloroacetate. *World J Clin Oncol* 2017;**8**(4):371.
246. Abildgaard C, Dahl C, Basse AL, Ma T, Guldberg P. Bioenergetic modulation with dichloroacetate reduces the growth of melanoma cells and potentiates their response to BRAF V600E inhibition. *J Transl Med* 2014;**12**(1):247.
247. Zheng MF, Shen SY. DCA increases the antitumor effects of capecitabine in a mouse B16 melanoma allograft and a human non-small cell lung cancer A549 xenograft. *Cancer Chemother Pharmacol* 2013;**72**(5):1031–41.
248. Pópulo H, Caldas R, Lopes JM, Pardal J, Máximo V, Soares P. Overexpression of pyruvate dehydrogenase kinase supports dichloroacetate as a candidate for cutaneous melanoma therapy. *Expert Opin Ther Targets* 2015;**19**(6):733–45.
249. Franco-Molina MA, Mendoza-Gamboa E, Sierra-Rivera CA, Zapata-Benavides P, Hernandez DF, Chávez-Reyes A, et al. In vitro and in vivo antitumoral activitiy of sodium dichloroacetate (DCA-Na) against murine melanoma. *Afr J Microbiol Res* 2012;**6**(22):4782–96.
250. Chaube B, Malvi P, Singh SV, Mohammad N, Meena AS, Bhat MK. Targeting metabolic flexibility by simultaneously inhibiting respiratory complex I and lactate generation retards melanoma progression. *Oncotarget* 2015;**6**(35):37281.
251. Michelakis ED, Sutendra G, Dromparis P, Webster L, Haromy A, Niven E, et al. Metabolic modulation of glioblastoma with dichloroacetate. *Sci Transl Med* 2010;**2**(31):31–4.
252. Duan Y, Zhao X, Ren W, Wang X, Yu KF, Li D, et al. Antitumor activity of dichloroacetate on C6 glioma cell: in vitro and in vivo evaluation. *Onco Targets Ther* 2013;**6**:189.
253. Kumar K, Wigfield S, Gee HE, Devlin CM, Singleton D, Li JL, Ivan M. Dichloroacetate reverses the hypoxic adaptation to bevacizumab and enhances its antitumor effects in mouse xenografts. *J Mol Med (Berl)* 2013;**91**(6):749–58.
254. Kolesnik DL, Pyaskovskaya ON, Boichuk IV, Solyanik GI. Hypoxia enhances antitumor activity of dichloroacetate. *Exp Oncol* 2014;**36**(4):231–5.
255. Vella S, Conti M, Tasso R, Cancedda R, Pagano A. Dichloroacetate inhibits neuroblastoma growth by specifically acting against malignant undifferentiated cells. *Int J Cancer* 2012;**130**(7):1484–93.
256. Sradhanjali S, Tripathy D, Rath S, Mittal R, Reddy MM. Overexpression of pyruvate dehydrogenase kinase 1 in retinoblastoma: a potential therapeutic opportunity for targeting vitreous seeds and hypoxic regions. *PLoS One* 2017;**12**(5):e0177744.
257. Park JM, Recht LD, Josan S, Merchant M, Jang T, Yen YF, et al. Metabolic response of glioma to dichloroacetate measured in vivo by hyperpolarized 13C magnetic resonance spectroscopic imaging. *Neuro Oncol* 2013;**15**(4):433–41.
258. Fedorchuk AG, Pyaskovskaya ON, Gorbik GV, Prokhorova IV, Kolesnik DL, Solyanik GI. Effectiveness of sodium dichloroacetate against glioma C6 depends on administration schedule and dosage. *Exp Oncol* 2016;**38**:80–3.
259. Wicks RT, Azadi J, Mangraviti A, Zhang I, Hwang L, Joshi A, et al. Local delivery of cancer-cell glycolytic inhibitors in high-grade glioma. *Neuro Oncol* 2014;**17**(1):70–80.
260. Ward NP. *Therapeutic modulation of cancer metabolism with dichloroacetate and metformin*. Doctoral dissertation, University of South Florida; 2017. Available at: https://search.proquest.com/openview/a84ae60fe3dc10edca4bc13fdeb0abee/1?pq-origsite=gscholar&cbl=18750&diss=y.
261. Lu X, Zhou D, Hou B, Liu QX, Chen Q, Deng XF, et al. Dichloroacetate enhances the antitumor efficacy of chemotherapeutic agents via inhibiting autophagy in non-small-cell lung cancer. *Cancer Manag Res* 2018;**10**:1231.
262. Yang Z, Tam KY. Anti-cancer synergy of dichloroacetate and EGFR tyrosine kinase inhibitors in NSCLC cell lines. *Eur J Pharmacol* 2016;**789**:458–67.
263. Allen KT, Chin-Sinex H, DeLuca T, Pomerening JR, Sherer J, Watkins III JB, et al. Dichloroacetate alters Warburg metabolism, inhibits cell growth, and increases the X-ray sensitivity of human A549 and H1299 NSC lung cancer cells. *Free Radic Biol Med* 2015;**89**:263–73.
264. Sun H, Zhu A, Zhou X, Wang F. Suppression of pyruvate dehydrogenase kinase-2 re-sensitizes paclitaxel-resistant human lung cancer cells to paclitaxel. *Oncotarget* 2017;**8**(32):52642–50.
265. Kolesnik DL, Pyaskovskaya ON, Boychuk IV, Dasyukevich OI, Melnikov OR, Tarasov AS, et al. Effect of dichloroacetate on Lewis lung carcinoma growth and metastasis. *Exp Oncol* 2015;**37**(2):126–9.

266. Fiebiger W, Olszewski U, Ulsperger E, Geissler K, Hamilton G. In vitro cytotoxicity of novel platinum-based drugs and dichloroacetate against lung carcinoid cell lines. *Clin Transl Oncol* 2011;**13**(1):43–9.
267. Shan HG, Mao GX, Pan JQ, Zhou XF. Study on apoptosis of human lung adenocarcinoma cell line-A549 induced by sodium dichloroacetate combined with cisplantin. *Pract Pharm Clin Rem* 2013;**1**:006.
268. Li B, Li X, Ni Z, Zhang Y, Zeng Y, Yan X, et al. Dichloroacetate and metformin synergistically suppress the growth of ovarian cancer cells. *Oncotarget* 2016;**7**(37):59458.
269. Ward NP, Poff AM, Koutnik AP, D'Agostino DP. Complex I inhibition augments dichloroacetate cytotoxicity through enhancing oxidative stress in VM-M3 glioblastoma cells. *PLoS One* 2017; **12**(6):e0180061.
270. Reed LJ, De BB, Gunsalus IC, Hornberger Jr. CS. Crystalline alpha-lipoic acid; a catalytic agent associated with pyruvate dehydrogenase. *Science* 1951;**114**:93–4.
271. Korotchkina LG, Sidhu S, Patel MS. R-lipoic acid inhibits mammalian pyruvate dehydrogenase kinase. *Free Radic Res* 2004;**38**:1083–92.
272. Konrad T, Vicini P, Kusterer K, Höflich A, Assadkhani A, Böhles HJ, et al. Alpha-lipoic acid treatment decreases serum lactate and pyruvate concentrations and improves glucose effectiveness in lean and obese patients with type 2 diabetes. *Diabetes Care* 1999;**22**(2):280–7.
273. Rahimifard M, Navaei-Nigjeh M, Baeeri M, Maqbool F, Abdollahi M. Multiple protective mechanisms of alpha-lipoic acid in oxidation, apoptosis and inflammation against hydrogen peroxide induced toxicity in human lymphocytes. *Mol Cell Biochem* 2015;**403**(1–2):179.
274. Schwartz L, Abolhassani M, Guais A, Sanders E, Steyaert JM, Campion F, et al. A combination of alpha lipoic acid and calcium hydroxycitrate is efficient against mouse cancer models: preliminary results. *Oncol Rep* 2010;**23**(5):1407–16.
275. Wenzel U, Nickel A, Daniel H. α-lipoic acid induces apoptosis in human colon cancer cells by increasing mitochondrial respiration with a concomitant $O2^-$ generation. *Apoptosis* 2005;**10**(2): 359–68.
276. Moungjaroen J, Nimmannit U, Callery PS, Wang L, Azad N, Lipipun V, et al. Reactive oxygen species mediate caspase activation and apoptosis induced by lipoic acid in human lung epithelial cancer cells through Bcl-2 down-regulation. *J Pharmacol Exp Ther* 2006;**319**(3):1062–9.
277. Simbula G, Columbano A, Ledda-Columbano GM, Sanna L, Deidda M, Diana A, et al. Increased ROS generation and p53 activation in α-lipoic acid-induced apoptosis of hepatoma cells. *Apoptosis* 2007;**12**(1):113–23.
278. Feuerecker B, Pirsig S, Seidl C, Aichler M, Feuchtinger A, Bruchelt G, et al. Lipoic acid inhibits cell proliferation of tumor cells in vitro and in vivo. *Cancer Biol Ther* 2012;**13**(14):1425–35.
279. Pyaskovskaya ON, Kolesnik DL, Fedorchuk AG, Prokhorova IV, Solyanik GI. 2-Deoxy-D-glucose enhances dichloroacetate antitumor action against Lewis lung carcinoma. *Exp Oncol* 2016;**38**(3):176–80.
280. Ayyanathan K, Kesaraju S, Dawson-Scully K, Weissbach H. Combination of sulindac and dichloroacetate kills cancer cells via oxidative damage. *PLoS One* 2012;**7**(7):e39949.
281. Dhar S, Lippard SJ. Mitaplatin, a potent fusion of cisplatin and the orphan drug dichloroacetate. *Proc Natl Acad Sci U S A* 2009;**106**(52):22199–204.
282. Pathak RK, Marrache S, Harn DA, Dhar S. Mito-DCA: a mitochondria targeted molecular scaffold for efficacious delivery of metabolic modulator dichloroacetate. *ACS Chem Biol* 2014;**9**(5):1178–87.
283. Bradford SA, Khan A. Individualizing chemotherapy using the anti-diabetic drug, metformin, as an "adjuvant": an exploratory study. *J Cancer Sci Ther* 2013;**5**(6):120–5. https://doi.org/10.4172/1948-5956.1000197.
284. Abildgaard C, Dahl C, Abdul-Al A, Christensen A, Guldberg P. Inhibition of retinoic acid receptor β signaling confers glycolytic dependence and sensitization to dichloroacetate in melanoma cells. *Oncotarget* 2017;**8**(48):84210.
285. Kaufmann P, Engelstad K, Wei Y, Jhung S, Sano MC, Shungu DC, et al. Dichloroacetate causes toxic neuropathy in MELAS: a randomized, controlled clinical trial. *Neurology* 2006;**66**(3):324–30.
286. Shroads AL, Langaee T, Coats BS, Kurtz TL, Bullock JR, Weithorn D, et al. Human polymorphisms in the glutathione transferase zeta 1/maleylacetoacetate isomerase gene influence the toxicokinetics of dichloroacetate. *J Clin Pharmacol* 2012;**52**(6):837–49.

287. Washington JT, Quintyne NJ. Dichloroacetate induces different rates of cell death in cancer and noncancer cell lines in vitro. *Tumori J* 2012;**98**(1):142–51.
288. Ishiguro T, Ishiguro R, Ishiguro M, Iwai S. Co-treatment of dichloroacetate, omeprazole and tamoxifen exhibited synergistically antiproliferative effect on malignant tumors: in vivo experiments and a case report. *Hepatogastroenterology* 2012;**59**(116):994–6.
289. Lemmo W, Tan G. Prolonged survival after dichloroacetate treatment of non-small-cell lung carcinoma-related leptomeningeal carcinomatosis. *J Med Cases* 2016;**7**(4):136–42.
290. Garon EB, Christofk HR, Hosmer W, Britten CD, Bahng A, Crabtree MJ, et al. Dichloroacetate should be considered with platinum-based chemotherapy in hypoxic tumors rather than as a single agent in advanced non-small cell lung cancer. *J Cancer Res Clin Oncol* 2014;**140**(3):443–52.
291. Dunbar EM, Coats BS, Shroads AL, Langaee T, Lew A, Forder JR, et al. Phase 1 trial of dichloroacetate (DCA) in adults with recurrent malignant brain tumors. *Invest New Drugs* 2014;**32**(3):452–64.
292. Heshe D, Hoogestraat S, Brauckmann C, Karst U, Boos J, Lanvers-Kaminsky C. Dichloroacetate metabolically targeted therapy defeats cytotoxicity of standard anticancer drugs. *Cancer Chemother Pharmacol* 2011;**67**(3):647–55.
293. Shahrzad S, Lacombe K, Adamcic U, Minhas K, Coomber BL. Sodium dichloroacetate (DCA) reduces apoptosis in colorectal tumor hypoxia. *Cancer Lett* 2010;**297**(1):75–83.
294. Zwicker F, Kirsner A, Peschke P, Roeder F, Debus J, Huber PE, et al. Dichloroacetate induces tumor-specific radiosensitivity in vitro but attenuates radiation-induced tumor growth delay in vivo. *Strahlenther Onkol* 2013;**189**(8):684–92.
295. Feuerecker B, Seidl C, Pirsig S, Bruchelt G, Senekowitsch-Schmidtke R. DCA promotes progression of neuroblastoma tumors in nude mice. *Am J Cancer Res* 2015;**5**(2):812.
296. Niewisch MR, Kuçi Z, Wolburg H, Sautter M, Krampen L, Deubzer B, et al. Influence of dichloroacetate (DCA) on lactate production and oxygen consumption in neuroblastoma cells: is DCA a suitable drug for neuroblastoma therapy? *Cell Physiol Biochem* 2012;**29**(3–4):373–80.
297. Kailavasan M, Rehman I, Reynolds S, Bucur A, Tozer G, Paley M. NMR-based evaluation of the metabolic profile and response to dichloroacetate of human prostate cancer cells. *NMR Biomed* 2014;**27**(5):610–6.
298. Armitage JO, Sidner RD. Antitumour effect of cimetidine. *Lancet* 1979;**1**:882–3.
299. Adams WJ, Morris DL. Short-course cimetidine and survival with colorectal cancer. *Lancet* 1994;**344**:1768–9.
300. Matsumoto S. Cimetidine and survival with colorectal cancer. *Lancet* 1995;**346**:115.
301. Svendsen LB, Ross C, Knigge U, Frederiksen HJ, Graversen P, Kjaergard J, et al. Cimetidine as an adjuvant treatment in colorectal cancer. A double blind randomized pilot study. *Dis Colon Rectum* 1995;**38**:514–8.
302. Hellstrand K, Naredi P, Lindner P, Lundholm K, Rudenstam CM, Hermodsson S, et al. Histamine in immunotherapy of advanced melanoma. *Cancer Immunol Immunother* 1994;**39**:416–9.
303. Creagan ET, Ahman DL, Green SJ, Long HJ, Frytak S, Itri LM. Phase II study of recombinant leukocyte A interferon (IFN-RA) plus cimetidine in disseminated malignant melanoma. *J Clin Oncol* 1992;**3**:977–81.
304. Sagaster P, Micksche M, Flamm J, Ludwig H. Randomised study using IFN-α versus IFN-α plus coumarin and cimetidine for treatment of advanced renal cell cancer. *Ann Oncol* 1995;**6**:999–1003.
305. Morris DL, Adams WJ. Cimetidine and colorectal cancer—old drug, new use? *Nat Med* 1995;**1**:1243–4.
306. Kelly MD, King J, Cherian M, Dwerryhouse SJ, Finlay IG, Adams WJ, et al. Randomized trial of preoperative cimetidine in patients with colorectal carcinoma with quantitative assessment of tumor-associated lymphocytes. *Cancer* 1999;**85**:1658–63.
307. Sasson AR, Gamagami R, An Z, Wang X, Moossa AR, Hoffman RM. Cimetidine: an inhibitor or promoter of tumor growth? *Int J Cancer* 1999;**81**:835–8.
308. Lefranc F, Yeaton P, Brotchi J, Kiss R. Cimetidine, an unexpected antitumor agent, and its potencial for the treatment of glioblastoma. *Int J Oncol* 2006;**28**(5):1021–30.

309. Niwa K, Onogi K, Wu Y, et al. Pronostic implications of cimetidine on advanced serous ovarian carcinoma related too cyclooxygenase 2 expression. *Mol Med Rep* 2008;**1**:119–22.
310. Gifford RRM, Tilberg AF. Histamine type-2 receptor antagonist immune modulation II. Cimetidine and ranitidine increase interleukin-2 production. *Surgery* 1987;**102**:242–7.
311. Gifford RR, Tilberg AF. Cimetidine reduces cyclosporine inhibition of interleukin-2 production. *J Surg Res* 1988;**45**(3):276–80.
312. Tzathas C, Koilakou S. H2 receptor antagonists and gastrointestinal cancer. *Ann Gastroenterol* 2004; **17**(3):237–8.
313. Takahashi HK, Watanabe T, Yokoyama A, Iwagaki H, Yoshino T, Tanaka N, et al. Cimetidine induces interleukin-18 production through H2-agonist activity in monocytes. *Mol Pharmacol* 2006;**70**(2):450–3.
314. Osband ME, Shen Y-J, Shlesinger M, Brown A, Hamilton D, Cohen E, et al. Successful tumor immunotherapy with cimetidine in mice. *Lancet* 1981;**1**:636–8.
315. Sahasrabudhe DM, McCune CS, L'Donnell RW, et al. Inhibition of suppressor T lymphocytes (Ts) by cimetidine. *J Immunol* 1987;**138**:2760–3.
316. Adams WJ, Morris DL, Ross WB, Lubowski DZ, King DW, Peters L. Cimetidine preserves non-specific immune function after colonic resection for cancer. *ANZ J Surg* 1994;**64**:847–52.
317. Lin CY, Bai DJ, Yuan HY, Wang K, Yang GL, Hu MB, et al. Perioperative cimetidine administration promotes peripheral blood lymphocytes and tumor infiltrating lymphocytes in patients with gastrointestinal cancer. Results of a randomized controlled clinical trial. *World J Gastroenterol* 2004; **10**(1):136–42.
318. Dejito B, Guoliang Y, Hongying Y, Yan L, Kun W, Hua S. Perioperative cimetidine application modulates natural killer cells in patients with colorectal cancer: a randomized clinical study. *J Tongji Med Univ* 1999;**19**(4):300–3.
319. Adams WJ, Morris DL. Pilot study: cimetidine enhances lymphocyte infiltration of human colorectal carcinoma. Results of a small randomized control trial. *Cancer* 1997;**80**:15–21.
320. Li B, Cao F, Zhu Q, Li B, Gan M, Wang D. Perioperative cimetidine administration improves systematic immune response and tumor infiltrating lymphocytes in patients with colorectal cancer. *Hepatogastroenterology* 2013;**60**(122):244–7.
321. Kubota T, Fujiwara H, Ueda Y, Itoh T, Yamashita T, Yoshimura T, et al. Cimetidine modulates the antigen presenting capacity of dendritic cells from colorectal cancer patients. *Br J Cancer* 2002;**86**:1257–61.
322. Kannagi R, Izawa M, Koike T, Miyazaki K, Kimura N. Carbohydrate-mediated cell adhesion in cancer metastasis and angiogenesis. Review article. *Cancer Sci* 2004;**95**(5):377–84.
323. Tang NH, Chen YL, Wang XQ, Li XJ, Fin FZ, Wang XZ. Cooperative inhibitory effects of antisense oligonucleotide of cell adhesion molecules and cimetidine on cancer cell adhesion. *World J Gastroenterol* 2004;**10**(1):62–6.
324. Kawase J, Ozawa S, Kobayashi K, Imaeda Y, Umemoto S, Matsumoto S, et al. Increase in E-selectin expression in umbilical vein endothelial cells by anticancer drugs and inhibition by cimetidine. *Oncol Rep* 2009;**22**(6):1293–7.
325. Liu FR, Jiang CG, Li YS, Li JB, Li F. Cimetidine inhibits the adhesion of gastric cancer cells expressing high levels of sialyl Lewis x in human vascular endothelial cells by blocking E-selectin expression. *Int J Mol Med* 2011;**27**(4):537–44.
326. Laferriere J, Houle F, Taher MM, Valerie K, Huot J. Transendothelial migration of colon carcinoma cells requires expression of E-selectin by endothelial cells and activation of stress-activated protein kinase-2 (SAPK/p38) in the tumour cells. *J Biol Chem* 2001;**276**:33762–72.
327. Fukuda M, Tanaka S, Suzuki S, Kusama K, Kaneko T, Sakashita H. Cimetidine induces apoptosis of human salivary gland tumor cells. *Oncol Rep* 2007;**17**(3):673–8.
328. Tozawa K, Okamoto T, Kawai N, Hashimoto Y, et al. Effects of anti-nuclear factor kB reagents in blocking adhesion of human cancer cells to vascular endothelial cells. *Cancer Res* 1995;**55**:4162–7.
329. Kobayashi K, Matsumoto S, Morishima T, Kawabe T, Okamoto T. Cimetidine inhibits cancer cell adhesion to endothelial cells and prevents metastasis by blocking E-selectin expresion. *Cancer Res* 2000;**60**:3978–84.

330. Fujikawa T, Shiraha H, Nakanishi Y, et al. Cimetidine inhibits epidermal growth factor-induced cell signaling. *J Gastroenterol Hepatol* 2007;**22**(3):436–43.
331. Tomita K, Izumi K, Okabe S. Roxatidine- and cimetidine-induced angiogenesis inhibition suppresses growth of colon cancer implants in syngeneic mice. *J Pharmacol Sci* 2003;**93**(3):321–30.
332. Chihara Y, Fujimoto K, Miyake M, Hiasa Y, Hirao Y. Anti-tumor effect of cimetidine via inhibiting angiogenesis factors in N-butyl-N-(4-hydroxybutyl) nitrosamine-induced mouse and rat bladder carcinogenesis. *Oncol Rep* 2009;**22**(1):23–8.
333. Orlowski J. Heterologous expression and functional properties of amiloride high affinity (NHE1) and low affinity (NHE3) isoforms of the rat Na/H exchanger. *J Biol Chem* 1993;**268**(22):16369–77.
334. Nemeth ZH, Deitch EA, Lu Q, Szabo C, Hasko G. NHE blockade inhibits chemokine production and NF-kB activation in immunostimulated endothelial cells. *Am J Physiol Cell Physiol* 2002; **283**:396–403.
335. Kulanthaivel P, Furesz TC, Moe AJ, Smith CH, Mahesh VB, Leibach FH, et al. Human placental syncyotiotrophoblast expresses two pharmacologically distinguishable types of Na = -H^+ exchangers, NHE-1 in the maternal- facing (brush border) membrane and NHE2 in the fetal-facing (basal) membrane. *Biochem J* 1992;**284**:33–8.
336. Wang D, Balkovetz DF, Warnock DG. Mutational analysis of transmembrane histidines in the amiloride-sensitive Na^+/H^+ exchanger. *Am J Physiol Cell Physiol* 1995;**269**(38):392–402.
337. MacLeod J, Hamilton RJ. Activation of Na^+/H^+ exchange is required for regulatory volume decrease after "physiological" volume increases in jejunal villus epithelial cells. *J Biol Chem* 1996; **271**(38):23138–45.
338. Guerra L, Di Sole F, Valenti G, Ronco PM, Perlino E. Polarizad distribution of Na^+/H^+ exchanger isoforms in rabbit collecting duct cells. *Kidney Int* 1998;**53**(5):1269–77.
339. Orlowsky J, Kandasamy RA. Delineation of transmembrane domains of the Na^+/H^+ exchanger that confer sensitivity to pharmacological antagonists. *J Biol Chem* 1996;**271**(33):19922–7.
340. Pantziarka P, Bouche G, Meheus L, Sukhatme V, Sukhatme VP. Repurposing drugs in oncology (ReDO)-cimetidine as an anti-cancer agent. *Ecancermedicalscience* 2014;**8**:485. PMC4268104.

CHAPTER 17

New horizons on pH regulators as cancer biomarkers and targets for pharmacological intervention

Fátima Baltazar[a,b,]*, Diana Valente[a,b,c,]*, Julieta Afonso[a,b,]*, Odília Queirós[c,]*, Sara Granja[a,b,]*
[a]Life and Health Sciences Research Institute (ICVS), School of Medicine, University of Minho, Braga, Portugal
[b]ICVS/3B's— PT Government Associate Laboratory, Braga /Guimarães, Portugal
[c]IINFACTS—Institute of Research and Advanced Training in Health Sciences and Technologies, Department of Sciences, University Institute of Health Sciences (IUCS), CESPU, CRL, Gandra, Portugal

Introduction

Addiction to glucose is a feature common to invasive cancer cells (see Chapter 2). These cells display high rates of glycolysis, associated with increased glucose uptake and production of copious amounts of lactic acid. This phenomenon, that occurs independently of the oxygen levels, was first described by Otto Warburg the 1920s, and is currently known as the "Warburg effect."[1,2] Lactic acid is removed from the cancer cell by monocarboxylate transporters (MCTs), leading to the acidification of the microenvironment, linked to aggressiveness: angiogenesis, invasion, metastasis, immune escape and resistance to therapy.[1] To cope with this metabolic phenotype, by avoiding intracellular acidification, cancer cells also rely on the activity of other pH regulators, which include carbonic anhydrases (CAs), Na^+/H^+ exchangers (NHEs), Vacuolar-type H^+-ATPases (V-ATPases) and anion exchangers (AEs).[3] MCTs are a family of 14 members and MCT1-4 mediate the transmembrane transport of monocarboxylates, including pyruvate and lactate. These isoforms are proton symporters and play a dual role in cancer metabolism: they support the glycolytic metabolism by removing lactate from the cancer cell, and they also function as pH regulators.[4] The CA family comprises 15 members and is responsible for converting carbon dioxide into carbonic acid, which is rapidly dissociated into H^+ and HCO_3^-, acidifying the tumor microenvironment (TME).[5] Due to their abnormal expression in tumors, CAIX and CAXII are known as cancer-related isoforms. The nine-member human NHE family consists of membrane-associated sodium-proton antiporters. Of these nine, NHE1 is the most important isoform involved in pH regulation in cancer.[6] ATPases are enzymes that couple the transport of ions across membranes with ATP synthesis or hydrolysis. Of all the ATPase

*Equal contribution

An Innovative Approach to Understanding and Treating Cancer: Targeting pH
https://doi.org/10.1016/B978-0-12-819059-3.00017-4
© 2020 Elsevier Inc.
All rights reserved.

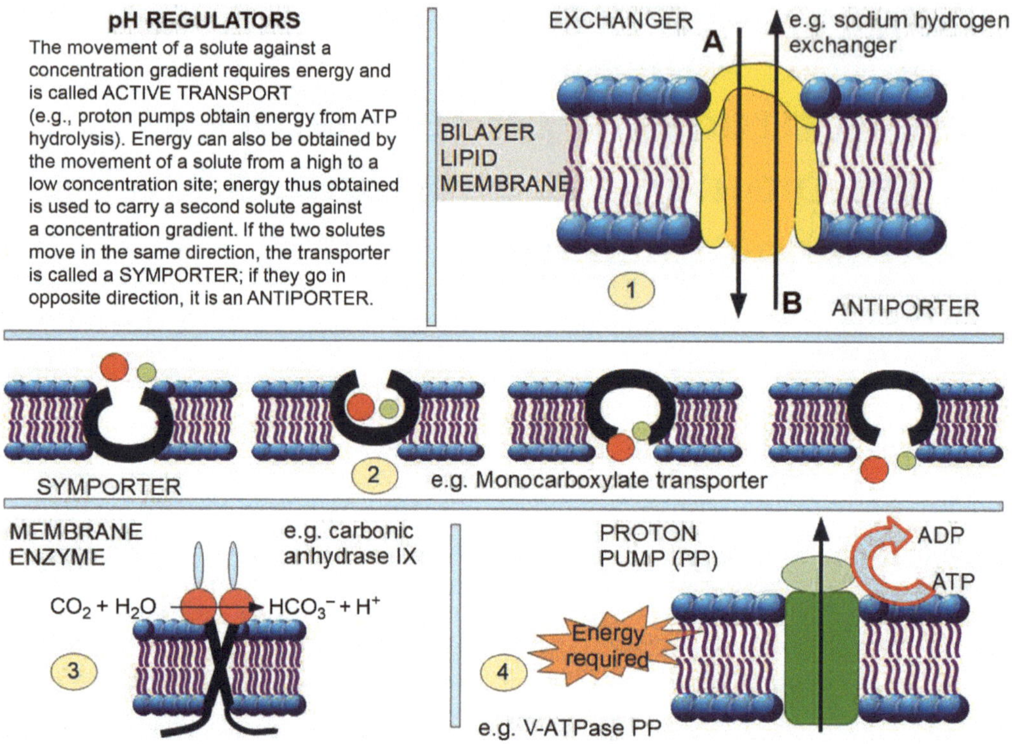

Fig. 1 pH Regulators occurring in cancer cells. Panel 1: the antiporter NHE1; Panel 2: MCTs co-transporting lactate and a proton; Panel 3: the membrane enzyme CAIX; Panel 4: the energy consuming proton pump V-ATPase. 1, 2 and 4 are examples of pH regulators that perform active transport.

members, V-ATPases are the most important in the cancer context, and they will be described below.[7,8] The AE family of transporters consists of 3 members, AE1-3, their main function is to exchange Cl^- for HCO_3^- across the plasma membrane.[9] However, the role of AEs in cancer has yet to be fully explored.

In this chapter we will focus on the role of these pH regulators in cancer (Fig. 1), as biomarkers of diagnosis and prognosis, and also on their exploitation as treatment targets (Table 1) and we will include current data from clinical studies (Fig. 2).

Monocarboxylate transporters

Monocarboxylate transporters (MCTs) comprise a family of proton-linked plasma membrane transporters that move monocarboxylates across cell membranes.[10] These transporter proteins are part of the larger SLC16 family of solute carriers, which has 14 members all, sharing similar topology. Their structure consists of 12 transmembrane (TM) α-helices domains with intracellular N- and C-termini and a large loop between

TM6 and 7.[11] However, only the first four isoforms (MCT1, MCT2, MCT3 and MCT4) are functionally confirmed to mediate the bidirectional proton-linked transport of monocarboxylates across the plasma membrane[10,11] (see also Chapter 4). Although lactate is the molecule that is transported by MCTs through the plasma membrane in highest quantities, MCTs also transport other metabolically important monocarboxylates such as pyruvate and ketone bodies. Therefore, MCTs play a central role in mammalian metabolism and are critical for the metabolic communication between cells.[12] Each isoform's expression pattern is correlated with its function, with the affinity for the substrate and with each tissue's metabolic requirements. MCT1 is ubiquitously expressed in normal tissues and due to its high affinity for the substrate, it is known to carry out both the influx and efflux. MCT2 is the isoform with the highest affinity for monocarboxylates. This isoform is mainly present in cells where rapid uptake of monocarboxylates is required. MCT3 is only expressed on the basal membrane of the retinal pigment epithelium (RPE) of the eye and on the choroid plexus epithelium, and MCT4 has lower affinity for most substrates. However, it has an important role in removing lactate from cells with high glycolytic activity.[10–12]

MCT1 and 4 are important carriers involved in the extrusion of lactate, in the maintenance of the hyperglycolytic phenotype and the metabolic symbiosis between cancer cells[4] (lactate shuttle). Thus, these will be the isoforms addressed in this chapter. However, the role of MCTs 1/4 is not confined to cell metabolism. MCTs also participate in the regulation of pH homeostasis by co-transporting monocarboxylate with a proton and by interacting with bicarbonate transport proteins and carbonic anhydrases.[13]

Expression and prognostic value

The most proliferative and aggressive tumor cells need huge amounts of glucose to meet their energy needs. This high glycolytic phenotype of cancer cells, translates into high production of lactate, which promotes extracellular acidosis and consequently an aggressive TME.[14] Lactate exported by MCT1/4 is essential for maintaining the hyperglycolytic acid-resistance phenotype of the TME.[4] MCT1 and MCT4 isoforms have been consistently described as proteins that are upregulated in cancer cells. Over the past years, our group and others have reported the expression of these transporters and their prognostic value in a wide range of human cancers.[15] MCT1 and/or MCT4 are up-regulated in several human cancers including lung,[16] breast,[17] head and neck squamous cell carcinomas (HNSCC),[18] renal cell carcinoma (RCC),[19] brain,[20] adrenocortical,[21] melanomas,[22] pancreatic,[23] colorectal,[24] ovarian,[25] cervical[26] and in testicular germ cell tumors.[27] However, the expression pattern of MCT1 and MCT4 varies due to the differences in tumor lactate content and how it is used by tumor cells. In this sense, while MCT4 is up-regulated, MCT1 was found to be down-regulated in prostate and liver cancer cells when compared with adjacent non-tumoral tissue.[28,29]

Moreover, the prognostic potential of MCT1 and MCT4 was found in various tumor types (for an extensive review see Refs. 3 and 15). MCT1 expression was associated with features of cancer aggressiveness in lymphomas[30] and testicular germ cell tumors,[27] and significantly correlated with reduced overall survival (OS) and/or reduced progression-free survival (PFS) in renal cell carcinoma (RCC)[31] and melanomas.[32]

The expression of MCT4 was associated with higher TMN stages in esophageal squamous cell carcinoma (ESCC),[33] urothelial carcinoma,[34] and testicular germ cell tumors and associated with shorter OS in urothelial carcinoma,[34] ESCC,[33] testicular germ cell tumors,[27] and RCC,[31] among others. However, a recent meta-analysis reported that only MCT4 expression was significantly associated with decreased OS and PFS across cancer types, whereas there was no clear association of MCT1 expression with OS or PFS.[35]

Exploitation as therapeutic targets

Due to the key role of MCTs in cancer development, it seems reasonable to consider these transporters as promising therapeutic targets in cancer. Encouraging results using in vitro and in vivo models showed the impact of inhibiting MCTs in different tumors. For example, genetic ablation of MCT1 and MCT4 has been found to reduce intracellular pH via intracellular lactate accumulation[23] and consequently tumor growth.[23,34,36–40] In the hypoxic TME, evidence points to a crucial role of MCT4 in the pH-regulating systems in vivo, promoting tumor progression.[41]

Several small molecules were found to efficiently inhibit MCTs. Quercetin is a flavonoid, and has been characterized as a specific but weak inhibitor of MCT1, like the toxic organomercurial *p*-chloromercuribenzenesulfonate.[42] Other inhibitors are the anion channel blockers, 4,4′-diisothiocyanatostilbene-2,2′-disulfonate (DIDS) and the widely studied α-cyano-4-hydroxycinnamate (CHC). Although all were reported as having anti-cancer effects[43,44] they are neither selective nor specific to the different MCT isoforms. Recently, Futagi et al., identified bindarit (2-[(1-benzyl-1H-indazol-3-yl)methoxy]-2-methylpropanoic acid) as the first selective inhibitor of MCT4 that worked in a non-competitive manner.[45] However further research on the biological effects in cancer cells is needed.

AstraZeneca has been doing interesting research on powerful MCT1-specific inhibitors with molecules such as AR-C117977[46] and AR-C155858.[47] These compounds, initially developed as immunosuppressive drugs, can increase intracellular lactate accumulation and impair in vivo tumor growth.[36] Recently, AZD-3965, an orally bioavailable MCT1 specific inhibitor, derived from AR-C155858, was reported to efficiently disrupt lactate export, reducing tumor growth and providing the rationale for clinical evaluation in cancer patients.[48,49] Curtis et al. explored its antitumor activity both as monotherapy and in combination with doxorubicin or rituximab. The work showed that

AZD-3965 caused lactate accumulation in Raji Burkitt's lymphoma model and decreased tumor growth.[50]

In the meantime, AstraZeneca succeeded in generating an MCT4 (AZ93) inhibitor that is selective and highly efficient in blocking growth of a wide range of cancer cells.[51] These compounds are powerful MCTs inhibitors, however, detailed information about possible side effects is lacking. Functional redundancy between MCT1 and MCT4, and tumor heterogeneity have been behind the lack of success in targeting only one isoform. Therefore, new approaches have to be developed to overcome these issues. Benjamin et al. described a reserpine derivative called syrosingopine, as a dual MCT1/4 inhibitor.[52] Using a mouse liver tumor model, they observed that syrosingopine-treated tumor cells sequester lactate.

Clinical studies

Although interest in targeting lactate transporters in cancer development has been widely described and studied in recent years, the inhibition of both MCT1 and MCT4 is still a challenge. AZD3965 is a potent and selective MCT1 inhibitor, developed by AstraZeneca, that went into a phase I clinical trial in the UK for patients with solid tumors such as prostate cancer and diffuse large-cell B lymphoma (clinicaltrials.gov/NCT01791595). There are still no results concerning the efficacy of this treatment.[53]

Carbonic anhydrases

Acidity in the TME is a hallmark of malignancy, resulting from the accumulation of protons in the extracellular milieu. Although acidosis in the tumor could be primarily attributed to increased lactic acid export to maintain tumor glycolytic phenotype, we cannot neglect the important contribution of respiration-derived CO_2 for TME acidosis.[54]

Carbonic anhydrases (CA) are among the most widely-studied proton flux regulators. These enzymes are metalloenzymes that catalyze the reversible hydration of carbon dioxide to bicarbonate and a proton ($CO_2 + H_2O \rightarrow HCO_3^- + H^+$), being key pH regulators[55] (see also Chapter 7). To date, 15 isoforms have been identified, which differ in sequence but share the same 3D tertiary structure. According to their location in the cell, CA enzymes in mammals are divided into four groups: the cytosolic (CA I-III, CAVII), mitochondrial (CAV), secreted (CAVI) and membrane associated (CAIV, CAIX, CAXII and CAXIV).[56] The different isoforms have different tissue distribution, cellular location and function (for an extensive review see Refs. 5,55,57). CAIX and CAXII are transmembrane proteins that contain an extracellular catalytic domain responsible for regulating the pH of the TME.[5] Given the established role of CAIX and CAXII in tumorigenesis, cancer cell signaling, tumor progression and acidification, these are the CA isoforms which will be detailed below.

Expression and prognostic value

CAs, and specifically CAIX and CAXII, are widely expressed in human malignancies. CAIX is a tumor-associated protein since it is rarely detected under normal physiological conditions. Robustly regulated by HIF-1α, high levels of CAIX expression have been described in several human cancers such as breast,[58,59] colon/rectum,[60,61] HNSCC,[18,62] lung,[63] brain,[64] bladder,[65] ovarian,[66] and RCC,[67] (for reviews see Refs. 18 and 55). The peculiar differential expression pattern makes CAIX a strong tumor hypoxia biomarker that could be useful in the clinic. A study conducted in human breast cancer provided definitive evidence of CAIX as an independent poor prognosis biomarker for distant metastases and survival.[68] This was also confirmed in pancreatic ductal adenocarcinoma[69] and hepatocellular carcinoma.[70] Similarly, in pancreatic neuroendocrine neoplasms, CAIX was correlated with high-grade staging, distant metastases, perineural invasion, and other malignant parameters. Importantly, patients with high CAIX expression had a shorter recurrence-free survival.[69] In contrast, a recent meta-analysis in RCC concluded that high CAIX expression was associated with better OS.[71] For a comprehensive review of the prognostic value of CAIX see Refs. 72 and 73. The predictive value of CAIX for clinical decision-making has also been investigated. Whereas in rectal adenocarcinomas the presence of CAIX indicates preoperative chemotherapy response,[74] in metastatic clear cell RCC it has been shown to be a predictive biomarker for response to sorafenib.[75]

Like CAIX, CAXII expression has also been found to be up-regulated in many tumor types including RCC,[76] colorectal,[60] breast,[58] bladder,[5] lung,[77] and others.[5,60,78] This correlation with tumor tissues could indicate CAXII as potential cancer biomarker. Indeed, as opposed to what has been found for CAIX, CAXII has been described as a marker of "good prognosis" in some cancers, including lung,[77,79] cervical,[80] and breast.[5] However, CAXII up-regulation has been associated with poor prognosis, in colorectal tumors, oral squamous carcinoma,[81] and some renal cancers.[76]

Exploitation as therapeutic targets

A growing body of evidence is supporting that inhibiting CAs has potent antiproliferative and anti-metastatic effects. An ever-increasing amount of experimental data recognizes that inhibition of CAIX and CAXII by gene knockdown has a strong impact on both extracellular and intracellular pH and consequently on tumor aggressiveness.[82] Silencing CAIX and/or CAXII reduces tumor growth,[83–86] promotes radio-sensitivity[84] and inhibits metastasis.[85]

Inhibitors that specifically target the tumor-associated isoforms CAIX and CAXII were developed and tested for their antitumor activity in recent years. One strategy is to use small molecule inhibitors that modulate and target the catalytic function of CAIX and CAXII. For instance, acetazolamide derivatives bind to CAIX and CAXII present in tumor cells and efficiently reduced tumor growth in a RCC xenograft mouse model.[87] Also, compounds belonging to the sulfamate class decreased tumor cell migration and spreading in vitro and reduced the metastatic tumor burden[88] in an orthotopic mouse model of breast carcinoma. The novel ureido-sulfonamide CAIX inhibitor (U104) depletes cancer stem cells and sensitizes breast tumor cells to the chemotherapeutic agent paclitaxel, leading to a delay in tumor growth and eradication of lung metastases.[85,89] This compound, also known as SLC-0111, was effective against glioblastoma xenografts and reversed tumor growth in combination with temozolomide,[90] and is being tested in phase I clinical trials.

Cancer-related distribution of extracellular CAIX and CAXII allows the use of antibodies to hit cancer cells. This second strategy is receiving considerable attention and efforts are underway to develop specific and efficient CAIX-CAXII-specific antibodies. In this sense, Zaticova et al. described a monoclonal mouse antibody VII/20 directed to the catalytic domain of CAIX,[91] 6A10, the first CAXII-specific inhibitory monoclonal antibody, that significantly delayed tumor outgrowth.[92,93] Petrul et al. reported that a potent and selective internalizing human antibody directed against CAIX, BAY 79-4620, decreased tumor growth in mice xenografted with several different human tumors.[94]

Interestingly, the tyrosine kinase inhibitors imatinib and nilotinib have been found to reduce the activity of CAIX,[95] however, further studies are needed to understand the mechanisms involved.

Clinical studies

Among the high number of small molecule inhibitors reported to date, only SLC-0111 (also known as WBI-5111), a selective CAIX inhibitor, proceeded to clinical trials. SLC-0111 entered Phase I clinical trials for the treatment of advanced, metastatic solid tumors in 2014 (clinicaltrials.gov/NCT02215850). Although the results of the Phase I clinical trial have not yet been published, the compound is scheduled for Phase II trials.[55]

Many clinical studies have been conducted on antibody-based therapies. The isolation of the monoclonal antibody G250, that specifically recognizes CAIX, led to the initiation of its clinical evaluation in patients with RCC.[96] Clinical studies confirmed the promising role of G250 as a diagnostic and therapeutic tool[55,97] (reviewed in Ref. 89). Also known as RENCAREX or GIRENTUXIMAB B, this was the first anti-CAIX monoclonal antibody used in a phase I/II clinical trial in mRCC patients. The treatment was shown to be safe and well tolerated in all the trials and, importantly, led to stable disease.[98] The ARISER phase III clinical trial (clinicaltrials.gov/NCT00087022) investigated Girentuximab as monotherapy for patients who have undergone surgery for primary non-metastatic kidney cancer. Although the treatment was safe and well tolerated, Girentuximab showed no clinical benefit as an adjuvant treatment.[99] Despite the potential of G250 to target CAIX, several clinical trials have failed, with G250 showing no benefit. In vitro studies demonstrated that the cytokine interferon gamma (IFN-γ) increased cG250-mediated cellular toxicity, suggesting a promising combination therapy for RCC.[100] In this line, a multicenter phase I/II trial combining radio-labeled cG250 with IFN-α 2a showed that patients with high CAIX-expressing tumors have a prolonged PFS of about 22 months.[98] cG250 was also tested as an imaging agent for diagnosing RCC. A phase II/III clinical trial to show the impact of the Zr-89-girentuximab in treatment is currently recruiting participants (clinicaltrials.gov/NCT02883153).

CAIX represents an ideal target for anti-tumor therapy and for imaging hypoxic solid tumors.

Sodium proton exchangers

As we know, sodium proton exchangers (NHEs) are membrane proteins involved in the maintenance of the tumor's inverted pH gradient. There are currently thirteen evolutionarily conserved NHE isoforms known in mammals, divided into three subgroups of the SLC9 (solute carrier family 9) gene family, and their activity consists in the counter-transport of Na^+ and H^+ across membranes.[6,101] NHE1 is one of the most important players; its protein structure comprises two domains: a hydrophobic N-terminal membrane domain, that exchanges intracellular H^+ with extracellular Na^+, and the cytoplasmic domain (C-terminus tail), essential for protein regulation. NHE1 is responsible for different cell functions, such as regulating intracellular pH, acid-base electrolyte control and cell migration.[102] Different studies have reported the involvement of NHE1 in pH regulation in many cancer models.[6,103,104] Both expression and activity of NHE1 are upregulated in cancer cells, increasing the efflux of protons towards the TME and contributing to the inverted pH gradient.[105,106] Additionally, NHE1 cooperates with bicarbonate-transporting systems such as, Na^+-HCO_3^- co-transporters, Na^+-dependent HCO_3^-/Cl^- exchangers and Cl^-/HCO_3^- exchangers.[107]

Expression and prognostic value

Many studies have shown the involvement of NHE1 in different types of cancers and distinct patterns of neoplastic progression.[108–110] NHE1 expression and activity have been associated with cell proliferation, motility, resistance to chemotherapy, and tumor growth, as well as with spontaneous regression of cancer in vivo.[111–113]

NHE1 was found to be highly expressed in metastatic human HNSCC cells. Cervical cancer samples expressed NHE1 protein in different amounts and the differential expression of NHE1 was associated with clinical outcomes.[106] Increased levels of NHE1, either at the mRNA or protein level, were found in hepatocellular carcinoma tissues when compared to normal counterparts. NHE1 expression was associated with increased tumor size, venous invasion and advanced tumor stages.[114] High levels of NHE1 protein were also described in primary human glioma cells, glioma xenografts and glioblastoma.[115] Ovarian cancer cells present high expression of NHE1 compared with normal ovarian cells, making this expression a negative prognostic factor.[116] However, different results were obtained in ESCC, where NHE1 expression was higher in well-differentiated tumors.[117] In addition, 5-year survival rates were poorer in the low NHE1 expression group and multivariate analyses showed that low NHE1 expression was associated with shorter postoperative survival.

Exploitation as therapeutic targets

Different pharmacological compounds can be used to target NHE1. The most common are pyrazine derivatives: amiloride and analogues, including ethylisopropylamiloride (EIPA), hexamethylamiloride (HMA) and dimethylamiloride (DMA),[102] and the benzoylguanidines cariporide and eniporide.[118]

Amiloride, an old potassium-sparing diuretic, was reported to decrease VEGF expression, the activity of urokinase-type plasminogen activator (μPA), metalloproteinases and other proteases involved in invasion and metastasis of breast cancer cells.[119] Furthermore, inhibition of NHE1 by amiloride was shown to have cytostatic effects in gliomas,[120] hepatocellular carcinoma[121] and breast cancer cells.[122] Recently, Rojas et al. showed the powerful anti-cancer effect of amiloride against myeloma cells in vitro and in a xenograft mouse model. Moreover, amiloride had a synergistic effect when combined with conventional drugs, such as dexamethasone, pomalidomide and guggulsterone, without significant toxicity to normal tissues in mice.[123,124] Another NHE1 inhibitor, zoniporide, that belongs to the amiloride family, reduced ovarian cancer cell proliferation.[116]

New amiloride analogues (EIPA, HMA and DMA), synthesized over the past few years, showed higher activity. It has been demonstrated that EIPA, a specific NHE1 inhibitor, is 200-times more potent than amiloride. EIPA prevented cancer cell proliferation and migration, reduced invasion and promoted apoptosis in a hepatocellular carcinoma model.[125] It has also been found that this compound acts synergically with doxorubicin in colon cancer cells.[126] HMA was reported to alter the inverted pH gradient in cancer cells, by decreasing pHi and increasing pHe, inducing cell death in leukemic cells[125] and to decrease hepatocarcinoma cell growth and viability.[127] DMA inhibits proliferation and induces apoptosis of acute promyelocytic leukemia cells through intracellular acidification.[128] Co-administration of paclitaxel with DMA showed synergistic effects in breast cancer cells.[129]

Regarding the benzoylguanidine class that includes cariporide, there are only few reports describing benefits in cancer.[118] This compound was tested in a cardiological setting and never entered clinical practice due to some reported cerebrovascular side effects,[118] and lack of benefits in ischemic heart disease.[130] In cancer, cariporide has been able to reduce tumor cell invasion in human tongue squamous cell carcinoma[131] and in melanoma cells.[110] Cariporide was used to induce intracellular acidification in glioblastoma allowing its localization by NMR. Cariporide showed that it can induce cell death and cancer regression through intracellular acidification which is toxic for tumor cells.[132]

More recently, other drugs targeting NHE1 were developed with more promising results, such as SL-591227, Phx-3, and compound 9t. For instance, Phx-3 was shown to promote apoptosis in different cancer cells, such as gastric cancer cells,[133] breast cancer, epidermoid carcinoma, lung adenocarcinoma, pancreatic cancer, RCC, colon adenocarcinoma and glioblastoma.[134] As to compound 9t, there are no studies either in vitro or in vivo and attempts to use it in clinical studies have been unsuccessful so far.[118]

In HNSCC cells, NHE1 knockdown led to reduction in migration and invasion and suppressed metastasis in vivo.[104] Flinck et al. demonstrated the role of NHE1 and HCO_3^- transporter NBCn1 on breast cancer cell cycle progression.[135] Knockdown of either NHE1 or NBCn1 reduced proliferation, increased cell cycle progression through S phase prolongation and delayed G2/M transition. A recent study showed that the combination of sonodynamic therapy and temozolomide decreases NHE1 expression which is involved in glioblastoma cell escape to Temozolomide-mediated toxicity. Thus, the combined treatment decreased glioma cell migration and increased the production of apoptotic proteins.[136]

Clinical studies

The only non-amiloride based compounds with NHE1 inhibitory activity that have undergone clinical trials are cariporide and eniporide. However, those trials were not in the cancer field but in a cardiology setting aimed at treating ischemic-reperfusion injury.[118,137–140] A recent phase 4 clinical trial studied the affect of amiloride in reducing urinary protein in patients with chronic kidney disease (NCT03170336). The results were disappointing, however, since the clinical trials were only conducted on patients with cardiac or kidney diseases, and the side effects may be related to those conditions. Therefore, studies of NHE1 inhibition in cancer patients with healthy hearts remain to be undertaken.[141]

Vacuolar-type H^+-ATPases

ATPases are a group of enzymes that either synthesize or hydrolyze ATP. A gradient of protons across the mitochondrial membranes is the driving force for the ATP synthesis process, whereas the hydrolysis is coupled to proton transport across intracellular or plasma membranes. Many different tissues express ATPases, which are involved in producing energy or in maintening cellular pH.[8,142] There are different types of ATPases, functionally classified as F-ATPases (primarily used in ATP synthesis), vacuolar-ATPases (V-ATPases), A-ATPases (from "Archaea"), P-ATPases (phosphorylated-ATPases) and

E-ATPases (extracellular-ATPases).[143] Of these, V-ATPase are the ones that have been more associated with cancer cells, i.e., involved in pH regulation.

V-ATPases are ubiquitous multimeric proteins consisting of several subunits organized in two functional domains: the peripheral V1 and the integral V0[144,145] (see Chapter 8). The V_1 domain (with eight different subunits: A, B, C, D, E, F, G and H) catalyzes ATP hydrolysis, whereas the V_0 domain (with five subunits: a,c,c″,d,e) is responsible for proton transport.[7] V-ATPase assembly can be regulated by many stimuli including glucose concentrations,[146] amino acid starvation,[147] dendritic cell maturation[148,149] and also during viral infection.[150]

Although initially described in a vacuolar system, V-ATPases are also present in the plasma membrane of specialized cells, such as osteoclasts, epididymal clear cells and renal epithelial intercalated cells, acting in physiological pH regulation.[151] V-ATPases are responsible for the acidification of intracellular compartments such as endosomes, lysosomes and secretory vesicles,[152–154] and thereby influence many processes associated with these organelles, including vesicular trafficking, endocytosis, autophagy, signaling pathways (e.g., NOTCH), receptor recycling, and protein degradation.[155]

In cancer cells, it has been reported that V-ATPases are found mainly at the plasma membrane, contributing to cell differentiation, angiogenesis, metastasis and resistance to treatment.[156] V-ATPases are the main proteins responsible for the efflux of protons from cancer cells to the extracellular milieu[157] and their expression is associated with metastatic processes in ovarian,[156] breast,[158,159] oral,[160] hepatocellular,[161] pancreatic,[162] and prostate cancers.[163] Furthermore, it was also reported that V-ATPase can be involved in vesicular trafficking, and in regulating other signaling pathways important for tumorigenesis such as Wnt, Notch and mTOR.[164,165]

Expression and prognostic value

A large number of studies show altered expression or subcellular localization of V-ATPases in many types of cancer. V-ATPases are overexpressed in human malignancies including pancreatic ductal adenocarcinoma,[162] ESCC,[164] oral squamous cell carcinoma,[166] hepatocellular carcinomas,[167] non-small cell lung carcinomas (NSCLC),[168] glioblastomas,[165] ovarian,[169] gastric,[170] and breast cancer.[158] In some of these tumors, V-ATPase over-expression is correlated with increased ability to invade the adjacent tissues.[171] For example, in breast, prostate, lung, ovarian, liver, pancreatic, melanoma and esophageal cancers, V-ATPase is localized on the leading edge of invasive cells.[172] Specifically, in ovarian cancer subunit a2 of the V1 domain is highly expressed in these cells and associated with metalloproteinase 9 activity.[169]

Moreover, the presence of V-ATPases in some types of cancer has been associated with different prognostic variables. For instance, in ESCC, V-ATPases are associated with lower disease-free survival, tumor invasion and lymph node metastasis.[164] Another study demonstrated that shortened disease free-survival in ESCC was associated with increased expression of V-ATPases, specifically subunit V1E1. Furthermore, a correlation has been also found between the expression of this subunit and the enzyme PKM2 (pyruvate kinase isoform M2), involved in the glycolytic phenotype, characteristic of these tumor types.[164] In glioma, V-ATPases are associated with shorter overall survival and high-grade glioblastoma tumors,[165] and in NSCLC their expression was correlated with the histopathological type (adenocarcinoma) and higher grade.[168] In cervical cancer, higher V-ATPase expression in adenocarcinomas was associated with shorter PFS and OS.[173] Similarly, in pancreatic cancer, higher expression of V-ATPase is associated with a higher cancer stage.[162] In acute myeloid leukemia, V-ATPase expression can vary among patients, but increased expression and activity are associated with leukemogenesis and development of resistance to therapy.[174] Furthermore, Liu et al. found that increased V-ATPase expression in gastric cancer tissues was associated with high histological grade, lymph node metastasis and vascular invasion.[175]

Exploitation as therapeutic targets

Different reports showed the involvement of V-ATPase over-expression in the treatment of resistant cancers. This can be due to both its expression at the plasma membrane, where it is responsible for the efflux of protons, or to the ion-trapping model.[153] In addition to being expressed at the plasma membrane, and thus involved in the acidic extracellular pH, V-ATPases are also present in intracellular compartments like lysosomes or endosomes, leading to their acidification and promoting entrapment of weak base antitumor drugs in these organelles, since they become protonated (ionized) in more acidic spaces.[176] This fact induces an alteration in drug distribution and accumulation, decreasing their therapeutic effects.[177] Thus, V-ATPase expression is associated with chemoresistance, but also with other aggressive features of cancer, namely invasion and metastasis.[155] For this reason, inhibition of this pH regulator is a promising new strategy for targeting tumor cells.

Plecomacrolide antibiotics, including bafilomycins and concanamycins are the most common V-ATPase inhibitors.[157,178,179] For example, Licon-Munoz *et al.* found an association between V-ATPase and androgen receptor activity in prostate cancer.[180] Inhibition of V-ATPase using concanamycin A, induced a decrease in androgen receptor expression linked to a defective endo-lysosomal pH regulation in prostate cancer cells. It was also observed that such inhibition decreased tumor growth and proliferation. A decrease in cancer proliferation through V-ATPase inhibition was also observed in hepatoblastoma, but in this case using bafilomycin.[181] Archazolid, another V-ATPase inhibitor, induced cell death in breast carcinoma cells.[182] Pharmacologic inhibition of V-ATPase by this compound interfered with epithelial-mesenchymal transition and inhibited tumor-initiating cell formation, decreasing cell motility and the number of mammospheres.[183] It also caused cellular depletion of free cholesterol in hepatocellular carcinoma cells, a strategy that reduced malignant proliferation in vitro and in vivo.[184] Similar effects were found using these compounds in several kinds of cancer cell lines.

Studies with in vivo models have also shown the potential antitumor effect of V-ATPase inhibitors. For instance, the use of archazolid decreased growth and increased cell death in acute lymphoblastic leukemia cell lines obtained with a xenograft mouse model, injected with primary leukemic cells obtained from patients.[185] In a mouse breast cancer xenograft model, archazolid treatment inhibited the development of lung metastasis, without apparent side effects.[186,187] In other xenograft mouse breast cancer (MCF-7 and MDA-MB-231) models, bafilomycin reduced tumor volume by 50%[188] whereas its use induced growth arrest in a hepatocarcinoma xenograft model in nude mice.[167] Treatment of a 4T1 breast cancer mouse model with archazolid, significantly reduced tumor growth, concomitantly increasing caspase-3 activity.[189] In a mouse model xenografted with the ductal breast adenocarcinoma JIMT-1 cell line, archazolid treatment decreased expression of the HER2 oncogene that accumulated in dysfunctional endosomal vesicles and was associated with reduced tumor growth. Therefore, it has been suggested as a potential treatment for trastuzumab resistant Her2 positive breast cancer.[190] In a JJN3 myeloma mouse xenograft, bafilomycin inhibited bone acidification and reduced bone pain.[191]

Several reports suggest that genetic manipulation of V-ATPases can be also a strategy for blocking cancer cell proliferation. The subunit c1 of the V_1 domain is essential for the assembly and activity of V-ATPase; the knockdown of this subunit reduced tumor growth, lung, liver and bone metastases and increased survival in a xenografted mouse 4T1 breast cancer model.[192] In a similar way, silencing the isoform a3 of the V_0 domain in a B16-F10 melanoma mouse model led to decreased lung and bone metastases. Likewise, the use of the V-ATPase a3 specific inhibitor FR167356 in the same model, decreased MMP9 expression, bone metastases and tumor size.[193] In a hepatocellular carcinoma mouse model, knockdown of V-ATPase's subunit ATP6L significantly reduced tumor size and metastatic capacity[161] (this subunit provides protons through a hydrophilic path across the membrane).

More recently, a new class of compounds, known as proton pump inhibitors (PPIs), traditionally used to treat acid-related gastro-intestinal disorders, showed excellent results as ATPase inhibitors.[194] It is important to note that PPIs are weak bases, and consequently accumulate in acidic environments, like the TME. For this reason, PPIs can selectively target cancer cells.[195,196] PPIs irreversibly inhibit H^+/K^+-ATPases in the gastric epithelial parietal cells.[197] Due to the similarity of both ATPases (V-ATPase and H^+/K^+-ATPase), some reports also showed the inhibition effect of PPIs in V-ATPases.[141] Indeed, there are some studies that demonstrated the efficacy of these compounds in different types of cancer. For instance, the PPIs omeprazole or pantoprazole decreased cell migration and increased cell death in gastric cancer,[198] melanoma cells[199] and in breast cancer[200]; omeprazole pretreatment improved conventional chemotherapy efficacy in a chemoresistant ovarian mouse model.[201] Furthermore, in the study of melanoma cells xenografted mice, esomeprazole induced a reversal of the inverted pH gradient[199] (see also Chapter 20).

PPIs are able to increase drug cytotoxicity due to the involvement of acidic pH and its relation with treatment resistance.[202,203] It has been shown that pre-treatment with omeprazole, before cisplatin and 5-FU treatment, induced a synergistic drug effect in esophageal cancer cells, but also on tumor growth in orthotopic and patient-derived xenograft mouse models.[204] In the same line, Azzarito et al. also demonstrated the synergistic effect of pretreatment with lansoprazole followed by paclitaxel in vitro and in vivo melanoma models.[205]

Other studies have proposed an association of V-ATPases and other proteins involved in drug resistance, namely ABC-transporters, such as glycoprotein P (Pgp) and multidrug resistance protein 1 (MRP1). Therefore, these compounds can overcome resistance to treatment by inhibiting different proteins and could become an important strategy with minor collateral effects for normal tissues.[206]

Different synthetic and natural plant compounds have been considered as potential anti- neoplastic drugs and have increasingly been used alone or in combination with conventional chemotherapeutics.[207,208] Recently, myrtenal, an essential oil found in plants (Artemisia, Coriandrum, Eucalyptus, etc.), showed antitumor activity against melanoma cells through V-ATPase inhibition, decreasing the migration and invasion abilities in melanomas cell lines and the number of metastases in mice.[209]

Clinical studies

There are few clinical trials regarding the use of V-ATPase inhibitors.

A Phase II Trial using omeprazole with the chemotherapeutic drugs carboplatin 5-fluorouracil, paclitaxel and pemetrexed, aimed to evaluate whether this combination could improve progression free survival in patients with HNSCC. The estimated primary completion date was December 2016, but no results have been published as yet (NCT02013453). Another phase II study completed in February 2012, aimed to evaluate the impact and compatibility of high-dose PPI esomeprazole treatment combined with chemotherapy in metastatic breast cancer (NCT01069081). The results showed that PPI improved the effect of conventional drugs. Finally, the combination of pantoprazole and docetaxel in metastatic castration-resistant prostate cancer patients was evaluated (NCT01748500). No results were yet published.

More recently, in an Icelandic population-based case-control study, with cases of breast cancer, prostate cancer, and malignant melanoma, the authors studied different levels of PPI exposure, and they reported that their results do not support the chemo-preventive effect of PPIs for the three cancer types.[210]

Anion exchangers

Na^+-independent anion (Cl^-/HCO_3^-) exchangers (AEs), AE1, AE2 and AE3, are part of the *Solute carrier 4* (*Slc4*) family of ten genes that includes five Na^+-coupled bicarbonate transporters (NBCe1, NBCe2, NBCn1, NBCn2 and NDCBE) and two other unusual members (AE4, BTR1). SLC4 proteins, also known as the bicarbonate-transporter (BT) family, are widely expressed N-glycosylated integral membrane proteins (see Chapter 9). They are important regulators of acid-base homeostasis by moving HCO_3^- (or CO_3^{2-}) to and of the cells. Coupled to their functions in pH regulation, BTs also promote the transepithelial movement of other electrolytes (e.g., Na^+ and Cl^-), contributing to central physiological processes like reabsorption of HCO_3^- in the kidney, CO_2 carriage facilitation (by chloride shift) in erythrocytes, modulation of digestive function and neuronal excitability, regulation of cellular volume, etc. Several pathological conditions like erythrocyte's disorders, diseases of the central nervous system, renal tubular acidosis and cancer are associated with dysfunctions of SLC4 proteins (reviewed in Refs. 9 and 211).

AE1 (SLC4A1), AE2 (SLC4A2) and AE3 (SLC4A3) are acid loaders, usually mediating electroneutral transmembrane exchange of one HCO_3^- anion by one Cl^- anion; reports exist on the sulfate/H^+ and oxalate/H^+ co-transport in exchange for Cl^- by AE1.[212,213] These proteins share 53–56% amino-acid composition. encompassing an N-terminal cytoplasmic domain (NCD), a transmembrane domain (TD) consisting of 10–14 segments, and a C-terminal cytoplasmic domain (CCD). The NCD acts as an anchorage site for multiple cytoskeleton proteins, some glycolytic enzymes and hemoglobin.[214,215] NCT is not necessary for the transport activity of AEs, being the TD responsible for that function, namely the formation of the ion translocation pore.[216] The CCD is essential for the functional expression and membrane trafficking of AEs, and for binding of protein partners, namely carbonic anhydrase II,[217] GADPH,[218] and p16.[219] Similarly to most SLC4 members, AEs are functionally inhibited by 4,4′-diisothiocyanatostilbene-2,2′-disulfonate (DIDS) (reviewed in Ref. 220).

AE1 was the first member of the SLC4 family to be cloned[221,222]; through its two promoters, SLC4A1 regulates the expression of the erythrocyte variant eAE1[221,222] and the kidney variant kAE1[223]; these two variants differ in their N-terminal cytoplasmic domain. In erythrocytes, AE1 accounts for 25–30% of the total transmembrane proteins, playing a crucial role in CO_2 carriage from systemic tissues to the respiratory alveoli by modulating the Jacobs-Stewart cycle.[224] In the kidney, AE1 promotes HCO_3^- reabsorption from the nephron tubules to the blood, with consequent urinary excretion.[225]

AE2 has three alternative promoters, originating three variants (AE2a, AE2b1, AE2b2) that are widely distributed in the basolateral membranes of most epithelial cells, where they are extremely sensitive to pH alterations.[226] AE3 has two alternative promoters, and their variants (bAE3 and cAE3) are mainly found in excitable tissues, like brain, retina, heart and smooth muscle, where it is stimulated, similarly to AE2, by rapid increases in intracellular pH.[227]

Expression and prognostic value

Tissue distribution and the physiological functions of AEs are largely known, but characterization of their expression levels and prognostic significance in malignancy is still in its infancy. Although AE1's expression in the normal kidney, is restricted to the collecting duct-intercalated cells of the nephron, it was detected in metanephric-derived cells of oncocytomas. It was not found in RCC or nephroblastoma. This led Bonsib et al. to suggest it as a specific oncocytoma marker.[228] AE1 expression was also observed in erythroleukemia cells,[229] in Epstein Barr Virus-transformed lymphocytes[230] and in high grade breast tumors.[231] There are reports of AE1 protein dysfunction in chronic myelogenous leukemia (CML).[232] In transformed erythroblasts, the v-erbA oncogene was described as a constitutive repressor of AE1 expression.[233] It has been suggested that increased erythrophagocytosis in CML is associated with AE1 clustering.[234] Chronic lymphocytic leukemia activated cells were able to bind, arrest and present AE1 to T cells, a process that may mediate the auto-aggression of erythrocytes in autoimmune hemolytic anemia.[235] In ESCC, an elevated cytoplasmic and membrane expression of AE1 was found in cancer tissues, that correlated to increased tumor stage, lack of differentiation and poor prognosis.[236] Similarly, AE1 expression was significantly associated with features of gastric cancer aggressiveness, such as increased tumor size, invasion and presence of non-lymphatic metastasis, additionally predicting a worse outcome.[237] Shen et al. reported AE1 expression in gastric and colon carcinoma tissues, concomitantly with p16 expression, describing a p16 sequestering mechanism at the cytoplasm by AE1, blocking p16 passage into the nucleus and, consequently, hampering its tumor-suppressive function.[238] This direct interaction had been previously described by Fu et al. in K562 cells.[239] Interestingly, Liu et al. reported that co-localization of p16 and AE1 in gastric cancer cells is a factor of good prognosis, indicating absence of lymph node metastasis. Recently, AE1 expression was identified as the causal mechanism of secondary autoimmune hemolytic anemia following colon adenocarcinoma.[240]

In the study by Yang et al., parietal cell-related expression of AE2 was found in normal human gastric tissues, being down-regulated in malignant lesions involving the body of the stomach; a weak diffuse profile was observed in the antrum.[241] Similarly, a single nucleotide polymorphism in AE2 gene was correlated with a reduction in the risk of developing bladder cancer.[242] On the other hand, a strong expression of AE2 was observed in Von Hippel-Lindau disease-associated clear-cell RCC, being described as a VHL target gene.[243] Over-expressed AE2 was also reported in poorly-differentiated hepatocellular carcinoma cells and tumor tissues.[244] Increased AE2 expression correlated with the concomitant expression of Ki67, and predicted a worse prognosis in colon cancer. EGR1 (early growth response protein 1) mediated AE2 up-regulation which, in turn, sequestered p16 in the cytoplasm of cancer cells—similarly to what has been described in the study by Shen et al.—a process that was inhibited by gastrin.[245] Later, it was shown that this EGR1-mediated AE2 up-regulation occurs in a cholecystokinin B receptor-dependent manner.[246] In ovarian cancer tissues, AE2 was found to be up-regulated when compared to adjacent normal tissue, and it was associated with a worse prognosis.[247] In ESCC, cytoplasmic or membrane expression of AE2 was observed, and its presence at the invasive front correlated with tumor stage and a worse prognosis.[248]

To the best of our knowledge, no studies on AE3 expression and prognostic value in cancer patients have been reported so far.

Exploitation as therapeutic targets

AEs have been poorly explored as therapeutic targets for cancer patients. Even so, a few interesting results have been reported to date. Tian et al. showed, in in vitro experiments, that degradation of AE1 in gastric cancer cells was due to p16 suppression by gastrin, and that disruption of either AE1 or p16 significantly impaired cellular proliferation. In a gastric cancer mouse model, administration of omeprazole induced systemic hypergastrinemia, but diminished AE1 and p16 expression and, consequently, tumor growth. This led the authors to hypothesize that moderate hypergastrinemia due to a slight increase in gastrin plasma levels may be beneficial for gastric cancer patients.[249] Another study suggested that miR-24 modulates AE1 expression through binding to the 3′UTR of AE1 mRNA. When transfecting miR-24 into gastric cancer cells, the authors showed a return of p16 into the nucleus by reduction of AE1 expression and loss of AE1-mediated p16 sequestration in the cytoplasm, with a consequent decrease in cell proliferation.[250] Suo et al. found that using two gastric cancer mouse models, targeting AE1 with siRNAs suppressed tumor growth,[251] similarly to what Tian et al. described.[249] The use of a miR-24 inhibitor led to the up-regulation of a functional AE1 in K562 leukemic cells.[250]

In highly aggressive human hepatocellular carcinoma (HHC) cell lines, treatment with DIDS inhibited cell proliferation and induced apoptosis, in a dose-dependent manner.[252] The same group investigated the effect of genetic ablation of AE2, showing a reduction in cell viability and induction of cell cycle arrest and apoptosis in the same poorly differentiated cell line. Absence of such effects in other less aggressive cell lines led the authors to suggest AE2 as a key player in HCC progression.[253] Knockdown of AE2 in ovarian cancer cells reduced cell proliferation and arrested the cell cycle at G1 phase. The results were corroborated in an in vivo mouse model. The authors additionally suggested activation of the mTOR/p70S6K1 pathway by upregulated AE2 as a possible mechanism of ovarian cancer progression.[247] Similarly, knockdown of AE2 by siRNAs markedly inhibited AE2 expression in ESCC cell lines while promoting variations in several matrix metalloproteinase (MMP) signaling pathway-related genes; this led to increased cell migration and decreased apoptosis.[248] Interestingly, in the study by Celai et al., AE2-deleted mice showed a progressive decrease in the number of regulatory T cells (Tregs), known for their role as suppressors of anti-tumor immune response. The authors designed specific peptides to target AE2 in mouse and human Tregs, and found apoptosis induction in these cells, concomitantly with activation of effective immune functions. Peptide p17AE2 also increased cell death in leukemia, lymphoma and multiple myeloma cells and primary malignant cultures. An in vivo stable peptide reduced tumor burden in B-cell lymphoma xenografted-mice.[254]

Clinical studies

To the best of our knowledge, there are no clinical trials listed at the ClinicalTrials.gov database involving inhibitors of AEs, either in the cancer field or in any other area. This is probably due to the lack of specific inhibitors. In the 1990s, Alper et al. described that an oxonol drug [diBA(5)C4] and two polyaminosterol analogs of the shark antibiotic squalamine (1360 and 1361) differentially inhibited AE1 and AE2 (respectively) and, thus, showed some isoform selectivity, but no similar studies have been reported since then.[255] The disulfonic stilbenes DIDS and SITS (4-acetamido-4′-isothiocyanostilbene 2′-disulfonate), as well as the cinnamic acid derivative CHC (α-cyano-4-hydroxycinnamate), are known to inhibit AEs activity,[220] but additionally they inhibit the function of monocarboxylate transporters.[11] Arakawa et al. recently described the crystal structure of AE1,[256] and although it is known that all AEs share a common structure, it will surely be possible to develop selective inhibitors in the near future.

Fig. 2 and Table 1 summarize the concepts discussed above.

Conclusions

The accelerated metabolism of cancer cells leads to the production of significant amounts of acidic compounds including lactic acid. In order to sustain high rates of cell proliferation, cancer cells expel the excess acids, acidifying the microenvironment. Tumor acidity contributes to many cancer cell aggressiveness features, including increased angiogenesis, invasion and metastasis, escape from the immune system, and resistance to treatment and is an attractive target for therapy.[257]

The pH dynamics of malignant tumors is maintained by a wide range of membrane-associated families of proteins, that in one way or another facilitate the acidification of the TME (Fig. 2): MCTs are responsible for the proton-coupled extrusion of lactic acid produced by the cancer glycolytic phenotype; CAs mediate the reversible hydration of carbon dioxide produced during respiration to carbonic acid; NHEs, promote the reversible electroneutral exchange of Na^+/H^+; ATPases couple ATP synthesis or hydrolysis for the transport of ions, including H^+, across membranes; and AEs mediate the electroneutral transmembrane exchange of HCO_3^- by Cl^-.

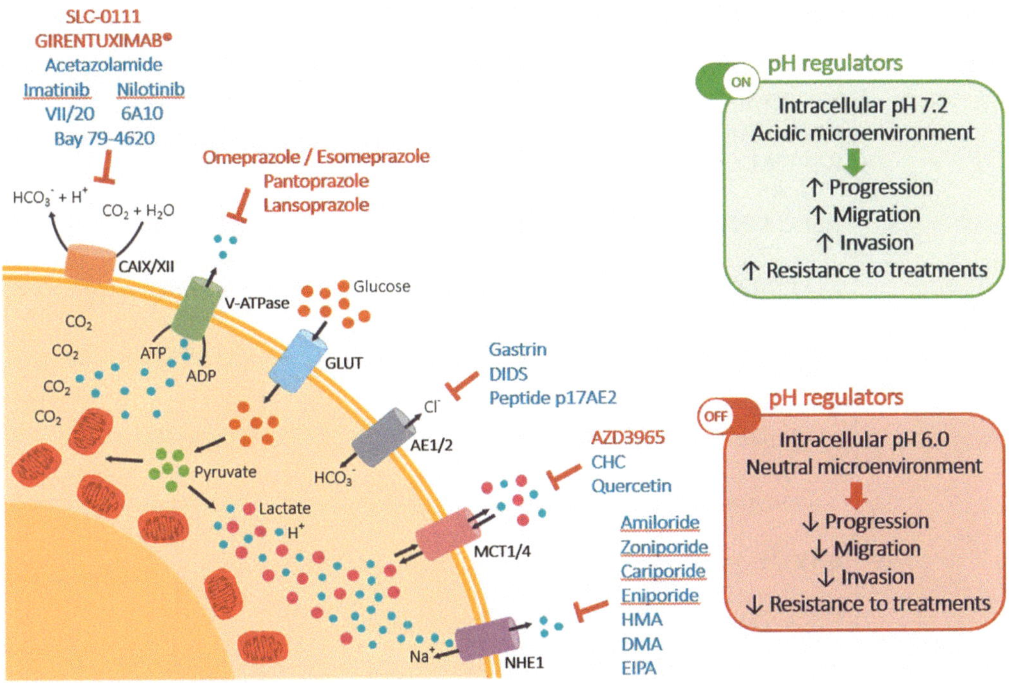

Fig. 2 Mechanisms of intracellular pH regulation in malignancy, and effects of inhibiting pH regulators as a therapeutic strategy to disrupt cancer growth. Compounds currently in clinical trials are shown in red, and compounds for which interesting results have been obtained in preclinical ("in vitro" and/or "in vivo") studies are shown in *blue*.

Table 1 In vivo studies targeting the different pH regulators in human cancer models.

Targets	Compounds	Type of cancer	Results
MCTs	KD MCT1/MCT4	Breast cancer[37]	↓ Tumor growth
	KD MCT1 + MCT4	Colon carcinoma[36]	↑ Mice survival
	KD MCT4	Bladder,[34] prostate[40] and pancreatic cancer[39]	↓ Tumor growth
	CHC	Glioma[44]	↓ Invasion Necrosis induction
		Colorectal cancer[43]	↓ Tumor growth
	Syrosingopine	Hepatocellular cancer[52]	↓ Serum lactate levels Intracellular lactate accumulation
	AZD3965	Lung and gastric cancer[49]	↓ Tumor growth ↑ Radiosensitivity
		Lung cancer[48]	↓ Tumor growth
		Raji Burkitt's lymphoma[50]	Intracellular lactate accumulation ↓ Tumor growth

Continued

Table 1 In vivo studies targeting the different pH regulators in human cancer models—cont'd

Targets	Compounds	Type of cancer	Results
CAs	VII/20 mAb	Colorectal cancer[91]	↓ Tumor growth
	6A10 mAb	Lung cancer[92]	↓ Tumor growth
	KD CAIX + CAXII	Colon carcinoma[83,84]	↓ Tumor growth ↑ Radiosensitivity
	KD CAIX	Breast cancer[85]	↓ Tumor growth ↓ Metastasis
	KD CAIX	Pancreatic cancer[86]	Prevented tumor initiation
	Acetazolamide derivatives	Renal cell carcinoma[87]	↓ Tumor growth
	Sulfamate class-CAs inhibitors	Breast cancer[88]	↓ Tumor growth ↓ Metastasis
	Ureidosulfonamide Inhibitor U104	Breast cancer[89]	↓ Tumor growth ↑ Chemosensitivity ↓ Metastasis
	BAY 79-4620	Lung, colorectal and cervical cancer[94]	↓ Tumor growth
	SLC-0111	Glioma[90] and breast cancer[85]	↓ Tumor growth ↑ Chemosensitivity ↓ Metastasis
NHEs	Amiloride	Esophageal cancer[124] Breast cancer[119] Myeloma[123]	↓ Tumor growth ↓ Metastasis ↓ Tumor growth ↑ Chemosensitization
V-ATPases	Concanamycin A/ Bafilomycin A	Hepatocellular cancer[167]	↓ Tumor growth
	Archazolid	Breast cancer[186,187] Hepatocellular cancer[184]	↓ Metastasis ↓ Membrane-related signaling pathways
	Omeprazole	Ovarian cancer[201] Leukemia[203]	↓ Tumor growth ↑ Chemosensitization
	Esomeprazole	Melanoma[199]	↓ Tumor growth Altered pH gradient
		Esophageal cancer[204]	↓ Tumor growth ↑ Chemosensitization
	Lansoprazole	Melanoma[205]	↓ Tumor growth ↑ Chemosensitization
	Pantoprazole	Gastric cancer[206]	↓ Tumor growth
	KD	Hepatocellular cancer[155]	↓ Tumor growth
	KD	Breast cancer[192]	↓ Metastasis
	Myrtenal	Melanoma[209]	↓ Metastasis

Table 1 In vivo studies targeting the different pH regulators in human cancer models—cont'd

Targets	Compounds	Type of cancer	Results
AEs	Omeprazole[a]	Gastric cancer[249]	↓ AE1 and p16 expression ↓ Tumor growth
	KD AE1	Gastric cancer[249,251]	↓ Tumor growth
	KD AE2	Ovarian cancer[247]	↓ Tumor growth
	p17AE2-HT[b]	B-cell lymphoma[254]	↓ Tumor growth

AEs, anion exchangers; *CAs*, carbonic anhydrases; *CHC*, α-cyano-4-hydroxycinnamic acid; *MCTs*, monocarboxylate transporters; *mAb*, monoclonal antibody; *NHEs*, Na^+/H^+-exchangers; *KD*, knockdown; *V-ATPases*, vacuolar-type H+-ATPases; ↓, decreased; ↑, increased.
[a]For induction of systemic hypergastrinemia.
[b]Macrocyclic peptide with a head-to-tail cyclization targeting AE2.

There is compelling evidence about the over-expression of these pH regulators in different cancer types. They have a role in cancer progression, and increased expression has been associated with poor prognosis. In addition, several in vitro and in vivo models have been used to confirm their role in cancer aggressiveness and demonstrate their potential as therapeutic targets (Table 1, Fig. 2). However, very few cancer pH regulator inhibitors have reached clinical trials. So far, only inhibitors of MCTs, CAs and V-ATPases have been tested in cancer patients and the results are still very limited (Fig. 2). One of the main factors limiting their exploitation in the clinical cancer treatment is the low availability of specific inhibitors. Thus, based on the encouraging preclinical data, most cancer pH regulators hold promise as anti-cancer agents and the researchers and clinical trial promoters should invest in the development of more specific inhibitors and innovative combinations of pharmaceuticals.

Acknowledgments

This manuscript has been developed under the scope of the project NORTE-01-0145-FEDER- 000013, supported by the Northern Portugal Regional Operational Programme (NORTE 2020) under the Portugal Partnership Agreement, through the European Regional Development Fund (FEDER), and through the Competitiveness Factors Operational Programme (COMPETE) and by National funds, through the Foundation for Science and Technology (FCT), under the scope of the project POCI-01-0145-FEDER-007038. JA and SG received fellowships from FCT, ref. SFRH/BPD/116784/2016 and SFRH/BPD/117858/2016, respectively.

References

1. Gatenby RA, Gillies RJ. Why do cancers have high aerobic glycolysis? *Nat Rev Cancer* 2004; **4**(11):891–9.
2. Warburg O. On the origin of cancer cells. *Science* 1956;**123**(3191):309–14.

3. Granja S, Tavares-Valente D, Queiros O, Baltazar F. Value of pH regulators in the diagnosis, prognosis and treatment of cancer. *Semin Cancer Biol* 2017;**43**:17–34.
4. Pinheiro C, Longatto-Filho A, Azevedo-Silva J, Casal M, Schmitt FC, Baltazar F. Role of monocarboxylate transporters in human cancers: state of the art. *J Bioenerg Biomembr* 2012;**44**(1):127–39.
5. Mboge MY, Mahon BP, McKenna R, Frost SC. Carbonic anhydrases: role in pH control and cancer. *Metabolites* 2018;**8**(1).
6. Donowitz M, Ming Tse C, Fuster D. SLC9/NHE gene family, a plasma membrane and organellar family of Na(+)/H(+) exchangers. *Mol Asp Med* 2013;**34**(2-3):236–51.
7. Stransky L, Cotter K, Forgac M. The function of V-ATPases in cancer. *Physiol Rev* 2016; **96**(3):1071–91.
8. Bublitz M, Morth JP, Nissen P. P-type ATPases at a glance. *J Cell Sci* 2011;**124**(Pt 15):2515–9.
9. Thornell IM, Bevensee MO. Regulators of Slc4 bicarbonate transporter activity. *Front Physiol* 2015;**6**:166.
10. Halestrap AP. The SLC16 gene family—structure, role and regulation in health and disease. *Mol Asp Med* 2013;**34**(2-3):337–49.
11. Halestrap AP. The monocarboxylate transporter family—structure and functional characterization. *IUBMB Life* 2012;**64**(1):1–9.
12. Halestrap AP, Price NT. The proton-linked monocarboxylate transporter (MCT) family: structure, function and regulation. *Biochem J* 1999;**343**(Pt 2):281–99.
13. Adijanto J, Philp NJ. The SLC16A family of monocarboxylate transporters (MCTs)—physiology and function in cellular metabolism, pH homeostasis, and fluid transport. *Curr Top Membr* 2012;**70**:275–311.
14. San-Millan I, Brooks GA. Reexamining cancer metabolism: lactate production for carcinogenesis could be the purpose and explanation of the Warburg Effect. *Carcinogenesis* 2017;**38**(2):119–33.
15. Pinheiro C, Morais-Santos F, Granja S, Miranda-Gonçalves V, Afonso J, Amorim R, Baltazar F. Targeting metabolic reprogramming as an anti- cancer strategy: aiming at monocarboxylate transporters. In: Atta-ur-Rahman F, Choudhary MI, editors. *Frontiers in anti-cancer drug discovery*. vol. 6. Bentham e Books; 2016. p. 3–65.
16. Granja S, Marchiq I, Le Floch R, Moura CS, Baltazar F, Pouyssegur J. Disruption of BASIGIN decreases lactic acid export and sensitizes non-small cell lung cancer to biguanides independently of the LKB1 status. *Oncotarget* 2015;**6**(9):6708–21.
17. Pinheiro C, Albergaria A, Paredes J, Sousa B, Dufloth R, Vieira D, et al. Monocarboxylate transporter 1 is up-regulated in basal-like breast carcinoma. *Histopathology* 2010;**56**(7):860–7.
18. Simoes-Sousa S, Granja S, Pinheiro C, Fernandes D, Longatto-Filho A, Laus AC, et al. Prognostic significance of monocarboxylate transporter expression in oral cavity tumors. *Cell Cycle (Georgetown, TX)* 2016;**15**(14):1865–73.
19. Kim Y, Choi JW, Lee JH, Kim YS. Expression of lactate/H(+) symporters MCT1 and MCT4 and their chaperone CD147 predicts tumor progression in clear cell renal cell carcinoma: immunohistochemical and The Cancer Genome Atlas data analyses. *Hum Pathol* 2015;**46**(1):104–12.
20. Miranda-Goncalves V, Honavar M, Pinheiro C, Martinho O, Pires MM, Pinheiro C, et al. Monocarboxylate transporters (MCTs) in gliomas: expression and exploitation as therapeutic targets. *Neuro-Oncology* 2013;**15**(2):172–88.
21. Pinheiro C, Granja S, Longatto-Filho A, Faria AM, Fragoso MC, Lovisolo SM, et al. Metabolic reprogramming: a new relevant pathway in adult adrenocortical tumors. *Oncotarget* 2015;**6**(42):44403–21.
22. Ho J, de Moura MB, Lin Y, Vincent G, Thorne S, Duncan LM, et al. Importance of glycolysis and oxidative phosphorylation in advanced melanoma. *Mol Cancer* 2012;**11**:76.
23. Schneiderhan W, Scheler M, Holzmann KH, Marx M, Gschwend JE, Bucholz M, et al. CD147 silencing inhibits lactate transport and reduces malignant potential of pancreatic cancer cells in in vivo and in vitro models. *Gut* 2009;**58**(10):1391–8.
24. Pinheiro C, Longatto-Filho A, Scapulatempo C, Ferreira L, Martins S, Pellerin L, et al. Increased expression of monocarboxylate transporters 1, 2, and 4 in colorectal carcinomas. *Virchows Arch* 2008;**452**(2):139–46.

25. Chen H, Wang L, Beretov J, Hao J, Xiao W, Li Y. Co-expression of CD147/EMMPRIN with monocarboxylate transporters and multiple drug resistance proteins is associated with epithelial ovarian cancer progression. *Clin Exp Metastasis* 2010;**27**(8):557–69.
26. Pinheiro C, Longatto-Filho A, Ferreira L, Pereira SM, Etlinger D, Moreira MA, et al. Increasing expression of monocarboxylate transporters 1 and 4 along progression to invasive cervical carcinoma. *Int J Gynecol Pathol* 2008;**27**(4):568–74.
27. Silva ECA, Carcano FM, Bonatelli M, Zaia MG, Morais-Santos F, Baltazar F, et al. The clinicopathological significance of monocarboxylate transporters in testicular germ cell tumors. *Oncotarget* 2018;**9**(29):20386–98.
28. Alves VA, Pinheiro C, Morais-Santos F, Felipe-Silva A, Longatto-Filho A, Baltazar F. Characterization of monocarboxylate transporter activity in hepatocellular carcinoma. *World J Gastroenterol* 2014;**20**(33):11780–7.
29. Pertega-Gomes N, Vizcaino JR, Miranda-Goncalves V, Pinheiro C, Silva J, Pereira H, et al. Monocarboxylate transporter 4 (MCT4) and CD147 overexpression is associated with poor prognosis in prostate cancer. *BMC Cancer* 2011;**11**:312.
30. Afonso J, Pinto T, Simoes-Sousa S, Schmitt F, Longatto-Filho A, Pinheiro C, et al. Clinical significance of metabolism-related biomarkers in non-Hodgkin lymphoma—MCT1 as potential target in diffuse large B cell lymphoma. *Cell Oncol (Dordr)* 2019;**42**:303–18.
31. Cao YW, Liu Y, Dong Z, Guo L, Kang EH, Wang YH, et al. Monocarboxylate transporters MCT1 and MCT4 are independent prognostic biomarkers for the survival of patients with clear cell renal cell carcinoma and those receiving therapy targeting angiogenesis. *Urol Oncol* 2018;**36**(6):311. e15–e25.
32. Pinheiro C, Miranda-Goncalves V, Longatto-Filho A, Vicente AL, Berardinelli GN, Scapulatempo-Neto C, et al. The metabolic microenvironment of melanomas: prognostic value of MCT1 and MCT4. *Cell Cycle (Georgetown, TX)* 2016;**15**(11):1462–70.
33. Cheng B, Chen X, Li Y, Huang X, Yu J. Prognostic value of monocarboxylate transporter 4 in patients with esophageal squamous cell carcinoma. *Oncol Rep* 2018;**40**(5):2906–15.
34. Todenhofer T, Seiler R. Selective inhibition of the lactate transporter MCT4 reduces growth of invasive bladder cancer. *Mol Cancer Ther* 2018;**17**(12):2746–55.
35. Bovenzi CD, Hamilton J, Tassone P, Johnson J, Cognetti DM, Luginbuhl A, et al. Prognostic indications of elevated MCT4 and CD147 across cancer types: a meta-analysis. *Biomed Res Int* 2015;**2015**:242437.
36. Le Floch R, Chiche J, Marchiq I, Naiken T, Ilc K, Murray CM, et al. CD147 subunit of lactate/H+ symporters MCT1 and hypoxia-inducible MCT4 is critical for energetics and growth of glycolytic tumors. *Proc Natl Acad Sci U S A* 2011;**108**(40):16663–8.
37. Morais-Santos F, Granja S, Miranda-Goncalves V, Moreira AH, Queiros S, Vilaca JL, et al. Targeting lactate transport suppresses in vivo breast tumor growth. *Oncotarget* 2015;**6**(22):19177–89.
38. Sanita P, Capulli M, Teti A, Galatioto GP, Vicentini C, Chiarugi P, et al. Tumor-stroma metabolic relationship based on lactate shuttle can sustain prostate cancer progression. *BMC Cancer* 2014;**14**:154.
39. Baek G, Tse YF, Hu Z, Cox D, Buboltz N, McCue P, et al. MCT4 defines a glycolytic subtype of pancreatic cancer with poor prognosis and unique metabolic dependencies. *Cell Rep* 2014;**9**(6):2233–49.
40. Choi SY, Xue H, Wu R, Fazli L, Lin D, Collins CC, et al. The MCT4 gene: a novel, potential target for therapy of advanced prostate cancer. *Clin Cancer Res* 2016;**22**(11):2721–33.
41. Chiche J, Le Fur Y, Vilmen C, Frassineti F, Daniel L, Halestrap AP, et al. In vivo pH in metabolic-defective Ras-transformed fibroblast tumors: key role of the monocarboxylate transporter, MCT4, for inducing an alkaline intracellular pH. *Int J Cancer* 2012;**130**(7):1511–20.
42. Izumi H, Takahashi M, Uramoto H, Nakayama Y, Oyama T, Wang KY, et al. Monocarboxylate transporters 1 and 4 are involved in the invasion activity of human lung cancer cells. *Cancer Sci* 2011;**102**(5):1007–13.
43. Sonveaux P, Vegran F, Schroeder T, Wergin MC, Verrax J, Rabbani ZN, et al. Targeting lactate-fueled respiration selectively kills hypoxic tumor cells in mice. *J Clin Invest* 2008;**118**(12):3930–42.

44. Colen CB, Shen Y, Ghoddoussi F, Yu P, Francis TB, Koch BJ, et al. Metabolic targeting of lactate efflux by malignant glioma inhibits invasiveness and induces necrosis: an in vivo study. *Neoplasia (New York)* 2011;**13**(7):620–32.
45. Futagi Y, Kobayashi M, Narumi K, Furugen A, Iseki K. Identification of a selective inhibitor of human monocarboxylate transporter 4. *Biochem Biophys Res Commun* 2018;**495**(1):427–32.
46. Bueno V, Binet I, Steger U, Bundick R, Ferguson D, Murray C, et al. The specific monocarboxylate transporter (MCT1) inhibitor, AR-C117977, a novel immunosuppressant, prolongs allograft survival in the mouse. *Transplantation* 2007;**84**(9):1204–7.
47. Ovens MJ, Davies AJ, Wilson MC, Murray CM, Halestrap AP. AR-C155858 is a potent inhibitor of monocarboxylate transporters MCT1 and MCT2 that binds to an intracellular site involving transmembrane helices 7-10. *Biochem J* 2010;**425**(3):523–30.
48. Polanski R, Hodgkinson CL, Fusi A, Nonaka D, Priest L, Kelly P, et al. Activity of the monocarboxylate transporter 1 inhibitor AZD3965 in small cell lung cancer. *Clin Cancer Res* 2014;**20**(4):926–37.
49. Bola BM, Chadwick AL, Michopoulos F, Blount KG, Telfer BA, Williams KJ, et al. Inhibition of monocarboxylate transporter-1 (MCT1) by AZD3965 enhances radiosensitivity by reducing lactate transport. *Mol Cancer Ther* 2014;**13**(12):2805–16.
50. Curtis NJ, Mooney L, Hopcroft L, Michopoulos F, Whalley N, Zhong H, et al. Pre-clinical pharmacology of AZD3965, a selective inhibitor of MCT1: DLBCL, NHL and Burkitt's lymphoma anti-tumor activity. *Oncotarget* 2017;**8**(41):69219–36.
51. Marchiq I, Pouyssegur J. Hypoxia, cancer metabolism and the therapeutic benefit of targeting lactate/H symporters. *J Mol Med (Berl)* 2015;**94**(2):155–71.
52. Benjamin D, Robay D, Hindupur SK, Pohlmann J, Colombi M, El-Shemerly MY, et al. Dual inhibition of the lactate transporters MCT1 and MCT4 is synthetic lethal with metformin due to NAD+ depletion in cancer cells. *Cell Rep* 2018;**25**(11):3047–58. e4.
53. Halford SER, Jones P, Wedge S, Hirschberg S, Katugampola S, Veal G, et al. A first-in-human first-in-class (FIC) trial of the monocarboxylate transporter 1 (MCT1) inhibitor AZD3965 in patients with advanced solid tumors. *J Clin Oncol* 2017;**35**(15_suppl):2516.
54. Parks SK, Cormerais Y, Pouyssegur J. Hypoxia and cellular metabolism in tumor pathophysiology. *J Physiol* 2017;**595**(8):2439–50.
55. Supuran CT, Alterio V, Di Fiore A, D'Ambrosio K, Carta F, Monti SM, et al. Inhibition of carbonic anhydrase IX targets primary tumors, metastases, and cancer stem cells: three for the price of one. *Med Res Rev* 2018;**38**(6):1799–836.
56. Supuran CT. Carbonic anhydrases—an overview. *Curr Pharm Des* 2008;**14**(7):603–14.
57. Supuran CT. Carbonic anhydrase inhibitors as emerging agents for the treatment and imaging of hypoxic tumors. *Expert Opin Investig Drugs* 2018;**27**(12):963–70.
58. Wykoff CC, Beasley N, Watson PH, Campo L, Chia SK, English R, et al. Expression of the hypoxia-inducible and tumor- associated carbonic anhydrases in ductal carcinoma in situ of the breast. *Am J Pathol* 2001;**158**(3):1011–9.
59. Pinheiro C, Sousa B, Albergaria A, Paredes J, Dufloth R, Vieira D, et al. GLUT1 and CAIX expression profiles in breast cancer correlate with adverse prognostic factors and MCT1 overexpression. *Histol Histopathol* 2011;**26**(10):1279–86.
60. Kivela A, Parkkila S, Saarnio J, Karttunen TJ, Kivela J, Parkkila AK, et al. Expression of a novel transmembrane carbonic anhydrase isozyme XII in normal human gut and colorectal tumors. *Am J Pathol* 2000;**156**(2):577–84.
61. Saarnio J, Parkkila S, Parkkila AK, Haukipuro K, Pastorekova S, Pastorek J, et al. Immunohistochemical study of colorectal tumors for expression of a novel transmembrane carbonic anhydrase, MN/CA IX, with potential value as a marker of cell proliferation. *Am J Pathol* 1998;**153**(1):279–85.
62. Yang JS, Chen MK, Yang SF, Chang YC, Su SC, Chiou HL, et al. Increased expression of carbonic anhydrase IX in oral submucous fibrosis and oral squamous cell carcinoma. *Clin Chem Lab Med* 2014;**52**(9):1367–77.
63. Ilie M, Mazure NM, Hofman V, Ammadi RE, Ortholan C, Bonnetaud C, et al. High levels of carbonic anhydrase IX in tumour tissue and plasma are biomarkers of poor prognostic in patients with non-small cell lung cancer. *Br J Cancer* 2010;**102**(11):1627–35.

64. Miranda-Goncalves V, Reis RM, Baltazar F. Lactate transporters and pH regulation: potential therapeutic targets in glioblastomas. *Curr Cancer Drug Targets* 2016;**16**(5):388–99.
65. de Martino M, Lucca I, Mbeutcha A, Wiener HG, Haitel A, Susani M, et al. Carbonic anhydrase IX as a diagnostic urinary marker for urothelial bladder cancer. *Eur Urol* 2015;**68**(4):552–4.
66. Hynninen P, Vaskivuo L, Saarnio J, Haapasalo H, Kivela J, Pastorekova S, et al. Expression of transmembrane carbonic anhydrases IX and XII in ovarian tumors. *Histopathology* 2006;**49**(6):594–602.
67. Wykoff CC, Beasley NJ, Watson PH, Turner KJ, Pastorek J, Sibtain A, et al. Hypoxia-inducible expression of tumor-associated carbonic anhydrases. *Cancer Res* 2000;**60**(24):7075–83.
68. Hussain SA, Ganesan R, Reynolds G, Gross L, Stevens A, Pastorek J, et al. Hypoxia-regulated carbonic anhydrase IX expression is associated with poor survival in patients with invasive breast cancer. *Br J Cancer* 2007;**96**(1):104–9.
69. Yang L, Yang Z, Li D, Liu Z, Zou Q, Yuan Y, et al. Overexpression of FZD1 and CAIX are associated with invasion, metastasis, and poor-prognosis of the pancreatic ductal adenocarcinoma. *Pathol Oncol Res* 2018;**24**(4):899–906.
70. Finkelmeier F, Canli O, Peiffer KH, Walter D, Tal A, Koch C, et al. Circulating hypoxia marker carbonic anhydrase IX (CA9) in patients with hepatocellular carcinoma and patients with cirrhosis. *PLoS ONE* 2018;**13**(7):e0200855.
71. Zhao Z, Liao G, Li Y, Zhou S, Zou H, Fernando S. Prognostic value of carbonic anhydrase IX immunohistochemical expression in renal cell carcinoma: a meta-analysis of the literature. *PLoS ONE* 2014;**9**(11):e114096.
72. Chiche J, Ilc K, Brahimi-Horn MC, Pouyssegur J. Membrane-bound carbonic anhydrases are key pH regulators controlling tumor growth and cell migration. *Adv Enzym Regul* 2010;**50**(1):20–33.
73. Tafreshi NK, Lloyd MC, Bui MM, Gillies RJ, Morse DL. Carbonic anhydrase IX as an imaging and therapeutic target for tumors and metastases. *Subcell Biochem* 2014;**75**:221–54.
74. Hong YS, Cho HJ, Kim SY, Jung KH, Park JW, Choi HS, et al. Carbonic anhydrase 9 is a predictive marker of survival benefit from lower dose of bevacizumab in patients with previously treated metastatic colorectal cancer. *BMC Cancer* 2009;**9**:246.
75. Choueiri TK, Regan MM, Rosenberg JE, Oh WK, Clement J, Amato AM, et al. Carbonic anhydrase IX and pathological features as predictors of outcome in patients with metastatic clear-cell renal cell carcinoma receiving vascular endothelial growth factor-targeted therapy. *BJU Int* 2010;**106**(6):772–8.
76. Parkkila S, Parkkila AK, Saarnio J, Kivela J, Karttunen TJ, Kaunisto K, et al. Expression of the membrane-associated carbonic anhydrase isozyme XII in the human kidney and renal tumors. *J Histochem Cytochem* 2000;**48**(12):1601–8.
77. Ilie MI, Hofman V, Ortholan C, Ammadi RE, Bonnetaud C, Havet K, et al. Overexpression of carbonic anhydrase XII in tissues from resectable non-small cell lung cancers is a biomarker of good prognosis. *Int J Cancer* 2011;**128**(7):1614–23.
78. Kivela AJ, Parkkila S, Saarnio J, Karttunen TJ, Kivela J, Parkkila AK, et al. Expression of transmembrane carbonic anhydrase isoenzymes IX and XII in normal human pancreas and pancreatic tumors. *Histochem Cell Biol* 2000;**114**(3):197–204.
79. Kobayashi M, Matsumoto T, Ryuge S, Yanagita K, Nagashio R, Kawakami Y, et al. CAXII Is a serodiagnostic marker for lung cancer. *PLoS ONE* 2012;**7**(3):e33952.
80. Yoo CW, Nam BH, Kim JY, Shin HJ, Lim H, Lee S, et al. Carbonic anhydrase XII expression is associated with histologic grade of cervical cancer and superior radiotherapy outcome. *Radiat Oncol (Lond)* 2010;**5**:101.
81. Chien MH, Ying TH, Hsieh YH, Lin CH, Shih CH, Wei LH, et al. Tumor-associated carbonic anhydrase XII is linked to the growth of primary oral squamous cell carcinoma and its poor prognosis. *Oral Oncol* 2012;**48**(5):417–23.
82. Iessi E, Logozzi M, Mizzoni D, Di Raimo R, Supuran CT, Fais S. Rethinking the combination of proton exchanger inhibitors in cancer therapy. *Metabolites* 2017;**8**(1).
83. Chiche J, Ilc K, Laferriere J, Trottier E, Dayan F, Mazure NM, et al. Hypoxia-inducible carbonic anhydrase IX and XII promote tumor cell growth by counteracting acidosis through the regulation of the intracellular pH. *Cancer Res* 2009;**69**(1):358–68.

84. Doyen J, Parks SK, Marcie S, Pouyssegur J, Chiche J. Knock-down of hypoxia-induced carbonic anhydrases IX and XII radiosensitizes tumor cells by increasing intracellular acidosis. *Front Oncol* 2012;**2**:199.
85. Lou Y, McDonald PC, Oloumi A, Chia S, Ostlund C, Ahmadi A, et al. Targeting tumor hypoxia: suppression of breast tumor growth and metastasis by novel carbonic anhydrase IX inhibitors. *Cancer Res* 2011;**71**(9):3364–76.
86. Pore N, Jalla S, Liu Z, Higgs B, Sorio C, Scarpa A, et al. In vivo loss of function screening reveals carbonic anhydrase IX as a key modulator of tumor initiating potential in primary pancreatic tumors. *Neoplasia (New York)* 2015;**17**(6):473–80.
87. Ahlskog JK, Dumelin CE, Trussel S, Marlind J, Neri D. In vivo targeting of tumor-associated carbonic anhydrases using acetazolamide derivatives. *Bioorg Med Chem Lett* 2009;**19**(16):4851–6.
88. Gieling RG, Babur M, Mamnani L, Burrows N, Telfer BA, Carta F, et al. Antimetastatic effect of sulfamate carbonic anhydrase IX inhibitors in breast carcinoma xenografts. *J Med Chem* 2012;**55**(11):5591–600.
89. Lock FE, McDonald PC, Lou Y, Serrano I, Chafe SC, Ostlund C, et al. Targeting carbonic anhydrase IX depletes breast cancer stem cells within the hypoxic niche. *Oncogene* 2013;**32**(44):5210–9.
90. Boyd NH, Walker K, Fried J, Hackney JR, McDonald PC, Benavides GA, et al. Addition of carbonic anhydrase 9 inhibitor SLC-0111 to temozolomide treatment delays glioblastoma growth in vivo. *JCI Insight* 2017;**2**(24).
91. Zatovicova M, Jelenska L, Hulikova A, Csaderova L, Ditte Z, Ditte P, et al. Carbonic anhydrase IX as an anticancer therapy target: preclinical evaluation of internalizing monoclonal antibody directed to catalytic domain. *Curr Pharm Des* 2010;**16**(29):3255–63.
92. Gondi G, Mysliwietz J, Hulikova A, Jen JP, Swietach P, Kremmer E, et al. Antitumor efficacy of a monoclonal antibody that inhibits the activity of cancer-associated carbonic anhydrase XII. *Cancer Res* 2013;**73**(21):6494–503.
93. Battke C, Kremmer E, Mysliwietz J, Gondi G, Dumitru C, Brandau S, et al. Generation and characterization of the first inhibitory antibody targeting tumor-associated carbonic anhydrase XII. *Cancer Immunol Immunother* 2011;**60**(5):649–58.
94. Petrul HM, Schatz CA, Kopitz CC, Adnane L, McCabe TJ, Trail P, et al. Therapeutic mechanism and efficacy of the antibody-drug conjugate BAY 79-4620 targeting human carbonic anhydrase 9. *Mol Cancer Ther* 2012;**11**(2):340–9.
95. Parkkila S, Innocenti A, Kallio H, Hilvo M, Scozzafava A, Supuran CT. The protein tyrosine kinase inhibitors imatinib and nilotinib strongly inhibit several mammalian alpha-carbonic anhydrase isoforms. *Bioorg Med Chem Lett* 2009;**19**(15):4102–6.
96. Oosterwijk E, Ruiter DJ, Hoedemaeker PJ, Pauwels EK, Jonas U, Zwartendijk J, et al. Monoclonal antibody G 250 recognizes a determinant present in renal-cell carcinoma and absent from normal kidney. *Int J Cancer* 1986;**38**(4):489–94.
97. Oosterwijk-Wakka JC, Boerman OC, Mulders PF, Oosterwijk E. Application of monoclonal antibody G250 recognizing carbonic anhydrase IX in renal cell carcinoma. *Int J Mol Sci* 2013;**14**(6):11402–23.
98. Siebels M, Rohrmann K, Oberneder R, Stahler M, Haseke N, Beck J, et al. A clinical phase I/II trial with the monoclonal antibody cG250 (RENCAREX(R)) and interferon-alpha-2a in metastatic renal cell carcinoma patients. *World J Urol* 2011;**29**(1):121–6.
99. Belldegrun ASCK, Kloepfer P, Fall B, Bevan P, Störkel S. ARISER: a randomized double blind phase III study to evaluate adjuvant cG250 treatment versus placebo in patients with high-risk ccRCC—results and implications for adjuvant clinical trials. *J Clin Oncol* 2013;**31**(31 Suppl). Abstr 4507.
100. Liu Z, Smyth FE, Renner C, Lee FT, Oosterwijk E, Scott AM. Anti-renal cell carcinoma chimeric antibody G250: cytokine enhancement of in vitro antibody-dependent cellular cytotoxicity. *Cancer Immunol Immunother* 2002;**51**(3):171–7.
101. Padan E, Landau M. Sodium-proton (Na^+/H^+) antiporters: properties and roles in health and disease. *Met Ions Life Sci* 2016;**16**:391–458.
102. Orlowski J, Grinstein S. Diversity of the mammalian sodium/proton exchanger SLC9 gene family. *Pflugers Arch* 2004;**447**(5):549–65.

103. Reshkin SJ, Cardone RA, Harguindey S. Na+-H+ exchanger, pH regulation and cancer. *Recent Pat Anticancer Drug Discov* 2013;**8**(1):85–99.
104. Kaminota T, Yano H, Shiota K, Nomura N, Yaguchi H, Kirino Y, et al. Elevated Na(+)/H(+) exchanger-1 expression enhances the metastatic collective migration of head and neck squamous cell carcinoma cells. *Biochem Biophys Res Commun* 2017;**486**(1):101–7.
105. Reshkin SJ, Bellizzi A, Caldeira S, Albarani V, Malanchi I, Poignee M, et al. Na+/H+ exchanger-dependent intracellular alkalinization is an early event in malignant transformation and plays an essential role in the development of subsequent transformation-associated phenotypes. *FASEB J* 2000;**14**(14):2185–97.
106. Chiang Y, Chou CY, Hsu KF, Huang YF, Shen MR. EGF upregulates Na+/H+ exchanger NHE1 by post-translational regulation that is important for cervical cancer cell invasiveness. *J Cell Physiol* 2008;**214**(3):810–9.
107. Asgharzadeh MR, Barar J, Pourseif MM, Eskandani M, Jafari Niya M, Mashayekhi MR, et al. Molecular machineries of pH dysregulation in tumor microenvironment: potential targets for cancer therapy. *Bioimpacts* 2017;**7**(2):115–33.
108. Amith SR, Fliegel L. Regulation of the Na+/H+ exchanger (NHE1) in breast cancer metastasis. *Cancer Res* 2013;**73**(4):1259–64.
109. Reshkin SJ, Bellizzi A, Albarani V, Guerra L, Tommasino M, Paradiso A, et al. Phosphoinositide 3-kinase is involved in the tumor-specific activation of human breast cancer cell Na(+)/H(+) exchange, motility, and invasion induced by serum deprivation. *J Biol Chem* 2000;**275**(8):5361–9.
110. Vahle AK, Domikowsky B, Schwoppe C, Krahling H, Mally S, Schafers M, et al. Extracellular matrix composition and interstitial pH modulate NHE1-mediated melanoma cell motility. *Int J Oncol* 2014;**44**(1):78–90.
111. Cardone RA, Casavola V, Reshkin SJ. The role of disturbed pH dynamics and the Na+/H+ exchanger in metastasis. *Nat Rev Cancer* 2005;**5**(10):786–95.
112. Rauch C. On the relationship between drug's size, cell membrane mechanical properties and high levels of multi drug resistance: a comparison to published data. *Eur Biophys J* 2009;**38**(4):537–46.
113. Harguindey S, Cragoe Jr. EJ. The Na+/H+ antiporter in oncology in the light of the spontaneous regression of cancer and cell metabolism. *Med Hypotheses* 1992;**39**(3):229–37.
114. Yang X, Wang D, Dong W, Song Z, Dou K. Over-expression of Na+/H+ exchanger 1 and its clinicopathologic significance in hepatocellular carcinoma. *Med Oncol* 2010;**27**(4):1109–13.
115. Cong D, Zhu W, Shi Y, Pointer KB, Clark PA, Shen H, et al. Upregulation of NHE1 protein expression enables glioblastoma cells to escape TMZ-mediated toxicity via increased H(+) extrusion, cell migration and survival. *Carcinogenesis* 2014;**35**(9):2014–24.
116. Sanhueza C, Araos J, Naranjo L, Toledo F, Beltran AR, Ramirez MA, et al. Sodium/proton exchanger isoform 1 regulates intracellular pH and cell proliferation in human ovarian cancer. *Biochim Biophys Acta Mol basis Dis* 2017;**1863**(1):81–91.
117. Ariyoshi Y, Shiozaki A, Ichikawa D, Shimizu H, et al. Na+/H+ exchanger 1 has tumor suppressive activity and prognostic value in esophageal squamous cell carcinoma. *Oncotarget* 2017;**8**(2):2209–23.
118. Harguindey S, Arranz JL, Polo Orozco JD, Rauch C, Fais S, Cardone RA, et al. Cariporide and other new and powerful NHE1 inhibitors as potentially selective anticancer drugs—an integral molecular/biochemical/metabolic/clinical approach after one hundred years of cancer research. *J Transl Med* 2013;**11**:282.
119. Giusti I, D'Ascenzo S, Millimaggi D, Taraboletti G, Carta G, Franceschini N, et al. Cathepsin B mediates the pH-dependent proinvasive activity of tumor-shed microvesicles. *Neoplasia* 2008;**10**(5):481–8.
120. Harley W, Floyd C, Dunn T, Zhang XD, Chen TY, Hegde M, et al. Dual inhibition of sodium-mediated proton and calcium efflux triggers non-apoptotic cell death in malignant gliomas. *Brain Res* 2010;**1363**:159–69.
121. Yang X, Wang D, Dong W, Song Z, Dou K. Inhibition of Na(+)/H(+) exchanger 1 by 5-(N-ethyl-N-isopropyl) amiloride reduces hypoxia-induced hepatocellular carcinoma invasion and motility. *Cancer Lett* 2010;**295**(2):198–204.
122. Wong P, Kleemann HW, Tannock IF. Cytostatic potential of novel agents that inhibit the regulation of intracellular pH. *Br J Cancer* 2002;**87**(2):238–45.

123. Rojas EA, Corchete LA, San-Segundo L, Martinez-Blanch JF, Codoner FM, Paino T, et al. Amiloride, an old diuretic drug, is a potential therapeutic agent for multiple myeloma. *Clin Cancer Res* 2017; **23**(21):6602–15.
124. Guan B, Hoque A, Xu X. Amiloride and guggulsterone suppression of esophageal cancer cell growth in vitro and in nude mouse xenografts. *Front Biol (Beijing)* 2014;**9**(1):75–81.
125. Rich IN, Worthington-White D, Garden OA, Musk P. Apoptosis of leukemic cells accompanies reduction in intracellular pH after targeted inhibition of the Na(+)/H(+) exchanger. *Blood* 2000; **95**(4):1427–34.
126. Matthews H, Ranson M, Kelso MJ. Anti-tumor/metastasis effects of the potassium-sparing diuretic amiloride: an orally active anti-cancer drug waiting for its call-of-duty? *Int J Cancer* 2011; **129**(9):2051–61.
127. Garcia-Canero R, Trilla C, Perez de Diego J, Diaz-Gil JJ, Cobo JM. Na+ :H+ exchange inhibition induces intracellular acidosis and differentially impairs cell growth and viability of human and rat hepatocarcinoma cells. *Toxicol Lett* 1999;**106**(2-3):215–28.
128. Chang C, Kong PY, Chen XH, Peng XG, Liu L, Liu H, et al. Effect of the NHE-1-specific inhibitor DMA on pHi, proliferation and apoptosis of HL-60/ADM cells in vitro. *Zhongguo Shi Yan Xue Ye Xue Za Zhi* 2006;**14**(3):488–91.
129. Reshkin SJ, Bellizzi A, Cardone RA, Tommasino M, Casavola V, Paradiso A. Paclitaxel induces apoptosis via protein kinase A- and p38 mitogen-activated protein-dependent inhibition of the Na+/H+ exchanger (NHE) NHE isoform 1 in human breast cancer cells. *Clin Cancer Res* 2003;**9**(6):2366–73.
130. Karmazyn M. NHE-1: still a viable therapeutic target. *J Mol Cell Cardiol* 2013;**61**:77–82.
131. Lv C, Yang X, Yu B, Ma Q, Liu B, Liu Y. Blocking the Na+/H+ exchanger 1 with cariporide (HOE642) reduces the hypoxia-induced invasion of human tongue squamous cell carcinoma. *Int J Oral Maxillofac Surg* 2012;**41**(10):1206–10.
132. Albatany M, Li A, Meakin S, Bartha R. In vivo detection of acute intracellular acidification in glioblastoma multiforme following a single dose of cariporide. *Int J Clin Oncol* 2018;**23**(5):812–9.
133. Nagata H, Che XF, Miyazawa K, Tomoda A, Konishi M, Ubukata H, et al. Rapid decrease of intracellular pH associated with inhibition of Na+/H+ exchanger precedes apoptotic events in the MNK45 and MNK74 gastric cancer cell lines treated with 2-aminophenoxazine-3-one. *Oncol Rep* 2011;**25**(2):341–6.
134. Che XF, Zheng CL, Akiyama S, Tomoda A. 2-Aminophenoxazine-3-one and 2-amino-4,4alpha-dihydro-4alpha,7-dimethyl-3H-phenoxazine-3-one cause cellular apoptosis by reducing higher intracellular pH in cancer cells. *Proc Jpn Acad Ser B Phys Biol Sci* 2011;**87**(4):199–213.
135. Flinck M, Kramer SH, Schnipper J, Andersen AP, Pedersen SF. The acid-base transport proteins NHE1 and NBCn1 regulate cell cycle progression in human breast cancer cells. *Cell Cycle* 2018; **17**(9):1056–67.
136. Chen L, Cong D, Li Y, Wang D, Li Q, Hu S. Combination of sonodynamic with temozolomide inhibits C6 glioma migration and promotes mitochondrial pathway apoptosis via suppressing NHE-1 expression. *Ultrason Sonochem* 2017;**39**:654–61.
137. Rupprecht HJ, vom Dahl J, Terres W, Seyfarth KM, Richardt G, Schultheibeta HP, et al. Cardioprotective effects of the Na(+)/H(+) exchange inhibitor cariporide in patients with acute anterior myocardial infarction undergoing direct PTCA. *Circulation* 2000;**101**(25):2902–8.
138. Mentzer Jr. RM, Bartels C, Bolli R, Boyce S, Buckberg GD, Chaitman B, et al. Sodium-hydrogen exchange inhibition by cariporide to reduce the risk of ischemic cardiac events in patients undergoing coronary artery bypass grafting: results of the EXPEDITION study. *Ann Thorac Surg* 2008; **85**(4):1261–70.
139. Zeymer U, Suryapranata H, Monassier JP, Opolski G, Davies J, Rasmanis G, et al. The Na(+)/H(+) exchange inhibitor eniporide as an adjunct to early reperfusion therapy for acute myocardial infarction. Results of the evaluation of the safety and cardioprotective effects of eniporide in acute myocardial infarction (ESCAMI) trial. *J Am Coll Cardiol* 2001;**38**(6):1644–50.
140. Avkiran M, Cook AR, Cuello F. Targeting Na+/H+ exchanger regulation for cardiac protection: a RSKy approach? *Curr Opin Pharmacol* 2008;**8**(2):133–40.

141. Parks SK, Chiche J, Pouyssegur J. Disrupting proton dynamics and energy metabolism for cancer therapy. *Nat Rev Cancer* 2013;**13**(9):611–23.
142. Spugnini EP, Citro G, Fais S. Proton pump inhibitors as anti vacuolar-ATPases drugs: a novel anticancer strategy. *J Exp Clin Cancer Res* 2010;**29**:44.
143. Sakthivel S. ATP-ase as a potential drug target for cancer, tumor growth and cellular functions. *Int J Hum Genet* 2012;**12**(3):151–6.
144. Toei M, Saum R, Forgac M. Regulation and isoform function of the V-ATPases. *Biochemistry* 2010;**49**(23):4715–23.
145. Ma Y, Yang F, Wang Y, Du Z, Liu D, Guo H, et al. CaMKKbeta is involved in AMP-activated protein kinase activation by baicalin in LKB1 deficient cell lines. *PLoS ONE* 2012;**7**(10):e47900.
146. Sautin YY, Lu M, Gaugler A, Zhang L, Gluck SL. Phosphatidylinositol 3-kinase-mediated effects of glucose on vacuolar H+-ATPase assembly, translocation, and acidification of intracellular compartments in renal epithelial cells. *Mol Cell Biol* 2005;**25**(2):575–89.
147. Stransky LA, Forgac M. Amino acid availability modulates vacuolar H+-ATPase assembly. *J Biol Chem* 2015;**290**(45):27360–9.
148. Liberman R, Bond S, Shainheit MG, Stadecker MJ, Forgac M. Regulated assembly of vacuolar ATPase is increased during cluster disruption-induced maturation of dendritic cells through a phosphatidylinositol 3-kinase/mTOR-dependent pathway. *J Biol Chem* 2014;**289**(3):1355–63.
149. Trombetta ES, Ebersold M, Garrett W, Pypaert M, Mellman I. Activation of lysosomal function during dendritic cell maturation. *Science* 2003;**299**(5611):1400–3.
150. Marjuki H, Gornitzky A, Marathe BM, Ilyushina NA, Aldridge JR, Desai G, et al. Influenza A virus-induced early activation of ERK and PI3K mediates V-ATPase-dependent intracellular pH change required for fusion. *Cell Microbiol* 2011;**13**(4):587–601.
151. Ma B, Xiang Y, An L. Structural bases of physiological functions and roles of the vacuolar H(+)-ATPase. *Cell Signal* 2011;**23**(8):1244–56.
152. Breton S, Brown D. New insights into the regulation of V-ATPase-dependent proton secretion. *Am J Physiol Ren Physiol* 2007;**292**(1):F1–10.
153. Sennoune SR, Luo D, Martinez-Zaguilan R. Plasmalemmal vacuolar-type H+-ATPase in cancer biology. *Cell Biochem Biophys* 2004;**40**(2):185–206.
154. Jefferies KC, Cipriano DJ, Forgac M. Function, structure and regulation of the vacuolar (H+)-ATPases. *Arch Biochem Biophys* 2008;**476**(1):33–42.
155. Whitton B, Okamoto H, Packham G, Crabb SJ. Vacuolar ATPase as a potential therapeutic target and mediator of treatment resistance in cancer. *Cancer Med* 2018;**7**(8):3800–11.
156. Lu X, Chen L, Chen Y, Shao Q, Qin W. Bafilomycin A1 inhibits the growth and metastatic potential of the BEL-7402 liver cancer and HO-8910 ovarian cancer cell lines and induces alterations in their microRNA expression. *Exp Ther Med* 2015;**10**(5):1829–34.
157. Perez-Sayans M, Somoza-Martin JM, Barros-Angueira F, Rey JM, Garcia-Garcia A. V-ATPase inhibitors and implication in cancer treatment. *Cancer Treat Rev* 2009;**35**(8):707–13.
158. Cotter K, Liberman R, Sun-Wada G, Wada Y, Sgroi D, Naber S, et al. The a3 isoform of subunit a of the vacuolar ATPase localizes to the plasma membrane of invasive breast tumor cells and is overexpressed in human breast cancer. *Oncotarget* 2016;**7**(29):46142–57.
159. Hinton A, Sennoune SR, Bond S, Fang M, Reuveni M, Sahagian GG, et al. Function of a subunit isoforms of the V-ATPase in pH homeostasis and in vitro invasion of MDA-MB231 human breast cancer cells. *J Biol Chem* 2009;**284**(24):16400–8.
160. Otero-Rey EM, Somoza-Martin M, Barros-Angueira F, Garcia-Garcia A. Intracellular pH regulation in oral squamous cell carcinoma is mediated by increased V-ATPase activity via over-expression of the ATP6V1C1 gene. *Oral Oncol* 2008;**44**(2):193–9.
161. Lu X, Qin W, Li J, Tan N, Pan D, Zhang H, et al. The growth and metastasis of human hepatocellular carcinoma xenografts are inhibited by small interfering RNA targeting to the subunit ATP6L of proton pump. *Cancer Res* 2005;**65**(15):6843–9.
162. Chung C, Mader CC, Schmitz JC, Atladottir J, Fitchev P, Cornwell ML, et al. The vacuolar-ATPase modulates matrix metalloproteinase isoforms in human pancreatic cancer. *Lab Investig* 2011;**91**(5):732–43.

163. Xu X, You J, Pei F. Silencing of a novel tumor metastasis suppressor gene LASS2/TMSG1 promotes invasion of prostate cancer cell in vitro through increase of vacuolar ATPase activity. *J Cell Biochem* 2012;**113**(7):2356–63.
164. Son SW, Kim SH, Moon EY, Kim DH, Pyo S, Um SH. Prognostic significance and function of the vacuolar H+-ATPase subunit V1E1 in esophageal squamous cell carcinoma. *Oncotarget* 2016;**7**(31):49334–48.
165. Di Cristofori A, Ferrero S, Bertolini I, Gaudioso G, Russo MV, Berno V, et al. The vacuolar H+ ATPase is a novel therapeutic target for glioblastoma. *Oncotarget* 2015;**6**(19):17514–31.
166. Garcia-Garcia A, Perez-Sayans Garcia M, Rodriguez MJ, Antunez-Lopez J, Barros-Angueira F, Somoza-Martin M, et al. Immunohistochemical localization of C1 subunit of V-ATPase (ATPase C1) in oral squamous cell cancer and normal oral mucosa. *Biotech Histochem* 2012;**87**(2):133–9.
167. Xu J, Xie R, Liu X, Wen G, Jin H, Yu Z, et al. Expression and functional role of vacuolar H(+)-ATPase in human hepatocellular carcinoma. *Carcinogenesis* 2012;**33**(12):2432–40.
168. Lu Q, Lu S, Huang L, Wang T, Wan Y, Zhou CX, et al. The expression of V-ATPase is associated with drug resistance and pathology of non-small-cell lung cancer. *Diagn Pathol* 2013;**8**:145.
169. Kulshrestha A, Katara GK, Ibrahim S, Pamarthy S, Jaiswal MK, Gilman Sachs A, et al. Vacuolar ATPase 'a2' isoform exhibits distinct cell surface accumulation and modulates matrix metalloproteinase activity in ovarian cancer. *Oncotarget* 2015;**6**(6):3797–810.
170. Wang P, Wang L, Sha J, Lou G, Lu N, Hang B, et al. Expression and transcriptional regulation of human ATP6V1A gene in gastric cancers. *Sci Rep* 2017;**7**(1):3015.
171. Sennoune SR, Martinez-Zaguilan R. Plasmalemmal vacuolar H+-ATPases in angiogenesis, diabetes and cancer. *J Bioenerg Biomembr* 2007;**39**(5-6):427–33.
172. Cotter K, Stransky L, McGuire C, Forgac M. Recent insights into the structure, regulation, and function of the V-ATPases. *Trends Biochem Sci* 2015;**40**(10):611–22.
173. Song T, Jeon HK, Hong JE, Choi JJ, Kim TJ, Choi CH, et al. Proton pump inhibition enhances the cytotoxicity of paclitaxel in cervical cancer. *Cancer Res Treat* 2016;**49**(3):595–606.
174. Aasebo E, Bartaula-Brevik S, Hernandez-Valladares M, Bruserud O. Vacuolar ATPase as a possible therapeutic target in human acute myeloid leukemia. *Expert Rev Hematol* 2018;**11**(1):13–24.
175. Liu P, Chen H, Han L, Zou X, Shen W. Expression and role of V1A subunit of V-ATPases in gastric cancer cells. *Int J Clin Oncol* 2015;**20**(4):725–35.
176. Daniel C, Bell C, Burton C, Harguindey S, Reshkin SJ, Rauch C. The role of proton dynamics in the development and maintenance of multidrug resistance in cancer. *Biochim Biophys Acta* 2013;**1832**(5):606–17.
177. Larsen AK, Escargueil AE, Skladanowski A. Resistance mechanisms associated with altered intracellular distribution of anticancer agents. *Pharmacol Ther* 2000;**85**(3):217–29.
178. Hernandez A, Serrano-Bueno G, Perez-Castineira JR, Serrano A. Intracellular proton pumps as targets in chemotherapy: V-ATPases and cancer. *Curr Pharm Des* 2012;**18**(10):1383–94.
179. Drose S, Altendorf K. Bafilomycins and concanamycins as inhibitors of V-ATPases and P-ATPases. *J Exp Biol* 1997;**200**(Pt 1):1–8.
180. Licon-Munoz Y, Fordyce CA, Hayek SR, Parra KJ. V-ATPase-dependent repression of androgen receptor in prostate cancer cells. *Oncotarget* 2018;**9**(48):28921–34.
181. Morimura T, Fujita K, Akita M, Nagashima M, Satomi A. The proton pump inhibitor inhibits cell growth and induces apoptosis in human hepatoblastoma. *Pediatr Surg Int* 2008;**24**(10):1087–94.
182. von Schwarzenberg K, Wiedmann RM, Oak P, Schulz S, Zischka H, Wanner G, et al. Mode of cell death induction by pharmacological vacuolar H+-ATPase (V-ATPase) inhibition. *J Biol Chem* 2013;**288**(2):1385–96.
183. Merk H, Messer P, Ardelt MA, Lamb DC, Zahler S, Muller R, et al. Inhibition of the V-ATPase by archazolid A: a new strategy to inhibit EMT. *Mol Cancer Ther* 2017;**16**(11):2329–39.
184. Bartel K, Winzi M, Ulrich M, Koeberle A, Menche D, Werz O, et al. V-ATPase inhibition increases cancer cell stiffness and blocks membrane related Ras signaling—a new option for HCC therapy. *Oncotarget* 2017;**8**(6):9476–87.

185. Zhang S, Schneider LS, Vick B, Grunert M, Jeremias I, Menche D, et al. Anti-leukemic effects of the V-ATPase inhibitor archazolid A. *Oncotarget* 2015;**6**(41):43508–28.
186. Wiedmann RM, von Schwarzenberg K, Palamidessi A, Schreiner L, Kubisch R, Liebl J, et al. The V-ATPase-inhibitor archazolid abrogates tumor metastasis via inhibition of endocytic activation of the Rho-GTPase Rac1. *Cancer Res* 2012;**72**(22):5976–87.
187. Kubisch R, Frohlich T, Arnold GJ, Schreiner L, von Schwarzenberg K, Roidl A, et al. V-ATPase inhibition by archazolid leads to lysosomal dysfunction resulting in impaired cathepsin B activation in vivo. *Int J Cancer* 2014;**134**(10):2478–88.
188. Alavian KN, Beutner G, Lazrove E, Sacchetti S, Park HA, Licznerski P, et al. An uncoupling channel within the c-subunit ring of the F1FO ATP synthase is the mitochondrial permeability transition pore. *Proc Natl Acad Sci U S A* 2014;**111**(29):10580–5.
189. Schneider LS, von Schwarzenberg K, Lehr T, Ulrich M, Kubisch-Dohmen R, Liebl J, et al. Vacuolar-ATPase inhibition blocks iron metabolism to mediate therapeutic effects in breast cancer. *Cancer Res* 2015;**75**(14):2863–74.
190. von Schwarzenberg K, Lajtos T, Simon L, Muller R, Vereb G, Vollmar AM. V-ATPase inhibition overcomes trastuzumab resistance in breast cancer. *Mol Oncol* 2014;**8**(1):9–19.
191. Hiasa M, Okui T, Allette YM, Ripsch MS, Sun-Wada GH, Wakabayashi H, et al. Bone pain induced by multiple myeloma is reduced by targeting V-ATPase and ASIC3. *Cancer Res* 2017;**77**(6):1283–95.
192. Feng S, Zhu G, McConnell M, Deng L, Zhao Q, Wu M, et al. Silencing of atp6v1c1 prevents breast cancer growth and bone metastasis. *Int J Biol Sci* 2013;**9**(8):853–62.
193. Nishisho T, Hata K, Nakanishi M, Morita Y, Sun-Wada GH, Wada Y, et al. The a3 isoform vacuolar type H(+)-ATPase promotes distant metastasis in the mouse B16 melanoma cells. *Mol Cancer Res* 2011;**9**(7):845–55.
194. Schwartz MD. Dyspepsia, peptic ulcer disease, and esophageal reflux disease. *West J Med* 2002;**176** (2):98–103.
195. Luciani F, Spada M, De Milito A, Molinari A, Rivoltini L, Montinaro A, et al. Effect of proton pump inhibitor pretreatment on resistance of solid tumors to cytotoxic drugs. *J Natl Cancer Inst* 2004;**96** (22):1702–13.
196. De Milito A, Fais S. Proton pump inhibitors may reduce tumor resistance. *Expert Opin Pharmacother* 2005;**6**(7):1049–54.
197. Ward RM, Kearns GL. Proton pump inhibitors in pediatrics: mechanism of action, pharmacokinetics, pharmacogenetics, and pharmacodynamics. *Paediatr Drugs* 2013;**15**(2):119–31.
198. Yeo M, Kim DK, Kim YB, Oh TY, Lee JE, Cho SW, et al. Selective induction of apoptosis with proton pump inhibitor in gastric cancer cells. *Clin Cancer Res* 2004;**10**(24):8687–96.
199. De Milito A, Canese R, Marino ML, Borghi M, Iero M, Villa A, et al. pH-dependent antitumor activity of proton pump inhibitors against human melanoma is mediated by inhibition of tumor acidity. *Int J Cancer* 2010;**127**(1):207–19.
200. Goh W, Sleptsova-Freidrich I, Petrovic N. Use of proton pump inhibitors as adjunct treatment for triple-negative breast cancers. An introductory study. *J Pharm Pharm Sci* 2014;**17**(3):439–46.
201. Lee YY, Jeon HK, Hong JE, Cho YJ, Ryu JY, Choi JJ, et al. Proton pump inhibitors enhance the effects of cytotoxic agents in chemoresistant epithelial ovarian carcinoma. *Oncotarget* 2015;**6** (33):35040–50.
202. Perez-Sayans M, Somoza-Martin JM, Barros-Angueira F, Diz PG, Rey JM, Garcia-Garcia A. Multidrug resistance in oral squamous cell carcinoma: the role of vacuolar ATPases. *Cancer Lett* 2010;**295** (2):135–43.
203. De Milito A, Iessi E, Logozzi M, Lozupone F, Spada M, Marino ML, et al. Proton pump inhibitors induce apoptosis of human B-cell tumors through a caspase-independent mechanism involving reactive oxygen species. *Cancer Res* 2007;**67**(11):5408–17.
204. Lindner K, Borchardt C, Schopp M, Burgers A, Stock C, Hussey DJ, et al. Proton pump inhibitors (PPIs) impact on tumor cell survival, metastatic potential and chemotherapy resistance, and affect expression of resistance-relevant miRNAs in esophageal cancer. *J Exp Clin Cancer Res* 2014;**33**:73.
205. Azzarito T, Venturi G, Cesolini A, Fais S. Lansoprazole induces sensitivity to suboptimal doses of paclitaxel in human melanoma. *Cancer Lett* 2015;**356**(2 Pt B):697–703.

206. Chen M, Lu J, Wei W, Lv Y, Zhang X, Yao Y, et al. Effects of proton pump inhibitors on reversing multidrug resistance via downregulating V-ATPases/PI3K/Akt/mTOR/HIF-1alpha signaling pathway through TSC1/2 complex and Rheb in human gastric adenocarcinoma cells in vitro and in vivo. *Onco Targets Ther* 2018;**11**:6705–22.
207. Bakkali F, Averbeck S, Averbeck D, Idaomar M. Biological effects of essential oils—a review. *Food Chem Toxicol* 2008;**46**(2):446–75.
208. Ferreira de Oliveira JMP, Pacheco AR, Coutinho L, Oliveira H, Pinho S, Almeida L, et al. Combination of etoposide and fisetin results in anti-cancer efficiency against osteosarcoma cell models. *Arch Toxicol* 2018;**92**(3):1205–14.
209. Martins BX, Arruda RF, Costa GA, Jerdy H, de Souza SB, Santos JM, et al. Myrtenal-induced V-ATPase inhibition—a toxicity mechanism behind tumor cell death and suppressed migration and invasion in melanoma. *Biochim Biophys Acta Gen Subj* 2019;**1863**(1):1–12.
210. Halfdanarson OO, Fall K, Ogmundsdottir MH, Lund SH, Steingrimsson E, Ogmundsdottir HM, et al. Proton pump inhibitor use and risk of breast cancer, prostate cancer, and malignant melanoma: an Icelandic population-based case-control study. *Pharmacoepidemiol Drug Saf* 2019;**28**(4):471–8.
211. Liu Y, Yang J, Chen LM. Structure and function of SLC4 family [formula: see text] transporters. *Front Physiol* 2015;**6**:355.
212. Milanick MA, Gunn RB. Proton-sulfate cotransport: external proton activation of sulfate influx into human red blood cells. *Am J Phys* 1984;**247**(3 Pt 1):C247–59.
213. Jennings ML, Adame MF. Characterization of oxalate transport by the human erythrocyte band 3 protein. *J Gen Physiol* 1996;**107**(1):145–59.
214. Alper SL. Molecular physiology and genetics of Na^+-independent SLC4 anion exchangers. *J Exp Biol* 2009;**212**(Pt 11):1672–83.
215. Zhang D, Kiyatkin A, Bolin JT, Low PS. Crystallographic structure and functional interpretation of the cytoplasmic domain of erythrocyte membrane band 3. *Blood* 2000;**96**(9):2925–33.
216. Barneaud-Rocca D, Etchebest C, Guizouarn H. Structural model of the anion exchanger 1 (SLC4A1) and identification of transmembrane segments forming the transport site. *J Biol Chem* 2013;**288**(37):26372–84.
217. Becker HM, Klier M, Deitmer JW. Carbonic anhydrases and their interplay with acid/base-coupled membrane transporters. *Subcell Biochem* 2014;**75**:105–34.
218. Su Y, Blake-Palmer KG, Fry AC, Best A, Brown AC, Hiemstra TF, et al. Glyceraldehyde 3-phosphate dehydrogenase is required for band 3 (anion exchanger 1) membrane residency in the mammalian kidney. *Am J Physiol Renal Physiol* 2011;**300**(1):F157–66.
219. Sterling D, Alvarez BV, Casey JR. The extracellular component of a transport metabolon. Extracellular loop 4 of the human AE1 Cl^-/HCO_3-exchanger binds carbonic anhydrase IV. *J Biol Chem* 2002;**277**(28):25239–46.
220. Romero MF, Chen AP, Parker MD, Boron WF. The SLC4 family of bicarbonate (HCO(3)(-)) transporters. *Mol Asp Med* 2013;**34**(2-3):159–82.
221. Kopito RR, Lodish HF. Primary structure and transmembrane orientation of the murine anion exchange protein. *Nature* 1985;**316**(6025):234–8.
222. Lux SE, John KM, Kopito RR, Lodish HF. Cloning and characterization of band 3, the human erythrocyte anion- exchange protein (AE1). *Proc Natl Acad Sci U S A* 1989;**86**(23):9089–93.
223. Kollert-Jons A, Wagner S, Hubner S, Appelhans H, Drenckhahn D. Anion exchanger 1 in human kidney and oncocytoma differs from erythroid AE1 in its NH2 terminus. *Am J Phys* 1993;**265**(6 Pt 2):F813–21.
224. Aliberti G, Proietta M, Pulignano I, Di Giovanni C, Tritapepe L, Vercillo G. Respiratory changes in human red cells. *Clin Lab Haematol* 2001;**23**(6):361–3.
225. van Adelsberg JS, Edwards JC, al-Awqati Q. The apical Cl/HCO3 exchanger of beta intercalated cells. *J Biol Chem* 1993;**268**(15):11283–9.
226. Stewart AK, Kerr N, Chernova MN, Alper SL, Vaughan-Jones RD. Acute pH-dependent regulation of AE2-mediated anion exchange involves discrete local surfaces of the NH2-terminal cytoplasmic domain. *J Biol Chem* 2004;**279**(50):52664–76.

227. Brosius 3rd FC, Pisoni RL, Cao X, Deshmukh G, Yannoukakos D, Stuart-Tilley AK, et al. AE anion exchanger mRNA and protein expression in vascular smooth muscle cells, aorta, and renal microvessels. *Am J Phys* 1997;**273**(6 Pt 2):F1039–47.
228. Bonsib SM, Bromley C. Immunocytochemical analysis of band 3 protein in renal cell carcinoma, nephroblastoma, and oncocytoma. *Arch Pathol Lab Med* 1994;**118**(7):702–4.
229. Nehls V, Zeitler-Zapf P, Drenckhahn D. Different sequences of expression of band 3, spectrin, and ankyrin during normal erythropoiesis and erythroleukemia. *Am J Pathol* 1993;**142**(5):1565–73.
230. Kay MM, Vollard CH. Human erythroid band 3 "anion exchanger 1" is expressed in transformed lymphocytes. *Cell Mol Biol (Noisy-le-Grand)* 1996;**42**(7):945–52.
231. Zhao J, Lang R, Guo X, Chen L, Gu F, Fan Y, et al. Clinicopathologic characteristics of pleomorphic carcinoma of the breast. *Virchows Arch* 2010;**456**(1):31–7.
232. Kundu M, Basu J, Rakshit MM, Chakrabarti P. Abnormalities in erythrocyte membrane band 3 in chronic myelogenous leukemia. *Biochim Biophys Acta* 1989;**985**(1):97–100.
233. Fuerstenberg S, Leitner I, Schroeder C, Schwarz H, Vennstrom B, Beug H. Transcriptional repression of band 3 and CAII in v-erbA transformed erythroblasts accounts for an important part of the leukaemic phenotype. *EMBO J* 1992;**11**(9):3355–65.
234. Das AK, Sinha AK, Basu J. Erythrophagocytosis and its relation to band 3 clustering in chronic myelogenous leukemia. *Biochem Mol Biol Int* 1997;**41**(3):539–45.
235. Galletti J, Canones C, Morande P, Borge M, Oppezzo P, Geffner J, et al. Chronic lymphocytic leukemia cells bind and present the erythrocyte protein band 3: possible role as initiators of autoimmune hemolytic anemia. *J Immunol* 2008;**181**(5):3674–83.
236. Shiozaki A, Kudou M, Ichikawa D, Shimizu H, Arita T, Kosuga T, et al. Expression and role of anion exchanger 1 in esophageal squamous cell carcinoma. *Oncotarget* 2017;**8**(11):17921–35.
237. Xu WQ, Song LJ, Liu Q, Zhao L, Zheng L, Yan ZW, et al. Expression of anion exchanger 1 is associated with tumor progress in human gastric cancer. *J Cancer Res Clin Oncol* 2009;**135**(10):1323–30.
238. Shen WW, Wu J, Cai L, Liu BY, Gao Y, Chen GQ, et al. Expression of anion exchanger 1 sequestrates p16 in the cytoplasm in gastric and colonic adenocarcinoma. *Neoplasia (New York)* 2007;**9**(10):812–9.
239. Fu GH, Wang Y, Xi YH, Shen WW, Pan XY, Shen WZ, et al. Direct interaction and cooperative role of tumor suppressor p16 with band 3 (AE1). *FEBS Lett* 2005;**579**(10):2105–10.
240. Kawamoto S, Kamesaki T, Masutani R, Kitao A, Hatanaka K, Imakita M, et al. Ectopic expression of band 3 anion transport protein in colorectal cancer revealed in an autoimmune hemolytic anemia patient. *Hum Pathol* 2019;**83**:193–8.
241. Yang Y, Wu PP, Wu J, Shen WW, Wu YL, Fu AF, et al. Expression of anion exchanger 2 in human gastric cancer. *Exp Oncol* 2008;**30**(1):81–7.
242. Andrew AS, Gui J, Sanderson AC, Mason RA, Morlock EV, Schned AR, et al. Bladder cancer SNP panel predicts susceptibility and survival. *Hum Genet* 2009;**125**(5-6):527–39.
243. Karumanchi SA, Jiang L, Knebelmann B, Stuart-Tilley AK, Alper SL, Sukhatme VP. VHL tumor suppressor regulates Cl-/HCO3- exchange and Na+/H+ exchange activities in renal carcinoma cells. *Physiol Genomics* 2001;**5**(3):119–28.
244. Wu TT, Hsieh YH, Wu CC, Tsai JH, Hsieh YS, Huang CY, et al. Overexpression of anion exchanger 2 in human hepatocellular carcinoma. *Chin J Physiol* 2006;**49**(4):192–8.
245. Song LJ, Liu RJ, Zeng Z, Alper SL, Cui HJ, Lu Y, et al. Gastrin inhibits a novel, pathological colon cancer signaling pathway involving EGR1, AE2, and P-ERK. *J Mol Med (Berl)* 2012;**90**(6):707–18.
246. Wang T, Zhao L, Yang Y, Tian H, Suo WH, Yan M, et al. EGR1 is critical for gastrin-dependent upregulation of anion exchanger 2 in gastric cancer cells. *FEBS J* 2013;**280**(1):174–83.
247. Zhang LJ, Lu R, Song YN, Zhu JY, Xia W, Zhang M, et al. Knockdown of anion exchanger 2 suppressed the growth of ovarian cancer cells via mTOR/p70S6K1 signaling. *Sci Rep* 2017;**7**(1):6362.
248. Shiozaki A, Hikami S, Ichikawa D, Kosuga T, Shimizu H, Kudou M, et al. Anion exchanger 2 suppresses cellular movement and has prognostic significance in esophageal squamous cell carcinoma. *Oncotarget* 2018;**9**(40):25993–6006.
249. Tian H, Zhang N, Suo WH, Wang T, Song LJ, Wu J, et al. Gastrin suppresses the interdependent expression of p16 and anion exchanger 1 favoring growth inhibition of gastric cancer cells. *Int J Cancer* 2010;**127**(6):1462–74.

250. Wu J, Zhang YC, Suo WH, Liu XB, Shen WW, Tian H, et al. Induction of anion exchanger-1 translation and its opposite roles in the carcinogenesis of gastric cancer cells and differentiation of K562 cells. *Oncogene* 2010;**29**(13):1987–96.
251. Suo WH, Zhang N, Wu PP, Zhao L, Song LJ, Shen WW, et al. Anti-tumor effects of small interfering RNA targeting anion exchanger 1 in experimental gastric cancer. *Br J Pharmacol* 2012;**165**(1):135–47.
252. Liu CJ, Hwang JM, Wu TT, Hsieh YH, Wu CC, Hsieh YS, et al. Anion exchanger inhibitor DIDS induces human poorly-differentiated malignant hepatocellular carcinoma HA22T cell apoptosis. *Mol Cell Biochem* 2008;**308**(1-2):117–25.
253. Hwang JM, Kao SH, Hsieh YH, Li KL, Wang PH, Hsu LS, et al. Reduction of anion exchanger 2 expression induces apoptosis of human hepatocellular carcinoma cells. *Mol Cell Biochem* 2009;**327**(1-2):135–44.
254. Celay J, Lozano T, Concepcion AR, Beltran E, Rudilla F, Garcia-Barchino MJ, et al. Targeting the anion exchanger 2 with specific peptides as a new therapeutic approach in B lymphoid neoplasms. *Haematologica* 2018;**103**(6):1065–72.
255. Alper SL, Chernova MN, Williams J, Zasloff M, Law FY, Knauf PA. Differential inhibition of AE1 and AE2 anion exchangers by oxonol dyes and by novel polyaminosterol analogs of the shark antibiotic squalamine. Biochemistry and cell biology. *Biochim Biol Cell* 1998;**76**(5):799–806.
256. Arakawa T, Kobayashi-Yurugi T, Alguel Y, Iwanari H, Hatae H, Iwata M, et al. Crystal structure of the anion exchanger domain of human erythrocyte band 3. *Science* 2015;**350**(6261):680–4.
257. Martinez-Outschoorn UE, Peiris-Pages M, Pestell RG, Sotgia F, Lisanti MP. Cancer metabolism: a therapeutic perspective. *Nat Rev Clin Oncol* 2017;**14**(1):11–31.

CHAPTER 18

Treating the pH gradient inversion

Introduction

This book started a journey on page one. The journey's main goal was to get to this chapter. It went from metabolism of the cancer cell and the intricacies of pH homeostasis that shed some light on how pH is deregulated and co-opted for cancer's development. All the preceding information showed that this abnormal proton (hydrogen ion, the essence of pH) handling was not just a by-product of cancer, but it is at the root of the entire malignant transformation process. Then, a lengthy pharmacological description followed explaining how drugs can modify the proton handling. Now the time has come to put together all the pieces of the puzzle and discuss the main issue: cancer treatment.

While, the increased production of lactic acid and extracellular acidification in malignant tumors has been a known fact since the days of Otto Warburg,[1] little if anything, was known about acid-base status inside the cell. This was due to technical problems: it was easier to insert a probe into the tumor itself (registering pHe) than to introduce a probe inside the cell because old probes were relatively big and were not very accurate. pH indicator dyes that were also used, were quite inaccurate. In 1920 Merkel H. Jacobs[2] found that the flowers of *Symphytum peregrinum* changed color according to pH and he used this indicator to show that intracellular acidification was produced when the cell was incubated in a highly saturated CO_2 medium. In the 1960s, the distribution of a weak radio-labeled acid was frequently used as a means for determining pHi.

Up to the 1950s, there was a generalized belief that the intracellular pH (pHi) was mainly dependent on the extracellular pH (pHe)[3]; which implied a free distribution of ions across the cell membrane. This mistaken concept was popular in the scientific world, in spite of Jacobs's findings in 1920, showing that intracellular and extracellular pH were not fully interdependent. In 1959, using the distribution of oxazolidinedione, Waddell et al.[4] measured the intracellular pH of dog muscle and found that it was in the range of 6.96–7.10 with an average of 7.04.

Things started to change as new techniques, such as microelectrodes and better indicator dyes were developed. NMR was first introduced by Moon and Richards to measure pHi in 1973[5] and it allowed more accurate and reliable measurements. Before this innovation, the study of equilibrium of radiolabeled weak acids on both sides of the cell membrane gave only approximate data about pHi. Since pH is constructed on a logarithmic scale, small methodological errors were significant enough to make results unreliable. NMR changed this.

An Innovative Approach to Understanding and Treating Cancer: Targeting pH
https://doi.org/10.1016/B978-0-12-819059-3.00018-6
© 2020 Elsevier Inc.
All rights reserved.

It took almost 50 years from the first discoveries of Warburg to find that intracellular pH (pHi) is strictly regulated in normal eukaryotic cells. In 1988, measurements of intracellular pH (pHi) showed a neutral to slightly alkaline cytosol, 7.0–7.2 in the normal cell.[6]

In 1989 it was found that respiratory-impaired mutated fermentative yeasts decreased fermentation if the pHi was lowered with acetic acid. The decrease of fermentation was proportional to the lowering of pHi.[7] In 1993, Karuri et al.[8] suggested that weak organic acids can produce intracellular acidification when the microenvironment is acidic, and this intracellular acidification might be cytotoxic to cancer cells.

In the 1970s by recording internal Na^+ and pHi, the importance of Na^+ exchange through the cell membrane was found to deeply influence the pHi. Na^+ ions appeared to enter as pHi increased, presumably in exchange for the H^+ ions leaving the cell.[9]

The inversion of the pH gradient in cancer has been the "leitmotif" of this book. Now it is time to discuss an integral treatment strategy targeting this pH gradient inversion. This strategy must keep in mind that:

(1) Many different exchangers, channels, transporters, metabolic pathways, and lactate transporters participate in the process (they will be called "participants").
(2) The activity of all these participants is usually up-regulated in cancer, but they are also present in normal cells (although with a lower activity rate).
(3) This means that a total shutdown of the participants is not possible because it would be highly toxic to normal cells.
(4) Therefore, the strategy to implement has to aim at reducing the participants' overactivity/overexpression without fully inhibiting them.
(5) The inhibition of one or only two participants would not be enough to achieve good results because cancer cells can compensate through increasing expression/activity by the other participants. Cancer cells have repeatedly proven to be very versatile when their survival is at stake.
(6) Therefore, the strategy must seek to partially inhibit the activity of as many participants as possible without reaching toxicity levels for normal cells.
(7) An order of precedence of the different participants must be established, based on the importance of their contribution to the pH gradient inversion as well as the commercial availability of inhibitory drugs.

This order of precedence is shown in Table 1.

(1) The pH reversal treatment should not interfere with standard treatments.
(2) The pH reversal can be enhanced by other ancillary drugs that do not act directly on the pH.

With an eye toward this chapter, the preceding chapters described all the pharmaceuticals that will be included in this proposed treatment schedule.

In 2009, for the first time, an integrated anti-pH gradient inversion was proposed for cancer treatment,[10] that has been updated and reviewed in 2017.[11] The scheme presented here is the offspring of those two publications.

Table 1 Order of precedence of pH participants

Order	Participant	Importance	Drugs
1	NHE1	+++++	Available
2	NBC	+++++	Not available
3	Glycolysis	+++++	Available
4	MCTs	+++++	Available
5	CAIX AND CAXII	++++	Available
6	Proton pumps	++++	Available
7	VGSCs	++++	Available
8	Lactate	++++	Available
9	Sp1	++	Available
10	I.C. buffers	++	Available
11	E.C. buffers	++	Available

NHE1, sodium/hydrogen exchanger 1; *NBC*, sodium/bicarbonate cotransporter; *CAIX*, carbonic anhydrase 9; *CAXII*, carbonic anhydrase 12; *VGSCs*, voltage gated sodium channels; *MCTs*, monocarboxylate transporters; *Sp1*, transcription factor specificity protein 1; *I.C.*, intracellular; *E.C.*, extracellular.

The pH reversal drugs

(1) NHE1: Amiloride is the only NHE inhibitor available on the market. It is not a strong nor specific inhibitor, and there are others, which are more powerful such as cariporide. Unfortunately, cariporide is not FDA approved and it is not available. The dose of amiloride should be between 20 and 30 mg a day. However, as described in the chapter on NHE1 there are a number of drugs that inhibit NHE1 as part of their action.

(2) NBC: There is no NBC inhibitor that has been tested on patients. Existing drugs are for experimental use only.

(3) Glycolysis: There are some products for "pushing" the cell away from the glycolytic pathway and into the oxidative phosphorylation pathway, but none are FDA approved. Dichloroacetate (DCA) is probably the best known and used, although not in mainstream scientific circles. It still remains an experimental drug in cancer. It is sold as a nutritional supplement in scientifically unaccredited markets. The usual dose is 500 mg twice a day. Alpha lipoic acid (ALA) has been proposed instead. It is also known as tioctic acid and it is on the market for the treatment of hepatic diseases at dosages of 400 mg twice a day.

(4) MCTs: The only inhibitor of monocarboxylate transporters available on the market is the flavonoid quercetin. It is not FDA approved and it is sold as a food supplement. The dose has never been established, however amounts around 1000 mg are well tolerated. There is no human toxicity described in the literature.[12] Benign tumors were found in rat kidneys at very high doses (above 1000 mg/kg/day).[13] However, it may reduce cisplatin cytotoxicity.[14]

(5) **CAIX and CAXII**: These membrane-bound carbonic anhydrases can be inhibited with the nonspecific blocker acetazolamide (AZA). It is marketed as a diuretic, for the treatment of glaucoma, increased intracranial pressure and mountain sickness. The dose of AZA should be 125 mg three times a day.

(6) **Proton pumps**: Proton pump inhibitors are widely available and are FDA approved for the treatment of gastric/duodenal ulcer, gastritis, and esophagitis. From the long list of V-ATPase proton pump inhibitors, two have shown certain superiority in cancer treatment: esomeprazol and lansoprazole. The doses necessary for cancer treatment are higher than those used for acid-sensitive disease. The doses of esomeprazol are from 20 to 40 mg three times a day and lansoprazol 60 mg three times a day.

(7) **VGSCs**: There is a long list of inhibitors of VGSCs. Some of them like phenytoin have shown medium range anti-cancer activity. The problem with phenytoin is that it interferes with the metabolism of different chemotherapeutic drugs. A substitute is topiramate, a cell acidifier that is compatible with almost all chemotherapeutic drugs. Full dosages of topiramate are 200 mg twice a day starting with 50 mg a day with progressive weekly increments. It also inhibits Aquaporin 1 and is a mild inhibitor of carbonic anhydrases.

(8) **Lactate**: In order to curb lactate production it is necessary to "return" cells to the oxidative pathway. The drugs are the same as described for glycolysis.

(9) **Sp1**: This transcription factor that promotes many of the genes related to pH inversion is downregulated by tolfenamic acid. This NSAID, which was used for many years for the treatment of migraine, is not available in many countries for human use. It is still used in veterinary medicine. An alternative is celecoxib at 200–400 mg twice a day.

(10) **Extracellular buffers**: Can be managed with bicarbonate, but actually this is not necessary since the core treatment lowers intracellular and increases extracellular pH levels.

(11) **Intracellular buffers**: there are some buffers that can modify the intracellular buffering system, but they are only useful for experimental purposes.

The pH gradient inversion reversal scheme

Table 2 summarizes the basic schedule for the reversal of the cancer cell inversion of the pH gradient. This scheme has shown very low, or no toxicity at all and does not interfere with either standard chemotherapy drugs or radiotherapy. The main adverse events reported were constipation and asthenia. At the indicated dose, acetazolamide requires increased fluid intake to prevent kidney stone formation. Paresthesia due to acetazolamide is not infrequent and sometimes requires dosage adjustment.

Table 2 The pH reversal scheme

Order	Participant	Drugs
1	NHE1	Amiloride 15–30 mg daily and cimetidine 400 mg twice a day.
2	NBC	–
3	Glycolysis	Alpha lipoic acid 800 mg daily
4	CAIX AND CAXII	Acetazolamide 375 mg daily
5	MCTs	Quercetin 300–500 mg daily[15,16]
6	proton pumps	Lansoprazole 180 mg daily or esomeprazole 80 mg daily
7	VGSCs	Topiramate 400 mg daily
8	Lactate	–
9	Sp1	Celecoxib 400 mg daily
10	I.C. buffers	–
11	E.C. buffers	–

Then, we must add aspirin to this basic scheme for two reasons:

(1) it reduces the risk of vascular events due to celecoxib[17,18] and

(2) it inhibits platelet activation[19,20] that forms a protective shield around circulating cluster malignant cells.

The logic behind the scheme

The main therapeutic target (participant) of each component of the scheme is shown in Table 2. However, each component has additional effects that have been considered for the inclusion of the drug in the scheme. For example, quercetin, which is a weak inhibitor of MCTs, also plays a role in reducing cellular motility and inhibiting the glycolytic pathway and producing a decrease of lactic acid production.[21] Quercetin increases doxorubicin's effects on the tumor while reducing its toxicity.[22] It down-regulates the chemokine receptor, CXCR4, and other cell adhesion molecules that play a role in metastasis.[23] However, it should be carefully considered the association of quercetin with cisplatin. There is evidence indicating that this association may increase genotoxicity in normal cells.[24] Quercetin at low concentrations decreases the therapeutic effects of cisplatin through reduction of ROS generation.[25] On the other hand, there are publications indicating beneficial effects of the association.[26,27] The issue remains controversial.

Topiramate is a cell acidifier and blocks VGSCs, but it also inhibits carbonic anhydrases[28] and aquaporin1.[29] Its antiepileptic effects make topiramate very useful in the treatment of glioblastoma.[30–32]

The proton pump inhibitor esomeprazole has many anti-tumor activities not related to proton pump inhibition: it reduces the concentration of prostaglandin E2,[33] increases the effects of doxorubicin and attenuates its adverse events,[34] inhibits tumor cell survival, adhesion and migration and it synergizes with cisplatin and 5-FU, reducing metastatic potential.[35] It increases the cytotoxicity of paclitaxel[36] and, in general, esomeprazole increases the cytotoxicity of standard cancer treatments.[37] Lansoprazole induces apoptosis in breast cancer cells.[38] Regarding intracellular acidification, lansoprazole seems to be a stronger acidifier than esomeprazole, although there is a report in which lansoprazole increased intracellular pH.[28] Therefore, we have restricted esomeprazole to esophageal cancer and use lansoprazole in other tumors.

The potassium-sparing diuretic amiloride is a weak inhibitor of NHE when compared with cariporide (40-fold more potent), but it is the only NHE1 inhibitor available on the market. Amiloride's antitumoral effects, however, are remarkable,[39] because in addition to inhibiting NHE1 it also inhibits urokinase-type plasminogen activator (uPA), an important molecule for migration and invasion. Amiloride decreases metastasis[40,41] and this anti-metastatic activity is further increased by celecoxib.[42]

Cimetidine acts as an inhibitor of NHE1 and at the same time decreases metastatic potential through E-selectine inhibition. In addition it has antiangiogenic and immunomodulatory effects.

Celecoxib, a NSAID, COX2 inhibitor has also shown remarkable anti-tumoral effects, specifically: (1) Sp1 inhibitor[43]; (2) anti-angiogenesis; (3) anti-metastasis; (4) inducer of apoptosis in cancer cells[44–46]; (5) Akt inhibitor[47,48]; (6) synergic activity with paclitaxel and carboplatin[49]; (7) increases radiosensitivity[50]; (8) increases doubling time of PSA in prostate cancer[51]; (9) modulates nuclear factor kB[52]; (10) enhances mitochondrial ROS production; and (11) decreases respiration followed by apoptosis in metastatic cells[53] (Fig. 1 and Table 3).

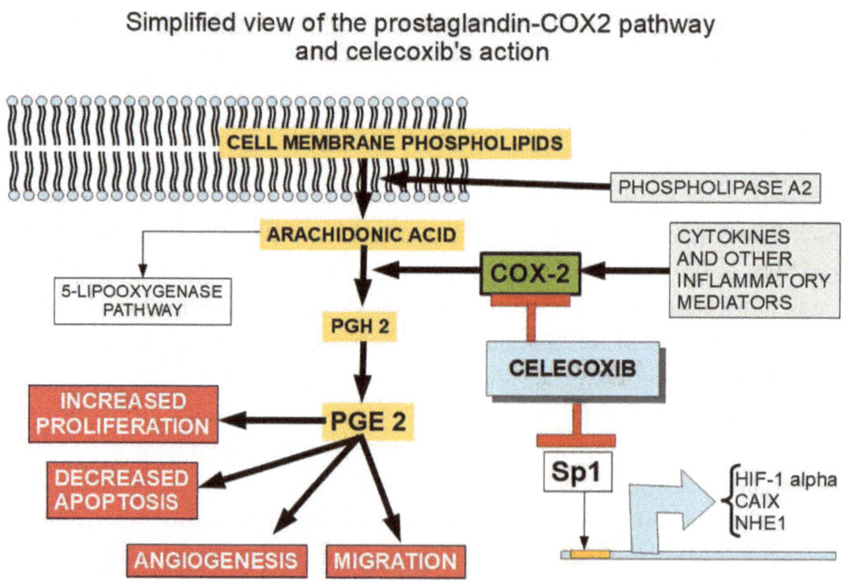

Fig. 1 Celecoxib is an indirect cell acidifier, because by down-regulating the Sp1 transcription factor it decreases the transcription of HIF-1α, carbonic anhydrase 9 (CAIX) and NHE1. The interruption of the COX2 pathway by celecoxib, contributes to the pH reversal treatment independently of its inhibition of the Sp1 transcription factor.

Table 3 Clinical evidence of celecoxib effects in cancer

Reference	Findings
Coombes et al. (2018)[54]	As a stand alone preventive treatment in a phase III multicenter double blind randomized trial of celecoxib versus placebo in primary breast cancer, patients did not show improvement in either overall survival or disease free survival.
Guylas et al. (2018)[55]	In a randomized trial between celecoxib and placebo in advanced non-small cell lung cancer (NSCLC) treated with chemotherapy, no differences were found in the two cohorts.
Araujo-Mino et al. (2018)[56]	In rectal cancer the addition of celecoxib to neoadjuvant chemotherapy showed a significant improvement in complete response.
Arber et al. (2011)[57]	In a 5 year follow up study of colonic sporadic adenoma formation, celecoxib significantly decreased the rate of new adenomas.
Pruthi et al. (2006)[58]	PSA elevation after treatment of prostate cancer showed significant flattening of slope when celecoxib was added vs. without celecoxib.

Analysis of Table 3: The results reported by Coombes et al. and Guylas et al. show that celecoxib is not useful either as a stand alone treatment or associated with chemotherapy in breast cancer and NSCLC. However, in rectal cancer the higher complete response (Araujo-Mino) and the decrease in new colonic adenoma rate in the study by Arber et al. suggest that the beneficial effects of celecoxib are tissue- or cancer-dependent. This last trial also showed that after 5 years, those receiving placebo had fewer adenomas than those on celecoxib. This probably means that after some time there is an escape phenomenon that makes COX2 inhibition ineffective. The reduction of PSA in the study by Pruthi et al. further suggests the tissue specificity of celecoxib in cancer.

Acetazolamide is a non-specific carbonic anhydrase inhibitor and it is also an Aquaporin1 inhibitor.[59] In cancer its main effect is the reduction of intracellular pH, reducing migration and invasion[60] and, therefore, it decreases metastasis.[61] As a collateral benefit, it reduces the nephrotoxicity of cisplatin.[62] Teicher et al.[63] have shown that acetazolamide as a stand-alone drug was able to significantly delay growth in a fibrosarcoma murine model. Its effects are further enhanced by chemotherapeutic drugs. Acetazolamide also modifies a protein that is related to CAII and intervenes in intracellular pH regulation: the chloride-bicarbonate exchanger AE1.[64] Inhibiting AE1 decreases bicarbonate transport into the cell, thereby decreasing the intracellular pH.

Alpha lipoic acid (ALA), that can be found in any pharmacy, is an alternative to the difficult-to-obtain, quality-certified dichloroacetate. Although it is not FDA approved, it is elaborated by pharmaceutical companies in Europe and South America and sold over the counter as a nutraceutical in the United States. ALA has shown anti-cancer effects such as cell cycle arrest and apoptosis in breast,[65] colon,[66] and hepatoma cancer cells,[67,68] and leukemia cells.[69] These effects are related to increased ROS (reactive oxygen species) due to restoration of oxidative phosphorylation. ALA reduced matrix metalloprotease activity in breast cancer cells.[70] Collateral benefits of ALA are: (1) possible risk-reduction of neuropathy in certain chemotherapeutic treatments[71] although a clinical trial did not confirm this[72]; (2) ALA attenuates myocardial damage caused by doxorubicin[73]; and (3) reduces the risk of nephropathy with cisplatin.[74] It also inhibits the activation of NF-kB in cell cultures.[75,76]

Associating all these compounds acidifies the intracellular milieu, induces apoptosis, decreases angiogenesis, migration, invasion and metastasis and also improves the cytotoxic effects of some chemotherapeutic drugs.

With this treatment there is a reduction in activity of all the proton-extruding systems, with the exception of NBC, that we have called the pH-tome, including the lactic acid production while ROS generation is increased (Fig. 2).

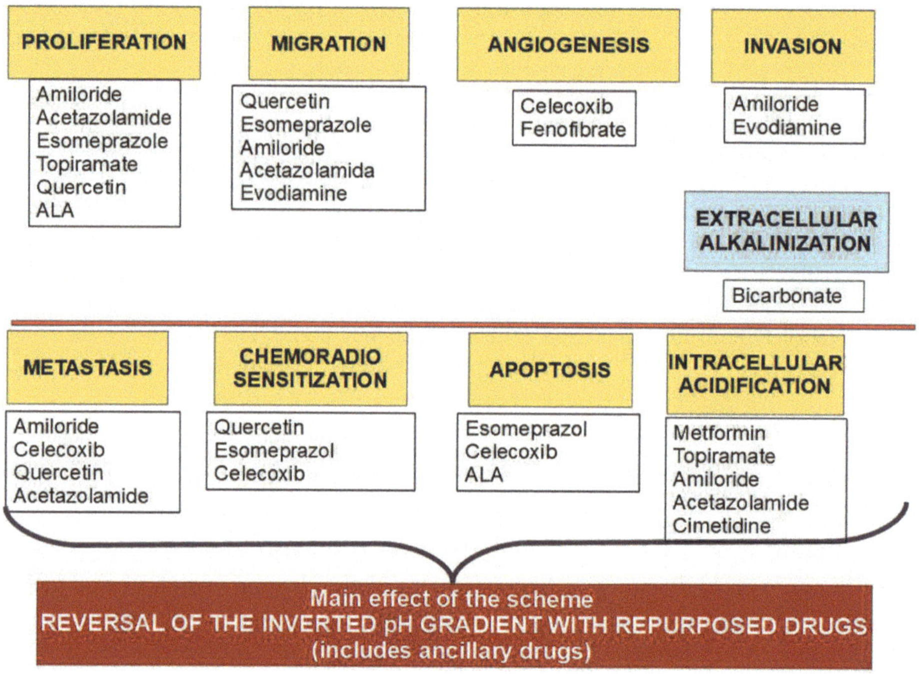

Fig. 2 A summary of the pH reversal treatment.

Some practical issues with the scheme

The central figure of the scheme is the inhibition of NHE1. Until newer inhibitors like cariporide[77] or compound 9t[78] are approved by FDA or come on the market, amiloride is the only available substance. The main problem with amiloride when used at high doses (more than 30 mg a day) is that hyperkalemia may develop. The usual symptoms are loss of appetite, nausea and vomiting. This is a frequent adverse event in older populations with decreased renal function. At the beginning of the treatment, the maximum tolerated dose should be carefully titrated. The association with another diuretic such as hydrochlorothiazide diminishes the risk of hyperkalemia. In most countries, the marketed product is an association of both in variable proportions: 5 mg of amiloride with 25 or 50 mg of hydrochlorothiazide. The third diuretic, acetazolamide contributes to the hydroelectrolytic imbalance. This means some paresthesias and muscle cramps. A dose schedule of 5 mg amiloride with 25 mg hydrochlorothiazide three times a day plus 125 mg of acetazolamide three times a day is usually well tolerated and does not cause hyperkalemia.

The rest of the components of the scheme, do not usually present any other problems.

The patients' main complaints were asthenia, constipation, paresthesias in face and limbs, muscular cramps and dysgeusia. None of these symptoms lead to interruption of the treatment.

When metformin was added to the scheme, constipation became a mounting problem that was difficult to handle. Finally, metformin was excluded from the treatment.

Cimetidine was originally part of the treatment for two reasons: (1) it works as an immunological modulator decreasing T regulatory cells[79–84] and (2) it decreases the activity of adhesion molecules[85–89] reducing the risk of metastasis. NHE1 inhibition was another asset of this drug. Unfortunately, it is no longer available on the market. In those countries, where cimetidine can be obtained, it should be used as part of the scheme.

This scheme has been tested in 11 patients with very advanced disease and multidrug resistance by one of us (TK) with encouraging results.

Ancillary drugs

The above scheme decreases intracellular pH and increases extracellular pH. However, cancer is a complicated process and many other factors must be dealt with: migration and angiogenesis, among others. The chapter on invadopodia discusses treating migration with evodiamine and dasatinib. The pH reversal scheme decreases migration. Adding evodiamine, which is a nutraceutical that lacks toxicity should decrease it further. However, this needs experimental testing.

Angiogenesis requires separate considerations. The association of pH inversion with metronomic chemotherapy seems to act synergistically against tumor proliferation and invasion, according to the experience with spontaneous cancer in dogs (unpublished). For the metronomic chemotherapy, the following scheme can be used: cyclophosphamide (50 mg a day for 6 days to a week) with 1 day of interruption in humans. In dogs weighing less than 10 kg, 12.5 mg were used and between 10 and 20 kg, 17.5 mg on a daily schedule. When metronomic chemotherapy was added to the pH reversal scheme, the cases achieved stable disease and partial or complete remission. Further, the progress free interval was much longer with important variations according to the type of tumor.

At this point we cannot determine if cyclophosphamide is the best choice for metronomic antiangiogenesis[90] but it is available for oral administration and this is a big advantage regarding our patients' comfort (see next chapter).

Metformin, as described in Chapter 16, is a cell acidifier, and this is the characteristic that makes it an anti-cancer drug. However, as with almost all the pharmaceuticals we have been discussing, it is not a stand-alone drug. Metformin "needs" something else to be fully active against cancer and shows synergies with DCA and 2-deoxy-glucose, two other cell acidifiers. Therefore, we believe that it would also synergize with the pH reversal scheme outlined above. Unfortunately, we lack experimental evidence. DCA and metformin are on the waiting list for admission to the scheme. The ideal companion to metformin would be a MCT inhibitor. In this case, lactic acid would accumulate in malignant cells without the possibility of being extruded. Intracellular acidity would increase to the point of inducing apoptosis, without affecting normal cells.

Fenofibrate should also be a good companion of the pH reversal scheme. Fenofibrate acts as an anti-angiogenic as it down-regulates the Vascular Endothelial Growth Factor (VEGF), Vascular Endothelial Growth Factor Receptor (VEGFR) and Hypoxia Inducible factor-1α (HIF-1α), and it inhibits endothelial cell migration.[91]

In general anti-angiogenics synergize with the pH reversal scheme (see next chapter).

We have no experience with bicarbonate as an extracellular buffer. However, there is some evidence (not in the context of cancer) showing that extracellular buffering with bicarbonate can reduce intracellular pH for a very short time but then is followed by intracellular alkalinization.[92]

Conclusions

The multidrug scheme for reducing intracellular pH described in this chapter is not a stand-alone therapy. However, one of us, has used it in many cases as a compassionate treatment and survival was extended well beyond expectations. Furthermore, the scheme improved quality of life. The number of cases treated was small and therefore no statistical analysis was possible.

The scheme should be tested in clinical trials as an adjunct to standard chemotherapeutic schedules as well as a compassionate treatment in patients that are beyond chemotherapy.

Anti-angiogenesis, whether with metronomic chemotherapy or fenofibrate would improve the results. This issue will be considered further in the next chapter.

References

1. Warburg O, Posener K, Negelein E. Uber den Stoffwechsel der Carcinomzelle. *Biochem Z* 1924;**152**:309–44.
2. Jacobs MH. The production of intracellular acidity by neutral and alkaline solutions containing carbon dioxide. *Am J Physiol* 1920;**53**(3):457–63.
3. Thomas RC. Review lecture: experimental displacement of intracellular pH and the mechanism of its subsequent recovery. *J Physiol* 1984;**354**(suppl):3–22.

4. Waddell WJ, Butler TC. Calculation of intracellular pH from the distribution of 5,5-dimethyl-2,4-oxazolidinedione (DMO). Application to skeletal muscle of the dog. *J Clin Invest* 1959;**38**(5):720–9.
5. Moon RB, Richards JH. Determination of intracellular pH by 31P magnetic resonance. *J Biol Chem* 1973;**248**(20):7276–8.
6. Madshus IH. Regulation of intracellular pH in eukaryotic cells. *Biochem J* 1988;**250**(1):1.
7. Pampulha ME, Loureiro-Dias MC. Combined effect of acetic acid, pH and ethanol on intracellular pH of fermenting yeast. *Appl Microbiol Biotechnol* 1989;**31**(5–6):547–50.
8. Karuri AR, Dobrowsky E, Tannock IF. Selective cellular acidification and toxicity of weak organic acids in an acidic microenvironment. *Br J Cancer* 1993;**68**(6):1080.
9. Murer H, Hopfer U, Kinne R. Sodium/proton antiport in brush-border-membrane vesicles isolated from rat small intestine and kidney. *Biochem J* 1976;**154**(3):597–604.
10. Harguindey S, Arranz JL, Wahl ML, Orive G, Reshkin SJ. Proton transport inhibitors as potentially selective anticancer drugs. *Anticancer Res* 2009;**29**(6):2127–36.
11. Harguindey S, Stanciu D, Devesa J, Alfarouk K, Cardone RA, Orozco JDP, et al. Cellular acidification as a new approach to cancer treatment and to the understanding and therapeutics of neurodegenerative diseases. In: *Seminars in cancer biology*. vol. 43. Academic Press; 2017. p. 157–79.
12. Harwood M, Danielewska-Nikiel B, Borzelleca JF, Flamm GW, Williams GM, Lines TC. A critical review of the data related to the safety of quercetin and lack of evidence of in vivo toxicity, including lack of genotoxic/carcinogenic properties. *Food Chem Toxicol* 2007;**45**(11):2179–205.
13. Dunnick JK, Hailey JR. Toxicity and carcinogenicity studies of quercetin, a natural component of foods. *Toxicol Sci* 1992;**19**(3):423–31.
14. Kuhlmann MK, Horsch E, Burkhardt G, Wagner M, Köhler H. Reduction of cisplatin toxicity in cultured renal tubular cells by the bioflavonoid quercetin. *Arch Toxicol* 1998;**72**(8):536–40.
15. Wu Q, Kroon PA, Shao H, Needs PW, Yang X. Differential effects of quercetin and its two derivatives (isorhamnetin and isorhamnetin-3-glucuronide) in inhibiting proliferation of human breast cancer MCF-7 cells. *J Agric Food Chem* 2018;**66**(27):7181–9.
16. Li S, Zhao Q, Wang B, Yuan S, Wang X, Li K. Quercetin reversed MDR in breast cancer cells through down-regulating P-gp expression and eliminating cancer stem cells mediated by YB-1 nuclear translocation. *Phytother Res* 2018;**32**(8):1530–6.
17. Mukherjee D, Nissen SE, Topol EJ. Risk of cardiovascular events associated with selective COX-2 inhibitors. *JAMA* 2001;**286**(8):954–9.
18. Graham DJ, Campen D, Hui R, Spence M, Cheetham C, Levy G, et al. Risk of acute myocardial infarction and sudden cardiac death in patients treated with cyclo-oxygenase 2 selective and non-selective non-steroidal anti-inflammatory drugs: nested case-control study. *Lancet* 2005;**365**(9458):475–81.
19. Schrör K. Aspirin and platelets: the antiplatelet action of aspirin and its role in thrombosis treatment and prophylaxis. In: *Seminars in thrombosis and hemostasis*. vol. 23(04). Copyright© 1997 by. Thieme Medical Publishers, Inc.; 1997. p. 349–56.
20. Malinin AI, Atar D, Callahan KP, McKenzie ME, Serebruany VL. Effect of a single dose aspirin on platelets in humans with multiple risk factors for coronary artery disease. *Eur J Pharmacol* 2003;**462**(1–3):139–43.
21. Jia L, Huang S, Yin X, Zan Y, Guo Y, Han L. Quercetin suppresses the mobility of breast cancer by suppressing glycolysis through Akt-mTOR pathway mediated autophagy induction. *Life Sci* 2018;**208**:123–30.
22. Li S, Yuan S, Zhao Q, Wang B, Wang X, Li K. Quercetin enhances chemotherapeutic effect of doxorubicin against human breast cancer cells while reducing toxic side effects of it. *Biomed Pharmacother* 2018;**100**:441–7.
23. Wang R, Yang L, Li S, Ye D, Yang L, Liu Q, et al. Quercetin inhibits breast cancer stem cells via down-regulation of aldehyde dehydrogenase 1A1 (ALDH1A1), chemokine receptor type 4 (CXCR4), mucin 1 (MUC1), and epithelial cell adhesion molecule (EpCAM). *Med Sci Monit* 2018;**24**:412.
24. Cross HJ, Tilby M, Chipman JK, Ferry DR, Gescher A. Effect of quercetin on the genotoxic potential of cisplatin. *Int J Cancer* 1996;**66**(3):404–8.
25. Li N, Sun C, Zhou B, Xing H, Ma D, Chen G, et al. Low concentration of quercetin antagonizes the cytotoxic effects of anti-neoplastic drugs in ovarian cancer. *PLoS One* 2014;**9**(7).

26. Wang Y, Han A, Chen E, Singh RK, Chichester CO, Moore RG, et al. The cranberry flavonoids PAC DP-9 and quercetin aglycone induce cytotoxicity and cell cycle arrest and increase cisplatin sensitivity in ovarian cancer cells. *Int J Oncol* 2015;**46**(5):1924–34.
27. Sánchez-González PD, López-Hernández FJ, Dueñas M, Prieto M, Sánchez-López E, Thomale J, et al. Differential effect of quercetin on cisplatin-induced toxicity in kidney and tumor tissues. *Food Chem Toxicol* 2017;**107**:226–36.
28. Supuran CT. Carbonic anhydrase inhibitors and their potential in a range of therapeutic areas. *Expert Opin Ther Pat* 2018;**28**(10):709–12.
29. Ma B, Xiang Y, Li T, Yu H-M, Li X-J. Inhibitory effect of topiramate on Lewis lung carcinoma metastasis and its relation with AQP1 water channel. *Acta Pharmacol Sin* 2004;**25**:54–60.
30. Marathe K, McVicar N, Li A, Bellyou M, Meakin S, Bartha R. Topiramate induces acute intracellular acidification in glioblastoma. *J Neurooncol* 2016;**130**(3):465–72.
31. Weller M, van den Bent M, Hopkins K, Tonn JC, Stupp R, Falini A, et al. EANO guideline for the diagnosis and treatment of anaplastic gliomas and glioblastoma. *Lancet Oncol* 2014;**15**(9):e395–403.
32. Maschio M, Dinapoli L, Zarabla A, Pompili A, Carapella CM, Pace A, et al. Outcome and tolerability of topiramate in brain tumor associated epilepsy. *J Neurooncol* 2008;**86**(1):61–70.
33. Falk GW, Buttar NS, Foster NR, Ziegler KL, DeMars CJ, Romero Y, et al. A combination of esomeprazole and aspirin reduces tissue concentrations of prostaglandin E2 in patients with Barrett's esophagus. *Gastroenterology* 2012;**143**(4):917–26.
34. Goh W, Sleptsova-Freidrich I, Petrovic N. Use of proton pump inhibitors as adjunct treatment for triple-negative breast cancers. An introductory study. *J Pharm Pharm Sci* 2014;**17**(3):439–46.
35. Lindner K, Borchardt C, Schöpp M, Bürgers A, Stock C, Hussey DJ, et al. Proton pump inhibitors (PPIs) impact on tumour cell survival, metastatic potential and chemotherapy resistance, and affect expression of resistance-relevant miRNAs in esophageal cancer. *J Exp Clin Cancer Res* 2014;**33**(1):73.
36. Song T, Jeon HK, Hong JE, Choi JJ, Kim TJ, Choi CH, et al. Proton pump inhibition enhances the cytotoxicity of paclitaxel in cervical cancer. *Cancer Res Treat* 2017;**49**(3):595.
37. Wang BY, Zhang J, Wang JL, Sun S, Wang ZH, Wang LP, et al. Intermittent high dose proton pump inhibitor enhances the antitumor effects of chemotherapy in metastatic breast cancer. *J Exp Clin Cancer Res* 2015;**34**(1):85.
38. Zhang S, Wang Y, Li SJ. Lansoprazole induces apoptosis of breast cancer cells through inhibition of intracellular proton extrusion. *Biochem Biophys Res Commun* 2014;**448**(4):424–9.
39. Matthews H, Ranson M, Kelso MJ. Anti-tumour/metastasis effects of the potassium-sparing diuretic amiloride: an orally active anti-cancer drug waiting for its call-of-duty? *Int J Cancer* 2011;**129**(9):2051–61.
40. Kellen JA, Mirakian A, Kolin A. Antimetastatic effect of amiloride in an animal tumour model. *Anticancer Res* 1988;**8**(6):1373–6.
41. Evans DM, Sloan-Stakleff K, Arvan M, Guyton DP. Time and dose dependency of the suppression of pulmonary metastases of rat mammary cancer by amiloride. *Clin Exp Metastasis* 1998;**16**(4):353–7.
42. Evans DM, Stakleff KD. Control of pulmonary metastases of rat mammary cancer by inhibition of uPA and COX-2, singly and in combination. *Clin Exp Metastasis* 2004;**21**(4):339–46.
43. Wei D, Wang L, He Y, Xiong HQ, Abbruzzese JL, Xie K. Celecoxib inhibits vascular endothelial growth factor expression in and reduces angiogenesis and metastasis of human pancreatic cancer via suppression of Sp1 transcription factor activity. *Cancer Res* 2004;**64**(6):2030–8.
44. Jendrossek V. Targeting apoptosis pathways by celecoxib in cancer. *Cancer Lett* 2013;**332**(2):313–24.
45. GroSch S, Tegeder I, Niederberger E, Bräutigam LU, Geisslinger G. COX-2 independent induction of cell cycle arrest and apoptosis in colon cancer cells by the selective COX-2 inhibitor celecoxib. *FASEB J* 2001;**15**(14):2742–4.
46. Arico S, Pattingre S, Bauvy C, Gane P, Barbat A, Codogno P, et al. Celecoxib induces apoptosis by inhibiting 3-phosphoinositide-dependent protein kinase-1 activity in the human colon cancer cell line HT-29. *J Biol Chem* 2002;**277**(31):27613–21.
47. Hsu AL, Ching TT, Wang DS, Song X, Rangnekar VM, Chen CS. The cyclooxygenase-2 inhibitor celecoxib induces apoptosis by blocking Akt activation in human prostate cancer cells independently of Bcl-2. *J Biol Chem* 2000;**275**(15):11397–403.

48. Kulp SK, Yang YT, Hung CC, Chen KF, Lai JP, Tseng PH, et al. 3-Phosphoinositide-dependent protein kinase-1/Akt signaling represents a major cyclooxygenase-2-independent target for celecoxib in prostate cancer cells. *Cancer Res* 2004;**64**(4):1444–51.
49. Altorki NK, Keresztes RS, Port JL, Libby DM, Korst RJ, Flieder DB, et al. Celecoxib, a selective cyclooxygenase-2 inhibitor, enhances the response to preoperative paclitaxel and carboplatin in early-stage non-small-cell lung cancer. *J Clin Oncol* 2003;**21**(14):2645–50.
50. Shin YK, Park JS, Kim HS, Jun HJ, Kim GE, Suh CO, et al. Radiosensitivity enhancement by celecoxib, a cyclooxygenase (COX)-2 selective inhibitor, via COX-2-dependent cell cycle regulation on human cancer cells expressing differential COX-2 levels. *Cancer Res* 2005;**65**(20):9501–9.
51. Smith MR, Manola J, Kaufman DS, Oh WK, Bubley GJ, Kantoff PW. Celecoxib versus placebo for men with prostate cancer and a rising serum prostate-specific antigen after radical prostatectomy and/or radiation therapy. *J Clin Oncol* 2006;**24**(18):2723–8.
52. Narayanan NK, Narayanan BA, Reddy BS. A combination of docosahexaenoic acid and celecoxib prevents prostate cancer cell growth in vitro and is associated with modulation of nuclear factor-κB, and steroid hormone receptors. *Int J Oncol* 2005;**26**(3):785–92.
53. Pritchard R, Rodríguez-Enríquez S, Pacheco-Velázquez SC, Bortnik V, Moreno-Sánchez R, Ralph S. Celecoxib inhibits mitochondrial O_2 consumption, promoting ROS dependent death of murine and human metastatic cancer cells via the apoptotic signalling pathway. *Biochem Pharmacol* 2018;**154**: 318–34.
54. Coombes RC, Tovey H, Kilburn L, Mansi J, Palmieri C, Bartlett J, et al. Abstract GS3-03: a phase III multicentre double blind randomised trial of celecoxib versus placebo in primary breast cancer patients (REACT—Randomised EuropeAn celecoxib trial). *Cancer Res* 2018;**78**(4). https://doi.org/10.1158/1538-7445.SABCS17-GS3-03.
55. Gulyas M, Mattsson JS, Lindgren A, Ek L, Lamberg Lundström K, Behndig A, et al. Swedish lung Cancer study group. COX-2 expression and effects of celecoxib in addition to standard chemotherapy in advanced non-small cell lung cancer. *Acta Oncol* 2018;**57**(2):244–50.
56. Araujo-Mino EP, Patt YZ, Murray-Krezan C, Hanson JA, Bansal P, Liem BJ, et al. Phase II trial using a combination of oxaliplatin, capecitabine, and celecoxib with concurrent radiation for newly diagnosed resectable rectal cancer. *Oncologist* 2018;**23**(1):2–5.
57. Arber N, Spicak J, Rácz I, Zavoral M, Breazna A, Gerletti P, et al. Five-year analysis of the prevention of colorectal sporadic adenomatous polyps trial. *Am J Gastroenterol* 2011;**106**(6):1135.
58. Pruthi RS, Derksen JE, Moore D, Carson CC, Grigson G, Watkins C, et al. Phase II trial of celecoxib in prostate-specific antigen recurrent prostate cancer after definitive radiation therapy or radical prostatectomy. *Clin Cancer Res* 2006;**12**(7):2172–7.
59. Bin K, Shi-Peng Z. Acetazolamide inhibits aquaporin-1 expression and colon cancer xenograft tumor growth. *Hepatogastroenterology* 2011;**58**(110 – 111):1502–6.
60. Parkkila S, Rajaniemi H, Parkkila AK, Kivelä J, Waheed A, Pastoreková S, et al. Carbonic anhydrase inhibitor suppresses invasion of renal cancer cells in vitro. *Proc Natl Acad Sci U S A* 2000;**97**(5):2220–4.
61. Yang XI, Ma B, Tao LI. Acetazolamide suppresses tumor metastasis and related protein expression in mice bearing Lewis lung carcinoma. *Acta Pharmacol Sin* 2002;**23**(8):745–51.
62. Osman NM, Copley MP, Litterst CL. Amelioration of cisplatin-induced nephrotoxicity by the diuretic acetazolamide in F344 rats. *Cancer Treat Rep* 1984;**68**(7–8):999–1004.
63. Teicher BA, Liu SD, Liu JT, Holden SA, Herman TS. A carbonic anhydrase inhibitor as a potential modulator of cancer therapies. *Anticancer Res* 1993;**13**:1549–56.
64. Sterling D, Reithmeier RA, Casey JR. A transport metabolon. Functional interaction of carbonic anhydrase II and chloride/bicarbonate exchangers. *J Biol Chem* 2001;**276**:47886–94.
65. Dozio E, Ruscica M, Passafaro L, Dogliotti G, Steffani L, Pagani A, et al. The natural antioxidant alpha-lipoic acid induces p27Kip1-dependent cell cycle arrest and apoptosis in MCF-7 human breast cancer cells. *Eur J Pharmacol* 2010;**641**(1):29–34.
66. Wenzel U, Nickel A, Daniel H. α-Lipoic acid induces apoptosis in human colon cancer cells by increasing mitochondrial respiration with a concomitant $O_2^{-}\cdot$-generation. *Apoptosis* 2005;**10**(2):359–68.
67. Shi DY, Liu HL, Stern JS, Yu PZ, Liu SL. Alpha-lipoic acid induces apoptosis in hepatoma cells via the PTEN/Akt pathway. *FEBS Lett* 2008;**582**(12):1667–71.

68. Simbula G, Columbano A, Ledda-Columbano GM, Sanna L, Deidda M, Diana A, et al. Increased ROS generation and p53 activation in α-lipoic acid-induced apoptosis of hepatoma cells. *Apoptosis* 2007;**12**(1):113–23.
69. Selvakumar E, Hsieh TC. Regulation of cell cycle transition and induction of apoptosis in HL-60 leukemia cells by lipoic acid: role in cancer prevention and therapy. *J Hematol Oncol* 2008;**1**(1):4.
70. Lee HS, Na MH, Kim WK. α-Lipoic acid reduces matrix metalloproteinase activity in MDA-MB-231 human breast cancer cells. *Nutr Res* 2010;**30**(6):403–9.
71. Melli G, Taiana M, Camozzi F, Triolo D, Podini P, Quattrini A, et al. Alpha-lipoic acid prevents mitochondrial damage and neurotoxicity in experimental chemotherapy neuropathy. *Exp Neurol* 2008;**214**(2):276–84.
72. Guo Y, Jones D, Palmer JL, Forman A, Dakhil SR, Velasco MR, et al. Oral alpha-lipoic acid to prevent chemotherapy-induced peripheral neuropathy: a randomized, double-blind, placebo-controlled trial. *Support Care Cancer* 2014;**22**(5):1223–31.
73. Al-Majed AA, Gado AM, Al-Shabanah OA, Mansour MA. Alpha-lipoic acid ameliorates myocardial toxicity induced by doxorubicin. *Pharmacol Res* 2002;**46**(6):499–503.
74. El-Beshbishy HA, Bahashwan SA, Aly HA, Fakher HA. Abrogation of cisplatin-induced nephrotoxicity in mice by alpha lipoic acid through ameliorating oxidative stress and enhancing gene expression of antioxidant enzymes. *Eur J Pharmacol* 2011;**668**(1–2):278–84.
75. Lee KY, D'Acquisto F, Hayden MS, Shim JH, Ghosh S. PDK1 nucleates T cell receptor-induced signaling complex for NFkappaB activation. *Science* 2005;**308**:114–8.
76. Suzuki YJ, Aggarwal BB, Packer L. Alpha-lipoic acid is a potent inhibitor of NF-kappa B activation in human T cells. *Biochem Biophys Res Commun* 1992;**189**:1709–15.
77. Harguindey S, Arranz JL, Orozco JD, Rauch C, Fais S, Cardone RA, et al. Cariporide and other new and powerful NHE1 inhibitors as potentially selective anticancer drugs—an integral molecular/biochemical/metabolic/clinical approach after one hundred years of cancer research. *J Transl Med* 2013;**11**(1):282.
78. Atwal KS, O'Neil SV, Ahmad S, Doweyko L, Kirby M, Dorso CR, et al. Synthesis and biological activity of 5-aryl-4-(4-(5-methyl-1Himidazol-4-yl) piperidin-1-yl) pyrimidine analogs as potent, highly selective, and orally bioavailable NHE-1 inhibitors. *Bioorg Med Chem Lett* 2006;**16**:4796–9.
79. Wang J, Su B, Ding Z, Du X, Wang B. Cimetidine enhances immune response of HBV DNA vaccination via impairment of the regulatory function of regulatory T cells. *Biochem Biophys Res Commun* 2008;**372**(3):491–6.
80. Zhang Y, Chen Z, Luo X, Wu B, Li B, Wang B. Cimetidine down-regulates stability of Foxp3 protein via Stub1 in Treg cells. *Hum Vaccin Immunother* 2016;**12**(10):2512–8.
81. Zheng Y, Xu M, Li X, Jia J, Fan K, Lai G. Cimetidine suppresses lung tumor growth in mice through proapoptosis of myeloid-derived suppressor cells. *Mol Immunol* 2013;**54**(1):74–83.
82. Pantziarka P, Bouche G, Meheus L, Sukhatme V, Sukhatme VP. Repurposing drugs in oncology (ReDO)—cimetidine as an anti-cancer agent. *Ecancermedicalscience* 2014;**8**.
83. Li B, Cao F, Zhu Q, Li B, Gan M, Wang D. Perioperative cimetidine administration improves systematic immune response and tumor infiltrating lymphocytes in patients with colorectal cancer. *Hepatogastroenterology* 2013;**60**(122):244–7.
84. Lefranc F, James S, Camby I, Gaussin JF, Darro F, Brotchi J, et al. Combined cimetidine and temozolomide, compared with temozolomide alone: significant increases in survival in nude mice bearing U373 human glioblastoma multiforme orthotopic xenografts. *J Neurosurg* 2005;**102**(4):706–14.
85. Kobayashi KI, Matsumoto S, Morishima T, Kawabe T, Okamoto T. Cimetidine inhibits cancer cell adhesion to endothelial cells and prevents metastasis by blocking E-selectin expression. *Cancer Res* 2000;**60**(14):3978–84.
86. Lefranc F, Yeaton P, Brotchi J, Kiss R. Cimetidine, an unexpected anti-tumor agent, and its potential for the treatment of glioblastoma. *Int J Oncol* 2006;**28**(5):1021–30.
87. Liu FR, Jiang CG, Li YS, Li JB, Li F. Cimetidine inhibits the adhesion of gastric cancer cells expressing high levels of sialyl Lewis x in human vascular endothelial cells by blocking E-selectin expression. *Int J Mol Med* 2011;**27**(4):537–44.

88. Takeuchi Y, Okayama N, Imaeda K, Okouchi M, Omi H, Imai S, et al. Effects of histamine 2 receptor antagonists on endothelial–neutrophil adhesion and surface expression of endothelial adhesion molecules induced by high glucose levels. *J Diabetes Complications* 2007;**21**(1):50–5.
89. Boueuf M, KovAkf J, TALrvov O. Inhibition of adhesion breast cancer cells by anticoagulant drugs and cimetidine. *Neoplasma* 2003;**50**:2.
90. Man S, Bocci G, Francia G, Green SK, Jothy S, Hanahan D, et al. Antitumor effects in mice of low dose (metronomic) cyclyophoshphamide administered continuously through the drinking water. *Cancer Res* 2002;**62**:2731–5.
91. Koltai T. Fenofibrate in cancer: mechanisms involved in anticancer activity. *F1000Research* 2015;**4**:55. https://doi.org/10.12688/f1000research.6153.2.
92. Goldsmith DJA, Forni LG, Hilton PJ. Bicarbonate therapy and intracellular acidosis. *Clin Sci* 1997;**93**(6):593–8.

CHAPTER 19

Metronomic anti-angiogenesis: The ideal companion of pH-centered treatments

Introduction

Maximum tolerated dose (MTD) chemotherapy: Classical conventional chemotherapeutic drugs damage DNA or the mitotic apparatus of proliferating cells whether benign or malignant, thus impeding replication. Therefore, standard cytotoxic chemotherapeutic drugs are used with the idea of treating cancer by directly killing, or inhibiting the growth of replicating cells.

Drug regimens have been designed to kill as many tumor cells as possible by treating them with "maximum tolerated doses" (MTDs) of these cytotoxic agents. Side effects such as neurotoxicity, cardiotoxicity, and damage to proliferating cells in healthy tissues (hematopoietic system, intestinal mucosa, etc.) pose serious problems and limitations to the use of chemotherapy.

Classical chemotherapy "kills the cells that can be killed," but not those that are resistant to their action or those that are "hidden" in very hypoxic areas that drugs are unable to reach. These resistant or hidden cells are the origin of relapse (through Darwinian selection) with new cells that do not respond to chemotherapy. Furthermore, the standard MTD regimens seriously impair quality of life and many patients are unable to complete the protocols.

Metronomic chemotherapy (MC) is defined as frequent uninterrupted administration of low dose cytotoxic chemotherapy for long periods of time. The doses should be low enough to avoid myelo-suppression, but at the same time, high enough to kill endothelial cells. This concept of targeting the tumor vasculature rather than the tumor cell itself, was first proposed by Folkman in 1971.[1] This treatment concept is based on the indispensable role of angiogenesis in tumor growth. The newly formed vessels of a tumor are selectively sensitive to low doses of various chemotherapeutic agents[2] when administered on a constant basis. These low doses of chemotherapeutics have a negligible effect on tumor proliferation. Tumors are unable to grow unless they can establish an adequate vascular infrastructure.[1] Therefore the effect of MC is suppression of tumor

An Innovative Approach to Understanding and Treating Cancer: Targeting pH
https://doi.org/10.1016/B978-0-12-819059-3.00019-8
© 2020 Elsevier Inc.
All rights reserved.

growth, by inhibiting angiogenesis[3,4] but without directly affecting the tumor mass itself. The goal of these types of treatments, which are successfully growing in veterinary medicine,[5] is to target the vascular endothelial cells, decreasing angiogenesis rather than killing tumor cells.[6–8] The basic idea is to use drugs that target not the transformed tumor cells themselves, but rather a genetically stable constituent cell type of tumors: the endothelial cells that form blood vessels. Angiogenesis, the process by which new blood vessels are formed, is a hallmark of cancer. "A considerable body of research spanning almost three decades has documented that tumor growth and metastasis require persistent new blood vessel growth."[9] And angiogenesis is not a simple event: "Endothelial cells must degrade their own basement membrane, develop sprouts from preexisting microvessels, invade the extracellular matrix, form tubes, and connect the tips of these tubes to create loops capable of handling blood flow."[10]

Endothelial cells that participate in the angiogenic process need migration and invasion abilities, so the pH inversion treatment should in theory also decrease angiogenesis.

Hanahan et al.[3] explained how to target these endothelial cells:

> *"It is well established that tumor-associated endothelial cells proliferate during chronic angiogenesis in tumors, albeit at lower frequencies than the tumor cells themselves. Apparently because of their lower rate of cell division, replication of these endothelial cells is only weakly disrupted by the episodic regimens of standard chemotherapeutic protocols. In these two new studies, however, cytotoxic drugs were administered routinely, to target the slowly proliferating tumor endothelial cells and abrogate their apparent capability to repair and recover during the usual rest periods. Both groups worked with mice bearing subcutaneous tumors, and each presents data suggesting that "metronomic" dosing regimens—either continuous infusion or frequent administration without extended rest periods—could have real value in the clinic. Both also demonstrated combinatorial effects of such altered cytotoxic drug regimens with newer, more specific angiogenesis inhibitors."*

What they proposed was the use of standard chemotherapeutic drugs in low doses (very far from MTD) using a continuous and permanent scheme (metronomic). This would impede the proliferation of endothelial cells on a chronic basis, without modifying the proliferation of tumor cells. The name "metronomic chemotherapy" meaning a regular and permanent administration of low dose of chemotherapeutic drugs targeting angiogenesis rather than the tumor itself was coined by Hanahan et al.

The concept of circulating endothelial progenitor cells as an important pro-tumoral factor goes a step beyond VEGF (vascular endothelial growth factor) as the main inducer of angiogenesis. The work by Bertolini et al.[11] showed that chemotherapy at MTD actually promoted circulating endothelial progenitors and therefore increased angiogenesis while the metronomic approach did exactly the opposite.

In metronomic chemotherapy, usually an "old" cancer chemotherapy drug is used but at a much lower dose. This is not a repurposed drug therapy but the use of an anticancer drug in a different way (for a review of MC read Mross et al.[12] and Bouche et al.[13]).

Most low dose metronomic protocols use the chemotherapic agent associated with a non-steroidal anti-inflammatory drug. In 2002, Colleoni et al.[14] published the first clinical experience with low dose cyclophosphamide and methotrexate in breast cancer patients. Their protocol was 2.5 mg of methotrexate twice a day the first and second day of each week and 50 mg/day of cyclophosphamide administered continuously. All the patients had been heavily pretreated and were in an advanced stage of the disease. The overall response rate in 63 patients was 19% with two complete and 10 partial remissions. All the patients showed a significant reduction in VEGF levels. In the experience of one of the authors (TK data not published,) with a small population of dogs with various natural cancers, the response rate was 50% (there were no heavily pretreated cases) with 25% complete remissions and 25% stable disease. The average remission period was 10 months. In this case the dogs received only 12.5 mg/10 kg wt. daily cyclophosphamide and 100 mg/10 kg wt. celecoxib (as COX2 inhibitor) Anti-angiogenic therapy used alone brings only modest benefits and for a limited time.[15–17] The reason is that even if they decrease the level of VEGF, modern anti-angiogenic drugs lack the ability of inhibiting other chemokines involved in the angiogenic process and resistance ensues after a short time. VEGFR tyrosine-kinase inhibitors, though successful in certain types of tumors (e.g., clear cell renal carcinoma) have very limited effects in many other cancers. These shortcomings of anti-angiogenic treatments gave us the idea to combine them with the pH inversion scheme. The results of this combination will be analyzed at the end of this chapter. There is an ongoing research for an ideal partner for anti-angiogenic therapies, starting from conventional chemotherapies to immunologic checkpoint inhibitors.[18, 19] However, this ideal partner has not been found yet.

Tumor-associated endothelial cells proliferate during chronic angiogenesis in tumors, albeit at a somewhat lower rate than the tumor cells. Due to this lower rate of mitosis, replication of tumor-associated endothelial cells is only weakly disrupted by the episodic administration of standard chemotherapeutic protocols at MTD. In metronomic schedules, cytotoxic drugs are administered routinely to target the slowly proliferating tumor endothelial cells and decrease their apparent ability to repair and recover during the usual rest periods. However, these tumor-associated endothelial cells (TAECs) are different from their normal counterpart. St Croix et al. identified 79 genes that were differently expressed in TAECs,[20] Lu et al.[21] found more than 400 genes differentially expressed in ovarian cancer TAECs and Akino et al.[22] identified frequent cytogenetic alterations in renal cell carcinoma TAECs. Endothelial cells of glioma are resistant to apoptosis by overexpressing GRP78 (a stress chaperone protein)[23] and survivin.[24] This creates resistance to conventional chemotherapies. Therefore, the classic idea that tumor endothelial cells are similar to normal endothelial cells is a mistaken concept. Actually, we do not know how anti-angiogenic schemes work with TAECs.

Cyclophosphamide (CTX)

We prefer cyclophosphamide as a chemotherapy drug for anti-angiogenesis due to many characteristics of this agent:

(1) It reduces angiogenesis.

(2) It improves the immune system attack on cancer cells. CTX has immunomodulatory effects, increasing anti-tumor immunologic activity; decreases T regulatory cells (Tregs), and immunosuppressive cytokines.[25–34] Even if the decrease of Tregs is a transient phenomenon, according to Ge et al.[35] it "induces stable tumor-specific T-cell responses, which correlate with improved clinical outcome in advanced-stage breast cancer."

(3) CTX can be administered orally.

(4) The low doses of CTX have very low risk of toxicity and are well tolerated.

(5) Low-dose metronomic CTX interacts with the different tumor microenvironment cells: endothelial cells, pericytes, circulating endothelial progenitor cells, and tumor-infiltrating lymphoid cells.[36]

(6) Enhanced effect in combination treatments with celecoxib[37, 38] and other chemotherapeutic drugs (paclitaxel, cisplatin, doxorubicin, sunitinib, everolimus, etc.).

(7) Enhances thrombospondin-1 expression.

(8) In dogs with natural tumors (splenic hemangiosarcoma), metronomic CTX seems to provide an effect similar to classic high-dose chemotherapy, but with much lower toxicity.[39]

(9) Orlando et al.[40] in 153 patients with metastatic breast cancer found that 15.7% of patients achieved prolonged clinical benefit with a median time to progression of 21 months.

(10) Metronomic CTX treatment can be associated with other chemotherapeutic drugs. Viale et al.[41] used cisplatin with CTX metronomic treatment in triple negative breast cancer. Bazzoli et al.[42] used letrozole, sorafenib and metronomic CTX in primary breast cancer and a complete response was observed in 6 out of 13 patients; a significant reduction in tumor size, was also found between baseline and 14 days of treatment in all 13 patients.

(11) CTX reduces VEGF expression and also decreases circulating endothelial progenitor cells. Both have pro-tumor activities in addition to angiogenesis.

(12) Oral CTX-based metronomic chemotherapy is the most widely studied metronomic regimen as a single-agent treatment or in combination and there is adequate experience with CTX in the clinical setting.[43]

Liyanage et al.[44] have shown that T regulatory cells co-expressing CD4 and CD25 are increased in peripheral blood and also in the tumor microenvironment of patients harboring breast or pancreatic carcinoma. These T regulatory cells also co-expressed CTLA-4 and CD45RO markers and secreted TGF-β and IL-10 but did not secrete IFN-γ. The direct consequence of this presence is a decreased immune response against cancer.

Other drugs commonly used in antiangiogenic protocols are methotrexate, capecitabine, etoposide, oral vinorelbin and vinblastine but by far the most used are cyclophosphamide and methotrexate, either alone or together.

There is now strong evidence to maintain that metronomic dosing protocols, with either continuous infusion or frequent administration without extended rest periods, could have real value in the clinic.

The MC approach has been slowly growing in the oncology armamentarium, however as Kareva et al.[45] wrote: "for most clinicians, metronomic chemotherapy remains a mostly palliative care tool rather than active, upfront therapy. This categorization of metronomic chemotherapy as palliative tool leads to a disregard of its synergism when used in combination with targeted biological agents and infrequent use in active treatment."

Low dose metronomic cyclophosphamide: Clinical and experimental evidence (Table 1)

Table 1 Evidence of low dose metronomic cyclophosphamide in cancer treatment

Reference	Study	Result
Man et al. (2002)[46]	Cyclophosphamide was administered to mice at low doses on a daily basis through the drinking water. Mice were xenotransplanted with human prostate, colon and melanoma cells.	Growth delay and complete regression in many cases that lasted for 60 days. Growth was restarted after that.
Hamano et al. (2002)[47]	Study of the effects of low dose metronomic CTX on endothelial cells and its action on thrombospondin-1 (TSP-1) production.	CTX induced selective expression of TSP-1 in tumor cells and perivascular cells and induced apoptosis of proliferating endothelial cells. TSP-1 contributed to tumor growth suppression induced by CTX.
Glode et al. (2003)[48]	Retrospective analysis of patients with hormone resistant prostate carcinoma who were treated with dexamethasone and cyclophosphamide.	70% of the patients showed PSA reduction. 30% had no benefits. Remission lasted for 8 months.
Bocci et al. (2003)[49]	Study performed on cells.	Metronomic CTX showed anti-angiogenic effects through increased TSP-1. Same findings with other chemotherapeutics used at low dose and in metronomic fashion.

Continued

Table 1 Evidence of low dose metronomic cyclophosphamide in cancer treatment—cont'd

Reference	Study	Result
Emmeneger et al. (2004)[50]	The study was aimed at discovering toxic effects of MC on normal tissues.	CTX at low metronomic dose may not only be at least as efficient as MTD therapy but also less toxic. Persistent lymphopenia was found.
Lord et al. (2007)[51]	Phase II hormone resistant prostate cancer in humans.	Response rate: 34.5% Median duration of response: 7.5 months
Samaritani et al. (2007)[52]	A case presentation of salvage therapy of an advanced ovarian cancer (stage III C) resistant to cisplatin treated with metronomic cyclophosphamide.	65 month progression free survival with no side effects.
Elmslie et al. (2008)[53]	85 Dogs with incompletely resected soft tissue sarcoma were treated with cyclophosphamide and piroxicam.	Tumor recurrence was significantly delayed when compared with untreated dogs.
Gebbia et al. (2012)[54]	Retrospective study in metastatic breast carcinoma in humans.	Tumor growth control in 35% of cases for 5 months.
Perroud et al. (2013)[55]	Phase I/II study of cyclophosphamide (50 mg) p.o. once a day and celecoxib 200 mg twice a day in advanced breast cancer. Population: 15 patients.	Partial response in 1 patient. Stable disease in 9 of 15 patients. Overall benefit in 15 patients: 60%. Stable disease for 37.5 weeks.
Ferrandina et al. (2014)[56]	Study of 54 patients with advanced ovarian cancer, refractory to standard treatments started treatment with metronomic cyclophosphamide.	20% Responded to treatment. The responders had an average PFS of 17 months. Median overall survival in responders was 30 months.
Montagna et al. (2014)[57]	Systematic review of results of phase I, II, and III studies testing metronomic therapy (MT) in breast cancer patients.	MT is a treatment option for breast cancer patients
Buckstein et al. (2016)[58]	High dose celecoxib and metronomic CTX were used in heavily pretreated patients with aggressive non Hodgkin lymphoma.	Results: response rate of 37% with PFS of 14.4 months in full response patients and PFS of 4.7 months in partial response cases.
Guercetti et al. (2018)[59]	Female mice were orthotopically injected with M234p triple negative cancer cells. The mice were divided into four groups: I control; II treated with cyclophosphamide; III treated with losartan and IV treated with cyclophosphamide and losartan.	Group IV showed tumor inhibition and higher survival with 60% of complete tumor regressions. They also found lower Ki67 and higher apoptosis. No toxicities were observed.

Metronomic cyclophosphamide-based treatment in dogs achieved better results than surgery and/or chemotherapy in many cases, plus a better quality of life.

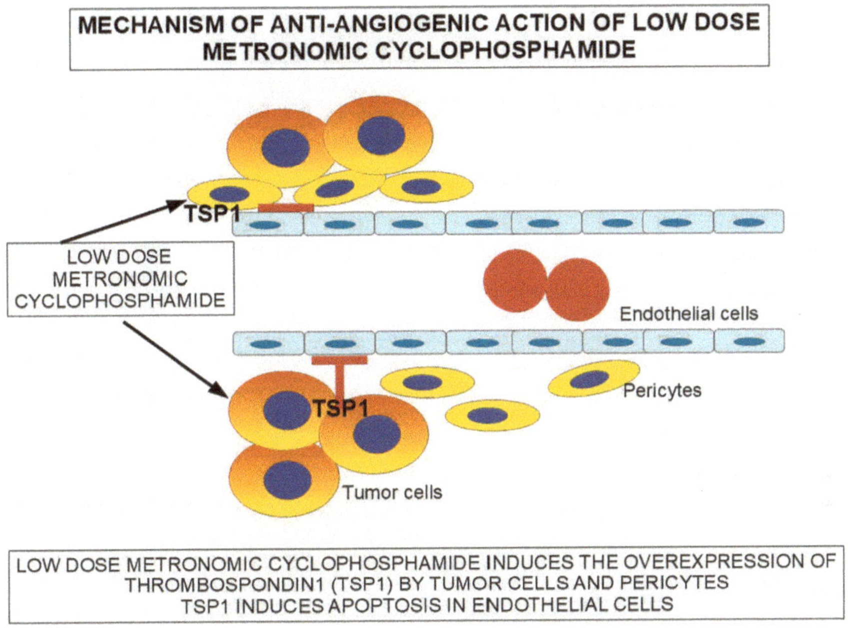

Fig. 1 Mechanism of action of low dose metronomic cyclophosphamide.[47]

Polton et al.[60] reported that in the case of advanced primary lung cancer in dogs the overall survival was significantly increased with MC as compared with surgery and chemotherapy.

Low dose metronomic cylophosphamide: Mechanism of action (Table 2)

One of the problems of anti-angiogenesis is that it decreases the uptake of other drugs used as part of the chemotherapy protocol[61] (Fig. 2). This may be partially compensated by increased drug retention. This also means that anti-angiogenesis does not normalize tumor vasculature in long-term treatment. Transient improvement of the vascular structure may be observed at the beginning of the treatment.[62]

Table 2 Mechanism of action of low dose metronomic cyclophosphamide in cancer

Mechanism of action (Fig. 1)
Antiangiogenesis
Depletion of T regulatory cells
Inhibition of T regulatory lymphocytes
Endothelial cells are more sensitive to low-dose chemotherapy compared with other types of normal cells
Increased production of thrombospondin-1

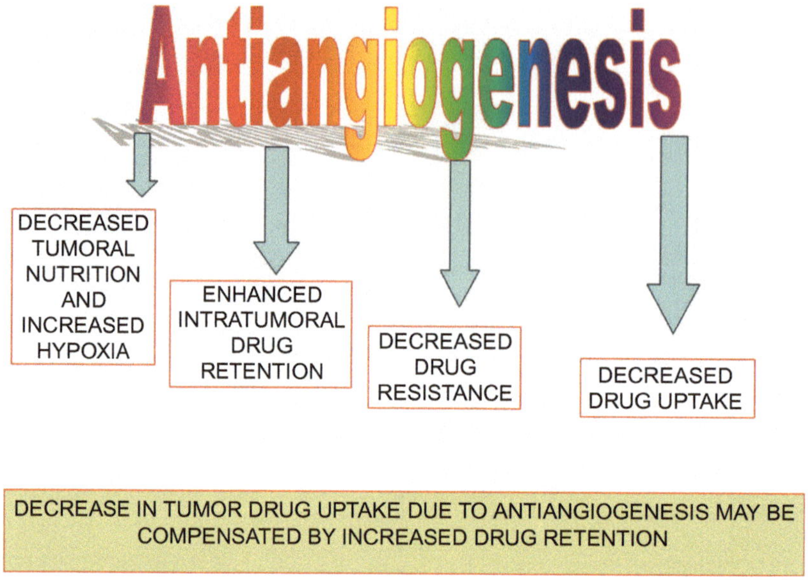

Fig. 2 Effects of anti-angiogenesis.

In addition to low dose metronomic cytotoxic drugs, other repurposed pharmaceuticals have been found to have anti-angiogenic activity. Two of them are:

(1) Ligands of the peroxisome proliferators-activated receptor α (PPARα) like fibrate hypolipidemic drugs.

(2) Ligands of the peroxisome proliferators-activated receptor γ (PPARγ) like thiazolidinedione antidiabetics.

Both PPARs are expressed in endothelial cells.

(1) Ligands of PPARα inhibit endothelial cell proliferation and tumor angiogenesis by down-regulating the expression of cytochrome *P*450 2C family epoxygenases, which convert arachidonic acid to epoxyeicosatrienoic acids (EET), which are potent angiogenic lipids.[63, 64] Fibrates decrease pro-angiogenic epoxygenase metabolites, hepatic EET biosynthesis, and Cyp2c epoxygenase expression.

This led us to incorporate fenofibrate in the anti-angiogenic schedule.

(2) Ligands of PPARγ increase expression of CD36, which is a receptor for thrombospondin-1 that works as an endogenous inhibitor of angiogenesis. (In cancer, the decrease of endogenous anti-angiogenic thrombospondin-1 (TSP1) often enables the angiogenic switch).[65]

Metronomic chemotherapy is not a stand-alone treatment, but a complement to other approved protocols. However, in cases of advanced disease and multidrug resistance, metronomic chemotherapy may achieve results by itself leading to stable disease for some time.

What is the best dose in metronomic chemotherapy?

The best dose is the one that achieves maximum anti-angiogenesis.[66] Shaked et al.[66] determined this optimal dose in mice corroborating anti-angiogenic activity by measuring the reduction of peripheral blood circulating VEGFR2 positive endothelial precursor cells (CEPs), and they proposed CEPs as a biomarker of anti-angiogenic activity.

In dogs with natural cancer, one of the authors (TK) determined that the best dose for cyclophosphamide was 0.8–1 mg/kg daily with no interruptions (unpublished material). This data was not checked with any biomarker, but through clinical tolerability and stable disease achieved.

Other possible components of the antiangiogenic scheme

The other two drugs employed in the anti-angiogenic scheme are celecoxib and fenofibrate. Importantly, Celecoxib was considered in the previous chapter, because it is also part of the anti-pH treatment.

Fenofibrate: is a fibric acid derivative that has been used for more than 40 years to reduce cholesterol and trygliceride levels in patients at risk of cardiovascular disease and for the treatment of atherosclerosis. Fenofibrate is one of the most commonly prescribed fibrates, and has a well known efficacy and safety profile.[67]

Fenofibrate is a PPARα (peroxisome proliferator activated receptor alpha) agonist. Peroxisome proliferator-activated receptors (PPARs) are widely distributed nuclear receptors. Ligand binding activates these receptors that play key roles in the transcription of genes related to cellular energy, modulating glucose and lipid metabolism. Fenofibrate is a ligand to these receptors.

Fenofibrate's main mechanism involved in anti-cancer activity is antiangiogenesis through downregulation of VEGF, VEGFR and HIF-1α, inhibition of endothelial cell migration, upregulation of endostatin and thrombospondin-1. But there are many other contributing mechanisms such as apoptosis, cell cycle arrest, down-regulation of NF-kB and Akt and cellular energy depletion by impairing mitochondrial function. Growth impairment is related to down-regulation of the PI3K/Akt axis and down-regulation of the p38 MAPK cascade. A possible role should be assigned to fenofibrate-stimulated overexpression of TRIB3 (tribbles homolog 3 is considered a down-regulator of NF-kB), which inhibits Akt phosphorylation. Important anti-cancer and metastatic activities due to down-regulation of MCP-1 (monocyte chemotactic protein-1), decreased MMP-9 production, weak downregulation of adhesion molecules such as E selectin, ICAM and VCAM and decreased secretion of chemokines like IL-6, and down-regulation of cyclin D-1. There is no direct link between fenofibrate's activity on lipid metabolism and anticancer activity, except for the fact that many anticancer effects are dependent on PPARα agonism. Fenofibrate exhibits also PPARα independent anti-cancer activities.

Fenofibrate's antiangiogenesis stems from PPARα agonism, inhibiting endothelial cell migration by targeting Akt phosphorylation[68] and inhibiting endothelial VEGFR2 expression.[69–71] Fenofibrate has a tissue-specific activity, because in the retina it does exactly the opposite, it protects retinal endothelial cells from apoptosis.[72–74]

A summary of fenofibrate's antitumoral effects is shown in Fig. 3.

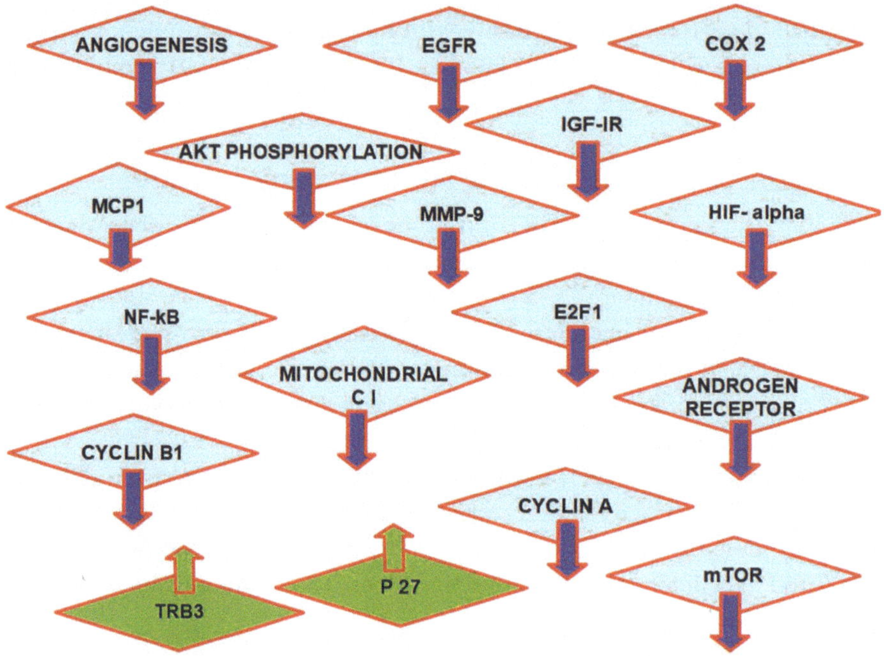

Fig. 3 A summary of fenofibrate's antitumoral activities: decreases angiogenesis[75]; decreases NF-kB signaling[76]; decreases Akt phosphorylation[77]; suppresses HIF-1α signaling[78]; reduces E2F2 and cyclin A activity[79]; impairs mitochondrial complex I activity[80, 81]; decreases COX2 expression[82, 83]; inhibits IGF-R[84]; down-regulates the androgen receptor in prostate[85]; increases the expression of p27[86]; down-regulates mTOR[87]; reduces cyclin B1 expression[88]; up-regulates TRB3[89]; and reduces the secretion of MMP9.

For all its antitumoral characteristics, fenofibrate is the ideal companion to celecoxib and metronomic chemotherapy. However, fenofibrate does not cross the blood brain barrier, which makes it useless for treating glioblastoma. The fact that it is highly cytotoxic to glioblastoma cells, led to the synthesis of a derivative which maintains all the cytotoxic abilities of fenofibrate and at the same time penetrates into the nervous system. The authors called it PP1.[90]

Resistance to antiangiogenic treatment

In theory, antiangiogenic treatment targets only endothelial cells. And endothelial cells are not supposed to be mutated, so that a Darwinian selection of resistant endothelial cells should not ensue.[91] Thus, there should not be resistance to antiangiogenic treatments.

This classical concept is incorrect. Endothelial cells in cancer are frequently over-expressing proteins that are not increased in the normal counterparts.

But this is only theory. In practice, there is resistance to angiogenic treatments and patients become unresponsive after an average of 1 year. What actually happens is that resistance comes from the malignant cells and the endothelial cells. Malignant cells find the way to escape from the extra hypoxia created by the inhibition of angiogenesis.

One of the escape mechanisms is probably through more glycolytic metabolism and therefore more proton production. The simultaneous use of anti-pH inversion drugs and anti-angiogenic drugs makes this escape less possible. That would explain the better results we obtained with the combination of both treatments.

The antiangiogenic schedule

Our treatment protocol consists of:

1. metronomic cyclophosphamide;
2. celecoxib;
3. fenofibrate.

Clinical experience

In a phase II clinical trial conducted on 37 patients over 60 years of age with advanced cancers of diverse origins, celecoxib (400 mg/b.i.d.) plus metronomic capecitabine (500 mg/b.i.d.) was administered continuously, and in one third of the cases stable disease was achieved. A reduction of tumor vessel permeability was found in this study.[92]

Dellapasqua et al.[93] used a combination of metronomic cyclophosphamide (50 mg/day), metronomic capecitabine (500 mg/t.i.d.) plus the anti-angiogenic drug bevacizumab (10 mg/kg every 2 weeks) in 46 patients with advanced breast cancer. They found one case of complete response, 21 partial responses and 19 cases of stable disease. The overall clinical benefit for more than 24 weeks was 68%. The median time for progression was 42 weeks.

In metastatic breast cancer, Soriano Gracia et al.[94] administered 50 mg of cyclophosphamide daily for 5 days a week and methotrexate 2.5 mg/b.i.d. for 2 days a week to 28 patients with complete and partial response in half of the cases and stabilization in 8 cases with an overall benefit (CR + PR + SD) of 78% with average response length of 8 months.

The experience of one of us (TK) in dogs with natural cancers had three stages. In the first stage, only metronomic chemotherapy was used in cases that had no previous treatments. One of the dogs was lost for follow up and from the remaining seven, four showed response to treatment with two dogs in complete remission and two with stable disease.

The average progression-free survival was 10 months among the responders and less than 4 months in non-responders. In a later experience with another seven dogs, the treatment consisted of a pH reversal approach. Five dogs showed stable disease and one dog partial response. This group included dogs that had received previous chemotherapy, and in general had a more advanced disease than those of the first group. The average survival was 11 months. Finally, a third group of dogs with natural cancers was treated with a combined anti-angiogenic/anti-pH treatment. This group had similarities with the second group in the sense that they had received chemotherapy before starting the treatment but the tumor had relapsed or metastasized. Six responded to treatment and the interesting finding was that the overall survival was 22 months. These unpublished data led to the idea that the two treatments should be associated. This research has a relatively low value, because there were many confounding factors: the different groups had different types of cancers, had been treated previously with different chemotherapeutic protocols and the stage of the disease was also quite diverse. No comparisons can be made among the three groups. However, the striking difference in overall survival and progression-free survival is a strong indicator that these two treatments should be administered together. Undoubtedly, these findings require further research with a larger population and similar characteristics in the three cohorts.

Discussion

Ever since Folkman's 1971[1] publication that pinpointed, for the first time, angiogenesis as a fundamental process in the tumor development a great deal of research has been dedicated to this issue. This led to the present "state of the art" with many FDA-approved anti-angiogenic drugs in use. However anti-angiogenesis has demonstrated important limitations to achieve sustainable results.[95, 96] That is the reason why anti-angiogenic treatments are usually combined with other classical chemotherapeutic protocols.

Even after initial success of anti-angiogenic treatments, resistance usually develops after approximately 6–10 months of treatment. The mechanisms involved in resistance are not clearly known.

The idea of associating anti-angiogenesis with pH-centered treatments goes toward avoiding resistance. At the present we do not know for sure, if this is achievable. What we do know is that angiogenesis would bring pH inversion one step closer to a complete reshuffling of the tumor environment.

In 2009, two publications reported that anti-angiogenic treatments may increase metastasis.[97, 98]

Paez-Ribes et al.[98] said: "...angiogenesis inhibitors targeting the VEGF pathway demonstrate antitumor effects in mouse models of pancreatic neuroendocrine carcinoma and glioblastoma but concomitantly elicit tumor adaptation and progression to stages of greater malignancy, with heightened invasiveness and in some cases increased lymphatic and distant metastasis. Increased invasiveness is also seen by genetic ablation of the Vegf-A gene."

In both cases, the research was done with drugs that target VEGF. This is not the case of the CTX-celecoxib association where the target is an increase in TSP-1, with a partial reduction in VEGF. While, the pro-metastatic effect of anti-angiogenesis has never been tested with CTX-celecoxib, CTX-celecoxib does not interfere directly with VEGF and VEGF expression never decreases beyond 40%. As pH inversion decreases migration, invasion and metastasis, we firmly believe that the association of a protocol based on metronomic CTX-celecoxib with pH inversion may probably avoid the risk of increased metastasis found with VEGF targeting alone.

We think that associating anti-angiogenesis with pH inversion may be an important step forward to avoid treatment resistance and increased invasion found with anti-angiogenesis alone.

As we know, the development of resistance to antiangiogenesis is one of the limiting factors of its benefits. The experience one of us (TK, discussed above) had with dogs with natural tumors showed that the simultaneous use of pH-centered treatment and anti-angiogenesis substantially delayed the onset of this phenomenon. On a speculative basis we think that this delay is related to the antagonism against known resistance factors.

Among more than 30 such factors we could identify a few that can be prevented or counteracted using pH-centered treatments, such as:

(1) **Increased invasion and metastasis with anti-angiogenic treatments.**[99] pH centered treatments reduce invasion through inhibition of NHE1, reduce motility, and reduce extracellular acidity with concomitant lower proteinase activation. Fenofibrate reduces MMPs expression while amiloride is an NHE1 inhibitor. The combined effect of amiloride, acetazolamide, and proton pump inhibitors increase extracellular pH.

(2) **Increased activity of other growth factors (growth factor redundancy).**[100] Fenofibrate that is an important ancillary drug in the pH-centered scheme and in anti-angiogenesis, has the ability to decrease EGFR and IGF-IR signaling, besides decreasing Akt phosphorylation[79, 101] which forms part of the pathway of many growth factor receptors.

(3) **Recruitment of bone marrow-derived cells like myeloid derived suppressor cells.**[102, 103] Cimetidine antagonizes the presence and the activity of these cells at the tumor level.[104, 105]

(4) **MCP1 (monocyte chemotactic protein-1)**[106] is involved in resistance to anti-angiogenesis. Fenofibrate downregulates MCP1 and other monocyte- derived inflammatory cytokines.[107, 108]
(5) **Increased expression of NF-kB** that recruits myeloid-derived tumor cells. After dexametasone, celecoxib is the most powerful NF-kB down-regulator.[109, 110]
(6) **Increased expression of aquaporin 1 is associated with increased angiogenesis.**[111] Acetazolamide inhibits AQP1.
(7) **Extracellular acidity is pro-angiogenic through up-regulation of angiogenic factors VEGF, cathepsin B, metalloproteases 2 and 9, and IL8.**[112, 113] Therefore, the pH reversal treatment has the ability to modify this situation by increasing extracellular pH.
(8) **V-ATPase proton pump inhibitors block endothelial cell migration.**[114] It has been suggested that V-ATPase proton pumps play an essential role in cancer-related angiogenesis.[115]
(9) **Tumor exosomes containing CD147 are angiogenic.**[116] Amiloride and the change in extracellular acidity decreases the release and internalization of exosomes.
(10) Carbonic anhydrase inhibition (that is part of the pH inversion scheme) heightens the activity of anti-angiogenic drugs like bevacizumab[117] in cholangiocarcinoma, and imatinib[118] and in other tumors,[119] and acetazolamide has anti-angiogenic effects.[120] On the other hand, cimetidine also has anti-angiogenic effects.[121]

Conclusions

The clinical outcome of anti-angiogenic treatments (e.g., metronomic chemotherapy) is extremely modest due to resistance, increased invasion and metastasis. The fundamentals of the anti-angiogenic approach need re-evaluation and probably analysis from a different angle. It is quite clear that anti-angiogenesis alone cannot do the job. Something else is needed.

There is evidence that anti-angiogenic metronomic chemotherapy and pH inversion reversal treatments may synergize and improve the results of cancer treatments. The exact mechanism that lies behind the synergy of these two therapies has not yet been fully clarified. However, it is possible that the pH-centered treatment decreases the invasiveness that seems to be increased by anti-angiogenic treatments. Fenofibrate plays a pivotal role between these two types of treatments. The reversion of the inverted pH gradient decreases angiogenesis and delays resistance to anti-angiogenic treatment. Therefore, we firmly believe that the association of pH-centered and anti-angiogenic treatments merits testing in clinical trials (Fig. 4).

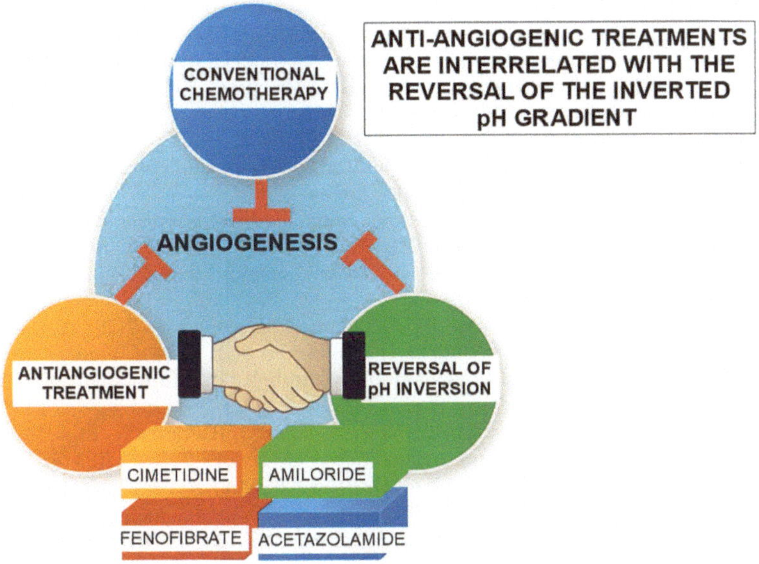

Fig. 4 There are many other possible interactions between these two treatment modalities but our knowledge is limited and this issue deserves further research. However, they seem to work like communicating vessels, which explains the need to address angiogenesis at the same time as the inverted pH.

References

1. Folkman J. Tumor angiogenesis: therapeutic implications. *N Engl J Med* 1971;**285**:1182–6.
2. Hirata S, Matsubara T, Saura R, Tateishi H, Hirokata K. Inhibition of in vitro vascular endothelial cell proliferation and in vivo neovascularization by low dose methotrexate. *Arthritis Rheum* 1989; **32**: 1065–73.
3. Hanahan D, Bergers G, Bergsland E. Less is more, regularly: metronomic dosing of cytotoxic drugs can target tumor angiogenesis in mice. *J Clin Invest* 2000;**105**(8):1045–7.
4. Kerbel RS, Kamen BA. The anti-angiogenic basis of metronomic chemotherapy. *Nat Rev Cancer* 2004;**4**:423–36.
5. Chun R, Thamm DH. Targeting angiogenesis and tumor vasculature. In: Withrow SJ, Vail DM, editors. *Withrow and MacEwen's small animal clinical oncology*. 4th ed. Saunders-Elsevier; 2007. p. 259.
6. Biller B. Metronomic chemotherapy in veterinary patients with cancer: rethinking the targets and strategies of chemotherapy. *Vet Clin North Am Small Anim Pract* 2014;**44**(5):817–29.
7. Marchetti V, Giorgi M, Fioravanti A, Finotello R, Citi S, Canu B, et al. First-line metronomic chemotherapy in a metastatic model of spontaneous canine tumours: a pilot study. *Invest New Drugs* 2012;**30**(4):1725–30.
8. Gaspar TB, Henriques J, Marconato L, Queiroga FL. The use of low-dose metronomic chemotherapy in dogs—insight into a modern cancer field. *Vet Comp Oncol* 2018;**16**(1):2–11.
9. Hanahan D, Folkman J. Patterns and emerging mechanisms of the angiogenic switch during tumorigenesis. *Cell* 1996;**86**:353–64.
10. Folkman J, Kalluri R. Tumor angiogenesis. In: Holland JF, et al., editors. *Cancer medicine*. Hamilton, Ontario, Canada: B.C. Decker Inc.; 2000 [Chapter 11].
11. Bertolini F, Paul S, Mancuso P, Monestiroli S, Gobbi A, Shaked Y, et al. Maximum tolerable dose and low-dose metronomic chemotherapy have opposite effects on the mobilization and viability of circulating endothelial progenitor cells. *Cancer Res* 2003;**63**(15):4342–6.

12. Mross K, Steinbild S. *Metronomic anticancer therapy—an ongoing treatment option for advanced cancer patients. J Cancer Res Ther* 2012;**1**:32. http://www.hoajonline.com/jctr/2049-7962/1/32.
13. Bouche G, André N, Banavali S, Berthold F, Berruti A, Bocci G, et al. Lessons from the fourth metronomic and anti-angiogenic therapy meeting, 24-25 June 2014, Milan. *Ecancermedicalscience* 2014;**8**:463. https://doi.org/10.3332/ecancer.2014.463.
14. Colleoni M, Rocca A, Sandri MT, Zorzino L, Masci G, Nole F, et al. Low-dose oral methotrexate and cyclophosphamide in metastatic breast cancer: antitumor activity and correlation with vascular endothelial growth factor levels. *Ann Oncol* 2002;**13**:73–80.
15. Jayson GC, Kerbel R, Ellis LE, Harris AL. Antiangiogenic therapy in oncology : current status and future directions. *Lancet* 2016;**388**(10043):518–29.
16. Zhao Y, Adjei AA. Targeting angiogenesis in cancer therapy: moving beyond vascular endothelial growth factor. *Oncologist* 2015;**20**(6):660–73.
17. Ribatti D. Tumor refractoriness to anti-VEGF therapy. *Oncotarget* 2016;**7**(29):46668–77.
18. Allen E, Jabouille A, Rivera LB, Lodewijckx I, Missiaen R, Steri V, et al. Combined antiangiogenic and anti-PD-L1 therapy stimulates tumor immunity through HEV formation. *Sci Transl Med* 2017; **9**(385):eaak9679.
19. Wang Z, Dabrosin C, Yin X, Fuster MM, Arreola A, Kimrym Rathmel W, et al. Broad targeting of angiogénesis for cancer prevention and therapy. *Semin Cancer Biol* 2015;**35**:224–43.
20. Croix BS, Rago C, Velculescu V, Traverso G, Romans KE, Montgomery E, et al. Genes expressed in human tumor endothelium. *Science* 2000;**289**(5482):1197–202.
21. Lu C, Bonome T, Li Y, Kamat AA, Han LY, Schmandt R, et al. Gene alterations identified by expression profiling in tumor-associated endothelial cells from invasive ovarian carcinoma. *Cancer Res* 2007;**67**(4):1757–68.
22. Akino T, Hida K, Hida Y, Tsuchiya K, Freedman D, Muraki C, et al. Cytogenetic abnormalities of tumor-associated endothelial cells in human malignant tumors. *Am J Pathol* 2009;**175**(6): 2657–67.
23. Virrey JJ, Dong D, Stiles C, Patterson JB, Pen L. Stress chaperone GRP78/BiP confers chemoresistance to tumor-associated endothelial cells. *Mol Cancer* 2008;**6**(8):1268–75.
24. Virrey JJ, Guan S, Li W, Schönthal AH, Chen TC, Hofman FM. Increased survivin expression confers chemoresistance to tumor-associated endothelial cells. *Am J Pathol* 2008;**173**(2):575–85.
25. Sharabi A, Ghera NH. Breaking tolerance in a mouse model of multiple myeloma by chemoimmunotherapy. *Adv Cancer Res* 2010;**107**:1–37.
26. Rabinovich GA, Rubinstein N, Matar P, Rozados V, Gervasoni S, Scharovsky GO. The antimetastatic effect of a single low dose of cyclophosphamide involves modulation of galectin-1 and Bcl-2 expression. *Cancer Immunol Immunother* 2002;**50**:597–603.
27. Liu F, Fang C, Ye X, Lu Q, Yu DH, Zhang WW, et al. Anti-4-1BB scFv immunogene therapy and low dose cyclophosphamide exhibit a synergistic antitumor effect in established murine lung tumors. *Cancer Biol Ther* 2009;**8**:707–13.
28. Ghiringhelli F, Larmonier N, Schmitt E, Parcellier A, Cathelin D, Garrido C. CD4+ CD25+ regulatory T cells suppress tumor immunity but are sensitive to cyclophosphamide which allows immunotherapy of established tumors to be curative. *Eur J Immunol* 2004;**34**(2):336–44.
29. Ghiringhelli F, Menard C, Puig PE, Ladoire S, Roux S, Martin F, et al. Metronomic cyclophosphamide regimen selectively depletes CD4+ CD25+ regulatory T cells and restores T and NK effector functions in end stage cancer patients. *Cancer Immunol Immunother* 2007;**56**(5):641–8.
30. Zhao J, Cao Y, Lei Z, Yang Z, Zhang B, Huang B. Selective depletion of CD4+ CD25+ Foxp3+ regulatory T cells by low-dose cyclophosphamide is explained by reduced intracellular ATP levels. *Cancer Res* 2010;**70**(12):4850–8.
31. van der Most RG, Currie AJ, Mahendran S, Prosser A, Darabi A, Robinson BW, et al. Tumor eradication after cyclophosphamide depends on concurrent depletion of regulatory T cells: a role for cycling TNFR2-expressing effector-suppressor T cells in limiting effective chemotherapy. *Cancer Immunol Immunother* 2009;**58**(8):1219.
32. Burton JH, Mitchell L, Thamm DH, Dow SW, Biller BJ. Low-dose cyclophosphamide selectively decreases regulatory T cells and inhibits angiogenesis in dogs with soft tissue sarcoma. *J Vet Intern Med* 2011;**25**(4):920–6.

33. Heylmann D, Bauer M, Becker H, Van Gool S, Bacher N, Steinbrink K, et al. Human CD4+ CD25+ regulatory T cells are sensitive to low dose cyclophosphamide: implications for the immune response. *PLoS One* 2013;**8**(12).
34. Madondo MT, Quinn M, Plebanski M. Low dose cyclophosphamide: mechanisms of T cell modulation. *Cancer Treat Rev* 2016;**42**:3–9.
35. Ge Y, Domschke C, Stoiber N, Schott S, Heil J, Rom J, et al. Metronomic cyclophosphamide treatment in metastasized breast cancer patients: immunological effects and clinical outcome. *Cancer Immunol Immunother* 2012;**61**(3):353–62.
36. Piedras K, Hanahan D. A multitargeted metronomic and maximum-tolerated dose "chemo-switch" regimen is antiangiogenic, producing objective responses and survival benefits in a mouse model of cancer. *J Clin Oncol* 2005;**23**(5):939–52.
37. El Bary NA, Hashem T, Metwally H, Ghany AA, El Mageed HA. A phase II study of high-dose celecoxib and metronomic 'low-dose' cyclophosphamide and methotrexate in patients with relapsed and refractory lymphoma. *Hematol Oncol Stem Cell Ther* 2010;**3**(1):13–7.
38. Krzyzanowska MK, Tannock IF, Lockwood G, Knox J, Moore M, Bjarnason GA. A phase II trial of continuous low-dose oral cyclophosphamide and celecoxib in patients with renal cell carcinoma. *Cancer Chemother Pharmacol* 2007;**60**(1):135–41.
39. Lana S, U'ren L, Plaza S, Elmslie R, Gustafson D, Morley P, et al. Continuous low-dose oral chemotherapy for adjuvant therapy of splenic hemangiosarcoma in dogs. *J Vet Intern Med* 2007;**21**:764–9.
40. Orlando L, Cardillo A, Rocca A, Balduzzi A, Ghisini R, Peruzzotti G, et al. Prolonged clinical benefit with metronomic chemotherapy in patients with metastatic breast cancer. *Anticancer Drugs* 2006;**17**: 961–7.
41. Viale G, Locatelli M, Criscitiello C, Trapani D, Curigliano G. Efficacy and safety of platinum and metronomic cyclophosphamide in triple negative breast cancer. *Breast* 2017;**36**:S47.
42. Bazzola L, Foroni C, Andreis D, Zanoni V, Cappelletti MR, Allevi G, et al. Combination of letrozole, metronomic cyclophosphamide and sorafenib is well-tolerated and shows activity in patients with primary breast cancer. *Br J Cancer* 2015;**112**(1):52.
43. Penel N, Adenis A, Bocci G. Cyclophosphamide-based metronomic chemotherapy: after 10 years of experience, where do we stand and where are we going? *Crit Rev Oncol Hematol* 2012;**82**(1):40–50.
44. Liyanage UK, Moore TT, Joo HG, Tanaka Y, Hermann V, Doherty G, et al. Prevalence of regulatory T cells is increased in peripheral blood and tumor microenvironment of patients with pancreas or breast adenocarcinoma. *J Immunol* 2002;**169**:2756–61.
45. Kareva I, Waxman DJ, Klement GL. Metronomic chemotherapy: an attractive alternative to maximum tolerated dose therapy that can activate anti-tumor immunity and minimize therapeutic resistance. *Cancer Lett* 2015;**358**(2):100–6.
46. Man S, Bocci G, Francia G, Green SK, Jothy S, Hanahan D, et al. Antitumor effects in mice of low dose (metronomic) cyclyophoshphamide administered continuously through the drinking water. *Cancer Res* 2002;**62**:2731–5.
47. Hamano Y, Sugimoto H, Soubasakos MA, Kieran M, Olsen BR, Lawler J, et al. Thrombospondin-1 associated with tumor microenvironment contributes to low-dose cyclophosphamide-mediated endothelial cell apoptosis and tumor growth suppression. *Cancer Res* 2004;**64**(5):1570–4.
48. Glode LM, Barqawi A, Crighton F, Crawford ED, Kerbel R. Metronomic therapy with cyclophosphamide and dexamethasone for prostate carcinoma. *Cancer* 2003;**98**(8):1643–8.
49. Bocci G, Francia G, Man S, Lawler J, Kerbel RS. Thrombospondin 1, a mediator of the antiangiogenic effects of low-dose metronomic chemotherapy. *Proc Natl Acad Sci U S A* 2003;**100**(22):12917–22.
50. Emmenegger U, Man S, Shaked Y, Francia G, Wong JW, Hicklin DJ, et al. A comparative analysis of low-dose metronomic cyclophosphamide reveals absent or low-grade toxicity on tissues highly sensitive to the toxic effects of maximum tolerated dose regimens. *Cancer Res* 2004;**64**(11):3994–4000.
51. Lord R, Nair S, Schache A, Spicer J, Somaihah N, Khoo V, et al. Low dose metronomic oral cyclophosphamide for hormone resistant prostate cancer: a phase II study. *J Urol* 2007;**177**(6): 2136–40 [discussion 2140].
52. Samaritani R, Corrado G, Vizza E, Sbiroli C. Cyclophosphamide "metronomic" chemotherapy for palliative treatment of a young patient with advanced epithelial ovarian cancer. *BMC Cancer* 2007;**7**(1):65.

53. Elmslie RE, Glawe P, Dow SW. Metronomic therapy with cyclophosphamide and piroxicam effectively delays tumor recurrence in dogs with incompletely resected soft tissue sarcomas. *J Vet Intern Med* 2008;**22**(6):1373–9.
54. Gebbia V, Boussen H, Valerio MR. Oral metronomic cyclophosphamide with and without methotrexate as palliative treatment for patients with metastatic breast carcinoma. *Anticancer Res* 2012;**32**(2): 529–36.
55. Perroud HA, Rico MJ, Alassino CA, Queralt F, Mainetti LE, Pezzotto MS, et al. Safety and therapeutic effect of metronomic chemotherapy with cyclophosphamide and celecoxib in advanced breast cancer patients. *Future Med* 2013;**9**(3):451–62.
56. Ferrandina G, Corrado G, Mascilini F, Malaguti P, Samaritani R, Distefano M, et al. Metronomic oral cyclophosphamide (MOC) in the salvage therapy of heavily treated recurrent ovarian cancer patients: a retrospective, multicenter study. *BMC Cancer* 2014;**14**(1):947.
57. Montagna E, Cancello G, Dellapasqua S, Munzone E, Colleoni M. Metronomic therapy and breast cancer: a systematic review. *Cancer Treat Rev* 2014;**40**(8):942–50.
58. Buckstein R, Kerbel RS, Shaked Y, Nayar R, Foden C, Turner R, et al. High-dose celecoxib and metronomic "low-dose" cyclophosphamide is an effective and safe therapy in patients with relapsed and refractory aggressive histology non-Hodgkin's lymphoma. *Clin Cancer Res* 2006;**12**(17):5190–8.
59. Guercetti J, Mainetti LE, Grillo MC, Del Giudice A, Baglioni MV, Caceres JM, et al. Metronomic chemotherapy with cyclophosphamide (Cy) and the repositioned drug losartan (Los) for the treatment of M-23p trople negative murine mammary adenocarcinoma. *Cancer Res* 2018;**78**(13) [Supplement: abstract 13].
60. Polton G, Finotello R, Sabattini S, Rossi F, Laganga P, Vasconi ME, et al. Survival analysis of dogs with advanced primary lung carcinoma treated by metronomic cyclophosphamide, piroxicam and thalidomide. *Vet Comp Oncol* 2018;**6**. Downloaded from: https://doi.org/10.1111/vco.12393.
61. Ma J, Chen CS, Blute T, Waxman DJ. Antiangiogenesis enhances intratumoral drug retention. *Cancer Res* 2011;**71**(7):2675–85.
62. Ma J, Waxman DJ. Combination of antiangiogenesis with chemotherapy for more effective cancer treatment. *Mol Cancer Ther* 2008;**7**(12):3670–84.
63. Pozzi A, Ibanez MR, Gatica AE, Yang S, Wei S, Mei S, et al. Peroxisomal proliferator-activated receptor-α-dependent inhibition of endothelial cell proliferation and tumorigenesis. *J Biol Chem* 2007;**282**:17685–95.
64. Pozzi A, Popescu V, Yang S, Mei S, Shi M, Poulitaival SM, et al. The anti-tumorigenic properties of peroxisomal proliferator-activated receptor {alpha} are arachidonic acid epoxygenase-mediated. *J Biol Chem* 2010;**285**:12840–50.
65. Huang H, Campbell SC, Bedford DF, Nelius T, Veliceasa D, Emelyn H, Shroff EH, et al. Peroxisome proliferator-activated receptor γ ligands improve the antitumor efficacy of thrombospondin peptide ABT510. *Mol Cancer Res* 2004;**2**:541–50.
66. Shaked Y, Emmenegger U, Man S, Cervi D, Bertolini F, et al. Optimal biologic dose of metronomic chemotherapy regimens is associated with maximum antiangiogenic activity. *Blood* 2005;**106**(9): 3058–71.
67. Yang L, Keating GM. Fenofibric acid: in combination therapy in the treatment of mixed dsylipidemia. *Am J Cardiovasc Drugs* 2009;**9**(6):401–9.
68. Goetze S, Eilers F, Bungenstock A, Kintscher U, Stawowy P, Blaschke F, et al. PPAR activators inhibit endothelial cell migration by targeting Akt. *Biochem Biophys Res Commun* 2002;**293**(5):1431–7.
69. Meissner M, Stein M, Urbich C, Reisinger K, Suske G, et al. PPARα activators inhibit vascular endothelial growth factor receptor-2 expression by repressing Sp1 dependent DNA binding and transactivation. *Circ Res* 2004;**94**:324–32.
70. Panigraphy D, Kaipainen A, Huang S, Butterfield CE, Barnes CM, et al. PPARα agonist fenofibrate suppresses tumor growth through direct and indirect angiogenesis inhibition. *PNAS* 2007;**105**(3): 985–90.
71. Dana N, Javanmard SH, Fazilati M, Pilehvarian AA. A comparison of peroxisome proliferator-activated receptor-α agonist and antagonist on human umbilical vein endothelial cells angiogenesis. *Adv Biomed Res* 2013;**2**(1):54 [e Collection].

72. Kim J, Ahn JH, Kim JH, Yu YS, Kim HS, et al. Fenofibrate regulates retinal endothelial cell survival through the AMPK signal transduction pathway. *Exp Eye Res* 2007;**84**(5):886–93.
73. Noonan JE, Jenkins AJ, Ma JX, Keech AC, Wang JJ, et al. An update on the molecular actions of fenofibrate and its clinical effects on diabetic retinopathy and other microvascular end points in patients with diabetes. *Diabetes* 2013;**62**(12):3968–75.
74. Chew EY, Ambrosius WT, Davis MD, Danis RP, Gangaputra S, Greven CM, et al. Effects of medical therapies on retinopathy progression in type 2 diabetes. *N Engl J Med* 2010;**363**:233–44.
75. Blann AD, Belgore FM, Constans J, Conri C, Lip GY. Plasma vascular endothelial growth factor and its receptor Flt-1 in patients with hyperlipidemia and atherosclerosis and the effects of fluvastatin or fenofibrate. *Am J Cardiol* 2001;**87**(10):1160–3.
76. Okamoto H, Iwamoto T, Kotake S, Momohara S, Yamanaka H, Kamatani N. Inhibition of NF-kappaB signaling by fenofibrate, a peroxisome proliferator-activated receptor-alpha ligand, presents a therapeutic strategy for rheumatoid arthritis. *Clin Exp Rheumatol* 2005;**23**(3):323–30.
77. Kubota T, Yano T, Fujisaki K, Itoh Y, Oishi R. Fenofibrate induces apoptotic injury in cultured human hepatocytes by inhibiting phosphorylation of Akt. *Apoptosis* 2005;**10**(2):349–58.
78. Zhou J, Zhang S, Xue J, Avery J, Wu J. Activation of peroxisome proliferators activator α suppresses hypoxia inducible factor 1α (HIF1α) signaling in cancer cells. *J Biol Chem* 2012;**287**:35161–9.
79. Yamasaki D, Kawabe N, Nakamura H, Tachibana K, Ishimoto K, Tanaka T, et al. Fenofibrate suppresses growth of the human hepatocellular carcinoma cell via PPARα-independent mechanisms. *Eur J Cell Biol* 2011;**90**(8):657–64.
80. Brunmair B, Lest A, Staniek K, Gras F, Sharf N, et al. Fenofibrate impairs rat mitochondrial function by inhibition of respiratory complex I. *JPET* 2004;**311**:109–14.
81. Nadanaciva S, Dykens JA, Bernal A, Capaldi RA, Will Y. Mitochondrial impaitment by PPAR agonists and statins identified via immunocaptured OXPHOS complex activities and respiration. *Toxicol Appl Pharmacol* 2007;**223**(3):277–87.
82. Lee DL, Wilson JL, Duan R, Hudson T, El-Marakby A. Peroxisome proliferator-activated receptor-α activation decreases mean arterial pressure, plasma interleukin-6, and COX-2 while increasing renal CYP4A expression in an acute model of DOCA-salt hypertension. *PPAR Res* 2011;**2011**:502631. https://doi.org/10.1155/2011/502631.
83. Grau R, Punzón C, Fresno M, Iñiguez MA. Peroxisome-proliferator-activated receptor α agonists inhibit cyclo-oxygenase 2 and vascular endothelial growth factor transcriptional activation in human colorectal carcinoma cells via inhibition of activator protein-1. *Biochem J* 2006;**395**(1):81–8.
84. Drukala J, Urbanska K, Wilk A, Grabaka M, Wybieralska E, et al. ROS accumumulation and IGF-IR inhibition contribute to fenofibrate/PPARα mediated inhibition of glioma cell motility in vitro. *Mol Cancer* 2010;**9**:159.
85. Zhao H, Zhu C, Qin C, Tao T, Li J, Cheng G, et al. Fenofibrate down-regulates the expressions of androgen receptor (AR) and AR target genes and induces oxidative stress in the prostate cancer cell line LNCaP. *Biochem Biophys Res Commun* 2013;**432**(2):320–5.
86. Han DF, Zhang J, Wei W, Tao T, Hu Q, Wang Y, et al. Fenofibrate induces G_0/G_1 phase arrest by modulating the PPARα/FoxO1/p27^{kip} pathway in human glioblastoma cells. *Tumor Biol* 2015;**36**(5): 3823–9.
87. Lian X, Gu J, Gao B, Li Y, Damodaran C, Wei W, et al. Fenofibrate inhibits mTOR-p70S6K signaling and simultaneously induces cell death in human prostate cancer cells. *Biochem Biophys Res Commun* 2018;**496**(1):70–5.
88. Morse E, Selim E, Cunard R. PPARαligands cause lymphocyte depletion and cell cycle block and this is associated with augmented TRB3 and reduced cyclin B1 expression. *Mol Immunol* 2009;**46**: 3454–61.
89. Selim E, Frkanec JT. Cunard R fibrates upregulates TRB3 in lymphocytes independent of PPAR alpha by augmenting CCAAT/enhancer binding protein beta (C/EBP beta) expression. *Mol Immunol* 2007;**44**(6):1218–29.
90. Stalinska J, Zimolag E, Pianovich NA, Zapata A, Lassak A, Rak M, et al. Chemically modified variants of fenofibrate with antiglioblastoma potential. *Transl Oncol* 2019;**12**(7):895–907.
91. Blagosklonny MV. Antiangiogenic therapy and tumor progression. *Cancer Cell* 2004;**5**(1):13–7.

92. Steinbild S, Arends J, Medinger M, Haring B, Frost A, Drevs J, et al. Metronomic antiangiogenic therapy with capecitabine and celecoxib in advanced tumor patients. Results of a phase II study. *Onkologie* 2007;**30**:629–35.
93. Dellapasqua S, Bertolini F, Bagnardi V, Campagnoli E, Scarano E, et al. Metronomic cyclophosphamide and capecitabine combined with bevacizumab in advanced breast cancer. *J Clin Oncol* 2008; **26**(39): 4899–905.
94. Soriano Gracia JL, Lima Perez M, Gonzalez Gonzalez J, Batista Albuerne N, Lopez Soto MV, et al. *Quimioterapia metronomica con ciclofosfamida y metotrexato en pacientes con cancer de mama metastasico en progresion. Rev Cubana Med* 2009;**48**(2). http://bvs.sld.cu/revistas/med/vol48_2_09/med02209.htm.
95. Bergersand G, Hanahan D. Modes of resistance to anti-angiogenic therapy. *Nat Rev Cancer* 2008;**8**: 592–603.
96. Senninoand B, McDonald DM. Controlling escape from angiogenesis inhibitors. *Nat Rev Cancer* 2012;**12**:699–709.
97. Ebos JML, Lee CR, Cruz-Munoz W, Bjarnason GA, Christensen JG, Kerbel RS. Accelerated metastasis after short-term treatment with a potent inhibitor of tumor angiogenesis. *Cancer Cell* 2009;**15**(3):232–9.
98. Pàez-Ribes M, Allen E, Hudock J, Takeda T, Okuyama H, Viñals F, et al. Antiangiogenic therapy elicits malignant progression of tumors to increased local invasion and distant metastasis. *Cancer Cell* 2009;**15**(3):220–31.
99. Ebos JM, Kerbel RS. Antiangiogenic therapy: impact on invasion, disease progression, and metastasis. *Nat Rev Clin Oncol* 2011;**8**:210–21.
100. Brooks AN, Kilgour E, Smith PD. Molecular pathways: fibroblast growth factor signaling: a new therapeutic opportunity in cancer. *Clin Cancer Res* 2012;**18**:1855–62.
101. Grabacka M, Plonka PM, Urbanska K, Reiss K. Peroxisome proliferator-activated receptor α activation decreases metastatic potential of melanoma cells in vitro via down-regulation of Akt. *Clin Cancer Res* 2006;**12**(10):3028–36.
102. Grunewald M, Avraham I, Dor Y, Bachar-Lustig E, Itin A, Jung S, et al. VEGF-induced adult neovascularization: recruitment, retention, and role of accessory cells. *Cell* 2006;**124**:175–89.
103. Crawford Y, Ferrara N. Tumor and stromal pathways mediating refractoriness/ resistance to anti-angiogenic therapies. *Trends Pharmacol Sci* 2009;**30**:624–30.
104. Zheng Y, Xu M, Li X, Jia J, Fan K, Lai G. Cimetidine suppresses lung tumor growth in mice through proapoptosis of myeloid-derived suppressor cells. *Mol Immunol* 2013;**54**(1):74–83.
105. Martin RK, Saleem SJ, Folgosa L, Zellner HB, Damle SR, Nguyen GKT, et al. Mast cell histamine promotes the immunoregulatory activity of myeloid-derived suppressor cells. *J Leukoc Biol* 2014;**96**(1):151–9.
106. Solinas G, Germano G, Mantovani A, Allavena P. Tumor-associated macrophages (TAM) as major players of the cancer-related inflammation. *J Leukoc Biol* 2009;**86**:1065–73.
107. Okopień B, Kowalski J, Krysiak R, Łabuzek K, Stachura-Kułach A, Kułach A, et al. Monocyte suppressing action of fenofibrate. *Pharmacol Rep* 2005;**57**(3):367–72.
108. Duez H, Chao YS, Hernandez M, Torpier G, Poulain P, Mundt S, et al. Reduction of atherosclerosis by the peroxisome proliferator-activated receptor α agonist fenofibrate in mice. *J Biol Chem* 2002;**277**(50):48051–7.
109. Funakoshi-Tago M, Shimizu T, Tago K, Nakamura M, Itoh H, Sonoda Y, et al. Celecoxib potently inhibits TNFα-induced nuclear translocation and activation of NF-κB. *Biochem Pharmacol* 2008;**76**(5):662–71.
110. Shishodia S, Koul D, Aggarwal BB. Cyclooxygenase (COX)-2 inhibitor celecoxib abrogates TNF-induced NF-κB activation through inhibition of activation of IκBα kinase and Akt in human non-small cell lung carcinoma: correlation with suppression of COX-2 synthesis. *J Immunol* 2004;**173**(3):2011–22.
111. Pan H, Sun CC, Zhou CY, Huang HF. Expression of aquaporin1 in normal, hyperplasic, and carcinomatous endometria. *Int J Gynecol Obstet* 2008;**101**(3):239–44.
112. Spugnini EP, Citro G, Fais S. Proton pump inhibitors as anti vacuolar-ATPases drugs: a novel anticancer strategy. *J Exp Clin Cancer Res* 2010;**29**:44.

113. Goh W, Sleptsova-Freidrich I, Petrovic N. Use of proton pump inhibitors as adjunct treatment for triple-negative breast cancers. An introductory study. *J Pharm Pharm Sci* 2014;**17**(3):439–46.
114. Rojas JD, Sanka SC, Luo D, Busch C, Martinez GM, Hendrix MJ, et al. Proton pumps, angiogenesis, and metastatic breast cancer. In: *Molecular imaging: reporters, dyes, markers, and instrumentation*. vol. 3924. International Society for Optics and Photonics; 2000. p. 67–79.
115. Sennoune SR, Martinez-Zaguilan R. Plasmalemmal vacuolar H^+-ATPases in angiogenesis, diabetes and cancer. *J Bioenerg Biomembr* 2007;**39**(5-6):427–33.
116. Millimaggi D, Mari M, D'Ascenzo S, Carosa E, Jannini EA, Zucker S, et al. Tumor vesicle-associated CD147 modulates the angiogenic capability of endothelial cells. *Neoplasia* 2007;**9**(4):349–57.
117. Vaeteewoottacharn K, Kariya R, Dana P, Fujikawa S, Matsuda K, Ohkuma K, et al. Inhibition of carbonic anhydrase potentiates bevacizumab treatment in cholangiocarcinoma. *Tumor Biol* 2016;**37**(7):9023–35.
118. Abd-El Fattah AA, Darwish HA, Fathy N, Shouman SA. Carbonic anhydrase inhibition boosts the antitumor effects of imatinib mesylate via potentiating the antiangiogenic and antimetastatic machineries. *Toxicol Appl Pharmacol* 2004;**316**:123–38.
119. McIntyre A, Patiar S, Wigfield S, Li JL, Ledaki I, Turley H, et al. Carbonic anhydrase IX promotes tumor growth and necrosis in vivo and inhibition enhances anti-VEGF therapy. *Clin Cancer Res* 2012;**18**(11):3100–11.
120. Xiang Y, Ma B, Li T, Gao JW, Yu HM, Li XJ. Acetazolamide inhibits aquaporin-1 protein expression and angiogenesis. *Acta Pharmacol Sin* 2004;**25**:812–6.
121. Natori T, Sata M, Nagai R, Makuuchi M. Cimetidine inhibits angiogenesis and suppresses tumor growth. *Biomed Pharmacother* 2005;**59**(1-2):56–60.

CHAPTER 20

Preventing metastasis with pH regulation

Introduction

When a malignant tumor is discovered in a patient, for example a breast carcinoma, the main concern is to find out if it has already metastasized. According to whether or not the tumor has metastasized we are going to treat the patient on the basis of proven protocols: surgery, radiotherapy and chemotherapy following well known-steps that usually vary only slightly among hospitals, oncology centers and countries. When this treatment is finished and the patient has no disease beyond loco regional lymph nodes, the next usual step is to prescribe tamoxifen or exemestane or some other anti-estrogenic for the next 5 or 10 years. A medical control is done every 6 months. This is the standard of care. Should this be all? Is it enough?

In most cases it is enough, but in many others, our original concern about metastasis will come up again, because eventually the metastases do show up. This means that not enough was done, at least in these cases or in those where there is a relapse.

Metastasis is the end product of an evolutionary process,[1] and one of the hallmarks of cancer.[2] Usually cancer mortality is not caused by the primary tumor but by the metastatic spread.[3] The lack of an effective methodology for preventing metastasis (or at least insufficient prevention) results in a high mortality rate in oncology patients. Therefore, it seems reasonable that if this risk can be reduced, the outlook for the patients may significantly improve survival, quality of life, and decrease medical costs. Solving the metastasis problem is almost solving the cancer problem.[4] And the way to solve it is by preventing metastasis, or at least reducing the risk.

We agree with Ghajar[5] when he says: "I argue that rather than waiting for metastasis to emerge before initiating treatment, it would be more effective to target metastatic seeds before they sprout."

There are simple, effective, non-toxic, and low cost resources that can be implemented in order to reduce the metastatic risk. In this chapter metastasis prevention will be focused on pH homeostatic drugs, however other pharmaceuticals with potential in this matter will also be mentioned.

Starting with aspirin, probably the most "non prescribed" drug used in the world. Aspirin is a NSAID (non-steroidal anti-inflammatory drug). Currently, aspirin is considered a preventive drug for colonic adenomas[6] and colon cancer.[7] The mechanism

An Innovative Approach to Understanding and Treating Cancer: Targeting pH
https://doi.org/10.1016/B978-0-12-819059-3.00020-4
© 2020 Elsevier Inc.
All rights reserved.

involved in this prevention is considered to be its ability to inhibit the three known isoforms of COX (cyclooxygenase): COX1, COX2, and COX3.[8] This inhibition impedes the production of prostaglandin E2 (PGE2) through transcriptional and non transcriptional mechanisms.[9] Another less known effect of aspirin in intestinal mucosa is the ability of salicylates to uncouple oxidative phosphorylation in mitochondria.[10] This uncoupling was also found in hepatocytes,[11] gastric mucosa,[12] cartilage[13] and the brain. Aspirin seems to be the weakest "uncoupler" of the salicylate derivatives. It also releases inorganic phosphate from ATP.[14] Aspirin and its metabolite, salicylate, are proton transporters through phospholipid bilayered membranes. Theoretical calculations indicate that salicylates increase proton uptake in the mitochondria.[15]

Aspirin induces apoptosis in certain cells by mitochondrial release of caspases.[16] Aspirin is an inhibitor of transcription factor AP-1.[17] The interesting issue is that AP1 inhibition is achieved through decrease of intracellular pH (pHi). Finally aspirin inhibits platelet activation. Platelets are an important component of metastasis. Circulating groups of malignant cells are protected by platelets gathered around them (Fig. 1) and aspirin has various activities against cancer including blocking this interaction (Fig. 2).

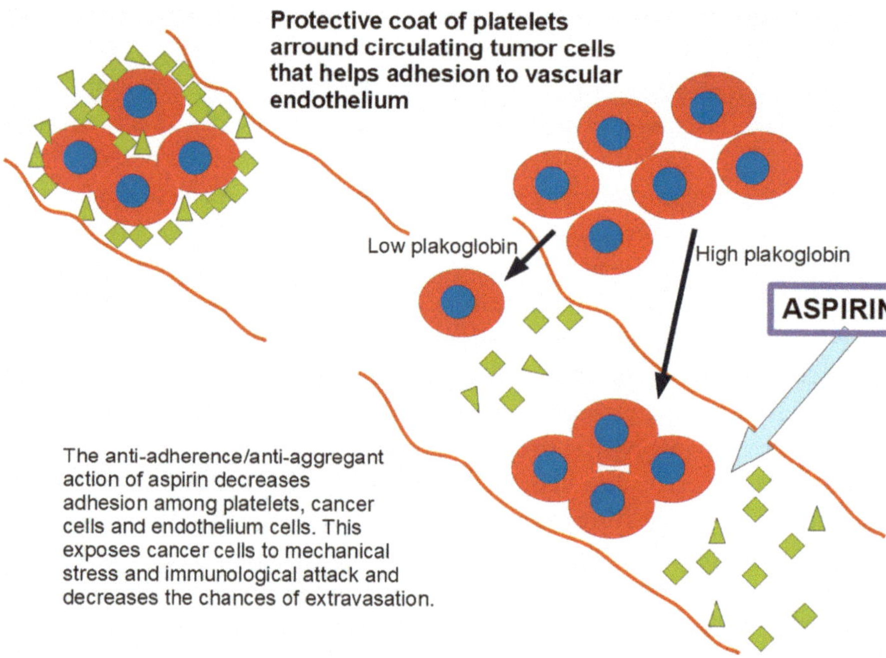

Fig. 1 Reducing adherence by aspirin. Plakoglobin-poor cells are shed into the circulation individually, while plakoglobin-rich cells are shed and travel the circulation as clusters.[18] Aspirin impedes the platelet protection of these clusters by reducing adhesiveness.

Aspirin is probably the simplest and most ancient drug for metastasis prevention, and furthermore, it is a cell acidifier.

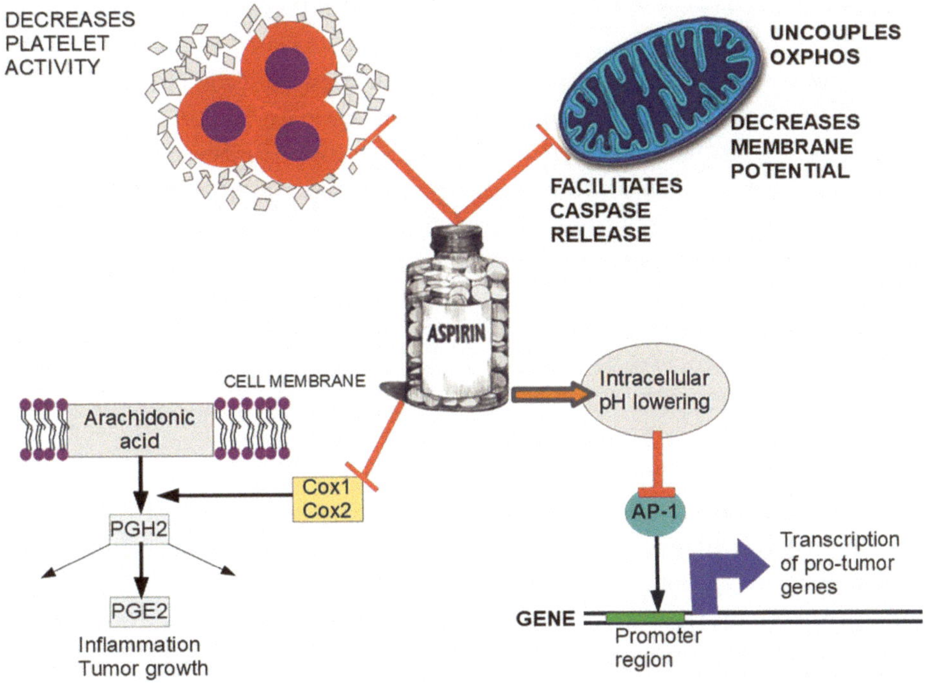

Fig. 2 The different activities of aspirin against cancer are explained in the text. AP-1 is a transcription factor that promotes the activity of many pro-tumoral genes related to growth and proliferation. It is formed by the union of other two transcription factors cFos and cJun (it is a heterodimer) and is stimulated by growth factors and cytokines.

However, very few oncologists regularly prescribe aspirin to their patients, and aspirin is not part of any protocol to prevent metastasis because there is no such protocol. That is precisely the aim of what comes next: building a protocol to reduce the risk of metastasis. Of course, aspirin will form part of this protocol.

The metastatic risk can be reduced

Many natural products, like genistein,[19] resveratrol[20] and curcumin[21] have shown interesting anti-metastatic activity. The same happens with pharmaceuticals like NSAIDs, aspirin[22] and celecoxib[23] as the best examples. Preventing cancer cell migration is probably the most effective way to reduce the metastatic risk. In this sense in the chapter on invadopodia (Chapter 11), the action of drugs like dasatinib and evodiamine plus inversion of the pH gradient are described. They are all effective anti-migratory agents. Many chemicals have been identified with anti-metastatic effects, such as increasing NO,[24] cimetidine, doxycycline, heparin, low molecular heparins, metapristone,[25] etc.

Ardiani et al.[26] developed a vaccine against Twist, a transcription factor involved in invasion and metastasis.

Many known pharmaceuticals that are, or have been used for purposes other than cancer are revealing anti-metastatic effects. This is the case of thiabendazole, an antifungal and anti-parasitic drug that has been in use for over 40 years. Thiabendazole is an inhibitor of migration,[27] in addition to having anti-angiogenic and immuno-modulating properties. Every day new drugs are found with potential preventive activity in metastasis. Old drugs are also emerging as potentially anti-metastatic (the repurposed drugs). The anti-metastatic armamentarium has been enriched with pharmaceuticals that decrease the risk of bone metastasis. Among them, we can mention zoledronic acid[28] that decreases invasion, and denosumab,[29] a monoclonal antibody directed against the receptor activator of nuclear factor kappa B ligand (RANKL). Denosumab diminishes the number of circulating cancer cells and prevents bone metastasis (clinical trials NCT01952054, NCT01951586, NCT02129699).

Finally, reversing the abnormal pH gradient in cancer, is probably the simplest way to decrease the risk of metastasis. There is no metastasis without invasion, and there is no invasion without migration. This concept takes us back to invadopodia (Chapter 11). Targeting invadopodia is probably the more efficient method to reduce the risk of metastasis. And invadopodia are highly sensitive to a reversal of the inverted pH gradient. If migration and invasion are blocked, metastasis should not occur.

The metastatic cascade

Paget's seed and soil model[30] in which the cancer cell is the seed that colonizes the soil that is represented by the organ in which the metastatic tissue grows, is still an oversimplified representation of the basic knowledge of what we think happens in this complex process. Metastasis is a multi-step development, in which each step depends on completion of the previous one. Each step is amenable to pharmacological targeting.

In a more detailed and amplified view on the subject it can be described as various sequential and interrelated steps[31]:

1. Primary tumor growth and angiogenesis.
2. Cells adopt a mesenchymal phenotype (epithelial-mesenchymal transition).
3. Future metastatic cells free themselves from the primary tumor.
4. Cellular protrusions into the matrix are created: invadopodia (see Chapter 11).
5. These protrusions are the engines that degrade surrounding matrix and polymerize actin fibers allowing migration.
6. Migrating cells reach endothelium of vessels.
7. They enter the blood vessels (intravasation).
8. Cancer cells circulate and survive in circulation. Platelet protection of these cells facilitates smooth and uneventful trafficking.

9. These cells reach the target organ where a pre-metastatic niche has been prepared.
10. They adhere to endothelial cells.
11. They migrate through the vessel wall (extravasation).
12. The newly arrived cells may stay dormant or start to grow in the colonized site inducing new vessels' formation (angiogenesis).
13. And finally they may produce new metastatic cells reinitiating the whole process.

This is a purely mechanical way of looking at this complex process which involves the activity of many molecules including enzymes, growth factors, cytokines and chemokines. The length of the chain of events necessary for metastasis, as shown in the metastatic cascade, and the amount of molecules involved, makes it a targetable objective.

The other important point to consider from the prevention angle, is the moment in which metastasis is produced, because there is evidence suggesting that it can be an early event,[32] and that at the time of diagnosis 60–70% of patients have initiated the metastatic process.

Metastasis is a very inefficient process: millions of cancer cells are permanently reaching the circulation, but very few have the ability to colonize and develop. It is possible that many of these cancer cells reaching the circulation have not fully gone through all the changes necessary for fulfilling all the steps of the cascade. So, not all the circulating tumor cells have the *metastatic signature* and finding circulating cancer cells is by no means a synonym of metastasis. The failure of many circulating cancer cells to colonize may also occur because the secondary tissue sites, which they reach have not developed the pre-metastatic niche (see below).

A difference should be made between single circulating cancer cells and oligoclonal tumor cell groupings. These groups have 23- to 50-fold greater metastatic potential.[33–35]

The cancer cell must go through all the steps of the cascade. It cannot skip a step. Failure to complete any one step results in failure of the metastatic process. So, that weakness may become our strength: we have to block one step successfully and metastasis will be aborted.

Circulating cancer cells with metastatic potential benefit greatly from platelets, not only from a mechanical point of view, but also through Autaxin (AUX) that is a lysophospholipase produced by activated platelets and controls levels of lysophosphatidic acid (LPA) in blood. As Leblanc et al.[34] demonstrated, AUX is released upon tumor cell-induced platelet aggregation. AUX is strongly related to bone metastasis in breast cancer.

We suppose that avoiding platelet aggregation with aspirin or dipiridamol may prevent this step. This needs experimental confirmation.

Metastasis prevention: When and how

Extracellular acidity promotes metastasis.[36–38] This issue has been extensively discussed in the first part of the book. For a review see Ref. 39.

The logical assumption should be that by reversing extracellular acidity, the risk of metastasis should decrease.

But, when should prevention start?

Our answer is: as soon as we know our patient has a malignant tumor, or even before, when we suspect that our patient may have a tumor and we start the medical workup to reach a diagnosis.

We do not know the exact moment when metastasis happens, but the nihilistic position of assuming that it is too late by the time the diagnosis is made, finds no justification in oncological practice. It is quite frequent to find patients with big tumor masses and elevated mitosis numbers that never develop a metastasis.

The pre-metastatic niche

The formation of tumor-promoting pre-metastatic niches in close or distant organs is a second new chapter of Paget's "seed and soil" hypothesis. Malignant tumors have the ability to induce pre-metastatic niche formation where metastatic cells will find an adequate environment for colonization. The primary tumor achieves this distant process through the release of exosomes containing a variety of cytokines, growth factors, coding and non-coding short RNAs. When the exosomes reach the pre-metastatic niche they use these cytokines that recruit bone marrow-derived cells.[40]

Costa Silva et al.[41] showed that pancreatic cancer cells secreted exosomes that prepared the liver to host the future metastasis: the pre-metastatic niches. These exosomes induced the host to:

(1) produce β-transforming growth factor;
(2) upregulate fibronectin;
(3) recruit bone marrow derived macrophages;
(4) express macrophage migration inhibitory factor (MIF). The inhibition of this factor inhibited metastatic formation in the liver.

Indeed, tumor-derived exosomes (TDEs) are essential players in the preparation of the terrain for metastatic colonization.[42, 43] Under non-cancer conditions exosomes are extracellular bilayer vesicles that work as an intercellular communication system. In cancer this system is used to promote cancer cell survival, deregulate the host's immunosurveillance, and prepare the pre-metastatic niche. Their integrin expression determines the selection of the organ to be metastasized.

There is clear evidence of the relation between exosomes and extracellular pH:

(1) Invadopodia show an increased secretion of exosomes, that regulates the aggressiveness of the invasion.[44] Invadopodia is the site where the pH gradient inversion is more noticeable and where there is a preferential localization of NHE1.
(2) Monesin, a Na^+/H^+ exchanger, stimulates exosome secretion through a mechanism of increased Ca^{2+}.[45]

(3) Microenvironmental pH is an important regulator of exosome activity; there is increased uptake and secretion with low extracellular pH. Treating cells with proton inhibitors decreases exosome uptake.[46, 47]

(4) Amiloride decreases the release of exosomes in malignant cells.[48] This could be through decrease in extracellular acidity or a yet undefined relation between NHE1 and exosomes.

There is also evidence of a relation between COX2 and exosome trafficking because COX2 inhibitors decrease it. This is the case of indometacine,[49] ketotifen[50] and celecoxib.[51] Aspirin reduces the protein content of platelet derived exosomes without affecting the total level of exosomes.[52] Platelet derived exosomes show pro-invasive effects on tumors.[53, 54] However, the exact nature of aspirin effects on these exosomes has not been clarified.

Therefore, the associated use of inhibitors of proton pumps, NHE1 and COX2 may decrease the modulation of the pre-metastatic niche.

Another important factor in metastasis is the SDF1/CXCR4 axis. Mesenchymal and marrow-derived stromal cells constitutively secrete a HIF-dependent compound known as SDF1 (stromal cell derived factor 1) or CXCL12. SDF1's expression is upregulated in cancer, where it attracts cells expressing the receptor, CXCR4, which is also expressed in many types of cancer cells, particularly in breast cancer. The binding of SDF1/CXCR4 has pro-proliferative, and pro-migratory effects, increasing local invasion and metastasis. For a review of the SDF/CXCL4 axis see Ref. 55.

Plerixafor, (trade name in USA Mozobil, FDA-approved) a small synthetic molecule, antagonizes the SDF1/CXCR4 axis reducing the number of metastasis.[56] Plerixafor also decreases the recurrence of glioma post-irradiation in a mice model.[57, 58] Anti-metastatic activity of plerixafor was found in experimental models with metastasis of melanoma,[59] cervical,[60] prostate,[61] and colon[62] cancer.

However, plerixafor is not FDA approved as a metastatic prevention drug, but for increasing the mobilization of hematopoietic stem cells in cancer patients. These cells are harvested and then re-injected.

Cancer dormancy

After surgery, chemoradiotherapy or any other type of treatment, residual malignant cells may remain, whether at the tumor site or at metastatic niches. These cells are silent and basically undetectable with available diagnostic tools. After a variable length of time, these cells "wake up" and relapse or metastasis appear. The dormant cells are not dormant any more.[63] This is a very well known and striking situation to any physician who treats patients with cancer. There are only two ways to prevent the comeback of the dormant cells: kill them or keep them dormant forever. Actually, killing them is not an option: they produce no symptoms, they are not findable, and, as they are not in active reproduction, they are resistant to conventional chemotherapeutic therapies. Therefore, only the second option is available: keep them dormant. The mechanism for achieving this is to create an hostile environment for their return to "life." Why and exactly how dormancy develops are matters of intensive research and have not been established yet.

It has been suggested that these dormant cells are usually in very acidic niches.[64] This has not been fully proved experimentally. However, keeping an inadequate microenvironment should create poor conditions for their comeback.

Metastasis prevention treatment based on pH

The association of six or seven drugs used in four patients by one of us (TK) with cancer with and without known metastasis showed no metastasis or new metastases, after more than 2 years of follow up (unpublished data).

The drugs:

(1) Acetazolamide 125 mg twice a day.
(2) Lansoprazol 60 mg twice a day.
(3) Amiloride 10 mg twice a day.
(4) Celecoxib 200 mg twice a day.
(5) Fenofibrate 120 mg once a day.
(6) Evodiamine 500 mg a day.
(7) Aspirin 500 mg daily.
(8) Topiramate 50 mg a day

This scheme has different objectives: the first three drugs (acetazolamide, lansoprazole and amiloride decrease intracellular alkalinity and at the same time increase extracellular pH; celecoxib works against angiogenesis; fenofibrate besides its VEGF antagonizing action, down-regulates MCP-1 (monocyte chemotactic protein-1), and decreases metalloprotease-9 (MMP-9) production; evodiamine decreases migration by inhibiting Src phosphokinase; aspirin decreases the adhesion between platelets and CCC (circulating cancer cells).

This multi-sided attack on different parts of the metastatic cascade should be able to reduce the metastatic risk. However, we must note that we have no personal experience with plerixafor.

The triple association acetazolamide-lansoprazole-amiloride targets the pH gradient inversion modifying the pre-metastatic niche conditions through decreased exosome release. In the event that the metastatic cells fulfill their destiny by colonizing the secondary site, it reduces invasion and proliferation.

Celecoxib and fenofibrate target angiogenesis at the niche and evodiamine targets the migration process. Finally, aspirin decreases the platelet protection of the circulating cancer cell clusters. In many cases adding flunarizine may give an extra advantage. Flunarizine is a voltage gated-sodium channel inhibitor and as such cooperates with the triple association in cellular acidification. Flunarizine is also a vasodilator. Better oxygenation of the pre-metastatic niche should decrease the activity of HIF-1α. This last step should decrease SDF1 expression, which is HIF-dependent.

Individual consideration of each of these drugs may seem "weak" to face the metastatic risk. It is using all of them together that gives strength to this approach.

Fig. 3 shows also some other drugs that can be used in a metastasis-prevention schedule.

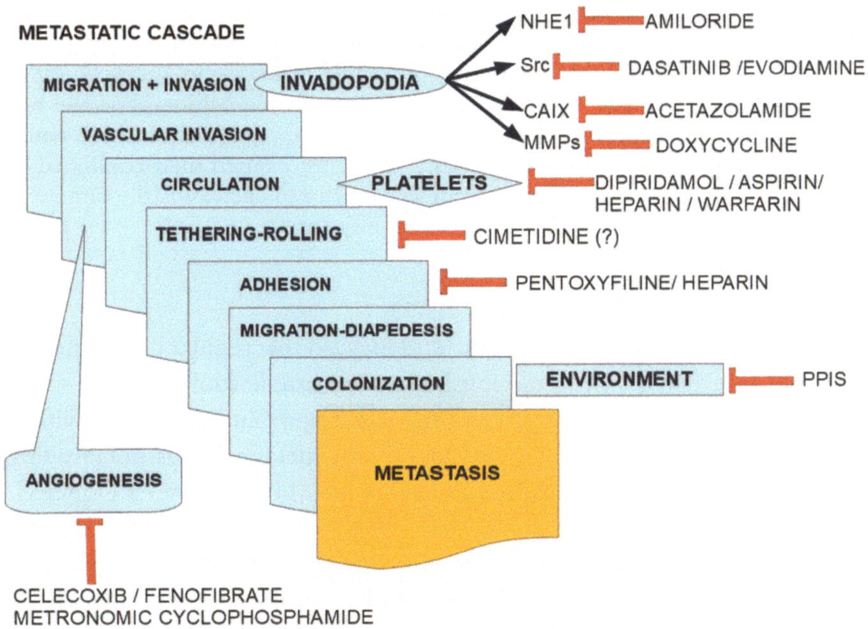

Fig. 3 A group of repurposed drugs with possible action against diverse steps of the metastatic cascade. *CAIX*, carbonic anhydrase 9; *MMPs*, metalloproteases; *PPIS*, proton pump inhibitors; *NHE1*, sodium hydrogen exchanger 1; *Src*, Src phosphokinase. Very little is known about the conditioning of the host niche. Alkalinization of the niche may decrease survival of metastatic cells.

Celecoxib was introduced in the metastatic risk reduction scheme according to the evidence shown in Table 1.

Table 1 Evidence for anti-metastatic activity of celecoxib

Reference	Findings
Zhang et al. (2014)[65]	Celecoxib inhibited metastasis of lung tumor cells in the circulation enhanced by PGE2 after surgery by blocking endogenous PGE2 expression and by suppression of the GSK-3β-catenin pathway.
Barlow et al. (2012)[66]	Celecoxib inhibited invasion of Ewing sarcoma cells in vitro probably through a COX-2 independent mechanism.
Yoshinaka et al. (2006)[67]	In a murine model of breast adenocarcinoma lung metastasis, metastasis occurred in 100% of controls, and in 90% of the animals given 15 mg/kg celecoxib, but the metastatic foci tended to be smaller in the celecoxib-treated group. Microvessel density was decreased in the celecoxib group.

Continued

Table 1 Evidence for anti-metastatic activity of celecoxib—cont'd

Reference	Findings
Roh et al. (2004)[68]	In the postoperative period there is a tendency toward tumor growth and metastasis. Celecoxib had a significant suppressive effect on tumor growth, angiogenesis, and metastasis in tumor-implanted mice with surgical wounds.
Ninomiya et al. (2012)[69]	Celecoxib reduced growth and metastatic potential of human colorectal carcinoma xenografts in mice through COX-2 inhibition, anti-angiogenesis and induction of apoptosis.
Xu et al. (2014)[70]	COX-2 inhibition prevented metastasis in a pre-clinical model.
Niu et al. (2008)[71]	The combined use of Celecoxib and Minocycline had better inhibitory effects on breast cancer bone metastasis compared to any of the drugs alone. They exerted their combined effects by increasing tumor-cell death and decreasing the tumor expression of MMP-9 and VEGF systems.

Cancer patients have a pro-thrombotic tendency, so that regular use of anticoagulants is quite frequent. In 1973,[72] Brown published a very simple tool to decrease metastasis: anticoagulation with warfarin. Brown used mice that were anticoagulated with warfarin and injected with tumor cells. The number of lung metastasis was notably less in this group as compared with non anti-coagulated controls. He observed a higher clearance of tumor cells in the lung when warfarin was administered and hypothesized that anti-coagulants prevented the formation of adequate thrombi permitting the activity of normal defense mechanisms. Ivarsson and Rudenstam[73] found similar results with heparin. The strong relation between metastasis and platelets is sustained by many authors[74–78] who have assigned a central role to platelets in the development of inflammation and cancer.

Another mechanism involved in metastasis prevention is the ability of heparins to block P-selectin-mediated intercellular adhesion between tumor cells and platelets. The coat of platelets that frequently envelopes tumor cells, is supposed to protect them from shear forces in the vascular bed and immunologic reconnaissance and attack.[79]

Cimetidine, a drug that reduces E-selectin expression, can diminish the tethering of circulating cancer cells to the endothelia expressing the sialyl Lewis epitope and therefore should decrease colonization.[80, 81] This issue gave way to some controversies in the clinical setting. While Ali et al.[82] found that with perioperative administration of cimetidine, time to recurrence and death was prolonged, Jameson et al.[83] reported no differences. Both studies have a conceptual bias: they studied the effects of perioperative administration of cimetidine for a period of not more than 5 weeks. Cimetidine should be evaluated by administering it in an uninterrupted and permanent manner before and after surgery. This is the only reliable way to find out its effects over the long term.

Radiotherapy as a source of increased metastatic risk

Radiotherapy is used very frequently in many cancers, and especially breast cancer. In the last 20 years its efficiency has increased thanks to better methods and instruments, delivering more radiation to the tumor and less to adjacent tissues. However, minor damage to other tissues cannot be fully avoided. In the paradigmatic case of breast cancer, lungs always receive some radiation. Irradiated tissues have a higher production of pro-tumoral exosomes and SDF1 expression, thus a certain modulation of a pre-metastatic niche may result. The importance of this modulation in the case of breast radiotherapy and lung metastasis has not been fully clarified in clinical practice. On the other hand, there is experimental evidence that irradiation of the lung and other organs predisposes to metastasis[84–87] and other complications that can be diminished through inhibition of the SDF1/CXCR4 axis.[88]

Fundamentals of the preventive schedule

Amiloride

Evans et al.[89] showed that amiloride had the ability to reduce pulmonary metastasis in a rat mammary cancer model in a dose-dependent manner. They considered uPA (urokinase-type plasminogen activator) inhibition as the reason for this effect. Many other studies have confirmed this finding.[90–97] This has also been validated on clinical grounds.[98] We think that uPA inhibition is only one of the reasons for the preventive action, and the main reason is NHE1 inhibition.

The down-regulation of E-selectin resulted in the reduction of hepatocarcinoma growth and amiloride resulted in a higher inhibition of E-selectin than cimetidine.[99] This would represent a third plus for amiloride.

The inhibition of NHE1 is an effective way of reducing metastatic abilities.[100]

As said above, amiloride reduces exosome release (the fourth plus).

Celecoxib

Many tumors, particularly colon cancer, over-express COX2. Tumors from patients with hepatic metastasis expressed high levels of COX 2.[101] Celecoxib reduced metastasis in many experimental situations and different cancers[67, 68, 102–105] including breast,[67, 106] lung,[65] bone,[107] prostate,[108] melanoma,[109] colon cancer,[69, 110] Ewing sarcoma,[66] Lewis lung carcinoma,[111] etc. Celecoxib also decreases lymphangiogenesis and lymph node metastasis.[112]

Cancer-derived extracellular vesicles are an important factor in preparing distant niches.

Nichida-Aoki et al.[113] using anti-CD9 or anti-CD63 serums that target these vesicles have shown a significant reduction in experimental metastasis models.

VEGF participates in pre-metastatic niche preparation inducing endothelial prostaglandin E2 that enhances the adhesiveness of metastatic cells. Celecoxib antagonizes this activity.[114]

Aspirin

The preventive effects of aspirin have been confirmed in adenomas and colon cancer,[115] and also in other types of tumors.[116, 117] The reduction of metastatic risk was confirmed by large population studies.[118]

Extracellular acidity treatment

The importance of extracellular acidity in the processes of migration and invasion has been explained in detail in many chapters of this book. We are not going to repeat those concepts here. In order to happen, metastasis requires those same abilities: motility and invasiveness. There is no metastasis without them. Therefore, extracellular acidity must be targeted if metastasis is to be prevented. We shall discuss some other aspects of acidity closely related to the metastatic process.

There is evidence showing that the "minimal residual disease" cells and the dormant cells "hide" in acidic niches where they resist therapeutic attacks.[61] However, all this evidence is indirect because up to now, no one has had the opportunity to make direct measurements in vivo in the pre-metastatic niche. In spite of this lack of direct knowledge of the pre-metastatic niche, there are certain issues that are well known and they suggest acidity as an important factor in the metastatic process.

Exosomes produced by the tumor play a key role in communication with other cells[119] and in preparing the pre-metastatic niche,[120] which is not a random process.[121] Microenvironmental acidity facilitates the secretion of pro-oncogenic exosomes. Parolini et al.[46] showed that acidity increased the release and uptake of exosomes in cancer and Logozzi et al.[122] showed that CD63 positive exosomes were significantly increased in the blood of melanoma patients (threefold) compared to normal donors. Another publication by Logozzi et al.[123] confirmed increased exosomal trafficking related to acidity. This concept applies to a large variety of cancer cells, including osteosarcoma, melanoma, breast, prostate and colon cancers.[124] Buffering that increases pH diminishes exosome trafficking.[43]

This means that acidity is a facilitator but not indispensable condition for tumor- associated exosome (TAE) release and uptake. Normal cells at the pre-metastatic niche that incorporate these TAEs modify their metabolism that switches from OXPHOS to glycolytic.[125, 126] This is also strong evidence that the pre-metastatic niche will become acidic.

Therefore, the pharmacological interventions addressing the inverted pH have many mechanisms that prevent metastatic spreading:

(a) decreased migration and invasion;

(b) reduced exosomes trafficking;

(c) decreased pre-metastatic niche acidity.

Topiramate

Topiramate has a two-pronged effect: it acidifies the cell and inhibits VGSCs. The inhibition of VGSCs reduces the metastatic risk.[127, 128]

Clinical cases

Four patients were treated with the above mentioned preventive scheme by one of us (TK). The unpublished data on these patients follow.

The patients:

(1) Melanoma of the face with deep penetration into the subcutaneous tissue without node invasion. Treated with surgery involving ample excision and skin graft. No metastasis after 22 month of follow up with the scheme. No chemotherapy administered.

(2) Right side colon adenocarcinoma Dukes C. Treated with surgery. No metastasis appeared after 48 month of follow up. No chemotherapy administered.

(3) Clear cell renal carcinoma of 5 cm diameter. Treated with surgery. Two lung and one mediastinal metastasis were removed 2 years after the first surgery. Spine and iliac bone metastasis appeared 1 year later, which were treated with sunitinib and radiotherapy. A rib metastasis and two lung metastasis appeared 1 year after the sunitib treatment. The patient was started on the scheme described above. The costal metastasis decreased its size, but did not disappear. The lung metastases did not increase their size. No new metastases appeared after 71 months follow up since the last metastatic finding. (No evodiamine in this treatment).

(4) Prostate carcinoma with invasion of the Gleason layer. A radical prostatectomy was performed. Neither chemotherapy nor hormonal treatment was administered due to patient refusal. No metastasis after 58 months of treatment with the metastasis prevention scheme without evodiamine.

With such a small number of patients we cannot reach any conclusion and this short list of cases is not a proof of concept. We are only describing certain facts that can provide justification for a case-control study involving a broader range of cancers and patients and a longer follow up.

Conclusions

There is no accepted protocol treatment for reducing metastatic risk. As a matter of fact, standard care does not even consider the problem in depth. With the exception of hormonal treatments in the aftermath of surgery and chemoradiotherapy in prostate and breast cancer, nothing else is done to prevent metastasis. The metastasis prevention scheme should start as soon as it is strongly suspected that a patient has a malignant tumor. There are simple measures that can be put in place to prevent metastasis. These measures should be harmless and devoid of toxicity and implemented until a definitive diagnosis and treatment is decided. In certain health systems this may take some time due to bureaucratic delays, resource shortages, or patient negligence.

The prevention scheme described above deserves well-planned clinical trials because it lacks toxicity, side effects are negligible, does not affect quality of life, the cost is minimal and it does not interfere with the main treatment.

References

1. Chiang AC, Massagué J. Molecular basis of metastasis. *N Engl J Med* 2008;**359**(26):2814–23.
2. Hanahan D, Weinberg RA. Hallmarks of cancer: the next generation. *Cell* 2011;**144**(5):646–74. 4.
3. Mehlen P, Puisieux A. Metastasis: a question of life or death. *Nat Rev Cancer* 2006;**6**:449–58.
4. Taketo MM. Reflections on the spread of metastasis to cancer prevention. *Cancer Prev Res* 2011;**4**:324–8.
5. Ghajar CM. Metastasis prevention by targeting the dormant niche. *Nat Rev Cancer* 2015;**15**(4):238–47.
6. Baron JA, Cole BF, Sandler RS, et al. A randomized trial of aspirin to prevent colorectal adenomas. *N Engl J Med* 2003;**348**:891–9.
7. Rayyan Y, Williams J, Rigas B. The role of NSAIDs in the prevention of colon cancer. *Cancer Invest* 2002;**20**:1002–11.
8. Paez Espinosa EV, Murad JP, Khasawneh FT. Aspirin: pharmacology and clinical applications. *Thrombosis* 2011;**2012**.
9. Schrör K. Pharmacology and cellular/molecular mechanisms of action of aspirin and non-aspirin NSAIDs in colorectal cancer. *Best Pract Res Clin Gastroenterol* 2011;**25**(4-5):473–84.
10. Somasundaram S, Sigthorsson G, Simpson RJ, Watts J, Jacob M, Tavares IA, et al. Uncoupling of intestinal mitochondrial oxidative phosphorylation and inhibition of cyclooxygenase are required for the development of NSAID-enteropathy in the rat. *Aliment Pharmacol Ther* 2000;**14**(5):639–50.
11. Yukiko T, Sokpong L, Michio U. In vitro effects of nonsteroidal anti-inflammatory drugs on oxidative phosphorylation in rat liver mitochondria. *Biochem Pharmacol* 1977;**26**(22):2101–6.
12. Jörgensen TG, Weis-Fogh US, Nielsen HH, Olesen HP. Salicylate-and aspirin-induced uncoupling of oxidative phosphorylation in mitochondria isolated from the mucosal membrane of the stomach. *Scand J Clin Lab Invest* 1976;**36**(7):649–54.
13. Whitehouse MW. Biochemical properties of anti-inflammatory drugs—III: uncoupling of oxidative phosphorylation in a connective tissue (cartilage) and liver mitochondria by salicylate analogues: relationship of structure to activity. *Biochem Pharmacol* 1964;**13**(3):319–36.
14. Brody TM, Fouts W. Action of sodium salicylate and related compounds on tissue metabolism in vitro. *J Pharmacol Exp Ther* 1956;**117**(1):39–51.
15. Gutknecht J. Aspirin, acetaminophen and proton transport through phospholipid bilayers and mitochondrial membranes. *Mol Cell Biochem* 1992;**114**(1-2):3–8.
16. Zimmermann KC, Waterhouse NJ, Goldstein JC, Schuler M, Green DR. Aspirin induces apoptosis through release of cytochrome c from mitochondria. *Neoplasia* 2000;**2**(6):505–13.

17. Dong Z, Huang C, Brown RE, Ma WY. Inhibition of activator protein 1 activity and neoplastic transformation by aspirin. *J Biol Chem* 1997;**272**(15):9962–70.
18. Aktary Z, Pasdar M. Plakoglobin: role in tumorigenesis and metastasis. *Int J Cell Biol* 2012;**2012**.
19. Pavese JM, Krishna SN, Bergan RC. Genistein inhibits human prostate cancer cell detachment, invasion, and metastasis. *Am J Clin Nutr* 2014;**100**(Suppl. 1):431S–6S.
20. Yang X, Li X, Ren J. From French Paradox to cancer treatment: anti-cancer activities and mechanisms of resveratrol. *Anticancer Agents Med Chem* 2014;**14**(6):806–25.
21. Kronski E, Fiori ME, Barbieri O, Astigliano S, Mirisola V, Killian PH, et al. miR181b is induced by the chemopreventive polyphenol curcumin and inhibits breast cancermetastasisvia down-regulation of the inflammatory cytokines CXCL1 and -2. *Mol Oncol* 2014;**8**(3):581–95.
22. Langley RE, Rothwell PM. Aspirin in gastrointestinal oncology: new data on an old friend. *Curr Opin Oncol* 2014;**26**(4):441–7.
23. Li WW, Long GX, Liu DB, Mei Q, Wang JF, Hu GY, et al. Cyclooxygenase-2 inhibitor celecoxib suppresses invasion and migration of nasopharyngeal carcinoma cell lines through a decrease in matrix metalloproteinase-2 and -9 activity. *Pharmazie* 2014;**69**(2):132–7.
24. Lu Y, Yu T, Liang H, Wang J, Xie J, Shao J, et al. Nitric oxide inhibits hetero-adhesion of cancer cells to endothelial cells: restraining circulating tumor cells from initiating metastatic cascade. *Sci Rep* 2014;**4**:4344.
25. Wang J, Chen J, Wan L, Shao J, Lu Y, Zhu Y, et al. Synthesis, spectral characterization, and in vitro cellular activities of metapristone, a potential cancer metastatic chemopreventive agent derived from mifepristone (RU486). *AAPS J* 2014;**16**(2):289–98.
26. Ardiani A, Gameiro SR, Palena C, Hamilton DH, Kwilas A, King TH, et al. Vaccine-mediated immunotherapy directed against a transcription factor driving the metastatic process. *Cancer Res* 2014; **74**(7):1945–57.
27. Zhang J, Zhao C, Gao Y, Jiang Y, Liang H, Zhao G. Thiabendazole, a well-known antifungal drug, exhibits anti-metastatic melanoma B16F10 activity via inhibiting VEGF expression and inducing apoptosis. *Pharmazie* 2013;**68**(12):962–8.
28. Rietkötter E, Menck K, Bleckmann A, Farhat K, Schaffrinski M, Schultz M, et al. Zoledronic acid-inhibits macrophage/microglia-assisted breast cancer cell invasion. *Oncotarget* 2013;**4**(9):1449–60.
29. Rolfo C, Raez LE, Russo A, Reguart N, Campelo RG, Bronte G, et al. Molecular target therapy for bonemetastasis: starting a new era with denosumab, a RANKL inhibitor. *Expert Opin Biol Ther* 2014; **14**(1):15–26.
30. Ribatti D, Mangialardi G, Vacca A. Stephen Paget and the 'seed and soil' theory of metastatic dissemination. *Clin Exp Med* 2006;**6**:145–9.
31. Samatov TR, Tonevitsky AG, Schumacher U. Epithelial-mesenchymal transition: focus on metastatic cascade, alternative splicing, non-coding RNAs and modulating compounds. *Mol Cancer* 2013;**12**:107.
32. Hunter KW, Crawford NPS, Alsarraj J. Mechanisms of metastasis. *Breast Cancer Res* 2008;**10**(Suppl 1): S2.
33. Aceto N, Bardia A, Miyamoto DT, Donaldson MC, Wittner BS, et al. Circulating tumor cell clusters are oligoclonal precursors of breast cancer metastasis. *Cell* 2014;**158**(5):1110–22.
34. Leblanc R, Lee SC, David M, Bordet JC, Norman DD, Patil R. Interaction of platelet-derived autotaxin with tumor integrin αVβ3 controls metastasis of breast cancer cells to bone. *Blood* 2014; **124**(20):3141–50.
35. Azvolinski A. Beyond counting: new way to use circulating tumor cells. *J Natl Cancer Inst* 2014; **106**(10):dju343.
36. Rofstad EK, Mathiesen B, Kindem K, Galappathi K. Acidic extracellular pH promotes experimental metastasis of human melanoma cells in athymic nude mice. *Cancer Res* 2006;**66**(13):6699–707.
37. Abaza M, Luqmani YA. The influence of pH and hypoxia on tumor metastasis. *Expert Rev Anticancer Ther* 2013;**13**(10):1229–42.
38. Calorini L, Peppicelli S, Bianchini F. Extracellular acidity as favouring factor of tumor progression and metastatic dissemination. *Exp Oncol* 2012;**34**(2):79–84.
39. Thews O, Riemann A. Tumor pH and metastasis: a malignant process beyond hypoxia. *Cancer Metastasis Rev* 2019;1–7.

40. Kaplan RN, Riba RD, Zacharoulis S, Bramley AH, Vincent L, Costa C, et al. VEGFR1-positive haematopoietic bone marrow progenitors initiate the pre-metastatic niche. *Nature* 2005; **438**(7069):820.
41. Costa-Silva B, Aiello NM, Ocean AJ, Singh S, Zhang H, Thakur BK, et al. Pancreatic cancer exosomes initiate pre-metastatic niche formation in the liver. *Nat Cell Biol* 2015;(6):816.
42. Hood JL, San RS, Wickline SA. Exosomes released by melanoma cells prepare sentinel lymph nodes for tumor metastasis. *Cancer Res* 2011;**71**:3792–801.
43. Zhao H, Achreja A, Iessi E, Logozzi M, Mizzoni D, Di Raimo R, et al. The key role of extracellular vesicles in the metastatic process. *Biochim Biophys Acta, Rev Cancer* 2018;**1869**(1):64–77.
44. Hoshino D, Kirkbride KC, Costello K, Clark ES, Sinha S, Grega-Larson N, et al. Exosome secretion is enhanced by invadopodia and drives invasive behavior. *Cell Rep* 2013;**5**(5):1159–68.
45. Savina A, Furlán M, Vidal M, Colombo MI. Exosome release is regulated by a calcium-dependent mechanism in K562 cells. *J Biol Chem* 2003;**278**(22):20083–90.
46. Parolini I, Federici C, Raggi C, Lugini L, Palleschi S, De Milito A, et al. Microenvironmental pH is a key factor for exosome traffic in tumor cells. *J Biol Chem* 2009. https://doi.org/10.1074/jbc.M109.041152.
47. Federici C, Petrucci F, Caimi S, Cesolini A, Logozzi M, Borghi M, et al. Exosome release and low pH belong to a framework of resistance of human melanoma cells to cisplatin. *PLoS One* 2014;**9**(2).
48. Chalmin F, Ladoire S, Mignot G, Vincent J, Bruchard M, Remy-Martin JP, et al. Membrane-associated Hsp72 from tumor-derived exosomes mediates STAT3-dependent immunosuppressive function of mouse and human myeloid-derived suppressor cells. *J Clin Invest* 2010;**120**:457–71.
49. Koch R, Aung T, Vogel D, Chapuy B, Wenzel D, Becker S, et al.Nuclear trapping through inhibition of exosomal export by indomethacin increases cytostatic efficacy of doxorubicin and pixantrone. *Clin Cancer Res* 2016;**22**(2):395–404.
50. Khan FM, Saleh E, Alawadhi H, Harati R, Zimmermann WH, El-Awady R. Inhibition of exosome release by ketotifen enhances sensitivity of cancer cells to doxorubicin. *Cancer Biol Ther* 2018; **19**(1):25–33.
51. Schlaepfer IR, Nambiar DK, Ramteke A, Kumar R, Dhar D, Agarwal C, et al. Hypoxia induces triglycerides accumulation in prostate cancer cells and extracellular vesicles supporting growth and invasiveness following reoxygenation. *Oncotarget* 2015;**6**:22836–56.
52. Goetzl EJ, Goetzl L, Karliner JS, Tang N, Pulliam L. Human plasma platelet-derived exosomes: effects of aspirin. *FASEB J* 2016;**30**(5):2058–63.
53. Janowska-Wieczorek A, Marquez-Curtis LA, Wysoczynski M, Ratajczak MZ. Enhancing effect of platelet-derived microvesicles on the invasive potential of breast cancer cells. *Transfusion* 2006; **46**(7):1199–209.
54. Dashevsky O, Varon D, Brill A. Platelet-derived microparticles promote invasiveness of prostate cancer cells via upregulation of MMP-2 production. *Int J Cancer* 2009;**124**(8):1773–7.
55. Teicher B, Fricker S. CXCL12 (SDF-1)/CXCR4 pathway in cancer. *Clin Cancer Res* 2010;**16**:2927–31.
56. Smith MC, Luker KE, Garbow JR, Prior JL, Jackson E, Piwnica-Worms D, et al. CXCR4 regulates growth of both primary and metastatic breast cancer. *Cancer Res* 2004;**64**(23):8604–12.
57. Kioi M, Vogel H, Schultz G, Hoffman RM, Harsh GR, Brown JM. Inhibition of vasculogenesis, but not angiogenesis, prevents the recurrence of glioblastoma after irradiation in mice. *J Clin Invest* 2010;**120**(3):694–705.
58. Liu S-C, Alomran R, Chernikova S, Lartey F, Stafford J, Jang T, et al. Blockade of SDF-1 after irradiation inhibits tumor recurrences of autochthonous brain tumors in rats. *Neuro Oncol* 2014;**16**:21–8.
59. D'Alterio C, Barbieri A, Portella L, Palma G, Polimeno M, Riccio A, et al. Inhibition of stromal CXCR4 impairs development of lung metastases. *Cancer Immunol Immunother* 2012;**61**(10): 1713–20.
60. Chaudary N, Pintilie M, Jelveh S, Lindsay P, Hill RP, Milosevic M. Plerixafor improves primary tumor response and reduces metastases in cervical cancer treated with radio-chemotherapy. *Clin Cancer Res* 2017;**23**(5):1242–9.

61. Conley-LaComb MK, Semaan L, Singareddy R, Li Y, Heath EI, Kim S, et al. Pharmacological targeting of CXCL12/CXCR4 signaling in prostate cancer bone metastasis. *Mol Cancer* 2016;**15**(1):68.
62. Heckmann D, Maier P, Laufs S, Wenz F, Zeller WJ, Fruehauf S, et al. CXCR4 expression and treatment with SDF-1α or Plerixafor modulate proliferation and chemosensitivity of colon cancer cells. *Transl Oncol* 2013;**6**(2):124–32.
63. Aguirre-Ghiso JA. Models, mechanisms and clinical evidence for cancer dormancy. *Nat Rev Cancer* 2007;**7**(11):834.
64. Peppicelli S, Andreucci E, Ruzzolini J, Laurenzana A, Margheri F, Fibbi G, et al. The acidic microenvironment as a possible niche of dormant tumor cells. *Cell Mol Life Sci* 2017;**74**(15):2761–71.
65. Zhang S, Da L, Yang X, Feng D, Yin R, Li M, et al. Celecoxib potentially inhibits metastasis of lung cancer promoted by surgery in mice, via suppression of the PGE2-modulated β-catenin pathway. *Toxicol Lett* 2014;**225**(2):201–7.
66. Barlow M, Edelman M, Glick RD, Steinberg BM, Soffer SZ. Celecoxib inhibits invasion and metastasis via a cyclooxygenase 2-independent mechanism in an in vitro model of Ewing sarcoma. *J Pediatr Surg* 2012;**47**(6):1223–7.
67. Yoshinaka R, Shibata MA, Morimoto J, Tanigawa N, Otsuki Y. COX-2 inhibitor celecoxib suppresses tumor growth and lung metastasis of a murine mammary cancer. *Anticancer Res* 2006;**26**(6B):4245–54.
68. Roh J, Sung M, Park SW, Heo DS, Lee DW, Kim KH. Celecoxib can prevent tumor growth and distant metastasis in postoperative setting. *Cancer Res* 2004;**64**(9):3230–5.
69. Ninomiya I, Nagai N, Oyama K, Hayashi H, Tajima H, Kitagawa H, et al. Antitumor and anti-metastatic effects of cyclooxygenase-2 inhibition by celecoxib on human colorectal carcinoma xenografts in nude mouse rectum. *Oncol Rep* 2012;**28**(3):777–84.
70. Xu L, Stevens J, Hilton MB, Seaman S, Conrads TP, Venstra TD, et al. COX-2 inhibition potentiates antiangiogenic cancer therapy and prevents metastasis in preclinical models. *Sci Transl Med* 2014;**6**(242):242ra84.
71. Niu G, Liao Z, Cai L, Wei R, Sun L. The combined effects of celecoxib and minocycline hydrochloride on inhibiting the osseous metastasis of breast cancer in nude mice. *Cancer Biother Radiopharm* 2008;**23**(4):469–76.
72. Brown MJ. A study of the mechanism by which anticoagulation with warfarin inhibits blood-borne metastases. *Cancer Res* 1973;**33**:1217–24.
73. Ivarsson L, Rudenstam CM. Heparin, dextran 1000 and metastasis formation after i.v. tumour cell injection in dextran non-sensitive rats. *Br J Cancer* 1975;**32**:502.
74. Guillem-Llobat P, Dovizio M, Alberti S, Bruno A, Patrignani P. Platelets, cyclooxygenases, and colon cancer. *Semin Oncol* 2014;**41**(3):385–96.
75. Bambace NM, Holmes CE. The platelet contribution to cancer progression. *J Thromb Haemost* 2011;**9**(2):237–49.
76. Labelle M, Begum S, Hynes RO. Direct signaling between platelets and cancer cells induces an epithelial-mesenchymal-like transition and promotes metastasis. *Cancer Cell* 2011;**20**(5):576–90.
77. Gay LJ, Felding-Habermann B. Contribution of platelets to tumour metastasis. *Nat Rev Cancer* 2011;**11**(2):123.
78. Sharma D, Brummel-Ziedins KE, Bouchard BA, Holmes CE. Platelets in tumor progression: a host factor that offers multiple potential targets in the treatment of cancer. *J Cell Physiol* 2014;**229**(8):1005–15.
79. Borsig L, Wong R, Feramisco J, Nadeau DR, Varki NM, Varki A. Heparin and cancer revisited: mechanistic connections involving platelets, P-selectin, carcinoma mucins, and tumor metastasis. *Proc Natl Acad Sci U S A* 2001;**98**(6):3352–7.
80. Kobayashi KI, Matsumoto S, Morishima T, Kawabe T, Okamoto T. Cimetidine inhibits cancer cell adhesion to endothelial cells and prevents metastasis by blocking E-selectin expression. *Cancer Cell* 2000;**60**(14):3978–84.
81. Liu FR, Jiang CG, Li YS, Li JB, Li F. Cimetidine inhibits the adhesion of gastric cancer cells expressing high levels of sialyl Lewis x in human vascular endothelial cells by blocking E-selectin expression. *Int J Mol Med* 2011;**27**(4):537–44.

82. Ali AH, Hale L, Yalamanchilli B, Ahmed M, Ahmed M, Zhou R, et al.The effect of perioperative cimetidine administration on time to colorectal cancer recurrence. *Am J Ther* 2018;**25**(4):e405–11.
83. Jameson MB, Arendse M, Pillai A, Warren J, Lolohea S, van Dalen R, et al. Final analysis of a ransomized placebo-controlled double-blind phase II trial of perioperative cimetidine (CIM) in early colorectal cancer (CRC). *J Clin Oncol* 2018;**36**(15).
84. Feys L, Descamps B, Vanhove C, Vral A, Veldeman L, Verneulen S, et al. Radiation-induced lung damage promotes breast cancer lung-metastasis through CXCR4 signaling. *Oncotarget* 2015; **6**(29):26615–32.
85. Müller A, Homey B, Soto H, Ge N, Catron D, Buchanan ME, et al. Involvement of chemokine receptors in breast cancer metastasis. *Nature* 2001;**410**:50–6.
86. Gunjal P, Schneider G, Ismail A, Kakar S, Kucia M, Ratajczak M. Evidence for induction of a tumor metastasis-receptive microenvironment for ovarian cancer cells in bone marrow and other organs as an unwanted and underestimated side effect of chemotherapy/radiotherapy. *J Ovarian Res* 2015;**8**:20.
87. Huang Q, Liu X, Shi W, O'Sullivan B, He Z, Peng Y, et al. Caspase 3-mediated stimulation of tumor cell repopulation during cancer radiotherapy. *Nat Med* 2011;**17**:860–6.
88. Shu H-K, Yoon Y, Hong S, Xu K, Gao H, Hao C, et al. Inhibition of the CXCL12/CXCR4-axis as preventive therapy for radiation-induced pulmonary fibrosis. *PLoS One* 2013;**8**.
89. Evans DM, Sloan-Stakleff K, Arvan M, Guyton DP. Time and dose dependency of the suppression of pulmonary metastases of rat mammary cancer by amiloride. *Clin Exp Metastasis* 1998;**16**(4):353–7.
90. Kellen JA, Mirakian A, Kolin A. Antimetastatic effect of amiloride in an animal tumour model. *Anticancer Res* 1988;**8**(6):1373–6.
91. Matthews H, Ranson M, Kelso MJ. Anti-tumour/metastasis effects of the potassium-sparing diuretic amiloride: An orally active anti-cancer drug waiting for its call-of-duty? *Int J Cancer* 2011; **129**(9):2051–61.
92. Evans DM, Sloan-Stakleff K. Suppression of the invasive capacity of human breast cancer cells by inhibition of urokinase plasminogen activator via amiloride and B428. *Am Surg* 2000;**66**:460–4.
93. Evans DM, Sloan-Stakleff K. Control of pulmonary metastases of rat mammary cancer by inhibition of uPA and COX-2, singly and in combination. *Clin Exp Metastasis* 2004;**21**:339–46.
94. Iishi H, Tatsuta M, Baba M, Yano H, Uehara H, Nakaizumi A. Suppression by amiloride of bombesin-enhanced peritoneal metastasis of intestinal adenocarcinomas induced by azoxymethane. *Int J Cancer* 1995;**63**:716–9.
95. Pilat MJ, Lehr JE, Quigley MM, Pienta KJ. The effect of amiloride on the metastatic properties of prostate cancer in the Dunning rat model. *Oncol Rep* 1998;**5**(4):889–91.
96. Yang X, Wang D, Dong W, Song Z, Dou K. Suppression of Na^+/H^+ exchanger 1 by RNA interference or amiloride inhibits human hepatoma cell line SMMC-7721 cell invasion. *Med Oncol* 2011;
28(1):385–90.
97. Ray P, Bhatti R, Gadarowski J, Bell N, Nasruddin S. Inhibitory effect of amiloride on the urokinase plasminogen activators in prostatic cancer. *Tumor Biol* 1998;**19**(1):60.
98. Harguindey S, Orive G, Pedraz JL, Bello G, Arranz JL, Samaniego JM. Long term remission of metastatic ovarian cancer after chronic treatment with the Na^+–H^+ antiport inhibitor amiloride. *Oncologia* 2002;**25**:472–6.
99. Borentain P, Carmona S, Mathieu S, Jouve E, El-Battari A, Gérolami R. Inhibition of E-selectin expression on the surface of endothelial cells inhibits hepatocellular carcinoma growth by preventing tumor angiogenesis. *Cancer Chemother Pharmacol* 2016;**77**(4):847–56.
100. Amith SR, Wilkinson JM, Fliegel L. KR-33028, a potent inhibitor of the Na^+/H^+ exchanger NHE1, suppresses metastatic potential of triple-negative breast cancer cells. *Biochem Pharmacol* 2016;**118**:31–9.
101. Chen WS, Wei SJ, Liu JM, Hsiao M, Kou-Lin J, Yang WK. Tumor invasiveness and liver metastasis of colon cancer cells correlated with cyclooxygenase-2 (COX-2) expression and inhibited by a COX-2-selective inhibitor, etodolac. *Int J Cancer* 2001;**91**(6):894–9.
102. Farooqui M, Li Y, Rogers T, Poonawala T, Griffin RJ, Song CW, et al. COX-2 inhibitor celecoxib prevents chronic morphine-induced promotion of angiogenesis, tumour growth, metastasis and mortality, without compromising analgesia. *Br J Cancer* 2007;**97**(11):1523.

103. Wei D, Wang L, He Y, Xiong HQ, Abbruzzese JL, Xie K. Celecoxib inhibits vascular endothelial growth factor expression in and reduces angiogenesis and metastasis of human pancreatic cancer via suppression of Sp1 transcription factor activity. *Cancer Res* 2004;**64**(6):2030–8.
104. Kundu N, Walser TC, Ma X, Fulton AM. Cyclooxygenase inhibitors modulate NK activities that control metastatic disease. *Cancer Immunol Immunother* 2005;**54**(10):981–7.
105. Ho VW, Hamilton MJ, Dang NH, Hsu BE, Adomat HH, Guns ES, et al. A low carbohydrate, high protein diet combined with celecoxib markedly reduces metastasis. *Carcinogenesis* 2014;**35**(10):2291–9.
106. Zhang S, Lawson KA, Simmons-Menchaca M, Sun L, Sanders BG, Kline K. Vitamin E analog α-TEA and celecoxib alone and together reduce human MDA-MB-435-FL-GFP breast cancer burden and metastasis in nude mice. *Breast Cancer Res Treat* 2004;**87**(2):111–21.
107. Klenke FM, Gebhard MM, Ewerbeck V, Abdollahi A, Huber PE, Sckell A. The selective Cox-2 inhibitor celecoxib suppresses angiogenesis and growth of secondary bone tumors: an intravital microscopy study in mice. *BMC Cancer* 2006;**6**(1):9.
108. Smith MR, Manola J, Kaufman DS, Oh WK, Bubley GJ, Kantoff PW. Celecoxib versus placebo for men with prostate cancer and a rising serum prostate-specific antigen after radical prostatectomy and/or radiation therapy. *J Clin Oncol* 2006;**24**(18):2723–8.
109. Bhatt RS, Merchan J, Parker R, Wu HK, Zhang L, Seery V, et al. A phase 2 pilot trial of low-dose, continuous infusion, or "metronomic" paclitaxel and oral celecoxib in patients with metastatic melanoma. *Cancer* 2010;**116**(7):1751–6.
110. Sade A, Tunçay S, Çimen İ, Severcan F, Banerjee S. Celecoxib reduces fluidity and decreases metastatic potential of colon cancer cell lines irrespective of COX-2 expression. *Biosci Rep* 2012; **32**(1):35–44.
111. Park W, Oh YT, Han JH, Pyo H. Antitumor enhancement of celecoxib, a selective cyclooxygenase-2 inhibitor, in a Lewis lung carcinoma expressing cyclooxygenase-2. *J Exp Clin Cancer Res* 2008; **27**(1):66.
112. Liu H, Yang Y, Xiao J, Lv Y, Liu Y, Yang H, et al.Inhibition of cyclooxygenase-2 suppresses lymph node metastasis via VEGF-C. *Anat Rec* 2009;**292**(10):1577–83.
113. Nishida-Aoki N, Tominaga N, Takeshita F, Sonoda H, Yoshioka Y, Ochiya T. Disruption of circulating extracellular vesicles as a novel therapeutic strategy against cancer metastasis. *Mol Ther* 2017; **25**(1):181–91.
114. Liu S, Jiang M, Zhao Q, Li S, Peng Y, Zhang P, et al.Vascular endothelial growth factor plays a critical role in the formation of the pre-metastatic niche via prostaglandin E2. *Oncol Rep* 2014;**32**(6):2477–84.
115. Burn J, Gerdes AM, Macrae F, Mecklin JP, Moeslein G, Olschwang S, et al. Long-term effect of aspirin on cancer risk in carriers of hereditary colorectal cancer: an analysis from the CAPP2 randomised controlled trial. *Lancet* 2011;**378**(9809):2081–7.
116. Thun MJ, Jacobs EJ, Patrono C. The role of aspirin in cancer prevention. *Nat Rev Clin Oncol* 2012; **9**(5):259.
117. Rothwell PM, Fowkes FGR, Belch JF, Ogawa H, Warlow CP, Meade TW. Effect of daily aspirin on long-term risk of death due to cancer: analysis of individual patient data from randomised trials. *Lancet* 2011;**377**(9759):31–41.
118. Rothwell PM, Wilson M, Price JF, Belch JF, Meade TW, Mehta Z. Effect of daily aspirin on risk of cancer metastasis: a study of incident cancers during randomised controlled trials. *Lancet* 2012; **379**(9826):1591–601.
119. Valadi H, Ekström K, Bossios A, Sjöstrand M, Lee JJ, Lötvall JO. Exosome-mediated transfer of mRNAs and microRNAs is a novel mechanism of genetic exchange between cells. *Nat Cell Biol* 2007;(6):654.
120. Wu M, Wang G, Hu W, Yao Y, Yu XF. Emerging roles and therapeutic value of exosomes in cancer metastasis. *Mol Cancer* 2019;**18**(1):53.
121. Hoshino A, Costa-Silva B, Shen TL, Rodrigues G, Hashimoto A, Tesic Mark M, et al. Tumour exosome integrins determine organotropic metastasis. *Nature* 2015;**527**:329–35.
122. Logozzi M, De Milito A, Lugini L, Borghi M, Calabro L, Spada M, et al. High levels of exosomes expressing CD63 and caveolin-1 in plasma of melanoma patients. *PLoS One* 2009;**4**(4).

123. Logozzi M, Angelini DF, Iessi E, Mizzoni D, Di Raimo R, Federici C, et al. Increased PSA expression on prostate cancer exosomes in in vitro condition and in cancer patients. *Cancer Lett* 2017;**403**:318–29.
124. Logozzi M, Mizzoni D, Angelini DF, Di Raimo R, Falchi M, Battistini L, et al. Microenvironmental pH and exosome levels interplay in human cancer cell lines of different histotypes. *Cancers (Basel)* 2018;**10**(10):370.
125. Fong MY, Zhou W, Liu L, Alontaga AY, Chandra M, Ashby J, Chow A, O'Connor ST, Li S, Chin AR, et al. Breast-cancer-secreted miR-122 reprograms glucose metabolism in premetastatic niche to promote metastasis. *Nat Cell Biol* 2015;**17**:183–94.
126. Yogev O, Henderson S, Hayes MJ, Marelli SS, Ofir-Birin Y, Regev-Rudzki N, et al. Herpesviruses shape tumour microenvironment through exosomal transfer of viral microRNAs. *PLoS Pathog* 2017;**13**.
127. Yildirim S, Altun S, Gumushan H, Patel A, Djamgoz M. Voltage-gated sodium channel activity promotes prostate cancer metastasis in vivo. *Cancer Lett* 2012;**323**:58–61.
128. Gumushan Aktas H, Akgun T. Naringenin inhibits prostate cancer metastasis by blocking voltage-gated sodium channels. *Biomed Pharmacother* 2018;**106**:770–5.

CHAPTER 21

Vacuolar-ATPase proton pump inhibition in cancer therapy: Veterinary and human experience

Enrico Pierluigi Spugnini[a], Stefano Fais[b]
[a]UOSD SAFU, Department of Research, Diagnosis and Innovative Technologies, IRCCS Regina Elena National Cancer Institute, Rome, Italy
[b]Department of Oncology and Molecular Medicine, Istituto Superiore di Sanità, Rome, Italy

Introduction

Very early in their life, tumor cells switch from aerobic to anaerobic glycolysis, a phenomenon that contributes to the induction of a hypoxic microenvironment.[1–5] This happens because uncontrolled proliferating tumor cells produce large volumes of H^+ through glycolysis and glucose consumption, amino acid metabolism, and ATP hydrolysis, all combined with transmembrane proton release and extracellular acidification.[1–5] That said, a singular property of cancer cells is their frequent perseverance in using glycolysis even in the areas with ample oxygen supply capable of supporting oxidative phosphorylation, a phenomenon known as the Warburg effect.[1–5] While apparently being inefficient in an energy-wise way, this metabolic switch actually provides cancer cells the ability to synthesize various macromolecules needed to sustain high proliferation rates. On the other hand, by adopting this energy strategy, cancer cells can also compete with the immune population on their own terms, as actively proliferating cells of the immune system and bypass oxidative phosphorylation during glucose metabolism even in areas with ample oxygen source.[6–8] In this regard, some researchers postulated that the hostile microenvironment created within tumors gradually culls the least competitive cell clones that are unsuited for survival in this high selective milieu. The tumor milieu has proven to be extremely toxic for either normal or more differentiated cells. This environmental pressure selects those clones capable of circumventing intracellular accumulation of toxic molecules, through the over-expression of detoxifying mechanisms including the expression and activation of several proton extruders.[1–8] The identified tumor proton extruders include V-ATPase, Na^+/H^+ exchanger (NHE), monocarboxylate transporters (MCTs) and carbonic anhydrase 9 (CAIX).[1–4] Their over-expression and increased metabolic function lead to a derangement of pH gradients typical of neoplastic cells, that is characterized by a reversed pH gradient between acidic extracellular and alkaline intracellular compartments.[1–6] Several advantages offered by the Warburg

An Innovative Approach to Understanding and Treating Cancer: Targeting pH
https://doi.org/10.1016/B978-0-12-819059-3.00021-6
© 2020 Elsevier Inc.
All rights reserved.

effect to cancer cells have been proposed such as enhanced biosynthetic activity, accelerated adenosine triphosphate (ATP) production, altered cell signaling, and a reduced risk of reactive oxygen species (ROS) mediated damage to malignant cells.[9–11] Another interpretation of the potential "benefit" of the Warburg effect in tumor metabolism, is correlated with the altered physiology of cancer cells where the glycolysis is pivotal for increasing their acid production and invading the neighboring healthy tissue.[12] Several experiments have shown that peritumoral regions with greater acidity generally express stronger invasive potential than regions with normal extracellular pH.[13] In addition, as mentioned above, it has been shown that the microenvironmental acidity helps tumors induce a state of anergy in the immunocompetent cells.[6–8] Finally, a second major mechanism of tumor impairment of the immune response is the depletion of microenvironmental glucose.[14, 15] Several mathematical models of the Warburg effect have been proposed over the years.[16, 17] Indeed, proton detoxifying mechanisms efficiently remove H^+ from the cell or sequester H^+ inside cytoplasmic vesicles, thus sidestepping potentially highly cytotoxic acidification of the tumor cell cytoplasm. Many features of tumor malignancy seem to be associated with increased acid milieu such as invasiveness, metastatic behavior and resistance to cytotoxic agents.[18–22] Other investigations postulate that an altered pH may be critical for several tumor-associated cellular functions, such as exosome synthesis and release, acidic vesicular trafficking, drug resistance, lytic enzyme activation and aberrant phagocytic activity.[23, 24] Proton pumps may act as detoxifying mechanisms by sequestering chemotherapy agents in acid liposomes and extruding them out of the cytosol, a feature that tumors share with several microorganisms.[23, 25] Here, we propose that an efficient anticancer approach could be to block or inhibit mechanisms involved in cell detoxification in order to deprive cancer cells of a key survival option, thereby triggering tumor cell death.

Contribution of V-ATPases to cancer progression and chemoresistance

The peculiar way cancer cells use glucose results in the buildup of acid metabolites that promote an acid milieu that intensively manipulates tumor cells and their host.[26–28] A primary causative agent of tumor unresponsiveness to the majority of cytotoxic drugs, is acidity, that acts through the protonation of the chemotherapeutic agents causing the salification of the agent outside the cells and the impossibility of reaching the intracellular targets.[28, 29] Acidification of neoplasms is induced by the release of byproducts of tumor metabolism—specifically protons—combined with local hypoxia. Over time, this competitive environment gradually selects cells able to thrive in the acidic milieu, which happens through the over-expression and activation of membrane-bound pH-regulating systems that protect tumor cells from intracellular acidification. Of these, vacuolar-type H^+ ATPases seem to be major contributors to the acidification of the tumor microenvironment.[26–29] Vacuolar H^+ ATPase (V-ATPase) is a complex multi-subunit protein that couples the energy of ATP hydrolysis to proton transport across intracellular and plasma membranes in particular from the cytoplasm toward intracellular compartments and from inside to outside the cell through the cytoplasmic membrane.[29–31] V-ATPases are made of a transmembrane subunit, called V_0 complex, devoted to proton transfer and a cytoplasmic portion, known as V_1 complex, that provides the necessary energy for proton translocation.[30] In lieu of its pivotal function in the regulation of cellular pH homeostasis, V-ATPase is involved in several cellular functions including endocytosis and activation of proteases,[25, 30, 31] angiogenesis,[32, 33] autophagy,[34] and amino acid sensing via interaction with mTOR.[35] Tumor cells growing at the periphery of the neoplasm are forced to survive in an hypoxic environment created by the distance from tumor blood vessels, that results in scarce oxygen and nutrient supply.[36–38] Over-expression of V-ATPase is an elegant compensatory apparatus that actually promotes endurance and growth advantages to cancer cells.[36–41] There are several investigations enlightening the crucial part of V-ATPase in tumor invasion and chemoresistance in breast cancer,[42–47] oral and head and neck carcinoma,[48–51] esophageal carcinoma,[52, 53] hepatocellular and pancreatic carcinoma,[54–56] prostatic carcinoma,[57] neuroblastoma,[58] myeloid leukemia,[59] lung carcinoma,[60, 61] sarcoma,[62, 63] and solid tumors in general.[64, 65] Therefore, inhibition of V-ATPase has become a novel therapeutic avenue for thwarting proton metabolism in cancer, which has been extensively investigated in vitro and in vivo, in both preclinical and clinical settings.[1–4]

Anti-V-ATPase compounds

The number of V-ATPase inhibitors is still relatively limited but they are currently being widely investigated to reveal their binding properties and their mode of inhibition.[64, 65] Most of the known V-ATPase inhibitors are natural compounds of microbial origin such as bafilomycin, the first specific inhibitor isolated from *Streptomyces griseus* in the 1980s, or Concanamycin A isolated from *Streptomyces neyagawaensis*.[66, 67] Archazolid, like bafilomycin A1 or concanamycin, binds to the V0 subunit c of V-ATPase and has also been recently investigated as an anti-cancer agent in vitro and in vivo.[68] Regrettably, their use is confined to preclinical investigations due to their elevated toxicity to normal tissues.[66] In this context, omeprazole and its offshoots have elicited great interest and enthusiasm within the scientific community for their ability to inhibit V-ATPase by linking to subunit A of the nucleotide binding domain.[69] An amazing property of these drugs is that they need acidic conditions to be activated, therefore creating the possibility of tumor selectivity.

V-ATPase and acidic tumor pH in autophagy

Autophagy is a highly conserved evolutionary, catabolic, and quality control process which involves the formation of vesicles called autophagosomes that surround cytoplasmic proteins and organelles for their degradation in the lysosome where they find optimal pH conditions for degradation activity.[70] Basal levels of autophagy are normally low but increase following exposure to starvation or diverse types of stress as a defense mechanism that recycles cytosolic material in order to counteract the lack of nutrients.[71] As to the role of autophagy in cancer, it is now widely accepted that said role is dependent on the stage of tumorigenesis.[71, 72] In general, before the initiation of a tumor, autophagy serves a tumor suppressor function in normal cells, removing damaged organelles and protein aggregates which could support genomic instability and tumorigenesis. Conversely, once a tumor is established, autophagy acts as a survival strategy in cancer cells, helping them to cope with hypoxia, nutrient deprivation, and anoikis.[71, 72] It is well known that the inhibition of V-ATPase by Bafilomycin A1 is a standard mechanism for blocking autophagy in the terminal stages.[73] Furthermore, some human diseases induced by mutations in V-ATPase subunits are associated with defective autophagy.[74, 75] Many studies have proved that highly metastatic cells express higher and/or increased levels of V-ATPase activity,[76, 77] and it is plausible that elevated V-ATPase activity in also involved in the upregulated autophagy characteristic of metastatic cancers.[78] Additionally, it has been shown that the exposure of melanoma cells to esomeprazole prompts the inhibition of mTOR signaling, induces the formation of autophagosomes, and concurrently reduces the autophagic flux, probably by inhibiting lysosome acidification.[79] An alternate inhibitory pathway of autophagy by proton pump inhibitors (PPIs) is the limitation of the regulatory functions of V-ATPase without inhibiting its proton pump activity.[80] In fact, after activation by E2F1, V-ATPase becomes a major modulator of mTOR[78] which is in turn, a major regulator of protein synthesis, lysosomal function and autophagy. In this paradigm, it is reasonable to assume that PPIs down-grade the V-ATPase mediated activation of mTOR. Autophagy has been reported as an indispensable survival mechanism used by melanoma, breast, and gastric carcinoma cells in highly acid environments.[81–83] These observations led to the therapeutic approach to cancer of inhibiting autophagy thus depriving cancer cells that are chronically exposed to acidosis of their defensive mechanism. The only available drug tested as autophagy inhibitor in clinical trials in cancer patients is the antimalaric compound chloroquine (CQ).[84] A recent meta-analysis showed that anti-autophagy therapy could improve the outcome of standard anticancer therapies.[84] Given the multifaceted function of V-ATPase in both lysosome acidification and membrane trafficking, further investigation is required to understand how inhibition of V-ATPase could modulate autophagy and tumor growth.[79–85]

Studies in murine models

Confirmatory studies in laboratory animals were performed in the wake of the promising in vitro observations of the antitumor action produced by PPI. Italian researchers validated that pretreatment with PPI omeprazole strongly enhanced the in vitro efficacy of chemotherapy drugs against tumor cell lines.[86] Additionally, the same schedule was effective at decreasing tumor load in SCID mice carrying orthotopic melanoma xenografts.[86] A different strategy that yielded promising results has been RNA interference with subunit ATP6L of V-ATPase that led to growth inhibition and metastatic delay in murine xenografts of hepatocellular carcinoma.[87] The antitumor properties of omeprazole were even more clearly defined by another study that evaluated this activity both in vitro and in vivo.[88] In this study, SCID mice carrying B-cell tumor xenografts were randomized to receive either high dose oral omeprazole or placebo, and the PPI cohort showed an impressive tumor response. This research proved that PPIs not only have chemosensitizing properties but also a direct anticancer activity.[88] Contemporarily, Niikura showed that oral administration of a V-ATPase inhibitor to SCID mice carrying orthotopic breast cancer xenografts resulted in delayed tumor growth as well as decreased bone metastasis.[89] In the wake of these results, other tumor histotypes have been treated with protocols involving PPIs pretreatment followed by chemotherapy. The tumors investigated included melanoma,[90, 91] gastric and colonic carcinoma,[92–95] breast[96] and ovarian carcinoma,[97] osteosarcoma,[98] solid tumors[99] in general and multiple myeloma.[100] All these studies showed the high tolerability and efficacy of this combination, even in notoriously chemo-resistant tumor histotypes. More recently, studies in rodents evidenced the involvement of V-ATPases in tumor-induced immunosuppression and in tumor resistance to biological therapy with targeted antibodies.[101] The study reported by Calcinotto and colleagues[101] is highly significant because it showed that tumor-infiltrating lymphocytes experience a state of anergy when exposed to the acid tumor milieu. Additionally, they proved the reversibility of this condition by countering tumor acidity. Hence, the possibility of downgrading tumor acidity could hypothetically sensitize tumors to conventional and engineered drugs, as well as promote the recovery of the immune system from acidity-induced anergy.[102]

Clinical evidence supporting the efficacy of PPIs as anticancer drugs

The value of comparative oncology as a bridge between mice and humans has been clearly demonstrated and its value as a strategy for speeding up the transition of new therapies from the laboratory to the bedside cannot be denied.[103, 104] Two studies have been conducted so far in pets with spontaneous neoplasms using PPIs in combination with standard chemotherapy. The first one evaluated the ability of high dose lansoprazole to reverse chemoresistance in dogs and cats with cancers unresponsive to chemotherapy.[105] Lansoprazole is a PPI that is converted into its active form upon exposure to an acid environment. Its application in veterinary medicine was confined at low dosage (1 mg/kg) for the treatment of gastric hyperacidity or at higher doses (2 mg/kg) for the therapy of gastrinoma.[105] In this study, the drug was used off-label with a 3 day loading dosage of 5 mg/kg followed by a 4 day maintenance regimen at 1 mg/kg as a chemosensitizer combined with standard veterinary chemotherapy protocols.[105] This strategy resulted in reversal of chemoresistance in 23 out of 34 treated animal patients (67% response rate). The therapy was well tolerated with side effects limited to transient gastrointestinal toxicity (grade 1–2 accordingly to current literature).[105, 106] In the last veterinary study, PPI combined with diet alkalization improved the outcome of metronomic chemotherapy, compared to the control group, independently of the tumor histotype or the animal species with cancer.[107] The cohort receiving alkalization showed enhanced tumor response (both in terms of the number of responders and the quality of response) when compared to the group receiving metronomic chemotherapy alone. Interestingly, alkalization was able to control histotypes unresponsive to chemotherapy, such as hepatic carcinoma, hemangiosarcoma and inflammatory mammary carcinoma. Fig. 1 shows the outcome of a dog with "linitis variant" gastric carcinoma treated with alkalization and metronomic chemotherapy: the dog achieved a partial remission with excellent quality of life for more than 17 months and is still being monitored. Alkalization could result in tumor control even in the case of two cancers simultaneously affecting the same individual, providing that the chemotherapy can be harmonized. Fig. 2 shows a paradigmatic case: a 16-year-old cat with gastrointestinal lymphoma treated with alkalization and chemotherapy. After 1 year, the cat, while still in good partial remission, developed a transitional-cell bladder carcinoma. The protocol was modified replacing vincristine with vinblastine, a drug effective against both histotypes. After 5 months the cat achieved complete remission of both histotypes and chemotherapy has been discontinued after 5 additional months of maintenance.

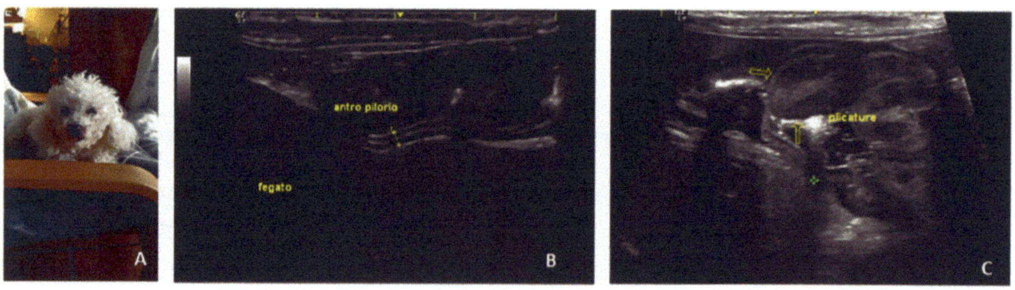

Fig. 1 A dog with "linitis variant" gastric carcinoma treated with alkalization and metronomic chemotherapy: the dog achieved a partial remission with excellent quality of life for more than 17 months and is still being monitored. (A) Twelve-year-old female spayed poodle with gastric carcinoma. (B) Ultrasonography at presentation. (C) Ultrasonography at 15 months of therapy.

So far, except for the occasional case report relating unexpected extended responses in patients treated with PPI,[108] the application of this strategy to humans has led to the publication of two clinical trials. These studies evaluated the tolerability and effectiveness of high-dose PPI in patients with osteosarcoma or metastatic breast cancer.[109, 110] In the first study it was shown that adding PPI, increased the effectiveness of chemotherapy in

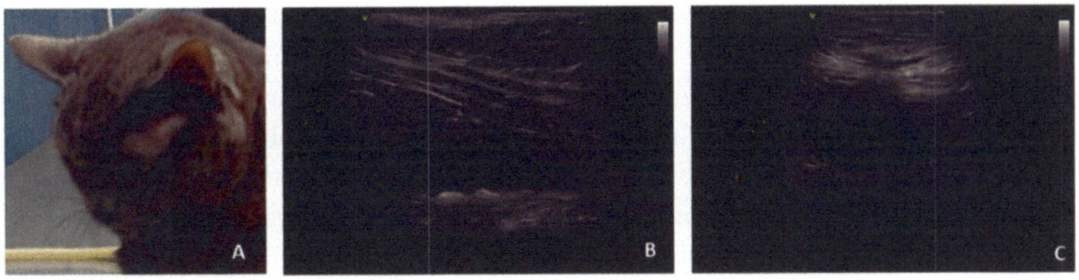

Fig. 2 A 16-year-old cat with gastrointestinal lymphoma treated with alkalization and chemotherapy. After 1 year, the cat, while still in good partial remission, developed a transitional-cell bladder carcinoma. The protocol was modified replacing vincristine with vinblastine, a drug effective against both histotypes. After 5 months the cat achieved complete remission of both histotypes and chemotherapy has been discontinued after 5 additional months of maintenance. (A) Sixteen-year-old male castrated cat with abdominal lymphoma and bladder cancer. (B) Bladder ultranography at the time of transitional cell carcinoma. (C) Bladder ultranography after 5 months of alkalization and chemotherapy.

osteosarcoma patients.[109] This improved the likelihood of successful limb-sparing procedure as well as increasing the overall response rate. Not surprisingly, this response was strongly dependent on the acidity status, as evidenced by the fact that the patients who obtained a statistically significant advantage from this strategy were those with the chondroblastic variant.[109] Indeed, this variant of osteosarcoma has been proven to be the most acidic of all osteosarcoma subtypes, and is a poor responder to chemotherapy.[109] The second trial recruited women with metastatic breast cancer, that were randomized to receive either conventional chemotherapy or chemotherapy with alkalization.[110] In this study, alkalized patients obtained the highest response rates and the longest survivals. Furthermore, there was a statistically significant survival advantage for women who continued their alkaline therapy after the completion of chemotherapy, compared to those who had alkalization limited to the period of chemotherapy.[110] Hence, it is likely that long term alkalization might become the more advantageous approach to the problem of maintenance therapy in cancer patients, allowing clinicians to decrease the duration of their protocols while preserving tumor control. Brana et al. investigated the feasibility of high dose i.v. pantoprazole prior to the administration of doxorubicin to patients with advanced solid tumors. The dose of 240 mg was recognized as the baseline for phase II studies, demonstrating the lack of interaction between the two drugs and confirming the increased uptake of anthracyclines by the tumors.[111] An indirect confirmation of the validity of PPIs as antitumor agents and alkaline therapy in general was provided by a published meta-analysis in head and neck tumor patients that found a better outcome for patients receiving anti-acidic drugs, in particular those treated with PPI.[112] A relatively recent study describes the outcome of 3 patients with metastatic colorectal cancer, refractory to standard chemotherapy and targeted therapy. The combination of high-dose rabeprazole (PPI inhibitor) and metronomic capecitabine[113] overcame drug resistance. This association resulted in a long-lasting stable disease, with good quality of life and relatively minor side effects.[113] These results led to the approval of a clinical trial (EudraCT Number: 2013-001096-20) aimed at testing the effectiveness of rescue alkalized metronomic chemotherapy.[114, 115] The results of this trial are expected to be published shortly and will be an advance in metronomic chemotherapy, the latest approved anticancer therapy. We hope that the same recognition will be granted to cancer alkalization as a standalone, non-mainstream anticancer therapy.

Conclusions

A common feature of malignant tumors is the anaerobic consumption of large amounts of glucose that results in low ATP production and a considerable amount of lactate as a terminal metabolic product. Otto Warburg proposed that these two phenomena were hallmarks of cancer.[116] The proposed mechanism for justifying this metabolic anomaly was an adaptation to the hypoxic tumor micro-environment in a fashion resembling anaerobic glycolysis in under-perfused tissue. However, a Warburg-like metabolism actually occurs early in the life of tumors, even in well-perfused districts.[1–4, 51] Conversion of glucose to lactate in aerobic conditions apparently appears to be a wasteful way of generating energy (roughly 18 times less efficient than mitochondrial oxidative phosphorylation), resulting in large amounts of non-metabolized reduced carbon, even for "selfish" tumor cells.[51] This metabolic pathway, however, supplies large amounts of carbon which in turn are diverted to proliferation.[51] In other words, as it is seen today, Warburg-like metabolism consists of the transformation of glucose in ribose-5-phosphate for nucleotide synthesis (pentose-phosphate-shunt). At the same time, large amounts of glutamine are transformed into mitochondrial tricarboxylic acid (TCA) cycle intermediates, a nitrogen source for nucleotide synthesis. Moreover, this anaplerosis acts as a balance for the migration of mitochondrial citrate to the cytosol, where it is used to generate acetyl-CoA, and for membrane lipid synthesis.[51] All these metabolic perturbations lead to the accumulation of enormous amounts of protons that are released beyond the cytoplasmic membrane by dedicated transporters and exchangers. These laboratory findings plus the preclinical and clinical results of proton pump inhibition and alkalizing therapy in general should lead to a revision of current approaches to cancer treatment.

References

1. Tredan O, Galmarini CM, Patel K, Tannock IF. Drug resistance and the solid tumor microenvironment. *J Natl Cancer Inst* 2007;**99**(19):1441–54.
2. Spugnini EP, Sonveaux P, Stock C, Perez-Sayans M, De Milito A, Avnet S, et al. Proton channels and exchangers in cancer. *Biochim Biophys Acta* 2015;**1848**(10 Pt B):2715–26.
3. Taylor S, Spugnini EP, Assaraf YG, Azzarito T, Rauch C, Fais S. Microenvironment acidity as a major determinant of tumor chemoresistance: proton pump inhibitors (PPIs) as a novel therapeutic approach. *Drug Resist Updat* 2015;**23**:69–78.

4. Spugnini E, Fais S. Proton pump inhibition and cancer therapeutics: a specific tumor targeting or it is a phenomenon secondary to a systemic buffering? *Semin Cancer Biol* 2017;**43**:111–8.
5. Izumi H, Torigoe T, Ishiguchi H, Uramoto H, Yoshida Y, Tanabe M, et al. Cellular pH regulators: potentially promising molecular targets for cancer chemotherapy. *Cancer Treat Rev* 2003;**29**(6):541–9.
6. Kareva I, Hahnfeldt P. The emerging "hallmarks" of metabolic reprogramming and immune evasion: distinct or linked? *Cancer Res* 2013;**73**(9):2737–42.
7. Pilon-Thomas S, Kodumudi KN, El-Kenawi AE, Russell S, Weber AM, Luddy K, et al. Neutralization of tumor acidity improves antitumor responses to immunotherapy. *Cancer Res* 2016;**76**(6):1381–90.
8. Huber V, Camisaschi C, Berzi A, Ferro S, Lugini L, Triulzi T, et al. Cancer acidity: an ultimate frontier of tumor immune escape and a novel target of immunomodulation. *Semin Cancer Biol* 2017;**43**:74–89.
9. Liberti MV, Locasale JW. The Warburg effect: how does it benefit cancer cells? *Trends Biochem Sci* 2016;**41**:211–8.
10. Vander Heiden MG, Cantley LC, Thompson CB. Understanding the Warburg effect: the metabolic requirements of cell proliferation. *Science* 2009;**324**:1029–33.
11. Stark H, Fichtner M, König R, Lorkowski S, Schuster S. Causes of upregulation of glycolysis in lymphocytes upon stimulation. A comparison with other cell types. *Biochimie* 2015;**118**:185–94.
12. Gatenby RA, Gillies RJ. Why do cancers have high aerobic glycolysis? *Nat Rev Cancer* 2004;**4**:891–9.
13. Estrella V, Chen T, Lloyd M, Wojtkowiak J, Cornnell HH, Ibrahim-Hashim A, et al. Acidity generated by the tumor microenvironment drives local invasion. *Cancer Res* 2013;**73**(5):1524–35.
14. Chang CH, Qiu J, O'Sullivan D, Buck MD, Noguchi T, Curtis JD, et al. Metabolic competition in the tumor microenvironment is a driver of cancer progression. *Cell* 2015;**162**(6):1229–41.
15. Lyssiotis CA, Kimmelman AC. Metabolic interactions in the tumor microenvironment. *Trends Cell Biol* 2017;**27**:863–75.
16. Schuster S, Boley D, Möller P, Stark H, Kaleta C. Mathematical models for explaining the Warburg effect: a review focussed on ATP and biomass production. *Biochem Soc Trans* 2015;**43**:1187–94.
17. Shamsi M, Saghafian M, Dejam M, Sanati-Nezhad A. Mathematical modeling of the function of Warburg effect in tumor microenvironment. *Sci Rep* 2018;**8**(1):8903.
18. Lee AH, Tannock IF. Heterogeneity of intracellular pH and of mechanisms that regulate intracellular pH in populations of cultured cells. *Cancer Res* 1998;**58**(9):1901–8.
19. Smallbone K, Gavaghan DJ, Gatenby RA, Maini PK. The role of acidity in solid tumour growth and invasion. *J Theor Biol* 2005;**235**:476–84.
20. Rofstad EK, Mathiesen B, Kindem K, Galappathi K. Acidic extracellular pH promotes experimental metastasis of human melanoma cells in athymic nude mice. *Cancer Res* 2006;**66**(13):6699–707.
21. Robertson-Tessi M, Gillies RJ, Gatenby RA, Anderson AR. Impact of metabolic heterogeneity on tumor growth, invasion, and treatment outcomes. *Cancer Res* 2015;**75**:1567–79.
22. Raghunand N, Mahoney B, van Sluis R, Baggett B, Gillies RJ. Acute metabolic alkalosis enhances response of C3H mouse mammary tumors to the weak base mitoxantrone. *Neoplasia* 2001;**3**(3):227–35.
23. Fais S, Fauvarque MO. TM9 and cannibalism: how to learn more about cancer by studying amoebae and invertebrates. *Trends Mol Med* 2012;**18**(1):4–5.
24. Spugnini EP, Logozzi M, Di Raimo R, Mizzoni D, Fais SA. Role of tumor-released exosomes in paracrine dissemination and metastasis. *Int J Mol Sci* 2018;**19**(12). pii: E3968.
25. Barar J, Omidi Y. Dysregulated pH in tumor microenvironment checkmates cancer therapy. *Bioimpacts* 2013;**3**(4):149–62.
26. Daniel C, Bell C, Burton C, Harguindey S, Reshkin SJ, Rauch C. The role of proton dynamics in the development and maintenance of multidrug resistance in cancer. *Biochim Biophys Acta* 2013;**1832**(5):606–17.
27. da Silva VP, Mesquita CB, Nunes JS, de Bem Prunes B, Rados PV, Visioli F. Effects of extracellular acidity on resistance to chemotherapy treatment: a systematic review. *Med Oncol* 2018;**35**(12):161.
28. Spugnini EP, Citro G, Fais S. Proton pump inhibitors as anti vacuolar-ATPases drugs: a novel anticancer strategy. *J Exp Clin Cancer Res* 2010;**29**(1):44.
29. McCarty MF, Whitaker J. Manipulating tumor acidification as a cancer treatment strategy. *Altern Med Rev* 2010;**15**(3):264–72.

30. Forgac M. Vacuolar ATPases: rotary proton pumps in physiology and pathophysiology. *Nat Rev Mol Cell Biol* 2007;**8**(11):917–29.
31. Lozupone F, Borghi M, Marzoli F, Azzarito T, Matarrese P, Iessi E, et al. TM9SF4 is a novel V-ATPase-interacting protein that modulates tumor pH alterations associated with drug resistance and invasiveness of colon cancer cells. *Oncogene* 2015;**34**(40):5163–74.
32. Rath S, Liebl J, Furst R, Vollmar AM, Zahler S. Regulation of endothelial signaling and migration by V-ATPase. *Angiogenesis* 2014;**17**(3):587–601.
33. Sennoune SR, Arutunyan A, del Rosario C, Castro-Marin R, Hussain F, Martinez-Zaguilan R. V-ATPase regulates communication between microvascular endothelial cells and metastatic cells. *Cell Mol Biol (Noisy-le-Grand)* 2014;**60**(1):19–25.
34. Mijaljica D, Prescott M, Devenish RJ. V-ATPase engagement in autophagic processes. *Autophagy* 2011;**7**(6):666–8.
35. Zoncu R, Bar-Peled L, Efeyan A, Wang S, Sancak Y, Sabatini DM. mTORC1 senses lysosomal amino acids through an inside-out mechanism that requires the vacuolar H(+)-ATPase. *Science* 2011; **334**(6056):678–83.
36. von Schwarzenberg K, Wiedmann RM, Oak P, Schulz S, Zischka H, Wanner G, et al. Mode of cell death induction by pharmacological vacuolar H+-ATPase (V-ATPase) inhibition. *J Biol Chem* 2013; **288**(2):1385–96.
37. Schempp CM, von Schwarzenberg K, Schreiner L, Kubisch R, Muller R, Wagner E, et al. V-ATPase inhibition regulates anoikis resistance and metastasis of cancer cells. *Mol Cancer Ther* 2014;**13**(4): 926–37.
38. Asgharzadeh MR, Barar J, Pourseif MM, Eskandani M, Jafari Niya M, Mashayekhi MR, et al. Molecular machineries of pH dysregulation in tumor microenvironment: potential targets for cancer therapy. *Bioimpacts* 2017;**7**(2):115–33.
39. Pérez-Sayáns M, Somoza-Martín JM, Barros-Angueira F, Rey JM, García-García A. V-ATPase inhibitors and implication in cancer treatment. *Cancer Treat Rev* 2009;**35**(8):707–13.
40. Martinez-Zaguilan R, Raghunand N, Lynch RM, Bellamy W, Martinez GM, Rojas B, et al. pH and drug resistance. I. Functional expression of plasmalemmal V-type H+-ATPase in drug-resistant human breast carcinoma cell lines. *Biochem Pharmacol* 1999;**57**(9):1037–46.
41. Raghunand N, Martinez-Zaguilan R, Wright SH, Gillies RJ. pH and drug resistance. II. Turnover of acidic vesicles and resistance to weakly basic chemotherapeutic drugs. *Biochem Pharmacol* 1999;**57**(9): 1047–58.
42. You H, Jin J, Shu H, Yu B, De Milito A, Lozupone F, et al. Small interfering RNA targeting the subunit ATP6L of proton pump V-ATPase overcomes chemoresistance of breast cancer cells. *Cancer Lett* 2009;**280**(1):110–9.
43. Feng S, Zhu G, McConnell M, Deng L, Zhao Q, Wu M, et al. Silencing of atp6v1c1 prevents breast cancer growth and bone metastasis. *Int J Biol Sci* 2013;**9**(8):853–62.
44. Fan S, Niu Y, Tan N, Wu Z, Wang Y, You H, et al. LASS2 enhances chemosensitivity of breast cancer by counteracting acidic tumor microenvironment through inhibiting activity of V-ATPase proton pump. *Oncogene* 2013;**32**(13):1682–90.
45. Feng S, Cai M, Liu P, Wei L, Wang J, Qi J, et al. Atp6v1c1 may regulate filament actin arrangement in breast cancer cells. *PLoS One* 2014;**9**(1):e84833.
46. von Schwarzenberg K, Lajtos T, Simon L, Muller R, Vereb G, Vollmar AM. V-ATPase inhibition overcomes trastuzumab resistance in breast cancer. *Mol Oncol* 2014;**8**(1):9–19.
47. Kuang Y, Wang S, Tang L, Hai J, Yan G, Liao L. Cluster of differentiation 147 mediates chemoresistance in breast cancer by affecting vacuolar H(+)-ATPase expression and activity. *Oncol Lett* 2018;**15** (5): 7279–90.
48. Garcia-Garcia A, Perez-Sayans Garcia M, Rodriguez MJ, Antunez-Lopez J, Barros-Angueira F, Somoza-Martin M, et al. Immunohistochemical localization of C1 subunit of V-ATPase (ATPase C1) in oral squamous cell cancer and normal oral mucosa. *Biotech Histochem* 2012;**87**(2):133–9.
49. Perez-Sayans M, Somoza-Martin JM, Barros-Angueira F, Diz PG, Rey JM, Garcia-Garcia A. Multidrug resistance in oral squamous cell carcinoma: the role of vacuolar ATPases. *Cancer Lett* 2010;**295**(2): 135–43.

50. Oliveira Alves MG, Carta CF, Padín-Iruegas ME, Pérez-Sayáns M, Suarez-Peñaranda JM, Issa JS, et al. Expression of ATP6V1C1 during oral carcinogenesis. *Biotech Histochem* 2016;**91**(4):263–8.
51. Huebbers CU, Adam AC, Preuss SF, Schiffer T, Schilder S, Guntinas-Lichius O, et al. High glucose uptake unexpectedly is accompanied by high levels of the mitochondrial ß-F1-ATPase subunit in head and neck squamous cell carcinoma. *Oncotarget* 2015;**6**(34):36172–84.
52. Huang L, Lu Q, Han Y, Li Z, Zhang Z, Li X. ABCG2/V-ATPase was associated with the drug resistance and tumor metastasis of esophageal squamous cancer cells. *Diagn Pathol* 2012;**7**:180.
53. Son SW, Kim SH, Moon EY, Kim DH, Pyo S, Um SH. Prognostic significance and function of the vacuolar H+-ATPase subunit V1E1 in esophageal squamous cell carcinoma. *Oncotarget* 2016;**7**(31):49334–48.
54. Bartel K, Winzi M, Ulrich M, Koeberle A, Menche D, Werz O, et al. V-ATPase inhibition increases cancer cell stiffness and blocks membrane related Ras signaling—a new option for HCC therapy. *Oncotarget* 2017;**8**(6):9476–87.
55. Kuchuk O, Tuccitto A, Citterio D, Huber V, Camisaschi C, Milione M, et al. pH regulators to target the tumor immune microenvironment in human hepatocellular carcinoma. *Oncoimmunology* 2018;**7**(7):e1445452.
56. Chung C, Mader CC, Schmitz JC, Atladottir J, Fitchev P, Cornwell ML, et al. The vacuolar-ATPase modulates matrix metalloproteinase isoforms in human pancreatic cancer. *Lab Invest* 2011;**91**(5):732–43.
57. Guedes JP, Pereira CS, Rodrigues LR, Côrte-Real M. Bovine milk lactoferrin selectively kills highly metastatic prostate cancer PC-3 and osteosarcoma MG-63 cells in vitro. *Front Oncol* 2018;**8**:200.
58. Hrabeta J, Groh T, Khalil MA, Poljakova J, Adam V, Kizek R, et al. Vacuolar-ATPase-mediated intracellular sequestration of ellipticine contributes to drug resistance inneuroblastoma cells. *Int J Oncol* 2015;**47**(3):971–80.
59. Aasebø E, Bartaula-Brevik S, Hernandez-Valladares M, Bruserud Ø. Vacuolar ATPase as a possible therapeutic target in human acute myeloid leukemia. *Expert Rev Hematol* 2018;**11**(1):13–24.
60. Lu Q, Lu S, Huang L, Wang T, Wan Y, Zhou CX, et al. The expression of V-ATPase is associated with drug resistance and pathology of non-small-cell lung cancer. *Diagn Pathol* 2013;**8**(145).
61. Kim YS, Jin HO, Hong SE, Song JY, Hwang CS, Park IC. Silencing of secretory clusterin sensitizes NSCLC cells to V-ATPase inhibitors by downregulating survivin. *Biochem Biophys Res Commun* 2018;**495**(2):2004–9.
62. Perut F, Avnet S, Fotia C, Baglio SR, Salerno M, Hosogi S, et al. V-ATPase as an effective therapeutic target for sarcomas. *Exp Cell Res* 2014;**320**(1):21–32.
63. Avnet S, Di Pompo G, Lemma S, Salerno M, Perut F, Bonuccelli G, et al. V-ATPase is a candidate therapeutic target for ewing sarcoma. *Biochim Biophys Acta* 2013;**1832**(8):1105–16.
64. Perez-Sayans M, Garcia-Garcia A, Reboiras-Lopez MD, Gandara-Vila P. Role of V-ATPases in solid tumors: importance of the subunit C (review). *Int J Oncol* 2009;**34**(6):1513–20.
65. Tay Z, Koo SH, Nguyen TT, Tan TS, Chen ML, Chin CF, et al. P-glycoprotein and vacuolar ATPase synergistically confer anthracycline resistance to fission yeast and human cells. *Curr Med Chem* 2014;**21**(2):251–60.
66. Bowman EJ, Siebers A, Altendorf K. Bafilomycins: a class of inhibitors of membrane ATPases from microorganisms, animal cells, and plant cells. *Proc Natl Acad Sci U S A* 1988;**85**(21):7972–6.
67. Muroi M, Shiragami N, Nagao K, Yamasaki M, Takatsuki A. Folimycin (concanamycin A), a specific inhibitor of V-ATPase, blocks intracellular translocation of the glycoprotein of vesicular stomatitis virus before arrival to the Golgi apparatus. *Cell Struct Funct* 1993;**18**(3):139–49.
68. Merk H, Messer P, Ardelt MA, Lamb DC, Zahler S, Müller R, et al. Inhibition of the V-ATPase by archazolid A: a new strategy to inhibit EMT. *Mol Cancer Ther* 2017;**16**(11):2329–39.
69. Mattsson JP, Vaananen K, Wallmark B, Lorentzon P. Omeprazole and bafilomycin, two proton pump inhibitors: differentiation of their effects on gastric, kidney and bone H(+)-translocating ATPases. *Biochim Biophys Acta* 1991;**1065**(2):261–8.
70. Levy JMM, Towers CG, Thorburn A. Targeting autophagy in cancer. *Nat Rev Cancer* 2017;**17**(9):528–42.

71. Cotzomi-Ortega I, Aguilar-Alonso P, Reyes-Leyva J, Maycotte P. Autophagy and its role in protein secretion: implications for cancer therapy. *Mediators Inflamm* 2018;**2018**:4231591. https://doi.org/10.1155/2018/4231591.
72. Galluzzi L, Pietrocola F, Bravo-San Pedro JM, Amaravadi RK, Baehrecke EH, Cecconi F, et al. Autophagy in malignant transformation and cancer progression. *EMBO J* 2015;**34**(7):856–80.
73. Klionsky DJ, Abdalla FC, Abeliovich H, Abraham RT, Acevedo-Arozena A, Adeli K, et al. Guidelines for the use and interpretation of assays for monitoring autophagy. *Autophagy* 2012;**8**(4):445–544.
74. Ochotny N, Voronov I, Owen C, Aubin JE, Manolson MF. The R740S mutation in the V-ATPase a3 subunit results in osteoclast apoptosis and defective early-stage autophagy. *J Cell Biochem* 2013; **114**(12):2823–33.
75. Ramachandran N, Munteanu I, Wang P, Ruggieri A, Rilstone JJ, Israelian N, et al. VMA21 deficiency prevents vacuolar ATPase assembly and causes autophagic vacuolar myopathy. *Acta Neuropathol* 2013;**125**(3):439–57.
76. Pamarthy S, Kulshrestha A, Katara GK, Beaman KD. The curious case of vacuolar ATPase: regulation of signaling pathways. *Mol Cancer* 2018;**17**(1):41.
77. De Milito A, Canese R, Marino ML, Borghi M, Iero M, Villa A, et al. pH-dependent antitumor activity of proton pump inhibitors against human melanoma is mediated by inhibition of tumor acidity. *Int J Cancer* 2010;**127**(1):207–19.
78. Meo-Evoli N, Almacellas E, Massucci FA, Gentilella A, Ambrosio S, Kozma SC, et al. V-ATPase: a master effector of E2F1-mediated lysosomal trafficking, mTORC1 activation and autophagy. *Oncotarget* 2015;**6**(29):28057–70.
79. Marino ML, Fais S, Djavaheri-Mergny M, Villa A, Meschini S, Lozupone F, et al. Proton pump inhibition induces autophagy as a survival mechanism following oxidative stress in human melanoma cells. *Cell Death Dis* 2010;**1**:e87.
80. Udelnow A, Kreyes A, Ellinger S, Landfester K, Walther P, Klapperstueck T, et al. Omeprazole inhibits proliferation and modulates autophagy in pancreatic cancer cells. *PLoS One* 2011;**6**(5):e20143.
81. Marino ML, Pellegrini P, Di Lernia G, Djavaheri-Mergny M, Brnjic S, Zhang X, et al. Autophagy is a protective mechanism for human melanoma cells under acidic stress. *J Biol Chem* 2012;**287**(36): 30664–76.
82. Wojtkowiak JW, Rothberg JM, Kumar V, Schramm KJ, Haller E, Proemsey JB, et al. Chronic autophagy is a cellular adaptation to tumor acidic pH microenvironments. *Cancer Res* 2012;**72**(16):3938–47.
83. Sheng J, Sun LB, Zhao SF, Qi WW, Lv J, Zhang ZG, et al. Acidic stress induces protective autophagy in SGC7901 cells. *J Int Med Res* 2018;**46**(8):3285–95.
84. Xu R, Ji Z, Xu C, Zhu J. The clinical value of using chloroquine or hydroxychloroquine as autophagy inhibitors in the treatment of cancers: a systematic review and meta-analysis. *Medicine (Baltimore)* 2018;**97**(46).
85. Kitazawa S, Nishizawa S, Nakagawa H, Funata M, Nishimura K, Soga T, et al. Cancer with low cathepsin D levels is susceptible to vacuolar (H(+))-ATPase inhibition. *Cancer Sci* 2017;**108** (6):1185–93.
86. Luciani F, Spada M, De Milito A, Molinari A, Rivoltini L, Montinaro A, et al. Effect of proton pump inhibitor pretreatment on resistance of solid tumors to cytotoxic drugs. *J Natl Cancer Inst* 2004;**96**(22): 1702–13.
87. Lu X, Qin W, Li J, Tan N, Pan D, Zhang H, et al. The growth and metastasis of human hepatocellular carcinoma xenografts are inhibited by small interfering RNA targeting to the subunit ATP6L of proton pump. *Cancer Res* 2005;**65**(15):6843–9.
88. De Milito A, Iessi E, Logozzi M, Lozupone F, Spada M, Marino ML, et al. Proton pump inhibitors induce apoptosis of human B-cell tumors through a caspase-independent mechanism involving reactive oxygen species. *Cancer Res* 2007;**67**(11):5408–17.
89. Niikura K. Effect of a V-ATPase inhibitor, FR202126, in syngeneic mouse model of experimental bone metastasis. *Cancer Chemother Pharmacol* 2007;**60**(4):555–62.
90. Nishisho T, Hata K, Nakanishi M, Morita Y, Sun-Wada GH, Wada Y, et al. The a3 isoform vacuolar type H(+)-ATPase promotes distant metastasis in the mouse B16 melanoma cells. *Mol Cancer Res* 2011;**9**(7):845–55.

91. Azzarito T, Venturi G, Cesolini A, Fais S. Lansoprazole induces sensitivity to suboptimal doses of paclitaxel in human melanoma. *Cancer Lett* 2015;**356**(2 Pt B):697–703.
92. Chen M, Lu J, Wei W, Lv Y, Zhang X, Yao Y, et al. Effects of proton pump inhibitors on reversing multidrug resistance via downregulating V-ATPases/PI3K/Akt/mTOR/HIF-1α signaling pathway through TSC1/2 complex and Rheb in human gastric adenocarcinoma cells in vitro and in vivo. *Onco Targets Ther* 2018;**11**:6705–22.
93. Wang X, Liu C, Wang J, Fan Y, Wang Z, Wang Y. Proton pump inhibitors increase the chemosensitivity of patients with advanced colorectal cancer. *Oncotarget* 2017;**8**(35):58801–8.
94. Zheng M, Luan S, Gao S, Cheng L, Hao B, Li J, et al. Proton pump inhibitor ilaprazole suppresses cancer growth by targeting T-cell-originated protein kinase. *Oncotarget* 2017;**8**(24):39143–53.
95. Zeng X, Liu L, Zheng M, Sun H, Xiao J, Lu T, et al. Pantoprazole, an FDA-approved proton-pump inhibitor, suppresses colorectal cancer growth by targeting T-cell-originated protein kinase. *Oncotarget* 2016;**7**(16):22460–73.
96. Schneider LS, von Schwarzenberg K, Lehr T, Ulrich M, Kubisch-Dohmen R, Liebl J, Trauner D, et al. Vacuolar-ATPase inhibition blocks iron metabolism to mediate therapeutic effects in breast cancer. *Cancer Res* 2015;**75**(14):2863–74.
97. Lee YY, Jeon HK, Hong JE, Cho YJ, Ryu JY, Choi JJ, et al. Proton pump inhibitors enhance the effects of cytotoxic agents in chemoresistant epithelial ovarian carcinoma. *Oncotarget* 2015;**6**(33): 35040–50.
98. Avnet S, Lemma S, Cortini M, Pellegrini P, Perut F, Zini N, et al. Altered pH gradient at the plasma membrane of osteosarcoma cells is a key mechanism of drug resistance. *Oncotarget* 2016;**7**(39): 63408–23.
99. Yu M, Lee C, Wang M, Tannock IF. Influence of the proton pump inhibitor lansoprazole on distribution and activity of doxorubicin in solid tumors. *Cancer Sci* 2015;**106**(10):1438–47.
100. Hiasa M, Okui T, Allette YM, Ripsch MS, Sun-Wada GH, Wakabayashi H, et al. Bone pain induced by multiple myeloma is reduced by targeting V-ATPase and ASIC3. *Cancer Res* 2017;**77** (6):1283–95.
101. Calcinotto A, Filipazzi P, Grioni M, Iero M, De Milito A, Ricupito A, et al. Modulation of microenvironment acidity reverses anergy in human and murine tumor-infiltrating T lymphocytes. *Cancer Res* 2012;**72**(11):2746–56.
102. Li W, Sun X. Recent advances in developing novel anti-cancer drugs targeting tumor hypoxic and acidic microenvironments. *Recent Pat Anticancer Drug Discov* 2018;**13**(4):455–68.
103. Park JS, Withers SS, Modiano JF, Kent MS, Chen M, Luna JI, et al. Canine cancer immunotherapy studies: linking mouse and human. *J Immunother Cancer* 2016;**4**:97.
104. Bentley RT, Ahmed AU, Yanke AB, Cohen-Gadol AA, Dey M. Dogs are man's best friend: in sickness and in health. *Neuro Oncol* 2017;**19**(3):312–22.
105. Spugnini EP, Baldi A, Buglioni S, Carocci F, de Bazzichini GM, Betti G, et al. Lansoprazole as a rescue agent in chemoresistant tumors: a phase I/II study in companion animals with spontaneously occurring tumors. *J Transl Med* 2011;**9**:221.
106. Veterinary cooperative oncology group—common terminology criteria for adverse events (VCOG-CTCAE) following chemotherapy or biological antineoplastic therapy in dogs and cats. *Vet Comp Oncol* 2016;**14**(4):417–46.
107. Spugnini EP, Buglioni S, Carocci F, Francesco M, Vincenzi B, Fanciulli M, et al. High dose lansoprazole combined with metronomic chemotherapy: a phase I/II study in companion animals with spontaneously occurring tumors. *J Transl Med* 2014;**12**:225.
108. Sreevathsa MR, Choudhury A. Long-term survival in a patient with carcinoid syndrome receiving treatment for Zollinger-Ellison syndrome. *Int Surg* 2009;**94**(1):20–2.
109. Ferrari S, Perut F, Fagioli F, Brach Del Prever A, Meazza C, Parafioriti A, et al. Proton pump inhibitor chemosensitization in human osteosarcoma: from the bench to the patients' bed. *J Transl Med* 2013;**11**:268.
110. Wang BY, Zhang J, Wang JL, Sun S, Wang ZH, Wang LP, et al. Intermittent high dose proton pump inhibitor enhances the antitumor effects of chemotherapy in metastatic breast cancer. *J Exp Clin Cancer Res* 2015;**34**:109 [Erratum].

111. Brana I, Ocana A, Chen EX, Razak AR, Haines C, Lee C, et al. A phase I trial of pantoprazole in combination with doxorubicin in patients with advanced solid tumors: evaluation of pharmacokinetics of both drugs and tissue penetration of doxorubicin. *Invest New Drugs* 2014;**32**(6):1269–77.
112. Papagerakis S, Bellile E, Peterson LA, Pliakas M, Balaskas K, Selman S, et al. Proton pump inhibitors and histamine 2 blockers are associated with improved overall survival in patients with head and neck squamous carcinoma. *Cancer Prev Res (Phila)* 2014;**7**(12):1258–69.
113. Falcone R, Roberto M, D'Antonio C, Romiti A, Milano A, Onesti CE, et al. High-doses of proton pumps inhibitors in refractory gastro-intestinal cancer: a case series and the state of art. *Dig Liver Dis* 2016;**48**(12):1503–5.
114. Roberto M, Romiti A, Onesti CE, D'Antonio C, Milano A, Falcone R, et al. A metronomic schedule as salvage chemotherapy for upper gastrointestinal tract cancer. *Anticancer Drugs* 2016;**27**(2):106–11.
115. Marchetti P, Milano A, D'Antonio C, Romiti A, Falcone R, Roberto M, et al. Association between proton pump inhibitors and metronomic capecitabine as salvage treatment for patients with advanced gastrointestinal tumors: a randomized phase II trial. *Clin Colorectal Cancer* 2016;**15**(4):377–80.
116. Warburg O. On respiratory impairment in cancer cells. *Science* 1956;**124**(3215):269–70.

Further reading

117. Huss M, Vitavska O, Albertmelcher A, Bockelmann S, Nardmann C, Tabke K, et al. Vacuolar H(+)-ATPases: intra- and intermolecular interactions. *Eur J Cell Biol* 2011;**90**(9):688–95.

CHAPTER 22

Taking advantage of cancer's pH alterations: pH-sensitive nanoparticles

Introduction

Extracellular acidity is a constant finding in all kind of cancers and its causes have been extensively discussed in Chapters 2–4. It represents an important instrument in the matrix degradation/migration/invasion process and at the same time a problem and an opportunity for the oncologist. The opportunity stems from the fact that this phenomenon is mainly restricted to malignant tissues.

Nano-systems, especially those that are activated by pH, makes it possible to target cancer while preserving normal cells and permitting the delivery of drugs directly to the tumor. At the same time, these systems can by-pass biological barriers and avoid the extrusion capabilities of multidrug resistance (MDR). The extracellular acidity of tumors represents the perfect opportunity for the development of pH-sensitive nanoparticules (NPs) that release the drug in the extracellular substance of the tumor or directly inside the malignant cells.[1]

The problems of chemotherapy

Among the many problems of chemotherapy, there are three that we shall discuss here:
(1) The bioavailability of the chemo drug at the tumor site;
(2) An adequate penetration into the malignant cell;
(3) Toxicity to normal cells.

The bioavailability of the chemo drug (CD) at the tumor site

Once injected or ingested, the drug enters the circulatory system where it is distributed throughout the entire organism. For this, the drug must be hydrosoluble in order to be transported by the blood.[2] For example, curcumin a very interesting nutraceutical for cancer treatment is not soluble in any kind of non-toxic solvents and thus, it cannot be used as a therapeutic substance.

Once in the blood, the CD may be denaturalized by normal blood components or may be bound to proteins that decrease its activity.

The goal of CDs is to achieve the highest possible concentration at the tumor site and the lowest possible concentration at normal tissues. If the drug can be "protected" from

An Innovative Approach to Understanding and Treating Cancer: Targeting pH
https://doi.org/10.1016/B978-0-12-819059-3.00022-8
© 2020 Elsevier Inc.
All rights reserved.

destruction (by macrophages) or inactivation (by other substances) during its widespread distribution throughout the organism, a higher concentration of active drug can be achieved at the required site of action. Nanoparticles (NPs) attached to CDs or CDs inside NPs can do this job. Another objective of treatment is the intracellular acidification of malignant cells with the goal of inducing acid stress and apoptosis, certain specialized NPs can do this too.

But, how does this NP-CD complex know where to release the CD?

This is the point where pH plays the capital role: NPs can be constructed in such a way that they release the drug specifically in sites with low pH, such as the extracellular space (ECS) of tumors. This are called the pH-sensitive NPs.

An adequate penetration into the malignant cell

The NP "protects" the drug and releases it specifically at the tumor's ECS. A second problem may ensue now: the acidic pH may inactivate the CD. Extracellular acidity has been found to be an important factor of chemoresistance. This is particularly evident in weakly basic anti-cancer drugs.[3,4] Proton pump inhibitors[5,6] and buffers like bicarbonate[7] were proposed to modify this situation. But the relation between chemoresistance and pH is not limited to the extracellular acidity of tumors: the acidity of intracellular vesicles may play a role too. One of the mechanisms used by cancer cells to decrease the activity of chemotherapeutic drugs consists in sequestering them in endocytic vesicles.[8]

With certain CDs (e.g., doxorubicin) is not enough to carry them to the specific site of action, it is also necessary to avoid sequestration by protonation due to the acidic conditions of the tumor's ECS.

Using doxorubicin (DOX), Mellor and Calaghan found that P-gp (P-glycoprotein) and extracellular acidity represented similar and equivalent problems for an adequate anti-cancer effect. The effects of doxorubicin were limited to the outer layer of tumor cells and failed to reach the more hypoxic areas.[9,10] Daunorubicin, doxorubicin and mitoxantrone are weak bases and therefore activity in a highly acidic medium is substantially reduced. Paclitaxel on the other hand is not affected by pH.

Extracellular acidity is highly heterogeneous, with very acid areas coexisting with less acidic ones. This means that acid-sensitive chemotherapeutic drugs will achieve good results in less acidic areas, but will fail in the more acidic ones. The very acidic niches that remain unharmed by the treatment will become the core for relapse and/or clonal evolution toward resistance.

Here again, pH-sensitive NPs can do the job by facilitating penetration into the cell and releasing the CD inside the cell. In this case, the design of the NP is different from the first situation.

Up to now, we have two different kinds of NPs:

(a) NPs that release the CD in acidic sites, and

(b) NPs that penetrate the malignant cell with its cargo releasing the drug inside the cell.

There are also five other types of NPs in which we are interested:

(c) an NP that can acidify the cytoplasm;

(d) an NP that deposits its load to the cell surface but does not let it enter to the cell;

(e) the double releasing NP;

(f) the multidrug NP;

(g) the multitarget NP (Fig. 1).

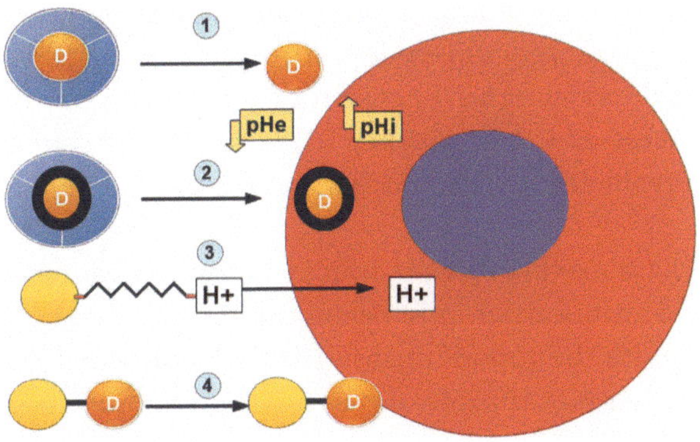

Fig. 1 Different types of NPs related to the pH gradient inversion: (1) pH-sensitive NP that releases the drug in acidic extracellular matrix; (2) pH-sensitive NP that facilitates drug penetration inside the cell; (3) pH-sensitive NP that acidifies the intracellular milieu; (4) pH-sensitive NP that deposits the drug on the cell surface but does not allow drug penetration inside the cell.

Toxicity to normal cells

The above mentioned NPs are designed in such a way that drug release mainly takes place in the malignant tissues and by-passing the normal ones. And this is possible due to cancer's altered pH homeostasis.

In this chapter we will focus on these four types of pH-sensitive NPs: the ones that release the drug in acidic sites, the NPs that release the drug inside the malignant cell, the NPs that acidify the intracellular compartment and the NPs that concentrate the drug on the cell surface without allowing penetration. We will also touch briefly the other types of NPs.

pH-sensitive NPs that release the drug in acidic sites

Many different nanocarriers have been tested in combination with chemotherapeutic drugs (CDs). These nanoparticles (NPs) are expected to increase the efficacy of therapeutic agents while reducing unwanted side effects. There are NPs that have been designed considering the advantage of the low pH of the extracellular space in tumors, being activated by this low pH and releasing the chemotherapeutic drug. This leads to a high accumulation of the drug at the tumor site.[11]

Hruby et al.[12] developed a polymeric micellar system for the delivery of doxorubicin that released 43% of the drug at pH5 and only 16% at pH7.4.

Meng et al.[13] created a mesoporous silica NP for doxorubicin release at low pH.

Many of these pH-activated delivery systems are engineered in such a way that they can reach lysosomes and nuclei, and be selectively released in the tumoral extracellular matrix.

Pegylated liposomal doxorubicin is doxorubicin protected by a liposomal layer reducing the uptake by the reticulo-endothelial system and stably retaining the drug as a result of liposomal entrapment. These features increase its permanence in circulation, leading to an increased tumor uptake.[14] It is not pH-activated, but it is a proof of concept that better vehicles for chemotherapeutic drugs can succeed in achieving improved results.

In order to produce a maximal tumor concentration of a cationic drug such as doxorubicin (Dox) a specific nanocarrier must be engineered. Dox enters the cell through diffusion and interferes with the topoisomerase II-DNA complex leading to cell death.[15]

The idea of delivering NPs capable of transporting Dox unchanged at normal pH to sites were pH is lower like the tumor extracellular matrix, where the drug is released of this protection, is particularly attractive and it has been the object of great amount of research.

The pH-activated drug release NPs are all based on the particular acid conditions of the pHe.[16] Box 1 and Fig. 2.

BOX 1 pH-sensitive NPs must fulfill three criteria

(1) Stability at physiologic pH.
(2) Destabilization/degradation or conformational changes at pH $<$ 6.8.
(3) Ability to release their load when pH falls below the trigger pH.

Fig. 3 shows the building of a pH-sensitive NP. Changes in pH introduce conformational changes in the NP that lead to disintegration or degradation of the NP thus releasing the drug.

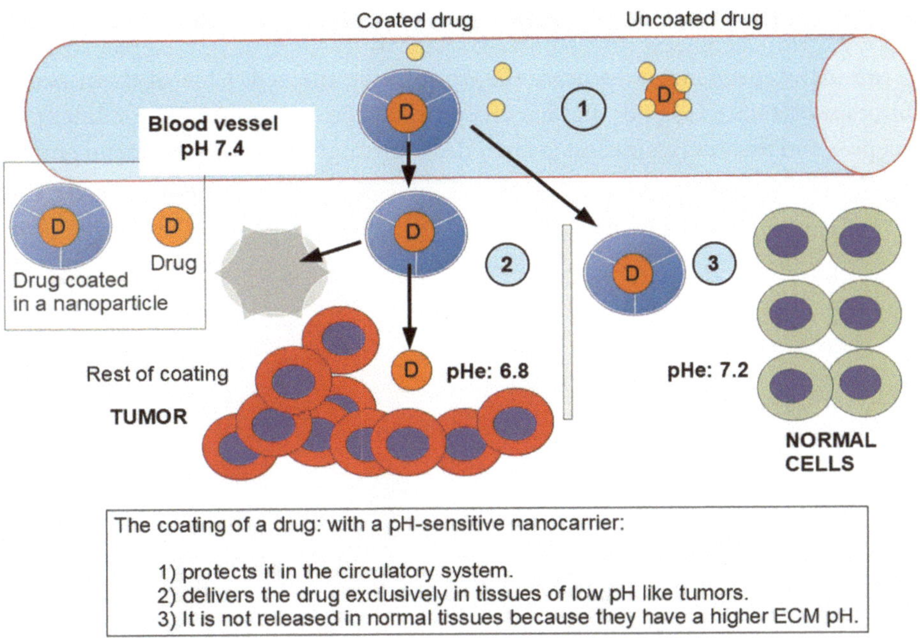

Fig. 2 The pH-sensitive NP protects the drug while in the circulatory system. When it reaches an area with low pH, such as in tumors, it releases the drug. There is no drug release in normal tissues because their extracellular pH is markedly higher than in tumors. The small *yellow circles* in the blood vessel represent possible substances that can decrease the chemotherapeutic activity. The NP "protects" the drug against macrophages, blood proteins, etc.

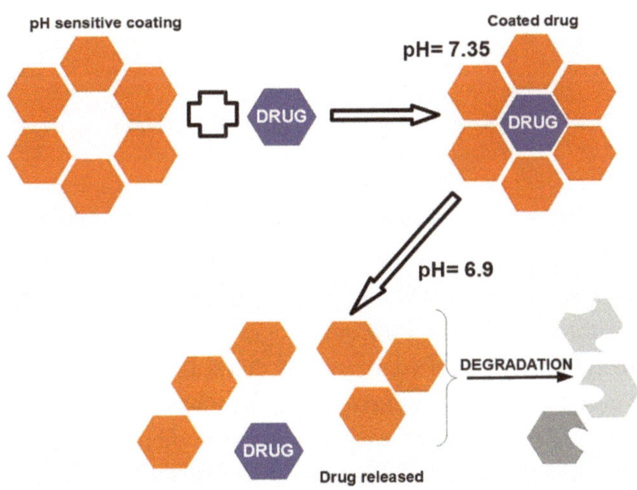

Fig. 3 Schematic representation of a system of pH-activated drug delivery. The drug is "protected" during its journey through blood vessels and tissues with normal pH. It is released when it encounters an acidic pH.

pH-sensitive NPs that release the drug inside the cell

There are many mechanisms to drive the drug inside the cell. One of them consists in attaching a substance to the NP that has receptors on the cell surface. The substance binds the receptor and the internalization is carried out through the formation of an endosome. The endosomal pH is low and it releases the drug from the nanoparticle. The NP can be a polymer or a micelle charged with the drug, and there is a linker or cleavable bond that attaches the substance to the NP (Fig. 4).

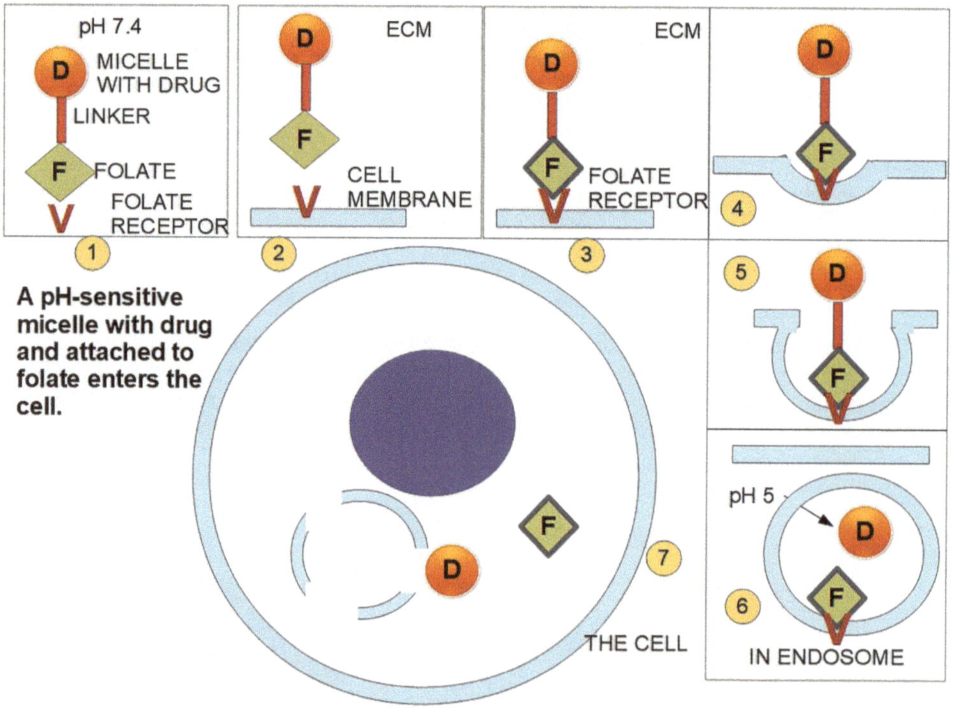

Fig. 4 A mechanism to deliver a chemotherapeutic drug inside the cell. See explanation in the text.

For example, ovarian cancer cells over-express folate receptors in more than 90% of the cases.[17] The inhibition of this folate receptors, essential for ovarian cancer metabolism, led to the development of Farletuzumab (a monoclonal antibody against folate receptor alpha). It failed in phase III clinical trials for relapsed ovarian cancer, but showed some minor benefits.[18] An NP loaded with a chemotherapeutic drug such as a vinca alkaloid can be bound either to farletuzumab or to folate through a cleavable linker (Fig. 4, panel 1). This NP Vinca-Folate or Vinca-Farletuzumab targets the folate receptor at the cell membrane (Fig. 4, panels 2 and 3) and induces the internalization of the complex through the formation of an endosome (Fig. 4, panels 4 and 5). The acidic condition of the endosome (pH5) breaks the cleavable linker and releases the drug (Fig. 4, panel 6).

Finally, endolysis (panel 7) releases the Vinca alkaloid in the cytoplasm where it acts against the microtubular system.[19] Vintafolide or EC145 is the name of an experimental product fabricated on these parameters. Actually, this compound does not incorporate the vinca alkaloid into a micelle. It is directly attached to the pH-sensitive linker. It is in phase II Clinical Trial.[20] In a phase I clinical trial Vintafolide showed low toxicity for the patients and very few side effects.

Incorporating a micelle to the structure of Vintafolin, as seen in Fig. 4 would allow the internalization of more than one drug simultaneously.

A step forward is the development of dual pH-activated nanoparticles, like the one proposed by Du et al.[21] This dual pH system has the ability of reversing the surface charge of the particle when meeting the acidic tumor environment. In this manner the internalization of the particle is facilitated. When the particle inside the cell enters the endosomes, which are even more acidic than the extracellular environment, doxorubicin is further released. These dual systems protect the drug from the extracellular acidity of tumors, and facilitate its penetration inside the cell (Fig. 5).

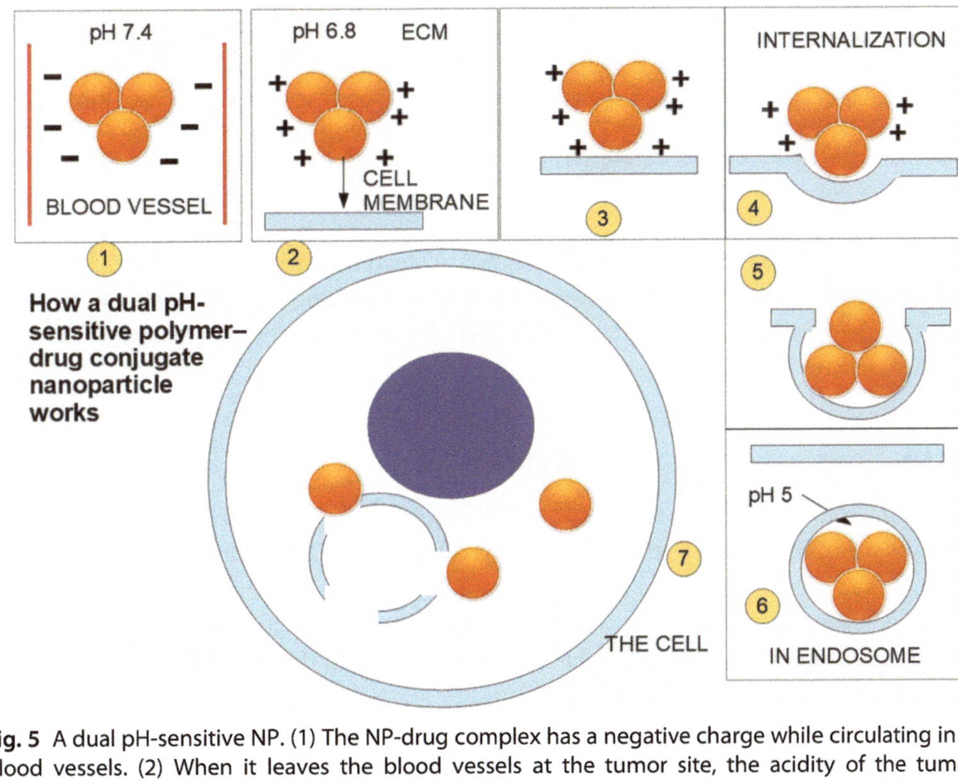

Fig. 5 A dual pH-sensitive NP. (1) The NP-drug complex has a negative charge while circulating in the blood vessels. (2) When it leaves the blood vessels at the tumor site, the acidity of the tumoral extracellular matrix produces an electrical inversion that facilitates the penetration (3)–(5) into the cell. The extreme acidity of the endosomes further releases the drug (6) and (7).

The intracellular delivery of substances by pH-activated nanoparticles is not exclusive for drugs, they can also be used to deliver other anti-cancer tools such as SiRNAs.[22] Fig. 6

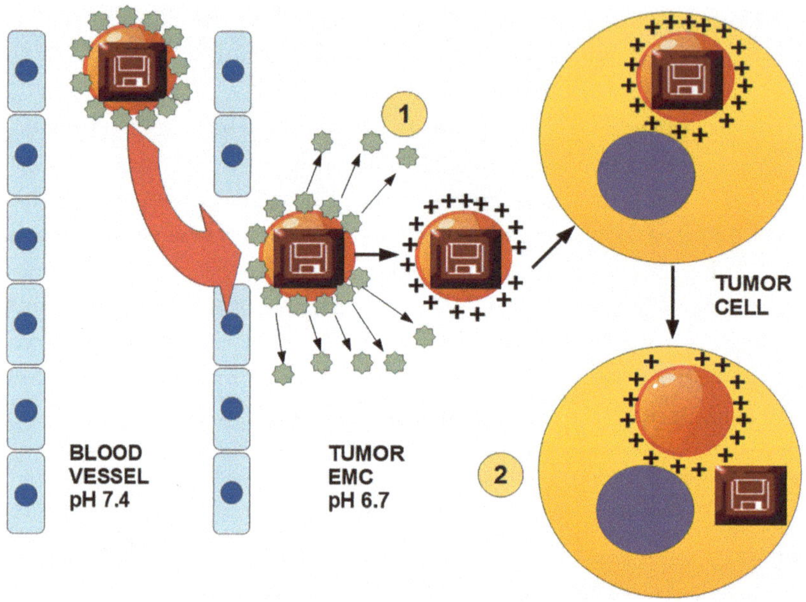

Fig. 6 A double coated NP delivery system for siRNA. (1) The first pH-sensitive coat drives the NP to the tumor's ECS where it loses its first coat. The second coat is revealed with a strong negative electrostatic charge that increases the cellular uptake. (2) Inside the cell, the second coat is shedded and the siRNA is released.

shows a delivery system that transports siRNA instead of drugs. It has a double coating and loses the first coat when it arrives to a low pH region. At this point the negative charge of the second coat facilitates NP-siRNA's entry to the cytoplasm. Then, the siRNA is released.

NPs that can acidify the cytoplasm

It has been established that cytoplasmic acidification is the main objective for the reversal of pH alterations in cancer. Carbone et al.[23] developed an NP in which they synthesized a chemical cage[24] that shields a proton (H^+) and coupled it to a gold nanoparticle. When this AuNP-caged H^+ is taken up by the cell, stimulation with ultraviolet light releases the proton from the cage[a] and the freed H^+ combines with CO_3H^- producing carbonic acid that decreases intracellular pH (Fig. 7).

[a] "The idea behind the caging technique is that a molecule of interest can be rendered biologically inert (or caged) by chemical modification with a photoremovable protecting group". From Ellis Davies.[24]

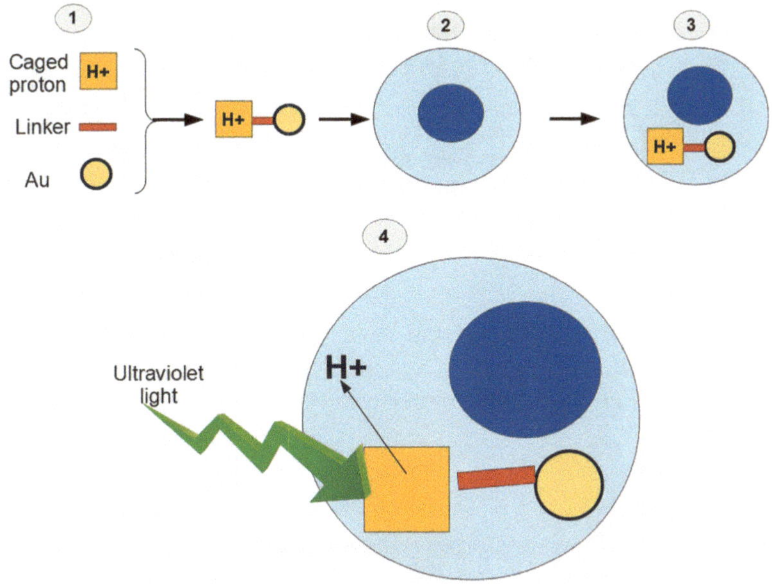

Fig. 7 (1) The NP consists of a chemically caged proton bound to a gold particle. (2) and (3) The NP is internalized into the cell. (4) Ultraviolet light stimulation "breaks" the cage releasing the proton that acidifies the cytoplasm.

These type of NPs have one drawback: their use would be limited to those tissues that can be irradiated with ultraviolet light. In the future, these NPs may become useful for treating oral cavity and oropharynx tumors. We also believe that it would be perfectly feasible to create endoscopes that can deliver UV light inside cavities.

NPs that deposit their load on the cell surface but do not let it enter the cell

Inhibition of carbonic anhydrase (CA) is an important step in the reversal of the inverted pH gradient. The problem is that CA is a family of different isoforms. Those related to cancer are the membrane CAs: CAIX and CAXII. The model inhibitor of CAs is acetazolamide that works as a pan-inhibitor. This means that acetazolamide inhibits all the isoforms of CA, including cytoplasmic and membranous CAs. CAs are housekeeping enzymes and therefore fully inhibiting them all would be highly toxic. A drug delivery system able to inhibit only the membranous CAs of tumors without affecting normal cells would be the goal.[25] There is advanced research with CA inhibitors attached to an NP that impedes cell penetration.[26,27] These mechanisms would preserve intracellular CAs and deliver the drug mainly to malignant tissues that over-express membrane CAs. Fig. 8.

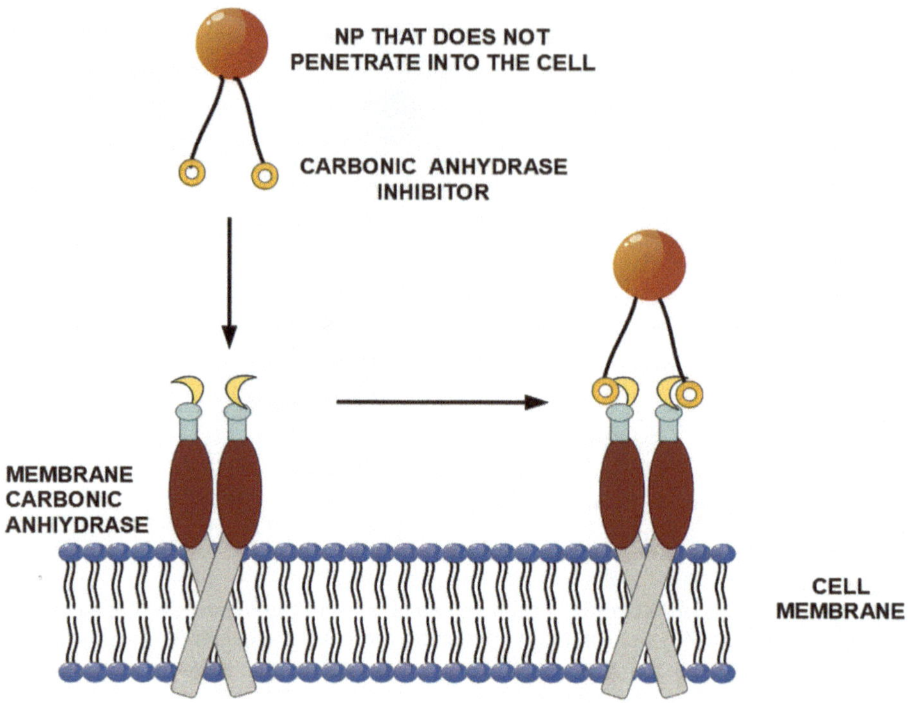

Fig. 8 A non-permeable carrier transports the carbonic anhydrase inhibitor to the cell surface.

Double release NPs

Fig. 5 is an example of a double release NP. When the NP reaches the tumor's ECS undergoes a change in polarity that increases its penetration into the cell. Inside the endosome, the very low pH induces an NP's conformational change with further release of the drug to the cytoplasm. This is just one example of many. Miao et al. developed a similar dual NP loaded with DOX.[28]

Kim et al.[29] synthesized an interesting system of double targeting micelles transporting doxorubicin (DOX). In the first step these micelles protected DOX and facilitated its arrival to the cell surface. Then a second component of the micelle containing folate engaged the folate receptors of the cell membrane inducing endocytosis. The ensuing endocytolysis, released the drug. This system could be applied successfully to cells with the MDR phenotype. Actually this is a system of double targeting more than double release.

Multidrug release NPs

NPs with the ability to deliver more than one drug have been developed recently.[30] Liao et al.[31] fabricated a system that delivers three chemotherapeutic drugs simultaneously:

doxorubicin, camptothecin and cisplatin. In this case the mechanism of release was not through pH-sensitive NPs but through light stimulation. In theory, a similar system could be developed with a pH-triggered release which would make the compound more focused on cancer cells.

Radiosensitizing NPs

Ma et al.[32] developed a NP based on the delivery of a carbonic anhydrase inhibitor and quercetin. This construction decreased the effects of hypoxia on tumors increasing the effects of radiotherapy. NP binding carbonic anhydrase is an interesting concept to deliver drug into hypoxic areas of the tumor.[33]

The making of an NP

A useful pH-sensitive NP for oncological use must combine at least two basic features:

(1) Ability to store the drug in stable conditions at normal pH.

(2) Release the drug when the pH decreases below a pre-established threshold.

These can be accomplished by adding chemical groups to the nanomaterial with the ability to:

(a) accept or donate protons;

(b) produce physical changes according to the pH.

Polymer-based drug delivery

Biopolymers (BPs) are the most frequently used nanocarriers for drug delivery. These compounds have some features that make them the best choice in the case of pH-sensitive systems:

(1) BPs are biodegradable.

(2) BPs are compatible with human tissues and body fluids.

(3) BPs are easy to manipulate in the laboratory.[34]

(4) BPs have the ability to respond to ionization with conformational changes.

Most of the pH-sensitive NPs are actually pH-sensitive polymers that can respond with changes in their physical properties when they encounter changes in the pH of their environment. In many of these NPs the polymer is dissolved when subjected to certain pH changes.[35] This is the case of NPs made of poly(β-amino) ester shown in Fig. 9.[36]

These polymers can transport the drug in two different ways[37]:

(1) the drug is physically entrapped in the NP;

(2) The drug is covalently bound to the NP (Fig. 10).

In the second case the drug is bound to the surface of a nanocarrier through a covalent acid-labile bond. The chemical bond is hydrolyzed when the pH decreases, releasing the drug.

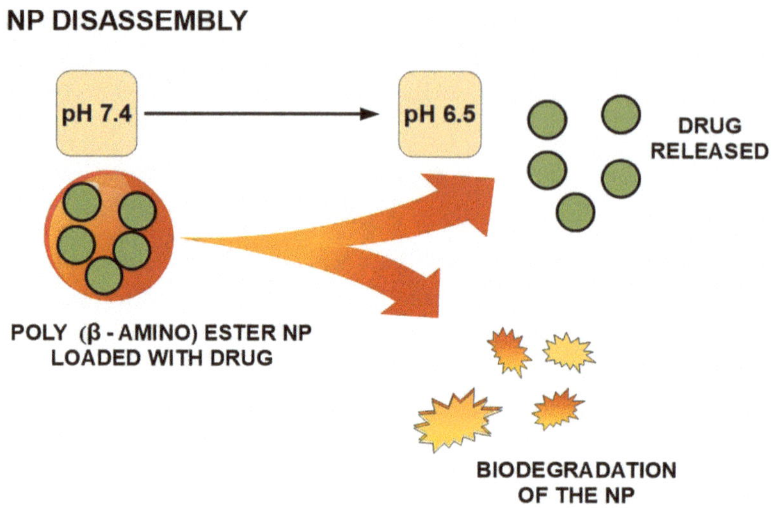

Fig. 9 Degradation or disassembly of the NP releases the drug when pH is lowered from the physiologic to the acidic range.

Micelle-based drug delivery

Another example of disassembly induced by low pH is the fabrication of NPs formed by micelles that are bound to hydrophobic drugs by a linker. The acidic pH dissolves the micelles releasing the drug (Fig. 10).

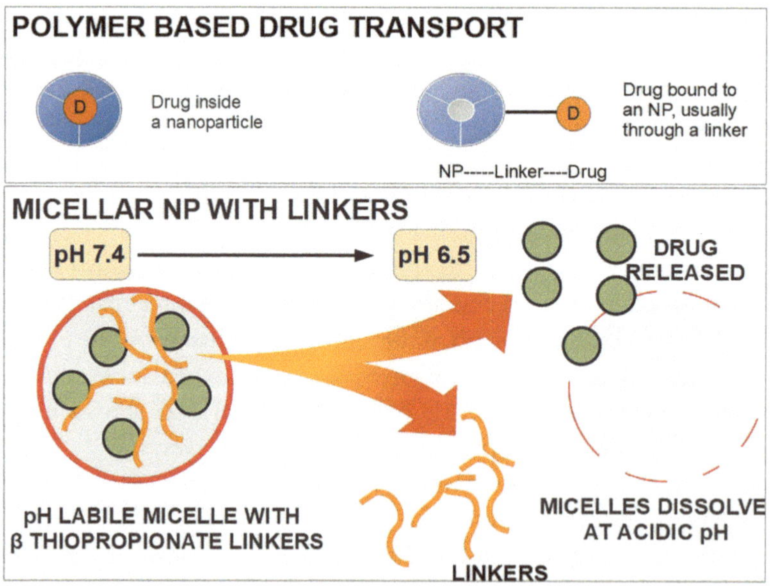

Fig. 10 Upper panel: two different mechanisms of drug transport by a polymer: inside the nanocarriers (left) or attached through a chemically labile bond (the linker) (right) Lower panel: a pH-sensitive micellar NP.

Polyester based NPs release acidic degradation products. This may represent a drawback if the degradation products remain in the ECS increasing its acidity. Hefferman et al.[38] developed an NP based on a new hydrophobic polymer called polyketal that avoids the problem.

The possibilities for the development of new site-specific pH-sensitive NPs for the delivery of drugs and other types of molecules are enormous and we are just at the beginning. Table 1 is only a small sample of pH-sensitive NPs.

Table 1 A summary of pH activated drug delivery systems for cancer treatment

Reference	Mechanism
Yang et al. (2018)[39]	Poly-L-lysine-lipoic acid modified by dimethylmaleic anhydride used as a carrier for doxorubicin with enhanced cell internalization and intracellular pH release.
Ju et al. (2017)[40]	A liposome prepared with a synthetic oligopeptide lipid, soy phosphatidylcholine, and cholesterol showed that this pH sensitive drug carrier enhanced tumor cell uptake and increase cytoplasmic distribution.
Ehi-Eromesele et al. (2017)[41]	Magnetic nanoparticles coated with silica, pH sensitive particles were synthesized as carriers of 5 FU. At pH 5 this NP released three times more 5 FU than at pH 7.4.
Guo et al. (2017)[42]	A gene transfection system was engineered in which the gene is part of a *pH*-responsive charge-convertible ternary complex.
Wu et al. (2017)[43]	A graft copolymer of poly(butylenes succinate)-*g*-cysteamine-bi-poly (ethylene glycol) was synthesized for doxorubicin transport. This micellar system could switch from less than 25% to 84% of doxorubicin release according to the media pH.
Li et al. (2016)[44]	A dual sensitive biodegradable micelle for drug transport was built with maximal release of the drug inside the tumor cell. The micelle was sensitive to pH and redox conditions. The micelles were quickly and fully internalized by cellular endosomes.
Jiang et al. (2017)[45]	A transactivator of transcription (TAT) nanoparticle was developed with increased intracellular retention and nuclear translocation of a platinum based chemotherapeutic drug. The intranuclear acid conditions triggered the TAT activity (not the extracellular acidity).
Piao et al. (2017)[46]	Gold nanocages with zwitterionic pH sensitive coating were used to increase delivery of chemotherapeutic drugs to the tumor.
Gisbert-Garzaran et al. (2017)[47]	pH-responsive mesoporous of silica and carbon was used to increase delivery of chemotherapeutic drugs into tumors.
Zignani et al. (2000)[48]	pH-sensitive release of drug was achieved with egg phosphatidylcholine liposomes adding *N*-isopropylacrylamide (NIPA) copolymers.

Clinical implications

Chemotherapeutic drugs are generally very toxic. Chemotherapy can be defined as a "controlled intoxication" where art and science join to keep the drug within a narrow therapeutic range. If that range is exceeded the patient will suffer serious adverse effects and if the range is not achieved the treatment will be ineffective.

Therefore, the way of administering the drug (drug delivery system) is of capital importance.[49]

The other main concern in oncology is to deliver the highest possible amount of drug to the tumor and ideally none to the rest of the organism.

Nano-systems that are pH-dependent will be a part of the new delivery systems that may eventually change oncological practice as we know it today.

Liu et al.[1] wrote: "Stimulus-responsive nanomaterials are termed as 'smart', 'intelligent', or 'environmentally sensitive', and have greater potential than traditional delivery systems."

A few interesting objectives can be achieved by using pH-sensitive nanocarriers:

(1) Specific and focalized delivery of drugs to the tumor without toxicity to normal cells.
(2) Intracellular acidification limited to cancer cells.
(3) Counteracting or by-passing multidrug resistance (MDR).
(4) Intracellular delivery of siRNAs that can block oncogenic pathways and proteins.
(5) Increased stability and permanency of the drug in circulation.
(6) Higher concentrations of the drug without generating intolerable toxicity.
(7) Increased retention of the drug inside the tumor.

These compounds are not yet in clinical use, but they will be soon.

The hand-foot syndrome (palmar-plantar erythrodysesthesia) is a well-known adverse event with some cytotoxic CDs[50,51] including the non pH-sensitive pegylated liposomal doxorubicin (PLD). The incidence with PLD is close to 50%.[52] This problem results in treatment interruption in many cases. However, it is less frequent when DOX is used alone.[53] This led to the conclusion that prolonged circulation due to the pegylated liposome NP increased leakage to non tumor-specific sites.[54] As a speculation, we may assume that using pH-sensitive NPs instead of PLD, the leakage probably would not be avoided. However the leaked NPs would not release DOX and therefore a foot and hand syndrome would be prevented. The advantages of a pH-sensitive NP chemotherapy drug delivery become very evident with the example described above.

Conclusions

We have oversimplified this chapter, sacrificing scientific details for the sake of clarity to offer a better understanding of this complex topic.

The first nanoparticles (NPs) were mainly designed for drug delivery improving its pharmacokinetics, prolonging its presence in the circulatory system and decreasing its degradation or inactivation before reaching its target. Then, new and more ambitious objectives were set for NPs: more focused delivery to the tumor, multidrug delivery, siRNA delivery, specific intracellular targeting, controlled drug release, environmental sensitive response, etc.

After the success of the pegylated liposomal doxorubicin (PLD), that has been used intensively in the last 10 years, a considerable effort has been dedicated to pH-sensitive drug carriers and very interesting developments have been achieved. The pH paradigm of cancer is a targetable soft spot of cancer that exposes malignant cells to an easier attack by pH-sensitive nanocarriers.

However, none of these new pH-sensitive carrier systems have yet been approved for human use. It is highly possible that we shall see breakthroughs on this issue in the near future. It opens the possibility to target certain components of the pH paradigm that, at present, lack efficient blockers. This would be the case of NBC (sodium/bicarbonate cotransporter), just to mention one example. As a membrane protein we think it may be targeted with nanocarriers that can protect normal cells from the attack.

The many broad and interesting possibilities that nanocarriers can offer to the world of oncology have not yet been fully explored. We trust that these tools will become part of clinical oncology very soon. This chapter is only a brief glance at this subject that would require—and deserves—many dedicated volumes. Peer et al.[55] wrote: "Nanotechnology has the potential to revolutionize cancer diagnosis and therapy. Advances in protein engineering and materials science have contributed to novel nanoscale targeting approaches that may bring new hope to cancer patients." We share this concept.

References

1. Liu J, Huang Y, Kumar A, Tan A, Jin S, Mozhi A, et al. pH-sensitive nano-systems for drug delivery in cancer therapy. *Biotechnol Adv* 2014;**32**(4):693–710.
2. Savjani KT, Gajjar AK, Savjani JK. Drug solubility: importance and enhancement techniques. *ISRN Pharmaceutics* 2012;**2012**. https://doi.org/10.5402/2012/195727.
3. De Milito A, Fais S. Tumor acidity, chemoresistance and proton pump inhibitors. *Future Med* 2005; **1**(6):779–86.
4. Taylor S, Spugnini EP, Assaraf YG, Azzarito T, Rauch C, Fais S. Microenvironment acidity as a major determinant of tumor chemoresistance: proton pump inhibitors (PPIs) as a novel therapeutic approach. *Drug Resist Updat* 2015;**23**:69–78.
5. Lee YY, Jeon HK, Hong JE, Cho YJ, Ryu JY, Choi JJ, et al. Proton pump inhibitors enhance the effects of cytotoxic agents in chemoresistant epithelial ovarian carcinoma. *Oncotarget* 2015;**6**(33):35040.
6. Fais S. Evidence-based support for the use of proton pump inhibitors in cancer therapy. *J Transl Med* 2015;**13**(1):368.
7. McCarty MF, Whitaker J. Manipulating tumor acidification as a cancer treatment strategy. *Altern Med Rev* 2010;**15**(3):264–72.
8. Weylandt KH, Nebrig M, Jansen-Rosseck N, Amey JS, Carmena D, Wiedenmann B, et al. ClC-3 expression enhances etoposide resistance by increasing acidification of the late endocytic compartment. *Mol Cancer Ther* 2007;**6**(3):979–86.
9. Mellor HR, Callaghan R. Accumulation and distribution of doxorubicin in tumour spheroids: the influence of acidity and expression of P-glycoprotein. *Cancer Chemother Pharmacol* 2011;**68**(5):1179–90.

10. Mahoney BP, Raghunand N, Baggett B, Gillies RJ. Tumor acidity, ion trapping and chemotherapeutics: I. Acid pH affects the distribution of chemotherapeutic agents in vitro. *Biochem Pharmacol* 2003; **66**(7):1207–18.
11. Lim EK, Huh YM, Yang J, Lee K, Suh JS, Haam S. pH-triggered drug-releasing magnetic nanoparticles for cancer therapy guided by molecular imaging by MRI. *Adv Mater* 2011;**23**(21):2436–42.
12. Hrubý M, Koňák Č, Ulbrich K. Polymeric micellar pH-sensitive drug delivery system for doxorubicin. *J Control Release* 2005;**103**(1):137–48.
13. Meng H, Xue M, Xia T, Zhao YL, Tamanoi F, Stoddart JF, et al. Autonomous in vitro anticancer drug release from mesoporous silica nanoparticles by pH-sensitive nanovalves. *J Am Chem Soc* 2010; **132**(36):12690–7.
14. Gabizon A, Shmeeda H, Barenholz Y. Pharmacokinetics of pegylated liposomal doxorubicin. *Clin Pharmacokinet* 2003;**42**(5):419–36.
15. Tacar O, Sriamornsak P, Dass CR. Doxorubicin: an update on anticancer molecular action, toxicity and novel drug delivery systems. *J Pharm Pharmacol* 2013;**65**:157–70.
16. Du JZ, Mao CQ, Yuan YY, Yang XZ, Wang J. Tumor extracellular acidity-activated nanoparticles as drug delivery systems for enhanced cancer therapy. *Biotechnol Adv* 2014;**32**(4):789–803.
17. Armstrong DK, White AJ, Weil SC, Phillips M, Coleman RL. Farletuzumab (a monoclonal antibody against folate receptor alpha) in relapsed platinum-sensitive ovarian cancer. *Gynecol Oncol* 2013; **129**(3):452–8.
18. Vergote I, Armstrong D, Scambia G, Teneriello M, Sehouli J, Schweizer C, et al. A randomized, double-blind, placebo-controlled, phase III study to assess efficacy and safety of weekly farletuzumab in combination with carboplatin and taxane in patients with ovarian cancer in first platinum-sensitive relapse. *J Clin Oncol* 2016;**34**(19):2271–8.
19. Marchetti C, Palaia I, Giorgini M, De Medici C, Iadarola R, Vertechy L, et al. Targeted drug delivery via folate receptors in recurrent ovarian cancer: a review. *Onco Targets Ther* 2014;**7**:1223.
20. Sharp M, Corp D. *Study of EC145 in patients with advanced ovarian and endometrial cancers*. In: *ClinicalTrials.gov*. Bethesda, MD: US National Library of Medicine; 2013. Available from:*http://clinicaltrials.gov/show/NCT00507741*. [NLM identifier: NCT00507741].
21. Du JZ, Du XJ, Mao CQ, Wang J. Tailor-made dual pH-sensitive polymer–doxorubicin nanoparticles for efficient anticancer drug delivery. *J Am Chem Soc* 2011;**133**(44):17560–3.
22. Sun CY, Shen S, Xu CF, Li HJ, Liu Y, Cao ZT, et al. Tumor acidity-sensitive polymeric vector for active targeted siRNA delivery. *J Am Chem Soc* 2015;**137**(48):15217–24.
23. Carbone M, Sabbatella G, Antonaroli S, Remita H, Orlando V, Biagioni S, et al. Exogenous control over intracellular acidification: enhancement via proton caged compounds coupled to gold nanoparticles. *Biochim Biophys Acta, Gen Subj* 2015;**1850**(11):2304–7.
24. Ellis-Davies GCR. Caged compounds: photorelease technology for control of cellular chemistry and physiology. *Nat Methods* 2007;**4**(8):619–28.
25. Liu S, Luo X, Liu S, Xu P, Wang J, Hu Y. Acetazolamide-loaded pH-responsive nanoparticles alleviating tumor acidosis to enhance chemotherapy effects. *Macromol Biosci* 2019; **19**(2):1800366.
26. Stiti M, Cecchi A, Rami M, Abdaoui M, Barragan-Montero V, Scozzafava A, et al. Carbonic anhydrase inhibitor coated gold nanoparticles selectively inhibit the tumor-associated isoform IX over the cytosolic isozymes I and II. *J Am Chem Soc* 2008;**130**:16130–1.
27. Sztandera K, Gorzkiewicz M, Klajnert-Maculewicz B. Gold nanoparticles in cancer treatment. *Mol Pharm* 2018;**16**(1):1–23.
28. Miao Y, Qiu Y, Yang W, Guo Y, Hou H, Liu Z, et al. Charge reversible and biodegradable nanocarriers showing dual pH-/reduction-sensitive disintegration for rapid site-specific drug delivery. *Colloids Surf B: Biointerfaces* 2018;**169**:313–20.
29. Kim D, Lee ES, Oh KT, Gao ZG, Bae YH. Doxorubicin-loaded polymeric micelle overcomes multidrug resistance of cancer by double-targeting folate receptor and early endosomal pH. *Small* 2008;**4**:2043–50.
30. Jhaveri A, Deshpande P, Torchilin V. Stimuli-sensitive nanopreparations for combination cancer therapy. *J Control Release* 2014;**190**:352–70.

31. Liao L, Liu J, Dreaden EC, Morton SW, Shopsowitz KE, Hammond PT, et al. A convergent synthetic platform for single-nanoparticle combination cancer therapy: ratiometric loading and controlled release of cisplatin, doxorubicin, and camptothecin. *J Am Chem Soc* 2014;**136**(16):5896–9.
32. Ma T, Liu Y, Wu Q, Luo L, Cui Y, Wang X, et al. Quercetin-modified metal–organic frameworks for dual sensitization of radiotherapy in tumor tissues by inhibiting the carbonic anhydrase IX. *ACS Nano* 2019;**13**(4):4209–19.
33. Shabana AM, Mondal UK, Alam MR, Spoon T, Ross CA, Madesh M, et al. pH-sensitive multiligand gold nanoplatform targeting carbonic anhydrase IX enhances the delivery of doxorubicin to hypoxic tumor spheroids and overcomes the hypoxia-induced chemoresistance. *ACS Appl Mater Interfaces* 2018;**10**(21):17792–808.
34. Prabaharan M, Mano JF. Stimuli-responsive hydrogels based on polysaccharides incorporated with thermo-responsive polymers as novel biomaterials. *Macromol Biosci* 2006;**6**:991–1008.
35. Schmaljohann D. Thermo- and pH-responsive polymers in drug delivery. *Adv Drug Deliv Rev* 2006;**58**:1655–70.
36. Shenoy D, Little S, Langer R, Amiji M. Poly(ethylene oxide)-modified poly(β-amino ester) nanoparticles as a pH-sensitive system for tumor-targeted delivery of hydrophobic drugs: part 2. In vivo distribution and tumor localization studies. *Pharm Res* 2005;**22**:2107–14.
37. Rawat M, Singh D, Saraf S, Saraf S. Nanocarriers: promising vehicle for bioactive drugs. *Biol Pharm Bull* 2006;**29**:1790–8.
38. Heffernan MJ, Murthy N. Polyketal nanoparticles: a new pH-sensitive biodegradable drug delivery vehicle. *Bioconjug Chem* 2005;**16**(6):1340–2.
39. Yang S, Wang Y, Ren Z, Chen M, Chen W, Zhang X. Stepwise pH/reduction-responsive polymeric conjugates for enhanced drug delivery to tumor. *Mater Sci Eng* 2018;**82**:234–43.
40. Ju C, Zhang C. Preparation and characterization of pH sensitive drug liposomes. In: Lu WL., Qi XR. (eds) Liposome-Based Drug Delivery Systems. *Biomaterial Engineering* 2018; Springer, Berlin, Heidelberg.
41. Ehi-Eromosele CO, Ita BI, Iweala EEJ. Silica coated LSMO magnetic nanoparticles for the pH-responsive delivery of 5-fluorouracil anticancer drug. *Colloids Surf A: Physicochem Eng Asp* 2017;**530**:164–71.
42. Guo A, Wang Y, Xu S, Zhang X, Li M, Liu Q, et al. Preparation and evaluation of pH-responsive charge-convertible ternary complex FA-PEI-CCA/PEI/DNA with low cytotoxicity and efficient gene delivery. *Colloids Surf B: Biointerfaces* 2017;**152**:58–67.
43. Wu S, Zheng L, Li C, Xiao Y, Huo S, Zhang B. Grafted copolymer micelles with pH triggered charge reversibility for efficient doxorubicin delivery. *J Polym Sci A: Polym Chem* 2017;**55**(12):2036–46.
44. Li D, Bu Y, Zhang L, Wang X, Yang Y, Zhuang Y, et al. Facile construction of pH-and redox-responsive micelles from a biodegradable poly (β-hydroxyl amine) for drug delivery. *Biomacromolecules* 2015;**17**(1):291–300.
45. Jiang W, Wang J, Yang J, He Z, Hou Z, Luo Y, et al. Acidity-triggered TAT-presenting nanocarriers augment tumor retention and nuclear translocation of drugs. *Nano Res* 2018;**11**(10):5716–34.
46. Piao JG, Gao F, Li Y, Yu L, Liu D, Tan ZB, et al. pH-sensitive zwitterionic coating of gold nanocages improves tumor targeting and photothermal treatment efficacy. *Nano Res* 2018;**11**(6):3193–204.
47. Gisbert-Garzarán M, Manzano M, Vallet-Regí M. pH-responsive mesoporous silica and carbon nanoparticles for drug delivery. *Bioengineering* 2017;**4**(1):3.
48. Zignani M, Drummond DC, Meyer O, Hong K, Leroux JC. In vitro characterization of a novel polymeric-based pH-sensitive liposome system. *BBA-Biomembranes* 2000;**1463**:383–94.
49. Langer R. Where a pill won't reach. *Sci Am* 2003;**288**:50–7.
50. Zuehle RL. Erythematous eruption of the palms and soles associated with mitotane therapy. *Dermatologia* 1974;**148**:90–2.
51. Nagore E, Insa A, Sanmartin O. Antineoplastic therapy-induced palmar plantar erythrodysesthesia ('hand-foot') syndrome: incidence, recognition and management. *Am J Clin Dermatol* 2000; **1**:225–34.
52. Lorusso D, Di Stefano A, Carone V, Fagotti A, Pisconti S, Scambia G. Pegylated liposomal doxorubicin-related palmar-plantar erythrodysesthesia ('hand-foot'syndrome). *Ann Oncol* 2007;**18**(7):1159–64.

53. O'Brien ME, Wigler N, Inbar M, Rosso R, Grischke E, Santoro A, et al. CAELYX Breast Cancer Study Group: reduced cardiotoxicity and comparable efficacy in a phase III trial of pegylated liposomal doxorubicin HCl (CAELYX/Doxil) versus conventional doxorubicin for first-line treatment of metastatic breast cancer. *Ann Oncol* 2004;**15**(3):440–9.
54. Yokomichi N, Nagasawa T, Coler-Reilly A, Suzuki H, Kubota Y, Yoshioka R, et al. Pathogenesis of hand-foot syndrome induced by PEG-modified liposomal doxorubicin. *Hum Cell* 2013;**26**(1):8–18.
55. Peer D, Karp JM, Hong S, Farokhzad OC, Margalit R, Langer R. Nanocarriers as an emerging platform for cancer therapy. *Nat Nanotechnol* 2007;**2**(12):751.

CHAPTER 23

Summary and final conclusions

A 16-year old shepherd, David defeated Goliath with one single stone. He impacted the head of the giant. Health professionals and researchers have been looking for ever for this unique stone to defeat cancer. This stone has been elusive to all the efforts. And there is a good reason for the failure. Cancer has many heads, it is not Goliath. In the meantime, many smaller stones had been thrown to cancer with variable results. Undoubtedly, many of these smaller stones had impacted cancer and made a difference. The effort was not worthless. However, cancer is still alive.

Fifty years ago, we only had three methods to combat cancer: surgery, chemotherapy and radiotherapy. Nowadays, we still have these three same methods with some new disciplines added like preventive medicine and early diagnosis. The three methods have evolved substantially. Fifty years ago we only had three targets to attack on the chemotherapeutic side: DNA with alkylating agents, essential molecules for cell proliferation with antimetabolites, and mitotic spindle with spindle poisons. Nowadays many new targets were found and taken care of: oncogenes, protumoral pathways, mitotic checkpoints, immunological checkpoints, immunological blockers, receptors, apoptosis blockers, and so on. These new developments made a difference in many types of cancers. Chronic myeloid leukemia was a fatal disease after a few years of progression. In the 1990s imatinib changed the history of the disease. This was a stone directed against a type of cancer that had only one head, like Goliath. However, some time later, cases of resistance to imatinib made their appearance on scene. This Goliath had the ability of growing a new head. This meant that new drugs were necessary and they were successfully developed. Unfortunately, most of the cancers have many heads. One stone was not enough, and this was the philosophy behind protocols of drug associations. New treatment schemes had also a history of success in many forms of cancer. Hodgkin's disease is one of the successful examples.

In the meantime, radiotherapy and surgery also went through positive changes. From the old cobalt based radiotherapy equipments, that killed cancer cells but also many normal cells in the surroundings, the newer equipments achieved high dose gamma radiation in the tumor with much less collateral damage. A practical result of this achievement is that radiotherapy now, shows the same outcome as radical surgery in not advanced prostate cancer.

Cancer surgery 50 years ago was quite mutilating with extensive resections trying to curb tumor's advance. The outcome was not better than with the more conservative approach that is employed nowadays.

An Innovative Approach to Understanding and Treating Cancer: Targeting pH
https://doi.org/10.1016/B978-0-12-819059-3.00023-X
© 2020 Elsevier Inc.
All rights reserved.

The first conclusion we can extract from this short history is that there have been important advances, but we should not wait for the silver bullet that will kill cancer.

The smaller stones addressed against the disease have an impact in many tumors, but not in all.

There is also a history of failures. Glioblastoma was uniformly considered a death sentence. This has not changed. Pancreatic cancer follows suit. Advanced metastatic cancer, once it developed chemoresistance, there is little more to do with the exception of palliative care which does not solve the problem.

Hematological malignancies as a whole have been benefited with the new drugs. This is not what happens with many solid tumors, particularly those that are in an advanced stage.

Kinase inhibitors, like sunitinib and sorafenib made a difference in overall survival in many solid tumors. But they are far from curing patients. The same happens with anti-angiogenic approaches. They are all small stones thrown against the giant. The giant feels the impact, withdraws for a short time and then comes back with a shield against the stones. We must find a way to disarticulate the shield.

We cannot miss any weakness of the giant in our quest.

We are flooded by scientific information and data on cancer. Every day a new MiRNA seems to be connected with proliferation, migration, invasion and metastasis. New pathways and cross-talks alike.

It is quite difficult to integrate all this new data in a comprehensive and meaningful way.

In the meantime old essential findings are simply ignored or bypassed in this obsessive search for the new. The pH alterations of cancer, now known for almost 100 years seem to belong to this neglected universe. And important as it is, it has no role at all in bedside medicine.

Cancer offers us a very weak point in the shield. Recognizing it, will give us a new target in order to really impact the disease. This weakness is the altered proton dynamics. Put it in other words, tumors have an altered pH gradient, and as we discussed it from Chapters 1–12, this represent an important weakness because tumors depend on it. Whenever we removed this cancer hallmark, proliferation, migration or invasion were hindered or the malignant cell simply died.

In the introductory words we said that we were going to climb the mountain of cancer from the pH side. When we arrived at the top of the mountain and had a panoramic view. What did we see?

What we saw from the mountain top

We saw a tumor that was growing, but before it could start proliferating it needed to modify its internal pH. It became more alkaline, and if we impeded alkalinization, there was no malignant transformation. The changes of one or more genes that originated cancer were not enough for the transformation. It was the gene plus the altered pH that gave way to transformation. In order to modify the pH the tumor activated a channel, that actually is not a channel but an exchanger which imported sodium and exported protons (ionized hydrogen or H^+). Getting rid of H^+ made the interior of the cell more alkaline. The extruded H^+ made the exterior of the cell more acid. This is the pH gradient inversion.

At the same time that protons were being exported, the exchanger imported Na^+. For each molecule of exported H^+ one molecule of Na^+ was imported. Supposing that only Na^+ would be imported without any trafficking of H^+, that would be enough to increase intracellular pH without modifying extracellular pH. This exchanger channel called NHE1 did both, imported Na^+ and exported H^+. Alkalinized the cell and acidified the surrounding exterior. The alkalinized interior sets the conditions necessary for proliferation. There is still an unsettled issue here: although cellular proliferation is initiated by alkalinization there are some authors that consider this event as a permissive rather than a necessary condition. However, if alkalinization is impeded, proliferation does not take place.

The acidified exterior creates the conditions necessary for migration and invasion and eventual metastasis. This makes NHE1 one of the main culprits of the malignant transformation. The malignant transformation and its intracellular alkalinization started the process toward cancer. At this initial moment in the life of the cancer cell, acidification of the extracellular surroundings or extracellular space (ECS) was not a determinant issue. It was only a consequence of what was happening in the cell. The acidic ECS will gain protagonism some time later. The cell started to proliferate. When it reached certain size, not very big indeed, like $0.4\,mm^3$, the cell found itself undernourished: lacked oxygen and nutrients because capillary vessels were too far from the growing cell. A normal capillary can provide for cellular needs up to a distance of $150\,\mu m$. However, this tumor that proliferated to $400\,\mu m^3$ surpassed the nourishment distance. The more dangerous point was the lack of oxygen because that would have killed the cell on short notice. The cell could handle the lack of glucose or other nutrients for some more time, but the lack of oxygen was life threatening immediately. No procrastination on the issue was possible. Normal cells and also the malignant cells, are prepared for contingencies like this. The lack of nutrients can be sustained for some time thanks to autophagy. But the lack of oxygen cannot be sustained. Therefore, eukaryotic cells that depend on aerobic metabolism have developed a very efficient instant response that is always in stand by: HIF-1α. This protein that is in a bound inactivated form in the cytoplasm, as soon as pO2 decreases is unfettered and free to bind another protein and thereof this dimeric protein activates the transcription of circa 150 genes that produce proteins related with adaptation to the new environmental conditions. This is precisely what our malignant cell did. One of the adaptation policies that the cell performed was the metabolic switch. This consisted in downregulating the oxidative metabolism of the mitochondria and using the glycolytic metabolism of the cytoplasm instead. This step solved one problem, the lack of oxygen, but created two new ones: (1) excessive production of lactic acid and CO2 that would have decreased the intracellular pH ending with the proliferative state and (2) an ineffective energy supply, because glycolysis produces 18-fold less energy than oxidative metabolism.

From the top of the mountain we could also watch how the malignant cell solved these new problems: (1) activated all the proton export mechanisms; (2) increased the uptake of nutrients, mainly (but not exclusively) glucose, the principal nutrient of normal and malignant cell.

On the mountain, we were passive observers. But if we would have had the goal to interrupt this process we would have blocked the proton export mechanisms or restrict the uptake of nutrients or restore the cell to the oxidative metabolism. Or even better, the three at the same time. What would have happened? The cell would have been submitted to an acidic stress, or to autophagy or to an oxidative stress. Independently of which kind of stress would have developed, what is logical to suppose that proliferation under these conditions would not have been possible.

But we did not intervene, therefore, the malignant cell continued with its proliferation unabated. Now, the cell was comfortably metabolizing great amounts of glucose and exporting protons at full speed. The alkaline intracellular pH, so necessary for the proliferative state was preserved and the overload of acid in the ECS created a very acidic environment. Normal cells in a very acidic environment would sooner or later acidify their interior and sooner or later they would die. Because after all, there is a relation between intracellular and extracellular pH. But this malignant cell we are scrutinizing neither died nor acidified its interior. Why?

The answer is simple: the very efficient system of proton export kept the balance. This also meant that it had to be overexpressed and overactive. The machinery that keeps the altered pH unabated we called the pH-tome. NHE1 is a fundamental player in the pH-tome team, but he is not alone. NHE1 works coordinately with VGSCs, NBC, carbonic anhydrases, V-ATPase proton pumps and monocarboxylate transporters. They all form part of the pH-tome. By now, our cell was strongly dependent on the pH-tome. Whenever we abandoned our passive behavior and perturbed one of the pH-tome participants, we could see that the malignant cell stopped its proliferation…for a short time. Because other participants of the pH-tome compensated for the lost companion and proliferation went on. But when we wanted to be very intrusive, and perturbed the whole proton extruding system things were more difficult for the cell. No adequate replacement could help out the cell. But we were not so intrusive and the cell went on.

Now that the pH-tome was fully functional, the pH of the ECS was low enough to mature proteolytic enzymes and the digestion of the structures around the cell was initiated. The invasion was on. Shortly before the invasion, we saw the cell changing its shape, it became like a mesenchymal cell. This change from an epithelial phenotype to a mesenchymal phenotype was called the EMT or epithelial mesenchymal transition.

EMT did not come alone. There were also biochemical changes that could be detected through specific indicators. From our pH observation post we could see that a pH polarization was going on in the cell. And it was precisely in the cell front end were the pH gradient was the highest where small protrusions into the ECS appeared. These protrusions had a sophisticated structure where many events were going on at the same time. As they invaded the ECS they were called invadopodia. Invadopodia had a very high expression of pH-tome structures: NHE1, VGSCs, aquaporins and CAs.

From our observation point, high in the mountain two things were evident:

(1) The cell needed to modify its pH homeostasis. The old system for keeping a balanced pH when the cell was normal, was unable to cope with the new requirements for endless proliferation. The malignant cell needed a different point of equilibrium.

(2) Therefore, here we had an essential "weakness" of the malignant cell's biology.

And while these events were well known by researchers in the field of molecular biology of the cancer for more than 20 years, no serious consideration was given to this soft spot at bedside therapy. This knowledge should have added one more stone in the armamentarium of the physicians involved in the treatment of patients with cancer. But, it did not.

The present

There is strong evidence showing that a pH oriented treatment for cancer has deep implications in the fight against the disease. Part of that evidence has been disclosed in this book. The basic fundamentals that sustain the pH centered approach have been also discussed. But Medicine being mainly an art is at the same time an experimental science. Therefore, this mixture of experience and art claims for more experience before bringing the pH approach to the bedside of the patients. What is lacking is a broad based clinical trial along the lines of a pH centered treatment. Not as a unique therapy, but as one more instrument in the fight against cancer. Treating the pH paradigm is not going to replace any of the multiple and well established therapies. It will probably improve their results, but definitely it will not cure cancer by itself.

When many basic and clinical researchers boarded this ship, they were not searching for the silver bullet that would solve the cancer problem. They were only trying to find out what, if anything, could be made with one evident weakness of malignant tumors: their dependence on the pH gradient inversion. Now, on board, they confirmed that pH alteration is a soft spot, and what is more important: a soft spot that can be targeted.

And what is of paramount interest, is that it can be attacked with existing drugs. This does not mean that more specific and powerful drugs will not improve the results. But at least, there are pharmaceuticals that can initiate the road. They are not perfect, they are not very specific, they are not very powerful...in a few words, they are not the panacea.

Pharmaceutical research may perfect many of them. But the pharmaceuticals are already there and can be tested. One drug is not enough. A rational combination of many of them is required in order to achieve a result. Many have toxicity, although low. Many lack adequate bioavailability, but this can be improved also. Many have been neglected, in spite of a promising future. That is the case of cariporide. Many are underappreciated. That is the case of tolfenamic acid and quercetin. Many were simply forgotten. And here the list is longer: betulinic acid, evodiamine, dichloroacetate, nelfinavir, sulindac, silymarin, etc. Many are enclosed in their initial objective and are hard to move them out from there. That is the case of proton pump inhibitors.

The drugs are here, the knowledge of how to use them is also here. This book only tried to put them together in a rational association targeting the weak spot: cancer's dependence on the altered pH.

The future

Large based clinical trials should confirm the benefits of targeting pH. This is the path to set in motion the research of better pharmaceuticals. When more specific carbonic anhydrase inhibitors will come to life, and there is a lot of research along this topic, higher inhibition of carbonic anhydrase will be possible.

When NHE1 inhibition is perfected, and there is also research on this issue, it will allow targeting this exchanger which is strongly associated with the initiation and progression of cancer.

We are only at the beginning of the road, but future envisions better compounds for an adequate targeting of the pH alteration.

However, this will not happen if the pH problem is not fully understood and keeps underestimated as a targetable soft spot. There is no more time for squander the pH approach. It is one powerful weapon that should be incorporated in the treatment of cancer.

What this book tried to show

Cells do not think. Cells have no strategy for developing cancer. Cells follow complex algorithms that may end up in cancer. And once achieving the malignant phenotype, new algorithms are set in place to keep this malignant machinery working indefinitely. Altering pH is part of these pro-oncogenic algorithms, not only at the beginning of the process but also during the progression of the disease. And whenever this algorithm is blocked, cancer cells find it quite difficult to continue their activity.

Lets analyze in detail what was said in this book.

Molecular level

(1) pH is altered in cancer.
(2) This alteration entails an inversion of the pH gradient.
(3) The pH gradient inversion is universal: all the malignant tumors have it.
(4) Cancer needs and depends on the altered pH.
(5) The pH alteration is a very early step in malignization.
(6) Blocking the pH alteration impedes malignization.
(7) Initially the pH alteration is not a consequence of tumor metabolism.
(8) Tumor metabolism increases the pH alteration but does not initiate it.
(9) Hypoxia can initiate a pH alteration.
(10) However, hypoxia appears a step after the initial pH alteration.
(11) The pH alteration is a consequence of one or more genetic changes.
(12) Many metabolic idiosyncrasies of cancer require an intracellular alkalinity.
(13) Progression of cancer through invasion and metastasis require a highly acidic microenvironment.
(14) The alteration that creates both, intracellular alkalinity and extracellular acidity, can be reversed with the adequate drugs.
(15) Reversing the inverted pH gradient is a hurdle for cancer development.
(16) The molecular mechanisms leading to the pH alteration are in great part known.
(17) Taking care of only one mechanism is insufficient to control cancer progression.
(18) There are drugs that target the mechanisms involved in the pH alteration.

Therapeutic level

(19) There is no unique drug that targets them all.
(20) The rational association of drugs can downregulate the altered pH mechanisms.
(21) Restoring a normal pH hinders tumor proliferation and invasion.
(22) Restoring a normal pH hinders metastatic colonization.
(23) The risk of metastasis can be reduced by targeting the pH alterations.
(24) The pH targeting is compatible with almost all the standard therapies.
(25) Treatments reversing the altered pH have low or no toxicity at all.
(26) Better drugs exist but they are not available on the market.
(27) Reversal of the altered pH is not a stand alone therapy.
(28) Association of metronomic chemotherapy and pH reversion improves results.
(29) Metronomic chemotherapy and pH reversion may be a stand alone compassionate treatment.
(30) pH reversion may reverse multidrug resistance.
(31) pH reversion treatments allow a good quality of life.
(32) pH reversion treatments decrease the risk of relapse and/or metastasis.

Future perspectives

(33) Large scale, well designed prospective clinical trials are needed.

(34) New and more specific drugs need to be developed.

(35) Old efficient drugs will return to the market.

Cancer has been confounding medical science for thousands of years. Hippocrates was the first to give a name to a "condition" that caused a pain "like the claws of a crab biting the flesh." We would also add that, like a crab, it moves insidiously—never following a straight path going forward.

We also know that patients treated for cancer often say that they feel as if they are living with a ticking bomb. Even if the tumor has been removed surgically or with chemo—or radiotherapy and they are in remission, as the date for their periodic control approaches they enter a phase of growing anxiety. A negative result is like a new lease on life that creates a sort of euphoria which gradually wanes until the next checkup.

Clearly the pH treatment described in the foregoing chapters is not a permanent cure. Its purpose is to slowdown the clock, and in some cases even stop the ticking. By blocking metastasis we can turn the disease into a chronic—for lack of a better word—condition the patient can live with, much like those suffering from HIV/AIDS. If prescribing our patients varying amounts of (inexpensive) drugs on a daily basis, for the rest of his or her life, can offer a longer and perhaps above all, a better, life with little or few side effects, we can say that we have made considerable progress in our climb. Or, to change metaphors, although the war is still being fought, we will have won an important battle.

We hope to have accomplished a landscape view of the pH problem in cancer and how to manage it in a context of low toxicity and compatibility with mainstay treatments.

If we have awakened oncologist's interest on the subject, we may say that the self imposed mission was accomplished.

Abbreviations

Akt	protein kinase B
AQP (s)	aquaporin (s)
CAII	carbonic anhydrase 2
CAIV	carbonic anhydrase 4
CAIX	carbonic anhydrase 9
CAs	carbonic anhydrases
CAXII	carbonic anhydrase 12
CRC	colorectal cancer
DCA	dichloroacetate
EGF	epidermal growth factor
EGFR	epidermal growth factor receptor
ER	estrogen receptor
ERER	estrogen receptor enhancer region
ESCC	esophagic squamous cell carcinoma
GBM	glioblastoma
GLUTs	glucose transporters
HIF	hypoxia inducible factor
HNSCC	head and neck squamous cell carcinoma
HRE	hypoxia responsive element
LA	lactic acid
LDH	lactate dehydrogenase
MCTs	monocarboxylate transporters
MMP	metalloproteases/metalloproteinases
mTOR	mammalian target of rapamycin
mTORC1	mammalian target of rapamycin complex 1
NBC	sodium/bicarbonate co-transporter
NF-kB	nuclear factor kappa B
NHE1	sodium/hydrogen exchanger 1
NSCLC	non-small cell lung cancer
OS	overall survival
OSCC	oral squamous cell carcinoma
OXPHOS	oxidative phosphorylation
PDGF	platelet derived growth factor
PDH	pyruvate dehydrogenase
PDK	pyruvate dehydrogenase kinase
PFS	progression free survival
PI3K	phospho-inositide-3-kinase
PKA	protein kinase A
PKC	protein kinase C
PPIs	proton pump inhibitors
PPP	pentose phosphate pathway
Pten	phosphatase and tensin homologue
PTI	proton transporter inhibitors

PTs	proton transporters
RCC	renal cell carcinoma
ROS	reactive oxygen species
SCC	squamous cell carcinoma
SCLC	small cell lung cancer
Sp1	specificity protein 1 (transcription factor)
TEA	tetraethylammonium
VEGF	vascular endothelial growth factor
VEGFR	vascular endothelial growth factor receptor
VGSC	voltage gated sodium channel
VHL	Von Hippel Lindau protein

Index

Note: Page numbers followed by *f* indicate figures, *t* indicate tables, and *b* indicate boxes.

D

E

T

U

V

FSC
www.fsc.org
MIX
Papier | Fördert
gute Waldnutzung
FSC® C083411

Kolibri 360 GmbH
Ferdinand-Jühlke-Straße 7
99095 Erfurt, Deutschland
produktsicherheit@kolibri360.de